Introduction to Mechanical Engineering

Introduction to Mechanical Engineering: Part 2 is the essential text for all second-year undergraduate students as well as those studying foundation degrees and Higher National Diplomas. Written by an experienced team of lecturers at the internationally renowned University of Nottingham, the text provides thorough coverage of the following core engineering topics, fully updated for the Second Edition:

- Fluid dynamics
- Thermodynamics
- Solid mechanics
- Electromechanical drive systems
- Feedback and control theory
- Structural vibration

As well as mechanical engineers, the text will be highly relevant to automotive, aeronautical/aerospace and general engineering students. All units include questions, with Units 4 and 5 including enhanced, detailed solutions online as a bonus feature.

Introduction to Mechanical Engineering

Part 2

Second Edition

Edited by
Michael Clifford

CRC Press
Taylor & Francis Group
Boca Raton London New York

CRC Press is an imprint of the
Taylor & Francis Group, an **informa** business

Designed cover image: www.shutterstock.com

Second edition published 2025
by CRC Press
2385 NW Executive Center Drive, Suite 320, Boca Raton FL 33431

and by CRC Press
4 Park Square, Milton Park, Abingdon, Oxon, OX14 4RN

CRC Press is an imprint of Taylor & Francis Group, LLC

© 2025 selection and editorial matter, Michael Clifford; individual chapters, the contributors

First edition published by CRC Press 2010

Library of Congress Cataloging-in-Publication Data
Names: Clifford, Michael, (Michael J.) editor.
Title: Introduction to mechanical engineering. Part 2 / edited by Michael Clifford.
Description: Second edition. | Boca Raton, FL : CRC Press, 2024. | Includes bibliographical references and index.
Identifiers: LCCN 2024026654 (print) | LCCN 2024026655 (ebook) | ISBN 9781032760216 (hbk) |
ISBN 9780367333775 (pbk) | ISBN 9780429319495 (ebk)
Subjects: LCSH: Mechanical engineering.
Classification: LCC TJ145 .I57 2024 (print) | LCC TJ145 (ebook) | DDC 620.1--dc23/eng/20240813
LC record available at https://lccn.loc.gov/2024026654
LC ebook record available at https://lccn.loc.gov/2024026655

ISBN: 978-1-032-76021-6 (hbk)
ISBN: 978-0-367-33377-5 (pbk)
ISBN: 978-0-429-31949-5 (ebk)

DOI: 10.1201/9780429319495

Typeset in Times
by SPi Technologies India Pvt Ltd (Straive)

Access the Instructor and Student Resources: www.routledge.com/cw/Clifford

Contents

Editor

Michael Clifford is a Senior Fellow of Advance HE and has lectured at the University of Nottingham since 1998. He has taught a wide range of subjects including professional studies, computational and numerical techniques, fibre-reinforced composites, design and engineering management. He has over 150 academic publications, including teaching case studies on the use of sustainable appropriate technologies in further education.

Contributors

Alastair Campbell Ritchie
University of Nottingham
Nottingham, United Kingdom

Kwing-So Choi
University of Nottingham
Nottingham, United Kingdom

Michael Clifford
University of Nottingham
Nottingham, United Kingdom

Donald Giddings
University of Nottingham
Nottingham, United Kingdom

Alan Howe
University of Nottingham
Nottingham, United Kingdom

Arthur Jones
University of Nottingham
Nottingham, United Kingdom

Introduction

'I think I should understand that better,' Alice said very politely, 'if I had it written down: but I can't quite follow it as you say it.'

Alice's Adventures in Wonderland by Lewis Carroll

This book builds on the experience and knowledge gained from *An Introduction to Mechanical Engineering, Part 1* and is written for undergraduate engineers and those who teach them. These textbooks are not intended to be a replacement for traditional lectures, but like Alice, we see the benefit of having things written down.

In this book, we introduce material to supplement the foundational units in Part 1 on solid mechanics, thermodynamics and fluid dynamics. In addition, the reader will encounter units on electromechanical drive systems, feedback and control and structural vibration.

This second edition contains additional material in Unit 1: fluid dynamics comparing the Navier–Stokes equations to the Bernoulli equation, vortex-shedding, the application of fluid dynamics to blood flow, the flight of golf balls and drag reduction using riblets.

Unit 4: Electromechanical Drive Systems contains a major new section on motors used in electric and hybrid vehicles. As 3D printers and related technologies are becoming increasingly prevalent, new material is included on the stepper motors often used within them, including the importance of logic circuits and power electronics in their control.

Unit 5: Feedback and Control Theory includes new material on applications of control including self-balancing transport devices and position control systems including a practical example involving the use of a microcontroller. There is also new material on the manipulation of block diagrams.

The material contained in this volume has been compiled from the authors' experience of teaching undergraduate engineers, mostly, but not exclusively, at the University of Nottingham. The knowledge contained within this textbook has been derived from lecture notes, research findings and personal experience from within the lecture theatre and from tutorial sessions.

We gratefully acknowledge the support and encouragement of Nicola Sharpe and Katya Porter at Taylor & Francis, without whom producing this second edition would still be just a good idea that needed to be put into action.

Dedicated to past, present and future engineering students at the University of Nottingham.

Michael Clifford, 2024

1 Fluid Dynamics

Kwing-So Choi
University of Nottingham, Nottingham, United Kingdom

1.1 INTRODUCTION

Fluid dynamics is the study of the dynamics of fluid flow. Here we learn how flows behave under different external forces and conditions. In a sense this is similar to rigid body dynamics in physics, where Newton's second law is used to describe the motion of rigid bodies. Here, we must apply Newton's second law to fluid flows in a different way since fluids do not behave exactly like rigid bodies. This will be discussed in Section 1.2, where basic equations to describe fluid motion are derived and explained. Some discussions on laminar and turbulent flows are also given there, paving the way for what will follow.

The fluid that we deal with in this unit is a viscous fluid, so the velocity of fluid flow becomes zero at solid surface. The consequence of this **no-slip condition** is that flow velocity changes from zero at the wall to the free-stream value sufficiently far away from the wall surface. This thin layer is called the **boundary layer**, an important concept in fluid dynamics, which explains how the fluid forces are generated. So, in Section 1.3, we learn the basic behaviour of boundary layers to be able to estimate the viscous drag acting on the solid surface.

The boundary layers over solid bodies behave differently depending on their shape. For example, the drag force acting on sports cars is much less than that on pickup trucks, where the boundary layer is separated from the body surface of the vehicle creating a strong flow disturbance. In Section 1.4, we study the streamlining strategy to reduce the drag force of immersed bodies. We also discuss how the drag of immersed bodies is affected by the Reynolds number as well as the wall roughness.

Pipes and ducts are important engineering components used in many fluid systems. It is important, therefore, that the flow resistance can be correctly estimated for different type of ducts and pipes. In general, there are two types of flow resistance. One is due to the friction drag, while the other relates to the loss of energy due to boundary layer separation. In Section 1.5, we study a method of minimizing the flow resistance similar to the streamlining strategy for immersed bodies. The discussion of pipes and ducts is extended to non-circular shapes by introducing the concept of hydraulic diameter.

The final section of this unit deals with the non-dimensional numbers of fluid dynamics. We are already familiar with the Reynolds number, but there are many other non-dimensional numbers in fluid dynamics. In Section 1.6, we learn how to identify relevant non-dimensional numbers for different types of fluid flow. We also study how these non-dimensional numbers are used to carry out model tests. Applications of the similarity principle to fluid machinery are given, emphasizing the importance of non-dimensional numbers in fluid dynamics.

1.2 BASIC CONCEPT IN FLUID DYNAMICS

1.2.1 NAVIER–STOKES EQUATIONS

The main aim of fluid dynamics is to understand the dynamic behaviour of fluid flows. Since all fluids are continuous, we can determine the velocities and pressure of flows as a function of space and time. To achieve this, we require the governing equations to represent the fluid flows: the Navier–Stokes equations.

DOI: 10.1201/9780429319495-1

For simplicity, we consider only two-dimensional, **isothermal** (no thermal input or output) and **Newtonian** (the shear stress is linearly proportional to the strain rate) flows with constant density and viscosity. Therefore, u, v and p (velocities and pressure in Cartesian coordinates) are functions of x, y and t. If the fluid flows are steady, they are only functions of x and y.

The Navier–Stokes equations can be derived by applying Newton's second law of motion to fluid flow. By considering a *small* control volume $dxdy$ with a unit depth, the fluid mass times acceleration is given in the vector form as:

$$m \cdot \vec{a} = m \cdot \frac{d\vec{V}}{dt} = \rho \cdot dxdy \cdot \frac{d\vec{V}}{dt} = \vec{F} \tag{1.1}$$

The x- and y-component forces, F_x and F_y, acting on the control volume are given by:

$$m \cdot a_x = \rho \cdot dxdy \cdot \frac{du}{dt} = F_x \tag{1.2}$$

$$m \cdot a_y = \rho \cdot dxdy \cdot \frac{dv}{dt} = F_y \tag{1.3}$$

Since, the total derivative (or material derivative) $\dfrac{d}{dt}$ represents

$$\frac{d}{dt} = \frac{\partial}{\partial t} + \frac{\partial x}{\partial t} \frac{\partial}{\partial x} + \frac{\partial y}{\partial t} \frac{\partial}{\partial y} = \frac{\partial}{\partial t} + u \frac{\partial}{\partial x} + v \frac{\partial}{\partial y} \tag{1.4}$$

Therefore,

$$m \cdot a_x = \rho dxdy \cdot \frac{du}{dt} = \rho \, dxdy \cdot \left[\frac{\partial u}{\partial t} + u \frac{\partial u}{\partial x} + v \frac{\partial u}{\partial y} \right] = F_x \tag{1.5}$$

$$m \cdot a_y = \rho \, dxdy \cdot \frac{dv}{dt} = \rho \, dxdy \cdot \left[\frac{\partial v}{\partial t} + u \frac{\partial v}{\partial x} + v \frac{\partial v}{\partial y} \right] = F_y \tag{1.6}$$

Here, we shall consider only those forces acting on the surfaces of the control volume, although a body force will be introduced later. Surface forces include hydrostatic pressure p, the normal stresses τ_{xx} and τ_{yy} and the shear stresses τ_{xy} and τ_{yx} (Figure 1.1).

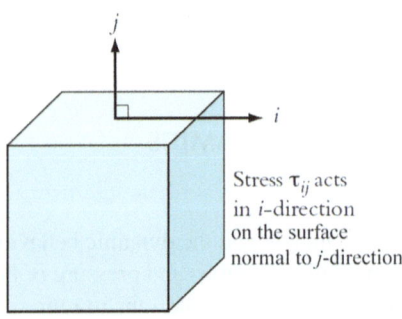

FIGURE 1.1 Stress tensor τ_{ij}. The first index i of the stress tensor τ_{ij} indicates the direction of the stress that is acting on the surface, whose normal direction is indicated by the second index j.

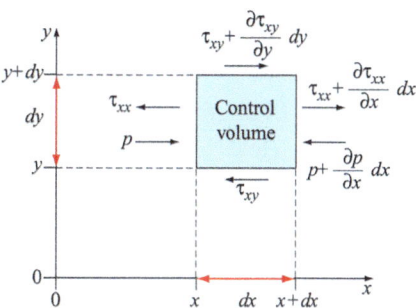

FIGURE 1.2 The balance of surface forces in the x-direction.

From Figure 1.2, the total surface forces in the x-direction can be obtained as follows:

$$\left[\left(\tau_{xx}+\frac{\partial \tau_{xx}}{\partial x}dx\right)dy-\tau_{xx}dy\right]+\left[\left(\tau_{xy}+\frac{\partial \tau_{xy}}{\partial y}dy\right)dx-\tau_{xy}dx\right]-\left[\left(p+\frac{\partial p}{\partial x}dx\right)dy-p\,dy\right] \quad (1.7)$$

Here, the forces acting on the surface at $(x + dx)$ can be obtained from the forces on the surface at x by using Taylor's expansion.

In a similar way, the total surface force in the y-direction can be obtained from Figure 1.3 as:

$$\left[\left(\tau_{yy}+\frac{\partial \tau_{yy}}{\partial y}dy\right)dx-\tau_{yy}dx\right]+\left[\left(\tau_{yx}+\frac{\partial \tau_{yx}}{\partial x}dx\right)dy-\tau_{yx}dy\right]-\left[\left(p+\frac{\partial p}{\partial y}dy\right)dx-p\,dx\right] \quad (1.8)$$

Therefore, the *total force* acting on the surfaces of control volume can be given by

$$F_x=\left(-\frac{\partial p}{\partial x}+\frac{\partial \tau_{xx}}{\partial x}+\frac{\partial \tau_{xy}}{\partial y}\right)dx\,dy \quad (1.9)$$

$$F_y=\left(-\frac{\partial p}{\partial y}+\frac{\partial \tau_{yx}}{\partial x}+\frac{\partial \tau_{yy}}{\partial y}\right)dx\,dy \quad (1.10)$$

So far, we have considered only the surface forces. There are flows, such as water waves around a ship, where the gravity force plays an important role. Therefore, we should also consider such a

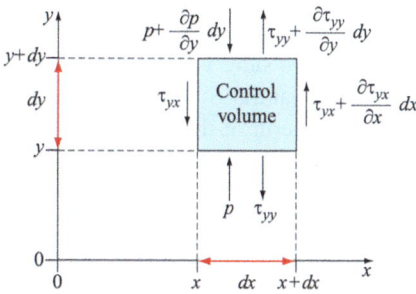

FIGURE 1.3 The balance of surface forces in the y-direction.

body force acting on the control volume together with the surface forces. In such a situation, the vertical force F_y in Equation (1.10) must be replaced by:

$$F_y = \left(-\frac{\partial p}{\partial y} + \frac{\partial \tau_{yx}}{\partial x} + \frac{\partial \tau_{yy}}{\partial y} \right) dx\,dy - \rho g\,dx\,dy \tag{1.11}$$

Here, the additional force due to gravity is $\rho g\,dy\,dx$ which always acts in the negative y-direction (downwards).

We finally obtained the following equations of fluid motion: Equations (1.12, 1.13):

$$\rho \left(\frac{\partial u}{\partial t} + u\frac{\partial u}{\partial x} + v\frac{\partial u}{\partial y} \right) = -\frac{\partial p}{\partial x} + \left(\frac{\partial \tau_{xx}}{\partial x} + \frac{\partial \tau_{xy}}{\partial y} \right) \tag{1.12}$$

$$\rho \left(\frac{\partial v}{\partial t} + u\frac{\partial v}{\partial x} + v\frac{\partial v}{\partial y} \right) = -\frac{\partial p}{\partial y} + \left(\frac{\partial \tau_{yx}}{\partial x} + \frac{\partial \tau_{yy}}{\partial y} \right) - \rho g \tag{1.13}$$

For the *Newtonian* flow, the stress is linearly proportional to the velocity gradient, we can write this relationship in a matrix form as follows:

$$\begin{pmatrix} \tau_{xx} & \tau_{xy} \\ \tau_{yx} & \tau_{yy} \end{pmatrix} = \begin{pmatrix} \mu\dfrac{\partial u}{\partial x} & \mu\dfrac{\partial u}{\partial y} \\ \mu\dfrac{\partial v}{\partial x} & \mu\dfrac{\partial v}{\partial y} \end{pmatrix} \tag{1.14}$$

By substituting this relationship into Equations (1.12) and (1.13), we have the final form of the Navier–Stokes equations after dividing all terms by the fluid density ρ.

$$\frac{\partial u}{\partial t} + u\frac{\partial u}{\partial x} + v\frac{\partial u}{\partial y} = -\frac{1}{\rho}\frac{\partial p}{\partial x} + v\left(\frac{\partial^2 u}{\partial x^2} + \frac{\partial^2 u}{\partial y^2} \right) \tag{1.15}$$

$$\frac{\partial v}{\partial t} + u\frac{\partial v}{\partial x} + v\frac{\partial v}{\partial y} = -\frac{1}{\rho}\frac{\partial p}{\partial y} + v\left(\frac{\partial^2 v}{\partial x^2} + \frac{\partial^2 v}{\partial y^2} \right) - g, \tag{1.16}$$

where $v = \mu/\rho$ is called the **kinematic viscosity** with the dimension $[L^2\,T^{-1}]$.

Now, the Navier–Stokes equations consist of four parts, which are:

$$\left(\text{Inertia force} \right) = \left(\text{Pressure force} \right) + \left(\text{Viscous force} \right) + \left(\text{Gravity force} \right) \tag{1.17}$$

In others words, the left-hand side of the Navier–Stokes equations represents the inertia force (i.e. mass times acceleration), while pressure force represented by the pressure gradient and viscous force led by viscosity are on the right-hand side. The gravity force can usually be omitted from the equations unless the surface wave or the natural convection is involved.

By examining the magnitude in each term of the Navier–Stokes equations, we can see that the left-hand side of the equations is of the order of u^2/L, while the right-hand side is of the order of vu/L^2. Here, u represents the velocity scale (either u or v), L represents the length scale (x or y) and v is the kinematic viscosity. The ratio of these two will give a *non-dimensional* value called the **Reynolds number** (Figure 1.4):

$$\text{Re}\left(\text{Reynolds number} \right) = \frac{u\boldsymbol{L}}{v} \propto \frac{\text{inertia force}}{\text{viscous force}} \tag{1.18}$$

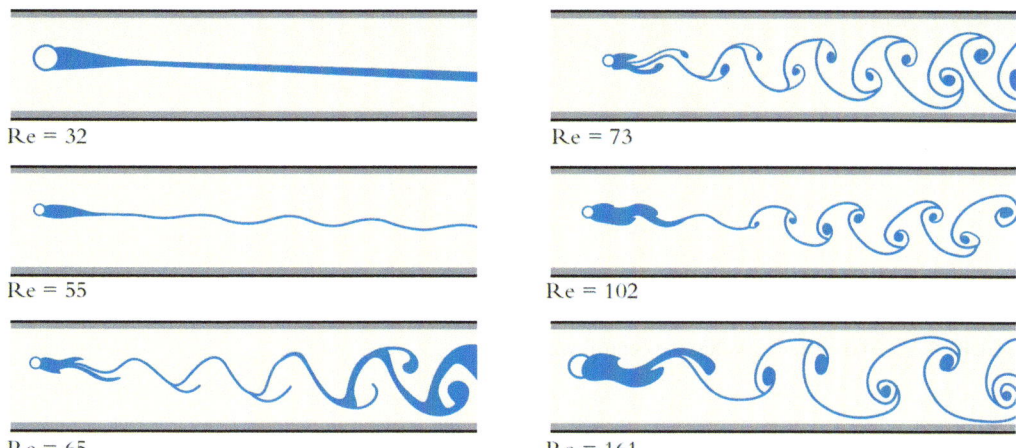

FIGURE 1.4 Effect of the Reynolds number on the vortex shedding from a circular cylinder. (H. Schlichting, 1968, *Boundary Layer Theory*, 6th edn, New York: McGraw Hill, reproduced with permission of the McGraw-Hill Companies.)

Similarly, the ratio between the magnitude of the inertia term $\left(\text{of the order of } \dfrac{u^2}{L}\right)$ and that of the gravity term (of the order of g) is called the **Froude number**:

$$\text{Fr}\left(\text{Froude number}\right) = \frac{u}{\sqrt{gL}} \propto \left(\frac{\text{inertia force}}{\text{gravity force}}\right)^{\frac{1}{2}} \tag{1.19}$$

Here, it is customary to take a square root of the ratio to define the Froude number. For example, a yacht with a length of $L = 20$ m travelling at $u = 10$ m/s would have Re = 1.3×10^8 and Fr = 0.7 assuming that $\nu = 1.5 \times 10^{-6}$ m²/s.

WORKED EXAMPLE

Find the acceleration of a fluid particle with the velocity field given by

$$\vec{V} = u\vec{i} + v\vec{j}$$

$$= \left(3t\right)\vec{i} + \left(2xy\right)\vec{j}$$

$$\frac{Du}{Dt} = \frac{\partial u}{\partial t} + u\frac{\partial u}{\partial x} + v\frac{\partial u}{\partial y} = 3$$

$$\frac{Dv}{Dt} = \frac{\partial v}{\partial t} + u\frac{\partial v}{\partial x} + v\frac{\partial v}{\partial y} = 0 + \left(3t\right)\left(2y\right) + \left(2xy\right)\left(2x\right)$$

$$\frac{D\vec{V}}{Dt} = \frac{Du}{Dt}\vec{i} + \frac{Dv}{Dt}\vec{j} = 3\vec{i} + \left(6ty + 4x^2y\right)\vec{j}$$

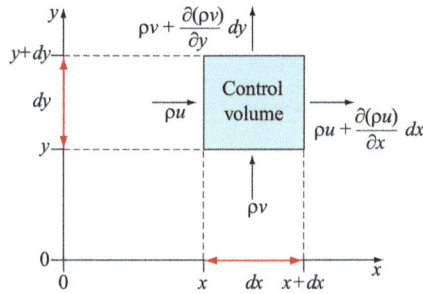

FIGURE 1.5   Mass balance within the control volume.

1.2.2 Continuity Equation

The **continuity equation** can be derived in a similar way as we have done to obtain the Navier–Stokes equation. Here, however, we should consider the *mass balance* in the control volume instead of the *force balance* (Figure 1.5).

Mass flux into the control volume in the *x*-direction is given by

$$\left[(\rho u) \cdot dy + \frac{\partial (\rho u)}{\partial x} dx \cdot dy \right] - (\rho u) \cdot dy = \frac{\partial (\rho u)}{\partial x} \cdot dx\, dy \qquad (1.20)$$

while the mass flux into the control volume in the *y*-direction is

$$\left[(\rho v) \cdot dx + \frac{\partial (\rho v)}{\partial y} dy \cdot dx \right] - (\rho v) \cdot dx = \frac{\partial (\rho v)}{\partial y} \cdot dx\, dy \qquad (1.21)$$

Here again, we have used Taylor's expansion to evaluate the mass flux out from the surface at $(x + dx)$ and $(y + dy)$ from the mass flux into the surface at *x* and *y*, respectively.

Since there must be no changes in mass within the control volume:

$$\frac{\partial (\rho u)}{\partial x} \cdot dx\, dy + \frac{\partial (\rho v)}{\partial y} \cdot dx\, dy = 0 \qquad (1.22)$$

Assuming that the fluid density ρ is constant, we have:

$$\frac{\partial u}{\partial x} + \frac{\partial v}{\partial y} = 0 \qquad (1.23)$$

This is called the continuity equation based on the conservation of mass.

WORKED EXAMPLE

When the velocity field is given by

$$\vec{V} = \left(x^2 - y^2 \right)\vec{i} + v\vec{j}$$

determine the ν-velocity from the two-dimensional continuity equation.

$$u = x^2 - y^2$$

$$\therefore \frac{\partial u}{\partial x} = 2x$$

Substituting into the continuity equation $\left(\dfrac{\partial u}{\partial x} + \dfrac{\partial v}{\partial y} = 0\right)$, we get

$$\frac{\partial v}{\partial y} = -2x$$

$$\therefore v = -2xy + C(x)$$

WORKED EXAMPLE

There is a two-dimensional steady flow whose velocity vector is given by

$$\vec{V} = 2xy\,\vec{i} - y^2\,\vec{j}$$

where $\vec{i}$ and $\vec{j}$ are unit vectors in the x- and y-directions, respectively. Obtain an expression for the pressure gradient of this flow if the fluid density is constant and the viscosity and gravity forces are negligible.

$$u = 2xy \quad v = -y^2$$

$$\frac{\partial u}{\partial x} = 2y \quad \frac{\partial u}{\partial y} = 2x$$

$$\frac{\partial v}{\partial x} = 0 \quad \frac{\partial v}{\partial y} = -2y$$

Substituting these into the Navier–Stokes equations, we get

$$(2xy)(2y) + (-y^2)(2x) = -\frac{1}{\rho}\frac{\partial p}{\partial x}$$

$$2xy^2 = -\frac{1}{\rho}\frac{\partial p}{\partial x} \quad \therefore \frac{\partial p}{\partial x} = -2\rho xy^2$$

$$(2xy)(0) + (-y^2)(-2y) = -\frac{1}{\rho}\frac{\partial p}{\partial y}$$

$$2y^3 = -\frac{1}{\rho}\frac{\partial p}{\partial y} \quad \therefore \frac{\partial p}{\partial y} = -2\rho y^3$$

TOPIC: NAVIER-STOKES EQUATIONS VS. BERNOULLI EQUATION

By neglecting the viscous term and gravity term on the right-hand side of the Navier–Stokes Equations (1.15) and (1.16) and assuming that the flow is steady ($\partial u/\partial t = \partial v/\partial t = 0$), we can derive Bernoulli's equation.

If we take a streamline along the y-axis, the Navier–Stokes equations will become

$$v\frac{\partial v}{\partial y} = -\frac{1}{\rho}\frac{\partial p}{\partial y} - g$$

because all other terms in the Navier–Stokes equations will vanish since $u = 0$ along this streamline. Also, partial derivatives in this equation will become total derivatives because v and p are only functions of x. Therefore, we have

$$v\frac{dv}{dy} = -\frac{1}{\rho}\frac{dp}{dy} - g$$

which can be rewritten as

$$\frac{d}{dy}\left[\frac{v^2}{2} + \frac{p}{\rho} + gy\right] = 0$$

Therefore,

$$\left[\frac{v^2}{2} + \frac{p}{\rho} + gy\right] = \text{constant}$$

along the streamline. This is Bernoulli's equation.

1.2.3 LAMINAR AND TURBULENT FLOWS

The flow through a pipe remains smooth and steady below the **critical Reynolds number**, given by

$$\text{Re} \equiv \frac{U \cdot d}{v} = 13,000 \tag{1.24}$$

However, the flow becomes fluctuating and random when the Reynolds number exceeds this value. Figure 1.6 shows this process where the transition from laminar to turbulent flow is demonstrated with dye injected into a pipe through a needle.

We can see the change in flow behaviour as a function of the Reynolds number. Figure 1.7(a) shows a laminar flow at subcritical Reynolds number, where the dye filament stays straight until the end of the pipe. As the Reynolds number is increased as shown in Figures 1.7(b) and (c), the flow develops patterns, signifying the flow transition that is taking place. Finally, turbulent flow is reached as shown in Figure 1.7(d) where the dye patterns seem to be random and chaotic.

Although the Reynolds number is an important parameter affecting the transition process to turbulence, there are other influential factors such as the wall roughness, initial disturbance and external disturbance. For example, a sharp inlet to the pipe will disturb the flow, and therefore reduce the critical Reynolds number. The vibrations of the experimental setup from the floor, as well as noise transmitted to the flow, will accelerate the process of transition.

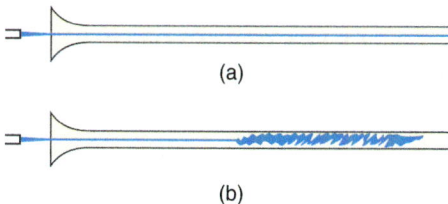

FIGURE 1.6 The flow in a pipe changes from laminar flow (a) to turbulent flow (b).

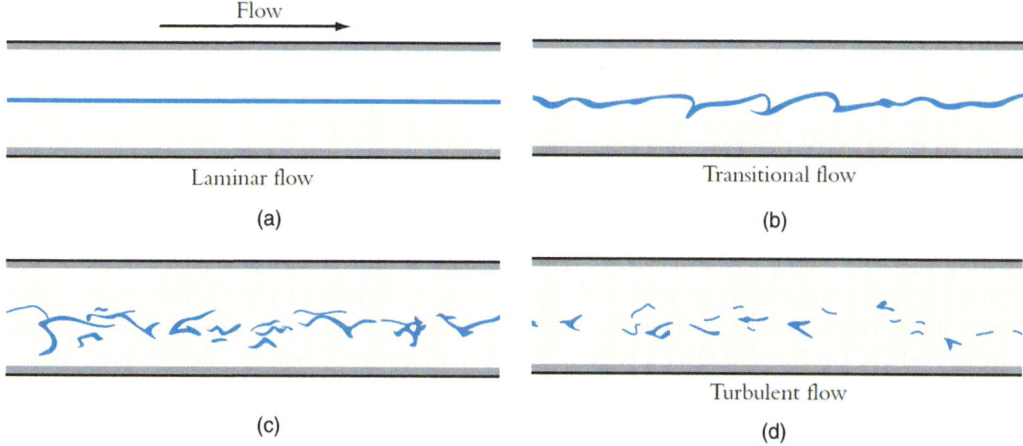

FIGURE 1.7 Flow transition to turbulence in a pipe flow. Laminar flow (a), where the injected dye stays straight through the pipe; transitional flow (b), where the flow develops certain patterns; turbulent flow (c, d), where the flow becomes random and chaotic.

There is, however, no precise definition of turbulence. Indeed, it is very difficult to determine whether a particular flow is turbulent or not. For example, random waves that can be observed over the surface of the water in a swimming pool are probably not turbulence. Therefore, we have to look at the symptoms of the flow to determine whether it is turbulent or not.

Symptoms of turbulence include:

- Irregularity
- High diffusivity
- High Reynolds number
- Three-dimensionality
- Dissipativeness
- Continuous flow.

Every turbulent flow is different, yet they have many common characteristics as listed above. We must, therefore, check whether a particular flow satisfies all of these characteristics before declaring that it is a turbulent flow

LEARNING SUMMARY

By the end of this section, you should have learnt:

✓ The Navier–Stokes equations are governing equations for fluid motion, which can be derived from Newton's second law of motion;

✓ The continuity equation guarantees the conservation of mass;

✓ The Reynolds number indicates the relative importance of inertial force in flow motion to viscous force;

✓ The Froude number signifies the importance of inertial force in flow motion against the gravity force;

✓ All flows become turbulent above the critical Reynolds number.

1.3   BOUNDARY LAYER

The boundary layer is a thin layer created over the surface of a body immersed in a fluid (Figure 1.8), where the viscosity plays a significant role. Due to the non-slip condition of viscous flows, the velocity at a solid surface is always zero. This means that the velocity gradually increases from zero at the wall to the freestream velocity U_o at the edge of the boundary layer. This creates a thin, highly sheared region called the boundary layer, over a body surface in a moving fluid (or over a moving body in a still fluid). Over a large commercial aircraft, for example, the boundary layer a few millimetre thick near the cockpit can grow to as much as half a metre thick towards the end of the fuselage.

1.3.1   REYNOLDS NUMBER

Similar to pipe flows, the boundary layer over the surface of a body is initially laminar, but will soon become turbulent as the Reynolds number increases. Here, the Reynolds number to describe the state of the boundary layer flow can take any of the following forms.

$$\mathrm{Re} \equiv \frac{U_o x}{v}, \frac{U_o \delta}{v}, \frac{U_o \delta^*}{v}, \frac{U_o \theta}{v} \tag{1.25}$$

While the pipe diameter is always used to define the Reynolds number of pipe flows, the length scale of boundary layer flows takes either the streamwise length x along the body surface or one of the boundary layer thicknesses such as δ, δ^* or θ. The boundary layer thickness δ is defined as the

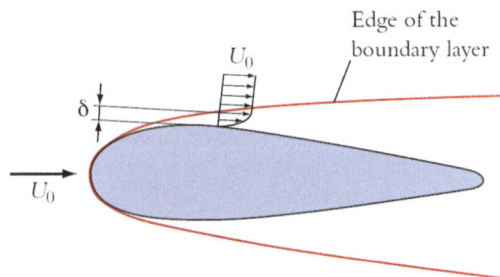

FIGURE 1.8   A boundary layer being developed over a wing.

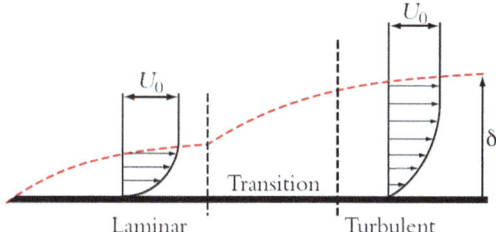

FIGURE 1.9 Development of the boundary layer over a flat plate parallel to the flow. Note that the boundary layer thickness grows faster when it is turbulent.

distance from the wall to the point where the boundary layer velocity reaches the freestream velocity. Discussions on other boundary layer thickness, such as δ^* or θ, will be given later in this section.

The boundary layer thickness over a flat plate at the streamwise length x from the leading edge can be obtained using the following formulae:

$$\delta = 5.0\,x\left(\frac{U_0 x}{\nu}\right)^{-\frac{1}{2}} \text{ for laminar flow}\,(\mathrm{Re}_x < 3\times10^6) \tag{1.26}$$

$$\delta = 0.37\,x\left(\frac{U_0 x}{\nu}\right)^{-\frac{1}{5}} \text{ for turbulent flow}\,(\mathrm{Re}_x > 3\times10^6) \tag{1.27}$$

The boundary layer growth depends on the flow condition, whether it is laminar or turbulent. The growth is much faster for turbulent boundary layers (Figure 1.9) as the diffusivity increases as a result of transition to turbulence (Section 1.1). The critical Reynolds number for the boundary layer over a flat plate is usually around 3×10^6 as indicated in Equations (1.26) and (1.27). It may take quite a different value, however, depending on the initial as well as boundary conditions of the flow. For the boundary layer over a rough surface, for example, the critical Reynolds number is less than 10^6. This means that the boundary layer transition takes place much earlier over a rough surface as compared to that over a smooth surface.

WORKED EXAMPLE

A laminar boundary layer is being developed over a flat plate with the freestream velocity of 1.5 m/s. Assuming that the pressure gradient along the plate is zero, obtain the distance x from the leading edge where the boundary layer thickness δ becomes 10 mm. Using the critical Reynolds number R_{xc} of 10^6, determine the transition point where the boundary layer becomes turbulent. The kinematic viscosity and density of the fluid (air) are 1.5×10^{-5} m²/s and 1.2 kg/m³, respectively.

$$\delta = 5.0\,x\left(\frac{U_0 x}{\nu}\right)^{-\frac{1}{2}} = 5.0\sqrt{\frac{\nu}{U_0}}\sqrt{x}$$

$$\therefore x = \frac{1}{25.0}\delta^2\frac{U_0}{\nu} = \frac{1}{25.0}(0.01)^2\frac{1.5}{1.5\times10^{-5}} = 0.40\left[\mathrm{m}\right]$$

The transition point x_t can be obtained from

$$R_{xc} = \frac{x_t U_o}{v} = 10^6$$

$$\therefore x_t = 10^6 \frac{v}{U_o} = 10^6 \times \frac{1.5 \times 10^{-5}}{1.5} = 10 \left[\text{m} \right]$$

1.3.2 REYNOLDS NUMBER OF LIVING THINGS

The Reynolds number of living things depends on their size, flight or swim speed and the medium they live in. The following are the Reynolds numbers of some familiar living things. Unless the body size is very small or the flight or swim speed is very low, as with butterflies, the flow around the body of living things is most likely turbulent.

Butterflies: $\text{Re} \approx \dfrac{\left(0.3\,\text{m/s} \times \left(0.08\,\text{m}\right)\right)}{\left(1.5 \times 10^{-5}\,\text{m}^2\,/\,\text{s}\right)} = 1600$

Cranes: $\text{Re} \approx \dfrac{\left(15\,\text{m/s} \times \left(1.0\,\text{m}\right)\right)}{\left(1.5 \times 10^{-5}\,\text{m}^2\,/\,\text{s}\right)} = 1,000,000$

Dolphins: $\text{Re} \approx \dfrac{\left(10\,\text{m/s} \times \left(2\,\text{m}\right)\right)}{\left(1.0 \times 10^{-6}\,\text{m}^2\,/\,\text{s}\right)} = 20,000,000$

Whales: $\text{Re} \approx \dfrac{\left(15\,\text{m/s} \times \left(25\,\text{m}\right)\right)}{\left(1.0 \times 10^{-6}\,\text{m}^2\,/\,\text{s}\right)} = 375,000,000$

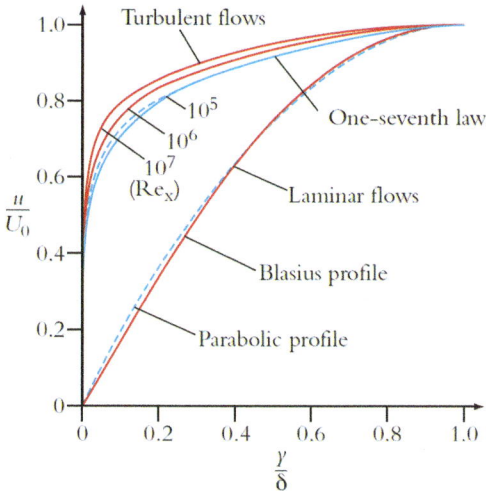

FIGURE 1.10 Velocity profiles of laminar and turbulent boundary layers.

1.3.3 VELOCITY PROFILES OF BOUNDARY LAYERS

The **Blasius profile** is the theoretical velocity profile of the **laminar boundary layer** over a flat plate where the static pressure does not change along the plate. Here, the Blasius profile is independent of the Reynolds number of the boundary layer as shown in Figure 1.10. The Blasius profile can be approximated by the parabolic velocity profile given by Equation (1.28).

$$\frac{u}{U_o} = \frac{y}{\delta}\left(2 - \frac{y}{\delta}\right) \tag{1.28}$$

For the **turbulent boundary layer**, however, there is no theoretical solution to represent the velocity profile. The turbulent boundary layer profile can be approximated by the **one-seventh law** given by

$$\frac{u}{U_o} = \left(\frac{y}{\delta}\right)^{\frac{1}{7}} \tag{1.29}$$

The atmospheric boundary layer is often simulated by the one-seventh law in a wind tunnel, where building or bridge models are tested. Indeed, this empirical law has a reasonable agreement with the actual profile of the turbulent boundary layer. However, the velocity gradient of the one-seventh law at the wall is always incorrect, since

$$\frac{d}{dy}\left(\frac{u}{U_o}\right)_{y=0} = \infty \tag{1.30}$$

Therefore, it cannot be used to investigate the turbulent boundary layer close to the wall surface. Here, we should use the logarithmic velocity profile instead (Figure 1.11). With the logarithmic velocity profile (the log law, for short), a large part of the turbulent boundary layer can be represented by

$$\frac{u}{u^*} = 5.75\log_{10}\frac{u^*y}{\nu} + 5.5 \tag{1.31}$$

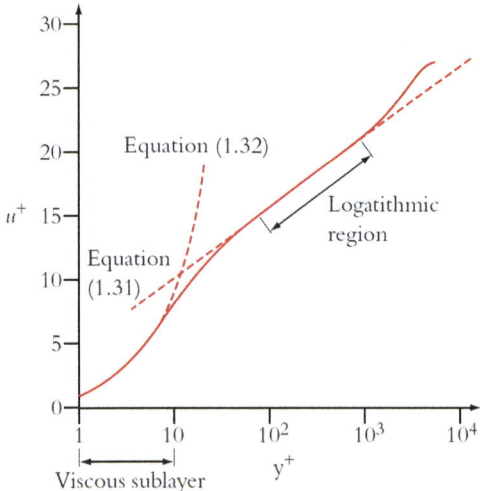

FIGURE 1.11   Logarithmic velocity profile of the turbulent boundary layer.

except for a very thin region near the wall (the viscous sublayer) where the velocity is given by the following linear profile.

$$\frac{u}{u^*} = \frac{u^* y}{v} \tag{1.32}$$

Here, the friction velocity u^* is given by $u^* = \sqrt{\dfrac{\tau_w}{\rho}}$. It should be noted that the turbulent velocity profile is dependent on the Reynolds number (Figure 1.10).

WORKED EXAMPLE

An atmospheric boundary layer ($\delta = 80$ m) with the freestream velocity of $U_o = 10$ m/s has the wall-shear stress of $\tau_w = 0.077$ Pa. The kinematic viscosity and density of air are $v = 1.5 \times 10^{-5}$ m²/s and $\rho = 1.2$ kg/m³, respectively.

1. **Estimate the thickness of the viscous sublayer.**
2. **Obtain the velocity at $y = 1$ m from the ground.**

ANSWERS

1. The thickness of the viscous sublayer is given by

$$y^+ = \frac{yu^*}{v} = 10 \quad \therefore y = 10\frac{v}{u^*}$$

$$\text{where, } u^* = \sqrt{\frac{\tau_w}{\rho}} = \sqrt{\frac{0.077}{1.2}} = 0.25\,\text{m/s}$$

$$\therefore y = 10 \times \frac{1.5 \times 10^{-5}}{0.25} = 0.6 \times 10^{-3}\,\text{m} = 0.6\,\text{mm}$$

2. Using the logarithmic velocity profile,

$$\frac{u}{u^*} = 5.5 + 5.75 \ \log_{10} \frac{u^* y}{\nu}$$

$$\therefore u = u^* \left(5.5 + 5.75 \log_{10} \frac{u^* y}{\nu} \right)$$

$$= 0.25 \left(5.5 + 5.75 \log_{10} \frac{0.25 \times 1}{1.5 \times 10^{-5}} \right) = 7.4 \, \text{m/s}$$

If we use the one-seventh law instead, we have

$$\frac{u}{U_o} = \left(\frac{y}{\delta} \right)^{\frac{1}{7}} \quad \therefore u = U_o \left(\frac{y}{\delta} \right)^{\frac{1}{7}}$$

$$u = 10 \left(\frac{1}{80} \right)^{\frac{1}{7}} = 5.3 \, \text{m/s}$$

1.3.4 FRICTION VELOCITY IN TURBULENT BOUNDARY LAYERS

Logarithmic velocity profile of turbulent boundary layer is given by

$$\frac{u}{u^*} = 5.75 \log_{10} \frac{u^* y}{\nu} + 5.5$$

Where $\dfrac{u}{u^*}$ and $\dfrac{u^* y}{\nu}$ are non-dimensional variables for the local velocity and the distance from wall, respectively.

$$u^+ \equiv \frac{u}{u^*} = \left[\frac{LT^{-1}}{LT^{-1}} \right] \quad y^+ \equiv \frac{u^* y}{\nu} = \left[\frac{\left(LT^{-1} \right) L}{L^2 T^{-1}} \right]$$

Using these notations, we can write the logarithmic profiles as

$$u^+ = 5.75 \log_{10} y^+ + 5.5$$

A logarithmic velocity profile can represent a large part of the turbulent boundary layers and turbulent pipe flows except in the viscous sublayer ($y^+ < 10$), where we should use

$$\frac{u}{u^*} = \frac{u^* y}{\nu} \quad \text{or} \quad u^+ = y^+$$

Since $u = \dfrac{u^{*2}}{\nu} y$ in the viscous sublayer, we have

$$\frac{du}{dy} = \frac{u^{*2}}{\nu}$$

$$\therefore \mu \frac{du}{dy} = \mu \frac{u^{*2}}{\nu} = \frac{\mu u^{*2}}{\left(\mu / \rho \right)} = \rho u^{*2} = \tau_w$$

We see that the velocity gradient within the viscous sublayer is constant.

1.3.5 Effect of Wall Roughness

The wall roughness can affect the laminar boundary layer by promoting an early transition to turbulence. For the turbulent boundary layer, the wall roughness enhances the fluid mixing to increase the near-wall velocity gradient, which leads to an increase in skin-friction drag. Here, it is important to know how rough the wall should be before it starts affecting the skin-friction drag of the turbulent boundary layer.

The wall surface of the turbulent boundary layer is **hydraulically smooth** when $\varepsilon^+ \equiv \dfrac{\varepsilon u^*}{\nu} < 5$. Here, ε is the roughness height, $u^* = \sqrt{\dfrac{\tau_w}{\rho}}$ is the friction velocity and ν is the kinematic viscosity of the fluid. In other words, the wall surface is smooth as far as the fluid dynamics of the turbulent boundary layer is concerned if the Reynolds number ε^+, based on the roughness height ε and the friction velocity u^*, is less than 5. This suggests that the skin friction drag of the turbulent boundary layer will be increased only when $\varepsilon^+ > 5$. Recalling that the thickness of viscous sublayer is given by $y^+ = 10$, we can say that the wall surface is hydraulically smooth if the roughness is completely submerged in the viscous sublayer.

If we know the roughness height ε, the **Moody chart** for pipe flows (and the analogous chart for the boundary layers) will give us the effects of roughness on the skin friction drag in the turbulent boundary layer. Wall roughness should be considered always relative to the boundary layer thickness, so we must use the non-dimensional roughness height in studying its effect on the skin friction drag (Figure 1.12).

1.3.6 Momentum Integral Equation

It is possible to estimate the skin-friction drag of the boundary layer using the momentum integral equation. To derive the equation, we consider a balance of forces on a control volume within the boundary layer (Figure 1.13), which is being developed over a flat plate. Here, lines 1 and 3 indicate the entry and exit of the control volume, while the streamline just above the boundary layer (line 2) and the flat plate (line 4) indicate the upper and bottom surfaces, respectively.

Since there is no flow across the streamline or the plate wall, the mass flow rate to the control volume through line 1 must be equal to the mass flow rate out of line 3. This gives

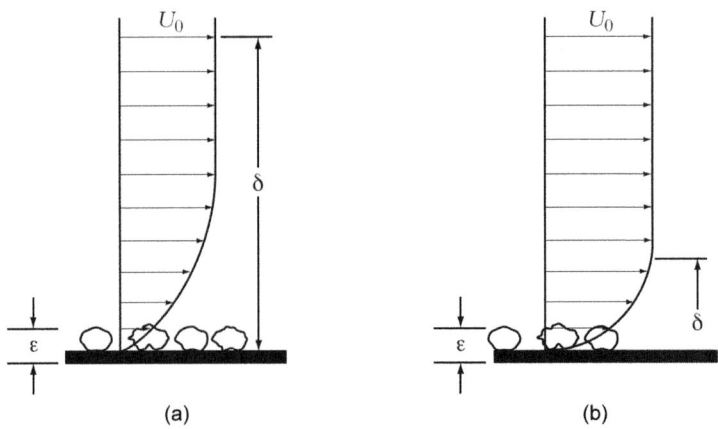

(a)   (b)

FIGURE 1.12 Schematic view of the boundary layer. The wall surface is considered smooth (a); it should be considered rough (b) despite the identical physical size of roughness.

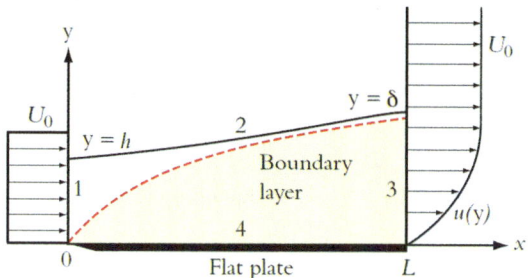

FIGURE 1.13 Control volume used in the derivation of the momentum integral equation.

$$\rho U_o h = \rho \int_0^{\delta(x)} u\, dy \tag{1.33}$$

The only force acting on the control volume is the skin friction drag over the flat plate since the pressure gradient over a flat plate is zero. Accordingly, the skin friction drag must be equal to the change in the momentum flux within the control volume. Therefore, the skin friction drag over the flat plate is given by

$$D(x) = \rho U_o^2 h - \rho \int_0^{\delta(x)} u^2 dy \tag{1.34}$$

By substituting Equation (1.33) into (1.34), we obtain

$$D(x) = \rho \int_0^{\delta(x)} U_o u\, dy - \rho \int_0^{\delta(x)} u^2 dy$$

$$= \rho \int_0^{\delta(x)} u(U_o - u)\, dy \tag{1.35}$$

Introducing the *momentum thickness* θ, a measure of the momentum loss as a result of the boundary layer growth, which is given by:

$$\theta = \int_0^{\delta} \frac{u}{U_o}\left(1 - \frac{u}{U_o}\right) dy \tag{1.36}$$

Equation (1.35) can be written as:

$$D(x) = \rho U_o^2 \theta \tag{1.37}$$

Therefore, the momentum thickness θ represents the skin friction drag D over a flat plate. A differential form of this equation can be given by:

$$\tau_w(x) = \rho U_o^2 \frac{d\theta}{dx} \tag{1.38}$$

if we note that

$$D(x) = \int_0^x \tau_{\mathrm{w}}(x)\,dx \tag{1.39}$$

Using the **skin friction coefficient**, $C_{\mathrm{f}} = \dfrac{\tau_{\mathrm{w}}}{\dfrac{1}{2}\rho U_{\mathrm{o}}^2}$, Equation (1.38) can be written as

$$C_{\mathrm{f}} = 2\frac{d\theta}{dx} \tag{1.40}$$

This is the **Kármán's momentum integral equation**, which is valid for both laminar and turbulent boundary layers as long as the pressure gradient is zero. The Kármán's momentum integral equation indicates that the local skin friction coefficient is exactly twice the streamwise change of the momentum thickness.

Another important parameter in the boundary-layer theory is the **displacement thickness δ^*** defined as

$$\delta^* = \int_0^\delta \left(1 - \frac{u}{U_{\mathrm{o}}}\right) dy \quad \text{or} \quad U_{\mathrm{o}}\delta^* = \int_0^\delta \left(U_{\mathrm{o}} - u\right) dy \tag{1.41}$$

The displacement thickness is a measure of mass-flow deficit in the boundary layer due to the non-slip condition of viscous flows. With this concept, we can treat the development of the boundary layer over a wall surface as if the on-coming flow was shifted by the amount of the displacement thickness δ^* (Figure 1.14). Therefore, we can estimate the change in flow rate though a duct, for example, without considering the change in velocity profile.

The ratio of the displacement thickness δ^* to the momentum thickness θ is called the **shape factor H**

$$H = \frac{\delta^*}{\theta} \tag{1.42}$$

This is a good indicator for the flow status, which can be used to check whether the flow is laminar or turbulent. Over a flat plate with zero pressure gradient, for example, the shape factor of the laminar

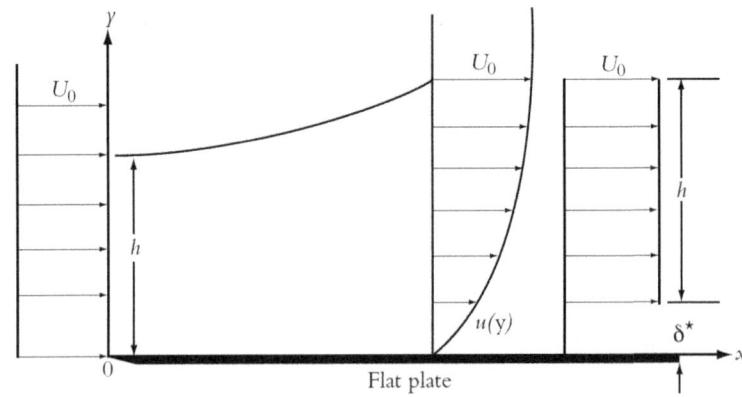

FIGURE 1.14   Simulated effect in the boundary layer using the displacement thickness concept.

boundary layer is $H \approx 2.6$, while the shape factor of the turbulent boundary layer is $H \approx 1.4$. For the boundary layer under transition, the shape factor takes a value between 1.4 and 2.6.

The shape factor can also be used to find out if it is close to flow separation (to be discussed next in this section). Since the velocity profile will become tall and thin as the flow separation is approached, the shape factor of the boundary layer will increase in value whether it is laminar or turbulent. Usually, this increase in the H value is quite rapid, giving warning that the boundary layer flow is about to detach from the wall.

WORKED EXAMPLE

Obtain the displacement thickness, momentum thickness and the shape factor of the boundary layer when the velocity profile is given by $\dfrac{u}{U_o} = \dfrac{y}{\delta}$ (Figure 1.15)

Changing the variables by setting $\eta = \dfrac{y}{\delta}$, we have

$$\delta^* = \int_0^{\delta}\left(1 - \frac{u}{U_o}\right)dy = \delta\int_0^1\left(1 - \frac{u}{U_o}\right)d\left(\frac{y}{\delta}\right) = \delta\int_0^1\left(1 - \frac{y}{\delta}\right)d\left(\frac{y}{\delta}\right)$$

$$= \delta\int_0^1(1-\eta)d\eta = \frac{1}{2}\delta$$

$$\theta = \int_0^{\delta}\frac{u}{U_o}\left(1 - \frac{u}{U_o}\right)dy = \delta\int_0^1\frac{u}{U_o}\left(1 - \frac{u}{U_o}\right)d\left(\frac{y}{\delta}\right) = \delta\int_0^1\left(\frac{y}{\delta}\right)\left(1 - \frac{y}{\delta}\right)d\left(\frac{y}{\delta}\right)$$

$$= \delta\int_0^1\eta(1-\eta)d\eta = \frac{1}{6}\delta$$

$$H = \frac{\delta^*}{\theta} = 3$$

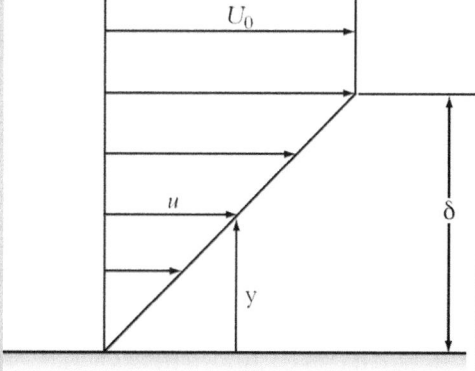

FIGURE 1.15 Linear velocity profile of the boundary layer.

1.3.7   Formulae for the Boundary Layer Development

Laminar:

$$\left.\begin{array}{l} \delta/x = 5.0 \\ \delta^*/x = 1.721 \\ \theta/x = 0.664 \\ C_f = 0.664 \\ C_D = 1.328 \end{array}\right\} \left(U_o x/v\right)^{-1/2} \tag{1.43}$$

Turbulent:

$$\left.\begin{array}{l} \delta/x = 0.37 \\ \delta^*/x = 0.046 \\ \theta/x = 0.036 \\ C_f = 0.058 \\ C_D = 0.0725 \end{array}\right\} \left(U_o x/v\right)^{-1/5} \tag{1.44}$$

The drag coefficient C_D is a non-dimensional drag of a flat plate of length x as defined by $C_D = \dfrac{D}{\dfrac{1}{2}\rho U_o^2 x}$, while the skin–friction coefficient C_f is a non–dimensional, local wall-share stress at x given by $C_f = \dfrac{\tau_w}{\dfrac{1}{2}\rho U_o^2}$.

WORKED EXAMPLE

Obtain the skin friction drag and the corresponding drag coefficient up to the transition point of the boundary layer in the worked example in section 1.3.1.

$$D = \int_0^x \tau_w(x)\,dx$$

$$= \int_0^x \frac{\rho}{2} U_o^2 \cdot C_f(x)\,dx$$

$$= 0.664 \cdot \frac{\rho}{2} U_o^2 \int_0^x \left(\frac{U_0 x}{v}\right)^{-\frac{1}{2}} dx$$

$$= 0.664 \times \frac{1.2}{2} \times 1.5^2 \times \left(\frac{1.5}{1.5 \times 10^{-5}}\right)^{-\frac{1}{2}} \int_0^{10} x^{-\frac{1}{2}}\,dx$$

$$= 0.00275 \times 2\sqrt{10} = 0.0179\,[\text{N}/\text{m}]$$

$$\therefore C_D = \frac{D}{\dfrac{1}{2}\rho U_o^2 L} = \frac{0.0179}{\dfrac{1.2}{2} \times 1.5^2 \times 10} = 0.00133$$

Alternatively, we can use $C_D = 1.328\ R_x^{-\frac{1}{2}}$ to give $C_D = 0.00133$, where $R_x = 10^6$. Therefore, $D = C_D \dfrac{1}{2} \rho U_o^2 L = 0.0180 \left[\text{N}/\text{m}\right]$

We could also compute the skin friction drag from Kármán's momentum integral equation,

$$\theta = 0.664 x\, R_x^{-\frac{1}{2}} = 0.664 \times 10 \times \left(10^6\right)^{-\frac{1}{2}} = 0.00664 \left[\text{m}\right]$$

Therefore,

$$D = \rho U_o^2 \theta = 1.2 \times 1.5^2 \times 0.00664 = 0.0179 \left[\text{N}/\text{m}\right]$$

This gives the drag coefficient of $C_D = 0.00133$.

1.3.8 BOUNDARY-LAYER EQUATIONS

Assuming that the boundary-layer thickness is small as compared to the streamwise length of development, we can derive a special form of the Navier–Stokes equations, called **Prandtl's boundary-layer equations**. In order to do that, we must make a number of assumptions.

1. The *length scale* in the vertical (normal) direction of the boundary layer is much smaller than that of the longitudinal (streamwise) scale. In other words,

$$\Delta y \ll \Delta x \quad \text{or} \quad \frac{\partial}{\partial y} \gg \frac{\partial}{\partial x}$$

2. The *velocity scale* in the vertical direction of the boundary layer is much smaller than that in the longitudinal direction:

$$v \ll u$$

3. The Reynolds number is very large, so that

$$\text{Re} = \frac{uL}{v} \sim \left(\frac{L}{\delta}\right)^2$$

Then, the Navier–Stokes Equations (1.15) and (1.16) will take a very simple form

$$u \frac{\partial u}{\partial x} + v \frac{\partial u}{\partial y} = -\frac{1}{\rho} \frac{dp}{dx} + v \frac{\partial^2 u}{\partial y^2} \tag{1.45}$$

$$0 = -\frac{1}{\rho} \frac{\partial p}{\partial y} \tag{1.46}$$

Equation (1.46) suggests that $p = p(x)$; therefore, the pressure is constant across the boundary layer. Since the pressure gradient of the boundary layer over a flat plate is always zero, that is $\partial p/\partial x = 0$. This means that the pressure is constant everywhere in the boundary layer over a flat plate.

The derivation of the boundary-layer equations can be done using the *order of magnitude* analysis. Here, we set

$x \sim L$ (x is of the same order of magnitude as the plate length L)

$y \sim \delta$ (y is of the same order of magnitude as the boundary layer thickness δ)

$p \sim \rho u^2$ (p is of the same order of magnitude as the dynamic pressure ρu^2)

$u \sim u$

$v \sim (\delta/L) u$ (using the continuity equation, we find that v is of the same order of magnitude as $(\delta/L) u$)

After replacing x, y, p, u and v with L, δ and u, we find that the first viscous term in the Navier–Stokes equations is much smaller than the second viscous term. It should also be noted that all terms in the y-equation become zero, except for the pressure gradient term.

1.3.9 Effect of Pressure Gradient

So far, we have studied only the boundary layers with zero pressure gradient. However, we can extend the boundary-layer theory to cover situations with non-zero pressure gradient. Figure 1.16 shows a typical development of the boundary layer over a curved surface, where a dramatic change in the velocity profile is taking place. Where the pressure gradient is negative, or favourable, that is $\partial p/\partial x < 0$ (from point A to C in Figure 1.16), the lost energy of the boundary layer due to skin friction drag can easily be replenished by the pressure force acting in the flow direction. Therefore, flow separation does not generally take place easily in such a pressure gradient condition. However, when the pressure gradient is positive, or adverse, $\partial p/\partial x > 0$ (from point C onward in Figure 1.16), the pressure force cannot easily replenish the lost momentum. This is because the pressure force acts *against* the boundary layer under an adverse pressure gradient. This leads to a separation of the boundary layer, or simply a **flow separation**, creating a region of **flow reversal** (Figure 1.16).

The flow separation point is defined as a location where the wall-shear stress becomes zero, that is

$$\tau_w \sim \left(\frac{\partial u}{\partial y}\right)_{y=0} = 0 \tag{1.47}$$

Therefore, the velocity gradient at $y = 0$ (at the wall) also becomes zero.

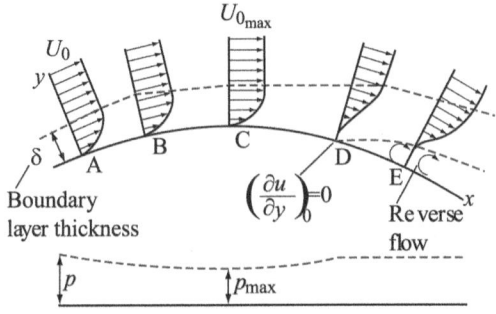

FIGURE 1.16 Development of the boundary layer over a convex surface, showing a retardation that leads to a flow separation with reverse flow. The flow is from left to right.

WORKED EXAMPLE

A sports equipment company is required to design a new swimming costume for the next Olympic Games to help swimmers break the world record in the 200 m freestyle. Answer the following questions assuming that the typical swim speed is 2 m/s.

1. **Suggest how you might design a new swim cap. In doing so, you need to explain the design concept based on fluid mechanical principles.**

2. **A new swimming costume must also cover the entire arms and legs. How should the design concept for these parts of the costume differ from that of a cap? Again, your answer must be based on fluid mechanical principles.**

3. **What considerations should be given to the choice of fabric for the swim cap and the swimming costume? Your answer must be accompanied by clear and sound reasons.**

ANSWERS

1. The Reynolds number of the boundary layer over a swim cap is estimated as

$$R_x = \frac{U_o x}{\nu} = \frac{2.0 \times 0.1}{1.0 \times 10^{-6}} = 2.0 \times 10^5$$

   which is a subcritical Reynolds number. Therefore, the flow is laminar. In this case, there are two possible strategies in designing the swim cap.

   (a) Maintain the laminar flow by making it as smooth as possible, since the skin–friction drag of laminar flows is much lower than that of turbulent flows.

   (b) Promote turbulent flow by *tripping* the boundary layer, which can reduce the pressure drag of the swimmer's head by moving the separation point further downstream.

2. To increase the thrust, the drag on arms and legs must be increased by

   (a) Making the surface of the swimming costume covering the arms and legs rough, thereby increasing the skin friction drag, or

   (b) Making the surface smooth as possible, thereby allowing the laminar flow to separate early.

3. We must choose the fabric of the swimming costume carefully to achieve these objectives: a smooth fabric where we want laminar flow, and a rough fabric in certain parts of the costume to promote turbulence.

LEARNING SUMMARY

By the end of this section, you should have learnt that

- ✓ Viscous fluid does not slip at a solid wall surface. This is called the non-slip condition of flow motion;

- ✓ The boundary layer is a thin fluid layer near a solid wall surface, where the velocity is less than the freestream velocity;

- ✓ The momentum thickness signifies the loss of momentum in the boundary layer due to skin-friction drag;

✓ The displacement thickness is a measure of mass flow deficit in the boundary layer;

✓ The boundary layer equations are a simplified form of the Navier–Stokes equations;

✓ Flow separation occurs over a curved surface when the static pressure increases in the flow direction.

1.4 DRAG ON IMMERSED BODIES

1.4.1 PRESSURE DRAG

While the friction drag D_{fric} results from the *viscous* action of fluids on the body surface, the pressure drag D_{pres} comes from the static pressure distribution around the body, mainly due to boundary-layer separation. The total drag acting on immersed bodies in incompressible flows, therefore, consists of the friction drag and the pressure drag. We can write

$$D_{\text{tot}}\left(\text{total drag}\right) = D_{\text{fric}}\left(\text{friction drag}\right) + D_{\text{pres}}\left(\text{pressure drag}\right) \tag{1.48}$$

The relative importance of D_{pres} to D_{fric} depends on the body shape as well as the Reynolds number. When the immersed bodies are streamlined, the friction drag dominates the total drag. When the non-streamlined bodies (bluff bodies) are placed in a fluid flow, however, the total drag is dominated by the pressure drag, and the contribution of the friction drag is usually negligible.

Drags can be expressed in terms of the non-dimensional drag coefficient, which is given by

$$C_{\text{D}} \equiv \frac{D}{\frac{1}{2}\rho V^2 A} \tag{1.49}$$

where A (which must be specified in quoting C_{D} values) is either the frontal area (for thick bodies, such as motor cars) or the planform area (for long bodies, such as aircraft wings).

In terms of drag coefficient, Equation (1.49) can be written as:

$$C_{\text{Dtot}} = C_{\text{Dfric}} + C_{\text{Dpres}} \tag{1.50}$$

1.4.2 FLOW AROUND A CIRCULAR CYLINDER

For a circular cylinder with radius a and length b, the pressure drag is given by

$$D_{\text{pres}} = \int_0^{2\pi} ab\left(p - p_\infty\right)\cos\theta \, d\theta \tag{1.51}$$

whose drag coefficient is given by

$$\begin{aligned}
C_{\text{Dpres}} &= \int_0^{2\pi} \frac{ab\left(p - p_\infty\right)\cos\theta \, d\theta}{\left(2ab\right)\frac{1}{2}\rho V^2} \\
&= \frac{1}{2}\int_0^{2\pi} \frac{\left(p - p_\infty\right)}{\frac{1}{2}\rho V^2}\cos\theta \, d\theta = \frac{1}{2}\int_0^{2\pi} C_{\text{p}}\cos\theta \, d\theta
\end{aligned} \tag{1.52}$$

where the pressure coefficient C_p is given by

$$C_p = \frac{p - p_\infty}{\frac{1}{2}\rho V^2} \tag{1.53}$$

It should be noted that the frontal area of a circular cylinder ($2ab$) is used to non-dimensionalize the drag to give the pressure drag coefficient. This equation suggests that C_{Dpress} can be obtained by integrating the streamwise component of C_p over the circular cylinder. The cylindrical coordinate system being used in the computation is shown in Figure 1.17, where the angle θ is measured clockwise from the frontal stagnation point.

Figure 1.18 compares the distribution of pressure coefficient C_p over a circular cylinder between the laminar flow and the turbulent flow. It should be noted that both C_p curves are asymmetric with respect to $\theta = 90°$, indicating that the static pressure over the front of the circular cylinder is much higher than that in the rear. The integrated pressure difference between the front and rear surfaces gives the pressure drag acting on the circular cylinder.

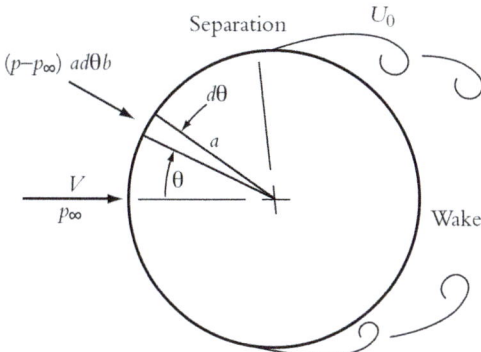

FIGURE 1.17 The coordinate system is used for the integral of static pressure around a circular cylinder to give the pressure drag. Here, p is the static pressure over the cylinder surface and p_∞ is the freestream pressure.

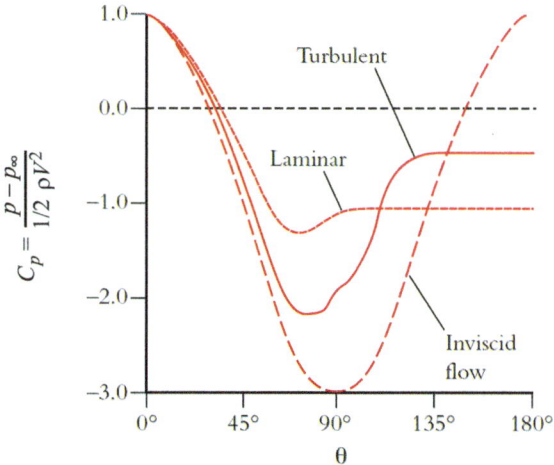

FIGURE 1.18 Non-dimensional pressure distribution over a circular cylinder, where the C_p curve for the laminar and turbulent flows are compared with the solution of inviscid flow.

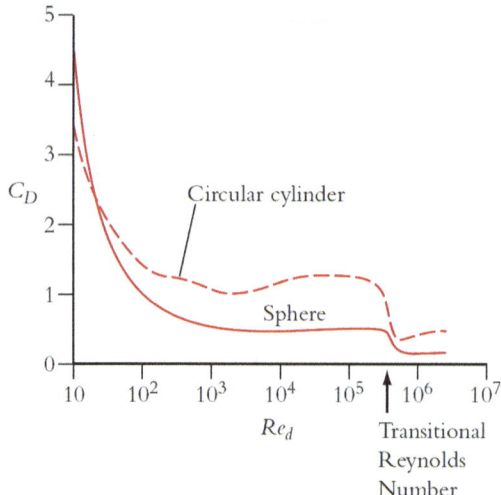

FIGURE 1.19 The drag coefficient of a circular cylinder and sphere as a function of the Reynolds number, showing that the C_D value reduces as the transition takes place.

Figure 1.18 also shows a significant difference in the static pressure distribution between the turbulent flow and the laminar flow. While the static pressure for the laminar flow stays near the minimum value of $C_p = -1.0$ in the rear of the circular cylinder, the turbulent flow recovers to a much greater value of $C_p = -0.4$ after reaching the minimum value of $C_p = -2.1$ at around $\theta = 75°$. This reflects a small C_D value of 0.3 for the turbulent flow as compared to $C_D = 1.2$ for the laminar flow (Figure 1.19). The main reason for the smaller C_D value for turbulent flow is that the flow separation takes place much further downstream due to the greater mixing capability of the turbulent flow. As a result, the wake region in the downstream of turbulent flow separation is narrower than for laminar flow.

However, the inviscid theory gives a symmetric C_p curve (Figure 1.18), suggesting that the drag on a circular cylinder is zero for zero-viscosity fluids. Certainly, this is not a realistic assumption in calculating the drag force on immersed bodies. Indeed, the inviscid theory cannot impose the non-slip condition on the wall; so, there will be no boundary-layer development or flow separation over the immersed bodies.

1.4.3 Drag of Bluff Bodies

As has been previously suggested, the drag coefficient C_D of immersed bodies is a function of the Reynolds number. The drag coefficient is gradually reduced with an increase in the Reynolds number as seen in Figure 1.19. Once the Reynolds number reaches the critical value (Re $\approx 5 \times 10^5$ for circular cylinders and spheres with smooth surface), the drag coefficient will drop suddenly. This is the transition point where the flow around the immersed bodies will become turbulent from laminar flow.

If the wall surface is rough, the transition to turbulences over a circular cylinder takes place at a lower Reynolds number (Figure 1.20), but the drag coefficient C_D for the turbulent flow (after transition) is greater than that with a smooth surface. Figure 1.20 also shows that the critical Reynolds number reduces with an increase in the roughness ratio ε/d. The flow over a sphere is qualitatively similar to that over a circular cylinder.

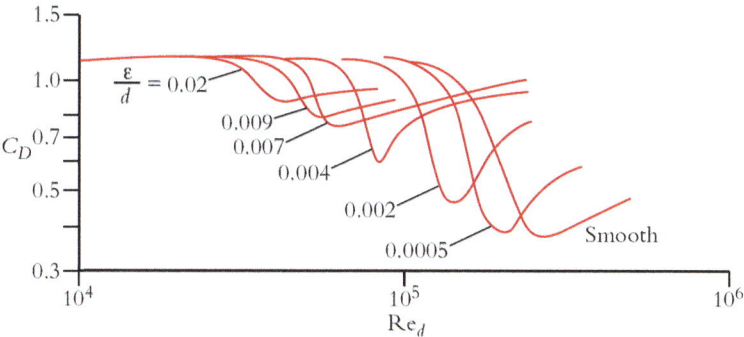

FIGURE 1.20 The drag coefficient of a circular cylinder, showing that the transition takes place at a lower Reynolds number with an increase in the surface roughness.

WORKED EXAMPLE

Obtain the drag force on a baseball of 73 mm diameter at the critical Reynolds number assuming that the flow around the ball is turbulent. The density and kinematic viscosity of air are 1.2 kg/m³ and 1.5 × 10⁻⁵ m²/s, respectively.

From Table 1.2, we find that $C_D \approx 0.2$ for a sphere when the flow is turbulent, where the critical Reynolds number is Re = 3 × 10⁵. Therefore, the drag force on the ball can be obtained by

$$D = C_D \frac{\rho}{2} U^2 A$$

where

$$A = \frac{\pi}{4} d^2 = \frac{\pi}{4} \times (0.073)^2 = 0.0042 \left[m^2 \right]$$

$$U = \text{Re} \frac{v}{d} = \left(3 \times 10^5 \right) \times \frac{1.5 \times 10^{-5}}{0.073} = 61.6 \left[m/s \right]$$

$$\therefore D = 0.2 \times \frac{1.2}{2} \times (61.6)^2 \times 0.0042 = 1.91 \left[N \right]$$

The dimples on a golf ball can reduce the pressure drag by making the ball surface rough. This reduces the transition Reynolds number by artificially forcing (tripping) the boundary layer to turbulent flow at low Reynolds numbers. As a result, the wake becomes narrower as can be seen in Figure 1.21. Although the friction drag is increased in this case, the total drag of the golf ball is reduced. This is because the golf balls are bluff (non-streamlined) bodies, where D_{pres} is much greater than D_{fric}.

Tables 1.1 and 1.2 give the drag coefficient of two-dimensional and three-dimensional bodies, respectively. It should be noted here that the drag coefficient of sharp-edged bodies, such as squares and cubes, is insensitive to the Reynolds number since the flow is always separated at the sharp edges. In other words, the C_D value of sharp-edged bodies remains constant whether the flow is laminar or turbulent as long as the Reynolds number is greater than 10⁴.

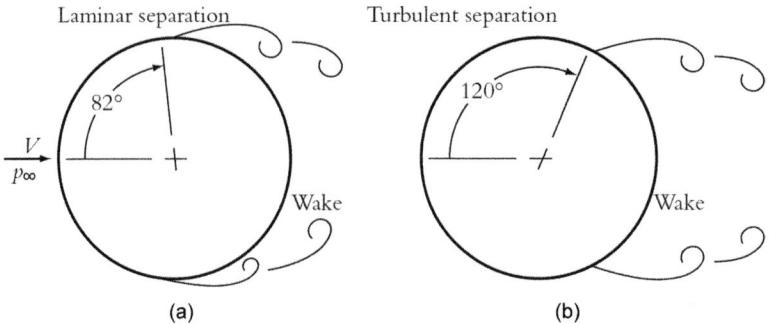

FIGURE 1.21 Comparison of laminar separation with turbulent separation showing that the separation point moves further downstream when the boundary layer becomes turbulent. This reduces the pressure drag by making the wake narrower.

TABLE 1.1
Drag Coefficient of Two-Dimensional Bodies at Re > 10⁴

Shape		C_D	Shape		C_D
Plate:	→│	2.0	Half-cylinder:	→◗	1.2
Square cylinder:	→▢	2.1		→◖	1.7
	→◇	1.6	Equilateral triangle:	→◁	1.6
Half tube:	→(	1.2		→▷	2.0
	→)	2.3		**Laminar**	**Turbulent**
Elliptical cylinder:	1:1 → ◯			1.2	0.3
	2:1 → ⬯			0.6	0.2
	4:1 → ⬭			0.35	0.15
	8:1 → ⬭			0.25	0.1

TABLE 1.2
Drag Coefficient of Three-Dimensional Bodies at Re > 10⁴

Body	L/D	C_D
Cube:		1.07
		0.81
60° cone:		0.5
Disk:		1.17
Cup:		1.4
		0.4
Parachute:		1.2
Circular cylinder:	0.5	1.15
	1	0.90
	2	0.85
	4	0.87
	8	0.99

	Laminar	Turbulent
	0.5	0.2
Ellipsoid:	0.47	0.2
	0.27	0.13
	0.25	0.1
	0.2	0.08

WORKED EXAMPLE

The fork ball is a baseball pitch thrown like a straight ball but with little or no rotation, where the ball initially travels straight but falls sharply as it gets closer to the batter who is standing about 18 m away from the pitcher.

1. Draw a figure showing the drag coefficient of the baseball as a function of the Reynolds number by considering the baseball as a smooth sphere.
2. Obtain the drag force on a baseball with 73 mm diameter at the critical Reynolds number assuming that the flow around the ball is turbulent. The density and kinematic viscosity of air are 1.2 kg/m³ and 1.5 × 10⁻⁵ m²/s, respectively.
3. By what percentage does the drag force change if the flow becomes laminar rather than turbulent?
4. Explain the behaviour of the fork ball as described above using the principles of fluid mechanics.

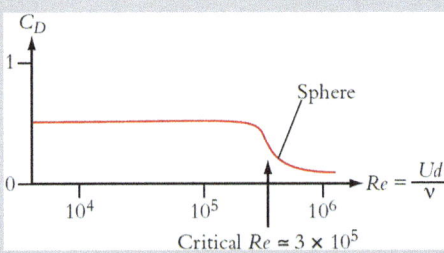

FIGURE 1.22 Drag coefficient of the baseball as a function of the Reynolds number.

5. **Discuss how the pitcher should adjust the delivery of the fork ball for it to remain effective if the ball surface becomes rough during a game.**

ANSWERS

1. see Figure 1.22

2. $C_D \approx 0.2$ for a sphere when the flow is turbulent, where

$$C_D = \frac{D}{\frac{\rho}{2} U^2 A}, A = \frac{\pi d^2}{4} = 0.0042\,\text{m}^2 \quad \text{and} \quad U = \text{Re}\frac{v}{d} = 61.6\,\text{m/s}$$

Therefore,

$$D = 0.2 \times \frac{1.2}{2} \times (61.6)^2 \times 0.0042 = 1.91\,\text{N}$$

3. When the flow around the sphere is laminar, $C_D \approx 0.5$. Therefore, the drag will be increased to 2.5 times (150% increase).

4. The fork ball is the result of reverse transition (from turbulent to laminar rather than usual laminar to turbulent route) of flow around the ball. During this transition, the ball will experience a 150% increase in drag, resulting in a sharp drop near the batter.

5. As shown in Figure 1.20 for a circular cylinder (similar for a sphere), the drag increase will be smaller when the surface is rough. Therefore, the amount of drop will be reduced as a result of ball roughness. The critical Reynolds number for flow transition will be lowered so that the pitcher must throw a fork ball with a lower initial speed for it to be effective.

WORKED EXAMPLE

A man jumped from an airplane with a parachute of 7.3 m in diameter. Assuming that the total mass of the man and parachute is 80 kg, calculate the speed of descent when he reaches terminal velocity.

The drag coefficient of parachute is $C_D = 1.2$ regardless of the Reynolds number as it is a "sharp-edged" body.

The terminal velocity will be reached when the drag of the parachute is balanced by the weight of the parachute and the man, that is $D = W$.

Here,

$$D = C_{\mathrm{D}}\frac{1}{2}\rho A V^2 = 1.2 \times (1.2/2) \times (\pi/4) \times (7.3)^2\, V^2\,[\mathrm{N}]$$

$$W = \mathrm{Mg} = 80 \times 9.8\,[\mathrm{N}]$$

$$\therefore V^2 = \frac{80 \times 9.8}{1.2 \times (1.2/2) \times (\pi/4) \times (7.3)^2} = 26.0\,[\mathrm{m}^2/\mathrm{s}^2]$$

to give $V = 5.1\,[\mathrm{m/s}]$

1.4.4 Streamlining Strategy

An important strategy in reducing pressure drag D_{pres} of immersed bodies is to **streamline** them by shaping the bodies in such a way as to move the flow separation point further downstream. This will effectively reduce the width of wake (the area in the downstream of flow separation), leading to a reduction of the low-pressure region in the rear of the immersed bodies.

Figure 1.23 shows a procedure for streamlining a rectangular cylinder that has sharp corners. The drag can be easily reduced to nearly a half by rounding the front corners of a cylinder, which reduces the drag coefficient C_{D} from 2.0 to 1.0. A more dramatic reduction in drag can be obtained by tapering the rear corners, resulting in a reduction of drag to nearly one seventh of its original. It is surprising to observe that a fully streamlined cylindrical body is equivalent in terms of the total drag with a circular cylinder one tenth of its width. This shows the effectiveness of streamlining strategy in reducing drag by tapering the rear of immersed bodies.

The width of the wake region can be reduced if the flow separation is moved back towards the rear of the body. In practice, however, it is often difficult to do this as it may reduce the capacity (volume) of the vehicle. Instead, trucks can benefit much by attaching a deflector on top of the cab (Figure 1.24), which reduces a large separation region in front of the trailer, leading to a drag reduction of up to 20%.

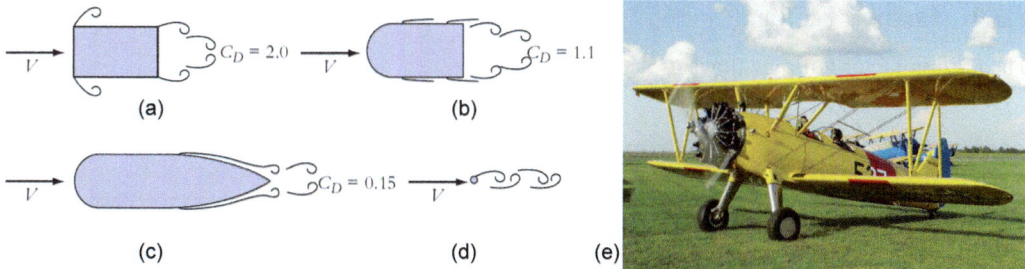

FIGURE 1.23 Reducing the drag coefficient through streamlining strategy. The drag coefficient of a rectangular cylinder (a) can be reduced by tapering the front (b) and rear (c) of the body. A circular cylinder, as shown in (d), has a diameter that is nearly one tenth of the width of the streamlined cylinder (c), yet both have the same drag. With its cylindrical wing struts, no wonder a biplane (e) cannot fly very fast.

FIGURE 1.24 A deflector can reduce the pressure drag of a truck by 20%, by steering the streamlines away from the frontal surface of the trailer. If there is no trailer attached to the cab, however, there will be a large increase in drag.

LEARNING SUMMARY

By the end of this section, you should have learnt:

✓ Pressure drag is a result of the boundary layer separation, where the static pressure difference is created between the front and rear of the bodies;

✓ Drag coefficient of immersed bodies is reduced with an increase in the Reynolds number when the flow is laminar;

✓ Drag coefficient of immersed bodies is suddenly reduced at the critical Reynolds number when the flow becomes turbulent;

✓ Surface roughness will reduce the critical Reynolds number of immersed bodies, thereby reducing their drag at lower Reynolds number;

✓ Streamlining is an effective strategy for reducing drag, where the immersed bodies are rounded at the front and tapered at the rear.

1.5 FLOW THROUGH PIPES AND DUCTS

1.5.1 FRICTION FACTOR

The friction factor f of pipe flows as defined by

$$f = \frac{h_f}{\dfrac{L}{d}\dfrac{V^2}{2g}} \tag{1.54}$$

is a non-dimensional form of the frictional head loss h_f. The friction factor is a function of the Reynolds number $R_d = \dfrac{Vd}{\nu}$ as well as the relative surface roughness $\dfrac{\varepsilon}{d}$, where V is the bulk velocity, d is the pipe diameter, ν is the kinematic viscosity of fluid and ε is the typical surface roughness height.

For laminar flows ($R_d < 2 \times 10^3$) it can be shown that the friction factor is the only function of the Reynolds number, where

$$f = \frac{64}{R_d} \tag{1.55}$$

This is called the Darcy–Weisbach equation for laminar pipe flows. It should be noted that the surface roughness does not affect the friction factor for laminar pipe flows.

For turbulent pipe flows, Colebrook gave the following formula for f, covering a wide range of Reynolds number and surface roughness

$$\frac{1}{\sqrt{f}} = -2.0 \log\left(\frac{\varepsilon/d}{3.7} + \frac{2.51}{R_d\sqrt{f}} \right) \tag{1.56}$$

Although accurate in presenting the friction factor for both transitional and fully turbulent pipe flows ($R_d > 4 \times 10^3$), this formula is difficult to use in practice since the friction factor is not given in a closed form. In other words, iteration is required to obtain the friction factor from this equation for a given Reynolds number and roughness ratio. It is for this reason Moody has presented a chart where the friction factor can be easily read. This is called the **Moody chart**, where the friction factor is given as a function of the Reynolds number R_d and the roughness ratio $\frac{\varepsilon}{d}$ (Figure 1.25a). Following a curve of constant $\frac{\varepsilon}{d}$ value (as shown on the right-hand side of the chart) to meet a constant Reynolds

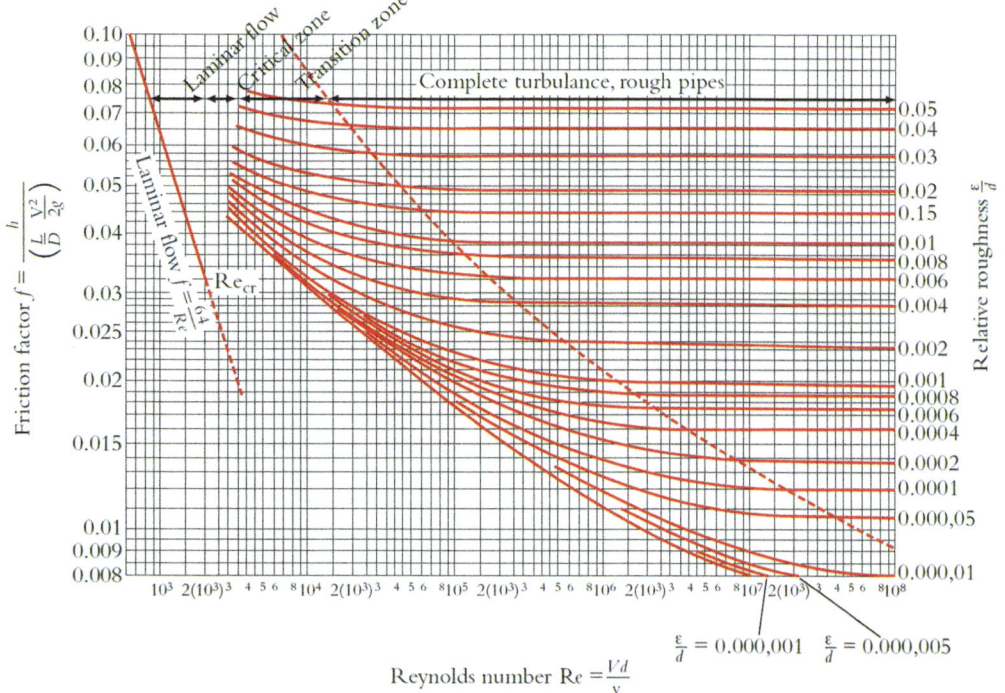

FIGURE 1.25A Moody chart. (Taken from F. M. White, 2008, *Fluid Mechanics*, New York: McGraw Hill and reproduced with permission of the McGraw-Hill Companies.)

number line, one can read off the friction factor on the left-hand side of the chart. Typical surface roughness ε for pipes and ducts, from iron to concrete, can be found in Table 1.3.

TOPIC: ATHEROSCLEROSIS

Atherosclerosis is a disease where fatty materials in blood circulation is built up to form plaque over the artery wall (see Figure 1.25b). Early stage of atherosclerosis can be considered as surface roughness which increases the friction factor (see the Moody chart in Figure 1.25a), thereby increasing the pressure drop in blood vessel. This requires the heart to work harder. As atherosclerosis develops further, there will be much greater pressure drop in blood vessels as the plaque becomes minor losses, such as sudden contraction and expansion (Figure 1.30). This will severely reduce the blood flow to the brain, causing stroke. Plaque can also break off the artery surface due to high shear stress and flow disturbance due to plaque formation. This will cut off the blood supply to the heart causing heart attack.

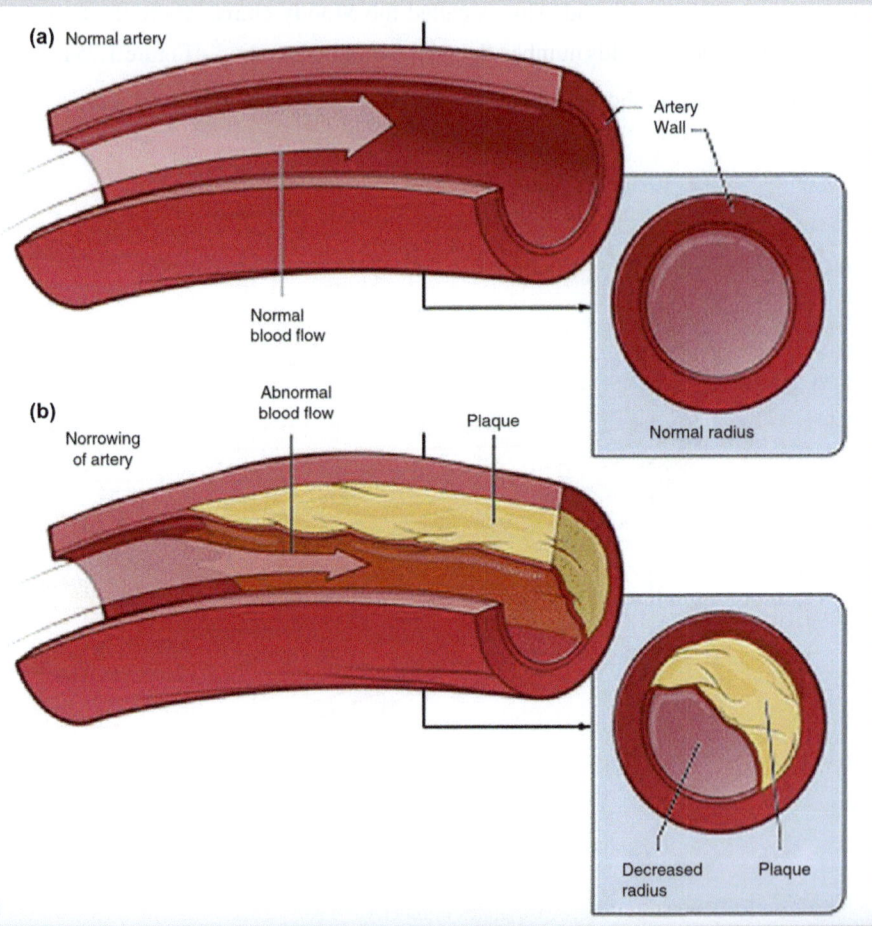

FIGURE 1.25B   Formation of atherosclerosis in which plaque is built up inside arteries restricting the blood flow. https://www.nhlbi.nih.gov/health-topics/atherosclerosis.

TABLE 1.3

Typical Surface Roughness in Pipe and Channel Flows

Material	Condition	ε mm	Uncertainty, %
Steel	Sheet metal, new	0.05	±60
	Stainless, new	0.002	±50
	Commercial, new	0.046	±30
Iron	Cast, new	0.26	±50
	Wrought, new	0.046	±20
	Galvanized, new	0.15	±40
Brass	Drawn, new	0.002	±50
Plastic	Drawn tubing	0.0015	±60
Glass	—	Smooth	
Concrete	Smoothed	0.04	±60
	Rough	2.0	±50
Rubber	Smoothed	0.01	±60
Wood	Stave	0.5	±40

There are two different definitions for the friction factor. The Darcy friction factor is used throughout this textbook, which is defined by

$$f = \frac{h_f}{\dfrac{L}{d}\dfrac{V^2}{2g}} \tag{1.57}$$

Meanwhile, the Fanning friction factor is given by

$$f_F = \frac{h_f}{4\dfrac{L}{d}\dfrac{V^2}{2g}} \tag{1.58}$$

which contains a factor of 4 in its definitions. Since the friction head is given by

$$h_f = \frac{\Delta p}{\rho g}$$

where,

$$\Delta p = \frac{\pi D L}{\dfrac{\pi}{4}d^2}.\tau_w = \frac{4L}{d}.\tau_w$$

Therefore,

$$f = \frac{\dfrac{1}{\rho g}.\dfrac{4L}{d}.\tau_w}{\dfrac{L}{d}\dfrac{V^2}{2g}} = 4.\frac{\tau_w}{\dfrac{\rho}{2}V^2} = 4C_f$$

Therefore, the Darcy friction factor is related to the skin-friction coefficient through

$$f = 4C_f \tag{1.59}$$

Using this relationship, the friction velocity u^* of the pipe flow can be given in terms of friction factor f as

$$f = 8\left(\frac{u^*}{V}\right)^2 \text{ or } u^* = \sqrt{\frac{f}{8}}V \qquad (1.60)$$

If we use the Fanning friction factor, we have $f_F = C_f$ instead.

1.5.2 Minor Losses

Whenever there are changes in velocity magnitude or direction in a pipe or duct system, there will be associated pressure drops, called minor losses. The minor losses are typically found at

- The entrance to the pipe or exit
- Sudden expansion or contraction
- Bends
- Valves.

The minor losses are caused by the internal flow separation as a result of changes in the magnitude or direction of the flow through the pipes or ducts. These are similar to the pressure reductions along the immersed bodies as a result of boundary layer separation. Although they are called minor losses, the pressure drops can be a significant part of the total pressure drop when the pipe or duct has a short straight section.

The minor head loss h_m in a duct or pipe system is expressed by

$$h_m = K \cdot \frac{V^2}{2g} \qquad (1.61)$$

where K is the minor loss coefficient or the K-factor. Therefore, the total head loss through the pipe system is given by summing the frictional head loss h_f and all the minor losses, to give

$$h_{tot} = h_f + \sum h_m \qquad (1.62)$$

The frictional head loss h_f through a pipe or duct is given from Equation (1.54) as

$$h_f = f\frac{L}{d}\frac{V^2}{2g} \qquad (1.63)$$

Examples of flow causing minor losses are shown in Figure 1.26, where the flow direction changes through a circular and square bend. Both flows are separated at the bend, but the degree of flow separation and the turbulence being produced are very different. The turbulence is much greater in a flow through a square bend than through a circular bend. This explains why the minor loss is much greater for the flow around a square bend. Therefore, the use of square bends should be avoided in a pipe or duct system.

The minor loss coefficient for mitre bends including that of the square bend is given in Figure 1.27. There is no Reynolds number dependency on this value since the flow is always separated at the sharp corner. Figure 1.28 shows the minor loss coefficient of circular bends at Re = 10^6 as a function of the bend-radius-to-pipe-diameter ratio $\frac{r}{d}$ and the bend angle θ_b. Here, the minor loss coefficient depends on the Reynolds number; so, a correction factor given in Figure 1.29 should be applied to this value.

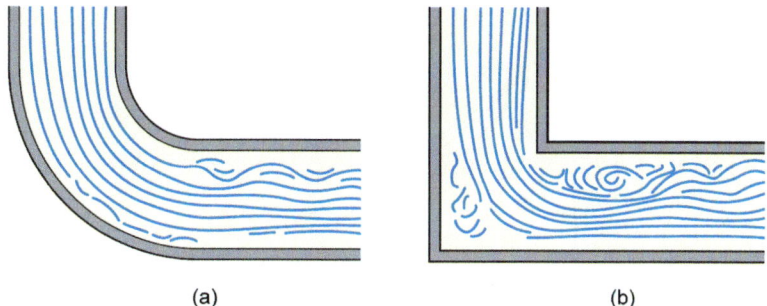

FIGURE 1.26 Flow through a circular (a) and square (b) bends.

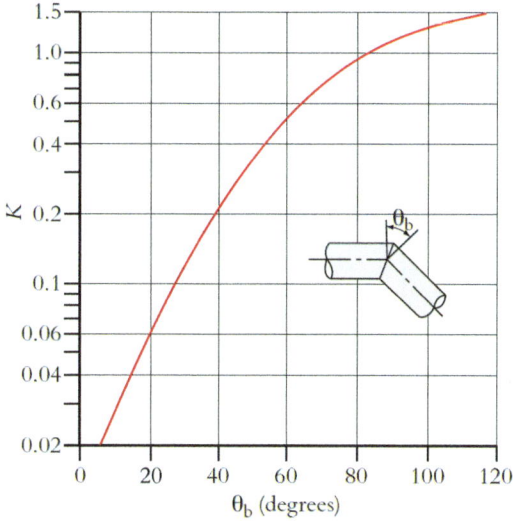

FIGURE 1.27 Mitre bend loss coefficient.

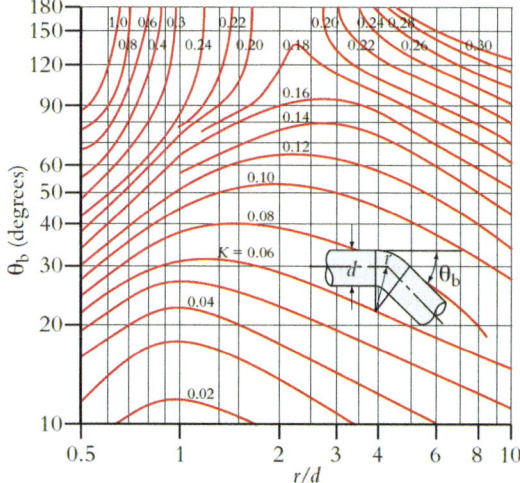

FIGURE 1.28 The loss coefficient of circular bends at Re = 10^6. For other Reynolds numbers, this coefficient must be multiplied by the correction factor given in Figure 1.29

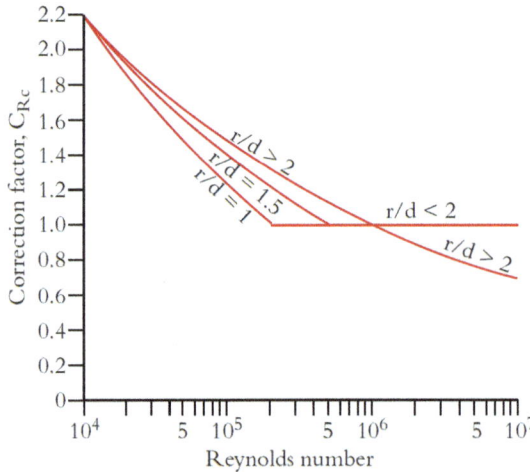

FIGURE 1.29 The Reynolds number correction factor for circular bend loss coefficient.

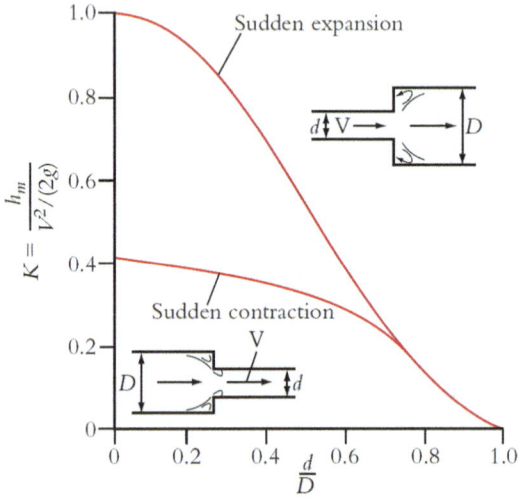

FIGURE 1.30 Minor losses for the flow through a sudden contraction or expansion as a function of $\dfrac{d}{D}$.

Other examples for large minor loss in a pipe and duct system include sudden contraction and expansion, where the flow separation takes place at the junctions (insets in Figure 1.30). The K-factors are a function of the rate of contraction or expansion.

Figure 1.31 illustrates some commercial valve geometries. Typical K-factor for the gate and disk valves is $K \approx 0.2$ when they are fully open, while it is $K \approx 4$ for the globe valve. Valves are the main source of minor losses in a pipe system as one can see in Figure 1.32, showing typical values of the K-factor when valves are partially open. It is shown that the head loss in the pipe or duct system will be increased by more than 100 times when a gate valve is closed by 75%. We must be careful, therefore, in the selection and use of the valves in a pipe and duct system.

There are further sources for minor losses in a pipe system. Figure 1.33 shows the entry losses for different entry geometries. As we would expect, the K-factor for the pipe entry is a function of the relative radius and length of the entry. Note that the K-factor is always unity for a sharp exit from the pipe.

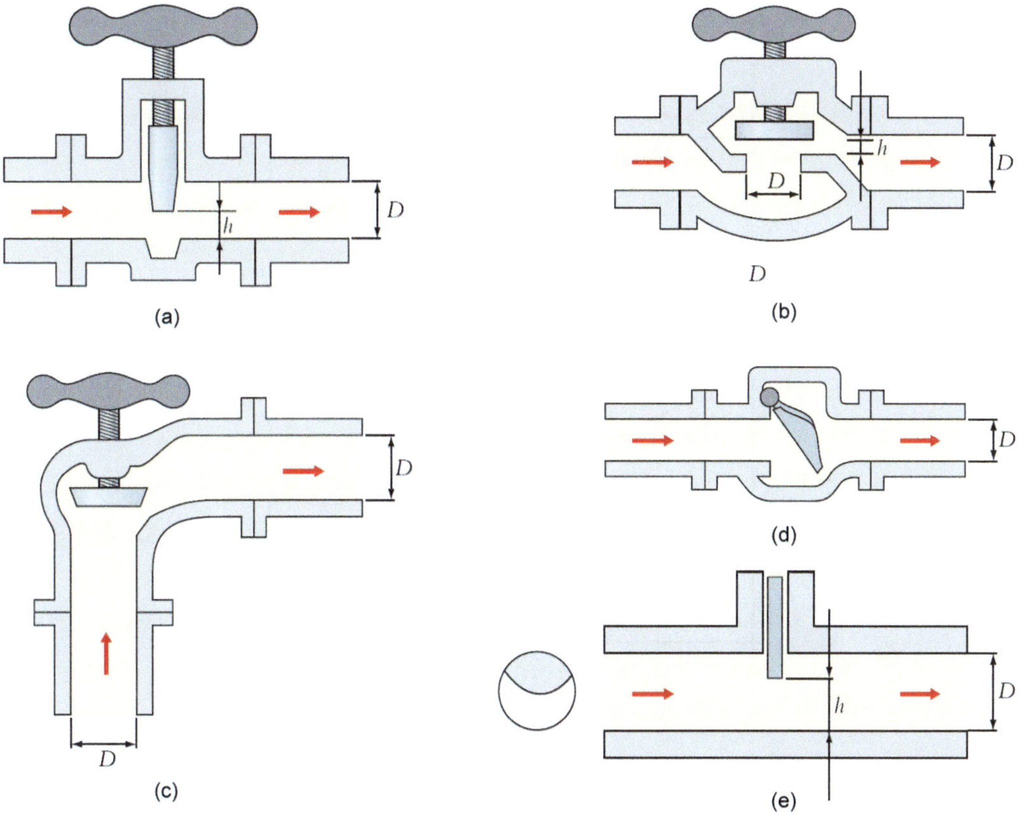

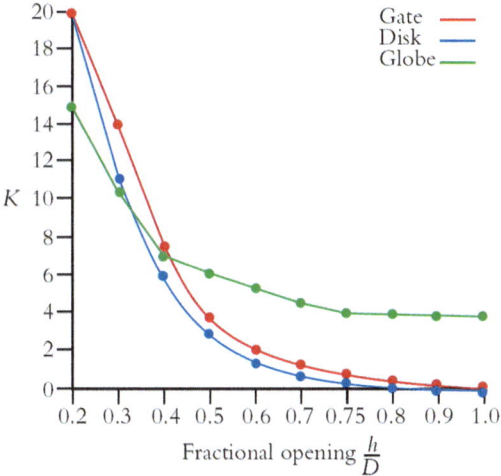

FIGURE 1.31 Commercial valve geometries. (a) Gate valve, (b) globe valve, (c) angle valve, (d) swing-check valve and (e) disk-type gate valve. (F. M. White, 2008, *Fluid Mechanics*, New York: McGraw Hill. Reproduced with permission of The McGraw-Hill Companies.)

FIGURE 1.32 Typical minor losses of valves when they are partially open. (F. M. White, 2008, *Fluid Mechanics*, New York: McGraw Hill. Reproduced with permission of The McGraw-Hill Companies.)

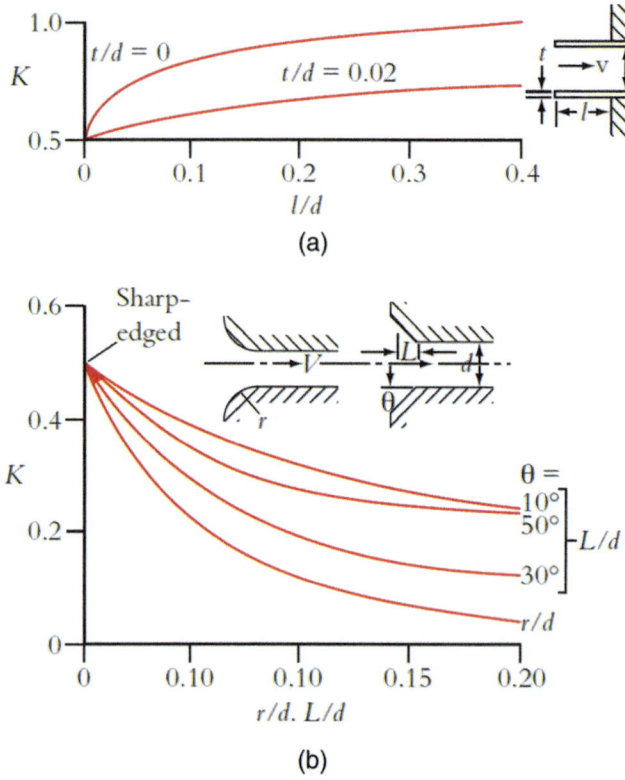

FIGURE 1.33 Entry losses to re-entrant inlets (a). Rounded inlets and bevelled inlets (b). Note that the exit losses are $K = 1$ for all exit shapes. (F. M. White, 2008, *Fluid Mechanics*, New York: McGraw Hill. Reproduced with permission of The McGraw-Hill Companies.)

1.5.3 HYDRAULIC DIAMETER

When the pipes and ducts are *not* circular, we can use the *hydraulic diameter* D_h in place for the diameter of the circular pipe to calculate pipe losses. The hydraulic diameter is defined by

$$D_h = 4 \times \frac{\text{cross-sectional areas}}{\text{wetted perimeter}} \qquad (1.64)$$

With this concept, we can obtain the friction factor of non-circular pipes and ducts using the Moody chart just as we have obtained the friction factor for a circular pipe from it. Here, the Reynolds number and the relative roughness can be defined by $\dfrac{VD_h}{v}$ and $\dfrac{\varepsilon}{D_h}$, respectively, using the hydraulic diameter D_h instead of d, while the frictional head loss is given by

$$h_f = f \frac{L}{D_h} \frac{V^2}{2g} \qquad (1.65)$$

The hydraulic diameter of a circular pipe is its physical diameter; that is $D_h = d$.

WORKED EXAMPLE

Calculate the hydraulic diameter D_h of the following pipes and channels.

Parallel plates, see Figure1.34.

$$D_h = 4 \times \frac{hl}{2l}$$
$$\therefore D_h = 2h$$

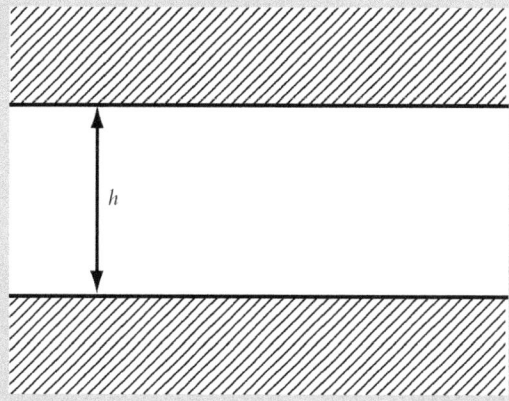

FIGURE 1.34 Parallel plates.

Concentric cylinders, see Figure 1.35.

$$D_h = 4 \times \frac{\frac{\pi}{4}\left(D^2 - d^2\right)}{\pi\left(D + d\right)}$$
$$\therefore D_h = D - d$$

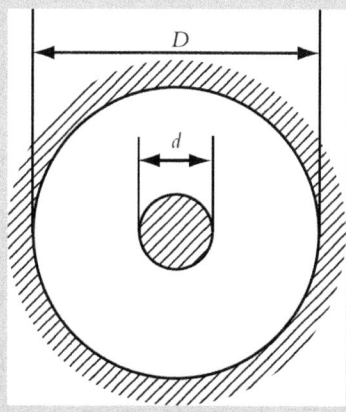

FIGURE 1.35 Concentric cylinders.

Rectangle, see Figure 1.36.

$$D_h = 4 \times \frac{lh}{2(l+h)}$$

$$\therefore D_h = \frac{2lh}{2(l+h)}$$

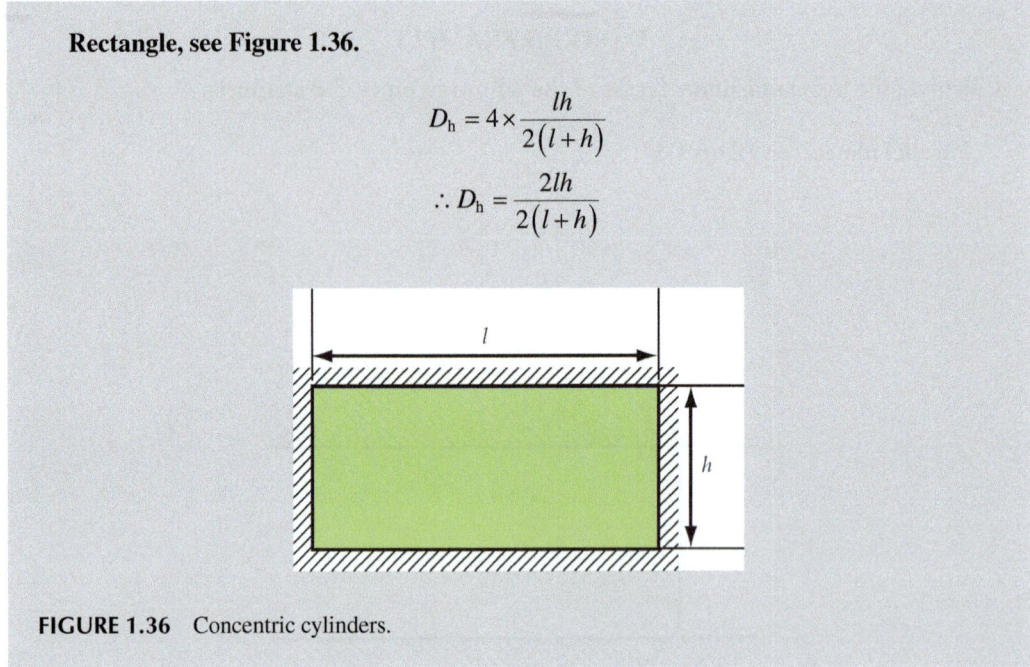

FIGURE 1.36   Concentric cylinders.

1.5.4   SECONDARY FLOWS

Secondary flows can be observed in non-circular pipes and ducts, where the fluid near the corners has a movement across the mean flow direction. The secondary flows are driven by the *turbulent shear stresses*, which act towards the corners of non-circular ducts. As a result, the isovelocity contours are similar in shape to the non-circular pipe or duct cross sections. This is why the hydraulic diameter concept works well for turbulent pipe or duct flows. Figure 1.37 shows the isovelocity contours (a) and secondary motions (b) across a triangular pipe section normal to the mean flow.

Secondary flows in non-circular pipes and ducts are turbulent flow phenomena, so there are no secondary motions for laminar flows. There are other types of secondary flows, which are caused

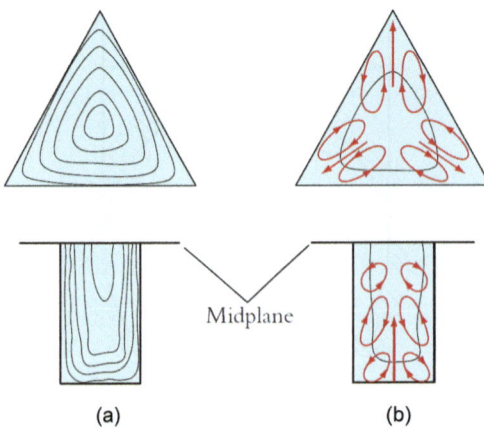

FIGURE 1.37   Isovelocity contours (a) and secondary motions (b) within non-circular ducts. Note that the isovelocity contours are similar to the duct shape. This is caused by the turbulent shear stresses, setting up the flow motion (secondary flow) normal to the mean flow direction.

by the centrifugal force acting on the flow around the circular bend. These secondary flows can be observed even for circular pipes or laminar flows.

WORKED EXAMPLE

A section of pipe is made of sheet steel, whose cross section has the shape of an equilateral triangle.

1. **If one side of the triangle is h, obtain the expression for the hydraulic diameter of this pipe.**

2. **Air of density $\rho = 1.2$ kg/m^3 and viscosity $\mu = 1.8 \times 10^{-5}$ kg/m s at atmospheric pressure is supplied by a blower through 30 m pipe with the cross section of an equilateral triangle of side 22.5 mm. Obtain an expression for the frictional head loss in terms of the volumetric flow rate and the friction factor. You can assume that the pipe section is straight and is placed horizontally.**

3. **If the blower is rated at 746W of output power, obtain the maximum flow rate through the pipe.**

ANSWERS

1. The wetted perimeter of a triangle of side h is $3h$, and its cross-sectional area is given by

$$A_c = \frac{1}{2} \times h \times \left(\frac{h}{2} \tan 60^\circ \right) = \frac{\sqrt{3}}{4} h^2$$

   Therefore,

$$D_h = 4 \times \frac{\dfrac{\sqrt{3}}{4} h^2}{3h} = \frac{\sqrt{3}}{3} h^2$$

2. Since $h = 22.5 \times 10^{-3}$ m, $A_c = 2.19 \times 10^{-4}$ m^2 and $D_h = 1.30 \times 10^{-2}$ m, the friction head loss is given by

$$h_f = f \frac{L}{D_h} \frac{V^2}{2g} = f \frac{L}{D_h} \frac{\left(\dfrac{Q}{A_c} \right)^2}{2g} = \frac{L}{D_h} \frac{A_c^{-2}}{2g} f Q^2$$

$$= \frac{30 \times \left(2.19 \times 10^{-4} \right)^{-2}}{\left(1.30 \times 10^{-2} \right) \times 2 \times 9.8} f Q^2$$

$$= 2.45 \times 10^9 \, f Q^2$$

3. The power is given by

$$P = 746 = \rho g h_f Q = 1.2 \times 9.8 \times \left(2.45 \times 10^9 \right) \left(f Q^2 \right) Q$$

$$f Q^3 = 2.59 \times 10^{-8} \tag{i}$$

Here, the pipe roughness ratio is given by

$$\frac{\varepsilon}{D_h} = \frac{0.05}{13.0} = 3.85 \times 10^{-3} \tag{ii}$$

The Reynolds number is

$$\text{Re} = \frac{\left(\dfrac{Q}{A_c}\right)D_h}{\left(\dfrac{\mu}{\rho}\right)} = \frac{\dfrac{\left(1.30\times10^{-2}\right)}{\left(2.19\times10^{-4}\right)}}{\left(\dfrac{1.8\times10^{-5}}{1.2}\right)}Q = 4.0\times10^6 Q \tag{iii}$$

The flow rate Q can be obtained by iteration through the following process.

1. Assume an appropriate flow rate Q.
2. Compute ε/D_h and Re using Equations (ii) and (iii), respectively.
3. Look up f against ε/D_h and Re using the Moody chart.
4. Check to see if f and Q satisfy Equation (i).
5. If not, change the flow rate Q and repeat the above until it converges.

You should get $Q = 0.0094$ m³/s.

LEARNING SUMMARY

By the end of this section, you should have learnt:

✓ The friction factor of a pipe flow is a function of the Reynolds number and the surface roughness ratio, which can be obtained from the Moody chart;
✓ Whenever there are changes in velocity magnitude or direction in a pipe or duct system, there will be associated pressure drops called minor losses;
✓ The total head loss through the pipe system is obtained by adding the frictional head loss and all the minor losses;
✓ When the pipes and ducts are not circular, we can use the hydraulic diameter D_h to calculate the pipe losses;
✓ The secondary flows in non-circular pipes and ducts are driven by the turbulent-shear stresses which act towards the corners of non-circular ducts.

1.6 DIMENSIONAL ANALYSIS IN FLUID DYNAMICS

1.6.1 NON-DIMENSIONAL NUMBERS

Non-dimensional numbers such as the Reynolds number, Re and the Froude number, Fr are important in understanding the characteristics of the flow as well as in comparing the flow behaviour with others. These non-dimensional quantities can be obtained as a ratio of two different physical quantities. For example, the Reynolds number represents the ratio of inertial force to viscous force,

TABLE 1.4

Non-Dimensional Numbers in Fluid Dynamics

Parameter	Definition	Ratio	Situation
Cavitation number (Euler number)	$Ca = \dfrac{p - p_v}{\rho U^2}$	$\dfrac{\text{Pressure}}{\text{Inertia}}$	Cavitation
Drag coefficient	$C_D = \dfrac{D}{\frac{1}{2}\rho U^2 A}$	$\dfrac{\text{Drag force}}{\text{Dynamic force}}$	Aerodynamics, hydrodynamics
Froude number	$Fr = \dfrac{U^2}{gL}$	$\dfrac{\text{Inertia}}{\text{Gravity}}$	Free-surface flow
Grashof number	$Gr = \dfrac{\beta \Delta T g L^3 \rho^2}{\mu^2}$	$\dfrac{\text{Buoyancy}}{\text{Viscosity}}$	Natural convection
Lift coefficient	$C_L = \dfrac{L}{\frac{1}{2}\rho U^2 A}$	$\dfrac{\text{Lift force}}{\text{Dynamic force}}$	Aerodynamics, hydrodynamics
Mach number	$Ma = \dfrac{U}{a}$	$\dfrac{\text{Flow speed}}{\text{Sound speed}}$	Compressible flow
Pressure coefficient	$C_P = \dfrac{p - p_\infty}{\frac{1}{2}\rho U^2}$	$\dfrac{\text{Stafic pressure}}{\text{Dynamic pressure}}$	Aerodynamics, hydrodynamics
Prandtl number	$Pr = \dfrac{\mu c_p}{k}$	$\dfrac{\text{Dissipation}}{\text{Comduction}}$	Heat convection
Reynolds number	$Re = \dfrac{\rho U L}{\mu}$	$\dfrac{\text{Inertia}}{\text{Viscosity}}$	Viscous flow
Roughness ratio	$\dfrac{\varepsilon}{L}$	$\dfrac{\text{Wall roughness}}{\text{Body length}}$	Turbulent, rough walls
Strouhal number	$St = \dfrac{\omega L}{U}$	$\dfrac{\text{Oscillation}}{\text{Mean speed}}$	Oscillating flow
Weber number	$We = \dfrac{\rho U^2 L}{Y}$	$\dfrac{\text{Inertia}}{\text{Surface tension}}$	Free-surface flow

while the Froude number is given as a ratio of inertial force to gravity force. The definition of non-dimensional numbers appearing in many fluid flows and their physical ratios are summarized in Table 1.4.

The **Strouhal number**, St, is a non-dimensional frequency of flow oscillation, appearing in such a situation as Kármán's vortex street (Figure 1.38). As seen in Figure 1.38, the Strouhal number of vortex shedding from a circular cylinder is nearly constant (St ≈ 0.2) over a large range of the Reynolds number.

TOPIC: VORTEX-SHEDDING FLOW METRE

Flow velocity through a pipe can be found by measuring the frequency of Kármán's vortex street since the Strouhal number of the flow around a bluff body (e.g. a circular cylinder) is nearly constant (St = 0.2) over a wider range of the Reynolds number (see Figure 1.38b). A typical vortex-shedding flow metre consists of a shedding bar spanning across the diameter of a pipe and a sensor to detect the frequency of vortex shedding (see Figure 1.38a).

From the definition of the Strouhal number, the flow velocity U through a pipe is given by

$$U = \frac{\omega L}{\text{St}} = \frac{\omega L}{0.2}$$

where ω is the vortex-shedding frequency and L is the diameter of the shedding bar. This gives the flow velocity U if we can measure the vortex-shedding frequency ω of the flow around a shedding bar.

A vortex-shedding flow metre is robust and is often maintenance free since there is no moving part involved. Therefore, it is used for industrial applications where the access to a pipe for maintenance is limited. On the other hand, the pressure drop in a pipe is quite significant due to the use of a shedding bar (see Table 1.2 for the drag coefficient of bluff bodies).

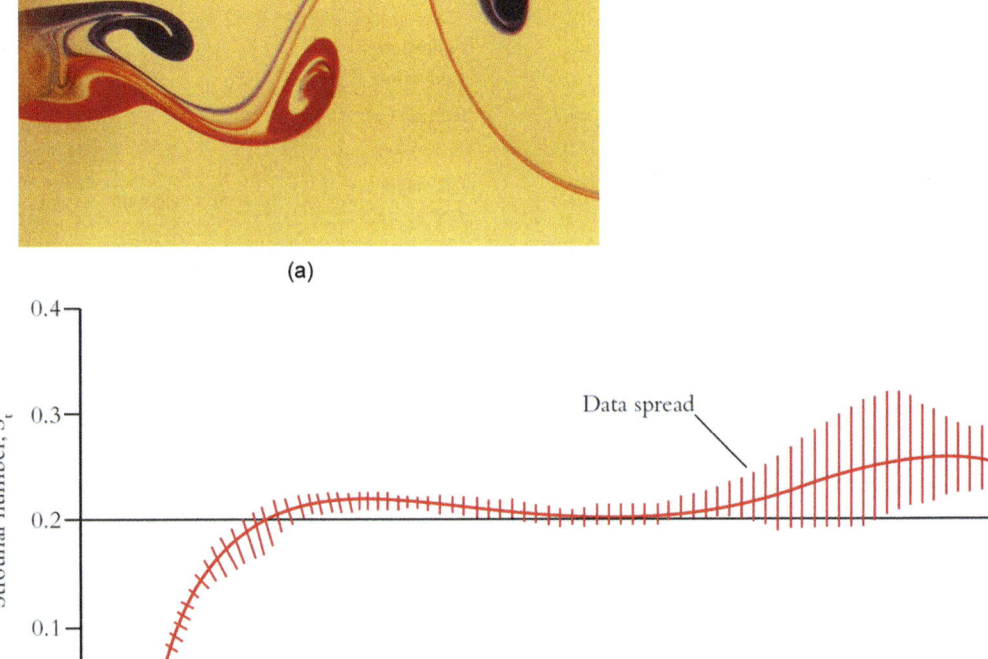

FIGURE 1.38 Vortex shedding from a circular cylinder. (a) Flow visualization; (b) Strouhal number (non-dimensional frequency for vortex shedding) as a function of the Reynolds number.

WORKED EXAMPLE

A chimney of 2 m in diameter and 40 m high is subjected to 22.35 m/s storm winds, where the density and dynamic viscosity of air are 1.225 kg/m³ and 1.78 × 10−5 Pa s, respectively.

1. Obtain the Reynolds number of the flow around the chimney, assuming that the velocity of the storm winds is uniform along the chimney. Is the flow laminar or turbulent at this Reynolds number?
2. Estimate the frequency of the vortex shedding from the chimney.

ANSWERS

1. The Reynolds number is given by

$$\text{Re} = \frac{Ud}{\mu/\rho} = \frac{22.35 \times 2}{1.78 \times 10^{-5}/1.225} = 3.08 \times 10^{6}$$

Therefore, the flow around the chimney is turbulent.

2. From Figure 1.38, we get the value for the Strouhal number as

$$\text{St} = \frac{fd}{U} \approx 0.25$$

Therefore,

$$f = \text{St}\frac{U}{d} = 0.25 \times \frac{22.35}{2} = 2.8\,\text{Hz}$$

Mach number, Ma, is a ratio of the flow speed to the speed of sound, indicating the compressibility effect of the fluid. Figure 1.39 presents the pictures of a flying bullet visualized by the shadowgraph technique, showing the shock waves over the bullet body at a high Mach number.

Weber number, We, indicates a relative magnitude of the inertial force to the surface tension. This non-dimensional number becomes important when there is significant effect of surface tension in the flow phenomena, such as the droplets and in capillary flows. Figure 1.40 shows a sequence of pictures to show the break-up of droplets, where the Weber number plays a significant role in determining its behaviour.

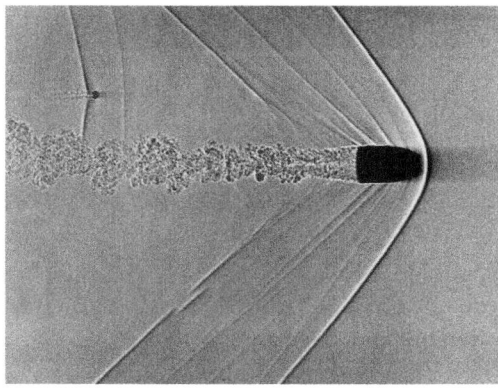

FIGURE 1.39 Effect of Mach number on the shockwave around a bullet.

FIGURE 1.40 The break-up of droplets is governed by the Weber number.

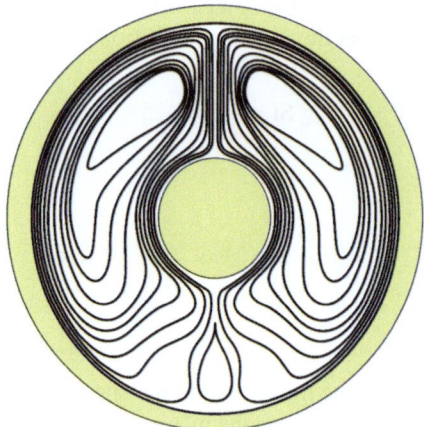

FIGURE 1.41 The streamlines in convection between concentric cylinders, which is governed by the Grashof number.

Grashof number, Gr, is defined as a ratio of the buoyancy force to the viscous force. The Grashof number becomes important when a strong buoyancy effect is considered in a situation such as natural convection. Figure 1.41 shows the pattern of thermal convection between two concentric cylinders, where the outer cylinder is cooled by 14.5 K. This gives the Grashof number of 120,000 based on the gap distance.

Drag coefficient C_D and **lift coefficient** C_L are non-dimensional numbers used for lift-generating bodies, such as wings and propellers. They represent the drag force and the lift force, respectively, acting on the bodies, relative to the dynamic pressure of the free stream. Figure 1.42 shows the lift coefficient (a) and drag coefficient (b) of the NACA 0012 airfoil. As the angle of attack is increased beyond the stall angle (around 15°), there will be a flow separation (c) to reduce the lift coefficient dramatically with an associated increase in the drag coefficient.

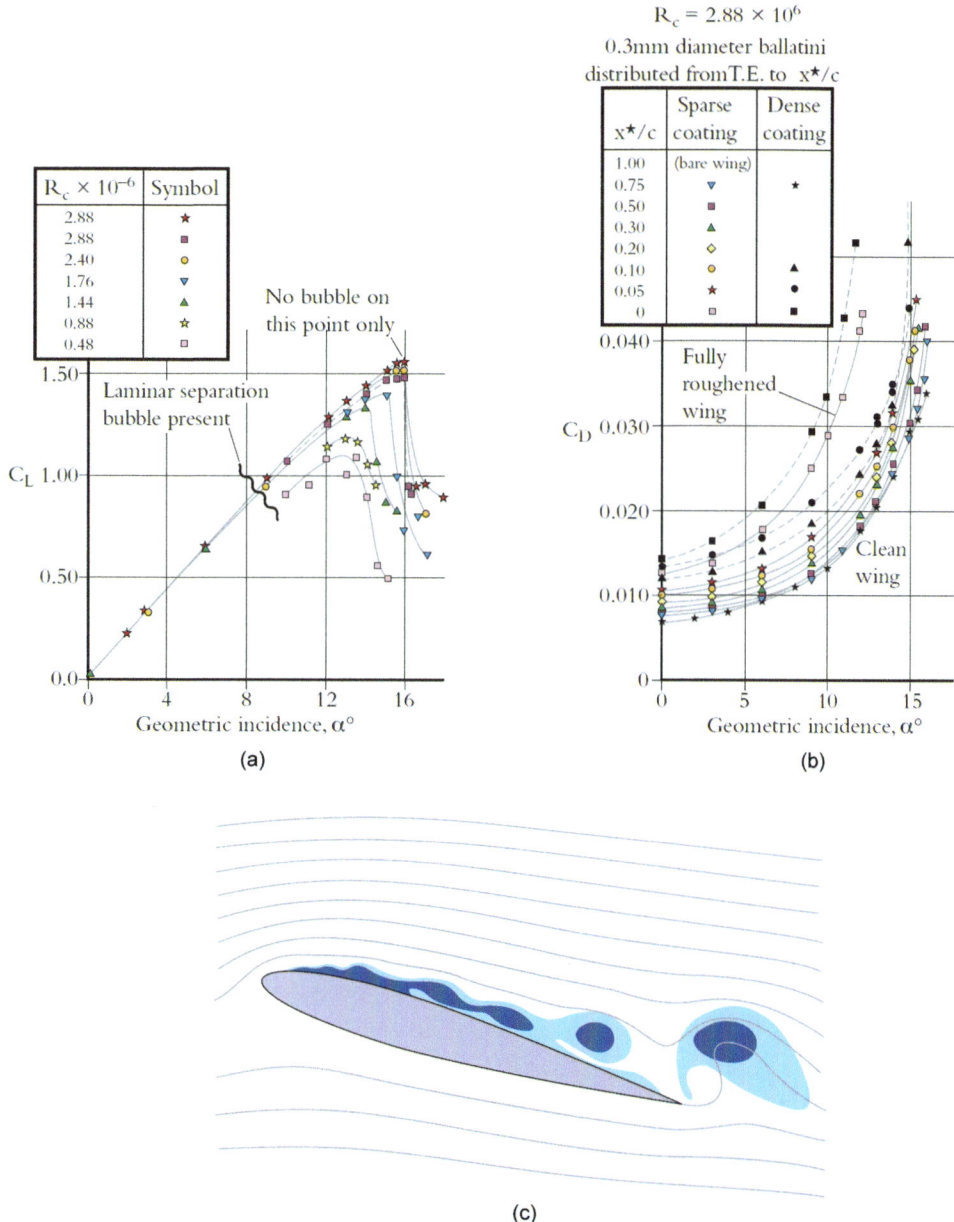

FIGURE 1.42   Lift (a) and drag coefficient (b) of the NACA 0012 airfoil (c).

TOPIC: SPINNING GOLF BALL

A golf ball can be launched at a speed of 80 m/s off the tee with a spin rate of 2,700 rpm when it is hit by a professional with a driver. Figure 1.43 shows the flow field around a spinning golf ball, suggesting that the flow separates asymmetrically at 100° on the top side and at 95° on the bottom. This difference in flow separation point is due to the spinning of the ball, which helps the turbulent boundary layer over the top of the ball to move further downstream despite

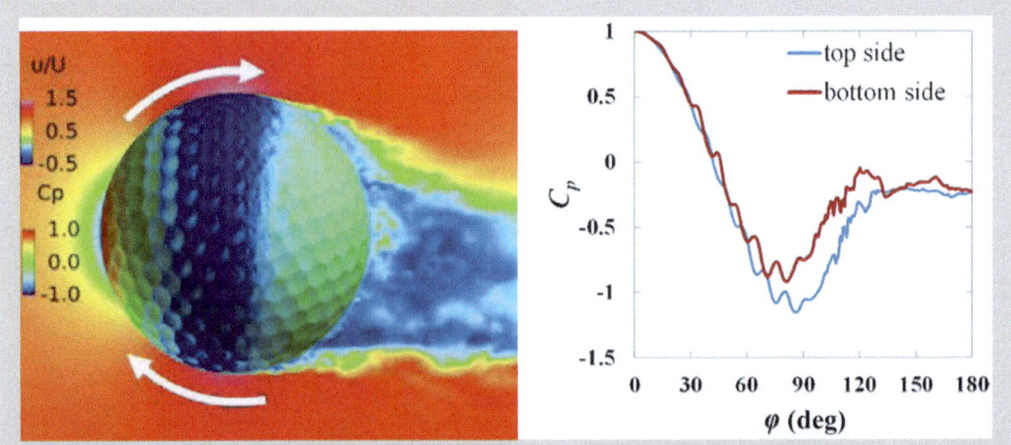

FIGURE 1.43   Flow around a spinning golf ball at Re = 1.1×10^5 and the static pressure distribution. (Jing Li et a. (2017) Numerical investigation of the flow past a rotating golf ball and its comparison with a rotating smooth sphere, *Flow Turbulence Combust*, 99: 837–864 https://doi.org/10.1007/s10494-017-9859-1.)

of the adverse pressure gradient. An opposite effect can be seen at the bottom of the ball, where the spinning effect is against the boundary-layer development. This asymmetric flow separation causes a downward shift of a wake region.

The non-dimensional pressure distribution of a spinning golf ball is also shown in Figure 1.43, which indicates that the static pressure is more negative on the top side of the ball than that on the bottom side. This pressure difference gives a lift force on the spinning golf ball. This is called the Magnus effect. The mechanism of lift generation on a spinning ball is essentially the same as that on an aircraft wing, where the lift force is a product of the flight velocity and the circulation around the body (which is called the Kutta–Joukowski theorem). Only difference is that necessary circulation around an aircraft wing is created naturally (without spinning) when it is set at a certain angle of attack.

TOPIC: RIBLETS

Marine biologists suspected for some time that the fine ridges of sharks may be contributing to their fast-swimming speed, as they found that shark scales form a streamline pattern along the body. The ridge spacing for fast swimming sharks is about 0.06 mm, corresponding to a non-dimensional value of $s^+ = s \cdot u^*/\nu \sim 24$, where s is the spacing, u^* is the friction velocity and ν is the kinematic viscosity of the fluid.

This discovery coincided with NASA's investigation into a new drag-reduction technique to improve fuel consumption of aircrafts, motor vehicles, trains and ships. As a result of this effort, they came up with a passive drag reduction device, called riblets – triangular-shaped microgrooves aligned in the flow direction, which are very much like sharks' scales. Test results indicated that turbulent skin friction can be reduced by 8% with riblets, confirming the expectation of marine biologists.

Although turbulence is associated with random and chaotic motions (see Figure 1.7), there are certain identifiable structures in some turbulent flows. These coherent structures are contributing to the self-sustainability of turbulent boundary layers. Figure 1.44 shows one of such

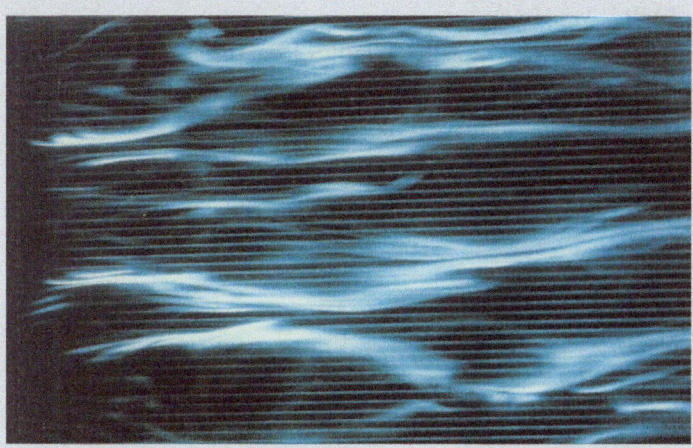

FIGURE 1.44 Quasi-streamwise vortices in the near-wall region of turbulent boundary layer over the riblet surface.

coherent structures in a near-wall region of turbulent boundary layers, called quasi-streamwise vortices (QSVs), which greatly contribute to turbulent energy production. Riblets are seen on the background of QSVs. It is found that riblets interact with QSVs to reduce the energy production, thereby reducing the skin–friction drag of turbulent boundary layers.

1.6.2   Buckingham's Theorem

The Buckingham π theorem, or the π theorem for short, is a systematic method for finding relevant non-dimensional quantities. It applies not only to the problems in fluid mechanics but also to many other areas of engineering and science. Buckingham's theorem consists of two parts. The first gives the number of non-dimensional quantities to account for a given flow problem, and the second determines each of the non-dimensional quantities involved.

1.6.2.1   Part 1

If there are n physical variables describing a physical process which contains m basic dimensions, where $n > m$. Then, this physical process can be described only with $(n - m)$ independent non-dimensional quantities:

$$\pi_1, \pi_{2,\cdots}, \pi_{n-m} \tag{1.66}$$

Here, the physical variables are those to indicate the characteristics of the physical quantity or status (e.g. velocity, pressure, density, etc.), while the basic dimensions are given by M (mass), L (length), T (time) or Θ (temperature).

1.6.2.2   Part 2

In order to find the non-dimensional quantities $\pi_1, \pi_2, \ldots, \pi_{n-m}$, we first pick m physical variables that do not form any of π_i themselves. Then, the product of these variables with one additional variable from $(n - m)$ remaining physical variables will give the non-dimensional quantities.

For example, if there are five physical variables v_1, v_2, v_3, v_4 and v_5 that contain three basic dimensions M, L and T, then there are two $(5 - 3 = 2)$ non-dimensional quantities to describe this physical process (Part 1). Picking v_1, v_2 and v_3, the non-dimensional quantities π_1 and π_2 can be given by

TABLE 1.5

Dimensions of Physical Variables Used in Fluid Dynamics

Quantity	Symbol	Dimensions
Acceleration	dV/dt	LT^{-2}
Angle	θ	None
Angular velocity	ω	T^{-1}
Area	A	L^2
Density	ρ	ML^{-3}
Force	F	MLT^{-2}
Kinematic viscosity	ν	L^2T^{-1}
Length	L	L
Mass flow	m	MT^{-1}
Moment, torque	M	ML^2T^{-2}
Power	P	ML^2T^{-3}
Pressure, stress	p, σ	$ML^{-1}T^{-2}$
Speed of sound	a	LT^{-1}
Strain rate	ε	T^{-1}
Surface tension	Y	MT^{-2}
Temperature	T	K
Velocity	V	LT^{-1}
Viscosity	μ	$ML^{-1}T^{-1}$
Volume	$\mathcal{V}$	L^3
Volume flow	Q	L^3T^{-1}
Work, energy	W, E	ML^2T^{-2}

multiplying v_1, v_2 and v_3 by v_4 and v_5, respectively, in such a way that π_1 and π_2 will become non-dimensional (Part 2) by choosing the powers a, b, c, d, e, f, g and h appropriately.

$$\pi_1 = \left(v_1\right)^a \left(v_2\right)^b \left(v_3\right)^c \left(v_4\right)^d$$

$$\pi_2 = \left(v_1\right)^e \left(v_2\right)^f \left(v_3\right)^g \left(v_5\right)^h$$

It is strongly advisable to go through worked examples carefully to understand Buckingham's theorem, where a detailed procedure to find the number of non-dimensional quantities as well as to determine their non-dimensional forms are given. To start the process, we need to understand the flow problem with a view to identifying relevant physical variables. Table 1.5 lists the physical variables appearing in many flow problems together with their dimensions, which can help apply Buckingham's theorem.

WORKED EXAMPLE

Obtain the non-dimensional quantities for a pipe flow, see Figure 1.45, by considering the following physical variables:

$$Q\left[L^3T^{-1}\right] \ a\left[L\right] \ l\left[L\right] \ \Delta p\left[ML^{-1}T^{-2}\right] \ \mu\left[ML^{-1}T^{-1}\right]$$

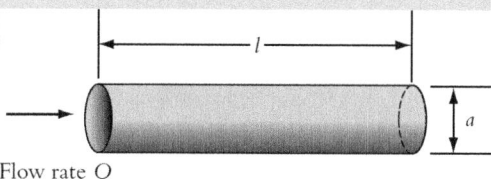

Flow rate Q

FIGURE 1.45 A pipe flow.

There are five ($n = 5$) physical variables Q, a, l, Δp and μ and three ($m = 3$) basic dimensions M, L and T. Therefore, the number of non-dimensional quantities in this flow problem will be two ($n - m = 2$).

In order to find these two non-dimensional quantities π_1 and π_2, we firstly pick the three ($m = 3$) physical variables Q, l and Δp. It is guaranteed that the products of these physical variables do not form any non-dimensional quantities, since only Δp contains the basic dimension M.

Therefore, we set $\pi_1 = Q^a \cdot l^b \cdot \Delta p^c \cdot \mu^d$, $\pi_2 = Q^e \cdot l^f \cdot \Delta p^g \cdot a^h$

We now have to determine a, b, c, d, e, f, g and h to make π_1 and π_2 non-dimensional. Substituting the basic dimensions into above,

$$\pi_1 = \left[L^{3a} \cdot T^{-a} \right] \cdot \left[L^b \right] \cdot \left[M^c L^{-c} T^{-2c} \right] \cdot \left[M^d L^{-d} T^{-d} \right]$$

$$\pi_2 = \left[L^{3e} \cdot T^{-e} \right] \cdot \left[L^f \right] \cdot \left[M^g L^{-g} T^{-2g} \right] \cdot \left[L^h \right]$$

In order for these quantities to be non-dimensional, we must solve the following:

$$\begin{array}{ll} L : 3a + b - c - d = 0 & 3e + f - g + h = 0 \\ T : -a - 2c - d = 0 & -e - 2g = 0 \\ M : c + d = 0 & g = 0 \end{array}$$

The solution is given by

$$g = 0, e = 0, f = -h, c = -d, a = d, b = -3d$$

Therefore, the non-dimensional quantities can be given by

$$\pi_1 = \left[Q \cdot l^{-3} \cdot \Delta p^{-1} \cdot \mu \right]^d \qquad \pi_2 = \left[l \cdot a^{-1} \right]^f$$

where d and f are arbitrary constants. We cannot determine these constants with the number of available equations. We set $d = f = 1$ for simplicity to get

$$\pi_1 = \frac{Q\mu}{\Delta p l^3} \qquad \pi_2 = \frac{l}{a}$$

In other words, π_1 is a non-dimensional flow rate and π_2 is the length to diameter ratio. The pipe flow problem can be described by only two non-dimensional numbers, so that

$$\pi_1 = f_n(\pi_2)$$

to give

$$Q = f_n\left(\frac{l}{a}\right) \cdot \frac{l^3}{\mu} \cdot \Delta p$$

Buckingham's theorem cannot determine the functional form of $\frac{l}{a}$.

1.6.3 DIMENSIONAL ANALYSIS AND MODEL TESTING

To carry out scaled model tests, it is important that we should follow the similarity principle. This basically consists of the geometric similarity and the dynamic similarity. First, it is vital that the geometric similarity is satisfied between the test model and the prototype. Only after that should we proceed with the scaled model test, ensuring that the flow around the prototype and that around a test model are dynamically similar.

To satisfy the geometric similarity, we must ensure that all body dimensions have the same linear scale in all directions. In other words, corresponding points of the prototype and the model must be related by the same linear scale ratio. For example, for a 1/10th scale wind tunnel test of a wing (Figure 1.46), the model must have

- Thickness, width and length that are 1/10th those of the prototype;
- Nose radius that is 1/10th of the prototype;
- 1/10th of the surface roughness as compared to that of the prototype;
- 1/10th the thickness of coating, if that is either painted or lacquered.

In addition, both the prototype and the model must have the same relative posture to the flow direction. This ensures that they have the same angle of attack. The dynamic similarity is guaranteed by making sure that all relevant non-dimensional numbers are the same during the scaled model tests.

When we carry out a wind tunnel test of an aircraft, for example, we need an exact scaled replica of the prototype aircraft in order to fulfil the geometric similarity. Dynamic similarity in the test will be satisfied by carrying out the model test at the same Reynolds number and Mach number as in the prototype flight. If the Mach number is small, and therefore the compressibility effect is negligible, then the only relevant non-dimensional number to consider will be the Reynolds number to satisfy the dynamic similarity (Figure 1.47).

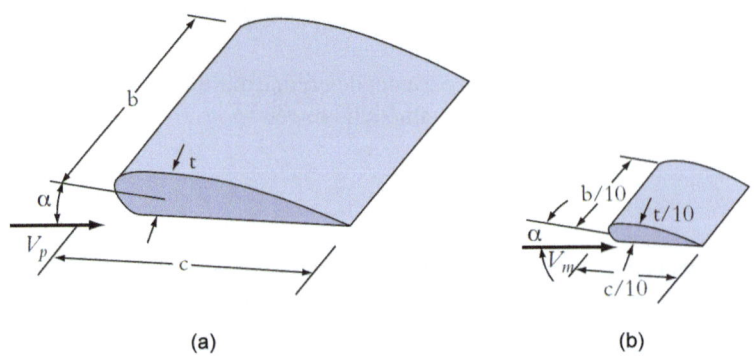

(a)   (b)

FIGURE 1.46 Prototype wing (a) and the 1/10th scale model (b) which are geometrically similar.

FIGURE 1.47   Wind tunnel test of the X-48 blended wing body at NASA Langley Research Center.

WORKED EXAMPLE

Derive the flow speed at which a scaled model experiment should be carried out, when the only relevant non-dimensional quantity is the Reynolds number.

If the Reynolds number Re is the only non-dimensional quantity for the flow being considered, we must ensure that

$$\mathrm{Re_p = Re_m}$$

Therefore,

$$\frac{V_p L_p}{\nu_p} = \frac{V_m L_m}{\nu_m}$$

So that, we have

$$\frac{V_m}{V_p} = \left(\frac{L_p}{L_m}\right)\left(\frac{\nu_m}{\nu_p}\right)$$

In order words, the velocity V_m in the model test must be determined by the scale of the model and the kinematic viscosity ratio.

For example, if the model size is a tenth of the prototype in the same working fluid ($\nu_p = \nu_m$), the test speed V_m must be 10 times faster than the prototype speed V_p in order to ensure the dynamic similarity. This is not always easy to achieve. There are special experimental facilities to attain high Reynolds numbers through a change in fluid density or viscosity.

WORKED EXAMPLE

A test for high-speed boat will be carried out in a towing tank. Here, we must consider the Froude number as the second non-dimensional quantity in addition to the Reynolds number. What are the conditions at which we should carry out the tank test to ensure dynamics similarity?

We must have the following relationships in order to ensure the dynamic similarity:

$$\text{Re}_p = \text{Re}_m \text{ and } \text{Fr}_p = \text{Fr}_m$$

Therefore,

$$\frac{V_p L_p}{v_p} = \text{Re} = \frac{V_m L_m}{v_m} \text{ and } \frac{V_p}{\sqrt{g L_p}} = \text{Fr} = \frac{V_m}{\sqrt{g L_m}}$$

The only way to satisfy these two conditions is

$$\frac{V_m}{V_p} = \sqrt{\frac{L_m}{L_p}} \text{ and } \frac{v_m}{v_p} = \frac{V_m}{V_p}\frac{L_m}{L_p} = \left(\frac{L_m}{L_p}\right)^{\frac{3}{2}}$$

In practice, the first condition can be met relatively easily, but the second condition is difficult to realize.

1.6.4 SIMILARITY RULES IN TURBOMACHINERY

Turbomachinery can be classified into two types – pumps and turbines. Pumps are used to carry out useful work, such as to circulate hot water in a central heating system, by adding energy to the fluid through the rotation of impellors. Turbines, on the other hand, extract energy from the moving fluid through shaft work as the blades or impellors rotate. In this section, we describe the performance of centrifugal pumps (Figure 1.48) through an application of similarity rules.

If we apply the Bernoulli equation at the inlet (1) and outlet (2) of the pump, we have

$$\left(\frac{p}{\rho g} + \frac{V^2}{2g} + z\right)_1 + h_s = \left(\frac{p}{\rho g} + \frac{V^2}{2g} + z\right)_2 + h_f$$

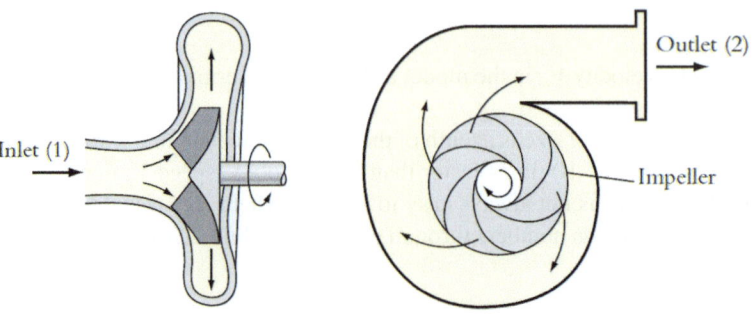

FIGURE 1.48 Centrifugal pump where the flow moves in (1) and exits from (2).

Therefore,

$$\left(\frac{p}{\rho g}+\frac{V^2}{2g}+z\right)_2 - \left(\frac{p}{\rho g}+\frac{V^2}{2g}+z\right)_1 = h_s - h_f \equiv H \tag{1.67}$$

Typically, the inlet and outlet areas of centrifugal pumps are similar ($A_1 \approx A_2$); therefore $V_1^2 \approx V_2^2$. Also, their locations are usually very close to each other ($z_1 \approx z_2$). Therefore,

$$H \approx \frac{\Delta p}{\rho g} = \frac{p_2 - p_1}{\rho g} \tag{1.68}$$

This is called the **net pump head**.

The power delivered to the fluid is called the **water horsepower**, which is given by

$$P_w = \rho g Q H \tag{1.69}$$

while the power required to drive the pump is given by

$$bhp = \omega T \tag{1.70}$$

This is called the **brake horsepower**. The efficiency of the pump is defined as a ratio of the water horsepower to the brake horsepower as

$$\eta = \frac{P_w}{bhp} = \frac{\rho g Q H}{\omega T} \tag{1.71}$$

WORKED EXAMPLE

Find out the non-dimensional quantities describing the performance of a centrifugal pump, see Figure 1.49, where we can assume that there are four ($n = 4$) physical quantities given by

pump head gH	$[L^2T^{-2}]$
flow discharge Q	$[L^3T^{-1}]$
impeller diameter D	$[L]$
Shaft speed n	$[T^{-1}]$

There are only two basic dimensions ($m = 2$) in this problem. From the π theorem, therefore, there must be only two ($n - m = 2$) non-dimensional quantities describing the pump performance.

In order to find these two quantities π_1 and π_2, we first pick two physical quantities D and n. We can easily see that these two *cannot* form a non-dimensional quantity.

Then,

$$\pi_1 = D^a \cdot n^b \cdot Q^c \qquad \pi_2 = D^d \cdot n^e \cdot (gH)^f$$

Substituting the basic dimensions into these, we have

$$\pi_1 = \left[L^a\right]\cdot\left[T^{-b}\right]\cdot\left[L^{3c}\cdot T^{-c}\right] \qquad \pi_2 = \left[L^d\right]\cdot\left[T^{-e}\right]\cdot\left[L^{2f}\cdot T^{-2f}\right]$$

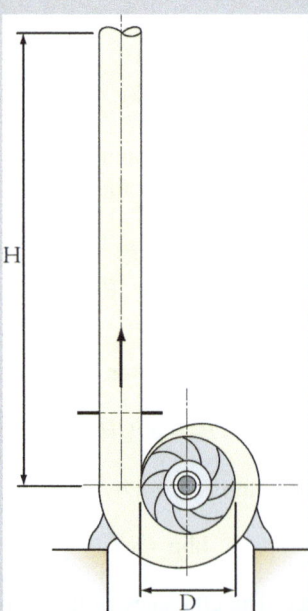

FIGURE 1.49 A centrifugal pump.

After some calculations, we find that $\pi_1 = \dfrac{nD^3}{Q}$ and $\pi_2 = \dfrac{\text{gH}}{n^2 D^2}$. Since the π theorem cannot give a functional form of a non-dimensional quantity, we take the liberty of making an assumption on its form. Conventionally, we use $\dfrac{Q}{nD^3}$ rather then $\dfrac{nD^3}{Q}$. Therefore, we set

$$\pi_1' = \frac{1}{\pi_1} = \frac{Q}{nD^3}$$

The performance of the centrifugal pump can be expressed as $\pi_2 = f_n(\pi_1')$:

$$\frac{\text{gH}}{n^2 D^2} = f_n\left(\frac{Q}{nD^3}\right)$$

or

$$C_\text{H} = f_n(C_\text{Q})$$

where C_H and C_Q are the head coefficient and the capacity coefficient, respectively.

1.6.5 CENTRIFUGAL PUMPS

In general, pump performance is dictated not only by the flow discharge, impeller diameter and shaft speed but also by the fluid density and viscosity as well as the surface roughness of the impellor. We should, therefore, write the pump head gH as a function of all of these variables.

$$\text{gH} = f_1(Q, D, n, \rho, \mu, \varepsilon)$$

from which we can find

$$C_H = g_1\left(C_Q, R_e, \frac{\varepsilon}{D}\right)$$

using Buckingham's theorem. Likewise, the brake horse power (bhp) has the following functional relationship:

$$bhp = f_2\left(Q, D, n, \rho, \mu, \varepsilon\right)$$

from which we can find

$$C_P = g_2\left(C_Q, R_e, \varepsilon / D\right)$$

Here,

$$\text{Capacity coefficient}\ldots C_Q = \frac{Q}{nD^3} \tag{1.72}$$

$$\text{Head coefficient}\ldots C_H = \frac{gH}{n^2 D^2} \tag{1.73}$$

$$\text{Power coefficient}\ldots C_P = \frac{bhp}{\rho n^3 D^5} \tag{1.74}$$

$$\text{Reynolds number}\ldots Re = \frac{\rho n D^2}{\mu} \tag{1.75}$$

$$\text{Roughness ratio}\ldots \frac{\varepsilon}{D} \tag{1.76}$$

And the efficiency of the centrifugal pump η is given by

$$\eta = \frac{\rho \cdot g \cdot H \cdot Q}{bhp} = \frac{C_H \cdot C_Q}{C_P} = g_3\left(C_Q\right)$$

If we assume that the effects of Re and the roughness ratio are small (i.e. the Reynolds number is sufficiently large and the internal pump surface is hydrodynamically smooth), we have

$$C_H = g_1\left(C_Q\right), C_P = g_2\left(C_Q\right)$$

which are exactly the same as in the previous worked example.

Unique performance curves exist for a family of pumps with similar geometries when the pump performance is plotted in terms of non-dimensional quantities such as C_Q, C_H and C_P (Figure 1.50). These save us much effort and time by using only one set of non-dimensional curves to describe the pump performance of an entire family.

For example, when the pumps from the same family are compared at the homologous points, such as at the best efficiency point (BEP), C_Q, C_H and C_P are equal. Therefore,

$$\frac{Q_2}{Q_1} = \frac{n_2}{n_1}\left(\frac{D_2}{D_1}\right)^3 \qquad \frac{H_2}{H_1} = \left(\frac{n_2}{n_1}\right)^2\left(\frac{D_2}{D_1}\right)^2$$

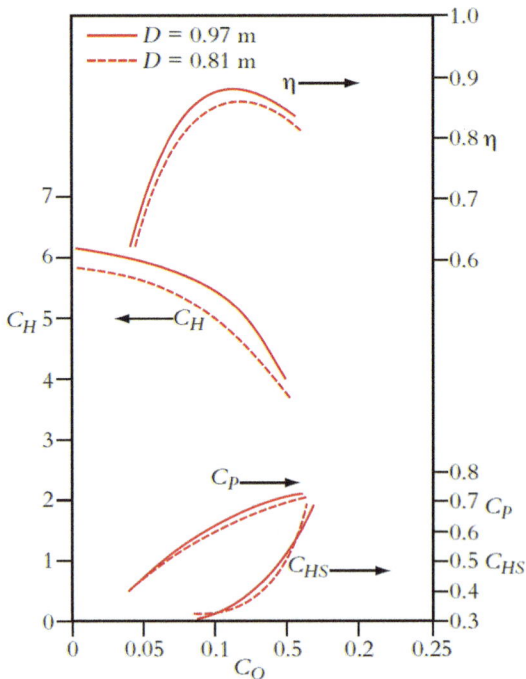

FIGURE 1.50 Performance of centrifugal pumps in non-dimensional quantities. (F. M. White, 2008, *Fluid Mechanics*, New York: McGraw Hill. Reproduced with permission of The McGraw-Hill Companies.)

$$\frac{P_2}{P_1} = \frac{\rho_2}{\rho_1}\left(\frac{n_2}{n_1}\right)^3\left(\frac{D_2}{D_1}\right)^5 \qquad \eta_1 = \eta_2$$

1.6.6 SPECIFIC SPEED

It is useful to know the type of pumps at the early stage of design when only the required flow discharge and pump head are known together with a likely shaft speed. For this purpose, we define the non-dimensional *specific speed* Ns′ as

$$\text{Ns}' = \frac{C_Q^{*\frac{1}{2}}}{C_H^{*\frac{3}{4}}} = \frac{n \cdot Q^{*\frac{1}{2}}}{\left(gH^*\right)^{\frac{3}{4}}}$$

$$\text{where} \quad C_Q^* = \frac{Q^*}{nD^3} \text{ and } C_H^* = \frac{\left(gH\right)^*}{n^2D^2} \qquad\qquad (1.77)$$

where (*) indicates the values at BEP. Pump designers can identify the type of required pumps by calculating the specific speed Ns′ from Q^*, H^* and n without requiring the size of the pump D.

As seen in Figure 1.51, centrifugal pumps with large pump head H and small flow discharge Q have relatively small specific speed Ns′ ≈ 0.10, while axial pumps (small H and large Q) tend to have larger specific speed Ns′ = 0.5 ~ 1.0.

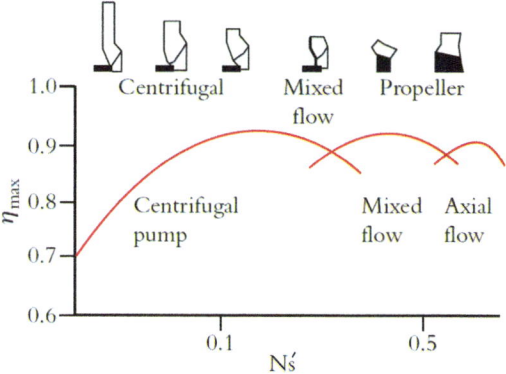

FIGURE 1.51   Non-dimensional specific speed of different type of pumps.

LEARNING SUMMARY

By the end of this section, you should have learnt:

✓ Non-dimensional numbers are important in understanding the characteristics of the flow as well as in comparing the type of flow with others;

✓ Buckingham's theorem gives not only the number of non-dimensional quantities involved but it also determines each non-dimensional quantity;

✓ To carry out model tests, we need to ensure both the geometric and dynamic similarities are satisfied;

✓ One can identify the shape of required pumps by calculating the specific speed without knowing the size of the pump.

REFERENCES

Colebrook, F., 1939, 'Turbulent flow in pipes, with particular reference to the transition between smooth and rough pipe laws', *Journal of the Institution of Civil Engineers*, vol. 11, no. 4, 133–156.

Moody, L.F., 1994, 'Friction factors for pipe flow', *Transactions of the ASME*, vol. 66, no. 8, 671–684.

2 Thermodynamics

Donald Giddings
University of Nottingham, Nottingham, United Kingdom

2.1 INTRODUCTION

2.1.1 TERMS USED

This unit continues from the foundation in thermodynamics laid in the first volume. Its aim is to provide an applied emphasis to the concepts already learned. With global warming, the energy crisis and the ever-increasing demand for heat and work to support human life, engineers have something very important to contribute in thermodynamics. The applications presented here concentrate mainly around power generation from liquid, gas, and solid fuels by combustion to provide heat transfer to produce work via steam power. The first section reviews and refines the information that is relevant from the thermodynamics in volume 1 and the subsequent sections provide analysis techniques for practical engineering applications.

The material covered will address the properties of working fluids as in the previous volume, perfect and semi-perfect gases and steam are of interest. Here, **refrigerants** are introduced along with **combustion reactions** and **product gas mixtures**, with a focus on application. Our attention is on what a **working fluid** can achieve. By manipulating the working fluid, practical ends can be met: **air conditioning** to affect the temperature and humidity of atmospheric air; producing electricity by heating steam with combustion reactions, which drives generators via **steam turbines**; **gas and vapour compressors** increase pressure at a given flow rate of fluid; **combustion** gases are used directly to drive engines; **transferring or insulating heat** is involved in all practical thermodynamic processes, and the modes of transfer and the basic techniques for calculation of heat transfer are presented. This unit considers how classical thermodynamic machines work and how to calculate the relationships between heat and work.

The first section considers *mixtures of gases* seeing how the volume and mass of each gas in a mixture relates to the others. Later sections progress on to the *chemistry of combustion reactions* and then on to the *energy release of the reactions* whether using gas, liquid, or solid fuels. *Moisture* is a life-supporting component of our environment and is contained in the atmosphere – rain, mist, and humidity – and in the reactants and products of combustion. Mixtures of gases with moisture have specific properties that require special treatment. The unit shows how to control atmospheric humidity.

Other characteristics of thermodynamic machines require heat transfer, either to produce work or to achieve desired temperatures. Situations are considered in which steady heat and work transfer occur, (steady meaning processes which do not vary over time). A particularly important thermodynamic system is the **steam power plant** shown in the schematic in Figure 2.1. This unit shows how steam can be used to produce power and how the devices within them relate to each other to produce useful power output. It relies not only on the internal steam cycle but also on the effective combustion energy release and heat transfer to the steam.

The law of conservation of energy states that energy can be neither created nor be destroyed but transformed from one type to another and this is the essence of the **First Law of Thermodynamics** met in Volume 1, and henceforward, referred to only as 'the First Law'. Heat and work are forms of energy – they have the same units as energy. The laws of thermodynamics consider the relationship of heat and work and the effect that the transfer has on the energy of the matter which undergoes the transfer.

It is important to be able to define the **system** and **control volume**. This means to have in mind what physical part of a thermodynamic device is under consideration. In order to deal with it in isolation, a sketch of the control volume within a process is often useful. Figure 2.2 shows a schematic

DOI: 10.1201/9780429319495-2

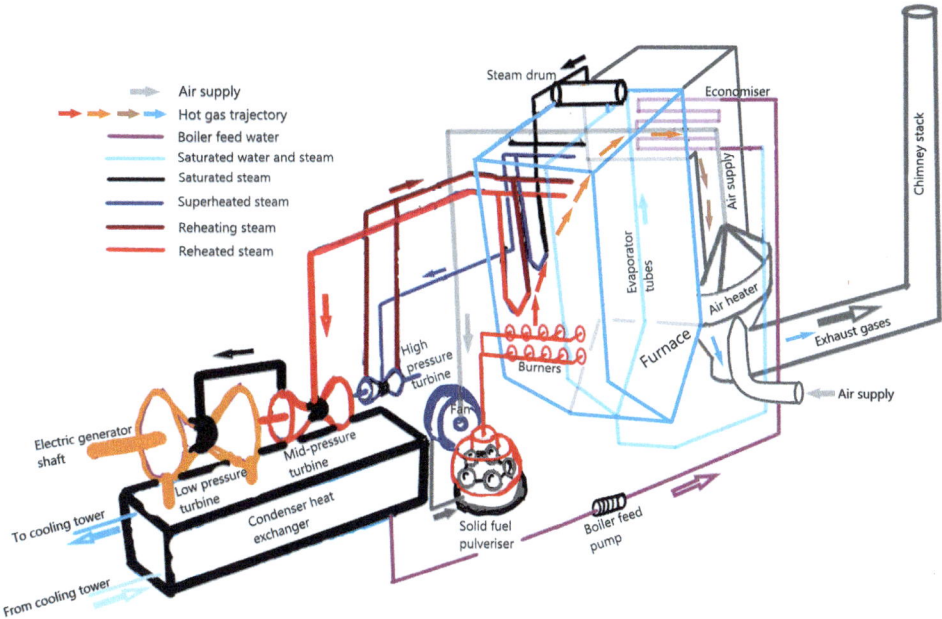

FIGURE 2.1 Schematic representation of the components used in a power generation plant. The furnace mixes air, which is warmed by the exhaust combustion gases in a heat exchanger, with a pulverised fuel that is then fed to burners. The water from the feed pump is heated to boiling point in an economiser heat exchanger prior to evaporating in tubes in the furnace wall. The steam is then separated in a steam drum before being superheated in an array of tubes suspended in the furnace. It then powers the high pressure turbine before being reheated in an array of tubes suspended in the furnace before powering the mid-pressure turbine and then the low pressure turbine. The steam is then condensed before returning to the cycle again.

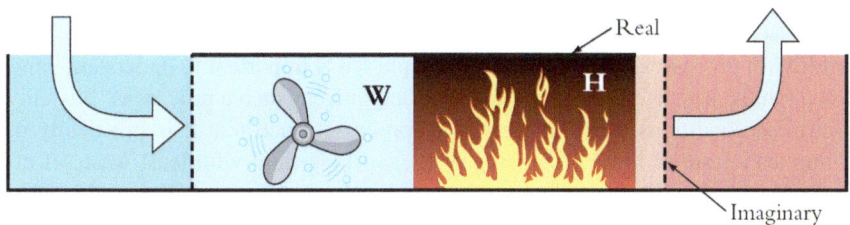

FIGURE 2.2 A control volume in a crude open system thermodynamic device having work and heat, real boundaries at physical walls and imaginary boundaries within the fluid marking the extent of the fluid under consideration.

diagram thermodynamic system in which a cool stream of fluid that enters a tube is accelerated by a propeller and receives energy in the form of work and proceeds to receive energy in the form of heat from a hot source (whether combustion or some other source) before passing out of the system. A control volume within the system describes what happens in a particular part of the system where some thermodynamic action occurs, and it is useful to separate out this space – breaking down the complex system into manageable parts. The **real** solid **boundaries** (perhaps moving or flexible but nonetheless impervious) are usually relatively easy to define. An **imaginary boundary** may be used to separate out internal, connected parts of the working fluid in process, by careful consideration of the separation point required to isolate the region of influence of heat and work and mass transfers. The thermodynamic system has to do with the fluid itself rather than the system containing it. The control volume is often specified by reference to an artefact, but it is important to note the distinction.

In Volume 1, both closed systems – such as pistons in closed cylinders – and open systems were considered, although no specific open systems were discussed. The open system is of significant importance in this Unit, and the First Law in the form of the **SFEE** (steady flow energy equation) dominates the analysis.

The direction of heat and work transfer is important, since maintaining an **energy audit** is vital. It is easy to lose track of an **energy budget** when the system is large and complex with many processes occurring. Even though the subsystems can be separated, for example, considering just one cylinder in a four-cylinder engine, it is important to maintain an **energy inventory** in the same sense. It is logical to consider the transfer of heat and work to the fluid as positive, and that is the convention adopted in this unit. This is the European convention. However, for argument's sake and for completeness it is useful to consider the alternative convention. The American method considers useful work output from the system as positive – a convention which has particular merit since it is usually work achieved by a machine that is of most interest. Note the terms above indicate a deliberate relationship to financial accounting – just like money, energy has to be accounted for properly.

Equilibrium is a concept that has been developed based on the understanding of kinetic energy (KE.) and potential energy (PE). If a system is in unstable equilibrium, then it is likely to lose PE. to KE. In the broader thermodynamic sense, there are several forms of available PE that can be unstable. However, when and only when all are balanced, a state of equilibrium is achieved. For example, after complete combustion when no further oxygen is available and the energy state of the products is extremely adverse to a reverse reaction, or when temperature is the same throughout, or a vapour is in contact with its liquid at constant volume and temperature. A state of equilibrium is one where measurable properties do not change with time.

In order to change from one state to another, a **process** must be followed. The idea of a sequence of processes in a **cycle** enables cyclic machines, the cycle implying a repeating sequence of events physically, but in the thermodynamics sense having a very specific meaning. A cycle is a sequence of processes which alters the state of the working fluid and returns it to its initial state. Cyclic processes are used to produce or make use of external work. **Para-cyclic** machines are very common, in which new materials of the same properties at a point in the cycle are induced from the environment each cycle and ejected to the environment to be recycled at some stage. These are open cyclic machines such as jet engines and internal combustion (IC) engines.

Reversibility of processes is an important principle – it is important to understand how a reversible process might be possible and what the useful output from such a process is. It is only possible in very controlled conditions and with very slow processes such that it is practically unachievable, but it implies a state of ideal processing of the working fluid with least waste of effort – the reason for this will be established when we look at the Second Law of Thermodynamics (henceforward, referred to as 'Second Law') in due course. On a typical **process diagram**, say pressure versus volume of a gas undergoing processes, we can indicate reversible and irreversible processes

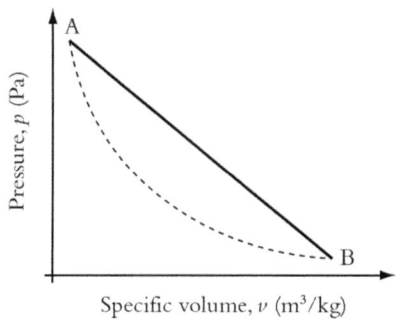

FIGURE 2.3 Schematic graph of pressure versus specific volume, illustrating a reversible process with a solid line, and an irreversible process with a dashed line, on a gas between two states A and B.

(Figure 2.3). This indication convention can be important and should be noted – dotted lines show irreversible processes and solid lines show reversible processes. Most lines drawn for real processes will therefore be dotted, but solid-lined reversible processes are often useful for comparison in order to calculate losses due to irreversibility.

2.1.2 The First Law of Thermodynamics

The **First Law** is very important – telling us about the exchange that happens between work and heat. In words:

> When any closed system is taken through a cycle, the net work delivered to the surroundings is proportional to the net heat taken from the surroundings.

The following formula is a statement of the First Law, and it represents the conservation of energy principle:

$$Q + W = U_2 - U_1 = m(u_2 - u_1)$$

Q [J] is the quantity of heat transferred in the process,
W [J] is the quantity of work produced by the process,
U [J] is the internal energy of the working fluid,
m [kg] is the mass of the working fluid in the system and
u [J/kg] is the specific internal energy of the working fluid.

States 1 and 2 are the initial and final states of the system. Internal energy is the amount of energy possessed by the molecules of the matter in the system in their kinetically energised states. The formula may be expressed in a differential form:

$$dq + dw = du$$

Note that the capital letters indicate bulk quantities, but importantly, lower case indicates that only 1 kg of the working fluid is being considered. As written here, it concerns the conservation of energy in a _closed system_ – that is where there is no new mass coming in. It is valid for both reversible and irreversible processes because it is just concerned with end states and how much heat and work has been transferred to the system. This law is unaffected by what the route between the states is.

The law in words actually talks about the work delivered to the surroundings and the heat taken from them _during a cycle_ – note that a cycle means that the state has returned to its starting point – that is $u_1 = u_2$. This is due to the route of the work, which can be displayed on a state diagram such as the p-v diagram, and there can be a variety of different routes. Similarly, there can be a variety of different routes that result in the same work. However, the route is usually defined by the engine that is acting on the system.

The SFEE is the first law applied to an open system, and is:

$$\dot{Q}_{12} + \dot{W}_{12} = \dot{m}\left(h_2 - h_1 + \frac{C_2^2 - C_1^2}{2} + g(z_2 - z_1) \right)$$

$\dot{Q}_{12}$ [kW] is the rate of heat addition.
$\dot{W}_{12}$ [kW] is the rate of work addition.
h [kJ/kg] is the specific enthalpy of the working fluid.
C [m/s] is the speed of the fluid.
g [m/s²] is gravitational acceleration.
z [m] is the height above a fixed datum point.

Here heat, work, enthalpy, KE and PE all have the same dimension of energy, the Joule, J. **Enthalpy** is the flow process equivalent of internal energy. Since $h = u + pv$, it consists of the internal energy plus the '**flow work**'; the pv part. Mostly in the machines that we consider, changes in velocity and hence KE and in height and hence PE are negligible compared to changes in enthalpy. For example, consider that the change in KE accelerating 1 kg of water from 0 to 10 m/s is 50 J and the PE for rising the same through 10 m is 98.1 J; the enthalpy change of water increasing in temperature from 10°C to 20°C – a relatively modest temperature change – is 41,900 J! Since KE and PE can often be neglected in reality, the SFEE can be reduced to a form which shows a close resemblance to the First Law for the closed system:

$$\dot{Q}_{12} + \dot{W}_{12} = \dot{m}\left(h_2 - h_1\right)$$

2.1.3 PROPERTIES

There are several **properties** to be aware of and the following table shows a selection.

Extensive	Intensive
Volume, V [m^3]	Density, ρ [m^3/kg]
Mass, m [kg]	Pressure, p [N/m^2]
Enthalpy, H [J]	Temperature, T [K]
Entropy, S [J/K]	Specific enthalpy, h [J/kg]

In the right-hand column are **intensive** properties. In the left are **extensive** properties. Intensive properties are the same for any quantity of material being considered. Extensive properties depend on the amount. Intensive properties are mostly expressed per kg of material; this is a **specific** measure of an extensive property. Extensive is indicated by capital symbols and intensive by lower case. In changing from extensive (the whole) to intensive (specific), H becomes h, S becomes s, V becomes v; T and p are inherently intensive and a specific measure cannot be stated. h, s and v are called specific enthalpy etc.

2.1.4 EQUATION OF STATE

The **equation of state** relating p, v and T is:

$$pv = RT,$$

or

$$pV = mRT,$$

where R is the specific gas constant having units J/kgK. Note the alternative use of specific and total volume, v and V. These equations hold for all gases.

Specific heat capacity is defined separately for constant pressure or for constant volume processes, and is directly related to the First Law since specific heat capacity at constant pressure is defined as:

$$c_\mathrm{p} = \left.\frac{\partial h}{\partial T}\right|_p \quad \text{which leads to} \quad h_2 - h_1 = c_\mathrm{p}\left(T_2 - T_1\right)$$

Specific heat capacity at constant volume is defined as:

$$c_v = \left.\frac{\partial u}{\partial T}\right|_v \text{ which leads to } u_2 - u_1 = c_v\left(T_2 - T_1\right).$$

Specific heat capacity is the amount of heat absorbed by 1 kg of gas when it is heated by 1 K at either constant pressure or constant volume. A **semi perfect gas** describes the behaviour of most gases except the very light ones (Helium and Hydrogen) – specific heat capacity varies with temperature, and it is useful to take the average from tables of thermodynamic properties when calculating heat transfer to or from a quantity of gas.

For a **perfect gas**, the relationship between the constants associated with the gas is important and useful. It is important to realise that since the specific gas constant R and the ratio of specific heats, γ (gamma) are usually known, it is possible to work out the specific heat capacities from the quantities above. This is illustrated in the following derivation:

Since $h = u + pv$, $dh = du + d(pv)$
So, $c_p dT = c_v dT + RdT$
And $c_p = c_v + R$
Also, $\gamma = c_p/c_v$

Since γ and R are usually known or determinable, useful relationships with c_p and c_v can be formed, e.g.

$$\gamma c_v = c_v + R \text{ so } c_v = R / \left(\gamma - 1\right)$$

2.1.5 PROCESSES

Processes describe how a working fluid is changed from one state to another. The following list describes processes met in Volume 1, in which c is a constant value:

Adiabatic = no heat transfer, $pv^\gamma = c$, $q = 0$
Isothermal = constant temperature, $pv = c$
Isobaric = constant pressure, $p = c$
Isochoric = constant specific volume, $v = c$
Polytropic = unspecified process type, $pv^n = c$
Isentropic = constant entropy, $pv^\gamma = c$, $q = 0$

Isentropic implies an adiabatic process which is reversible. This is commonly achieved approximately in open system devices such as axial flow turbines and compressors.

The p-v representation can be changed to a relation between p and T or v and T by using the equation of state.

e.g. in $pv^n = c$, substitute $p = mRT/v$ to give $mRT_1 v_1^n/v_1 = mRT_2 v_2^n/v_2$, i.e.

$$T_1 v_1^{n-1} = T_2 v_2^{n-1}$$

Work occurs when a boundary of a system moves against a force – this can be an external atmosphere, or a mechanical device. The formula

$$W = -\int_{1}^{2} pdV$$

W [J] – work done on the working fluid
p [N/m² or Pa] – instantaneous pressure of the working fluid
V [m³] – instantaneous volume of the working fluid

is the work done in a closed system per kg of working fluid, for a reversible process, and relies on a knowledge of pressure at every stage. A reversible process is ideal, so this formula tells us the ideal work assuming no friction, no turbulence and nothing unrecoverable. Work is negative because of the First Law sign convention that we have chosen to use, such that if the gas has expanded the energy has gone from the system in order to work against the exterior. It is useful to know how to calculate the ideal, reversible, work done for a process, to compare with the actual work and hence quantify the losses of energy from the system.

A *p-v* diagram, illustrated in Figure 2.4a, is a **state diagram**, which shows the relationship between pressure and specific volume. The limit of an individual process is bounded by constant pressure and constant volume process lines (horizontal and vertical, respectively) and with curves between representing polytropic processes, including isothermal ($n = 1$) and adiabatic ($n = \gamma$) processes. n is always positive. Other process lines may be possible if the pressure–volume relationship is prescribed by an external mechanical device, such as a spring on a piston. The area under the curve represents the work done for a reversible process. In irreversible processes, work is 'lost' to overcoming friction in the turbulent fluid and other such resistances.

For a **cyclic process** (Figure 2.4b), where the original state is returned to, the area contained in the loop of the cycle is the work output (or input) to the system. The sign depends on the sense of direction on the process path. Considering a clockwise process, net work is produced by the system; positive compression of the gas is indicated in the lower curve, and heat is added to the gas in order to make it work to expand against the external system in the upper curve. The area under the upper curve is greater than the area under the lower one and net work is negative (more work is removed from the gas than is put into it and the working fluid does work on the surrounding machinery). If the cycle

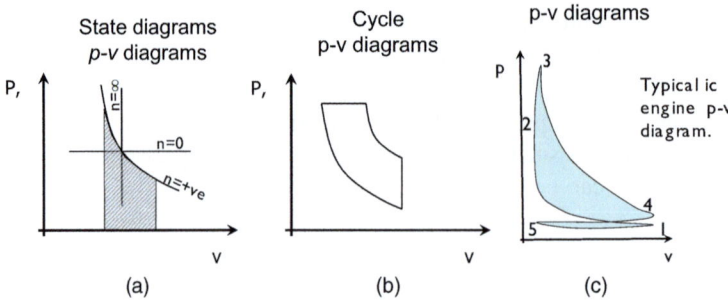

FIGURE 2.4   Schematic graphs of typical state diagrams using pressure and specific volume to indicate what is happening to the working fluid; (a) shows the effect of polytropic index on ideal processes from 0 (constant pressure process) to ∞ (constant volume process) and (b) shows a cyclic process involving four separate processes, (c) shows a real para-cyclic process, the Otto cycle for petrol engines.

was anticlockwise, then the positive work (to compress the gas) is the area under the upper curve; a larger amount of work is put into the system than is taken out and the system must lose heat to do this.

Figure 2.4c shows a typical p-v diagram for a four stroke engine – a polytropic compression from 1 to 2 as the crank drives the piston up. Combustion from 2 to 3 as the piston starts to be driven down by the expanding combustion gases. A polytropic expansion follows from 3 to 4, the pressure at 4 being still above the atmospheric indraft pressure at 1 – the gases still have expansion to do were there room to move. The combustion gases are exhausted from 4 to 5 and a new draught of fresh air is taken in as the crank drives the piston down from 5 to 1. The work done against the surroundings is the shaded area in the p-v graph. By controlling the form of the processes between states this way, a significant amount of work is produced. The upper area is the useful work out; the lower slim area is the pumping work to get exhaust out and fresh air in and is carried by the other cylinders in the cycle and by the flywheel. Note the direction round the cycle is clockwise – net work out.

2.1.6   THE SECOND LAW OF THERMODYNAMICS

It is important to consider the efficiency of processes. In the p-v diagram for a cycle, there is no indication about the effectiveness of the relationships between heat and work. Given that work cycle, how effective was the conversion of heat to work? How much heat was required to get that quantity of work done? The **Second Law** of thermodynamics addresses this issue; in words it can be stated as:

> It is impossible to construct a system, which will operate in a cycle, extract heat from a reservoir, and do an equivalent amount of work on the surroundings.

From the Second Law, there are eight **corollaries**, that is the facts that follow from it and can be proven by logical arguments as described in Volume 1, which can be summarised as follows:

1. Heat can't pass from cold to hot without work.
2. Reversibility implies maximum efficiency.
3. Temperatures of heat source and sink determine maximum efficiency.
4. There is a universal zero value of temperature.
5. If you exchange heat with more than two reservoirs, the efficiency reduces from the maximum possible.
6. $\int dQ/T$ for a complete cycle is zero if reversible, and less than zero if irreversible.
7. $\int dq/T$ between two states is the change of specific (i.e. per kg of the gas) entropy, s.
8. If $q = 0$ then s increases or remains constant.

The first corollary is part of everyday experience and is perhaps the best way to remember the second law for an engineer. The second corollary is not obvious unless the Second Law is considered, except that we intuitively know that friction causes energy to be lost to disorder. The third corollary is very interesting – the extreme temperatures available to a heat engine determine the best that it can perform. If it is possible to raise the upper temperature or lower the sink temperature, then efficiency will increase. The fourth is simple to remember, but was a new concept since most temperature scales before the Second Law depended on the known thermodynamic state points of fluids like water or alcohol. The fifth is reasonably obvious – increasing the number of devices that transfer heat in the system introduces more disorder. The sixth is surprising, and is where we find the difference between reversible and irreversible processes. This sets the scene for the most efficient cycle possible – which we will meet. The seventh is the definition of entropy – this affects the system and the immediate surroundings – that is the level of disorder in the external reservoirs of hot and cold. The eighth is a condition for a cycle – if a cycle involves no heat transfer, it is irreversible if entropy increases and reversible if it does not.

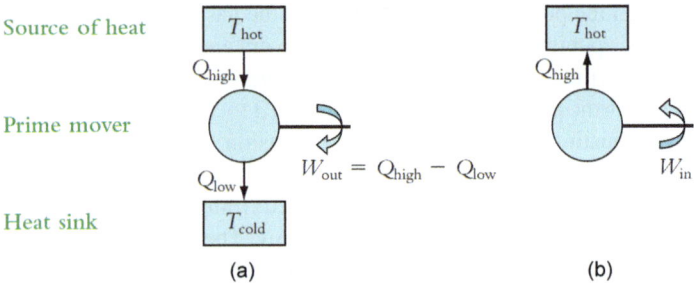

FIGURE 2.5 Schematic representation of heat engines; (a) shows a heat engine designed to produce work out of the working fluid, and (b) shows how the 2nd Law allows work to be entirely converted to heat.

Figure 2.5a shows a symbolic representation of a **heat engine** – having some undefined machinery in the circle, which by virtue of heat exchange with the hot and cold reservoirs, is able to develop work on the surroundings. We define a cycle efficiency as:

$$\eta = \text{Work done} / \text{Heat supplied} = W_{out} / Q_{high} = 1 - Q_{low} / Q_{high} < 1$$

The most important observation here is that in order to produce work output, heat has to be lost to a cold reservoir. To produce work, the engine must exchange heat with both hot and cold. With only one reservoir (Figure 2.5b), work can be done on the system, but not produced by it. Heat is a less useful form of heat transfer than work because all work can be converted to heat without necessarily having to lose any in the exchange, but not all heat can be converted to work; some heat supplied must be lost as heat.

2.1.7 REAL PROCESSES

Figure 2.6a shows various flow devices that are used in thermodynamic machines to relate the working fluid to the outside world. A **compressor** does work on gas by sucking gas axially into a series of disks with aerofoil blades on the periphery. As the gas is compressed, it passes through

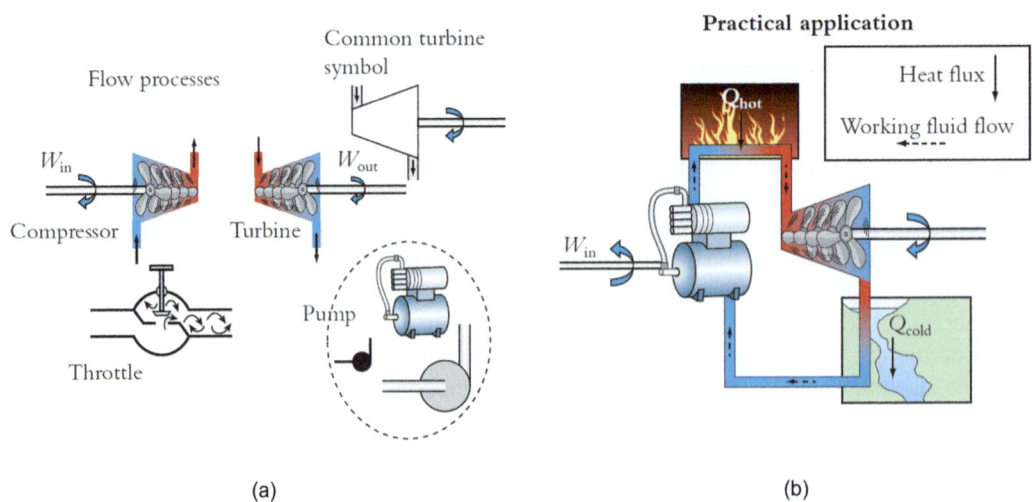

FIGURE 2.6 (a) Flow processes which alter the state of a working fluid in an open process; (b) a practical implementation of the heat engine.

even narrower passages with smaller blades, until it exits at high pressure (HP). These are commonly used in the inlet of a turbofan jet engine. The **turbine** is a compressor in reverse – it takes a hot, highly compressed fluid and expands it through ever widening passages to get work out while dropping the pressure and reducing the energy of the working fluid; a turbine is found at the rear of a jet engine. A common symbol for a compressor or turbine is the quadrilateral shape at the top right of the figure, which reflects the truncated cone shape of the machine due to the convergent tubular casing from the low-pressure (LP) end to the HP end. **Pumps** are used to deliver a volume of pressurised liquid, in which there is relatively little energy increase in the liquid because liquids are virtually incompressible. The **throttle** is a constriction on the flow, the most common example of which is a tap on a kitchen sink. A throttle involves a pressure drop but is usually at a flow rate sufficiently low so that the KE change is insignificant. In this case, the change of enthalpy is negligible – there is an exchange between u and pv internally.

Placing these flow processes in a cycle, as in Figure 2.6b, we see a schematic of a realistic heat engine, the **vapour power cycle**. The hot source of heat and the cold heat sink are external to the system – they transfer heat across pipes that contain the working fluid – a liquid being pushed through the system by a pump is energised by the heating process and converted into vapour, producing work when it passes through a turbine. The fluid is then condensed back to a liquid in the cold sink before repeating the cycle. The elements of this machine are a simplified description of what occurs in a power generation station.

Having a basic heat engine to analyse, we now consider its efficiency. The **Carnot efficiency** is a very important measure. Referring to the heat engine shown in Figure 2.5, the cycle efficiency is the ratio of the work output to the heat input. It can be seen from the heat engine schematic that work out is the difference between the heat in and heat out. This is determined by the temperature of each of the reservoirs, that is:

$$\eta_{\text{Carnot}} = \text{Work done / Heat supplied} = w_{\text{out}} / q_{\text{high}} = 1 - q_{\text{low}} / q_{\text{high}} = 1 - T_{\text{low}} / T_{\text{high}} < 1$$

q_{low} [J/kg] – rate of specific heat supplied to the low-temperature reservoir of heat energy.
q_{high} [J/kg] – rate of speicific heat supplied to the high-temperature reservoir of heat energy.
T [K] – temperature of a reservoir of heat energy.

This shows that a small temperature ratio is better for improving efficiency. The most efficient machine will exchange heat with a sink at absolute zero. For all real cycles, efficiency is lower than the Carnot efficiency because a small amount of work is required to drive the process (the pump in the vapour power cycle) and because heat is added at more than one temperature.

Entropy is a measure of disorder; the state of agitation of a system; the energy allocated to maintaining a state of disorder over the ordered state. This property is defined from the Second Law:

$$dq_{\text{reversible}} = T ds$$

q [J/kg] – specific heat transfer.
T [K] – instantaneous working fluid temperature.
s [J/kgK] – instantaneous specific entropy of the working fluid.

This is the heat transfer equivalent of $dw_{\text{reversible}} = -pdv$. It is the minimum heat transfer required to get from one state to another. This applies for any reversible process undergone by a closed system.

Isentropic efficiency is used to characterise the performance of many open-system machines. Since open-system machines (e.g. axial flow turbines and compressors as shown in Figure 2.7) operate under near adiabatic conditions, and a reversible adiabatic process is the most efficient, it is useful to compare the actual machine work with the ideal machine work. The **isentropic work** output of a turbine is greater than the energy transferred out of the fluid in practice, and the isentropic work

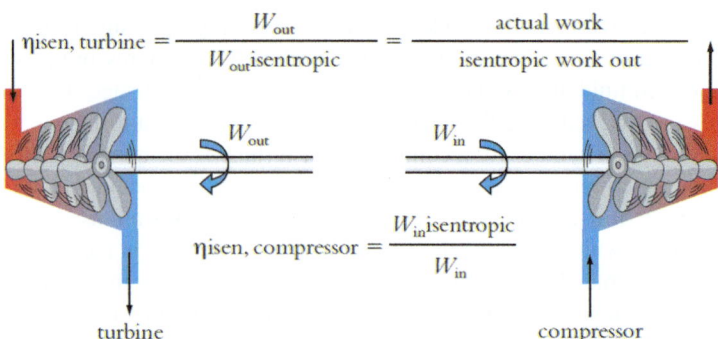

FIGURE 2.7 Isentropic efficiency is defined as the ratio of ideal work and actual work in an axial flow compressor or turbine.

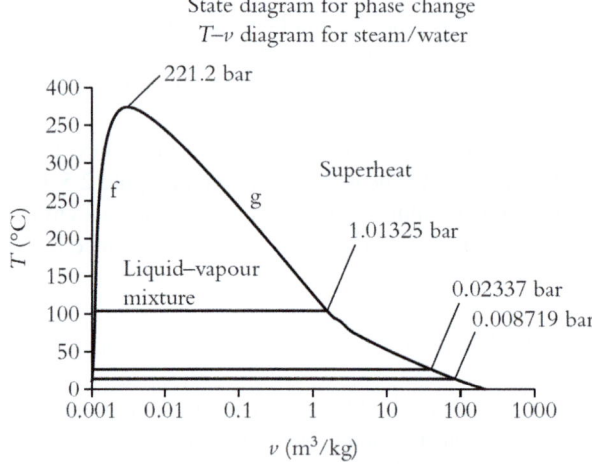

FIGURE 2.8 State diagram for water in terms of pressure and specific volume. The area under the curve is the region where vapour and liquid exist together, the left hand side labelled f indicates the line separating saturated liquid from the mixture region, and the right hand side labelled g indicates saturated vapour, separating superheated vapour from the mixture region. Note the constant pressure lines, which are horizontal in the mixture region, follow approximately the f line and rise sharply in the superheated steam region.

input to a compressor is less than the theoretical input power required to drive the compressor in practice, that is the useful increase in energy to the fluid is less than the isentropic amount. The work is measured in terms of change of working fluid enthalpy and can be determined from the SFEE.

The state diagram in Figure 2.8 describes the properties of steam. It is useful to notice the characteristics of the chart. The standard notation is to use 'f' to indicate saturated liquid and 'g' for saturated vapour. If the pressure is 1.01325 bar (i.e. 101325 Pa) and the temperature is 150°C, the line on the right of the domed region shows that the steam is 'superheated' above its boiling point, and if the temperature is 5°C, then the fluid is sub-cooled, and there will be no vapour. If we reduce the pressure slowly to 0.008719 bar, boiling will occur. Boiling happens on solid surfaces within the fluid and bubbles rise and expand within the fluid. 0.008719 bar is p_S, or saturation pressure, for 5°C; similarly, 0.02337 bar is p_S for 20°C, and 1.01325 bar is the saturation pressure of the commonly known boiling point of water at 100°C. Other points to note from this chart are that the saturated liquid region, to the left part of the curve labelled 'f' in Figure 2.8, is very narrow, that is liquid volume doesn't change a great deal with temperature, certainly up to 100°C at atmospheric pressure. The volume difference between f and g grows as the temperature is reduced. Phase change diagrams

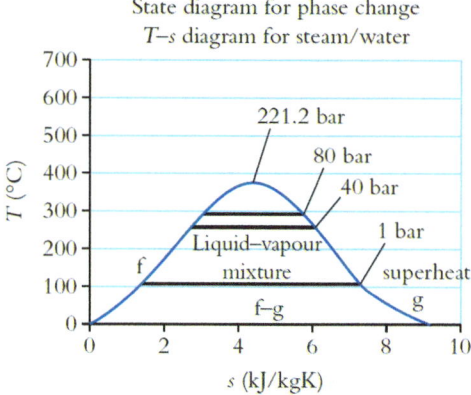

State diagram for phase change
T–s diagram for steam/water

FIGURE 2.9 State diagram for water in terms of temperature and specific entropy. The characteristics of this chart are similar to the *p-v* chart, except the dome is more spread out on the horizontal axis.

like this, regardless of the properties compared, all have this approximate form, with division into the three regions. The charts always have a domed line, underneath which is the vapour–liquid mixture region.

Just as the relation for work depends on *p* and *v*, and the *p-v* chart plot area under a plotted curve represents the work done for a reversible process following that curve, Figure 2.9 shows the chart of *T* and *s*, which represents the heat transfer for a reversible process in a similar way. The heat transfer for the reversible case is useful because it is the limiting, ideal case, heat transfer to which actual heat transfer can be compared. Although it cannot accurately represent the heat transfer for irreversible processes, it shows the ideal heat transfer for a particular type of process, and the characteristics of the real process points can be compared to the actual process points. It is particularly useful for liquid–vapour fluids such as water and refrigerants.

The T-s diagram for steam in Figure 2.9 has constant pressure lines drawn on it. The pressure lines are horizontal through the central vapour–gas mixture region under the dome, and then follow the contour of the saturated liquid side of the dome (the left-hand side from the critical point at the apex). The entropy for water is assigned as zero at 0°C for convenience. By the **Third Law of Thermodynamics**, the absolute value of entropy of any substance is zero at zero *K*. Despite this, use of a reference state is helpful. This diagram is particularly useful for the vapour power topic and for refrigeration. This is very useful regarding the Carnot efficiency, against which the efficiency of all power producing plant is compared. Ideal turbines are considered as constant entropy devices.

LEARNING SUMMARY

✓ There are principal descriptive terms used in the previous volume which are now to be used with an emphasis placed on how they find application in thermodynamic processes. In particular, the ideas of **cycles**, **processes** and **irreversibilities** are useful for considering practical thermodynamic systems.

✓ The **First Law of Thermodynamics** is directly represented by the **SFEE** in most engineering processes because the fluids move through the system. The effects of PE and KE terms are considered to be negligible in many practical thermodynamic processes. The equation then reduces to a form that considers the interchange between enthalpy of the fluid and the energy transferred to it in the forms of heat and work. *This is the most important and straightforward relationship to remember for applied thermodynamics – it is the cornerstone of applied thermodynamic machines.*

✓ The **properties** of thermodynamic working fluids are important in determining thermodynamic state, and the **equation of state** for **perfect** and **semi-perfect gases** is involved in most calculations involving machines where a gas is the working fluid.

✓ Thermodynamic processes change the state of a working fluid, and **cyclic processes** and heat and work transfers in a series of processes are commonly used to produce heat and work.

✓ The **work** done **on** the fluid in a closed process is $W = -\int_1^2 pdV$ for a *reversible* process.

✓ **State diagrams** and the representation of processes on them indicate what is going on in heat engines.

✓ The **Second Law of Thermodynamics** describes *how* heat and work can and cannot be converted in a thermodynamic machine.

✓ The eight **corollaries** of the Second Law are principles of real thermodynamic processes and demonstrate pragmatic observations from experience relating to the Second Law.

✓ **Cycle efficiency** describes how well a machine takes in-going energy and converts it to useful output.

✓ The practical components of real thermodynamic machines are **turbines** for work out, **pumps** and **compressors** for work in and **throttles** for reducing pressure.

✓ The **Carnot efficiency** is the measure of maximum achievable performance of a heat engine.

✓ **Isentropic efficiency** is a comparison of the real process power output to the ideal, reversible process power output.

2.2   AIR CONDITIONING

Air conditioning is mainly concerned with controlling the temperature and humidity of atmospheric air. It is important for controlling the comfort of living organisms and for keeping complex electronic circuitry at a suitable climatic condition:

Comfort air conditioning for people.

- People have limited comfort zone due to the requirement to maintain a steady core body temperature of 37°C.
- People produce heat and moisture into atmosphere.
- They heat at ~80 W resting, 120 W office work, up to 400 W physical working.
- They produce sweat at varying rates and 100% humid air during respiration.

Control conditioning for computers.

- Computers have a limited 'comfort' zone requirement for steady core temperature and dry conditions, which are cooler and dryer than for people.

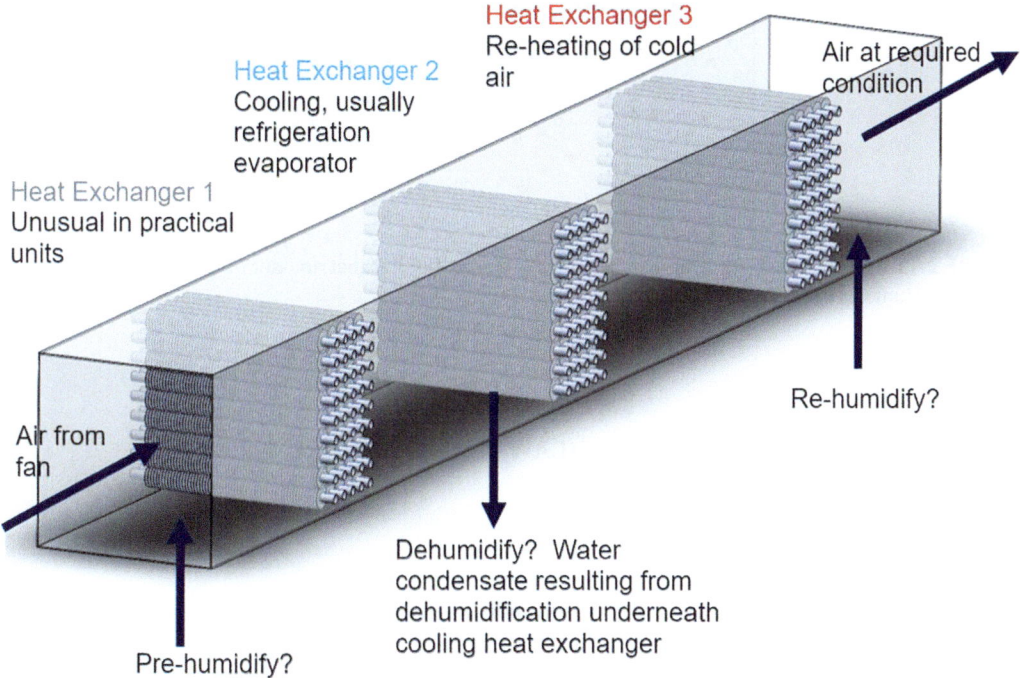

FIGURE 2.10 A basic air conditioning unit principle – fan draws air into a duct, there may be a pre-conditioning heat exchanger and pre-humidifying, followed by a heat exchanger cooled by a refrigerator circuit which cools to reduce humidity, and finally a heat exchanger to warm the air up and possibly re-humidifying to fine tune the humidity if necessary.

A so-called 'swamp cooler' is a good example of air conditioning – blowing air through a wet cloth in hot dry climates causes the air to emerge on the other side sensibly colder – typically in British indoor conditions, this will cause a 3 or 4°C drop in the air temperature. This is akin to the common experience of climbing out of a swimming pool and feeling cold in air that is actually warmer than the water. The temperature difference is due to evaporation – otherwise known as evaporative cooling. This difference increases with increasing atmospheric temperature and decreasing **humidity** – a term of primary importance in air conditioning.

Figure 2.10 shows a schematic apparatus for air conditioning processes. A duct has an in-draught fan, a pre-treatment heat exchanger and humidifier (optional in practical plants for fine tuning air condition), a cooling heat exchanger attached to a refrigeration circuit capable of cooling the air to a point where humidity is so high that water forms on the heat exchanger and dehumidifies the air and finally a heat exchanger to reheat the cold air and optional re-humidifier to fine-tune the air conditions. All of these components affect the **condition** of the air.

When it is very dilute, water vapour is considered to be a perfect gas. Figure 2.11 shows the principle of mixed gases in which gas molecules of different gases have different colours to illustrate that they are intimately mixed and are undergoing molecular motion and collisions with a level of KE (dependent on temperature) that gives internal energy to the system and results in an overall pressure. These gases can be imagined separately filling the entire space. They would then give a pressure in the space on their own. The **Law of Partial Pressures** observes that when the gases are mixed, their individual, separate pressures add up to make the total pressure of the combined gas mixture. The pressure ratio is directly proportional to the number of **moles** ratio as described below:

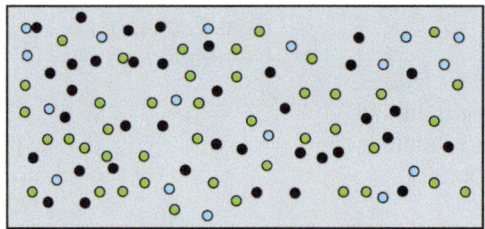

FIGURE 2.11 Perfect gases have partial pressures – the pressure of that amount of gas if it alone occupied the entire volume. Here molecules of 3 gases are inter-mixed.

GIBBS–DALTON LAW (LAW OF PARTIAL PRESSURES)

$$p = \sum p_i$$

p [Pa] mixture pressure
p_i [Pa] – the partial pressure due to one of the component gas species, i, in the mixture of gases (e.g. nitrogen)

We will see later that the partial pressure is related to the number of moles (n) of gas in the mixture:

$$\frac{p_i}{p} = \frac{n_i}{n} = \frac{V_i}{V}$$

n [moles] – number of moles of a species (subscript i) or of the mixture (no subscript);
V [m³] – volume of gas, and with subscript i volume of component gas, i.

Note that the number of moles indicates the quantity of a substance. Every molecular substance has a molecular mass; that is the mass of a specific number of molecules of the substance. The specific number of molecules in a mole is the **Avogadro number** which is **6.022×10^{23}** molecules per mole. Since molecules generally have differing sizes, they have different molecular masses. The molecular mass is defined in the units g/mol or kg/kmol.

Air is classified either with or without water vapour as **atmospheric air** or **dry air**, respectively. The properties of dry air are available in tables of data and are just that – air without any moisture whatsoever. Dry air is a mixture of gases which can be regarded to a good approximation as existing in a universally constant ratio across the globe at sea level, mainly nitrogen and oxygen. The partial pressures are determined by the number of moles of each gas present in a given volume, which is in the same ratio as the volume ratio. Water vapour can be supported in the air by virtue of its partial pressure – or its saturation pressure at the temperature of the atmospheric air.

WORKED EXAMPLE

Atmospheric air consists of dry air and water vapour

Dry air

- Composed of N_2 and O_2.
- By mass the proportions are, for approximate calculation, 76.7% nitrogen and 23.3% oxygen.
- By volume, 79% nitrogen and 21% oxygen.

What are the partial pressures, assuming atmospheric pressure is 1.01325 bar?
By using the formula relating mole and volume ratio to pressure ratio:

$$\text{nitrogen} - 0.79 \times 1.01325 = 0.80047 \, \text{bar}$$

$$\text{oxygen} - 0.21 \times 1.01325 = 0.21278 \, \text{bar}$$

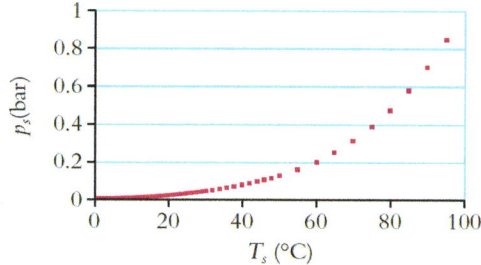

FIGURE 2.12 Water vapour and its saturation temperature and pressure.

The practice of boiling water is familiar, and it occurs when the **saturation pressure** of a fluid is equal to the atmospheric pressure. Water boils at 100°C (the boiling point) at a pressure of 1 atmosphere (1.01325 bar). This is the **saturation temperature**, T_s, and the corresponding saturation pressure, usually denoted p_s. The chart of saturation pressure *versus*. saturation temperature in Figure 2.12 shows that p_s drops sharply when the temperature falls below 100°C.

Saturation pressure shows the pressure water vapour exerts on a free water surface because of the energetic molecules it consists of. When the saturation pressure is equal to the local atmospheric pressure, boiling occurs; it can occur anywhere on a solid surface submerged in the liquid, the surface providing nucleation sites for vapour bubble growth. Initially, small bubbles grow as they rise through the liquid before bursting through the free surface.

Apart from boiling, vapour can evaporate from a free surface of water and condense at a free surface. At equilibrium, evaporation and condensation occur at equal rates. At equilibrium, the partial pressure of water vapour in contact with liquid water will be the saturation pressure corresponding to the temperature. When water vapour is not in contact with a free liquid water surface, the partial pressure will be less than or equal to the saturation pressure at that temperature. The level of **humidity** will be determined by the saturation pressure of the liquid at the prevailing temperature.

There are two measures of humidity, which are useful in different ways. **Absolute humidity, ω,** describes the mass of water vapour contained by a given mass of air.

$$\omega = \frac{m_s}{m_a} = \frac{m_s/V}{m_a/V} = \frac{v_a}{v_s}$$

where subscript 's' is for vapour and subscript 'a' is for dry air. Note here that $m_s/V = 1/v_s$ **also** $m_a/V = 1/v_a$ and hence the relationship in terms of specific volume also, where V is the volume of the gas mixture and v is the specific volume of each component gas type. Subscripts indicate air (a) and vapour (s).

The following derivation shows how this formula can be converted from mass fractions to partial pressures. Starting with the ideal gas law:

$$p_sV = m_sR_sT \rightarrow \frac{m_s}{V} = \frac{p_s}{R_sT}$$

And

$$p_aV = m_aR_aT \rightarrow \frac{m_a}{V} = \frac{p_a}{R_aT}$$

Therefore,

$$\omega = \frac{R_ap_s}{R_sp_a}$$

R_a (specific gas constant for dry air) is 287 J/kg.K and R_s (specific gas constant for water vapour) is 461 J/kg.K. The atmospheric pressure is the sum of the partial pressures, that is $P_{atmos} = P_a + P_s$. The equation becomes:

$$\omega = \frac{287p_s}{461\left(p_{atmos} - p_s\right)} = \frac{0.622p_s}{\left(p_{atmos} - p_s\right)}$$

This relationship is very useful in the analysis of air conditioning systems, as will become clear in the subsequent parts of this section.

The second measure is **relative humidity, φ,** which gives a measure of how far from maximum humidity that the atmospheric air is:

$$\phi = \frac{p_s}{p_g}$$

Note carefully the *re-definition of p_s for air conditioning only* – here p_s is the partial pressure of the water vapour in the atmospheric air, not the saturation pressure, and p_g is the partial pressure of vapour that has its maximum humidity for the temperature, T, of the air, that is p_g is the *saturation pressure* of water vapour corresponding to the temperature of the air. Note that this use of p_s is only to be used for air conditioning and not to be confused with the saturation pressure also commonly denoted p_s in the tables of steam properties. Here, and only for air conditioning, p_g means the saturation pressure from the tables, but the pressure corresponding to the temperature of the atmospheric air. This is the standard notation for air conditioning engineers.

A typical range of specific humidity, ω, is from zero (for totally dry air) to up to about 30 g H_2O per kg of dry air at normal sea level pressures. A typical range of relative humidity, φ, varies between 0 and 100% since it is the proportion of maximum water vapour pressure that can exist at the temperature of the air and so can be zero if no water vapour is present, and can be 100% when the vapour has a pressure corresponding to the boiling point pressure for the temperature of the air.

If the temperature of the air falls until the *saturated humidity point*, that is 100% relative humidity occurs, the air is at the **dewpoint** temperature; it is the temperature at which air becomes *saturated* with water vapour, that is it can contain no more water in the vapour state, when cooled at constant pressure. Since ω = constant before dewpoint is reached, p_s is constant during cooling down to the dewpoint temperature by which point $p_s = p_g$, because p_g reduces as temperature reduces.

EXAMPLES

1. **A sample of atmospheric air contains 12 g of water in 1.2 kg of dry air. What is the absolute humidity?**
   In this case, there is 12 g per 1.2 kg, which is a specific or absolute humidity, ω, 0.01. This is typical of the standard atmospheric condition.

2. **What is the specific volumetric ratio of moisture to air?**
   The inverse of ω is the volumetric ratio – that is partial volume ratio. The specific volume ratio of air to vapour follows, that is 0.01 m³/kg of air per m³/kg of vapour. This shows the significant difference in the specific volume of the water vapour and dry air, and that the vapour is dispersed thinly in the air

3. **If the dry air specific volume is 1 m³/kg and the partial pressure of water vapour is 0.016 bar and the air has 100% relative humidity, what is the specific volume ratio and, hence, what is the specific humidity?**
   Air specific volume v is 1 m³/kg, specific volume of vapour is (more accurately, from thermodynamic property tables) 83 m³/kg and specific volume ratio is 1/83 = 0.012, which is the specific humidity.

4. **Given a specific humidity of 0.02 and an atmospheric pressure of 1.01325 bar, what is the partial pressure of water vapour?**

$$\omega = 0.02, p_{\text{atmos}} = 1.01325 \, \text{bar, and}$$
$$p_s = \omega.p_{\text{atmos}} / (0.622 + \omega); \text{therefore}, p_s = 0.03156 \, \text{bar}.$$

5. **The humidity is 50% in atmospheric air at 20°C; what is the partial pressure of the water vapour in the air?**
   At 20°C, p_g is 0.02337 bar. Given the relative humidity is 50%, the vapour pressure, p_s must be 0.02337 × 0.5 = 0.0117 bar

6. **If the temperature of the atmospheric air is 35°C and the relative humidity is 50%, what is the dew point?**
   Find p_s from the relative humidity, ϕ, formula with p_g at 35°C. Then identify the temperature that corresponds to this, that is the dew point. p_g = 0.056 bar (from thermodynamic property tables). p_s = 0.5 × 0.056 = 0.028 bar. From the tables, the temperature that corresponds to this as its p_g is 23°C. This means that when the temperature falls to 23°C, moisture will condense out of the air, either as a cloud if the bulk air temperature decreases or more usually on surfaces which cool down due to losing heat to the night sky by thermal radiation.

2.2.1 Hygrometry or Psychrometry

The analysis of air condition is called **hygrometry** and the two measures just introduced provide the basis for the work of air conditioning engineers. In order to control air condition, we must be able to measure it easily. The basic tools of the trade are **wet and dry bulb thermometers** as shown in Figure 2.13. The wet bulb differs from the dry one only in that it has a muslin sock around the bulb that is soaked in water from a reservoir. If the local atmosphere has a humidity of less than 100%, then water will evaporate from the sock and cause the temperature of the bulb to drop due to the latent heat of evaporation, and therefore, the thermometer will register a lower temperature. This temperature enables us to calculate the relative humidity.

A spreadsheet can be calculated based on this theory to determine the relative and specific humidities from the wet and dry bulb temperatures. The chart in Figure 2.14 is called a **psychrometric chart** and it shows the condition of the air for a given pair of wet and dry bulb temperatures. Data represented includes relative and specific humidity, enthalpy of the mixture, and specific volume of the mixture.

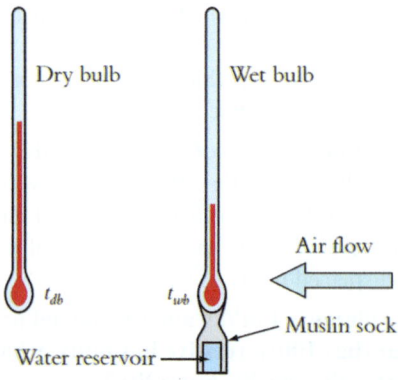

FIGURE 2.13 Wet and dry bulb thermometers.

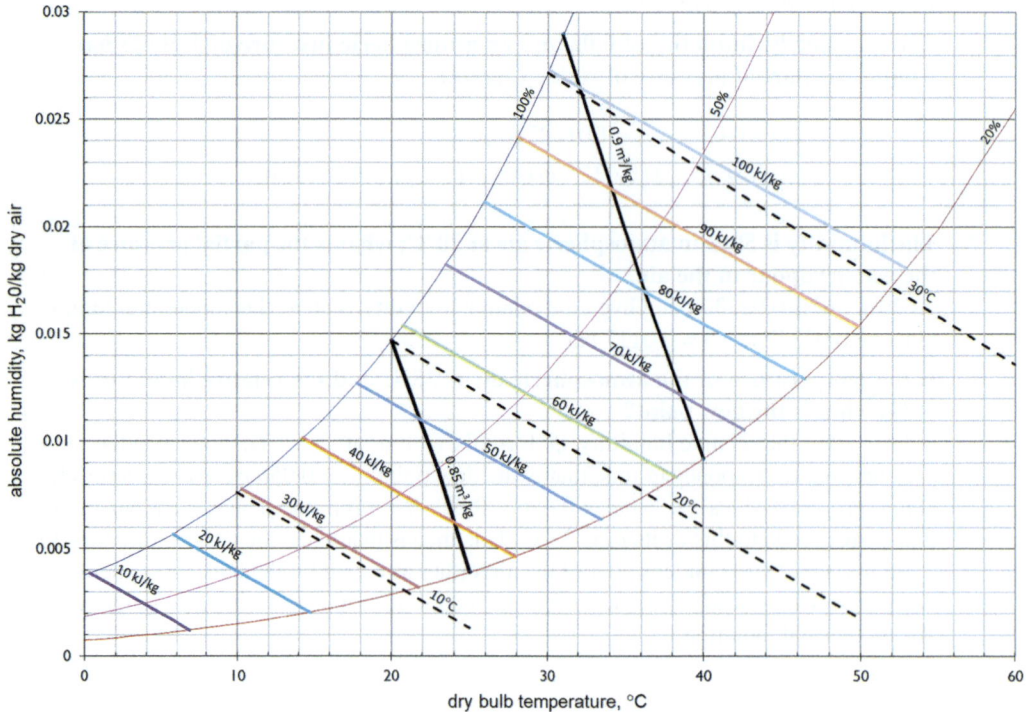

FIGURE 2.14 An approximation of the psychrometric chart at 1.01325 bar using perfect gas assumptions. For a more accurate chart, please refer to the CIBSE (Chartered Institute of Building Services Engineers). Dashed lines are of constant wet bulb temperature, colour lines are of constant air specific enthalpy, solid black lines are specific volume and curved colour lines are of constant relative humidity.

EXAMPLES OF READING THE PSYCHROMETRIC CHART

1. **What is the relative humidity if T_{DB} is 30°C and T_{WB} is 20°C?**
The dry bulb temperature is read from the bottom axis, and constant temperature lines rise vertically. The wet bulb temperature is read from the 100% relative humidity curve on the left-hand side. The constant wet bulb temperature lines are inclined to the horizontal. The intersection of the two lines defines the air condition and the relative humidity is indicated on the curves at 10% intervals. The approximate value for this particular condition is approximately 38%.

2. **What is the specific humidity if it is 100% humid at 30°C? What does this tell you about the relative masses of air and water at 30°C?**
Taking the position on the 100% relative humidity curve, and marking the point when the dry bulb, and wet bulb, temperature is 30°C defines the air condition. Now read horizontally across to the vertical axis on the right-hand side to show the absolute humidity is approximately 0.0274.

3. **What is the enthalpy for a 50% relative humidity at 20°C dry bulb?**
The intersection of the 50% relative humidity curve with the vertical 20°C dry bulb temperature defines the air condition. The enthalpy is found by placing a rule on the air condition point, and rotating about that point until the enthalpy on the upper enthalpy scale and the lower enthalpy scale are the same. In this case, it is 39 kJ/kg.

Figure 2.15 shows the basic principles of the operation of an air conditioning unit. Air to be conditioned is drawn in by a fan, cooled down either to simply reduce temperature or more commonly to also decrease humidity by reaching dew point and causing condensation. Air is heated up after the cooler when condensation has occurred in order to make it comfortable before being issued to the room. The power of heating and cooling is determined from the desired changes in the enthalpy of the dry air/water vapour mixture between each of the sections. This must be equal to the energy required to be drawn from the external source. The SFEE is used for applying the conservation of mass and energy. With reference to Figure 2.15,

Mass conservation:

$$m_{s2} = m_{s3} + m_{w4}$$

$$m_{a1} = m_{a2} = m_{a3} = m_{a4}$$

At stage 3, there is the same mass of dry air, but some moisture has fallen out of the cooling unit (condensing on the cold refrigerated coils of the heat exchanger in the air) to the drain. So we have

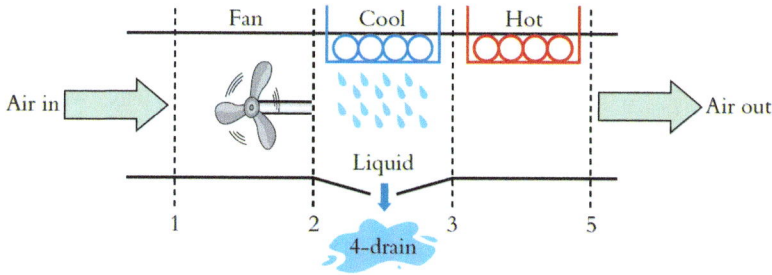

FIGURE 2.15 Enthalpy exchanges in the cooling section.

the same mass of dry air with different enthalpy (calculable by dry air temperature and c_p), and we have a smaller amount of water vapour with the air. We also have an amount of condensate – the condensed water vapour on the cooling coils.

Then $m_a h_{a1}$ is in-going air enthalpy, m_{s2} is from specific humidity and h_{s2} from enthalpy of water vapour at the temperature of the point 2, T_2; m_{w4} is the condensate at the temperature of the cooling coils with h_{w4} from tables as h_{fw4}.

WORKED EXAMPLE

An air conditioning unit has a fan which delivers a mass of atmospheric air. The mass flow rate of the dry air in this is m_a, which is 0.5 kg/s. The specific humidity is 0.01 kg/kg air, and it can be concluded that the mass flow rate of the water vapour, m_{s2}, is 0.005 kg/s. The in-going air temperature is 30°C.

The unit causes a drop in specific humidity to 0.006 kg/kg. What is the rate of heat taken out of the air in the cooler section of the air conditioning unit, $\dot{q}$? What is the pressure of the dry air, p_a, and of the water vapour, p_s, at exit?

a. Calculate p_{S2} and p_{S3} using specific humidity–vapour pressure formula.
b. Calculate the mass flow rated of dry air and water vapour at entry to and at exit from the cooler.
c. What is the dew point temperature corresponding to $p_g = p_{S3}$?
d. Find c_{pAIR} from the tables at the appropriate mean temperature.
e. Find, from the tables, the specific enthalpy of liquid water at the dew point temperature, and of the vapour at points 2 and 3.

Use the SFEE across the cooler to find $\dot{Q}$.

a. Using $\varpi = \dfrac{0.622 p_S}{p - p_S} = \dfrac{0.622 p_S}{1.01325 - p_S} = 0.01 \Rightarrow 0.01 \times 1.01325 - 0.01 p_{S2} = 0.622 p_{S2}$

   $\Rightarrow 0.01013 = 0.632 p_{S2} \Rightarrow p_{S2} = 0.016 \, \text{bar}$

   p_{S3} must be at 100% relative humidity condition, $\phi = 1, \therefore p_g = p_{SAT} = p_{S3}$

   $\varpi = \dfrac{0.622 p_{S3}}{p - p_{S3}} \Rightarrow 0.006 = \dfrac{0.622 p_{S3}}{1.01325 - p_{S3}} \Rightarrow 0.00608 - 0.006 p_{S3} = 0.622 p_{S3}$

   $\Rightarrow 0.00608 = 0.628 p_{S3} \rightarrow p_{S3} = 0.0097 \, \text{bar}$

b. Atmospheric air mass flow rate at entry is given as 0.5 kg/s. We know

   $\varpi = \dfrac{\dot{m}_W}{\dot{m}_{DRYAIR}} = 0.01$ and $\dot{m}_{DRYAIR} + \dot{m}_W = 0.5 \, \text{kg/s}$

   $\therefore \dot{m}_W = 0.005 \, \text{kg/s}$ and $\dot{m}_{DRYAIR} = 0.495 \, \text{kg/s}$.

   At exit $\dot{m}_{DRYAIR} = 0.495 \, \text{kg/s}$ still – it can only go out one way. We know that

   $\omega = 0.006$, and therefore we say $\varpi = \dfrac{\dot{m}_W}{0.495} = 0.006 \rightarrow \dot{m}_W = 0.003 \, \text{kg/s}$.

c. At p_{S3}, the saturation temperature, $T_{SAT} \sim 6.5°C$ from tables of saturated steam and water.

d. The mean temperature of 6.5°C and 30°C is 18.25°C or approximately 291 K. Using the data for dry air at low pressure,

$c_{p,air)291 K}$ could be interpolated for accuracy, but inspection shows it is approximately 1.0045 kJ/kgK.

e. Saturated water and steam table is required again,

From this it can be seen that h_f of liquid water at 6.5°C is approximately 27 kJ/kg, h_g at 6.5°C is 2485 kJ/kg and at 30°C, h_g at 30°C is 2555 kJ/kg.

f. Use the SFEE, which is the statement of the First Law (conservation of energy) for moving fluids and ignore the terms from KE and PE as negligible, $Q + W = \Delta H$. In the air conditioning unit between 2 and 3, there is no fan or other working device, so $W = 0$. We need the change in the enthalpy of the fluids in the section 2–3 in order to work out the heat transfer to cause it.

Referring to the schematic, the enthalpy flows in at 2 and out at 3 and through the condensate collection chute are all indicated in terms of mass flow rate and specific enthalpy. The flow of enthalpy can then be compared in the SFEE, using the convention of end state minus start state:

$$Q = \dot{m}_{DRYAIR}h_{DRYAIR,3} - \dot{m}_{DRYAIR}h_{DRYAIR,2} + \dot{m}_{W,3}h_{g,3} - \dot{m}_{W,2}h_{g23} + \dot{m}_{COND}h_{f,COND}$$

We can use $\Delta H = mC_p\Delta T$ for the enthalpy change of dry air, which is $0.495 \times 1.004 \times (6.5 - 30) = -11.68$ kW, just to cool the dry air.

The change in the enthalpy of the water vapour carried in the air is mainly due to the loss of mass as vapour which condenses. Therefore, with the data for h_g and mass flow rate of vapour going in at 2 and out at 3, we have $0.003 \times 2485 - 0.006 \times 2555 = -5.32$ kW. The condensate is assumed to leave at 6.5°C and it has a mass rate of 0.003 kg/s; therefore, the enthalpy flow rate is $0.003 \times 27 = 1.27$ kW. Putting all this in the SFEE, we have $Q = -11.68 - 5.32 + 1.27 = -15.73$ kW.

This heat is removed from the air and is therefore expressed as negative.

2.2.2 REFRIGERATION AND HEAT PUMPS

In considering the practical air conditioning plant, the cooling duty must be supplied by a refrigeration unit, and the application of refrigeration and heat pumps is therefore considered here. Figure 2.16 shows a heat pump in which work is converted into heat transfer to produce a flow of heat from cold to hot. It is a heat engine operating in reverse. The process is marked on – it's the reverse of a vapour power cycle with a **condensing hot reservoir**, an **evaporating cold reservoir** – a compressor and a throttle.

The reciprocating compressor pumps the **refrigerant** vapour to drive refrigerant round the circuit. The compressor maintains the evaporator heat exchanger at a low pressure by drawing evaporated vapour out of it; since the vapour pressure is low and the saturation temperature is correspondingly low – lower than the surrounding cold region – the liquid refrigerant in the evaporator tubes evaporates as heat is absorbed from the surroundings. The compressor drives the evaporated vapour through to the higher pressure condenser heat exchanger tubes at the top of the cycle. The HP means high saturation pressure and therefore high saturation temperature and the vapour is forced to condense since the saturation temperature is higher than the hot reservoir temperature, which is the surroundings of the condenser heat exchanger. This cycle is a refrigerator if the cool side is the useful output or a heat pump if the hot side is the useful output. Note the symbol for the throttle on the left, which reduces pressure at constant enthalpy and maintains the temperature difference between the hot and cold heat exchangers.

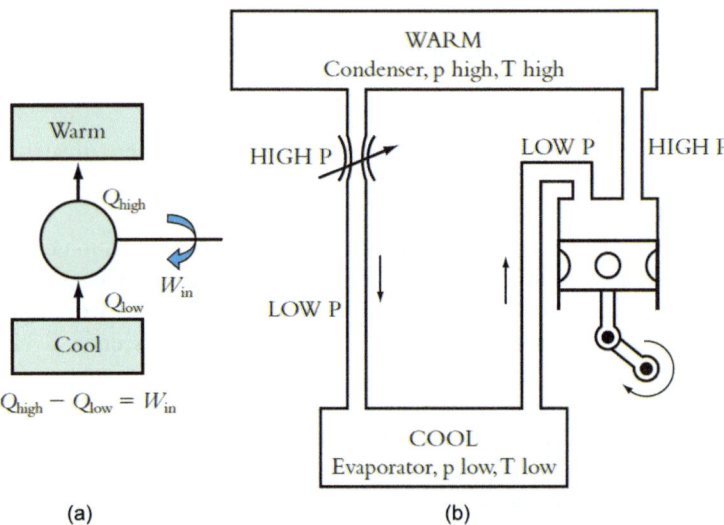

(a)   (b)

FIGURE 2.16 Refrigeration or heat pump circuit (a) schematic of the reversed heat engine, (b) schematic of the physical circuit required.

The pressure, p, vs specific enthalpy, h, properties chart is the easiest chart to deal with for refrigeration cycles since it yields the enthalpy change for a process directly. The enthalpy is the energy content of the working fluid, and hence energy balances can very easily be performed with operations occurring at two pressure levels. The **p-h diagram** for the commonly used refrigerant R134a is displayed in simplified form in Figure 2.17. Note the examples of constant temperature lines drawn on in blue and pink. On the superheated side of the chart (the right hand side of the domed line), the

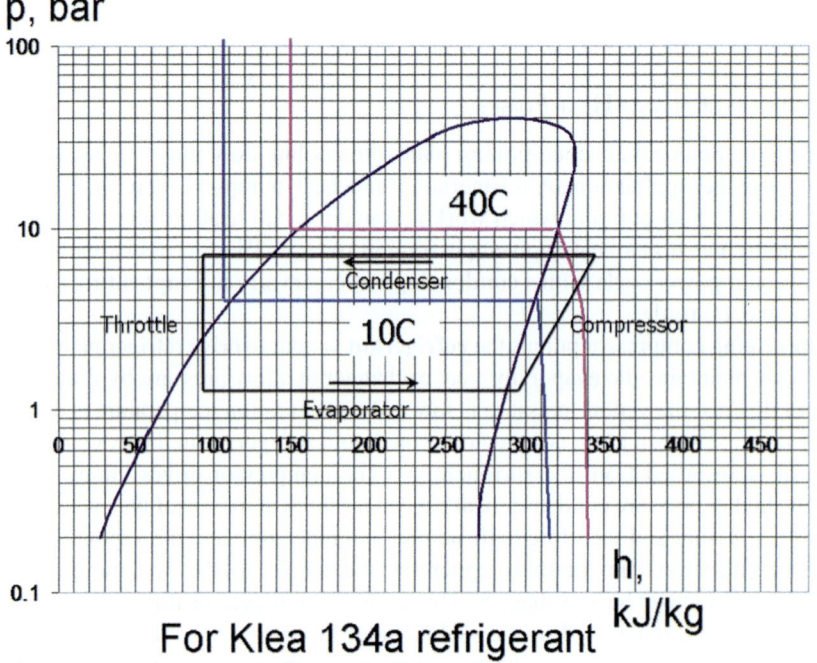

FIGURE 2.17 p-h diagram of a typical modern refrigerant.

temperature lines descend in a curve; on the sub-cooled liquid side, the lines rise vertically because the liquid is incompressible over the range of pressures represented. A typical heat pump process is plotted in Figure 2.17, showing a cycle which runs in the anti-clockwise direction. We know from the earlier description of cycles that this direction means that net work is being put into the cycle.

The process is plotted on the chart as follows:

- Horizontal and lines across at the high pressure and at the low pressure;
- Find the temperature of the evaporator exit at the low pressure in the evaporator in the superheated side of the chart;
- Find the temperature of compressor outlet, which is in the superheated region, at the high pressure of the condenser,
- Draw a line showing the process of the compressor between evaporator exit and compressor exit;
- Find the temperature of the outlet from the condenser, which is in the sub-cooled liquid region;
- A vertical line through the saturated liquid point corresponding to the condenser exit temperature defines the process in the throttle, assuming that it is a perfectly constant enthalpy device;
- The change in enthalpy across the three main components gives the heat/work transferred in that component.

The reason for the throttle having no change in enthalpy may not be immediately obvious, although in practice the reason is basically as follows: we know that the throttle is short and should be well insulated, and so Q is nearly zero. There is obviously no work in the throttle – no paddles or pistons to move the fluid. Use the SFEE $\dot{Q} + \dot{W} = \dot{m}\Delta h$, where Q and W are rates of heating and power input and $\dot{m}$ is the mass flow rate. It is easy to see that if Q and W are zero, then the enthalpy entering is the same as that leaving.

The heat pump circuit relies on vapour-liquid behaviour. Vapour enters the condenser tube (Figure 2.18a) at a temperature above T_{high} and exchanges heat, Q, out of condenser to cool the vapour. The vapour condenses when it reaches T_{sat} until all is condensed. T_{sat} needs to be a few degrees above T_{hot} to work properly. The condensed and slightly sub-cooled liquid passes through the throttle (Figure 2.18b) in which the SFEE causes no change in enthalpy as noted above. This reduces the pressure of the fluid and therefore the T_{sat} reduces to a point just below the cool reservoir (which contains the evaporator) temperature. As it enters the evaporator, liquid heats up to T_{sat} and starts to boil (Figure 2.18c). By the time it reaches the end of the evaporator, all is evaporated and the vapour gains a little superheat.

The heat pump is not only used for refrigeration. The hot reservoir may be the useful heat output of the unit. This is the case in **ground source heat pumps** used for domestic heating. This is an effective way of providing heating because the coefficient of performance ensures that a significantly higher amount of heat is transferred than the energy supplied to the compressor.

2.2.3 COEFFICIENT OF PERFORMANCE

The heat pump effectiveness is measured in terms of how much heat is transferred either out of the cold reservoir or into the hot reservoir, depending on whether it is used as a refrigeration unit or a heat pump. In either case, the cost of pumping the heat is the power into the compressor. A measure

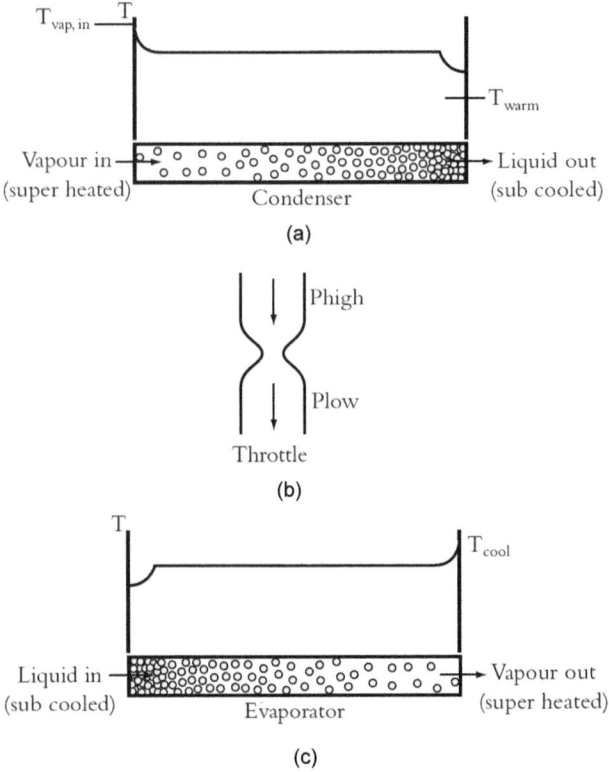

FIGURE 2.18 How the refrigeration circuit works.

of how well the heat pump works is therefore a ratio of the heat transferred to the work input, which is given by the enthalpy changes in the evaporator and condenser for the heat transfers and by the enthalpy change in the compressor. From Figure 2.17, the change in enthalpy is given on the horizontal axis in kJ/kg of refrigerant. The difference is obviously far greater in the evaporator and condenser than in the compressor. For the cycle shown, these enthalpy changes are approximately 200 kJ/kg, 250 kJ/kg and 50 kJ/kg, respectively. The ratio is therefore for refrigeration 200/50 = 4 and for the heat pump 250/50 = 5. These are far greater than unity and therefore cannot be stated as an efficiency. This ratio is known as the **coefficient of performance** and is the standard way of quantifying the effectiveness of heat pumps.

LEARNING OUTCOMES

✓ **Atmospheric air** is a mixture of **dry air** and **water vapour**.

✓ **Air condition** defines the **temperature** and **humidity** of atmospheric air.

✓ The **Law of Partial Pressures and Partial Internal Energy** determines what proportion of a gas pressure or internal energy is due to each individual gas component.

✓ There are two measures of humidity: **specific** (or **absolute**) humidity, which states the mass of water vapour per unit mass of dry air, and **relative** humidity which states the ratio of the partial pressure of water vapour in the air, denoted p_s, to the maximum partial pressure of the water vapour at that temperature, denoted p_g, which is the water saturation pressure.

✓ Gases may be quantified in **moles**, which is a specific number of molecules (the Avogadro number, 6.022×10^{26} per kmol).

✓ **Dew point** is the temperature at which atmospheric air becomes 100% saturated with water vapour (i.e. $p_s = p_g$), such that if the temperature is further reduced water condenses out of the atmosphere.

✓ The **psychrometric chart** is the key tool of air condition monitoring together with the **wet** and **dry bulb** thermometers, and the terms **psychrometry** and **hygrometry** are introduced to describe the study of atmospheric air.

✓ The principles of operation of an **air conditioning unit** require enthalpy balances to determine the heat power required by heating and cooling in the unit to produce a particular condition.

✓ The **heat pump** is the generic **refrigeration unit** which is required to produce cooling in the air conditioning unit, and which can also be used to provide heating energy from a cold source.

✓ The **p-h** diagram is used to plot heat pump processes, since it quickly yields enthalpy changes between points, which represent the heat and work transfers required to produce the changes of state.

✓ The **coefficient of performance** is introduced as the measure of heat pump effectiveness.

2.3 GAS MIXTURES

There are two ways of analysing gas mixtures – by mass fractions and by volume fractions. The law of partial pressures and internal energy leads to analysis by mass or **gravimetric analysis**. Using it enables the calculation of properties of a mixture, that is specific enthalpy, h, specific heat capacity at constant volume, c_v, specific heat capacity at constant pressure, c_p, specific gas constant, R, molar mass, M and specific entropy, s from the individual properties combined according to their mass fractions.

2.3.1 GRAVIMETRIC ANALYSIS

The proportions by mass are useful for determining the fluid properties of the mixture. The Gibbs-Dalton Law of Partial Pressures presented in Section 2.2 gives the method for doing this as follows:

DEDUCTIONS FROM THE GIBBS–DALTON LAW OF PARTIAL PRESSURES

Enthalpy

$$H = U + pV$$

$$H = \Sigma U_i + \Sigma p_i V$$

$$H = \Sigma H_i = \Sigma m_i h_i$$

in which m_i is the mass of the gas species identified as i.

Specific Heat at Constant Volume, c_v

From the summation (conservation) of internal energy

$$m\left(u_2 - u_1\right) = \sum m_i \left(u_{2i} - u_{1i}\right)$$

in which m is the total mass of the sample of gas

$$mc_v\left(T_2 - T_1\right) = \sum m_i\, c_{vi}\left(T_2 - T_1\right)$$

$$c_v = \sum_i \frac{m_i}{m} c_{vi}$$

Specific Heat at Constant Pressure, c_p

From the summation (conservation) of enthalpy

$$m\left(h_2 - h_1\right) = \sum m_i \left(h_{2i} - h_{1i}\right)$$

$$mc_p\left(T_2 - T_1\right) = \sum m_i\, c_{pi}\left(T_2 - T_1\right)$$

$$c_p = \sum_i \frac{m_i}{m} c_{pi}$$

Gas Constant R for the Mixture

From Joules law,

$$R = c_p - c_v$$

$$R = \sum_i \frac{m_i}{m} c_{pi} - \sum_i \frac{m_i}{m} c_{vi} = \sum_i \frac{m_i}{m}\left(c_{pi} - c_{vi}\right)$$

$$R = \sum_i \frac{m_i}{m} R_i$$

Apparent Molar Mass M

$$R = \sum_i \frac{m_i}{m} R_i = \frac{\tilde{R}}{\tilde{m}} = \sum_i \frac{m_i}{m}\frac{\tilde{R}}{\tilde{m}_i}$$

$$\frac{1}{\tilde{m}} = \sum_i \frac{m_i}{m}\frac{1}{\tilde{m}_i}$$

Entropy

This can be calculated by treating the mixture as a perfect gas.

From the first law, $dq = -dw + du$

Substituting from the Second Law and reversible work in a closed system,

$$Tds = pdv + du$$

$$ds = \frac{p}{T}dv + \frac{du}{T}$$

Now, $du = c_v dt$ and $p/T = R/v$

$$ds = R\frac{dv}{v} + c_v\frac{dt}{T}$$

Hence,

$$s_2 - s_1 = R\ln\left(\frac{v_2}{v_1}\right) + c_v\ln\left(\frac{T_2}{T_1}\right)$$

This can be converted to alternative forms – i.e. p-T and p-v forms – by using the equation of state.

Since $pv = mRT$, to get the p-T form: $v_2/v_1 = (p_1/p_2)(T_2/T_1)$

Substitute into the previous entropy formula to give:

$$s_2 - s_1 = R\ln\left(\frac{p_1}{p_2}\right) + R\ln\left(\frac{T_2}{T_1}\right) + c_v\ln\left(\frac{T_2}{T_1}\right)$$

$$s_2 - s_1 = R\ln\left(\frac{p_1}{p_2}\right) + \left(R + c_v\right)\ln\left(\frac{T_2}{T_1}\right)$$

$$s_2 - s_1 = R\ln\left(\frac{p_1}{p_2}\right) + c_p\ln\left(\frac{T_2}{T_1}\right)$$

Similar procedure for p-v form.

Therefore, for mixtures

$$m\left(s_2 - s_1\right) = \Sigma m_i\left(s_{i,2} - s_{i,1}\right)$$

$$= \Sigma m_i\left[c_{p_i}\ln\left(\frac{T_2}{T_1}\right) - R_i\ln\left(\frac{p_{i,2}}{p_{i,1}}\right)\right]$$

Note: Even for a perfect gas mixture, the entropy depends on state rather than temperature alone (see h & u which are just temperature dependent for a perfect gas). This means that entropy depends on temperature and another property, that is on the state, whereas h and u depend on temperature only.

EXAMPLE

Calculate c_p for a mixture of the following fuel and air mixture at 300 K: methane 0.1 kg, ethane 0.25 kg, nitrogen, 0.75 kg and oxygen 0.23 kg

Using the tables of thermodynamic properties, the c_p for each of the gases can be read at 300 K as methane, 2.26 kJ/kgK, ethane 1.766 kJ/kgK, nitrogen 1.040 kJ/kgK and oxygen 0.918 kJ/kgK. Using the formula from above the c_p of the mixture:

$$\frac{\left(0.1\times2.226+0.25\times1.766+0.75\times1.040+0.23\times0.918\right)}{0.1+0.25+0.75+0.23}=1.245\,kJ\,/\,kgK$$

2.3.2 VOLUMETRIC AND MOLAR ANALYSES

Volumetric analysis is directly related to **molar analysis**. So, if we have the proportions of the volumes of the gases reacting, we know the proportions of moles reacting. The analysis examined so far is **Gravimetric**, that is the analysis of gas mixtures by the separation of the constituents and their estimation by mass. However, in many cases the composition based on the volumes (or volumetric composition) of gases in the mixture is known.

EXAMPLE

The volumetric composition of air is approximately 79% nitrogen, N_2, 21% oxygen, O_2, by volume, that is in 1 m³ of air, the partial volumes of nitrogen, N_2 and oxygen, O_2 are 0.79 m³ and 0.21 m³.

Law of partial volumes Amagat's law of partial volumes.

The volume of a mixture of gases is equal to the sum of the volumes of the individual constituents when each exists alone at the pressure and temperature of the mixture, V = mixture volume at p,T

V_i = partial volume of gas component i at p,T and $V_{i,P,T}$ means the volume occupied by gas 'i' at the mixture pressure, p, and mixture temperature, T. It is important to understand the subscripts which identify the various components.

Formula for the relation between partial pressure and partial volume

Partial pressures and partial volumes are alternative ways of describing the mixture. Either a component occupies p_i at V or V_i at p but not V_i at p_i. Using the gas law, considering a perfect gas:

By partial pressure, $p_iV = m_iR_iT$
By partial volume, $pV_i = m_iR_iT$

Divide to give: $\dfrac{V_i}{V}=\dfrac{p_i}{p}$

This equation shows that proportions by volume are equal to proportions by pressure. So if a gas occupies 10% of the volume of a mixture, its partial pressure will be 10% of the mixture pressure.

Molar Analysis: By definition, this is the same as the volumetric analysis but using mole fractions. The mole was presented as a measure of fluids in the previous section on air conditioning.

Definitions:
a. 1 mole (1 mol) of gas has a mass in g equal to its molecular weight.
b. 1 mol of any perfect gas occupies *the same volume V* at a given p and T as 1 mol of any other perfect gas (e.g. the volume of 1 kmol of a gas at standard temperature and pressure, $T = 0°C$ and $p = 1$ atmosphere, is $V = 22.4$ m³).
c. 1 mol of a molecular species contains an Avogadro number of molecules, $N_A = 6.022 \times 10^{23}$ mol⁻¹.
d. Usually it is more convenient to quantify in kmol. Point b tells us that the volume occupied by the same number of molecules of any gas is the same.

The atomic masses of all elements can be found in the periodic table of the chemical elements. To get the molecular mass of a molecule made up of more than one atom, add the atomic masses of the atoms. For example, carbon dioxide is CO_2; for every single molecule of this, there is one atom of carbon with atomic mass 12 g/mol and two atoms of oxygen (indicated by subscript 2) of 16 g/mol each, thus total molecular mass of CO_2 is $12 + 16 + 16 = 44$ g/mol or for larger quantities, 44 kg/kmol.

Difference between mass and volume fractions

Consider 1 kmol of hydrogen and 2 kmol of oxygen mixing without reacting as shown in Figure 2.19, pictured by volume at constant pressure and temperature. It is known that at the same p and T, 1 kmol of any gas occupies the same volume – regardless of mass – as 1 kmol of any other gas. Since the molecular mass of hydrogen (2 g/mol for H_2) is so much less than oxygen (32 g/mol for O_2), this shows us the effect of molecular weight, that vastly different volume and mass fractions of gases can occur in a mixture.

The equation of state for perfect gas in terms of m is:

$$p_i V \left(\text{or } PV_i \right) = mRT$$

To calculate by number of moles instead of mass:

$n = \dfrac{m}{\tilde{m}}$ where n is the number of kmol, m is mass of sample [kg] and $\tilde{m}$ is molecular mass [kg/kmol]. Knowing that the specific gas constant, R, is the molar gas constant, $\tilde{R}$, 8.314 [J/mol.K], divided by the molecular mass of the gas:

$$mR = m \frac{\tilde{R}}{\tilde{m}} = n\tilde{R}$$

Substituting in gas law:

$$p_i V \left(\text{or } PV_i \right) = n_i \tilde{R} T$$

Molar analysis can be used to determine partial pressure and partial volumes and vice-versa.

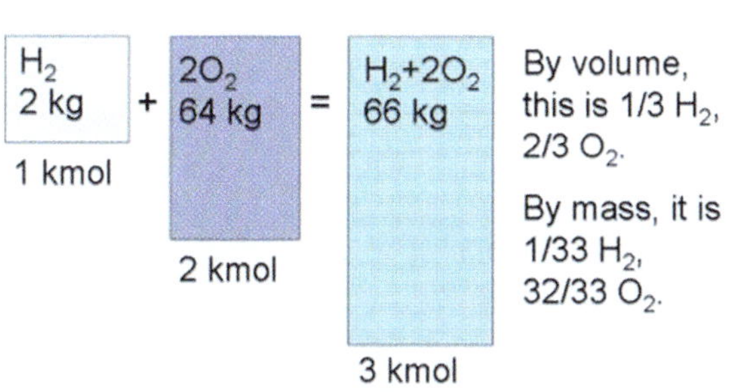

FIGURE 2.19 Proportions by mass compared to volume.

Derivations of relations for mixture properties in terms of molar or volumetric fractions:

Formula for the relation between partial pressure and partial volume and molar fraction.
For the mixture as a whole, considered as a perfect gas, we know the equation:

$$pV = n\tilde{R}T$$

(note: the molar form of the Gas Law). Use the equation of state for a perfect gas for the constituent gas with partial pressure p_i and the mixture at pressure p, and divide: $\dfrac{p_iV}{pV} = \dfrac{n_i\tilde{R}T}{n\tilde{R}T}$
Combine with the partial-pressure/partial-volume relation formula:
In which

$$\sum p_i = p, \sum V_i = V \text{ and } \sum n_i = n.$$

This equation together with the previous relation for V_i/V and p_i/p, shows us the important fact that volume fraction is equal to molar fraction. Consider a reaction $2H_2 + O_2 \rightarrow 2H_2O$. The molar fraction of H_2 in the reactants is 2/3. This is also its volume fraction. Usually, we have a volumetric analysis and a reaction to consider in combustion-related calculations.

Formula for apparent molar mass of mixture, $\tilde{m}$
This is calculated from

$$\tilde{m} = \frac{m}{n}$$

and therefore

$$m = \sum n_i\tilde{m}_i$$

by conservation of mass.
Therefore, the molar mass of the mixture is

$$\tilde{m} = \sum_i \frac{n_i\tilde{m}_i}{n}$$

Compare this with the relation derived by partial pressure analysis

$$\frac{1}{\tilde{m}} = \sum_i \frac{m_i}{m}\frac{1}{\tilde{m}_i}$$

Apparent gas constant of mixture, R
Since specific gas constant is defined as: $R = \dfrac{\tilde{R}}{\tilde{m}}$
Rearrange:

$$\tilde{m} = \frac{\tilde{R}}{R}$$

and

$$\tilde{m} = \sum_i \frac{n_i\tilde{m}_i}{n}$$

Therefore,

$$\frac{\tilde{R}}{R} = \sum \frac{n_i}{n} \frac{\tilde{R}}{R_i}$$

and

$$\frac{1}{R} = \sum \frac{n_i}{n} \frac{1}{R_i}.$$

Conversions between mass and volume fractions

To convert from gravimetric to volumetric proportions, consider volumetric analysis of constituent, i:

$$pV_i = \frac{m_i \tilde{R}}{\tilde{m}_i} T \tag{A}$$

of mixture:

$$pV = \frac{m \tilde{R}}{\tilde{m}} T \tag{B}$$

(A)/(B) gives $\dfrac{V_i}{V} = \dfrac{m_i}{\tilde{m}_i} \Big/ \dfrac{m}{\tilde{m}}$ or $\dfrac{m_i}{m} = \left(\dfrac{V_i}{V}\right) \dfrac{\tilde{m}_i}{\tilde{m}}$

Note that $pV_i = n_i \tilde{R} T$ and $n_i = m_i / \tilde{m}_i$.

The conversion can be done in tabular form:

An example of conversion from volume fraction to mass fraction. Typically, a volumetric analysis is stated as, for example, 10% methane (CH_4), 10% ethane (C_2H_6), 65% nitrogen (N_2) and 15% oxygen (O_2), that is $V_i/V = n_i/n = 0.1, 0.1, 0.65$ and 0.15, respectively for each gas.

Component	$V_i/V = n_i/n$	$\tilde{m}_i$	$(n_i/n)\ \tilde{m}_i$	$\dfrac{m_i}{m} = \dfrac{n_i}{n} \tilde{m}_i \cdot \dfrac{1}{\tilde{m}}$
A, methane, CH_4	0.1	16	1.6	0.058
B, ethane, C_2H_6	0.1	30	3.0	0.109
C, nitrogen, N_2	0.65	28	18.2	0.659
D, oxygen, O_2	0.15	32	4.8	0.174

$$\tilde{m} = \sum \frac{n_i \tilde{m}_i}{n} = 27.6 \,\text{kg} / \text{kmol}$$

Alternatively for mass to volume fraction:

Component	m_i/m	$\tilde{m}_i$	$(m_i/mm)(1/\tilde{m}_i)$	$\dfrac{n_i}{n} = \dfrac{m_i}{m}\dfrac{1}{\tilde{m}_i}\cdot M$
A carbon	0.09	12	0.0075	0.192
B O_2	0.21	32	0.0066	0.169
C N_2	0.70	28	0.0250	0.639

$$\frac{1}{\tilde{m}} = \Sigma \frac{m_i}{m}\frac{1}{\tilde{m}_i} = 0.0391\,\text{kmol}/\text{kg}$$

For 0.1 kg of carbon combusting in 1 kg of air, the mass ratio of air is $O_2 = 0.233$, $N_2 = 0.767$. Total mass is 1.1 kg. This is useful because $C + O_2 \rightarrow CO_2$ which we shall consider in the section on combustion.

2.3.2.1 Work and Heat Transferred

The equations given in this volume and in the first volume for the work done and heat transferred during various processes undergone by a single, perfect gas can be used for a mixture of perfect gases. That is, the appropriate values of specific heats and gas constants are calculated from the values applicable to individual constituents (c_v, c_p, R and γ). Therefore, knowing the mixture, all other equations can be used.

LEARNING OUTCOMES

✓ **Gas mixtures** have several gases intimately mixed, which can be quantified in proportion by mass or by volume.

✓ The **Gibbs–Dalton** Law of Partial Pressures leads to proportion analysis by mass, or **gravimetric analysis**.

✓ **Amagat's Law of Partial Volumes** leads to analysis by volume proportions, or **volumetric** or **molar analysis**.

✓ Combining the two with the Gas Law leads to a useful relationship between partial volumes, molar proportions, and partial pressures of the gases: $\dfrac{V_i}{V} = \dfrac{n_i}{n} = \dfrac{p_i}{p}$, which can be used to make conversions between gravimetric and volumetric analyses.

2.4 COMBUSTION

Fuels produce energy by reacting with **oxidisers**. The oxidiser is commonly the oxygen in atmospheric air but may be embedded in a compound with high oxygen content; for example, nitrates with NO_3 (i.e. one nitrogen atom per molecule with three oxygen atoms) as seen in fireworks in Figure 2.20, which contain gunpowder: potassium nitrate (75% mass), charcoal (15% mass) and sulphur (10% mass). The fuel may be a solid (e.g. coal or biomass), a liquid (e.g. petrol which may be considered as octane C_8H_{18}) or a gas (e.g. natural gas which may be considered as methane, CH_4, as shown burning in Figure 2.20b or hydrogen, H_2). Since fuels mainly comprise hydrogen and carbon, they are called **hydrocarbons**. If sufficient starting energy is provided to initiate a reaction, then the

FIGURE 2.20   Combustion characteristics.

oxygen will react with the hydrocarbon fuel in combustion. This section outlines how to calculate the amount of air required to burn a given amount of fuel and the composition of the products of combustion formed.

Solid and liquid fuels are analysed by mass (gravimetric analysis). Gaseous fuels are analysed by volume (volumetric analysis). In this section, all fuels are burned in air, and it is assumed that air contains 23.3% O_2, 76.7% N_2 by mass or 21% O_2, 79% N_2 by volume as described in the introductory part of Section 2.2 on air conditioning. The ratio of nitrogen to oxygen by volume is 79/21 = 3.76. The ratio by mass is 76.7/23.3 = 3.29. These analyses correspond to a mean molar mass, $\tilde{m}$, of 28.85 [kg/kmol] and a corresponding mean specific gas constant R of 0.287 kJ/kgK. If the small percentage of CO_2 and Ar are included, then $M = 28.96$ kg/kmol and $R = 0.287$ kJ/kg K. This assumes approximate air data from tables of thermodynamic and transport properties of fluids.

2.4.1   STOICHIOMETRIC COMBUSTION

Molar reaction equations are used for the analysis of a combustion process, which begins by the formulation of the chemical equation which shows how the atoms of the reactants are combined to form the products.

How to balance equations
  e.g. butane $C_4H_{10} + xO_2 \rightarrow aCO_2 + bH_2O$
  The unknown quantities are labelled by the unknown number of moles for each (x, a and b) and the combustion of only 1 mole (if working in grams) or 1 kmol (if working in kg) of the fuel is considered. We know that since matter cannot be created or destroyed, the atoms on the reactant side must be present on the products side and so a count of atoms reveals that:

  For carbon, C: $4 = a$
  For hydrogen, H: $10 = 2b$; $b = 5$
  For oxygen, O: $2x = 2a + b$; $x = 6.5$

  Therefore,

$$C_4H_{10} + 6.5O_2 \rightarrow 4CO_2 + 5H_2O$$

By the law of conservation of mass, the number of atoms of each element is the same at the beginning as at the end of any chemical reaction, that is,

the number of atoms of reactants = the number of atoms of products.

The reaction equations are **molar**, that is, in terms of molecules reacting, and therefore we can make use of molar masses to find the masses of each element involved in the reaction, and the molar proportions are equivalent to volume proportions for gases as described in Section 2.3 in the subsection on molar analysis. Equations conserve:

a. The number of atoms of each element from reactants to products of combustion.
b. Mass of reactants = mass of products.

Usually, the number of moles (alternatively called mols) of reactants will not equal the number of moles of products.

In order to convert to mass fractions from mole fractions, use $n = \dfrac{m}{\tilde{m}} \Rightarrow m = n\tilde{m}$ as shown in the example below in which:

n = number of kmol in sample
m = mass of the sample
$\tilde{m}$ = molecular mass of the sample

Example: forming water.

H_2	$+\frac{1}{2}O_2$	$\rightarrow$	H_2O
1 kmol	$+\frac{1}{2}$ kmol	$\rightarrow$	1 kmol
2 kg	+16 kg	$\rightarrow$	18 kg
1 kmol × 2 kg/kmol	$+\frac{1}{2}$kmol × 32 kg/kmol	$\rightarrow$	1 kmol × 18 kg/kmol

The proportions can then be scaled up or down depending on the quanitity of fuel used. Similarly, if we have the masses, we can convert to molar equations.

Note: Incidentally, at this point, it is worth noting that the naming of hydrocarbon fuels gives away their chemical formula. The prefixes, meth-, eth-, prop-, but-, pent-, hex-, hept-, oct- etc denote the number of carbon atoms in the molecules increasing from 1 to 8 in this list. The suffixes –ane and –ene denote the bonding of the carbon atoms to each other within the molecules. A carbon atom can be considered as having four links with which to connect to other atoms. Therefore, methane has one carbon atom with its four links connected one each to a hydrogen atom, hence, CH_4. Ethane has two carbon atoms with one link connected between them and then each one having three links connected to a hydrogen, hence C_2H_6. Ethene (Figure 2.21a) has two carbon atoms with two links connected between them and each with only two links each to hydrogen atoms, hence C_2H_4. The schematic representation of butane (Figure 2.21b) gives the idea of the single carbon links and the single hydrogen links.

Stoichiometric combustion describes a combustion reaction in which all of the fuel and oxidiser are used to make the products, such that there is no leftover fuel or oxidiser.

A combustion process can be:

Complete: where sufficient O_2 is available to convert all carbon and all hydrogen in the hydrocarbon fuel to CO_2 and H_2O. There may be excess oxygen in the products.
Incomplete: where not enough O_2 is available and other products such as CO appear. Oxygen prefers to react with hydrogen, so all hydrogen is consumed before carbon starts to be converted into CO or CO_2.

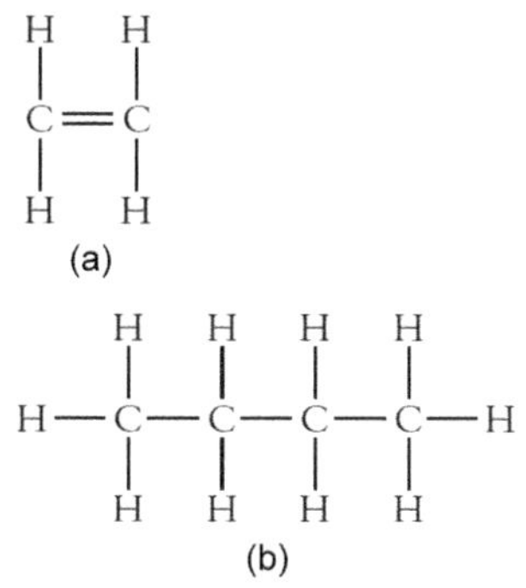

FIGURE 2.21   Schematic diagram of the structure of (a) ethene and (b) butane.

A stoichiometric reaction is one where all the oxygen is used up and all the fuel is burnt to the **ultimate** products (CO_2, H_2O), which are the most stable combinations of those atoms at atmospheric conditions. A stoichiometric combustion describes the most efficient use of oxygen and fuel, and it is useful to use the ratio of oxygen to fuel in the stoichiometric case in order to compare actual oxygen–fuel ratios. This can be done by volume or by mass. For example, if ethane is burned in oxygen, the equation is described by:

$$C_2H_6 + 3\tfrac{1}{2}O_2 \rightarrow 2CO_2 + 3H_2O$$

The stoichiometric oxygen/fuel ratio of C_2H_6 (by volume) is oxygen moles ÷ fuel moles = 3.5/1 = 3.5 – that is by number of moles. This can be done because of $V_i/V = n_i/n$, that is molar proportions represent volume proportions.

2.4.2   Air to Fuel Ratio

Most combustion applications use the oxygen in atmospheric air and therefore draw in significant amounts of nitrogen to the reaction. Nitrogen is not reactive at atmospheric conditions but can react at high temperatures, typically found in flames. Although there can be a reaction of nitrogen, it does not happen in significant quantity compared to the more aggressive carbon and hydrogen reactions and can be neglected with only a very small effect on the volume and mass fractions of the other reactions. It is a significant issue for air pollution since parts per million are sufficient to cause acid rain and respiratory problems which NOx pollutants can result in.

The stoichiometric air/fuel ratio (**AFR**) by volume is found from the stoichiometric oxygen:-fuel ratio equation but includes the nitrogen in the air as an inert gas. This is important not for the reaction but for the temperature of the flame – the more gas there is involved in the reaction which must be heated up from atmospheric conditions to the flame temperature, the lower will be the flame temperature.

Nitrogen is brought in with oxygen at a fixed ratio of:

$3.76 \times (O_2$ volume) – from approximate air data.

Total air volume includes both the N_2 and the O_2, that is air volume is $(1 + 3.76) \times (O_2$ volume). The stoichiometric AFR in the case of ethane combustion before is:

$$AFR_{volume} = (O_2 / fuel) \text{ volume ratio} \times 4.76$$
$$= 3.5 \times 4.76 = 16.7 \, (\text{byvolume})$$

This can be converted to stoichiometric AFR by mass by considering that volume fraction is equal to molar fraction and using the molar masses.

Since $n = m / \tilde{m}$,

$$\text{AFR}_{\text{mass}} = \frac{m_{\text{air}}}{m_{\text{fuel}}} = \text{AFR}_{\text{volume}} \frac{\tilde{m}_{\text{air}}}{\tilde{m}_{\text{fuel}}}$$

$$= 16.7 \frac{29}{30} = 16.14$$

in which the approximate molar mass of atmospheric air is 29 g/mol, and the molar mass of C_2H_6 is 30 g/mol. In practice, the complete reaction equation for the air and ethane stoichiometric case will include N_2 terms, where N_2 is treated as inert as in the following reaction equation:

$$C_2H_6 + 3\tfrac{1}{2}O_2 + 3\tfrac{1}{2} \times 3.76N_2 \rightarrow 2CO_2 + 3H_2O + 3\tfrac{1}{2} \times 3.76N_2$$

that is, $C_2H_6 + 3.5O_2 + 13.16N_2 \rightarrow 2CO_2 + 3H_2O + 13.16N_2$

2.4.3 NON-STOICHIOMETRIC COMBUSTION

This can either be **fuel rich** (insufficient O_2) or **fuel lean** (excess O_2). In these cases, combustion products can contain incomplete products such as CO and free oxygen as well as CO_2, H_2O, N_2 etc.

In the case of **excess fuel**, the incomplete combustion of the fuel leads to CO_2 and CO products (this is in the case of hydrocarbon fuels). The hydrogen is always the first to burn off because it is a much more aggressive reaction than the carbon reactions, so it seizes any oxygen present before the carbon can get to it. In real processes, there may be both oxygen (O_2) and CO in the products, because of incomplete mixing of fuel and air.

In the case of **excess O_2**, the products will include free oxygen, but the fuel is burnt to the ultimate products. The % excess air supplied is a useful indicator for mixture quality. This is defined as:

$$\% \text{ excess air} = \frac{\text{air supplied} - \text{minimum air for stoichiometric combustion}}{\text{minimum air for stoichiometric combustion}} \times 100 = \frac{V_a - V_{a,\text{stoich}}}{V_{a,\text{stoich}}}$$

In which V_a is the actual volume of air used and $V_{a,\text{stoich}}$ is the required stoichiometric volume of air to burn the fuel. This is the same by volume and by mass.

Dividing throughout by V_f, the volume of fuel used in the reaction:

$$\frac{\dfrac{V_a}{V_f} - \dfrac{V_{a,\text{stoich}}}{V_f}}{\dfrac{V_{a,\text{stoich}}}{V_f}}$$

But

$$\frac{V_a \tilde{m}_a}{V_f \tilde{m}_f} = \frac{m_a}{m_f}$$

Therefore,

$$\frac{\dfrac{V_a \tilde{m}_a}{V_f \tilde{m}_f} - \dfrac{V_{a,\text{stoich}}}{V_f}\dfrac{\tilde{m}_a}{\tilde{m}_f}}{\dfrac{V_{a,\text{stoich}}}{V_f}\dfrac{\tilde{m}_a}{\tilde{m}_f}} = \frac{\dfrac{m_a}{m_f} - \dfrac{m_{a,\text{stoich}}}{m_f}}{\dfrac{m_{a,\text{stoich}}}{m_f}} = \frac{m_a - m_{a,\text{stoich}}}{m_a}$$

2.4.4 WET AND DRY PRODUCTS

Since the products of hydrocarbon combustion contain H_2O, that is steam, which condenses when the product temperature is reduced to the atmospheric condition, it is useful to consider the products as wet when the water is contained as a vapour and dry when the vapour is considered to have condensed out of the products. This gives rise to the terms **wet products** and **dry products** of combustion. It also affects the energy value of the fuel, such that less energy is used for heating when the products escape to atmosphere containing the water as vapour without using the latent heat of condensation that it contains for a useful purpose. Notably, modern domestic *condensing* boilers make use of the energy in the water vapour by condensing it in the exhaust stream on the cold water entry pipes, and capturing some of its energy to initially warm the cold entry water. A gas analyser can be used to measure the component gases of a combustion exhaust stream to test how complete the combustion is, which informs the state of mixture of the incoming gases.

2.4.4.1 Example 1

Determine the stoichiometric AFR:

 (a) by mass
 (b) by volume

for propane, C_3H_8. Calculate the volumetric composition by wet and dry analyses of the products. If the fuel is burnt with 30% excess air, what are the volumetric and gravimetric compositions of the wet products?

Aim:

To realise stoichiometric AFR by mass and by volume.
To realise the importance of wet and dry analysis of product gases.

Stoichiometric AFR:

(a) by mass

Reaction equation of the fuel:

$$C_3H_8 + 5O_2 \rightarrow 3CO_2 + 4H_2O$$
$$44\,kg + 5(32\,kg) \rightarrow 3(44\,kg) + 4(18\,kg)$$

The oxygen to fuel ratio by mass is therefore 5 × 32 / 44 = 3.636. The AFR can be worked out using the approximate ratio of air to oxygen in atmospheric air, which is 1:0.233 by mass:

$$\frac{air[kg]}{fuel[kg]} = \frac{oxygen[kg]}{fuel[kg]} \times \frac{1}{0.233} = 3.636 \times 4.29$$

The AFR by mass is therefore 15.6.

(b) by volume

The oxygen to fuel ratio by volume is 5:1 from the reaction equation above, remembering that ratio by volume is the same as ratio by number of moles. To find the AFR by volume, use the approximate ratio of air to oxygen in atmospheric air, which is 1:0.21 by volume:

$$\frac{air[kmol]}{fuel[kmol]} = \frac{oxygen[kmol]}{fuel[kmol]} \times \frac{1}{0.21} = 5 \times 4.76$$

The AFR by volume is therefore 23.8.

The complete representative equation with all mixture components of the air–fuel stream includes nitrogen in the approximate ratio for atmospheric air. Including this in the reaction equation gives:

$$C_3H_8 + 5\left(O_2 + {}^{0.79}\!/_{0.21}N_2\right) \rightarrow 3CO_2 + 4H_2O + 5 \times {}^{0.79}\!/_{0.21}N_2$$

In order to get a volumetric analysis of wet and dry products, use a table for compactness. Wet includes the water in the reaction and dry does not include the water.

Component	Wet V_i	Wet V_i	Dry V_i	Dry V_i
CO_2	3	0.116	3	0.138
H_2O	4	0.155	0	0
N_2	18.81	0.729	18.81	0.862
Total	25.81	1	21.81	1

When there is 30% excess air in the reacting stream of air and fuel, there will be 30% more O_2 and N_2 in the representative equation:

$$C_3H_8 + 5 \times 1.3\left(O_2 + {}^{0.79}\!/_{0.21}N_2\right) \rightarrow 3CO_2 + 4H_2O + 0.3 \times 5O_2 + 5 \times 1.3 \times {}^{0.79}\!/_{0.21}N_2$$

The list of products can be made for gravimetric and volumetric analyses:

Component	v_i	V_i/v	$\tilde{m}_i$	$m_i = \tilde{m}_i \dfrac{v_i}{v}$	$\dfrac{m_i}{m} = \dfrac{\tilde{m}_i}{\tilde{m}} \dfrac{v_i}{v}$
	[kmol]	[–]	[kg/kmol]	[kg/kmol]	
CO_2	3	0.091	44	4	0.141
H_2O	4	0.122	18	2.187	0.077
O_2	1.5	0.046	32	1.456	0.051
N_2	24.45	0.742	28	20.78	0.731
Σ	32.95	1		$\Sigma = \tilde{m} = 28.4$	1

Note that the column of mass of individual mixture species components has units kg/kmol and not just kg. This happens because when multiplying by the volume fraction, we are considering per unit volume, and that is equivalent to per kmol, of the total product gases. The sum of this column, strictly speaking, makes up the molecular mass of the product gases as an incidental by-product of the table operations because we use kmol as a convenient total sample size of the product gases. The final column calculating the ratio of each gas to the total mass of product gases uses the molecular mass of the previous column by simply dividing each species component m_i by the sum of the m_i column.

2.4.4.2 Example 2

Benzene C_6H_6, shown in Figure 2.22, is burned in air and the dry products of combustion contain 2% CO by volume and no free oxygen. Determine the stoichiometric AFR by mass and the actual AFR by mass.

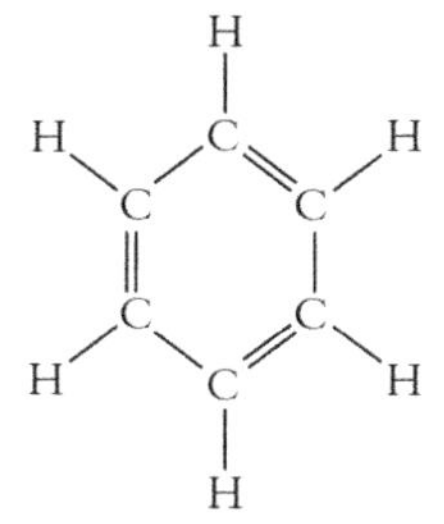

FIGURE 2.22   Schematic diagram of the structure of benzene.

Aim:

To calculate excess air supplied.

To show how to use the AFR to indicate sub-stoichiometric combustion. The benzene molecule is complex as shown in Figure 2.22, in the form of a ring with a mixture of double and
single links between the carbon.

For the stoichiometric combustion case, the reaction of the fuel with oxygen can be written as:

$$C_6H_6 + 7\tfrac{1}{2}O_2 \rightarrow 6CO_2 + 3H_2O$$

$$78\,kg + 7\tfrac{1}{2}(32\,kg) \rightarrow 6(44\,kg) + 3(18\,kg)$$

The oxygen to fuel ratio by mass is therefore 7½ × 32:78 = 3.077.

The AFR by mass can be calculated using the approximate ratio of air to oxygen in atmospheric
air by mass, which is 1:0.233 = 4.29:

$$\frac{air\,[kg]}{fuel\,[kg]} = \frac{oxygen\,[kg]}{fuel\,[kg]} \times \frac{air\,[kg]}{oxygen\,[kg]} = 3.077 \times 4.29$$

The AFR by mass is therefore 13.2.

For the actual combustion, insufficient oxygen is supplied to burn the fuel completely. We assume
the order of the reaction of the atoms in the fuel is hydrogen first completely as it is the fastest and
by far the most aggressive, and subsequent to this the carbon burns, which can be considered as
firstly all to carbon monoxide and further to this, some CO further reacts to form CO_2 to consume
the remaining oxygen in cases where there is sufficient O_2 to complete the reaction for all of the
CO. The lack of sufficient air therefore leads to some CO remaining in the exhaust gas stream.

We assume in the representative reaction that there is a proportion of the stoichiometric O_2
required and write:

$$C_6H_6 + y\left(O_2 + \frac{0.79}{0.21}N_2\right) \rightarrow aCO_2 + bCO + cH_2O + \frac{0.79y}{0.21}N_2$$

The first step is to conduct an atomic balance in terms of y, such that when proportion y is known,
the equation balances identical quantities of elements on each side.

Carbon: $6 = a + b$
Hydrogen: $6 = 2c$, that is $c = 3$.
Oxygen: $2y = 2a + b + c$ and substituting c, and $a = 6-b$, $2y = 12-2b + b + 3$ or $b = 15-2y$
   and $a = -9+2y$.

The second piece of information is the proportion of CO in the dry products by volume, which is 2%. We therefore create the proportion from the above equation and equate to 0.02:

$$\frac{b}{a+b+3.76y} = 0.02$$

Substituting $b = 2y-15$ and $a + b = 6$,

$$\frac{15-2y}{6+3.76y} = 0.02$$

Therefore, $y = (15 - 0.12) / (2 + 0.0752) = 7.17$.

The AFR by mass for the actual combustion process is:

$$\frac{air\left[kg\right]}{fuel\left[kg\right]} = \frac{7.17\times\left(32\,kg/kmol+3.76\times28\,kg/kmol\right)}{78\,kg} = \frac{983}{78} = 12.6$$

2.4.4.3 Example 3

A particular coal has a mass analysis of 81% C, 5% H, 5% O and 9% ash. The volumetric analysis of the dry products is:

10% CO_2, 1% CO, 8% O_2, 81% N_2

Determine the minimum air required for the complete combustion of 1 kg of coal and the percentage of excess air required.

Aim:

To realise the ash content effect on AFR.
To determine stoichiometric reaction.
To realise the effect of oxygen in coal on the stoichiometric reaction.
To calculate excess air supplied.

Solution

In this case, it is useful to determine the required oxygen by working out for each component how much oxygen is required to burn it individually. So, instead of a representative reaction equation, it is more obvious to create a table:

Constituent	Reaction	Mass of Constituent per kg of Coal	Mass of O_2 Required per kg of Constituent	Oxygen Required
Carbon	$C + O_2 \rightarrow CO_2$	0.81	32/12 = 2.67	2.16
Hydrogen	$H_2 + \frac{1}{2}O_2 \rightarrow H_2O$	0.05	16/2 = 8	0.4
Oxygen		0.05	−1	−0.05
Ash		0.09	0	0

Adding the oxygen required in the last column per kg of coal, 2.51 kg of O_2 required. For complete combustion, and the air required for 1 kg of coal is obtained by multiplying this mass of oxygen by the gravimetric (i.e. mass proportion) air:oxygen ratio by mass in the atmosphere, which is 1:0.233 = 4.29.

Therefore, the complete combustion of 1 kg of the coal requires $4.29 \times 2.51 = 10.77$ kg of air.

To find the actual air supplied, we have to have regard for the information presented by the constituents of the product gas stream. This contains nitrogen that can only have come from the air supplied since there is no nitrogen in the coal constituents. The carbon came in a fixed proportion of the mass of the coal (constituents stated in the question), and if we can work out the nitrogen to carbon ratio, we can infer the air–carbon ratio, and hence, the air–coal ratio. The other useful information is that the ratio of products is given by volume (in product gas constituents stated in the question), and we therefore know the molar proportions, which then give not only the mass proportions of each species molecule but also the mass proportions of the species atoms.

For 1 kmol of dry product gases, there is 0.81 kmol of N_2 (as indicated in the dry product gas constituents in the question), with mass $0.81 \times 28 = 22.68$ kg. This is carried in with a proportion of air by mass 1:0.767 or 22.68 / 0.767 = 29.57 kg.

There is 0.1 kmol of CO_2, containing 0.1 kmol of carbon, with mass $0.1 \times 12 = 1.2$ kg, and there is 0.01 kmol of CO, containing 0.01 kmol of carbon, with mass $0.01 \times 12 = 0.12$ kg. The total amount of carbon in 1 kmol of product gas is therefore 1.32 kg, which is carried in the proportion 1:0.81 of coal to carbon in the coal constituents. Therefore, the mass of coal corresponding to 1.32 kg of carbon in it is 1.32/0.81 = 1.63 kg.

The air to coal ratio is therefore 29.57:1.63 = 18.14 by mass.

Therefore, percentage excess air is by the difference of this and the stoichiometric AFR:

$$\text{excess air } \% = \frac{18.14 - 10.77}{10.77}$$

Or 68.4%.

2.4.5   How Solids Burn

Coal, plastics and wood all burn in an oxygen environment, and the pattern of combustion is very familiar if not widely recognised. The combustion has characteristic stages as follows: initial heating with evaporation of any moisture content up to the temperature at which volatile organic matter melts and begins to evaporate. This stage is called devolatilisation involving the evaporation of volatiles – which are low-boiling-point hydrocarbons in the fuel. The volatile molecules burn in the surrounding gas giving the familiar flame of airborne combustion. There then follows smouldering combustion of the char content that is left behind, until only the ash is left. Char is carbon, which is solid, does not evaporate and relies on diffusion of oxygen in through the pores of the solid matter to burn. As long as the diffusion rate through the solid is sufficiently fast to produce a reasonable amount of combustion, whilst insufficient cooling occurs, then the char remains at high temperature sufficient to initiate combustion at new sites penetrating deeper into the body. This type of solid-bound combustion has a characteristic glow dependent on the temperature of the char. Figure 2.23 shows the early stages of combustion of a piece of rubber, showing a sooty flame as volatile molecules are released, some are burned to gas products and some are cooled faster than they can be burned, forming carbon heavy particles which form soot.

2.4.6   Combustion Energy

This section describes how to calculate the energy released by combustion and the temperature of the products. The temperature affects the mechanical strengths of the materials of the combustion vessel as well as thermal efficiency. Combustion, just like other processes considered before, can be a closed system process, as in an IC engine when the valves are closed or an open process as in a jet engine or the furnace of a water boiler in power generation. As before, it is the First Law that we are interested in. The two forms are the direct First Law and the derived SFEE for each of the processes. Combustion releases chemical energy which is converted into thermal energy, internal energy in the closed system and enthalpy in the open system.

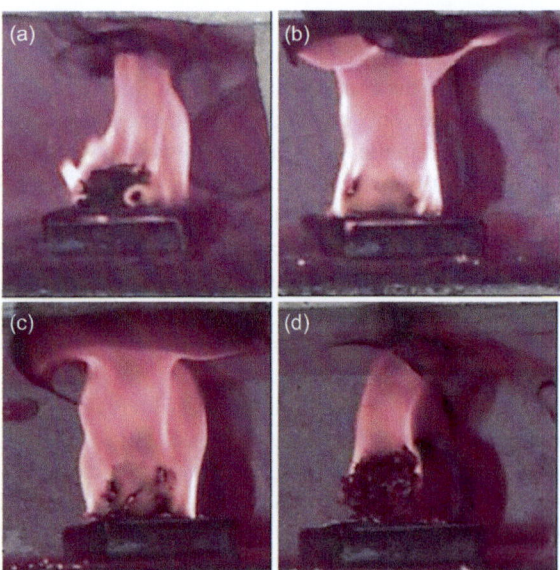

FIGURE 2.23   Initial stage of combustion of a piece of rubber, (a) initial combustion, (b) degradation of the structure as the flame penetrates and volatile components are released and burnt as the solid matter expands when its structure is released, (c) further combustion throughout of volatile matter with large amount of soot produced due to incomplete combustion, (d) final stages of the flame as volatile matter is exhausted and the solid combustion of char takes over as main form of combustion.

2.4.7   INTERNAL ENERGY (ΔU_0) OF COMBUSTION

The closed system is the easiest to consider to start with for a non-flow process. To use the First Law of Thermodynamics, $Q + W = \Delta U$, we need to know when the energy is released from combustion, how much is delivered into heat and work transfer and how much is taken to increase the internal energy of the working fluid. ΔU_0 is the **internal energy of combustion** at reference conditions of temperature T_0 (usually 25°C) and pressure p_Θ (1 bar). This is determined from the First Law applied to a constant volume, closed system process, as shown in the following.

For an analysis of closed system combustion, refer to the schematic diagram of the process in Figure 2.24.

The reactants are brought together; at V_0, T_0, the fuel and air mixture is ignited and burnt.

Then, combustion takes place; this releases chemical energy which is transformed into 'sensible' internal energy; that is KE of the molecules which dictates the temperature of the mixture.

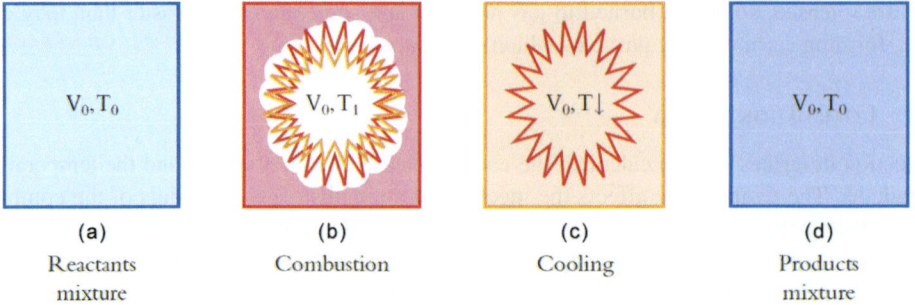

FIGURE 2.24   Schematic diagram of combustion in a closed vessel. (a) reactants mixture, (b) combustion, (c) cooling, (d) products mixture

Some changes to properties will also occur and the chemical species will change as the fuel reacts chemically with the oxygen in the air. Still at V_0, temperature would have increased.

The products are cooled back to T_0 and the amount of heat removed to do so is recorded. Notice that bringing the products back to temperature T_0 is a standardising procedure – it allows us to consider in a standard way the quantity of heat available from a combustion reaction.

Apply the First Law to the processes (a) to (c) $\dot{Q} = -\dot{W} + U_C - U_A$. But there is no work because it is a constant volume process. $\dot{Q} = U_{P0} - U_{R0}$ where subscript P indicates product species mixture and R indicates reactant species mixture and subscript zero indicates the start temperature. $\dot{Q} = U_{P0} - U_{R0} = \Delta U_0$, since Q is negative, because heat is transferred from the system to surroundings. Then, $U_{R0} > U_{P0}$ and the internal energy of combustion is negative. The internal energy of combustion tells us how much heat has to be removed when the reactants combust to the products and the heat removed brings the products back to the standard state temperature, subscript 0. So, the heat of this equation is the heat released by the reaction.

We can use this in the case where an alternative final state is arrived at – we will see how to do this in Section 2.4.10.

2.4.8 ENTHALPY (ΔH_0) OF COMBUSTION

For steady flow process combustion, the situation is similar except that the SFEE $Q + W = \Delta H$ (neglecting KE and PE) is used. In the same way as for the closed system, but where a standard state internal energy of combustion is used in the closed system, in the open system ΔH_0, the **enthalpy of combustion** at reference temperature T_0 (25°C) and pressure (1 bar) is used.

Analysis of open system combustion:

ΔH_0 can be determined from the SFEE for a flow process with no work and heat transfer to restore the temperature of the products to initial temperature of the reactants, T_0. Consider the situation shown in Figure 2.25, in which the combustion takes place from the standard state condition at point 1 (i.e. temperature is T_0), releasing energy. The combustion energy released in the flame is removed through the walls of the system as heat to reduce the hot combustion product gases temperature to the initial temperature of the reactants, T_0. Writing the SFEE including KE,

$$H_1 + \dot{m}\frac{C_1^2}{2} + \dot{Q}(-ve) = H_2 + \dot{m}\frac{C_2^2}{2}$$

$$Q = (H_2 - H_1) + \dot{m}\left(\frac{C_2^2}{2} - \frac{C_1^2}{2}\right)$$

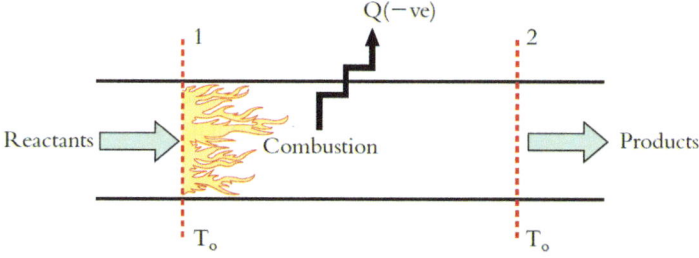

FIGURE 2.25 Open system combustion model. Note the beginning and end conditions are forced to be at the standard state.

The KE terms are often negligibly small compared to the enthalpy, and the equation simplifies to:

$$Q = H_{P0} - H_{R0} = \Delta H_0$$

For a combustion reaction, Q is negative, that is the system produces heat that is extracted from the working fluid via the containing system walls, to bring the products back to the standard condition temperature, T_0. Since the flow is in a duct of some kind, it is often the case that pressure does not change significantly across the flame.

2.4.9   CONVERTING ΔH_0 AND ΔU_0 FROM $H = U + pV$

The values of the enthalpy of a reaction are stated in data tables, but the internal energy of a reaction is not necessarily stated. If we want to know the internal energy of reaction for a closed process, then we must be able to work it out. It is therefore useful to be able to convert from one to the other, which is based on the formula for enthalpy as follows:

$$\Delta H_0 = H_{P0} - H_{R0} = \left(U_{P0} + p_{P0} V_{P0}\right) - \left(U_{R0} + p_{R0} V_{R0}\right)$$

$$\Delta H_0 = \left(U_{P0} - U_{R0}\right) + \left(p_{P0} V_{P0} - p_{R0} V_{R0}\right)$$

$$\Delta H_0 = \left(\Delta U_0\right) + \left(p_{P0} V_{P0} - p_{R0} V_{R0}\right)$$

- For solid and liquid components, pV is negligible
- For perfect gas components, $pV = nRT$

$$\Delta H_0 = \Delta U_0 + \left(n_P - n_R\right) R T_0$$

n_p = kmols of products per kmol of fuel burnt,
n_R = kmols of reactants per kmol of fuel burnt,
ΔH_0 and ΔU_0 are in kJ/kmol of fuel burnt.

Usually, energy contributions are considered in terms of mass, and so conversion from kJ/kmol to kJ/kg must be done before use in the SFEE. The enthalpy of a reaction is stated for common reactions in the data tables. There is a very easy way to work them out from the **enthalpy of formation** for the molecules involved in the reaction which are also given in the tables of data in order to determine the reaction enthalpy for a wider range of reactions. Formation enthalpies are stated in kJ/kmol for molecules, and represent the energy required to form the molecule from the constituent atoms. The formation enthalpies of molecules are dependent on their structure, but are often zero. All diatomic molecules have formation enthalpies of zero. To calculate the energy released by a chemical reaction, write the reaction down, write the formation enthalpy of each molecule below the reaction equation and multiply by the number of moles involved in the reaction. The difference between the sum of reactant formation enthalpies, h_{f0}^{θ} (the superscript θ indicates standard condition, 25°C), and the sum of the product formation enthalpies is the reaction energy. For example, consider the previous example of propane in Section 2.4.4.1 and writing firstly the mass of each constituent and then the formation enthalpy contributions from the specific formation enthalpy, h_{f0}^{θ} [kJ/kmol]:

$$C_3H_8 + 5O_2 \rightarrow 3CO_2 + 4H_2O$$
$$44\,\text{kg} + 5\left(32\,\text{kg}\right) \rightarrow 3\left(44\,\text{kg}\right) + 4\left(18\,\text{kg}\right)$$
$$-104{,}000 + 5 \times 0 \rightarrow 3 \times -393{,}520 + 4 \times -285{,}820$$

The enthalpy on the reactant side is −104,000 kJ and the enthalpy of the products is −2,323,840 kJ. Therefore, for the combustion of 1 kmol of propane, the reaction energy released is $-2,323,840 - (-104,000) = -2,219,840$ kJ. This is a lot of energy, and it is important to remember that it comes from 1 kmol of propane, which weighs $3 \times 12 \text{ kg} + 8 \times 1 \text{ kg} = 44$ kg, which is a substantial amount of fuel.

2.4.10 CALORIFIC VALUES

Δh_0 and Δu_0 are usually given at 25°C, per kg or kmol $\left(\tilde{h} \right)$ of burnt fuel. In the tables, Δh_0 is given and stated as $\Delta \tilde{h}^\theta$ kJ/kmol at $T = 25$°C. The phase of the reactants and products which could be vapour or liquid (e.g. H_2O) should be specified. Alternatively, calorific values are quoted instead of Δh_0 and Δu_0. These are defined according to British Standards.

> **Gross (or Higher) CV** at constant volume ($Q_{gr,v}$), $\Delta U_{25°C}$, with H_2O in products as liquid.
> **Net (or Lower) CV** at constant volume ($Q_{net,v}$), $\Delta U_{25°C}$, with H_2O in products as vapour.
> **Gross (or Higher) CV** at constant pressure ($Q_{gr,p}$), $\Delta H_{25°C}$, with H_2O in products as liquid.
> **Net (or Lower) CV** at constant pressure ($Q_{net,p}$), $\Delta H_{25°C}$, with H_2O in products as vapour.

By convention, calorific values are stated as positive values; however, we know that according to the convention, positive energy is energy into the system, that a combustion reaction produces negative heat for use in the First Law equations. The difference between gross and net is due to the state of the water in the product gases. Look at the tables of formation enthalpies: H_2O (liquid) is −285,820 kJ/kg and H_2O (vapour) is −241,830 kJ/kg. The difference is the enthalpy difference between liquid and vapour water at 25°C. This is the contribution of the H_2O content to the product gases overall enthalpy. Hence, the higher (more negative) enthalpy difference is with the water as liquid because it reclaims the energy of condensation before the gases are released to the atmosphere.

2.4.11 GAS MIXTURE CONDITIONS WHEN INLET AND OUTLET CONDITIONS ARE NOT AT STANDARD CONDITIONS

Consider the closed system combustion, like the situation in Figure 2.25, but this time with the beginning and end conditions that are different to the standard state. If initial and final states are not at the reference state temperature T_0, of 25°C, then artificial processes are considered in order to

---

EXAMPLE

What is the specific enthalpy change for carbon dioxide between 25°C and 327°C?

Solution:
For CO_2, molar enthalpy at 327°C or 600 K is 12,916 kJ/kmol. At 25°C or 298.15 K, it is zero.
So $\tilde{h}_{600K} - \tilde{h}_{29.15K} = 12,916 \text{ kJ / kmol}$ and since $\tilde{m}_{CO_2}$ is 44 kg/kmol, specific enthalpy change is $h_{600K} - h_{298K} = 293.5$ kJ/kg.

c.f. The alternative method of computation is to use specific heat capacity by using the average value of the maximum and minimum temperature. $c_p = 0.846$ kJ/kgK at 300 K and $c_p = 1.075$ kJ/kgK at 600 K.

$$h_{600K-298K} = mc_p\Delta T = 1 \times \left((0.846 + 1.075)/2 \right) \times 300 = 288.15 \text{ kJ / kg}.$$

> The first method with enthalpy is more accurate since it makes no assumptions and uses differences stated in tables of data that are empirically confirmed. The second method assumes that the change of c_p with temperature is linear – it is not quite accurate to do so.

adjust conditions such that combustion takes place at the reference temperature T_0. The reactants and products are assumed to be perfect gases (plus liquids and solids) so that h and u depend on T and only T. With these assumptions, we can use $\Delta u = c_v \Delta T$ and $\Delta h = c_p \Delta T$ for gas mixtures heating and cooling, or more accurately the absolute values of h and u directly from tables, to calculate the difference in enthalpy and internal energy between reference and actual states. The values of the molar enthalpy or energy content (*note now just enthalpy not formation enthalpy or combustion enthalpy*) and the molar internal energy are given for common gases in the tables of data over a large range of temperatures. The specific values, in order to consider fractions by mass rather than volume, are calculated by dividing by the molecular mass; for example, $\tilde{h}\left[kJ/kmol \right] / \tilde{m}\left[kg/kmol \right] = h\left[kJ/kg \right]$ where the wavy line accent indicates molar quantities.

2.4.12 Closed System Combustion

Having determined that the internal energy and enthalpy changes of gases between different temperatures can be determined, we can now return to the question at hand, which is to determine the enthalpy (or internal energy for the closed system) difference from the standard state for the beginning and end states of combustion. Figure 2.26 shows a schematic of a combustion in a closed cylinder with a moveable piston for extracting work, similar to an IC engine. In the analysis of the combustion, we know what the energy release is at standard conditions (1 bar, 298.15 K). So, our analysis breaks the process 1–2 into a number of artificial steps:

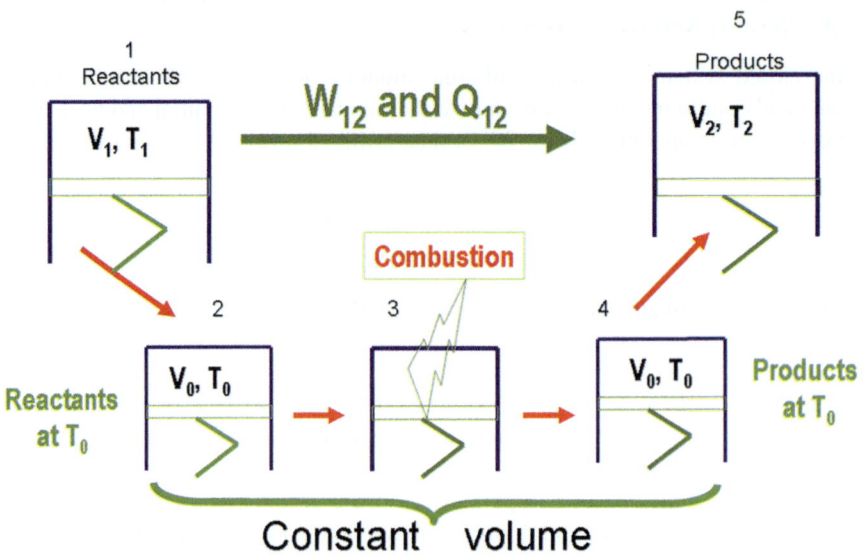

FIGURE 2.26 Closed system combustion case with conditions at beginning (1) and end (5) that are different to standard state. A cylinder is depicted similar to an internal combustion engine, and 3 intermediate artificial conditions are inserted to aid in the analysis of the process: (2) cool to T_0 and allow the volume to change to V_0 (3) combustion reaction at constant volume (4) cool to T_0 and V_0.

1. Start with the gas at the initial conditions v_1, T_1.
2. Make the gas go from the initial conditions v_1, T_1 to the standard conditions v_0, T_0 and note the heat and work to do this;
3. As in the analysis for Figure 2.24, combust and release the reaction energy;
4. As in the analysis for Figure 2.24, cool back to T_0 noting the energy release by heat transfer to do so;
5. Make v_0 T_0, the standard conditions go to final conditions v_2, T_2 noting the heat and work to do this.

Note that there are only two real states: at the beginning 1 and at the end 2. The three intermediate artificial conditions are inserted between the beginning and end states to simplify the analysis because we know what happens with energy release at standard conditions. This assumes that we know the start and end conditions and wish to know the energy released.

From applying the First Law to the processes (1) to (2),

$$Q_{12} + W_{12} = U_2 - U_1 \text{ but}$$

$$U_2 - U_1 = U_{P2} - U_{R1}$$

$$U_2 - U_1 = \left(U_{P2} - U_{P0}\right) - \left(U_{R1} - U_{R0}\right) + \left(U_{P0} - U_{R0}\right)$$

$$U_2 - U_1 = \underbrace{\sum m_i c_{vi} \left(T_2 - T_0\right)}_{\text{products}} - \underbrace{\sum m_i c_{vi} \left(T_1 - T_0\right)}_{\text{reactants}} + \Delta U_0$$

where U, with capital, indicates extensive internal energy for the whole quantity of gases considered, which comprises perhaps several gas species, each one indicated by subscript 'i' or with tables for values of internal energy.

$$U_2 - U_1 = \underbrace{\sum \dot{m}_i \left(u_{i2} - u_{i0}\right)}_{\text{PRODUCTS}} - \underbrace{\sum \dot{m}_i \left(u_{i1} - u_{i0}\right)}_{\text{REACTANTS}} + \dot{m}_{\text{fuel}} \Delta u_0$$

where subscripts 1 and 2 are the beginning and end conditions and subscript 0 is the standard state temperature condition (25°C) and i indicates a gas constituent and the sum operator is used to add the contributions of the various gas constituents. Note that internal energy does not depend on pressure for a perfect gas. Given the conditions that occur between 1 and 2, heat and work exchanges with the surroundings can then be calculated. For example, if we know that the system is perfectly insulated, then the energy change must be the work done by the surroundings on the working fluid; alternatively, if there is no machinery to extract work, then the energy must be exchanged as heat with the surroundings.

2.4.13 OPEN SYSTEM COMBUSTION

The open system can be handled in the same way as the closed system; apply the same idea as for the closed system of bringing reactants to the standard state T_0 and the products returned from T_0 to actual state 2. Figure 2.27 is a schematic representation of an open system, with the heat and work exchanges indicated.

Apply the SFEE:

$$H_1 + \dot{m}\frac{C_1^2}{2} + \dot{Q}_{12} = -\dot{W}_{12} + H_2 + \dot{m}\frac{C_2^2}{2}$$

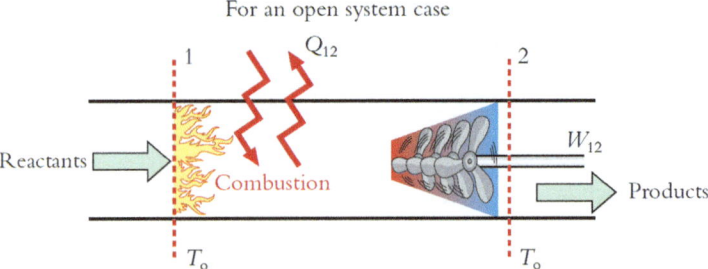

FIGURE 2.27   Open system combustion schematic in which the beginning and end states are at the standard condition.

$$\dot{Q}_{12} + \dot{W}_{12} = \left(H_2 - H_1\right) + \dot{m}\left(\frac{C_2^2}{2} - \frac{C_1^2}{2}\right)$$

Then, KE is negligible compared to the enthalpy terms:

$$\dot{Q}_{12} + \dot{W}_{12} = H_2 - H_1$$

Noting that condition 2 is for the products and that condition 1 is for the reactants:

$$\dot{Q}_{12} + \dot{W}_{12} = H_{P2} - H_{R1}$$

Now like the treatment of the closed system case, reduce the product condition to standard state, and the same for the reactants:

$$\dot{Q}_{12} + \dot{W}_{12} = \left(H_{P2} - H_{P0}\right) - \left(H_{R1} - H_{R0}\right) + \left(H_{P0} - H_{R0}\right)$$

Note that the last enthalpy difference term is the combustion enthalpy at standard conditions and using the expression for enthalpy from specific heat capacity at constant pressure:

$$\dot{Q}_{12} + \dot{W}_{12} = \sum m_i C_{pi}\left(T_2 - T_0\right) - \sum m_i C_{pi}\left(T_1 - T_0\right) + \Delta H_0$$

Alternatively, using the enthalpy of the gases from tables of data for a more accurate solution:

$$\dot{Q}_{12} + \dot{W}_{12} = \sum m_i\left(h_{i2} - h_{i0}\right) - \sum m_i\left(h_{i1} - h_{i0}\right) + \Delta H_0$$

2.4.13.1   Example 4

A mixture of CO and air is 10% fuel rich by volume and at 8.28 bar and 555 K. The mixture burns adiabatically and at constant volume. Calculate the product's temperature neglecting dissociation.

Solution:

The representative stoichiometric reaction equation gives the ideal reaction, which shows the required oxygen and nitrogen to thoroughly combust all fuel:

$$CO + \tfrac{1}{2}O_2 + \tfrac{1}{2} \times 3.76N_2 \rightarrow CO_2 + \tfrac{1}{2} \times 3.76N_2$$

We know that the reaction is not stoichiometric, because we are told that the mixture is 10% rich – that is, there is 10% more CO than can be burnt with the stoichiometric air in the above equation. Therefore, the true representative reaction equation will have 1.1CO. Thus,

$$1.1CO + \tfrac{1}{2}O_2 + \tfrac{1}{2} \times 3.76N_2 \rightarrow CO_2 + \tfrac{1}{2} \times 3.76N_2 + 0.1CO$$

Alternatively, divide by 1.1 to get the reaction in terms of combustion of 1 kmol of fuel:

$$CO + 0.91\left(\tfrac{1}{2}O_2 + \tfrac{1}{2}\times 3.76N_2\right) \rightarrow 0.91CO_2 + 1.71N_2 + 0.09CO$$

In order to determine the final temperature, we need to analyse the closed system with the method described above:

$$U_2 - U_1 = \sum_{\text{PRODUCTS}} m_i c_{v,i}\left(T_2 - T_0\right) - \sum_{\text{REACTANTS}} m_i c_{v,i}\left(T_1 - T_0\right) + \Delta U_0$$

The internal energy of combustion can be obtained from the tabulated values of enthalpy of combustion and the equation to convert between the two:

$$\Delta H_0 = \Delta U_0 + \left(n_P - n_R\right)RT_0$$

$\Delta \tilde{h}_0$ for CO is –283,000 kJ/kmol for the reaction $CO + \tfrac{1}{2}O_2 \rightarrow CO_2$

$$\Delta \tilde{u}_0 = \Delta \tilde{h}_0 - \left(n_P - n_R\right)\tilde{R}T_0$$

$$\Delta \tilde{u}_0 = -283,000 - \left(1 - 1.5\right)8.314 \times 298$$

Therefore, the internal energy of combustion per kmol of CO is –281,761 kJ/kmol.

To put this in specific terms, divide the molecular mass of CO, 28 kg/kmol, to give –10,063 kJ/kg. The First Law can now be applied, for adiabatic ($Q = 0$) and constant volume ($W = 0$) to give:

$$U_2 - U_1 = 0$$

where U_2 is the total internal energy of the products at the end of the process and U_1 is the total internal energy of the reactants before combustion. Knowing the temperature of the reactants, we can work out the internal energy of the reactants before combustion. We don't know the final temperature of the products – it is what is asked for. Using the equation that was developed in Section 2.4.12,

$$U_2 - U_1 = \sum_{\text{PRODUCTS}} m_i c_{v,i}\left(T_2 - T_0\right) - \sum_{\text{REACTANTS}} m_i c_{v,i}\left(T_1 - T_0\right) + \Delta U_0$$

Or more accurately, the other form of the equation with molar internal energy:

$$U_2 - U_1 = \sum_{\text{PRODUCTS}} n_i\left(\tilde{u}_{i2} - \tilde{u}_{i0}\right) - \sum_{\text{REACTANTS}} n_i\left(\tilde{u}_{i1} - \tilde{u}_{i0}\right) + n_{\text{fuel}}\Delta \tilde{u}_0$$

Reading values of molar internal energy from the tables to complete the following table of reactants:

Species	n_i kmol	$\tilde{u}_{i,555K}$ kJ/kmol	$n_i\tilde{u}_{i,555K}$ kJ
CO	1	2,984	2,984
O$_2$	0.455	3,233	1,471
N$_2$	1.71	2,944	5,034
Σ			9,489

where the internal energy at 555 K is found from the tables by interpolation. The final column is the internal energy per kmol of CO burned. The values are stated at 400 K and 600 K for a number of gaseous species. The interpolation is done as:

$$\tilde{u}_{i,555K} = \frac{555-400}{600-400}\left(\tilde{u}_{i,600K} - \tilde{u}_{i,400K}\right) + \tilde{u}_{i,400K}$$

The reader can easily confirm that the values in the table are correct.

There is no direct way to obtain the product temperature as it is determined by balancing the equation. The method used is to guess a temperature and see how far out of balance the equation is. In this case, the guess is 3000 K.

Making the same table as above for the products,

Species	n_i kmol	$\tilde{u}_{i,3000K}$ kJ/kmol	$n_i\tilde{u}_{i,3000K}$ kJ/kmol
CO_2	0.91	127,920	116,407
CO	0.09	68,598	6,174
N_2	1.71	67,795	115,929
Σ			238,510

Inserting the values in the equation:

$$U_2 - U_1 = 238,510 - 9,489 + \left(-281,761\right) = -52,740\,\text{kJ}$$

This is negative and shows that the products were not hot enough. Therefore, try the next temperature in the table which is 3200 K.

Species	n_i kmol	$\tilde{u}_{i,3200K}$ kJ/kmol	$\tilde{u}_{i,3200K}$ kJ/kmol
CO_2	0.91	138,720	126,235
CO	0.09	74,391	6,695
N_2	1.71	73,555	125,779
Σ			258,709

Now, the equation is $U_2 - U_1 = 258,709 - 9,489 + (-281,761) = -32,540$ kJ. This is still too cool; so, the process is repeated until the answer is acceptably close to zero. Using data for 3600 K provides a positive value of the internal energy sum of 8127 kJ indicating that the actual temperature is just below this point.

LEARNING OUTCOMES

✓ **Combustion** involves mixing a fuel (**hydrocarbon**) and an oxidiser in order to produce heat, whilst converting the chemicals involved in the reaction into reaction products.

✓ In this unit, the hydrocarbon fuels considered only consist of hydrogen and carbon and produce the ultimate product gases **carbon dioxide** and **water vapour**.

✓ **Molecular reaction equations** can be balanced by counting the atoms of all molecules on the reacting side of the equation and matching the number of atoms of all molecules on the product side of the equation.

✓ A **stoichiometric** combustion reaction is ideal; every carbon atom combines with exactly one **diatomic oxygen** molecule to produce a carbon dioxide molecule, and every hydrogen atom combines with one other hydrogen atom and one oxygen atom to produce water vapour.

✓ When there is either too much or too little oxygen, **incomplete products of reaction** appear including **carbon monoxide** and oxygen. The reaction is then non-stoichiometric.

✓ Most practical combustion installations use atmospheric air which carries oxygen as the primary oxidiser. The oxygen in the air brings with it a fixed proportion of **nitrogen**. The combined requirement for oxygen with its associated nitrogen gives an air requirement which leads to the AFR by mass or by volume.

✓ For non-stoichiometric combustion processes, the **excess air** ratio is defined as the ratio of excess air supplied to the air required for stoichiometric reaction.

✓ The product gases may be considered in proportions with or without moisture, since the moisture will condense out of the gases when they cool down. The alternatives are known as **wet** and **dry products of combustion**.

✓ Each reaction has an **enthalpy of reaction** at **standard conditions**, which is determined by the difference between the sum of **formation enthalpies** of the reactants and products.

✓ **The reference condition for combustion** is 1 bar pressure and 25°C, at which the enthalpy of all gases is set to zero by convention.

✓ The enthalpy of a reaction is referred to as the **calorific value** of the fuel, and may be considered as **gross** in which case, water is considered as condensed out of the gases, or as **net** in which case, the water is considered as vapour in the gases.

✓ Enthalpy of a reaction is directly used in flow process combustion but if required for closed process combustion it must be converted into **internal energy of reaction**.

✓ The final temperature of a combustion process can be determined by an **enthalpy balance** (or **internal energy balance** for closed processes).

2.5 RECIPROCATING COMPRESSORS

Reciprocating piston machines are typically needed to provide a supply of compressed air but are also used for other positive displacement requirements such as refrigerator pumps and vacuum pumps. They can be described as **single-** or **double-acting** (compression on one or both sides of piston) or **single-** or **multi-stage** (number of stages of compression before air is delivered). They are **positive displacement** (cylinder inlet and outlet valves are needed to prevent backflow) machines and usually have **self-acting valves** (opened and closed by pressure differentials across the valves). A typical example of compressor configurations is indicated in Figure 2.28. The range of sizes and pressures is from the order of millimetres of water gauge pressures (e.g. 5 mm water gauge is $p = \rho g h = 1,000$ [kg/m^3] × 9.81 [m/s^2] × 0.005 [m] = 49.05 Pa), as might be used for medical purposes, to several metres of water gauge pressure (e.g. 20 m water gauge is $p = \rho g h = 1,000$ [kg/m^3] × 9.81 [m/s^2] × 20 [m] = 196,200 Pa or 1.9 bar gauge or approximately 2.9 bar absolute) and hundreds of bar pressure, as might be used for compressing carbon dioxide for burying in disused oil and gas wells for carbon dioxide sequestration.

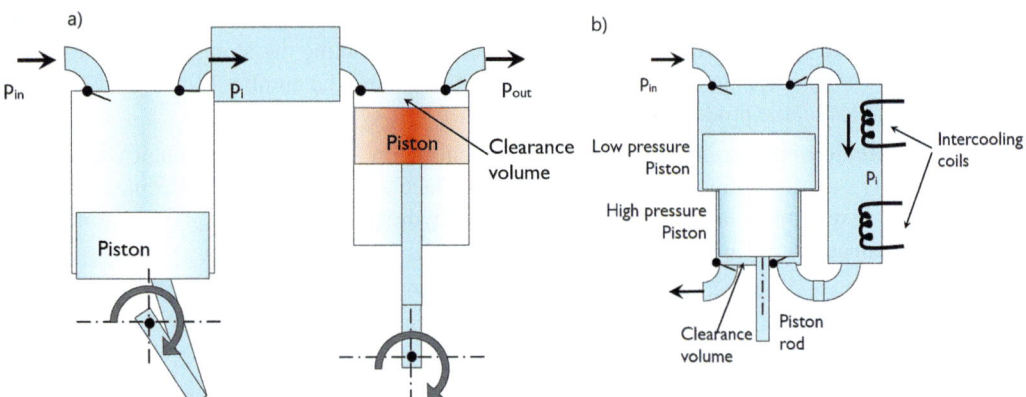

FIGURE 2.28 Schematic representation of a) two stages (low pressure feeds into high pressure), single acting (compression on up stroke of piston only) and b) two stage, double acting (compression on up and down stroke of piston).

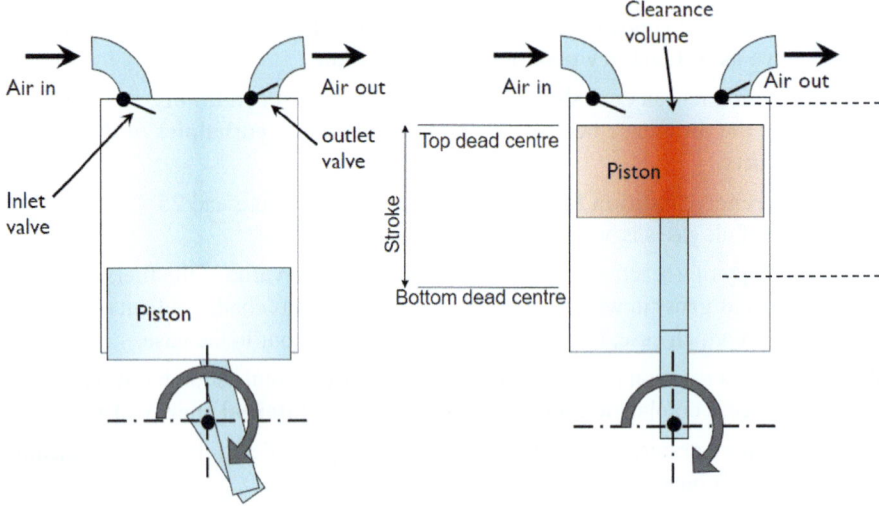

FIGURE 2.29 Schematic of reciprocating air compressor stage with inlet and outlet flap valves and piston with crank.

There is some basic terminology that must be remembered for a proper discussion of reciprocating compressors to describe their operation fully. These are related to Figures 2.29 and 2.30 and are described in the following:

Stroke, also called **swept volume** – distance from the Bottom Dead Centre (**BDC**) to Top Dead Centre (**TDC**) piston position.
Clearance volume – a small proportion of swept volume typically allows the valves to be clear of the piston at the top of its stroke.
Machine cycle – the gas goes through a machine cycle rather than a full thermodynamic cycle because it is drawn in and exhausted out of the compressor to be used.
Automatic valves – spring return valves.

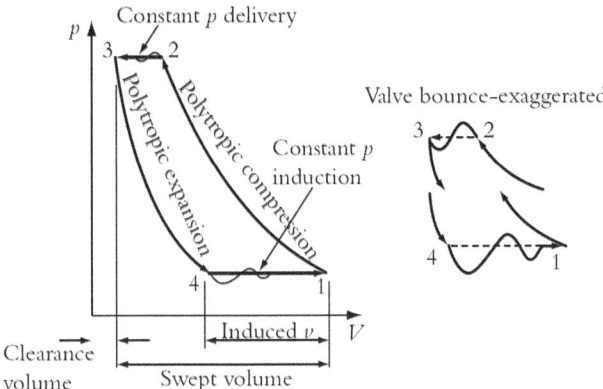

FIGURE 2.30 Machine cycle, alternatively called an indicator diagram, of a single stage of a reciprocating air compressor and valve bounce.

The TDC and BDC are related to the crank angle that drives the piston; at BDC, the volume above the piston in the cylinder is maximum and at TDC it is minimum. The gas in the compressor does not undergo a true thermodynamic cycle according to the definition of being taken around a series of states and returned to its initial condition. It is a **machine cycle**, since the same mass is not taken around the cycle, but rather has its state changed from state 1 to state 2 as shown in Figure 2.30, whereupon it is delivered to further processes separate to the compressor. The **indicator diagram** is a pressure–volume diagram of the processes occurring in the compressor (Figure 2.30) and it describes the progression of the gases through the compressor. These diagrams are commonly produced mechanically from compressors in order to establish the actual performance by comparing instantaneous pressure in the cylinder to the volume of the cylinder above the piston according to the position of the piston. There is a sequence of events operating on different quantities of the working fluid which can be described as follows (see Figure 2.30):

1. Some time after the TDC position of the piston, the intake valve opens at (4) when the pressure in the cylinder has dropped to p_1 and the air over the cylinder inlet is able to force its way through the flap valve;
2. As the piston continues travelling downwards (4 to 1), air is drawn in until the BDC position is at position 1;
3. The piston continues in its cycle now heading towards the top of the cylinder (1 to 2) with the gas inside the cylinder trapped by the two valves and compressing according to a polytropic process;
4. The delivery (exhaust) starts when the piston has moved some way towards the TDC at which time the pressure in the cylinder just overcomes pressure p_2 and the flap valve opens (2);
5. The piston continues to the TDC position (4) driving air out into the receiving chamber at the higher pressure;
6. The piston reaches the TDC, with the clearance volume remaining to allow the valve movements and a small volume of gas at the higher pressure trapped inside the cylinder;
7. Finally, the outlet valve closes and this volume is expanded in a polytropic process (4-1) to the lower pressure and the cycle repeats.

If the polytropic processes occur rapidly, then they are nearly adiabatic; if slowly, allowing heat transfer, then nearly isothermal. Between these two limits, the process is polytropic which is defined by pv^n = constant, and $1 < = n < = \gamma$. Typically, the value of n for air is 1.3 ($\gamma = 1.4$). The ideal (indicated in the diagram) and actual indicator diagrams differ because of imperfect valve operation,

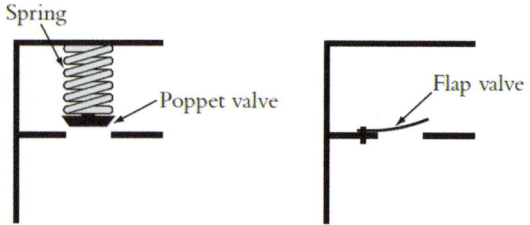

FIGURE 2.31 Typical arrangements of self operating valves.

leakage past the piston and non-constant polytropic index, n, during compression and expansion and valve bounce. Valve bounce can often cause pressure oscillations on the 'constant pressure' processes at the highest and lowest pressures, as the valve bounces on its springy mounting as indicated in Figure 2.31.

2.5.1 WORK DONE DURING COMPRESSION

The cost of running a compressor is in the work done, or rather the power requirement, to drive the machine cycle and deliver a required mass flow of the gas. There is the ideal case, and the actual work done, and further to these there is the case of the minimum that is thermodynamically possible, which are all covered in this section.

Reversible (ideal) work done (by compressor on gas +ve) for an equivalent steady flow open system is:

$$dw = \int_{1}^{2} vdp$$

This formula is derived from the SFEE: $q + w = dh$

And because $h = u + pv, dh = du + d(pv)$ so $du = dh - pdv - vdp$ by the chain rule

And $q_{rev} = Tds = du + pdv$ from the first law since $q_{rev} + w_{rev} = du$ and $w_{rev} = pdv$ and from the second law $q_{rev} = Tds$

Therefore, putting together these contributions, $du + pdv + w = du + pdv + vdp$ and cancelling terms produces $w = vdp$ by which we obtain the integral above.

Therefore, for a polytropic process,

$$w = \frac{n}{n-1}(p_2v_2 - p_1v_1) \text{ or } w = \frac{n}{n-1}R(T_2 - T_1)$$

w is specific work (work done per kg of working fluid) for polytropic steady flow process with polytropic index, n.

The overall work is therefore:

$$\dot{W} = \dot{m}p_1v_1\frac{n}{n-1}\left(\frac{p_2v_2}{p_1v_1} - 1\right)$$

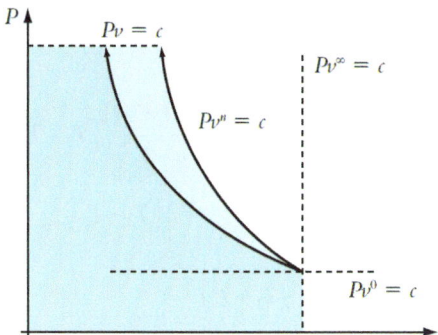

FIGURE 2.32   Ideal and actaul ideal compression processes.

Or

$$\dot{W} = \dot{m}R\frac{n}{n-1}\left(T_2 - T_1\right)$$

$\dot{W}$ dot is work rate, that is for total mass flow rate of air.

For calculation using pressures, using $\dfrac{V_2}{V_1} = \left(\dfrac{p_2}{p_1}\right)^{\frac{1}{n}}$, the formula in terms of p and V finally

becomes $\dot{W} = \dot{m}p_1 v_1 \dfrac{n}{1-n}\left(\left(\dfrac{p_2}{p_1}\right)^{\frac{n-1}{n}} - 1\right)$ or $\dot{W} = \dot{m}RT_1 \dfrac{n}{1-n}\left(\left(\dfrac{p_2}{p_1}\right)^{\frac{n-1}{n}} - 1\right)$

This is ideal in the sense that it takes no consideration of the practical workings of the air compressor, which will introduce mechanical work and viscous flow work and thus reduce the efficiency of the air compression, that is there will be mechanical efficiency.

Less work is better for a compressor because it is supplied by the work input from an external motor – it is the effort required to deliver the air at the higher pressure. The magnitude of the work decreases as n decreases – the limiting point is $n = 1$, that is isothermal as depicted in Figure 2.32. This is the minimum work to compress the air to the higher pressure, and therefore, would be the ideal if it were practically possible. The reversible polytropic index work is the actual ideal work done – with the polytropic index – somewhere between adiabatic and isothermal.

Figure 2.32 illustrates how isothermal compression requires less work than polytropic graphically. Work done for isothermal $pv = c$ and polytropic $pv^n = c$ is the area under the curve on the pressure–volume diagram which is less in the case of isothermal. The aim is to compress air with minimum work to save resources; therefore, closer to isothermal is better.

2.5.1.1   Heat Transfer to the Jacket Surrounding the Compression Chamber

For polytropic and isothermal processes, there must be heat transfer to the jacket around the compression chamber, which will always have some kind of cooling jacket in order to enhance cooling to help approach the isothermal conditions. This is due to the First Law which requires heat to be rejected if work is to be done. It is also the reason why the polytropic index is between the limiting conditions of 1 for isothermal (total heat loss) and γ for adiabatic (no heat loss). For a polytropic process, we can calculate the heat transfer using the SFEE:

$$Q + W = \dot{m}\Delta h$$

$$Q = \dot{m}c_P\left(T_2 - T_1\right) - \frac{n\dot{m}R}{n-1}\left[T_2 - T_1\right]$$

where n is the polytropic index. Using the previous expression for work and the definition of c_P and since $R = c_V(\gamma - 1)$

$$Q = \dot{m}c_v\gamma\left(T_2 - T_1\right) - \frac{n\dot{m}c_v\left(\gamma - 1\right)}{n-1}\left[T_2 - T_1\right]$$

$$Q = \left\{\dot{m}c_v\gamma - \frac{n\dot{m}c_v\left(\gamma - 1\right)}{n-1}\right\}\left[T_2 - T_1\right]$$

$$Q = \dot{m}\frac{\gamma - n}{1-n}c_v\left(T_2 - T_1\right)$$

Therefore, the work input and the enthalpy rise associated with it require that this heat is lost to the cooling jacket surrounding the cylinder.

2.5.2 Efficiency

There must be a standard method for describing how good a compressor is at doing its job, and there are several ways in which this can be done. The **isothermal efficiency**, first, describes how close the processes are to the unachievable (because it would require perfect heat transfer of all energy above the starting temperature of the compressed gas) ideal isothermal process. The isothermal work formula is:

$$\dot{W} = \dot{m}RT_1 \ln\frac{p_2}{p_1}$$

As obtained from the reversible work equation above together with the substitution of $pv = c$, where c is constant for isothermal processes, and by the gas law, $pv = RT$.

This is the minimum work required for compression by the pressure ratio with the given mass flow rate, $\dot{m}$. The ratio of this minimum theoretical work with the actual ideal work produces the isothermal efficiency.

$$\eta_{\text{isothermal}} = \frac{T_1 \ln\dfrac{p_2}{p_1}}{\dfrac{n}{n-1}\left(T_2 - T_1\right)}$$

Second, the **volumetric efficiency** describes how well the operation of the machine draws air into the cycle and delivers it to the HP reservoir. It results from the small clearance volume described in part 6 of the machine cycle description above, which is trapped in the cylinder when the piston starts its descent and is expanded until the lower pressure is reached, by which time it has taken a significant part of the potential inlet stroke of the piston.

$$\eta_{\text{vol}} = \frac{\text{volume induced at inlet pressure}}{\text{swept volume}}$$

Since $\eta_{\text{vol}} = \dfrac{V_1 - V_4}{V_s}$ and $V_1 = V_S + V_C$ (where subscript S indicates swept and C indicates clearance volume), then

$$\eta_{\text{vol}} = \frac{V_s - V_c - V_c\left(\dfrac{p_2}{p_1}\right)^{\frac{1}{n}}}{V_s} \quad \text{or } \eta_{\text{vol}} = 1 - \frac{V_c}{V_s}\left\{\left(\frac{p_2}{p_1}\right)^{\frac{1}{n}} - 1\right\}.$$

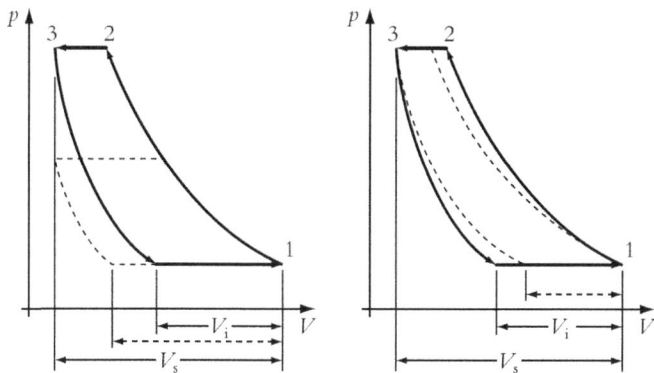

FIGURE 2.33 Effect of increasing pressure and index n on the volumetric efficiency.

Volumetric efficiency decreases as pressure ratio increases and as the polytropic index, n, is reduced as can be seen in Figure 2.33, there is less volume drawn as a part of the induction stroke when these variations happen. Dotted lines show the effects of decreasing pressure ratio and decreasing the polytropic index. In the first case, the swept volume is constant, and the expansion of the clearance volume takes up more of it as pressure increases. In the second case, reducing n causes the expansion and compression lines to lean more towards the horizontal (i.e. towards pv^0 – constant pressure process). Increasing isothermal efficiency reduces volumetric efficiency. Whatever the higher pressure is, the clearance volume is the same and the gas in the clearance volume expands during the same part of the induction downstroke of the piston, taking up an increasing part of the swept volume, V_S, as the HP is increased or the polytropic index is reduced.

Mass flow rate can be calculated using the volume induced per stroke of the piston, which is by volumetric efficiency, $\eta_{\text{volumetric}}$, the stroke volume per cycle, V_{swept} and the density of the gas at the inlet, ρ:

$$m_{\text{induced}} = \eta_{\text{volumetric}} V_{\text{swept}} \rho$$

and per second by including the number of cycles per second:

$$\dot{m}_{\text{induced}} = N \eta_{\text{volumetric}} V_{\text{swept}} \rho$$

where N will be the compressor speed that is r.p.m./60 for a single acting compressor and 2 × r.p.m./60 for a double acting compressor.

2.5.3 MULTISTAGING AND INTERCOOLING

Volumetric efficiency is affected by pressure ratio such that an absolute limit occurs, at which no delivery occurs and the entire stroke is taken up in expanding the clearance volume. For example, the formula for volumetric efficiency shows that delivery is zero when $p_{\text{HIGH}}/p_{\text{LOW}} = 22.6$ for $n = 1.3$ and $V_c/V_s = 10\%$. In practice, $(p_{\text{HIGH}}/p_{\text{LOW}})$ for a single stage is limited to approximately 4:1, and two or more stages are used for higher delivery pressures. The pressure between stages is the intermediate pressure p_i (which is now the inlet pressure for the second stage). This is referred to as **multistaging** in which the first stage of the compressor delivers to a further stage at the higher pressure, which in turn increases pressure further.

In the simplest case of two consecutive stages, as shown schematically in Figure 2.34, each stage can be treated as a separate subject, considering the conservation of mass between the two stages and the pressure from the outlet of the LP stage is the inlet pressure for the higher-pressure stage.

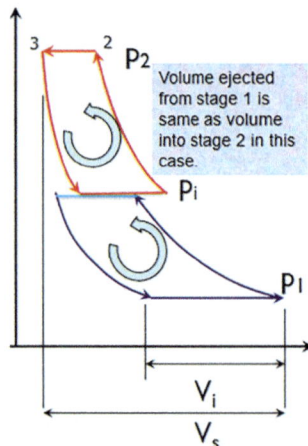

FIGURE 2.34   Schematic *p-V* diagram of a two stage compressor.

It is obviously desirable to reduce the work done by the compressor. The total work input is minimum when the intermediate pressure is $p_i = \sqrt{p_{high}p_{low}}$. This formula can be derived by expressing the formula for the combined power in terms of intermediate pressure, p_i, and differentiating with respect to p_i to give the point at which minimum work occurs when its value is zero. In this case, the work in the LP stage is equal to the work in the HP stage, $W_{LP} = W_{HP}$.

To maintain a good isothermal efficiency and thus keep the work done to a minimum, there is an opportunity in multistaging to reduce the temperature of the compressed gas at the intermediate point between stages. In air compressors, this is typically done by applying a water-cooling jacket around the intermediate receiver chamber as illustrated in Figure 2.35. This is referred to as **intercooling**. Ideally, the temperature in the intercooler would be reduced to the same temperature as the inlet temperature, and this is called **complete intercooling**. Figure 2.36 shows the benefit of this in a **single-acting two-stage compressor** on the indicator diagram. The dotted line shows the machine

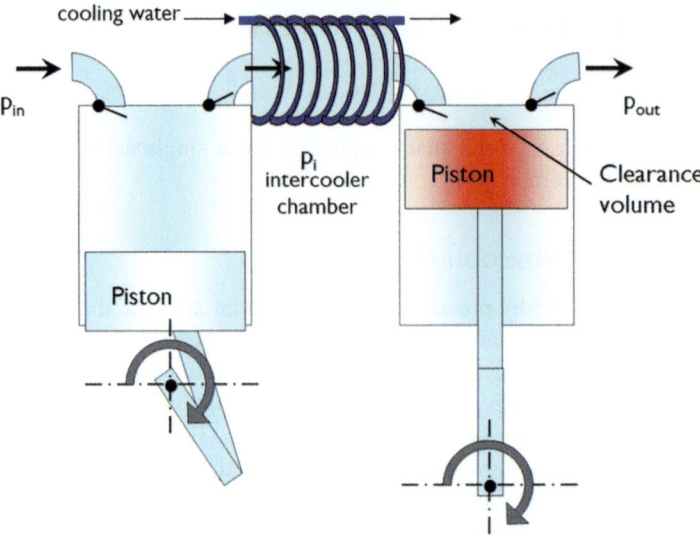

FIGURE 2.35   Schematic of a two stage compressor with an intermediate receiver chamber which can be cooled to apply intercooling to the working fluid.

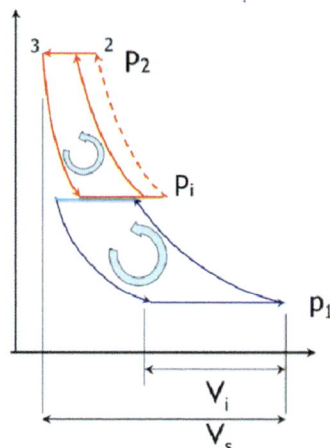

FIGURE 2.36   The effect of intercooling is to reduce the work in the higher pressure stage; this shows that the same mass can be compressed using a smaller high pressure compressor stage.

cycle for the second stage without intercooling for the same mass flow as the first stage. This area is reduced in the p-V diagram in the intercooled case, and a smaller (i.e. physically more compact) second stage is required for doing less work. The area between the dashed and solid lines is the work saved by intercooling.

A further benefit may be obtained from two-stage compression by making use of the downstroke of the piston as shown in Figure 2.37. This is described as a **double-acting two-stage compressor** as it acts in both directions of travel of the piston. Air drawn into the LP upper part of the machine is delivered to an intermediate-pressure chamber, which in turn provides the inlet fluid for the HP stage in the lower part of the machine. In practice, the sectional area of the lower cylinder must be smaller than that of the upper to maintain a balanced mass flow between the two stages. Incidentally, it is also possible to have a double-acting single-stage compressor to improve compactness of a particular design. The cost of this is in the complexity of the machine, which will make it more costly to make and therefore to sell. The figure shows that as with the case of a single-acting two-stage compressor, intercooling can be applied in an intermediate pressure chamber.

Two stage – double acting with intercooling

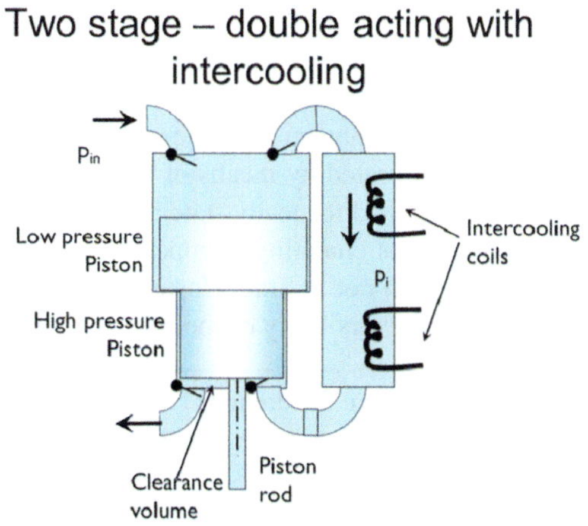

FIGURE 2.37   Schematic diagram of two stage, double acting compression, showing output of the low pressure side (top) goes into the high pressure inlet and the difference in size of the two compression chambers due to the change in specific volume.

LEARNING OUTCOMES

✓ **Reciprocating compressors** use a piston driven in a cylinder, with a volume of gas at low pressure induced on each cycle which is compressed and delivered at a higher pressure.

✓ The terminology for compressors describes how the machine is constructed and the method of operation. In particular, the **clearance volume** is significant for the operation of a compressor as it limits how much air can be drawn in during the **induction stroke**.

✓ Since a new volume of gas is induced at every induction, a state diagram does not represent a true cycle, but rather a **machine cycle**. The *p-v* diagram is known as the **indicator diagram** of the compressor.

✓ The compression requires work, and it is possible to calculate the **ideal work** assuming polytropic processes and no losses to friction in the machine. This work involves heat generation and heat which must be lost in order to comply with the First Law of Thermodynamics.

✓ The two measures of compressor efficiency are the **volumetric efficiency** which describes how much the volume drawn in is limited by the expansion of the clearance volume and the **isothermal efficiency** which describes how far from the theoretical minimum work of isothermal compression the machine cycle process is.

✓ **Multistaging** is used because of the limiting effect of volumetric efficiency on pressure ratio. An intermediate pressure is achieved in one stage, followed by another stage to the higher pressure.

✓ There is an **ideal intermediate pressure**, based on the least work done, which is found to be when each stage does equal work.

✓ **Intercooling** is used in multi-stage compression to more closely tend towards the isothermal condition and reduce the compression work done.

2.6   HEAT TRANSFER

Heat is energy in the process of transfer under the driving potential of a spatial temperature gradient. The subject of heat transfer deals with the manner and rate at which this transfer takes place. Heat appears at the boundary and may be generated by means of chemical and nuclear reactions within the body of a system. Our attention is limited to steady state heat transfer – that is, where there is an established temperature gradient that is not changing. Temperature gradient in this context refers to dT/dx, a variation in spatial dimension but not in time. Unsteady heat transfer is a little beyond our scope. Heat transfer contributes to the irreversibility of thermodynamic processes. This is due to:

1. Heat transfer through _finite_ temperature differences.
2. Combustion or chemical reactions with a finite temperature difference between the fluid and the surroundings.

We know two process types specified by the method of heat management in the process:

1. <u>Adiabatic</u> = no heat transfer $q = 0$, for which $pv^\gamma = c$
2. <u>Isothermal</u> = constant temperature, $pv = c$

In both, p is pressure, v is specific volume, γ is the ratio of specific heats (c_p/c_v) and c is a constant. Heat transfer can be seen to affect the efficiency of processes, and as shall be seen, this is a direct consequence of the Second Law. It is also important for the cooling of engineering components at all scales.

2.6.1 CONDUCTION

In conduction, energy is transferred from molecule to molecule of a substance by passing on vibrational KE. Under steady conditions, the molecule will pass on as much energy as it receives; under non-steady conditions, however, the mean energy level of the molecule will rise. The KE of the molecule is proportional to its absolute temperature. The most obvious example is transfer of heat through a solid, uniform material. Figure 2.38 shows a brick wall. If the surfaces of the brick wall are maintained at T_1 and T_2 with the brick initially at T_0, this simplified schematic shows that a transient development of temperature profile up to the steady condition will occur. The true profile of temperature development is similar to this, but is complicated by the true physical behaviour of the medium; transient heat transfer is beyond the scope of this volume; what is considered here is the end state of the transient, the **steady state** condition.

2.6.1.1 Fourier's Law for Conduction

The law which describes conduction is an experimentally determined law known as the **Fourier Rate Law**. For one-dimensional heat flow, it may be expressed as:

$$\dot{Q} = -kA\frac{dT}{dx} \text{ or } \dot{q}'' = -k\frac{dT}{dx}$$

$\dot{Q}$ is the heat flux density or the rate of heat flow through area A of the surface.
$\dot{q}''$ is the heat flux density or rate of heat flow through unit area indicated by the double prime.
k is the **thermal conductivity** of the material.
$\dfrac{dT}{dx}$ is the temperature gradient in the x-direction of the through-flow of heat.

Units of k are **[W/mK]**, that is the heat transferred per second, per unit length in the direction of heat conduction, per degree Kelvin temperature difference across that distance. 'One dimensional' means that the heat travels in only one direction, and the gradients of temperature in other directions are zero. Taking the example shown in Figure 2.38, the steady state condition has a linear gradient of temperature, and we may consider $\Delta T/\Delta x = (T_2 - T_1)/(x_2 - x_1)$. The area through which the heat flux flows is perpendicular to the page, and has area A, or we may usefully consider the heat flux per unit area of wall. The conductivity of the materials tends to alter with temperature, and certainly

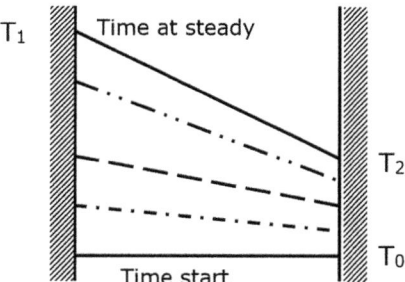

FIGURE 2.38 Temperature gradient through a brick wall, which develops from a change from the start temperature. On the left the start and end condition, the end condition remaining constant, and on the right graphs of the temperature as it changes from initial condition to the steady state.

TABLE 2.1

Thermal Conductivity of Some Gases, Liquids and Solids

Material	Thermal Conductivity, W/mK
Dry air at 20°C[a]	0.0257
Dry air at 100°C[a]	0.0335
Water at 20°C[a]	0.603
Water at 100°C[a]	0.681
Steam at 100°C[a]	0.0248
Stainless steel 304 at 20°C[b]	14.4
Aluminium Duralumin at 20°C[b]	164
Helium at 17°C and 1 bar[c]	0.148
Carbon dioxide at 17°C and 1 bar[c]	0.016
Hydrogen at 17°C and 1 bar[c]	0.178
Aviation oil MS—20 at 20°C[c]	0.134
Hydraulic oil AMG—10 at 20°C[c]	0.119
Fired clay brick, density 2000 kg/m³, protected[d]	0.7
Phenolic foam insulation[d]	0.035
Glass fibre wool at 10°C[d]	0.04

Notes:

[a] Rogers, G.F.C. and Mayhew, Y.R. Thermodynamic and Transport Properties of Fluids, fifth edition, 1995, Blackwell Publishing Ltd, Oxford UK.

[b] The Engineering Toolbox, online resource, https://www.engineeringtoolbox.com/thermal-conductivity-metals-d_858.html.html (accessed 4/4/23).

[c] Vargaftik, N.B. Touloukjan, Y.S, Tables on the thermophysical properties of liquids and gases in normal and dissociated states, 2nd ed. 1975, John Wiley and Sons Inc, New York.

[d] Environmental design – CIBSE Guide A (8th ed.) The Chartered Institution of Building Services Engineers (CIBSE). Retrieved from https://app.knovel.com/hotlink/toc/id:kpEDCIBSE1/environmental-design/environmental-design

varies according to the type of material – metallic solids are known to conduct well compared to non-metallic solids. A range of values for the thermal conductivity of some materials from various sources is shown in Table 2.1, which shows the variation in the conductivity of gases, liquids and conductive and insulating solids. There are therefore engineering materials suitable for differing heat transfer requirements, and every material has a conductivity which can be used to inform how well a body will conduct heat whether the purpose is to enhance or reduce heat transfer.

It is useful at this point to consider the Second Law consequences of heat transfer. All *heat transfer across finite temperature difference is irreversible*. Q transfers to increase the internal energy, u, [J/kgK] of the working fluid; the working fluid proceeds around the thermodynamic cycle and rejects the same heat, Q, in cycle at T_{COLD} to return to its initial state. The Second Law states that it is impossible to get the same heat to transfer back to the hot reservoir again without doing work – in other words, without adding extra energy to transfer the same heat back again. This demonstrates the irreversibility in a thermodynamic cycle.

Conduction is degradation of heat. Referring to Figure 2.39, which shows conduction through a solid from a hot source to a cold sink at steady state, the heat leaves hot at Q/T_{HOT} and enters the cold at Q/T_{COLD}. Since, $T_{HOT} > T_{COLD}$, $Q/T_{HOT} < Q/T_{COLD}$, i.e. $s_{COLD} > s_{HOT}$ because $ds = dq/T$. This shows that the entropy contained in the heat energy that is transferred has increased. The conductor at steady state has no change of state and its entropy has not changed by this heat transfer – it is the source and sink, that is, the surrounding 'universe' which has suffered the change of entropy resulting from this.

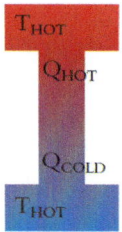

FIGURE 2.39 A solid conductor transferring heat from a hot source to a cold sink.

2.6.1.2 Thermal Resistance

A trick is used in order to make the calculation of heat transfer easier. It notes the similarity between the physical nature of heat transfer and electrical conduction. Starting with the Fourier Law for heat conduction, integrate in the direction of heat flow (x) to give:

$$q = \frac{-\Delta T}{\dfrac{x}{kA}}$$

$$q'' = \frac{-\Delta T}{x \, / \, k}$$

where ΔT is the overall temperature difference across a thickness x (and k is constant in the direction of heat flow, if the heat flow is unidirectional), and the double apostrophe indicates heat transferred per m^2 area of heat transfer area.

Compare this equation with $I = V/R$ - (Ohm's Law) and the analogy between heat flow and electric current flow becomes apparent.

Temperature difference ΔT is equivalent to voltage difference, ΔV.
Heat flow, Q is equivalent to current flow, I.
Thermal resistance x/kA is equivalent to electrical resistance, R.

Thermal resistance [K/W] is dependent on the material conductivity (k), the thickness through which heat is passing (x) and the cross-sectional area through which it travels (A).

The example in Figure 2.40 shows how thermal resistances can be used to simplify the analysis of a wall having several layers of different materials. This effect of thermal conductivity variation in a plane wall can be found in many household and engineering situations, for example, a wall of a house or a refrigerator compartment or an insulated pipe. At steady state, if heat, q, passes through the first wall surface, the same heat, q, must pass through all interior faces. Note that *in the steady state*, this does not work in the transient case. The relative temperature gradients shown in Figure 2.40 illustrate that materials with low k are more resistive to heat transfer – that is, they are good insulators, and therefore, maintain a high gradient of temperature. It is not necessary to know the interior face temperatures to know the overall heat transfer – all that is required is the material properties and its thickness. We know $\dot{q} = \dot{q}_1 = \dot{q}_2 = \dot{q}_3$ under steady conditions, hence applying Fourier's law to each layer gives

$$\dot{q} = \frac{-\Delta T_{2-1}}{R_{\text{Th}2-1}} = \frac{-\Delta T_{3-2}}{R_{\text{Th}3-2}} = \frac{-\Delta T_{4-3}}{R_{\text{Th}4-3}}$$

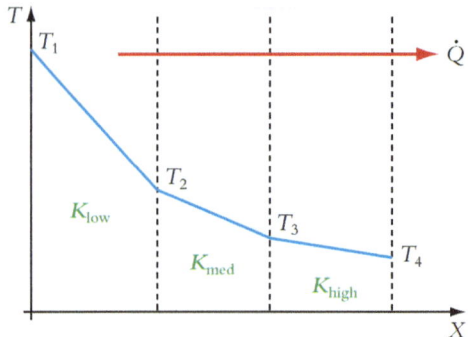

FIGURE 2.40 A plane wall with three layers of increasing conductivity; note that low conductivity results in high temperature gradients – useful in situations where insulation is required – and vice versa.

$$\dot{q}\left(R_{\text{Th}2-1} + R_{\text{Th}3-2} + R_{\text{Th}3-4}\right) = \left(-\Delta T_{1-2}\right) + \left(-\Delta T_{2-3}\right) + \left(-\Delta T_{3-4}\right)$$

$$\dot{q} = \frac{-\left(T_4 - T_1\right)}{\sum R_{\text{Th}}}$$

The negative is used to indicate the direction of heat loss from 1 to 2 to 3 to 4 in this case. It is clearly seen that by knowing the external temperatures and the property and thickness of the materials in between, the heat flux per metre squared of wall can be easily calculated. From this, the temperature at the interior boundaries can also be calculated if desired. The thermal resistance network can be used in more complex situations connecting complex heat flux paths between known temperature points; the method can be found in texts on heat transfer.

2.6.2 CONVECTION

Convection situations include both heat transfer from solid boundaries into fluids and between two mixing streams of fluids and the redistribution of internal energy within a fluid. Convection implies the movement of fluid to transmit energy, and so, the study of fluid motion characteristics of the situation is of particular importance. Fluid motion may be induced by buoyancy effects resulting from changes in temperatures within the fluid during the initial stages of heat transfer, which are then perpetuated as **natural** or **free convection**. Alternatively, the fluid motion may be 'forced' by an external source – pump or compressor – and **forced convection** of heat will develop. Fluid motion conveys (or convects, carries with it) thermal energy which can be exchanged with solids/fluids/gases at other temperatures.

2.6.2.1 Newton's Law of Cooling

The rate of heat transfer is defined using a **convective heat transfer coefficient** of the temperature gradient between the bulk part of the fluid and the surface of the wall by using **Newton's Law of Cooling**:

$$\dot{Q} = -hA\left(T_f - T_w\right)$$

where

h is the heat transfer coefficient (HTC) [W/m²K],
A is the area of the surface in contact with the fluid [m²],
T_w is the wall temperature [K, or °C] and
T_f is the bulk fluid temperature [K or °C], that is the temperature in the fluid far from the surface

TABLE 2.2

Ranges of Convection Heat Transfer Coefficient in Some Fluid Types according to Bejan[a]

Fluid Type	Typical Value for Convection Heat Transfer Coefficient, W/m²K
Gases, atmospheric pressure, natural convection	10
Gases, atmospheric pressure, forced convection	80
Water, forced convection	1000
Boiling or condensation of water	10000

[a] Bejan, A. Heat Transfer, 1993. John Wiley and Sons, Inc. New York.

This equation defines the HTC, h. The units of h are **[W/m²K]**, that is heat transferred per second, per m² of contacted surface area and per degree Kelvin temperature difference between bulk fluid and surface. The negative again indicates that in this case heat is lost and that T_f is less than T_w.

h is difficult to determine analytically for a particular situation as it depends on relatively 'complex' fluid flow behaviour which does not lend itself to analysis. The values are usually determined by experiment and the use of this empirical data is described in Section 6.2.2. The data in Table 2.2 shows typical values for various fluid types and flow conditions according to Bejan (1998).

Thermal resistance of combined convection/conduction can be calculated in a similar way to the conduction case, making use of the similarity with Ohm's Law. The thermal resistance of a fluid/solid boundary is defined from Newton's Law. Referring to Figure 2.41, which shows a plan view of a solid plane wall with a high-temperature gas on one side and a low-temperature gas on the other, the steady heat transfer when the temperature profile has become constant from the convection onto wall surface A from fluid at distance on side A ($T_{\infty A}$):

$$\dot{q}_A = -h_A A_A \left(T_{WA} - T_{\infty A}\right) = -\frac{\left(T_{WA} - T_{\infty A}\right)}{1/h_A A_A} = -\frac{\left(T_{WA} - T_{\infty A}\right)}{R_{TH,A}}$$

which is equal to the steady heat transfer through the wall from side A to side B:

$$\dot{q}_{A-B} = -\frac{kA}{\Delta x}\left(T_{WB} - T_{WA}\right) = -\frac{\left(T_{WB} - T_{WA}\right)}{\Delta x/kA} = -\frac{\left(T_{WB} - T_{WA}\right)}{R_{TH,A-B}}$$

which is equal to the steady heat transfer from the wall side B to the fluid distant from wall B ($T_{\infty B}$):

$$\dot{q}_B = -h_B A_B \left(T_{\infty B} - T_{WB}\right) = -\frac{\left(T_{\infty B} - T_{WB}\right)}{1/h_B A_B} = -\frac{\left(T_{\infty B} - T_{WB}\right)}{R_{TH,B}}$$

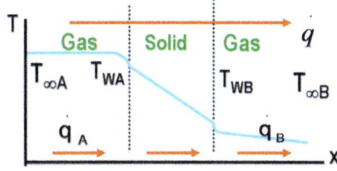

FIGURE 2.41   The effect of convective heat transfer at surfaces separating solid and gas regions.

The result of this in a combined convective–conductive steady state situation is shown in Figure 2.41.
These can be combined to give

$$\dot{q} = \frac{-\left(T_{\infty B} - T_{\infty A}\right)}{1/h_A A_A + \Delta x/kA + 1/h_B A_B}$$

In the same way as before, $\dot{q}$ is the same through both boundaries (i.e. in the three media that it passes through from ∞A to ∞B). There is no need to know the intermediate temperatures but only the difference in temperatures between the two distant sides, ΔT overall, need be known in order to calculate the overall heat transfer rate, $\dot{q}$. Note: a fluid layer could be contained within a structure, such as the air gap in a cavity wall, making multiple thermal resistances from the external to interior regions. The thermal resistance method is especially useful in such cases.

2.6.2.2   Overall Heat Transfer Coefficient

Conversely to thermal resistance, the **overall** HTC, U, is useful to measure how well heat is transferred in a combined mode of heat transfer case.

The overall HTC is defined similarly to the convective HTC:

$$q = UA\left(\Delta T_{\text{overall}}\right) = UA\left(T_{\infty A} - T_{\infty B}\right)$$

$$UA \equiv -\frac{1}{\sum R_{\text{TH}}}$$

The overall HTC is very useful in complex heat transfer situations such as that found in heat exchangers, where a large number of tubes and fins may be used, it is more convenient to select a suitable representative area and find the heat transfer for that surface taking into account all the thermal resistances that are involved. From the previous example, it can be seen that the thermal resistance is:

$$\sum R_{\text{Th}} = 1/h_A A_A + \Delta x/kA + 1/h_B A_B$$

The overall HTC, if the area in each part is the same, A, in the case of a plane wall:

$$UA = \frac{1}{\dfrac{1}{A}\left(1/h_A + \Delta x/k + 1/h_B\right)} = \frac{1}{1/h_A + \Delta x/k + 1/h_B}$$

The units are W/m²K.

2.6.2.3   Axisymmetric Problems

The case of a cylinder with a hot inside surface and cool outside surface is shown in Figure 2.42. The situation is commonly found in domestic central heating, and in thermodynamic cycles like the vapour power cycle and the refrigeration cycle. The temperature profile is not linear as we look radially from the centre to the outside, and this is because the area across which heat passes increases as the radius increases. We use Fourier's law of conduction, which still applies, but we must take into account the varying area:

$$\dot{q} = -kA\frac{\delta T}{\delta r}$$

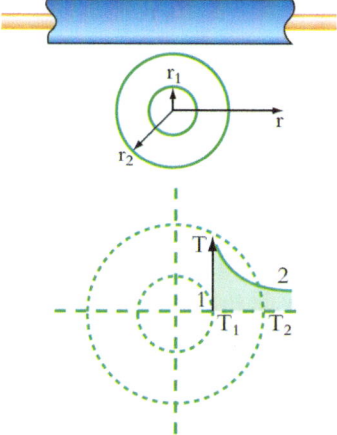

FIGURE 2.42 Axisymmetric case of conduction – temperature distribution across the wall of a cylinder.

$$\dot{q}'' = -k\frac{\delta T}{\delta r}$$

$$\dot{q}' = -k.2\pi r.\frac{\delta T}{\delta r}$$

The double prime and single prime marks on the heat transfer rate are used to indicate per unit surface area and per unit length, respectively. In this case, the temperature profile is no longer linear even if k is constant. The heat transfer rate per unit length is particularly useful in the case of pipes containing fluids exchanging heat.

Recasting the formula as

$$\frac{\dot{q}'}{2\pi rk}dr = -dT$$

and integrating between A and B gives

$$\frac{\dot{q}'}{2\pi k}\ln\frac{r_b}{r_a} = -\left(T_b - T_a\right)$$

So $\dot{q}' = -\dfrac{\left(T_b - T_a\right)}{\ln\left(\dfrac{r_b}{r_a}\right)\Big/2\pi k}$

For the total thermal resistance, comprising inner convection, then conduction, and then outer convection, R_{TH} is:

$$1/\left(R_{\text{TH,convection inner}} + R_{\text{TH conduction of wall}} + R_{\text{TH convection outer}}\right).$$

The $R_{\text{TH conduction}}$ of the annular wall is, from the above formula, $\ln\ (r_b/r_a)/2\pi k$. If there is convection heat transfer at the inner and outer surfaces, Newton's law of cooling still applies, but note that the inner surface is now smaller than the outer surface area.

$$\dot{q}' = \frac{-\Delta T_{\text{overall}}}{1/h_a 2\pi r_a + \ln\left(\frac{r_b}{r_a}\right)/2\pi k + 1/h_b 2\pi r_b}$$

where $\Delta T_{\text{overall}}$ is the temperature difference between the fluid outside and inside the pipe, and the areas are cylindrical surfaces of various radii having 1 m length to produce $\dot{q}'$, the heat transfer per unit length of the pipe.

2.6.2.4   Convection Mechanism

If a fluid moves near to a wall, at the wall, the velocity of the fluid is zero, and the heat transfer into the fluid takes place by conduction. Thus, the local heat flux per unit area q'' is

$$\dot{q}'' = -k\frac{\delta T}{\delta y}\bigg|_{\text{wall}}$$

The hot fluid close to the inner layer is then carried away by convection. From Newton's law of cooling, $\dot{q}'' = -h\left(T_{\text{wall}} - T_\infty\right)$. These equate at steady conditions to give:

$$h = \frac{-k\left(\delta T / \delta y\right)_{\text{wall}}}{T_{\text{wall}} - T_\infty}$$

This tells us how h can be determined experimentally. A heat flux probe having a known thickness is placed on the wall to determine conduction into the fluid. What we would prefer is a calculation. Something which relates experimental results with data about the flow – a **correlation** – is required. To understand how the fluid motion affects the heat transfer, it is necessary to have a basic understanding of the fluid mechanics involved and for this it may be useful to refer to Fluid Mechanics Section 1.3 which addresses boundary layers next to walls. The following sections describe the influence of the boundary layer on heat transfer.

2.6.2.5   Velocity Boundary Layer

In the case of a solid surface with fluid flowing over it, a boundary layer forms as indicated schematically in Figure 2.43. (as described previously in Figure 1.9) This is a layer in which, on the surface, the fluid is stationary at the solid/fluid interface, and as distance increases towards the free flowing fluid distant from the surface, grows to 99% of the main flow velocity. The laminar sublayer is very thin in the order of millimetres, and is the region in which the laminar motion of the fluid occurs, which is dominated by the viscous friction between the fluid molecules sliding over each other. In a fluid that is sufficiently slow, the entire boundary layer is occupied by laminar motion, and a linear increase of velocity is observed. In a faster moving fluid, the laminar flow cannot continue when the velocity gradient causes entrainment of the laminar layers into the faster moving fluid. This 'tripping up' of the flow causes a chaotic random motion known as turbulence, and the nature of the boundary

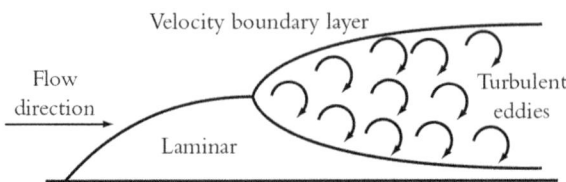

FIGURE 2.43   Development of the velocity boundary layer from the front edge of a plate placed stationary in the flowing liquid.

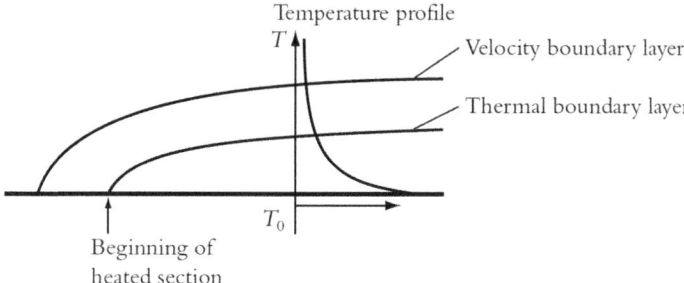

FIGURE 2.44 Development of the thermal boundary layer on a flat plate with a heated section.

layer then becomes turbulent, with a thin laminar sub-layer. Viscosity causes momentum to diffuse by enabling the redistribution of momentum by the dragging forces applied on the resulting viscous friction. The velocity boundary layer grows with distance in the flow direction to the order of several mm. The development depends on the viscosity of the fluid and the speed of the flow – a relationship that is expressed well in the non-dimensional Reynolds number.

2.6.2.6 Thermal Boundary Layer

Similar to the velocity boundary layer, the thermal boundary layer as illustrated schematically in Figure 2.44 describes the layer over the surface in which the temperature is less than 99% of the main fluid temperature difference from that of the wall surface. Thermal boundary layer thickness is usually less than the velocity boundary layer thickness. It depends on **thermal diffusivity**, which is dependent on the conductivity of the fluid, and on its **thermal capacity**, expressed as ρc_p.

2.6.2.7 Determination of Heat Transfer Coefficient from Boundary Layer Characteristics

The values of HTC can be measured for any particular geometry, flow condition and set of fluid physical properties. To extend the applicability of the data, dimensional analysis is used, which allows the comparison of similar situations by scaling the size of the particular situations such that the thermal transport character is compared according to the appropriate fluid properties. Convective heat transfer can be classified by the type of flow over the type of geometry. Dimensionless numbers are used to relate the scale of the flow.

There are three characteristic **dimensionless numbers** which are used to express the characteristics of convective heat transfer.

2.6.2.8 Nusselt Number

The Nusselt number, Nu, characterises the heat transfer itself, and is calculated from relationships with other dimensionless numbers as detailed in the subsequent sections. The HTC, h, is derived from it.

$$\mathrm{Nu_x} = \frac{h\,x}{k} \rightarrow h = \frac{k\mathrm{Nu_x}}{x}$$

Here, $\mathrm{Nu_x}$ is a **local Nusselt number** at a particular distance, x, in the geometry being considered. From this, if we know the Nusselt number, and the conductivity of the fluid, we have the convective heat transfer at any particular point on the geometry. Since the velocity and thermal boundary layers rely on the properties of the fluid and its flow, there must be a way to relate them to the heat transfer rate and hence Nu. There is fortunately an analysis to determine the Nusselt number for many different geometrical situations; these have been compiled by researchers, a very small sample of them will be considered in order to consider how they are used. Nusselt numbers are calculated using dimensionless numbers, which express the characteristics of the heat transfer situation, and they are the **Prandtl number**, which occurs in all Nusselt number correlations, the **Reynolds number**, which occurs in forced convection situations, and the **Grashof number**, which occurs in natural convection situations. These are described in the following:

Prandtl number, Pr, is the ratio of viscous diffusion to thermal diffusion.

It was mentioned earlier that the thickness of the velocity boundary layer depends on the viscosity, and that viscosity causes diffusion of momentum. It was also mentioned that the thermal boundary layer is determined by thermal diffusivity. The Prandtl number provides a comparison of these effects, and basically compares them in order to see how well a fluid diffuses viscously and thermally. It is expressed as

$$\mathrm{Pr} = \frac{\nu}{\alpha} = \frac{\mu / \rho}{k / \rho c_p}$$

$$\mathrm{Pr} = \frac{c_p \mu}{k}$$

in which, ν (pronounced 'nu') is the **kinematic viscosity**, which is the more familiar dynamic viscosity, μ, divided by density ρ, and α is the **thermal diffusivity**, which is conductivity, k, divided by after density insert Greek symbol rho × specific heat capacity, c_p.

Reynolds number, Re, is the ratio of momentum to viscous forces.

The Reynolds number is used in forced convection situations, where the fluid is driven mechanically through the geometry. It is familiar from fluid mechanics (see Section 1.2.3) and is given by:

$$\mathrm{Re} = \frac{\rho U D}{\mu}$$

in which density, velocity and dynamic viscosity have their usual symbols and D is the **characteristic length**, a length defining the scale of the fluid flow which must be carefully selected for each flow situation, being sometimes obvious; for instance, in a pipe it is the diameter, and sometimes not, and for instance, when a geometry has more than one scale, for example in a duct with a cross-sectional area change.

Grashof number, Gr, is the ratio of buoyancy to viscous forces.

The Grashof number is used for natural convection situations, those which are driven by thermally induced buoyancy. The idea of natural convection is familiar from the expression that heat rises. The convective motion results from the alteration of fluid density due to temperature gradients, and the hotter, less-dense, fluid rises in the cooler, more-dense, fluid surrounding. It is expressed as:

$$\mathrm{Gr} = \frac{g \beta l^3 \rho^2 \Delta T}{\mu^2}$$

g is gravitational acceleration (9.81 m/s²); $\beta = 1/T_f$ is the compressibility for perfect gases in which T_f is the film temperature which is the average of the wall surface temperature and the temperature of the fluid distant from the wall; l is the height, ρ is density, and μ is dynamic viscosity, the temperature difference between the free stream far from the surface, and the surface temperature is ΔT.

With these three dimensionless numbers, Pr, Re and Gr, the Nusselt number correlations can be constructed by observing experimental situations. The general form of the correlation is Nu = f(Pr, Gr), i.e. Nusselt number is a function of Pr and Gr, for natural convection and Nu = f(Pr,Re) for forced convection.

Example of the scale of the various quantities involved

Prandtl number for water is

$$\mathrm{Pr} = \frac{c_p \mu}{k}$$

which for water depends on $c_\mathrm{p} = 4.2$ kJ/kgK, k = 0.6 W/mK and dynamic viscosity $\mu = 0.001$ kg/ms; therefore, Pr = 4200 × 0.001/0.6 = 7.

The Grashof number for air in the case where the temperature of a wall is 300 K and the temperature of the nearby air (free-stream air) is 280 K. The difference $\Delta T = 20$°C and the average of the temperatures of the surface and the free stream is 290 K. Therefore, $\beta = 1/290$ and $\rho = 1.2$ kg/m³, and so the Grashof number for a vertical surface of 1 m height is $l = 1$ m, $\mu = 1.8 \times 10^{-5}$ kg/m.s. Therefore,

$$\mathrm{Gr} = \frac{9.81 \dfrac{1}{290} 1^3 1.2^2 \times 20}{1.8 \times 10^{-52}} = 2.506 \times 10^9$$

N.B. Gr has to do with buoyancy, Re has to do with forced flows and Pr has to do with all flows.

The calculation of Reynolds number should be familiar from fluid mechanics.

2.6.2.9   Nusselt Number Correlations

A great deal of work has been done to correlate physical heat transfer with Nusselt number.

For heat exchange on the inner surface of a tube,

$\mathrm{Nu}_\mathrm{d} = 3.66$ for laminar forced flow, with a constant surface temperature and

$$\mathrm{Nu}_\mathrm{d} = 0.023 \, \mathrm{Re}_\mathrm{d}^{0.8} \, \mathrm{Pr}^{0.4}$$

for turbulent forced flow.

For heat exchange on flat plate:

$$\mathrm{Nu}_\mathrm{x} = 0.332 \, \mathrm{Re}_\mathrm{x}^{0.5} \, \mathrm{Pr}^{0.33}$$

for laminar forced flow with constant surface temperature and for natural convection on a vertical plate, there is sensitivity to the particular range of Grashof and Prandtl number characteristics:

$$\mathrm{Nu}_\mathrm{x} = 0.59 \left(\mathrm{Gr}_\mathrm{x} \, \mathrm{Pr} \right)^{0.25}$$

for cases in which $10^3 < \mathrm{Gr}_\mathrm{x} \mathrm{Pr} < 10^9$.

$$\mathrm{Nu}_\mathrm{x} = 0.13 \left(\mathrm{Gr}_\mathrm{x} \, \mathrm{Pr} \right)^{0.33}$$

for cases in which $10^9 < \mathrm{Gr}_\mathrm{x} \mathrm{Pr} < 10^{12}$.

Once the Nusselt number is known, the HTC can be found directly from the Nusselt number relationship Nu = hL/k. More specifically, $h = k\mathrm{Nu}_\mathrm{d}/d$ for a pipe of diameter d and $h = k\mathrm{Nu}_\mathrm{x}/x$ for a vertical plate of length x.

2.6.3 RADIATION

When heat transfers by **thermal radiation** or by **radiant heat transfer**, energy is transferred via electromagnetic wave motion, which requires no intermediate material and which happens at the speed of light, 299,792,458 m/s. Radiation heat transfer will occur between any two separated bodies whenever there is a temperature difference between them and they have a direct line of sight of each other.

The bodies in Figure 2.45 both radiate heat between each other. It is not one way since both bodies are at temperatures above 0 K; however, the net heat transfer is from the hotter to the cooler. What they receive from each other depends on what they can directly see of each other. It is important to bear in mind that each body is radiating all the time regardless of what it can see, purely dependent on the temperature of its surface, and that net transfer occurs because one body emits more than the other.

2.6.3.1 Stefan–Boltzmann Law for Radiation

The basic radiation law (due to the work of Stefan (1879) and Boltzmann (1884)) is:

$$\dot{q}_b = \sigma A T^4$$

The rate of heat **emission** from a **'black body'** as indicated by the subscript, b, with a surface area A [m^2] at temperature T (must be in Kelvin) and σ is the Stefan–Boltzmann Radiation constant:

$$\sigma = 5.67 \times 10^{-8}\ \mathbf{Wm^{-2}K^{-4}}.$$

A 'Black Body' is a perfectly radiating body. It emits the maximum possible energy at all wavelengths at a given temperature. For real materials, radiation is less than the black body and depends on **emissivity, ε**, which is in the range 0 to 1. The formula is then:

$$\dot{q} = \sigma \varepsilon A \Delta T^4$$

The closer ε is to 1, the closer to black body it is. Table 2.3 shows a range of approximate emissivities for different materials.

From Table 2.3, it can be seen that there is a significant range of emissivity for a range of materials; for example, aluminium foil only emits 1% of the potential black body radiation at a particular temperature, whereas frost/snow is very close to being a black body.

The **Stefan–Boltzmann Law** shows us that if a body has a temperature above 0 K, then it will emit heat to the surroundings. However, it also receives heat from the surroundings by the same mechanism if there is anything in the surroundings with a temperature greater than 0 K. We have to consider what it will receive from other bodies.

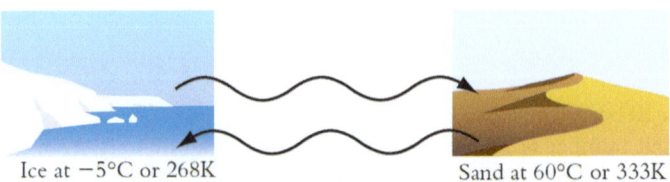

Ice at −5°C or 268K   Sand at 60°C or 333K

FIGURE 2.45 Radiant heat transfer occurs whenever there is a temperature difference between two bodies that can see each other.

TABLE 2.3

Typical Approximate Values of Emissivity (Bejan)

Aluminium foil	0.01
Copper, polished	0.04
Steel, cold rolled	0.08
Steel, rough plate	0.95
Zinc, oxidised	0.11
Fireclay brick	0.75
Asphalt pavement	0.9
Cotton cloth	0.77
Pyrex glass	0.9
Bituminous felt	0.9
Frost (snow)	0.98

2.6.3.2 Black Body Heat Transfer

The law of radiation heat transfer (for black bodies) is:

$$\dot{q}_{b,1} = \sigma A_1 F_{1-2} \left(T_2^4 - T_1^4 \right)$$

It depends on the surface area of the body that is being considered, A_1, and the view that it gets of the body it is exchanging heat with, the second body having area A_2. The amount of view is quantified by the **'black body view factor'**, F_{1-2}. For the simplest two body system, refer to Figure 2.46:

$$F_{1-2} = 1$$

$$F_{2-1} = A_1 / A_2$$

$$F_{11} = 0$$

$$F_{22} = 1 - \left(A_1 / A_2 \right)$$

Subscript 1–2 on the view factor means the area of object 2 that object 1 can see in order to receive radiation from it. This is the simplest case – a sphere within a hollow sphere. The calculation of other view factors is beyond the scope of this unit.

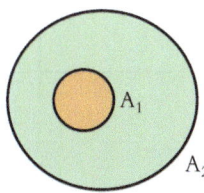

FIGURE 2.46 The simplest case of radiant heat transfer between two bodies – A sphere contained within a hollow sphere at different temperatures.

WORKED EXAMPLE

A man stands in a freezer room of a local fast food restaurant, with clothes having a surface temperature of 5°C and emissivity of 0.3. What is the rate of heat transfer from the surface of his clothes to the freezer walls if the temperature of the freezer walls is −20°C? Assume that the situation can be modelled as a small sphere of surface area 1 m² representing the man, inside a larger sphere. Use $\dot{q}_{b,1} = \sigma A_1 F_{1-2} \left(T_2^4 - T_1^4 \right)$ with $F_{1-2} = 1$ and $A_1 = 1$ m².

$$\dot{q}_{1-2} = 5.67 \times 10^{-8} \times 1 \times 1 \times \left(253_2^4 - 278_1^4 \right) = -106\,\text{W}.$$

It is interesting to note that 100 W is a reasonably maintainable heat generation rate from a healthy standing person, and that the thickness of cloth, y, of conductivity 0.08 W/mK, say, is required to maintain the surface temperature of 5°C if the skin temperature beneath is maintained at 22°C is from the Fourier Law:

$$\dot{q} = \frac{kA}{\Delta x} \left(T_{\text{inner}} - T_{\text{outer}} \right)$$

$106 = 0.08 \times 1 \times (22-5)/y$ and $y = 0.013$ m or 13 mm – the thickness of a thick fleece.

2.6.3.3 Solid/Fluid Boundary with Significant Radiation

Consider a 1-D heat flow in a solid as shown in Figure 2.47 in which there is convective heat transfer from the solid to the fluid and radiative heat transfer also from the same solid to the surroundings. This case is difficult to solve because it is non-linear with respect to temperature, T. So, the sum of thermal resistance, R_{TH}, approach cannot be used. The new procedure requires the solution of energy conservation at the surface. The heat transfer within the solid to the surface by conduction must all leave the surface by both convection and radiation in order to satisfy the steady state condition (with no change in temperature of the body), that is:

$$\dot{q}_{\text{cond}} = \dot{q}_{\text{conv}} + \dot{q}_{\text{rad}}$$

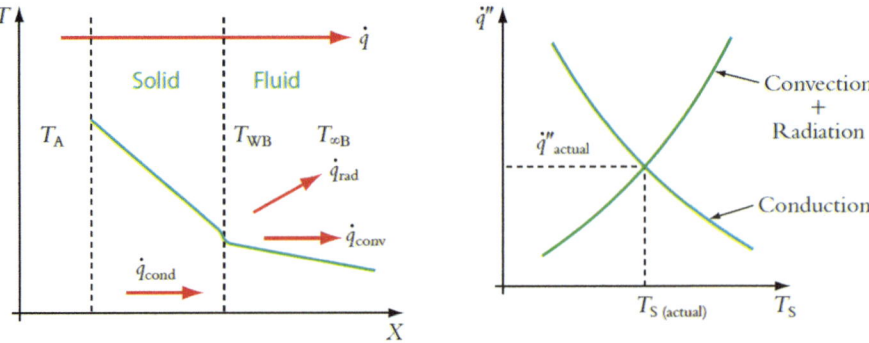

FIGURE 2.47 A solid body with a constant temperature on the left hand side and a fluid on the right hand side resulting in combined convective and radiative heat transfer on the right hand side.

and using Fourier's Law of thermal conduction heat transfer, Newton's Law of thermal convection heat transfer and Stefan-Boltzmann's formula for radiative heat transfer, for 1 m^2 area:

$$\frac{K\left(T_A - T_S\right)}{\Delta x} = h\left(T_S - T_\infty\right) + \sigma F_{S-\infty}\left(T_S^4 - T_\infty^4\right)$$

where subscript infinity, ∞, indicates temperature far from the surface.

Temperatures must all be in Kelvin in this formula in order to use the Stefan–Boltzmann part of it correctly. In order to calculate T_S, the surface temperature, we have to solve by a numerical iterative approach or plot on a graph as shown on the right in Figure 2.47.

Alternatively, in the case where the temperatures in Kelvin are reasonably close to each other (say within 10%), a linear relation may be contrived as follows by factorising the Stefan–Boltzmann contribution:

$$\dot{q}_{b,1} = \sigma A_S F_{S-\infty}\left(T_S^4 - T_\infty^4\right)$$

$$\dot{q}_{b,1} = \sigma A_S F_{S-\infty}\left(T_S^2 - T_\infty^2\right)\left(T_S^2 + T_\infty^2\right)$$

$$\dot{q}_{b,1} = \sigma A_S F_{S-2}\left(T_S + T_\infty\right)\left(T_S^2 + T_\infty^2\right)\left(T_S - T_\infty\right)$$

Considering this as comprising a thermal resistance, $\dot{q} = \dfrac{T_S - T_\infty}{R_{TH}}$, where

$$R_{TH} = \frac{1}{\sigma A_S F_{S-\infty}\left(T_S + T_\infty\right)\left(T_S^2 + T_\infty^2\right)}$$

Take the average of T_S and T_∞ and substitute in this formula for both as the average, T_{AVG}, and then $R_{TH} = 1/F_{s-\infty}\sigma 4 T_{AVG}^3$. The formula for R_{TH} is dependent on the mean temperature in order to use this. This method may be useful in complex thermal resistance networks having a radiative contribution.

LEARNING OUTCOMES

✓ There are three modes of heat transfer – **conduction**, **convection**, and **radiation**.

✓ Conduction heat transfer is determined by **Fourier's Law of conduction**. The conductivity used in the law depends on material properties and temperature and is defined for most materials.

✓ The linear relationship between heat energy transferred and the temperature difference through which it moves is analogous to Ohm's law of electrical resistance. In a similar manner, the heat transferred is analogous to the current transferred, the temperature difference to the potential voltage difference, and the remaining terms are analogous to the electrical resistance, and are termed the **thermal resistance**.

✓ Convective heat transfer results from fluid moving from place to place and conveying and mixing materials of differing temperatures. **Forced convection** describes situations in which the fluid is mechanically driven, by wind or machine, and **natural convection** describes transfer due to temperature gradients in the fluid causing buoyancy and hence naturally driven circulation.

✓ **Newton's law of cooling** assumes a known **convective** HTC which depends on the flow configuration. Once known, this heat transfer contribution can be treated as thermal resistance similar to the conductive case and combined overall thermal resistance can be used to calculate overall heat transfer.

✓ Conversely, the inverse of the overall thermal resistance divided by the overall surface area available for heat transfer is known as the **overall** HTC.

✓ In the case of heat flow radially, the surface area of conduction increases with radius and leads to a logarithmic expression for the thermal resistance due to conduction.

✓ Convection depends on fluid flow, and hence on **laminar** and **turbulent** flow characteristics.

✓ Similar to the velocity boundary layer, there is a **thermal boundary layer** in situations where heat transfer is taking place. It depends on the **thermal diffusivity** (the thermal equivalent of viscosity), and on **thermal capacity**, $m.c_p$.

✓ The **Nusselt number** is a dimensionless parameter, which represents convective HTC, and if known, provides the convective HTC directly. **Correlations** of experimental data with Nusselt number are made with other dimensionless numbers: the **Prandtl number** is used for all Nusselt number correlations and is the ratio of kinematic viscosity, ν, to thermal diffusivity, α; **Reynolds number** is used for Nusselt number correlations in forced convection situations; **Grashof number** is a dimensionless ratio, which represents buoyant and viscous forces, and it is used for Nusselt number correlations in natural convection situations.

✓ Nusselt number correlations are available in heat transfer texts from notable experimental evidence.

✓ **Radiant heat transfer** is by electromagnetic radiation due to the release of energy from molecules excited by heat energy. It is determined according to the **Stefan–Boltzmann law of radiant heat release**. For radiative heat transfer, two bodies at different temperatures having a direct sight of each other release and receive heat dependent on their temperatures.

✓ The **emissivity** of a surface limits the amount of radiant energy released and modifies the Stefan–Boltzmann Law for surfaces which are not thermally black. A truly thermally black surface is not necessarily coloured black but has the property that it releases all its potential for radiant energy due to its temperature.

✓ It is possible to calculate the **combined mode** heat transfer with conduction, convection and radiation, but in cases where radiant heat transfer is present, the calculation is not directly solvable due to the fourth power of T in the Stefan–Boltzmann Law. The calculation must be done iteratively or graphically.

2.7   HEAT EXCHANGERS

Heat exchangers allow exchange of heat between two fluid streams without mixing them – perhaps the most familiar is the car radiator which cools engine cooling water with a forced draught of cool air. There are two main types of heat exchanger – the **recuperator** and the **regenerator** as shown schematically in Figure 2.48.

The **recuperator** (Figure 2.48a) is the most familiar and common type of heat exchanger; it is used in condensers and evaporators and boilers, and for cooling liquids and gases. A hot stream is contained in tubes which pass through a vessel containing the cooler fluid; of course, the hot and

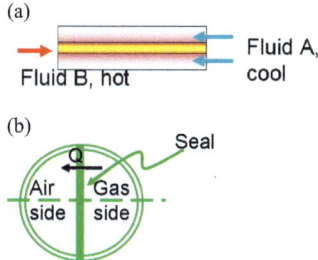

FIGURE 2.48 Schematic diagram of (a) a recuperator and (b) a regenerator.

cold streams can flow vice versa. The fluids do not make contact with each other. They exchange heat through the walls of the heat exchanger across a temperature gradient, which varies as progress is made through the recuperator, and as the temperatures of the two fluid streams change accordingly.

The **regenerator** (Figure 2.48b) is commonly used for preheating atmospheric air for combustion in the furnace of power stations using the exhaust gases at approximately 200°C as they travel to the chimney stack. As can be seen in the figure, there is a dividing wall centrally down the tube in which the regenerator is mounted, with the hot gases on one side of the wall and the cold gases on the other. The rotating mesh, or frame, of a material with a significant heat capacity (i.e. capacity to absorb heat) has elements which absorb heat in the hot side and release heat to the cold side, alternately taking heat from one side to the other as the mesh rotates. The mesh rotates steadily, continuously transferring heat between the two. In this way, in the power station, the exhaust gases leave with a lower temperature, and the combustion air has been warmed, which means that more heat has been used to generate power. Any heat loss in a power station is a corresponding loss in power generation, so any recovery of potentially lost heat is useful.

This section demonstrates how to calculate the heat exchange in recuperators and from this point they shall be referred to as just **heat exchangers**. The simplest type of heat exchanger is the **shell-and-tube** configuration, which implies an inner tube containing one of the two fluids, contained within an outer shell (which is often just a bigger tube) as shown in Figure 2.48a. This is the simplest type to analyse because we can estimate reasonably well the temperature profile from our knowledge of heat transfer in convective-conductive modes from Section 2.6, and hence the local and overall HTC. Figure 2.49 shows a simplified geometry of the shell and tube; there are obviously two directions that each fluid can pass in, and this gives rise to a very important distinguishing characteristic – whether the flow is **parallel flow**, in which case the fluids travel in the same direction along the geometry, or **counter-flow**, in which the fluids travel in opposite directions. From the figure, it can be seen that the corresponding temperature profiles are significantly affected by the flow directions.

2.7.1 HEAT TRANSFER ANALYSIS FOR SHELL AND TUBE HEAT EXCHANGER

2.7.1.1 Energy Balance

Referring to Figure 2.50, a schematic of a single pass shell and tube heat exchanger is seen which is the simplest possible configuration, in which an inner fluid is contained within the tube and an outer fluid is contained within the shell. By the principle of conservation of energy, heat transfer from the hot fluid = heat transfer to cold fluid, i.e.:

$$\dot{Q} = \left\{ \dot{m} c_p \left(T_{\text{in}} - T_{\text{out}} \right) \right\}_{\text{hot}} = \left\{ \dot{m} c_p \left(T_{\text{out}} - T_{\text{in}} \right) \right\}_{\text{cold}}$$

The term $\dot{m} c_p$ is the **thermal capacity rate** of the fluid, that is, its capacity for storing thermal energy per degree of temperature rise per second of flow. What is needed to solve this equation is $\dot{Q}$ or inlet temperatures and at least one outlet temperature.

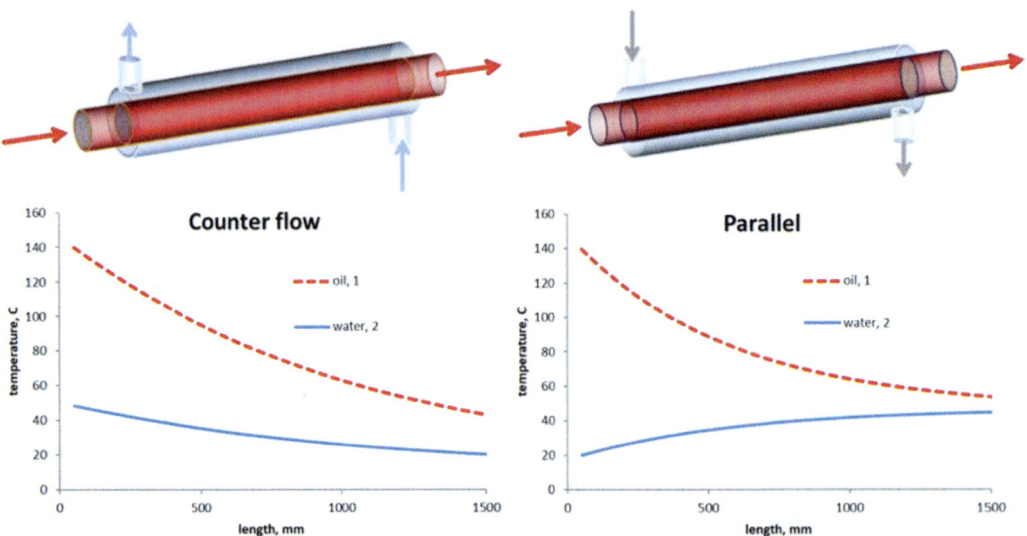

FIGURE 2.49 Schematic of the geometry and temperature profile in a shell and tube heat exchanger with counter and parallel flow. In this case oil, with specific heat capacity 2.4 kJ/kgK, density 760 kg/m³, heat transfer coefficient 1000 W/m²K and flow rate 10 g/s enters at 140°C and water, with specific heat capacity 4.18 kJ/kgK, density 1000 kg/m³, heat transfer coefficient 1000 W/m²K and flow rate 20 g/s enters at 20°C. The tube diameter is 20 mm and the shell diameter is 50 mm.

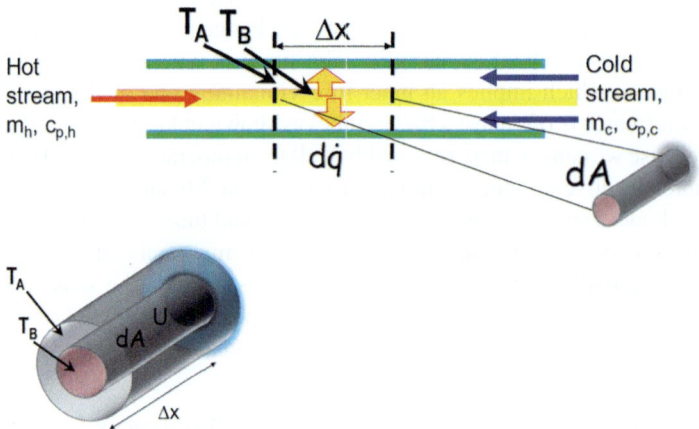

FIGURE 2.50 An elemental length of a shell and tube heat exchanger for the consideration of the heat transfer in it, with an inset of an elemental length of the inner tube to highlight the cylindrical area through which heat transfer occurs via the overall heat transfer coefficient, U.

2.7.1.2 Heat Transfer Calculation

With the applicable formulae, we know from heat transfer theory that we can, in the steady state, set up an 'Ohm's Law' analogical formula for the heat transfer in terms of the overall temperature difference between the hot and cold fluids at a particular position on the heat exchanger, and the thermal resistance at that point which is dependent on the convective heat transfer from the internal fluid to the wall of the tube, the conduction of heat from the inside of the wall to the outside of the wall and the convection from the outside of the wall to the external fluid. The formula becomes:

$$\dot{Q} = \frac{\Delta T_{\text{overall}}}{R_{\text{th}}} = \frac{\Delta T_{A-B}}{\dfrac{1}{h_i 2\pi r_i} + \dfrac{\ln \dfrac{r_o}{r_i}}{2\pi k L} + \dfrac{1}{h_o 2\pi r_o}}$$

in which subscript A represents the inner (tube bound) fluid and subscript B represents the outer (shell bound) fluid. Since the wall of the tube is *usually* thin compared to the pipe diameter, the pipe inner surface area is very nearly the same as the pipe outer surface area, so that $A_A = A_B = A$ and $r_i = r_o$, and the thermal resistance can be represented instead as an overall HTC acting on that area:

$$\dot{q} = UA\left(\Delta T_{\text{Overall}}\right)$$
$$= UA\left(T_{\infty A} - T_{\infty B}\right)$$

in which $UA \equiv \dfrac{1}{\sum R_{\text{TH}}}$ and the ∞ symbol indicates bulk temperature of the inner or outer fluid. That is the temperature of each fluid far enough from the surfaces are to be independent of radial temperature gradients. Considering the element shown in Figure 2.50, there is the contribution of heat transfer for the elemental length of the heat exchanger:

$$d\dot{q} = -dA \cdot U\left(T_B - T_A\right)$$

The total heat transfer is $\dot{q} = \sum d\dot{q}$, summing over all elements on length. This energy balance gives the overall *achievement* of the heat exchanger, the transfer of heat between the two fluids, by which the cold heats up and the hot cools down – it is the reason for the existence of the heat exchanger. The calculation depends on the performance of the heat exchanger, and the calculation of the heat exchanger performance is due to the heat transfer. Theory tells us how to get heat transfer across the surfaces, but it is complicated by the stream-wise gradients of temperature.

Note: Even if U is constant, $(T_B - T_A)$ varies along the heat exchanger. This must be taken into account. There are two approaches available for the analysis of heat exchangers, namely **logarithmic mean temperature difference (LMTD)** and **effectiveness-number of transfer units method (ε-NTU)** which will be elaborated on in the following sections.

2.7.2 LMTD Method for Sizing Heat Exchangers

In a parallel and counter-flow single-pass tubular heat exchanger, as shown in Figure 2.50, the temperature difference between the two fluid streams varies as the flow proceeds from inlet to outlet. The LMTD analysis gives the equivalent constant temperature difference for calculation of heat transfer, such that it is the same as the actual heat transfer with the temperature varying with progression through the heat exchanger. LMTD finds a representative temperature dependent on the overall temperature differences at inlets and outlets, which requires the heat exchange between the two separated fluids. It is used then to find the area of a heat exchanger that is required to achieve the desired heat transfer. Alternatively, if a heat flux is required and fluid entry or exit temperatures are not known, these can be found by iteration.

2.7.2.1 LMTD Method

Referring to Figure 2.51, the LMTD formula is expressed in terms of the temperature difference between the fluid in the tube and in the shell at position a and at position b:

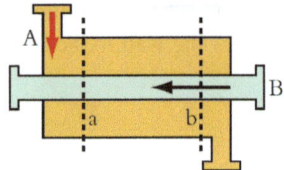

FIGURE 2.51 Single pass shell and tube heat exchanger.

$$\text{LMTD} = \Delta T_\mathrm{m} = \frac{\Delta T_\mathrm{a} - \Delta T_\mathrm{b}}{\ln\left(\Delta T_\mathrm{a} / \Delta T_\mathrm{b}\right)}$$

where a refers to inlet and b to outlet conditions or any two far apart points along the heat exchange surface. i.e.

$$\Delta T_\mathrm{a} = \left(T_\mathrm{B} - T_\mathrm{A}\right)_\mathrm{a}$$
$$\Delta T_\mathrm{b} = \left(T_\mathrm{B} - T_\mathrm{A}\right)_\mathrm{b}$$

These are the overall temperature differences between stream A and stream B. These relationships apply to, and are the same for, both parallel and counter-flow arrangements. The LMTD formula is derived from the consideration of elemental heat exchange and can be easily verified.

For **sizing heat exchangers** (i.e. working out required size for given flow rate and temperature change), the LMTD approach is used to give the area required from:

$$A = \frac{\dot{Q}}{U\Delta T_\mathrm{m}}$$

where $\dot{Q}$ is the total heat transfer rate between the two fluid streams, U is the overall HTC for the geometry and flow and ΔT_m is the LMTD.

The **rating of heat exchangers** (i.e. working out the temperature change in a given heat exchanger) can be done by iteration using the LMTD approach. This is what the LMTD is used for, when there is a known overall HTC, U. U must be determined by calculation or from tabulated data.

Example of how to use the LMTD method to rate and size a heat exchanger.

Calculate the LMTD for parallel and counter-flow cases which have the same inlet and outlet stream temperatures. The cold enters at 20°C and exits at 40°C and the hot enters at 100 C and exits at 50 C. Referring to Figure 2.52, the temperature profile is distinctly different in the counter-flow and parallel flow operations. In the parallel case, LMTD = (80 – 10)/[ln(80/10)], and LMTD = 33.7°C, and in the counter-flow case, LMTD = (60 – 30)/[ln(60/30)], and LMTD = 43.3°C. Note: The parallel arrangement has the lower LMTD – therefore it needs a larger surface area for the same heat transfer, that is a larger heat exchanger. This is one of the advantages of the counter-flow arrangement.

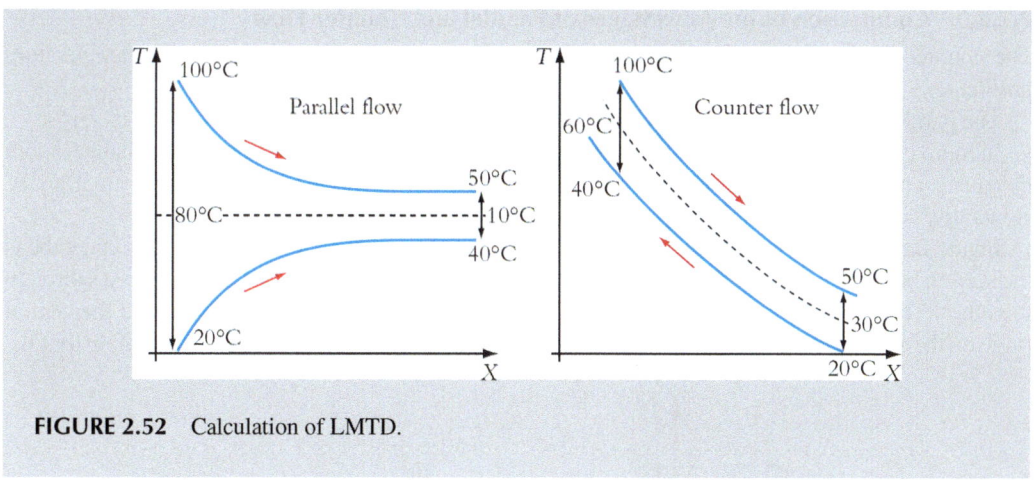

FIGURE 2.52 Calculation of LMTD.

2.7.2.2 Calculation of Heat Transfer Rate

$$Q = U \cdot A$$

2.7.2.2.1 LMTD

The overall HTC is determined from

$$UA = \cfrac{1}{\cfrac{1}{2\pi r_o h_o L} + \cfrac{\ln\dfrac{r_o}{r_i}}{2\pi k L} + \cfrac{1}{2\pi r_i h_i L}}$$

(i) and (o) are the inside and outside surfaces of the tube. If the HTC variations along the tube are large, then an average U value is used: $\bar{U} = (U_a + U_b)/2$. This is usually unnecessary. For a first approximation, evaluate U at $T = (T_a + T_b)/2$. The variation is because the HTC may vary with fluid temperatures since the fluid properties may vary with temperature. In a shell and tube heat exchanger, with the tube carrying hot fluid and the shell cold fluid, the conduction term can usually be neglected if the thickness of the wall is small and the conductivity high.

The LMTD analysis requires the inlet and outlet temperatures of the fluid streams to be known. This may necessitate an iterative solution involving guessed outlet values. This is the main disadvantage of the approach.

Procedure

1. Guess the outlet temperatures.
2. Calculate LMTD.
3. Calculate the heat transfer rate from step 2.
4. Calculate the outlet temperature for step 1.
5. Repeat until converged.

2.7.2.3 Comparison of the Advantages of Parallel and Counter Flow

The counter-flow type is the most compact for a given heat transfer rate due to the temperature profile.

The parallel flow type gives a lower T_{max} for the tube wall temperature.

In more complex designs of similar type, the 'tube' flow passes through the heat exchanger more than once, and the 'shell' side flow can be directed across the tube to increase turbulent heat transfer. Generally, heat exchangers of this type are called <u>shell and tube heat exchangers</u>.

Figure 2.53 shows a typical shell and tube heat exchanger. As shown in the schematic, the tube is the section between the two end caps and the end caps themselves contain the fluid that passes through the tubes. There are two passes of the tube fluid through the shell fluid in this case, and the central ridge on the opened end of the cap shows how the two halves of the cap are separated to allow for this.

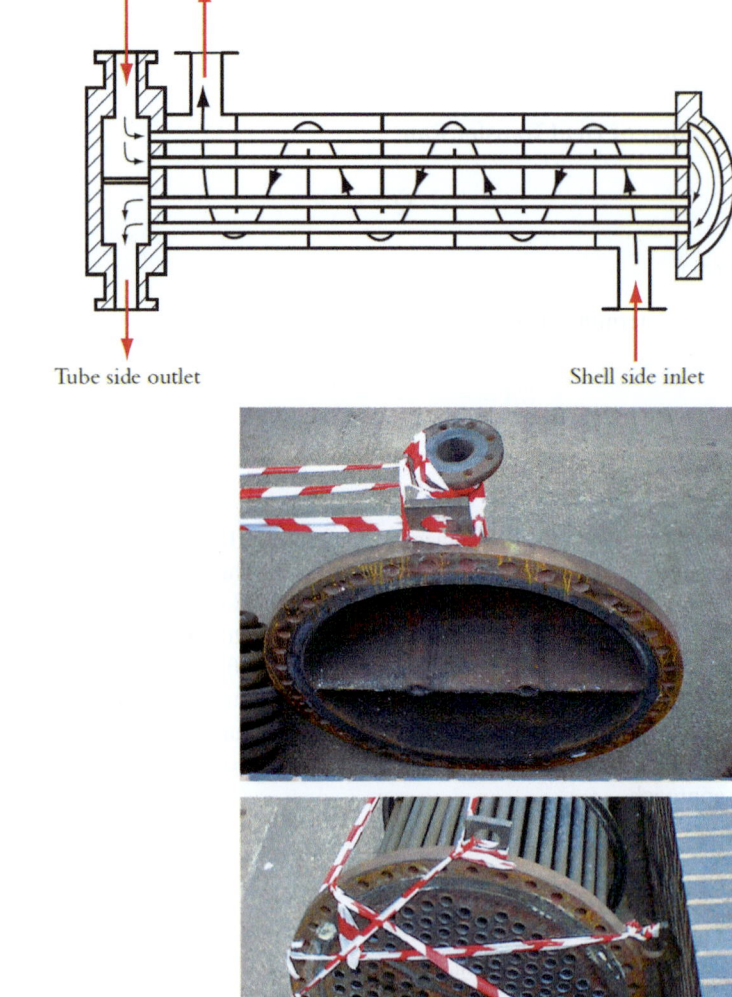

FIGURE 2.53 A typical shell and tube heat exchanger used in the hot water system at the University of Nottingham.

2.7.2.4   Crossflow and Multipass

The stream temperature differences are more complicated in cross-flow or multi-pass configurations. The effects of cross-flow and multi-pass are accounted for by introducing a **correction factor** F such that:

$$\dot{Q} = U\,A\,F\,\Delta T_m$$

Factor F is derived from calculations similar to the derivation of LMTD or read from graphs. Figure 2.54 shows a matrix heat exchanger with cross-flowing fluids and a counter–parallel flow shell and tube heat exchanger. The resulting temperature profiles are complex and a correction is required. The correction allows the simple counter-flow shell and tube heat exchanger calculation to be assumed, but corrects for the complexity of the flow with the correction factor. The LMTD method relies on prior experimental data.

A simpler and more direct approach for finding the temperatures is the ε-NTU method, which will follow after the treatment of the correction factor and is related to the correction factor method.

2.7.2.5   Correction Factor – LMTD

An excellent practical book on heat exchangers by Kays and London, which is the foundation of modern heat exchanger design and theory, presents methods for calculating the behaviour of many different heat exchanger types and configurations. The methods described here are described in detail in that text and for a full description the reader is referred thereto. This brief treatment is appropriate for the stage of this course in which it is a component of the course.

The correction factor allows us to use the LMTD type of calculation for more complex geometries. We must introduce two new variables:

$$R = \frac{T_1 - T_2}{t_2 - t_1} \text{ and } P = \frac{t_2 - t_1}{T_1 - t_1}$$

T_1 and T_2 are the entry and exit temperatures of one stream and t_1 and t_2 are the temperatures of the other stream. It is easy to see that R is the ratio of the temperature changes of the two streams and that P is the ratio of the temperature change of one stream to the maximum temperature difference between the hot inlet and the cold inlet.

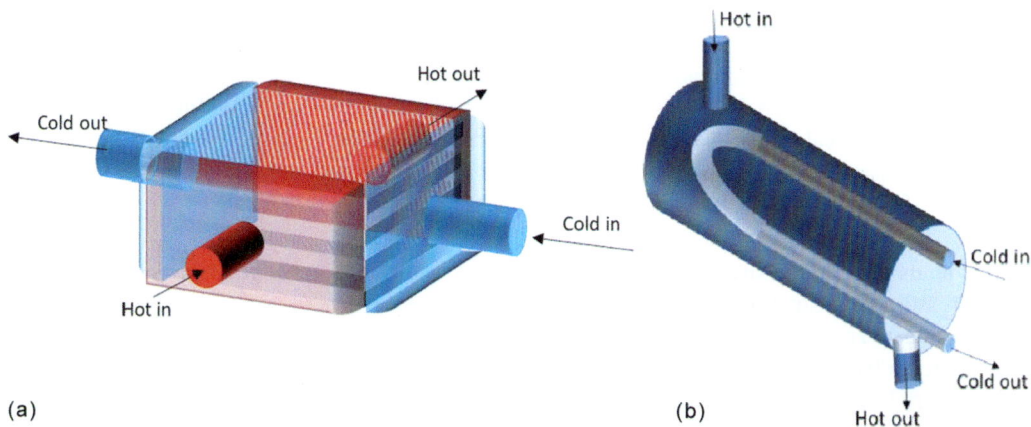

(a)   **(b)**

FIGURE 2.54   Two examples of cross flow heat exchanger types, (a) cross flow with plates separating each fluid, (b) cross flow with shell and tube arrangement having both counter-flow and parallel flow due to the reversal of the tube flow direction.

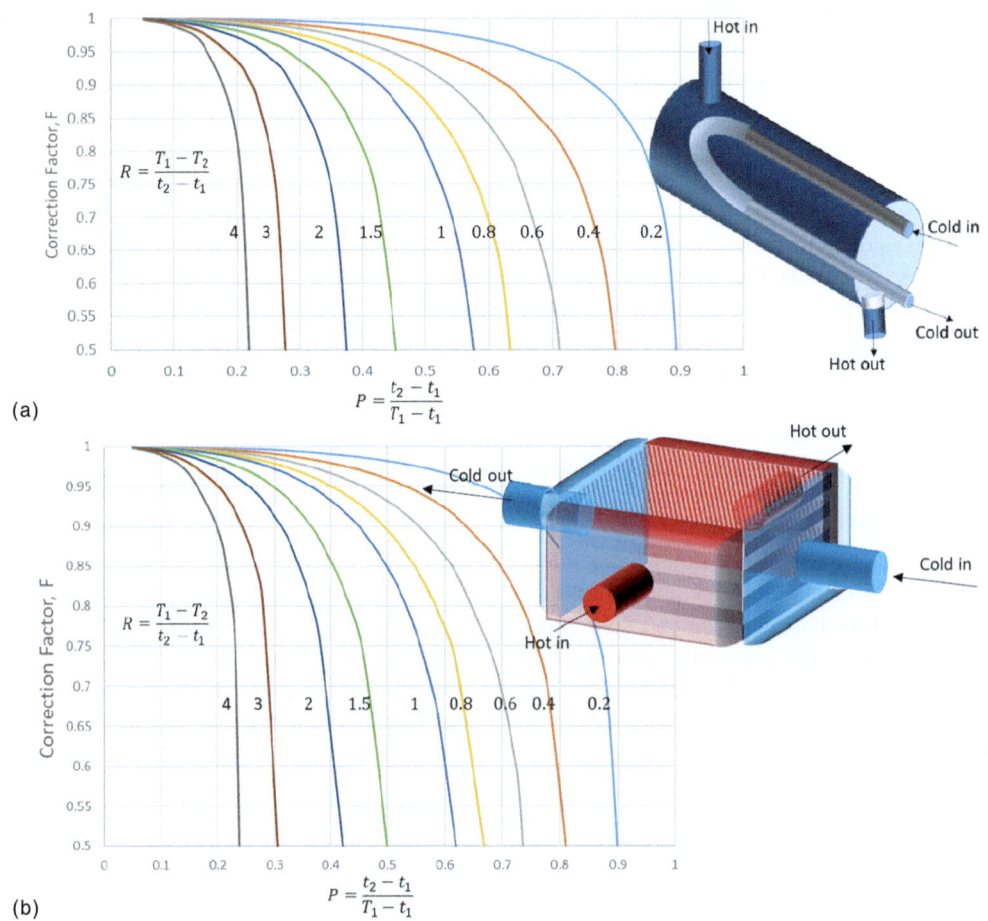

FIGURE 2.55 (a) One shell pass, two tube passes and multiples of them using formulae provided in Bowman et al (1940); (b) Cross flow, both streams mixed, alternate layers of the heat exchanger allow flow of the hot and then cold fluid, but inside the passages the fluids can migrate across the width of the passage and self-mix. Chart plotted using the formula of Bowman et al 1940 as detailed in Kays and London.

Using this method, we can read the correction factor from charts in the book, knowing P on the horizontal axis, a selection of curves for various values of R, and the corresponding correction factor, F from the vertical axis. Equations defining the correction factor, F, are derived by Bowman et al (1940) and presented in analysis comprehensively by Kays and London. The outcome is illustrated in Figure 2.55 for two particular configurations of heat exchangers.

If P tends to zero, the stream having temperatures t_1 and t_2 has a change of phase, that is if
   pressure drop is not too great $t_1 \rightarrow t_2$.
If R tends to zero, the stream having temperatures T_1 and T_2 has a change of phase.
If a stream has a phase change, the capacity rate is effectively infinite because the fluid will
   absorb heat without changing temperature. These are the two limits on the charts.

Figure 2.56 shows what is meant by the terms used with cross flow heat exchanger, the corrugations keep fluid moving as an **unmixed stream** as it passes through the channel, the open passage has no corrugations and the fluid is free to move across the full face of the channel and is called a **mixed stream**.

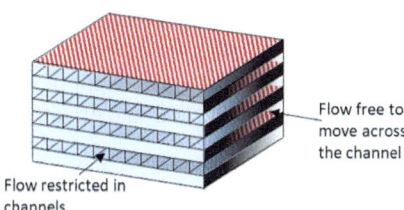

FIGURE 2.56   Examples of what is meant by cross flow heat exchangers from Bejan.

2.7.2.6   Care Using LMTD

It is important to understand the '**capacity rate**' of the fluids in the heat exchanger, that is

$$C_{\text{cold}} = \dot{m}_{\text{cold}} c_{p,\text{cold}} \text{ and } C_{\text{hot}} = \dot{m}_{\text{hot}} c_{p,\text{hot}}$$

These determine the magnitude of temperature changes for a certain heat transfer. It is important to always use the formula for LMTD:

$$\dot{q} = UA\,\Delta T_{\text{m}}$$

together with the heat capacity formulae:

$$\dot{q} = \left\{ \dot{m} c_p \left(T_{\text{in}} - T_{\text{out}} \right) \right\}_{\text{hot}} = \left\{ \dot{m} c_p \left(T_{\text{out}} - T_{\text{in}} \right) \right\}_{\text{cold}}$$

Using the former on its own suggests that the heat transfer does not depend on the capacity rate of the fluids. The limit of the heat exchanger capacity is when condensation of the hot fluid or evaporation of the cold fluid occurs across the heat exchanger, as in a boiler for example.

2.7.3   Effectiveness (ε)-NTU Method of Rating Heat Exchangers

This is an alternative method to LMTD for working out the temperature change in a heat exchanger which examines the thermal conductance, UA, as well as the capacity rates, C_{hot} and C_{cold}; this method introduces two dimensionless groups:

the **Number of heat Transfer Units**:

$$\text{NTU} = \frac{UA}{\left(\dot{m} C_p \right)_{\text{min}}}$$

where $(\dot{m} C_p)_{\text{min}}$ is the smaller capacity rate.

The effectiveness, ε:

$$\varepsilon = \frac{\text{actual heat transfer rate}}{\text{maximum heat transfer rate}} = \frac{\dot{q}}{\dot{q}_{\text{max}}}$$

in which $\dot{q}_{\text{max}} = C_{\text{min}} \Delta T_{\text{max}}$.

Equations defining the correction factor, F, in the LMTD method can be applied by noting that ratio R is equivalent to C_{min}/C_{max} and ratio P is equivalent to NTU (ref Kays and London) and using the further relationship $F = \text{NTU}_{counterflow}/\text{NTU}_{actual}$, (ref Wright in Kays and London). The formulae for F can be converted into a relationship between NTU and effectiveness making use of the formula for ε-NTU for the counter-flow single-pass case.

2.7.3.1 What Is the Maximum Possible Heat Transfer Rate?

This can best be considered in graphical terms as indicated in Figure 2.57, where the temperature of each fluid is depicted with progression through the heat exchanger; it asks the question if the UA could be increased to whatever we wanted, then what is the maximum achievable heat exchange between the two fluids?

As the size of the exchanger increases, the fluid with the smaller capacity rate, which has the steeper gradient of temperature change against stream direction (as shown in Figure 2.57 it is C_{cold}), exchanges heat with the hot fluid until the limit when the size of the heat exchanger is such that it leaves at the temperature of the hot fluid. This is the limit of the heat transfer. The minimum fluid may be either the hot or the cold fluid. For example, if the cold fluid has the minimum capacity rate then, $q_{max} = C_{cold}(T_{hot,in} - T_{cold,in})$.

Calculations usually entail evaluating effectiveness, ε from knowledge of the capacity rates, C, number of transfer units, NTU, values, and then evaluating q_{actual}. No iteration is required. Expressions for **NTU** in terms of ε can be derived by analysis of the rearranged **LMTD** heat transfer equation:

$$\ln\left(\Delta T_a / \Delta T_b\right) = \frac{1}{q}\left(\Delta T_a - \Delta T_b\right)\text{UA}$$

in terms of effectiveness and NTU by manipulation of the three equations. For example, a parallel flow gives the formula:

$$\text{NTU}_{parallel\ flow} = -\frac{\ln\left[1 - \varepsilon\left(1 + C_{min}/C_{max}\right)\right]}{1 + C_{min}/C_{max}}$$

which can be manipulated to give effectiveness:

$$\varepsilon = \frac{1 - \exp\left[-\text{NTU}\left(1 + C_{min}/C_{max}\right)\right]}{1 + C_{min}/C_{max}}$$

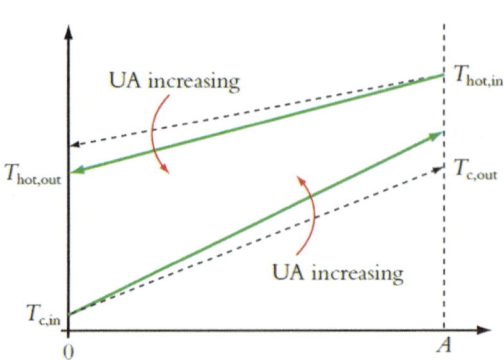

FIGURE 2.57 Demonstrating what the maximum possible heat transfer between two fluids in a heat exchanger is.

A formula is required for each configuration of heat exchanger. Fortunately, rather than using the formulae, data are presented in charts that can be read. If $C_{min}/C_{max} = 1$, the capacity rates are balanced – balanced heat exchanger. If $C_{min}/C_{max} = 0$, one stream, C_{max}, has phase change at nearly constant pressure.

EXAMPLE OF HOW TO USE THE ε-NTU METHOD

Knowing $UA/C_{min} = $ NTU, the chart gives effectiveness, which then yields q_{actual} from knowing q_{max}. Inspection of the chart in Figure 2.58a shows that if the area, C_{min} and overall HTC are known, then NTU follows for the horizontal axis point, and if C_{max} is known, then the ratio C_{min}/C_{max} indicates the curve to follow on the chart, and finally the effectiveness of the heat exchanger is given from the data represented by the chart on the vertical axis. This is for two shell passes, four tube passes and multiples of them. Figure 2.58b indicates the similar case for a cross-flow, both streams mixed heat exchanger configuration.

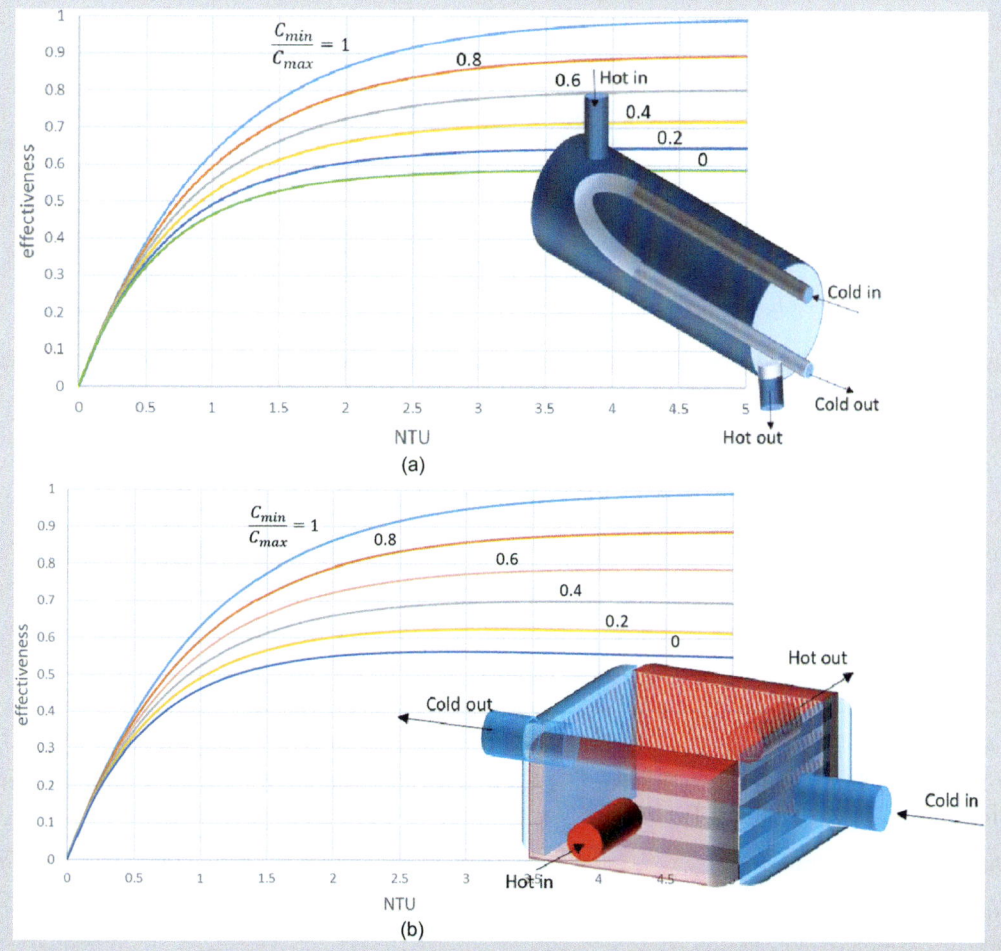

FIGURE 2.58   Effectiveness-NTU charts for the same geometries as the LMTD using formulae from Bowman et al. (1940) adapted to produce ε-NTU as described by Wright in Kays and London (1984).

2.7.3.2 Calculation of the Pressure Drop

Calculation of the pressure drop depends on detailed knowledge of the path that the fluid follows in flowing through the heat exchanger. Expansions, entries, and bends all need to be included.

There is a pump factor that needs to be taken into account. For a liquid it is:

$$\dot{W} = \frac{1}{\eta_{p}} \frac{\dot{m}}{\rho} \left(p_{in} - p_{out} \right)$$

in which η_{p} is the isentropic efficiency of the pump.

For a gas it is:

$$\dot{W} = \frac{1}{\eta_{c}} \mathrm{mc}_{p} T_{in} \left[\left(\frac{p_{out}}{p_{in}} \right)^{R/c_{p}} - 1 \right]$$

Intensification of heat transfer is usually accompanied by an increase in pressure drop. The Δp across the heat exchanger can be an issue with complex flow paths. Work must be supplied to deliver fluid through this pressure gradient by a pump.

2.7.3.3 Fouling Factors

Over time fluids tend to leave residual traces on the surfaces that they flow over. The thin layer of deposits from a particular fluid flow in the heat exchanger has a conductive resistance on the shell (s) and tube (t) sides. This contributes to overall U_{S} (HTC).

$$\dot{q} = \frac{\Delta T}{\dfrac{1}{h_{s}A_{s}} + \dfrac{1}{h_{t}A_{t}}} \text{ and } \dot{q} = U_{s}A_{s}\mathrm{LMTD} \text{ to give } U_{s}A_{s} = \frac{1}{\dfrac{1}{h_{s}A_{s}} + \dfrac{1}{h_{t}A_{t}}}$$

Over time the boundary surfaces corrode and acquire a scale coat, whilst the fluids gain impurities. So, fouling factors, r, are used for both shell and tube giving,

$$\frac{1}{U_{s}} = \left(\frac{1}{h_{s}} + \frac{1}{h_{t}} \frac{A_{s}}{A_{t}} \right) + \left(r_{s} + r_{t} \frac{A_{s}}{A_{t}} \right)$$

Fouling factor is additive to the thermal resistance components for the ideal HTCs, and the effect of the fouling depends on fluids involved.

TABLE 2.4

A Sample of Some Representative Fouling Factors [m²K/W] Related to Some Common Fluid Flows, from TEMA (Tubular Exchanger Manufacturers Association, 2007)

Fluid Type and Flow Characteristic	Fouling Factor, m²K/W
Sea water, >1 m/s	0.00009
Average river water, >1 m/s	0.00035
Refrigerants	0.00018
Natural gas flue gas	0.00088
Engine lubrication oil	0.00018

Some typical fouling factors are stated in Table 2.4 showing the effect of various fluids. The units are m²K/W, that is units of thermal resistance. Therefore, a value of 0.00018 for engine lubrication oil means 5678 W/m²K contribution, that is it is a small thermal resistance value, but comparable with the thermal resistance of thin conductive walls of the tubes for example.

LEARNING OUTCOMES

✓ Heat exchangers are classified first as **regenerators** and **recuperators**. The recuperator, which is the type involving tubes carrying one fluid surrounded by another fluid at a different temperature, is the concern of this unit.

✓ Recuperators are defined second by direction of flow as **counter-flow** or **parallel flow** of the two fluids involved.

✓ **Thermal capacity rate** is the specific heat capacity of a fluid multiplied by the amount of mass of the fluid considered, which is the mass flow rate of the fluid. It determines the temperature rise for a given heat input.

✓ Heat transfer theory shows how an elemental approach could be taken to the analysis of a particular heat exchanger, but there are two methods used with overall HTCs, which allow for simplified analysis in general cases.

✓ Since the temperature varies throughout a heat exchanger, the LMTD is used with a known overall HTC to work out either the heat transferred for a given sized heat exchanger or the size of a heat exchanger given the heat transfer required. It can be shown that the LMTD is the correct mean to calculate the overall heat transfer.

✓ The collection of experimental results by Kays and London provides a method for modifying the LMTD for more complex heat exchangers than the simple **shell-and-tube** configuration.

✓ It is important to remember that the capacity rate of each fluid determines its temperature and heat transfer.

✓ The **effectiveness (ε) – number of transfer units (NTU)** method relates the known data of overall HTC and the minimum of the two capacity rates in the NTU and relies on experimental data from heat transfer texts to relate this to the ratio of actual heat transfer to maximum possible heat transfer between the hottest and coldest temperatures available. Again, this relies on the catalogue of data by Kays and London.

✓ The cost of heat exchangers is the pressure drop for a given size. This must be considered in a practical heat exchanger and is determined from fluid mechanics calculations.

✓ **Fouling** often occurs in practical installations and the alteration to HTC must be considered.

2.8 VAPOUR POWER CYCLES

The idea of a vapour power cycle is to use condensable vapour to create power. Referring to Figure 2.59, we can see that it is similar to a heat pump running in reverse – that is

4-1 work is put in by a pump (at the position of the throttle in the heat pump) to drive the liquid fluid into a HP part of the circuit, (rather than expanding into the LP part of the circuit);

1-2 significant heat is added in the HP side to convert the liquid into vapour;

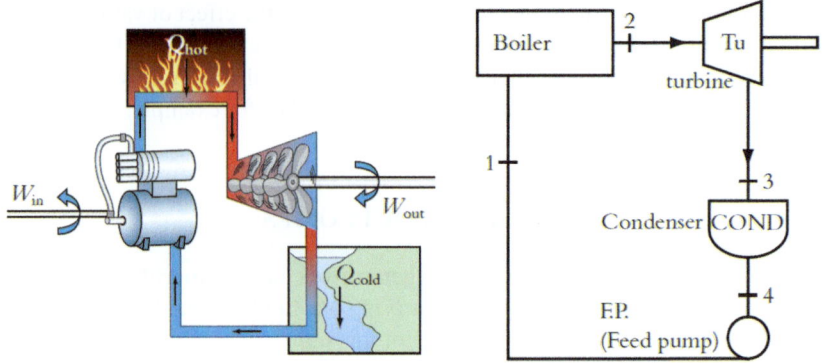

FIGURE 2.59   Schematic diagram of a simplified vapour power cycle.

2-3 the vapour passes through a turbine (at the position of the compressor in the heat pump), driving the blades round and producing work out of the turbine shaft, and in the process losing pressure back to the LP side of the circuit;

3-4 finally, before running around the circuit again, the vapour fluid is cooled in a heat exchanger to convert it fully back into a liquid.

The process is used for power production and requires that

    a. The working fluid is a condensable vapour
    b. The power cycle consists of a series of steady flow processes.

For the remainder of this section, steam will be considered as the condensable vapour. The analysis is simplified by the assumption that KE and PE changes are negligible compared with the change in enthalpy. The SFEE for change of state between points in the cycle is:

$$Q + W = H_2 - H_1$$

Since the purpose of the cycle is to produce usable work, it is useful to see where the work is coming from and where the expended energy used to create the work is being applied. The useful measures are **thermal efficiency, work ratio**, and **specific steam consumption**.

2.8.1 THERMAL EFFICIENCY

Closed cycle plants used for power generation have a thermal efficiency of between about 35% and 50%, where, with reference to Figure 2.59:

$$\eta_{th} = \frac{\text{net work output}}{\text{external heat supplied}} = \frac{\dot{w}_{out} - \dot{w}_{in}}{\dot{q}_{hot}} = \frac{\left(h_2 - h_3\right) - \left(h_4 - h_1\right)}{h_2 - h_1}$$

The net work output is measured in MW. It is useful to quantify the total energy output as electrical energy from the generator set that is mounted at the end of the turbine shaft, and this is often denoted as MW(e) to signify electrical power output. In practise, it is very close to the work output from the shaft of the turbine since the efficiency of electrical generators is very high (usually over 99%).

2.8.2 WORK RATIO (r_w)

The efficiency of the cycle is not useful on its own – we want to know how much we can get out of the process. Work occurs at two points in the cycle – driving the liquid fluid up to the HP, which is an energy cost in the system, and releasing energy from the HP vapour fluid on its passage through the turbine to deliver work to the electrical generator set. The ratio of these is expressed as the work ratio:

$$r_w = \frac{\text{net work output}}{\text{gross work output}} = \frac{\left(h_2 - h_3\right) - \left(h_1 - h_4\right)}{\left(h_2 - h_3\right)}$$

This tells us how much work we get out for work we put in – no consideration of heat here. Liquid compression requires less work due to (very nearly) no specific volume change as p increases. High r_w values are desirable, which indicates low sensitivity of efficiency to irreversibilities. Modern steam plants have $r_w \cong 0.98$, whilst for gas turbine cycles $r_w \cong 0.45$; the major difference is due to high compression work involved with gases.

2.8.3 SPECIFIC STEAM CONSUMPTION (SSC)

The efficiency and work ratio do not inform us of the amount of steam delivered to the turbine, which indicates the size of the plant. We use the specific steam consumption (SSC):

$$\text{S.S.C.} = \frac{\dot{m}_{\text{steam}}}{\dot{W}_{\text{net}}}$$

which is usually expressed in kilograms of steam required per kilo-watt-hour of power produced (i.e. the mass of steam required to maintain 1 kW of power output for 1 hour). A low SSC is good and implies the need for a smaller plant (that is lower capital cost).

2.8.4 CYCLES

The simplified cycle shown in Figure 2.59 can be modified to improve the performance on all three metrics by application of the aspects of the Second Law. The following applications are used routinely to improve power station performance and are regular parlance, so much so that the names of the inventors who first considered the implications of the process changes are synonymous with the processes. The technology applied in power stations increases the work produced for the size of plant and the amount of work that can be produced from the heat available. The theory explains Carnot cycle, the Rankine cycle, superheating, reheat and regeneration.

2.8.4.1 Carnot Cycle – The Most Efficient Thermodynamic Cycle

The Carnot cycle comprises four reversible processes: two isothermal and two adiabatic as indicated on a T-s chart of water and steam in Figure 2.60, and these are:

1-2 Isothermal expansion by adding heat, Q_h. In this part of the cycle, heat is added from the surroundings to convert the water to low-quality steam (i.e. high liquid droplet content or cloudy steam) to high-quality steam (low liquid content), and eventually to saturated steam (i.e. no water content, or dry steam) by evaporating the liquid and thus maintaining a constant temperature in the steam. The heat input is denoted as Q_H, since the heat is added in the HP or (relatively) hot side of the circuit.

2-3 Adiabatic expansion by passing the steam through a turbine, producing work. This assumes that the turbine is perfect, that is, it has no mechanical-friction or fluid-friction losses. The steam temperature drops from high on the HP side, T_H, to low on the low pressure side, T_L.

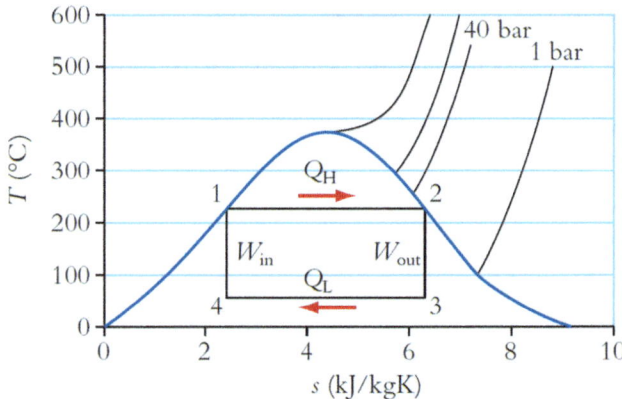

FIGURE 2.60   Representation of the Carnot cycle on the temperature-entropy diagram for steam.

3-4 Isothermal compression by removing heat, Q_L. Heat is removed from the system at this point in the cycle to the surroundings and the temperature is constant as steam gradually condenses as it travels, forming a film on the pipe walls of the system and eventually being entirely liquid. Note that a high water by mass content in the steam is not necessarily a high water by volume content since the density is high (about 1,000 kg/m³, compared to the steam at about 0.5 kg/m³).

4-1 Adiabatic compression by passing the low-quality steam through a pump, until the point at which the steam–water mixture in the system becomes saturated water with no steam content. Temperature increases from T_L to T_H.

The constancy of the temperature as heat is added in the HP side and heat is removed in the LP side is thermodynamically speaking a good thing because we know that the Carnot efficiency improves with increasing difference between the hottest and coldest reservoirs (surroundings), and we know from section 2.1.6 and point 5 in the list of corollaries of the 2nd law, that if heat is exchanged in a cycle with more than one reservoir at different temperatures, then the efficiency falls from the maximum available from the highest temperature and lowest temperature reservoirs to somewhere in between. The only way to add heat at constant temperature is by phase change and it is achieved in steam by working in the mixture region.

The usefulness of the *T-s* diagram is apparent from this process as shown in Figure 2.60 in which it can be seen that the constant entropy processes are vertical lines, and variation from the ideal will be seen obviously on this diagram if the vertical lines incline from the vertical.

The Carnot cycle is the basic cycle upon which improvements are built, since it is an impractical cycle, and although the thermal efficiency is highest for a given high temperature, higher temperatures are achievable by working outside the mixture region. The Carnot cycle is impractical for the following reasons: the turbine will be eroded on the surfaces as the steam expands in the mixture region since at high speeds the droplets of steam that form as the quality of the steam falls will have an impact on the blades and cause erosion which is very costly to repair; it is difficult to ensure that condensation reaches the exact point required for the saturated liquid at the point entering the compressor on the left side of the process, and the compressor work is quite high since it is not pumping a very dense liquid but a significantly less dense, and therefore higher specific volume, fluid and reducing efficiency which decreases the work ratio. As a point of interest, if the cycle is reversed it becomes the Carnot refrigeration cycle seen in section 2.2.2, in which all the arrows or processes are reversed.

WORKED EXAMPLE

Find the SSC, work ratio and thermal efficiency of the Carnot cycle operating between 100 bar and 0.3 bar.

For SSC, we require the work done to compress the fluid and the net work done. For the Carnot cycle, we need the enthalpy change between 4 and 1 on the diagram in Figure 2.60.

$$h_{f,100bar} 1408 kJ/kg$$

$$s_{f,100bar} 336 kJ/kgK$$

Dryness fraction at 4 is from entropy $s_4 = s_1$ and s_4 is in the mixture region with a dryness fraction, x, being x part vapour entropy and $(1-x)$ part liquid entropy:

$$s_{4f}(1-x) + s_{4g}x = s_1 0.944(1-x) + 7.767x = 3.36$$
$$x = 0.35$$

Therefore, enthalpy at 4 is:

$$h_4 = 0.35 h_{g0.3bar} + 0.65 h_{f0.3bar}$$
$$h_4 = 0.35 \times 2,625 + 0.65 \times 289 = 1,106 \, kJ/kg.$$

Therefore, $W_{4-1} = 1,408 - 1,106 = 302$ kJ/kg.
The work out of the turbine between 2 and 3 is:

$$\Delta h = h_{g100bar} - h_3$$

h_3 is found by first getting the dryness fraction as above for constant entropy, which is 0.685, and then using that to get the mixture region enthalpy, h_3 enthalpy, which is 1,889 kJ/kg. Therefore, $\Delta h = 2,725 - 1,889 = 836$ kJ/kg. Since the units of specific enthalpy are kJ/kg, then the SSC in kg/kWh is calculated for 3600 seconds as follows:

$$\text{SSC for Carnot is} \, 1 \times 3,600 / (836 - 302) = 6.742 \, kg/kWh.$$

$$r_W \text{ for Carnot is} \, (836 - 302) / 836 = 64\%.$$

For thermal efficiency, the external heat added is required. For the Carnot cycle, this is $h_{fg,100bar} = 1,317$ kJ/kg. Therefore, $\eta_{TH,Carnot} = (836 - 302)/1,317 = 40.5\%$.

2.8.4.2 Rankine Cycle – Reducing Compression Work

The Rankine cycle, shown on the *T-s* diagram in Figure 2.61, extends the condensation to the saturation line (process 3-4) before using a pump instead of a compressor to increase the pressure of liquid water (process 4-1) from the condenser pressure (typically a fraction of atmospheric pressure) to the boiler pressure (typically in the order of 100 bar). The mixture at outlet from the condenser is saturated liquid; so, the feed pump work is greatly reduced since compressing liquid requires less work than a liquid and vapour mixture. The work ratio is increased, because 1 and 4 are virtually coincident and, as a direct result of this, W_{in} is very small. For a flow process, the reversible process work is:

$$W = \int_4^1 v.dp$$

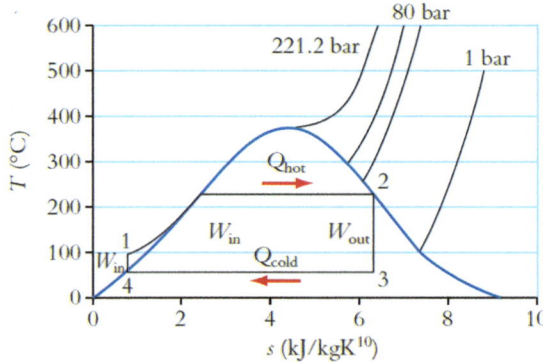

FIGURE 2.61 The Rankine cycle showing the small pump work in the liquid phase, which results in reduced loss.

Since v [m³/kg] is very nearly constant and not a function of pressure, the work is $v[p_1 - p_4]$. With v being the inverse of density, and knowing water density is approximately 1,000 kg/m³, an estimate of the work in changing pressure from 0.3 bar to 100 bar is $0.001 \times [100\text{-}0.3] \times 10^5 = 9,970$ J/kg or 10 kJ/kg. In the Rankine cycle, thermal efficiency is reduced because not all heat is added at the higher temperature, rather a significant amount is added at a temperature lower than the saturation temperature, but this is unimportant with regard to the work output increase. This can be illustrated by using typical enthalpy values. SSC is also lower (i.e. improved) due to the increased work out per kg of steam during the cycle. These effects are seen in the following example.

WORKED EXAMPLE

Compare the SSC, work ratio and thermal efficiency of the Rankine and Carnot cycles operating between 100 bar and 0.3 bar.

In the case of the Rankine cycle, the compression work is 10 kJ/kg of water. Using some of the results from the turbine work in the Carnot calculation above:

SSC for Rankine is $1 \times 3,600/(836 - 10) = 4.891$ kg/kWh. Carnot was 6.742 kg/kWh.
r_W for Rankine is $(836 - 10)/836 = 98.8\%$. For Carnot, it is 64%.

For thermal efficiency, the external heat added is required. For the Rankine cycle, it is $h_{g100bar} - (h_{f,0.3bar} + 10)$ kJ/kg $= 2,725 - (289 + 10) = 2,426$ kJ/kg.

Therefore, $\eta_{TH,Rankine} = (836 - 302)/2,426 = 22\%$.

The comparison shows that Rankine is a significant improvement over Carnot for SSC and r_W, but that the thermal efficiency is significantly affected.

2.8.4.3 Rankine Cycle with Superheat

To improve the cycle efficiency, the saturated steam leaving the boiler is **superheated**. Heat is applied at nearly constant pressure between 2 and 3 by passing the steam through pipes that pass through the furnace flue gases at the top of the furnace.

It is useful to note at this point, where a practical implementation is considered, that the boiling of water to saturated steam occurs in tubes built into the walls of the furnace which completely fill the wall surface area, receiving mostly radiant heat transfer from the hot flame in the vicinity of the

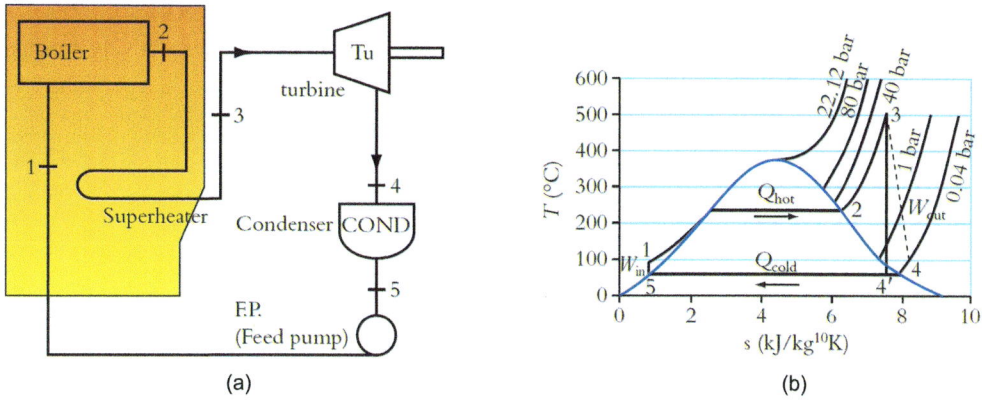

FIGURE 2.62   Schematic of the Rankine cycle with a superheater circuit.

burners, before rising to a steam drum, where any liquid is passed back to the furnace wall pipes and the saturated steam proceeds into the superheating tubes. This raises the steam state into the superheated region.

Figure 2.62a gives some indication of the layout in the furnace. The boiler is actually in the walls – a large number of vertical pipes line the entire wall surface, and in these pipes the water is heated from state 1 to state 2 in Figure 2.61. The superheater circuit consists of yet more pipes suspended above the flame of the furnace in the middle of the hot rising gases. The gases having a higher temperature than the steam in the pipes means that there is a temperature gradient between the hot gases and the steam in the pipes of a few hundred degrees and that significant heat transfer occurs mainly by forced convective heat transfer.

Figure 2.62b shows the process diagram on the T-s chart. The use of modern materials allows T_3 values up to 650°C. The Carnot cycle is limited to $T_{max} \leq 374.15°C$ (the critical temperature for H_2O), and superheating surpasses this temperature considerably. Some of the heat is supplied at $T > T_{sat}$, raising the average temperature above that of the purely Rankine cycle, and thus increasing the thermal efficiency. The other advantage of superheating is that the dryness fraction at exit from the turbine (4) is higher than in the Rankine cycle and consequently the turbine suffers less erosion by droplet impact.

The real process in the turbine becomes important at this point as indicated in Figure 2.62b, and the isentropic (constant entropy, reversible adiabatic) process 3-4 suffers an increase in entropy due to the Second Law and arrives at the same pressure line as 4, but with increased entropy and hence dryness fraction. The dotted line shows this irreversible process.

WORKED EXAMPLE

Calculate r_W, η_{TH} and SSC for a Rankine superheat cycle working between 100 bar and 0.3 bar, with 139°C of superheat.

The calculation of liquid work is the same as in the previous example. At 100 bar, $T_{SAT} = 311°C$; therefore, superheated temperature is 450°C. The isentropic work out of the turbine requires enthalpy in $h_{100bar,450°C} = 3{,}241$ kJ/kg, and the enthalpy at exhaust. Calculate the exhaust enthalpy using the mixture fraction at constant entropy. Entropy at 100 bar and

450°C is 6.419 kJ/kgK. Using h_f and h_g at 0.3 bar as 289 and 2,625 kJ/kg, respectively, results in dryness fraction as follows:

$$6.149 = 0.944(1-x) + 7.767x$$
$$x = 0.762$$

and hence exhaust enthalpy:

$$h_3 = 2,625x + 289(1-x)$$
$$h_3 = 2,069 \text{ kJ/kg}$$

Hence, work out is $3,241 - 2,069 = 1,172$ kJ/kg.

$$r_W = (1,172 - 10)/1,172 = 99.1\%.$$

External heat supplied is $h_{100bar,450°C} - (289+10) = 3,241 - 299 = 2,942$ kJ/kg.

$$\eta_{TH} = (1,172 - 10)/2,942 = 39.5\%.$$

$$SSC = 3,600/(1,172 - 10) = 3.098 \text{ kg}/\text{kWh}.$$

2.8.4.4 Use of Reheaters

A 'reheater' supplies heat at a point part way through the turbine expansion (between turbines). The reheater is a tube bank which passes steam back through the furnace flue gases in the same way as superheaters as shown in Figure 2.63(a). This avoids turbine expansion into low-dryness fraction conditions, which cause high turbine erosion by droplet impact.

The SSC is reduced, hence a small power station is required. Figure 2.63(b) shows the effect of the reheat, which is similar to the superheater, but on a lower pressure line, between points 4 and 5.

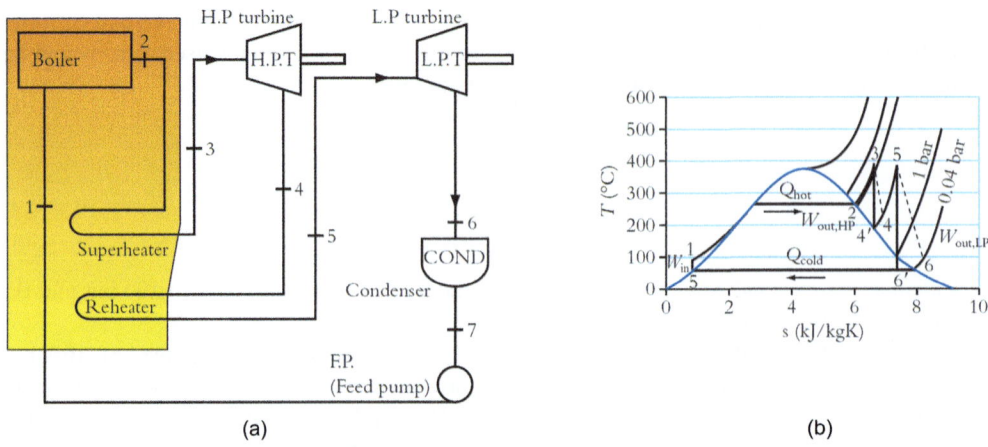

(a) (b)

FIGURE 2.63 Schematic diagram of the process with a reheater circuit.

2.8.4.5 The Mollier Chart or Enthalpy–Entropy Diagram for Steam

This chart, a section of which is shown in Figure 2.64, is used to find the state of steam at entry and exit from turbines. It is useful because, the ideal process is isentropic – constant entropy – which means there is a vertical line from one pressure and temperature state to a lower, known pressure. When the **Mollier diagram** is used with **isentropic efficiency**, the true exit state from the turbine can be easily determined by calculating the actual exit enthalpy and hence temperature. The change in specific enthalpy that is revealed by this is used to find the actual work produced by the turbines

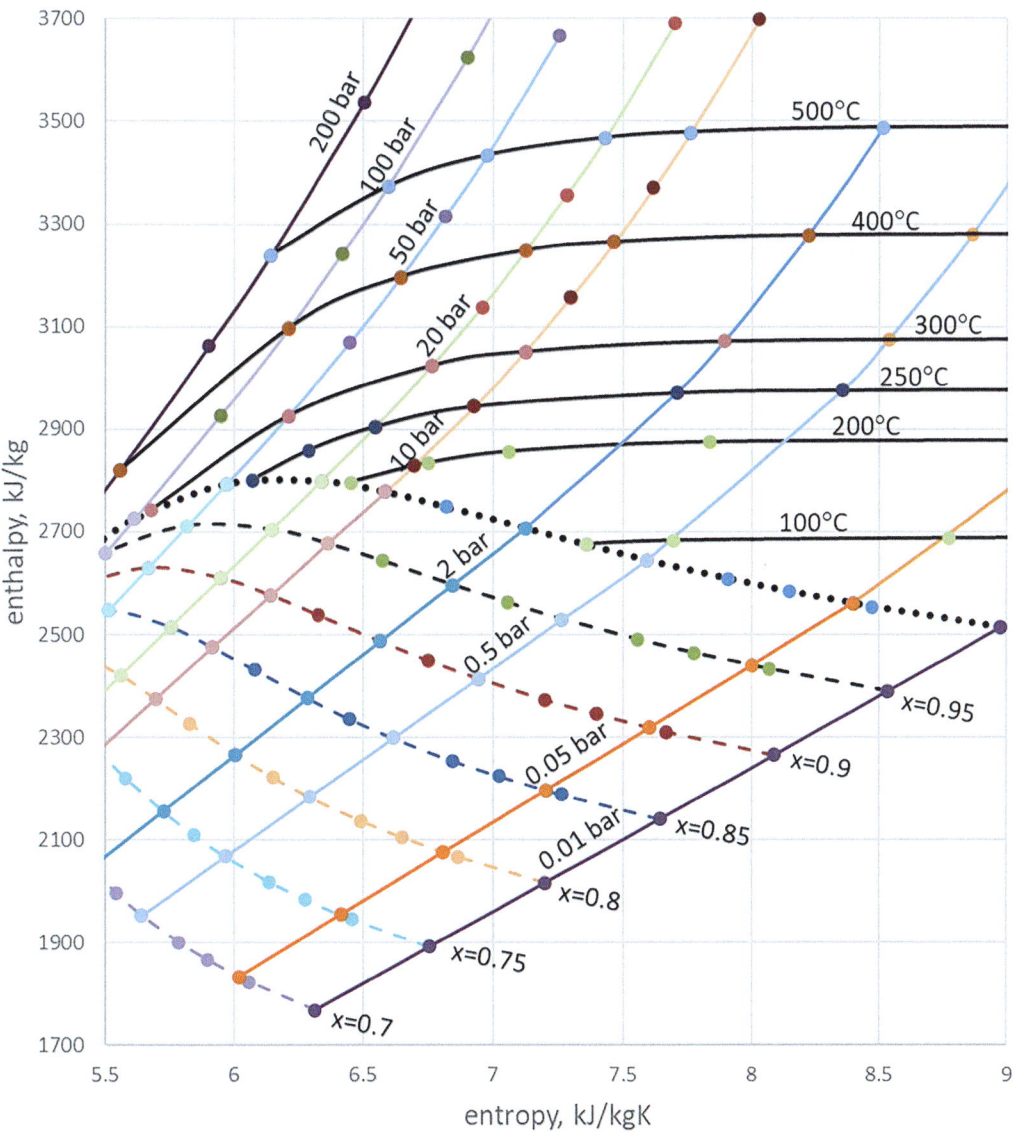

FIGURE 2.64 The Mollier enthalpy vs. entropy diagram for steam, with data points from Rogers, G.F.C. and Mayhew, Y.R., 1995, Thermodynamic and Transport Properties of Fluids, SI Units, Fifth Edition, Blackwell Publishing Ltd, Oxford, UK. Superheated and saturated conditions are separated by the dotted x = 1 plot.

in the steam cycle. The chart is a quick and reasonably accurate method for finding the change in properties, avoiding the need for interpolating values in the tables of steam data which is the alternative method.

2.8.4.6   Use of Feed Heaters – Regenerative Cycle

Feed heaters are used to reduce the external heat supplied (i.e. the heat drawn into the circuit external to the steam pipes, that is heat taken out of the furnace) at temperatures below T_{max}. This is known as **regenerative heating**. They therefore increase the thermal efficiency. The reduction is achieved by bleeding off steam part way through the turbine expansion and using it to heat the feed water flow to the boiler.

There are costs associated with this mode of operation since turbine work is reduced and specific steam consumption is increased. The plant is more complex (more expensive).

There is an advantage in regenerative heating in that the LP turbine doesn't have to be so big because it handles less steam.

Two types of feed heaters are used:

OPEN in which bleed steam and feedwater are mixed irreversibly and adiabatically in a constant pressure process.

CLOSED in which a recuperative heat exchanger is used with no mixing of the separate streams of steam.

In both cases, mass flow conservation and the SFEE are applicable. That is the inlet and outlet conditions are related by mass flow continuity and SFEE.

(N.B. recuperative means like a recuperative heat exchanger – where the fluids exchange heat through a wall with no mixing.)

Open Feed Heater

The mass flow of steam is split after the HP part of the turbine so that only a proportion of it goes on to the LP turbine.

The energy equation in the feed heater, assuming a perfectly insulated (i.e. adiabatic chamber) and referring to the state points indicated in Figure 2.65, becomes:

$$y\dot{m}h_4 + (1-y)\dot{m}h_7 = \dot{m}h_8$$

in which y is the diverted proportion of the total steam flow rate, $\dot{m}$, and h is the specific enthalpy at the indicated locations 4, 7 and 8 in the circuit. States 4 and 7 mix irreversibly, but at constant pressure to produce a fluid at state 8 which is usually in the saturated liquid state. The liquid reaching the boiler in the saturated state then has the requirement of receiving sufficient heat to change phase rather than having to heat up to saturation first. Since the pressure is below the boiler pressure, an extra feed pump is needed after the open feed heater.

$$p_4 = p_7 = p_8$$

$$m_4 + m_7 = m_8$$

The principle of energy conservation gives:

$$(mh)_4 + (mh)_7 = (mh)_8$$

The assumption of saturated liquid at the point before the pump is usually a reasonably good one to calculate the required bleed flow, m_4. It can be shown that for optimum efficiency, the pressure

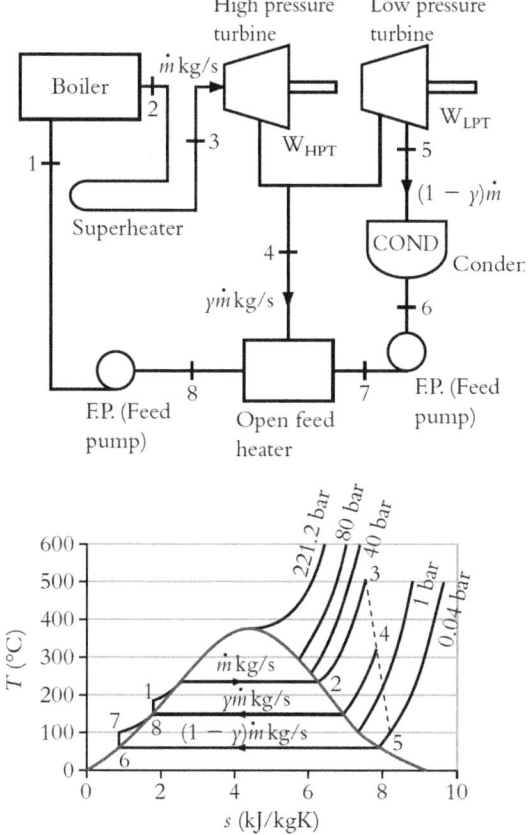

FIGURE 2.65 A steam circuit with open feed heater.

for bleeding is p_{SAT} for the temperature halfway between the average boiler and condenser temperatures, but the proof is beyond the scope of this section. With the assumption of reaching saturation temperature, the process diagram appears as shown schematically in Figure 2.65. Note that this is not a true process diagram because the mass carried around each part is *not* the same. The feedpump between 6 and 7 increases pressure almost at constant entropy, before the liquid reaches the feed heater chamber, where it mixes with the wet steam which has been bled out of the HP turbine and the combined water leaves at saturated condition before being pumped up to the boiler steam pressure at 1. When calculating the work out, it is important to remember that the LP turbine does not have the full flow rate of steam from the boiler but less due to the bled steam.

Closed Feed Heater

In this case, as indicated in Figure 2.66, some of the expanding steam is bled from the HP turbine and passed through the feed heater which raises the feed water temperature partially towards the saturation temperature at p_1. The bled steam at 4 is cooled by heat exchange to state 8 and then throttled down to the condenser pressure. During the throttling process, the expansion may lead to condensed water flashing off as steam at a lower pressure for which a flash chamber is provided. The feed water temperature is raised from T_7 to T_1 during the process of internal heat transfer. Note that 1 is still liquid (subcooled liquid) and state 8 is condensed liquid. The throttling process can form some steam which can lead to erosion.

Flash chambers are used to avert erosion by slowing down the large volume of steam. There is no mixing between streams.

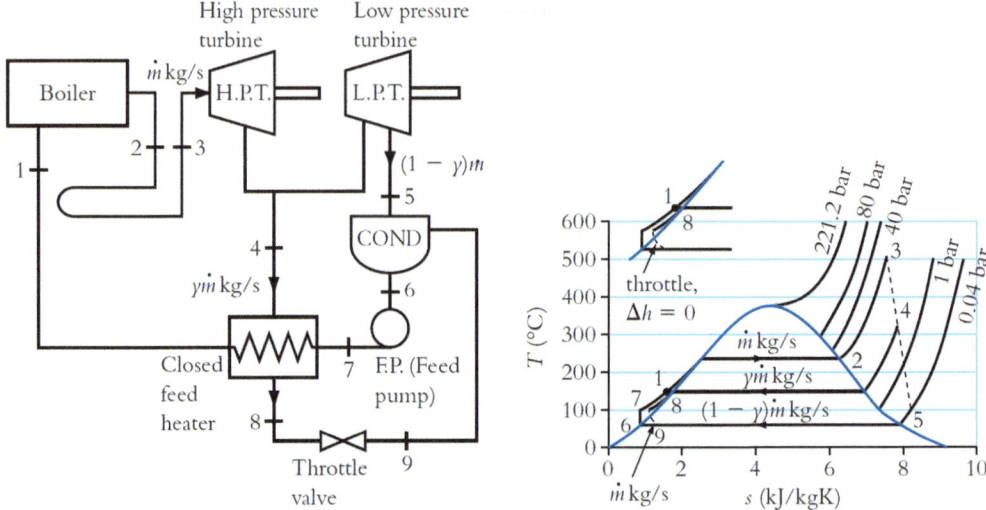

FIGURE 2.66   Schematic of regenerative feed water heating using a closed feed heater.

The energy equation gives (assuming adiabatic):

$$\left(mh\right)_4 - \left(mh\right)_8 = \left(mh\right)_1 - \left(mh\right)_7$$

Since enthalpy is constant from 8 to 9:

$$\left(mh\right)_8 = \left(mh\right)_9$$

The correct choice of bleed pressures is a matter of lengthy optimisation.

Note: As before, this is not a true process diagram because the mass carried around each part is *not* the same. The inset picture in Figure 2.66 shows the detail at point 8 – the bleed steam condenses as it heats up the feed water from state 7 to state 1 before it enters the boiler. The spent bleed steam then goes through a throttle to the pressure of the condenser and joins the feed water. The throttle process is indicated by a dashed line since it is an irreversible process.

2.8.4.7   Cycles with Power and Process Steam/Heat as Useful Outputs – Combined Heat and Power

On chemical process sites, there is often the requirement for steam and heat to be supplied in addition to the need for electrical power. Steam plants can be designed to meet these demands. Steam is required at a much lower temperature than power steam. It would be wasteful to have a heating only plant.

Efficiency is redefined as, for example:

$$\eta = \frac{\text{Net work} + \text{Process heat}}{\text{Heat supplied}}$$

Efficiencies are higher than for a conventional plant (70% typical) since the heat used for the process can be included in the benefit of the system.

It is convenient to use the plant for power and then extract heat from the condenser. The condenser is maintained at a relatively HP in order that $T_{SAT} = T_{REQUIRED}$. The turbine exhausting to the relatively HP condenser is called a Back Pressure Turbine, and operates as indicated schematically in Figure 2.67. This is practical when heat and power are fairly steady and well matched. When higher power is required, a Pass out Turbine can be used, as indicated schematically in Figure 2.68. This bleeds steam at an intermediate pressure to a heat exchanger before being fed back to the boiler. The amount of bled steam can be controlled.

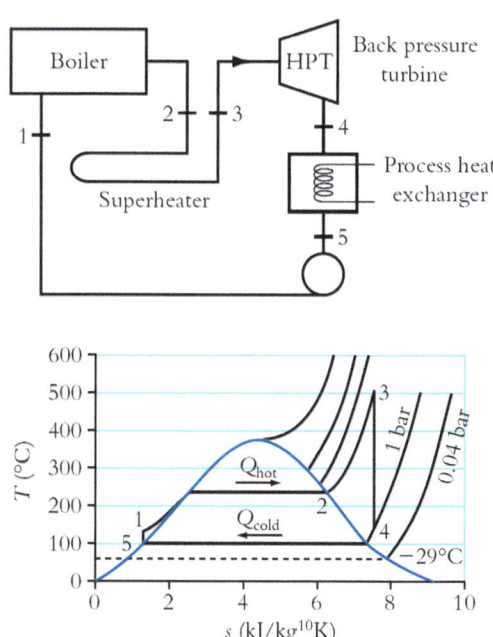

FIGURE 2.67 Back pressure turbine.

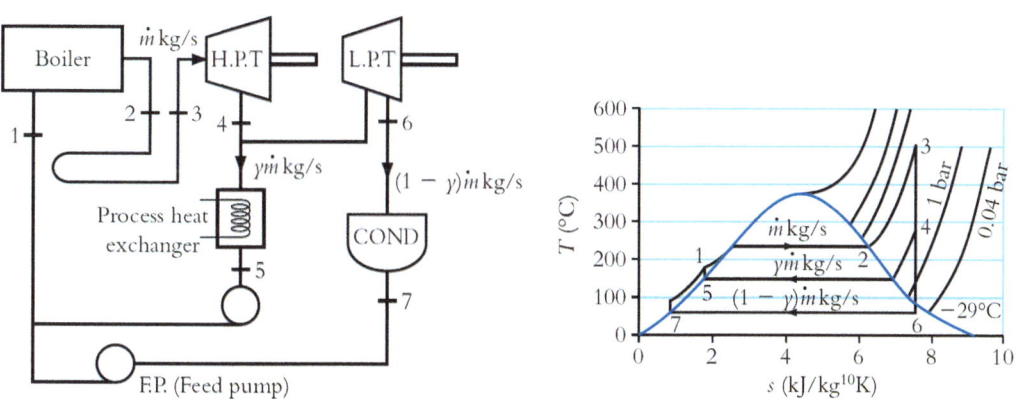

FIGURE 2.68 Pass out turbine.

LEARNING OUTCOMES

✓ A **vapour power cycle** operates on a **condensable** fluid in a **closed cycle**, receiving
heat from a hot reservoir, usually a **boiler**, to evaporate the fluid before expanding
through a **turbine** to a lower pressure in order to produce work; it is then condensed
to remove heat at a low pressure to a cold reservoir. A **pump** then increases the pressure to the HP side of the circuit prior to the heating part of the cycle.

✓ Enthalpy changes between state points on the cycle are used to calculate heat and work from the First Law in the form of the SFEE.

✓ Since vapour power cycles are widely used for power production, there are measures of the effectiveness of employment of the energy input to obtained output power. Thermal efficiency is the ratio of **work output** to **externally supplied heat**, that is the heat supplied from the hot reservoir. **Work ratio** is the ratio of **net work** (work output less work in) to work output, and gives an indication of the loss to compression. In the case of steam vapour power cycles, the **SSC** is the mass flow rate of steam per unit of power output, measured in kg per kW-hour.

✓ The basic cycle has been improved by altering it to improve the effectiveness. The basic cycle is the **Carnot cycle** which uses evaporation between saturated liquid and saturated vapour and which has expansion and compression within the vapour–liquid mixture region. This is the most thermally efficient cycle given any maximum and minimum temperature available and is based on the idea of the Carnot efficiency. The **Rankine cycle** is an improvement to the Carnot cycle, by simplifying the condensation part of the cycle such that all fluid is in the saturated liquid state and the **pumping work** is thus significantly reduced. The average upper temperature can be increased by employing a **superheat** part of the circuit after initial boiling, thus increasing the overall thermal efficiency; this also improves the turbine operation by reducing expansion into the mixture region. **Reheat** in the circuit is included after the initial turbine expansion, in order to avoid entering the mixture region significantly and is used in coal-fired power stations.

✓ The **Mollier chart** or **h-s** diagram is a convenient representation of the processes in the vapour power cycle since it directly yields enthalpy changes which are the heat and work exchanges with the working fluid. The process in the turbine is close to isentropic, and it is a simple exercise to compare the isentropic performance with the real performance by use of the isentropic efficiency.

✓ **Feed heating** is often employed, using some of the heat in the steam to heat the water being fed from the condenser to the pump. This improves the thermal efficiency by reducing the heat required from the boiler.

✓ Feed heaters can either mix the steam with the feed water, in an **open feed heater** or pass the fluids through a recuperator, in which the fluids are separate but exchange heat, in a **closed feed heater**.

✓ **Combined heat and power** is used where a quantity of heat is required for chemical processing in a factory or where domestic heating is required for long periods of time. The thermal efficiency is artificially improved by including some of the heat lost as useful output since it is put to good use. In this case, steam is taken at higher temperatures out of the final turbine, at a point where there is still useful heat in the steam.

2.9   RECIPROCATING INTERNAL COMBUSTION ENGINES

IC engines were developed in the 19th century, producing petroleum and diesel fuelled engines, which would subsequently find their most common and well-known application in the automotive industry, for power units for motor vehicles, as well as for small-scale power generation. In this section, an overview of reciprocating engines is given.

2.9.1 INTRODUCTION

The reciprocating engine mechanism, shown schematically in Figure 2.69, consists of a piston which moves in a cylinder and forms a movable gas-tight plug, a connecting rod and a crankshaft, similar to the construction of the reciprocating compressor described in section 2.6. If the engine has more than one cylinder, then the cylinders and pistons are identical, and all the connecting-rods are fastened to a common crankshaft. The angular positions of the crank-pins are such that the cylinders contribute their power strokes in a selected and regular sequence. By means of this arrangement the reciprocating motion of the piston is converted to a rotary motion at the crankshaft. There are many types and arrangements of engines and their classification defines them as described in the following.

2.9.2 OTTO AND DIESEL CYCLES

Petrol and **combustible gases** have high volatility and are appropriate for spark ignition (SI) engines and a thermodynamic cycle based on the **Otto cycle**. Petrol is a complex mixture of distilled organic oil, the closest chemical to which is **octane**, or **C_8H_{18}**, which leads to the definition of the **octane number**, defining the quality of the fuel. The combustion is started by **SI** and allows for controlling the timing of fuel detonation. The **air standard** Otto cycle is shown in Figure 2.70a. An air standard

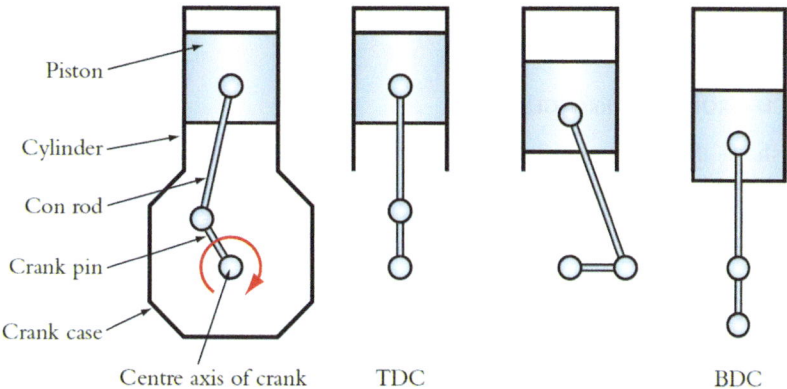

FIGURE 2.69   Schematic of a reciprocating engine.

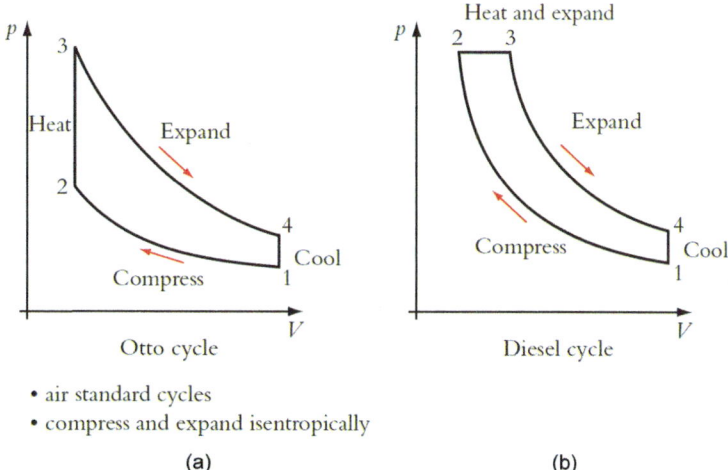

FIGURE 2.70   Air standard cycles representing the Otto and Diesel cycles.

cycle considers the processes in an ideal state, in which the heat is supplied and removed externally (rather than generated within the cylinder as in practice) and a single working fluid in a closed cycle is considered. The cycle consists of **isentropic compression** (1-2) assuming no heat transfer through the cylinder walls and a rapid process; this is followed by **constant volume heating** (2-3); the piston moves down and **isentropic expansion** occurs (3-4); the **heat is then removed** (4-1) and the cycle repeats. The power out of the cycle is the area in the cycle loop.

Diesel and **fuel oils** have relatively low volatility and are appropriate for compression ignition (CI) and a thermodynamic cycle based on the **diesel cycle**. Since the fuel is a complex mixture of distilled organic oil, the chemical composition is not precisely definable, but a close approximation is **cetane**, or $C_{16}H_{34}$, and this is the ideal diesel fuel leading to the definition of **cetane number** defining the quality of the fuel. The combustion is started by **CI** which works by increasing pressure to such a high level that the temperature is significantly raised, sufficient to commence a rapid deflagration or combustion, as compared to the explosive reaction occurring with more volatile fuels. The **air standard diesel cycle** is illustrated in Figure 2.70b. It consists of **isentropic compression** (1-2); **constant pressure heat addition** (2-3); **isentropic expansion** (3-4) and **constant volume heat removal** (4-1).

The practical implementation of these cycles are the **SI** engine, in which the air and the fuel are mixed before compression, and the **CI** engine, in which only the air is compressed and the fuel is injected into the compressed air which is then ignited at a sufficiently high temperature to initiate spontaneous combustion.

2.9.3 CYCLE PROCESSES ARRANGEMENT

Figure 2.71 shows the practical differences between three engine types, which between them represent the common engine cycles used. The stroke of the piston is the distance it moves from the position most extreme from the crankshaft to the nearest. This takes place over half a revolution of the crankshaft. In petrol engine practice, the extreme positions of the piston are referred to as **TDC**,

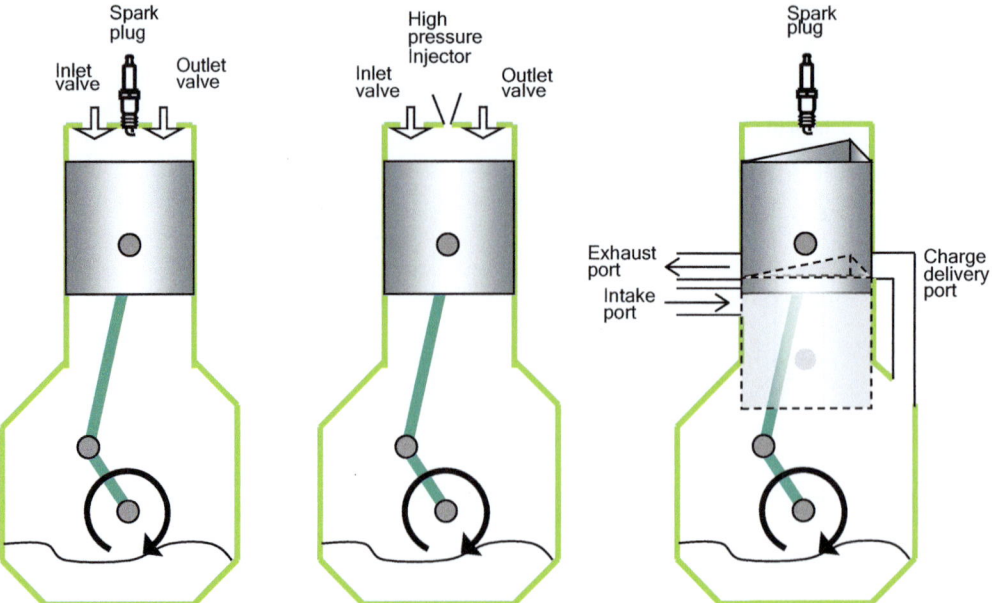

FIGURE 2.71 Schematic diagrams of engine types: (a) Spark ignition, 4-stroke engine, (b) compression ignition, 4 stroke engine, (c) spark ignition, 2-stroke engine with BDC position indicated to show clearance of charge delivery and exhaust ports.

and **BDC**. In oil-engine practice, they are referred to as **outer dead centre (ODC)** and **inner dead centre (IDC)**, respectively. An engine which requires four strokes of the piston (that is two revolutions of the crankshaft) to complete its cycle is called a **four-stroke** cycle engine. An engine which requires only two strokes of the piston (that is one crankshaft revolution) is called a **two-stroke** cycle engine. In all reciprocating **IC** engines, the gases are induced into and exhausted from the cylinder through ports, the opening and closing of which are related to the piston position. In a two-stroke engine (Figure 2.71c), the ports can be opened or closed by the piston itself, but in the four-stroke engine a separate shaft, called the **camshaft**, is required; this is driven from the crankshaft through a 2 to 1 speed reduction. The two-stroke engine has a piston head which is shaped to ensure that the charge delivery is directed upward into the cylinder away from the exhaust port to prevent a short circuit to the exhaust. The four-stroke petrol engine requires inlet and outlet valves. The diesel engine requires a HP injection port and outlet valves. There may be a number of valves and ports in each cylinder to improve fuel distribution and combustion.

2.9.4   CONTROL OF POWER OUTPUT

Spark ignition engines are **quantity governed**. Air and fuel are mixed outside the cylinder and the quantity of mixture induced is controlled by the throttle plate position. The throttle plate controls the air induced by restricting the inlet pipe, which may deliver to a distribution manifold or directly to each individual cylinder. Diesel engines are **quality governed**. They have minimum restriction to air flow into the engine (no throttle plate). Fuel is injected directly into the cylinder, under pressure, towards the end of the compression stroke. The amount of fuel injected controls the power output.

2.9.5   COMBUSTION INITIATION

In the case of the **SI engine**, the air and fuel mixture is **ignited by spark**, typically at 10° to 40° before the TDC during the compression stroke. The **compression ratio** is approximately 8:1, such that at the end of compression the gases in the piston are at eight times the atmospheric pressure. The mixture ratio must be near stoichiometric (~14.5:1 by mass) for ignition by spark to occur. Current engines run at stoichiometric under most conditions, for efficiency of the emissions control system. The AFR (described in section 2.4.2) affects the power and economy – slightly richer causes increased power and slightly leaner causes improved fuel economy. A **throttle** or restriction on the air flow adjusts the AFR.

   In the case of the **CI engine** fuel injection takes place near the end of the compression stroke. Combustion begins, as the fuel mixes with the air in the cylinder, by self-ignition when compression heating raises the mixture temperature to 800–900 K. The compression ratio in a CI engine is greater than 12:1 in order to achieve the required temperature. These engines normally run at AFR in the range 20:1–25:1. They can run out to approximately 40:1 AFR. The air flow is not restricted or otherwise controlled upstream; the mixture is controlled by injection of the fuel at an adjustable rate.

2.9.6   SPARK IGNITION ENGINE

Referring to Figure 2.72, the characteristics of a real SI engine cycle can be identified as follows:

   1-2 **Induction stroke**. Air/fuel mixture is drawn into the cylinder through the intake valve. The cylinder pressure is less than atmospheric due to losses across the valve, the intake manifold and the throttle plate.
   2-3 **Compression stroke**. The mixture is compressed and ignited by spark before TDC. Combustion proceeds at near-constant volume at approximately 50° of crank rotation. Peak cylinder pressure occurs just after TDC and is typically in the order of 25 bar for the **Wide Open Throttle** condition (WOT).
   3-4 **Power stroke**. The hot and HP combustion products do work on the piston, giving useful work output.

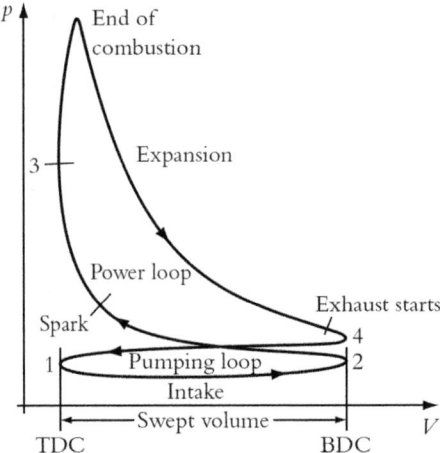

FIGURE 2.72   Spark ignition cycle on the *p-V* diagram.

4-1 **Exhaust stroke**. The energy depleted combustion products are expelled through the exhaust port into the exhaust manifold. Cylinder pressure is slightly above atmospheric, due to the restrictions on the gas flow of the valves and pipes downstream from the engine. Notice the energy lost when the exhaust starts as a free expansion from a pressure significantly higher than atmospheric.

Overall, the cycle is distinctly different to the ideal Otto cycle, mainly because of the combustion process used to produce the heat internal to the closed volume and the need to replenish the volume gases. The process is distinctly different to the standard air Otto cycle.

2.9.7   COMPRESSION IGNITION ENGINE

Referring to Figure 2.73, the Diesel cycle has characteristics as follows:

1-2 **Induction stroke**. Air is induced into the cylinder – no fuel is present in the air at this stage.

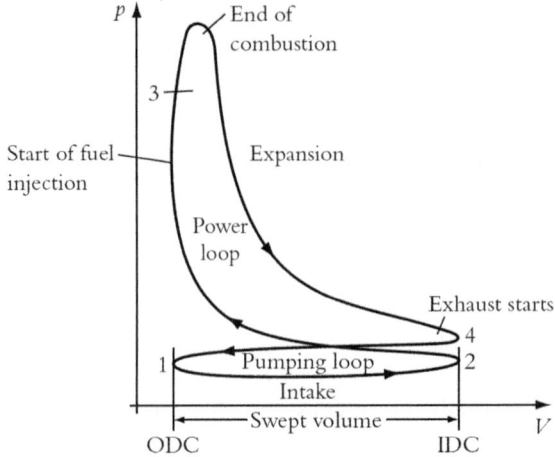

FIGURE 2.73   Compression ignition cycle on the *p-V* diagram.

2-3 **Compression**. The air only is compressed and fuel injection begins just before ODC and continues for ~20° to 30° of crank rotation. The injection is done at a very HP in order to form a fine and quick atomised spray. The rail pressure has increased to the point where most efficient diesel engines now have up to 2,000 bar. The combustion duration is longer than for a SI engine because it is nearer to constant pressure than constant volume. The rate of burn is controlled by the rate at which fuel evaporates and mixes with air in the cylinder. Peak cylinder pressure is in the order of 40 bar at full load.

3-4 **Power Stroke**. The combusting mixture expands against the load applied to the engine.

4-1 **Exhaust Stroke**. A stroke to remove the spent gases. Again notice the immediate loss of pressure at exhaust.

Notice that the cycle is again distinctly different to the ideal diesel cycle due to the real combustion processes used to provide the heat and due to the induction and exhaust strokes.

2.9.8 PERFORMANCE ASSESSMENT

For SI engines (similar for diesels), the main concerns are fuel economy at idle and part load, WOT power output and emission levels. Also important, but less easy to quantify, are the characteristics of the engine response to changes in demand (throttle opening or closing).

i. **Indicated power** or **ip**: Rate of work done on the piston by the gas as determined from the p-V diagram. The top loop on the diagram is the power loop (positive work done by gases) and the lower loop is the pumping loop (negative work done by gases). The indicated power is given by:

$$\text{ip} = \left(\text{imep}\right).\, LAN\left[\text{W}\right]$$

where L = stroke, A = piston area, N = cycles/sec, that is, (engine speed/2 for four-stroke engine), imep is the indicated mean effective pressure. From the diagram:

$$\text{imep} = \frac{\text{power loop area} - \text{pumping loop area}}{\text{diagram base length equivalent to swept volume}}$$

(imep values for SI engines are in the order of 5 bar).

The above is sometimes referred to as the net imep.

The gross imep is given by neglecting the pumping loop area.

ii. **Brake Power** or **bp**: Measured power output, $T\omega$, where T = brake torque and ω = engine speed (rad/s)

iii. **Mechanical Efficiency**: $\eta_m = \dfrac{\text{Brake Power}}{\text{Indicated Power}}$ This is typically ~90% at full-load. Losses are due to friction in the valve drive train, piston friction, bearings etc.

iv. **Brake Specific Fuel Consumption**: b.s.f.c. = m_f/Brake Power, where m_f is the fuel mass flow rate.

This is a measure of engine efficiency. It is more commonly used than **brake thermal efficiency**, $\eta_b = \dfrac{\text{BrakePower}}{m_f \cdot Q_{net}}$

where Q_{net} is the lower calorific value of the fuel. (η_b is the range ~35% for modern engines). $Q_{net} \equiv \Delta U_o$

v. **Volumetric Efficiency**: $\eta_v = \dfrac{\text{Volume induced / cycle}}{\text{Swept Volume}}$ where the volume induced is referred to conditions at inlet to the intake manifold. (WOT η_v is typically >80%).

2.9.9 COMPRESSION RATIO AND UNCONTROLLED COMBUSTION

Thermal efficiency increases with increasing compression ratio (r):

$$r = \frac{V_{max}}{V = \dfrac{V_{max}}{V_{min}} = \dfrac{V_{clearance} + V_{displaced}}{V_{clearance}}}$$

This reflects the increase in the upper temperature of the cycle, which increases thermal efficiency (recall that $\eta_{Carnot} = 1 - T_{MIN}/T_{MAX}$). In addition, there is less residual gases left in the cylinder at the end of the exhaust stroke to dilute the fresh change for the next cycle. SI engines have r values in the range of 9–10. These are unlikely to increase dramatically for the following reasons:

1. **Production tolerances**. High r values mean larger variations from build-to-build for given tolerances on dimensions. Variations of +½ a ratio are common.
2. **Spontaneous ignition**. If the r value is increased progressively, temperatures during compression will increase and eventually the charge will start to burn spontaneously before SI.
3. **Knock**. This is a problem even on current engine designs, and is more likely to occur when r is increased. This is an uncontrolled combustion phenomenon. Combustion starts normally by SI but unburnt gas ahead of the flame front is compressed. If the unburnt gas temperature is raised sufficiently, it will self-ignite as in (2), producing a characteristic knocking sound.

   In some circumstances, the flame front is accelerated and propagates across the combustion chamber with a shock wave. This detonation wave is reflected back and forth giving rise to high-frequency noise. Knock phenomena can cause serious overheating and loss of efficiency.
4. **Pre-ignition**. One effect of knock is to produce local hot spots on the combustion chamber wall which can initiate combustion before the spark on following cycles. Similarly, deposits on spark plugs and walls can also produce hot spots.

In CI engines, r is higher, but similar issues occur, due to the nature of the combustion. There is initial delay as fuel and air mix, followed by the initial spread of the flame and finally by combustion as the fuel is fully injected. **Diesel knock** may occur if the initial stages occur too rapidly.

2.9.10 SPARK TIMING

In SI engines, maximum work output is achieved when combustion occurs at TDC, at constant volume. In practice, this is not achieved. The best spark timing gives peak pressures which occur at approximately 15° after TDC. To achieve this, SI needs to occur before TDC. At a given nominal load, speed and AFR, the optimum timing is normally taken as the minimum advance for best torque (MBT timing). Usually, this is determined experimentally and timing requirements are mapped out as a calibration for engines of a given design. Flame speeds are approximately proportional to engine speed, though engine speed varies by an order of magnitude between idle and maximum engine speed, the duration of combustion requires a similar crank angle interval. Even so, timing is advanced from about 10° before TDC at idle and 30° before TDC at high speeds, full load.

In addition, timing is adjusted to compensate for changes in load. Timing is advanced under light load conditions by up to about 20°. This compensates for charge dilution by residual combustion products, and low cylinder pressures and temperatures which are adverse to good combustion.

Modern electronic ignition systems perform adjustments electronically, but still use engine speed and manifold vacuum (or air flow rate) to identify timing requirements.

2.9.11 Fuelling Systems

In SI engine, fuel injection systems have largely replaced carburetors, although the basic function is the same. The carburetor meters fuel into the throat of a venturi, such that the AFR is approximately constant for all running conditions. The air mass flow rate through the venturi depresses the throat gauge pressure and fuel flow has the same functional dependence on this pressure. This is because as the constriction occurs, the air flow speeds up and the static pressure drops correspondingly. The mixture strength is increased at idle (to compensate for poor combustion conditions due to residual dilution in particular) and at WOT (to give maximum power output, which is then limited by the availability of air).

Current fuel injection systems are multipoint (most expensive) or single point (more expensive than a carburetor). In multipoint systems, fuel is injected into each tract of the intake manifold near to the intake valve. One fuel injector per cylinder is required. Single-point system inject fuel into air flow before it is split into separate routes to individual cylinders.

Multipoint systems (either electronic (EFI), or mechanical (MFI)) are most common. Fuel is injected in discrete shots lasting approximately 10 msec. Fuel supply pressure (3 bar) is much lower than diesel fuel injection systems and the injection timing is not critical. For diesel engines, fuel injection is inherently necessary. Each cylinder has an injector which fires fuel into the combustion chamber directly (direct injection diesels) or into a pre-chamber connected to the combustion chamber (indirect injection diesels). The injection system is sequential (each cylinder receives the fuel charge at the same point in its cycle). Because the injection is into the cylinder, late in the compression stroke, the fuel supply pressure is much higher (200 bar and now much HPs are available) than for SI engine systems. The injection spray characteristics (droplet size, distribution, direction etc.) are also more critical since this directly influences the way in which combustion progresses.

2.9.12 Supercharging

The loss in the exhaust due to the sudden expansion when exhaust commences, can be used to benefit the charging part of the cycle. This is done generally on Diesel engines to very useful effect in improving engine efficiency. The exhaust gases are passed over a turbine, which is directly linked to a compressor which delivers the air to the inlet manifold. The compressor turns at speeds in the order of tens of thousands of r.p.m. This is called a **turbocharger**, and the basic idea is indicated on the p-V diagram in Figure 2.74. It improves the volumetric efficiency since the air in the cylinder is at a higher pressure.

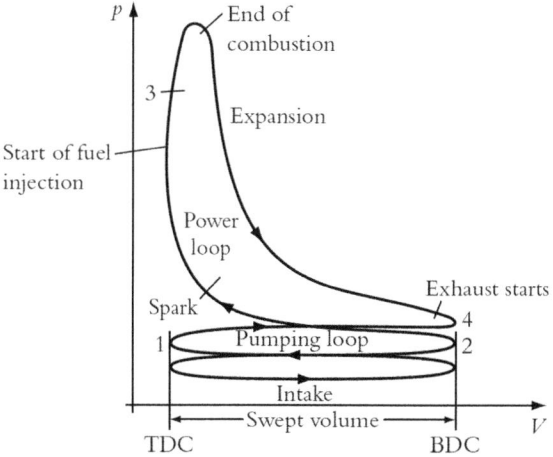

FIGURE 2.74 Principle of the use of a turbocharger to improve volumetric efficiency.

LEARNING OUTCOMES

✓ IC engines are either operated on the **Otto cycle**, in which case they work with petrol and are called **SI** engines or on the **Diesel** cycle, in which case they work with diesel fuel and are called **CI** engines.

✓ Diesel fuel may be mainly considered as **cetane, $C_{16}H_{34}$**. Petrol fuel may be mainly considered to be **octane, C_8H_{10}**.

✓ Both cycles are based on the ideal **standard air cycle** versions of the cycle, which assume external heat supply and ideally enacted processes. In reality, the cycles vary from this, mainly due to the nature of heat addition by internal combustion.

✓ SI engines have AFRs of approximately 14.5:1 (near stoichiometric) and a **compression ratio** of approximately 8. They are **quantity governed**, in that the rate of air flow is adjusted to alter power and economy.

✓ CI engines have AFRs between 20:1 and 25:1 and compression ratio in excess of 12. They are **quality governed**, in that the air flow is unrestricted and the injection of fuel controls the power and economy.

✓ There are several measures of engine performance: **indicated power** is the power directly from the p-V diagram; **brake power** is the measured power at the engine shaft; **mechanical efficiency** is the ratio of brake power to indicated power; **brake specific fuel consumption** which is the fuel required for a specific brake power; **volumetric efficiency** which is volume induced/swept volume, similar to that for air compressors.

✓ The compression ratio affects the combustion performance and can lead to poor combustion and hence poor mechanical behaviour if handled badly. In particular, it can lead to a noisy combustion behaviour known as **knock** or **diesel knock**.

✓ Spark timing is important for SI engines in order to provide the maximum pressure at the right moment in the cycle. The spark occurs a few degrees before the TDC position in order to create maximum pressure at the TDC.

✓ Fuelling systems for SI engines rely on injection of a premixed air and fuel mixture. Traditionally, **carburetors** were used, but recently these have been superseded by fuel injection. For CI engines, injection is necessary for the cycle to work, and the improvements have come from injection strategies due to electronic control and higher pressures available for fuel injection.

✓ The exhaust gases are ejected significantly above atmospheric pressure. This expansion can be used in a **turbocharger** to drive a turbine which in turn drives a compressor at a very high rotational speed. The compressor delivers air to the cylinder at significantly above atmospheric pressure, thus increasing the volumetric efficiency.

3 Solid Mechanics

Richard Brooks and Michael Clifford

3.1 INTRODUCTION

Part 1 of "An Introduction to Mechanical Engineering" covered the basic principles of Solid Mechanics (Unit 1) including basic mechanics and design analysis, stress and strain and the analysis of simple engineering loading situations such as uniaxial loading, beam bending, multi-axial stresses and torsion. This basic introduction assumed that all materials behaved in a linear elastic way and focussed on establishing the fundamental principles and simplified methods of analysis to solve for stresses and strains within components and structures.

In this unit, the application of solid mechanics to engineering problems is taken to a greater depth. Further elastic analysis is considered including aspects such as combined loading applied to components, further bending analyses, such as shear stresses in beams, the calculation of bending deflections and the bending of beams with asymmetric sections. Powerful methods of analysing elastic deformations in more complex shaped structures using the concept of stored strain energy are also included. The analysis of more detailed stress distributions, in thick cylinders, for instance, is also explained. The purpose of these further elastic analyses is to determine stresses, strains and deformations under more complex loading and where geometry varies in a more complex way.

Stresses and strains are calculated for a reason that is to design components fit for a purpose. In this sense, it is important that we consider how components might fail under loading. For this reason, this unit also includes several sections on failure, including yield criteria to assess when failure might occur, the analysis beyond the elastic region, where the elastic-plastic behaviour becomes important and the analysis of particular types of failure other than yielding including brittle fracture and fatigue (under cyclic loading). We learn how to analyse components under such conditions and how to design against such failures. Finally, failure of components is not necessarily restricted to excessive mechanical stress but may also be caused by other factors such as temperature or excessive deformations. The unit, therefore, also covers the analysis of thermal stresses and situations where elastic instability, that is buckling, might occur.

All of the above topics are covered in depth; first by introducing the basic concepts and theory, second by developing standard methods for analysis and finally by illustrating these methods with many worked examples. At the end of the unit, students will have learnt a wealth of easily usable analytical methods for solving many practical engineering problems.

3.2 COMBINED LOADING

Many engineering problems can be analysed as simple load situations; for example, uniaxial loading, beam bending, torsion etc. However, it is also very common in the real world for engineering components and structures to be subjected to several loads simultaneously. This is a combined loading situation and can be analysed by superposing the effects of the individual loads.

3.2.1 MOHR'S CIRCLE RECAP

Mohr's circle for plane stress was introduced in Part 1 of "Introduction to Mechanical Engineering" as a useful graphical technique for analysing plane stress acting on an element in a material or structure. For combined loading situations, it is common to reduce the problem to such a plane stress problem and analyse it using Mohr's circle. The analysis will give the principal stresses, the

DOI: 10.1201/9780429319495-3

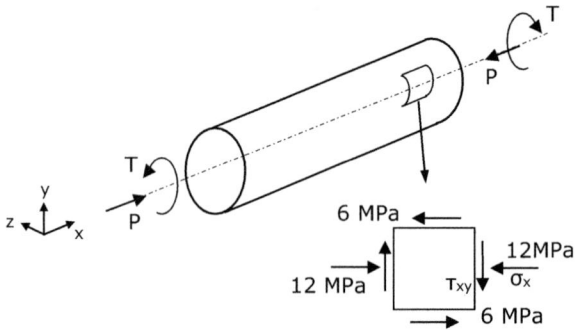

FIGURE 3.1   Shaft subjected to combined torque and compressive axial loading.

maximum shear stresses and the angles of the principal planes for the element. Figure 3.1 shows a shaft subjected to combined loading of a torque, T, and a compressive axial load, P. Let us assume that such loading gives rise to an axial stress of -12 MPa (i.e. compressive) and a shear stress of -6 MPa (i.e. causes the element to rotate clockwise) acting on a surface element as shown in the figure. The Mohr's circle analysis is then as follows:

The known stresses on the element are:

$\sigma_x = -12$ MPa
$\sigma_y = 0$ MPa
$\tau_{xy} = -6$ MPa
$\tau_{yx} = 6$ MPa

Figure 3.2 shows the Mohr's circle for this stress system. To draw the circle, first draw point B which represents stresses on the x-plane (coordinates: -12, -6). Next, draw point E which represents stresses on the y-plane (co-ordinates: 0, $+6$). Join the two points with the line BE which intersects the x-axis at the centre of the circle, C. The circle can now be drawn and the following quantities are measured:

$\sigma_1 = 2.5$ MPa
$\sigma_2 = -14.5$ MPa

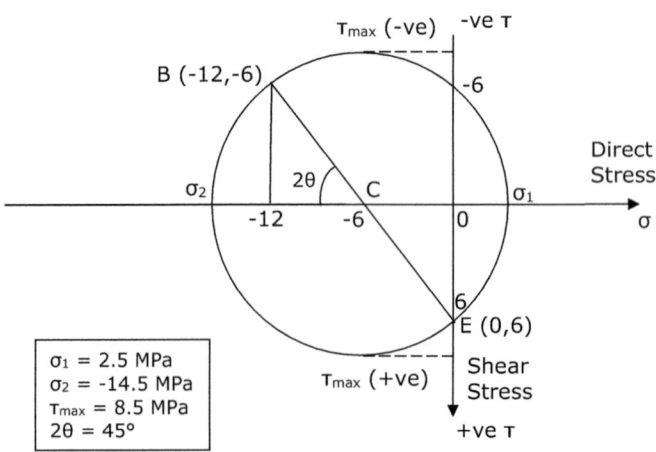

FIGURE 3.2   Mohr's circle for combined torsion and compression.

$\tau_{max} = 8.5$ MPa
$2\theta = 45°$

On the element, the angle of the principal plane (P1) from the y-plane is $\theta = 22.5°$ anticlockwise as shown in Figure 3.2.

Alternatively, the important parameters in the circle can be calculated analytically as follows:

The centre of the circle is given by $C = (\sigma_x + \sigma_y)/2 = -6$

The radius of the circle is given by $R = \sqrt{\left(\dfrac{\sigma_x - \sigma_y}{2}\right)^2 + \tau_{xy}^2} = 8.5$

The principal stresses are:

$$\sigma_1 = C + R = 2.5\,\text{MPa}$$

$$\sigma_2 = C - R = -14.5\,\text{MPa}$$

$$\tau_{max} = R = 8.5\,\text{MPa}$$

The angle of the principal planes: $\tan 2\theta = \dfrac{\tau_{xy}}{\left(\dfrac{\sigma_x - \sigma_y}{2}\right)} = 1$

$2\theta = 45°$
$\theta = 22.5°$

If the analytical approach is taken (which does give more accurate results), then it is always advisable to sketch Mohr's circle in order to gain a clear understanding of the orientation of the principal planes and the maximum shear planes with respect to the x- or y-planes.

3.2.2 SUPERPOSITION OF COMBINED LOADS

The **Principle of Superposition** states that:

$$\begin{bmatrix} \text{The total effect of } \underline{\text{combined}} \\ \text{loads applied to a body} \end{bmatrix} = \Sigma \begin{bmatrix} \text{The effects of the individual} \\ \text{loads applied } \underline{\text{separately}} \end{bmatrix}$$

[see Part 1 of "Introduction to Mechanical Engineering"]

Thus, when a body or structure is subjected to a combination of different types of loading simultaneously, we can consider the effect of each load on the local stress of an element separately. Stresses on the element can then be summed to determine the effect of the combined loading. A number of combined loading examples can be used to illustrate:

3.2.2.1 Combined Bending and Axial Loads

Figure 3.3 shows a beam carrying a uniformly distributed load (UDL) along its span, while simultaneously being subjected to an axial compressive force, F. Figure 3.3 shows how the effect of the combined loading, on the stress distribution through the thickness of the beam at the centre of its span, is determined. The effects of the UDL and the axial force are obtained separately and then summed to give the combined stress distribution on the beam. The symmetrical bending stress distribution about the neutral axis is essentially skewed to more compression by the effect of the axial stress.

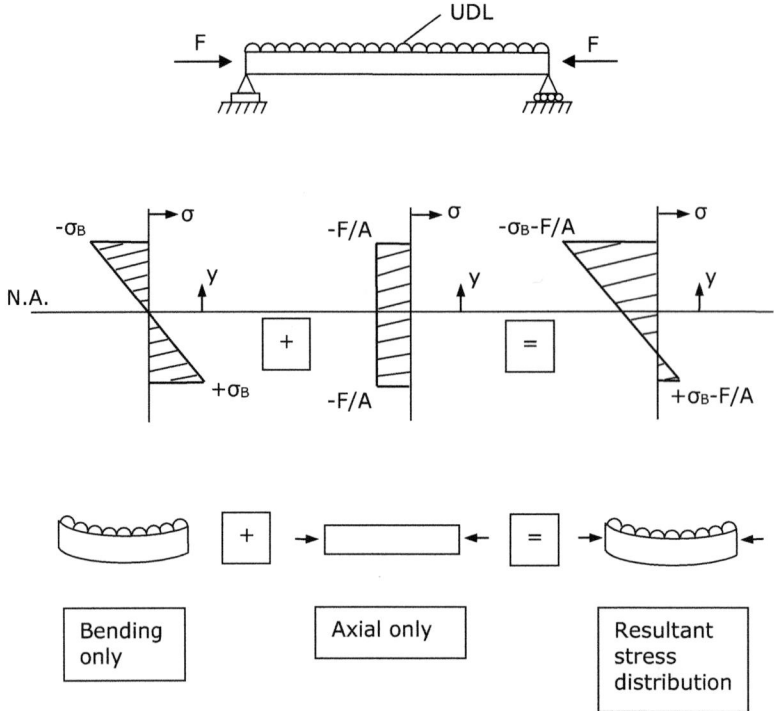

FIGURE 3.3   Combined bending and axial loading.

3.2.2.2   Combined Bending and Torsion

Figure 3.4 shows a similar beam to Figure 3.3, except now the beam carries a torque instead of the axial load. This loading situation is typical of a shaft with self-weight (UDL) transmitting a torque. In this case, the beam cross-section can be assumed to be a solid circle with diameter d. The stresses at the centre of the span, at the bottom surface of the beam, are given by the usual bending and torsion equations as follows:

Arising from the UDL:

$$\text{Bending stress}\left(\sigma_{B}\right) = \frac{My}{I} \text{ where } y = d\,/\,2$$

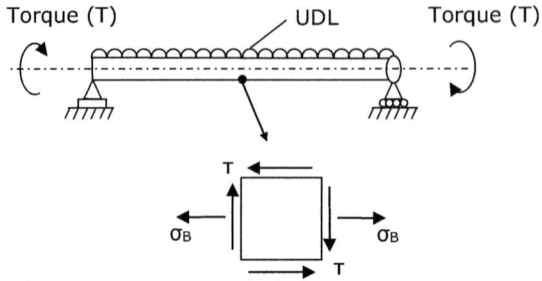

FIGURE 3.4   Combined bending and torsion.

Arising from the torque:

$$\text{Torsional shear stress}\,(\tau) = \frac{Tr}{J}\ \text{where}\ r = d/2$$

These two stresses can be superposed and illustrated acting on an element at the surface of the beam as shown in Figure 3.4.

Mohr's circle can now be used for this element to obtain the principal stresses and maximum shear stress at this position.

3.2.2.3 Combined Pressure and Axial and Torsional Loading

A combination of three loads can be illustrated by considering a thin-walled cylinder, as shown in Figure 3.5, subjected to internal pressure, P, an axial tensile force, F, and a twisting torque, T. Figure 3.5 shows the stresses, arising from each load separately, acting on a surface element in the plane of the cylinder wall. The superposition of these three stresses is also shown on the element. Mohr's circle can again be used to obtain the principal stresses and maximum shear stress for the element.

3.2.3 METHODOLOGY FOR COMBINED LOADING

The methodology for analysing components or structures under combined loading can now be summarised:

  i. Identify a 2D element at the location of interest in the component.
 ii. Determine the stresses acting on the element arising from each individual load.

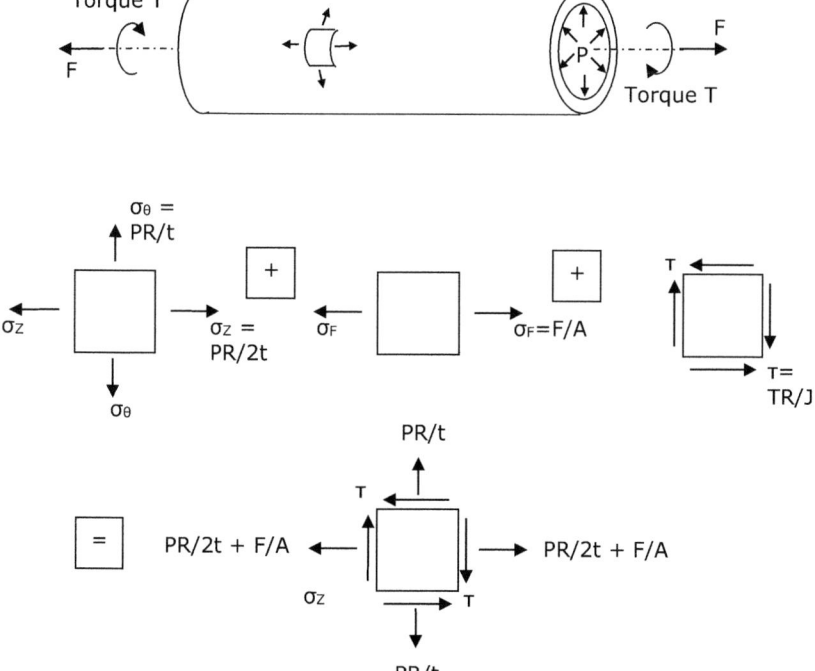

FIGURE 3.5   Combined internal pressure, axial tensile force and torque.

iii. Superpose the stresses from each individual load to obtain the combined stresses on the element.
iv. Use Mohr's circle to determine the principal stresses and the maximum shear stress on the element.

3.2.4   WORKED EXAMPLE

3.2.4.1   Combined Bending and Torsion – Offset Loading on a Cantilever

Figure 3.6 shows a solid circular cross-section cantilever beam, length, L, and diameter, d, fixed at one end. Attached at the free end of the beam is a crank arm which allows a vertical load, P, to be applied at an offset distance, a, from the axis of the cantilever.

Determine the maximum shear stress on the upper surface at the fixed support of the cantilever beam (position A).

The following load and dimensions apply:

$P = 1$ kN
$L = 200$ mm
$a = 120$ mm
$d = 30$ mm

We consider the stresses acting on a small surface element at position A. The load gives rise to a bending moment and torsional moment at the cross-section at position A as follows:

Bending moment $M = P \cdot L$
Torsional moment $T = P \cdot a$

These moments give rise to separate bending and shear stresses acting on the element at position A which can be superposed to give the total effect of the combined loading as shown in Figure 3.6. The stresses are:

$$\text{Bending stress } \sigma_B = \frac{My}{I} = \frac{PL\,d/2}{pd^4/64} = \frac{32PL}{pd^3} = \underline{75.45\,\text{MPa}}$$

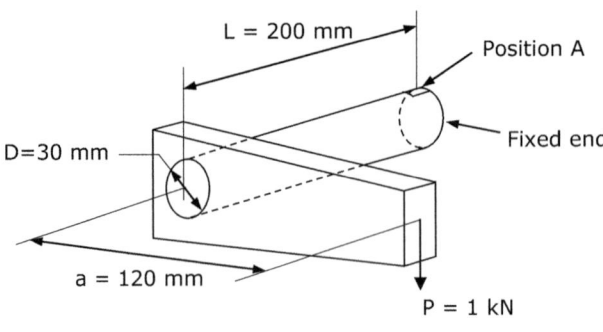

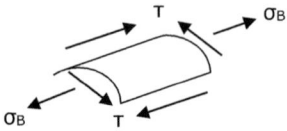

FIGURE 3.6   Combined bending and torque acting on a cantilever beam.

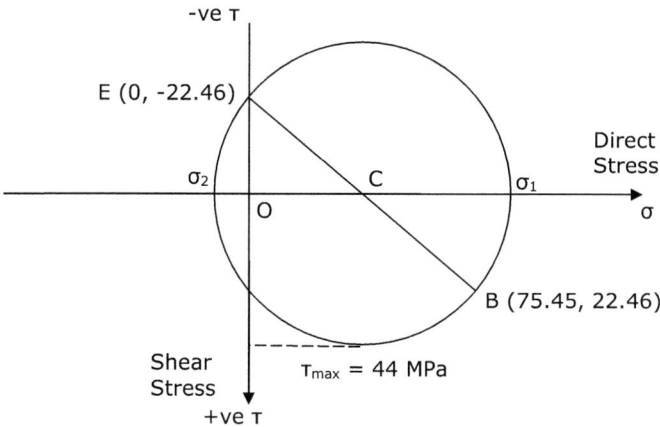

FIGURE 3.7 Mohr's circle for combined bending and torsion.

$$\text{Torsional shear stress } \tau = \frac{Tr}{J} = \frac{\text{Pa}\,d/2}{\pi d^4/32} = \frac{16\text{Pa}}{\pi d^3} = \underline{22.64\,\text{MPa}}$$

Mohr's circle can now be drawn for the element to determine the maximum shear stress as shown in Figure 3.7 The coordinates of Point B on the circle (σ_B, τ) correspond to the stresses on the element in the longitudinal direction, that is along the cantilever. Point E corresponds to the stresses $(0, -\tau)$ in the transverse direction to this. The line joining these two points defines the diameter of the circle and, where it crosses the σ-axis, the centre, C. The circle can now be drawn and its radius measured to give the maximum shear stress as follows:

$$T_{\text{max}} = \text{Radius} = \underline{44\,\text{MPa}}$$

Alternatively, by calculation:
Given the element stresses,

$\sigma_x = 75.45$ MPa
$\sigma_y = 0$ MPa
$\tau_{xy} = 22.64$ MPa

$$\tau_{\text{max}} = \frac{\sigma_y - 0}{2} = \frac{\sigma_y}{2}$$

LEARNING SUMMARY

By the end of this section, you should have learnt:

✓ The basic use of Mohr's circle for analysing the general state of plane stress;
✓ How the effect of combined loads on a component can be analysed by considering each load as initially having an independent effect;
✓ How to use the Principle of Superposition to determine the combined effect of these loads.

3.3 YIELD CRITERIA

3.3.1 ELASTIC-PLASTIC DEFORMATIONS

Uniaxial tensile tests (Figure 3.8) are used to obtain some important material properties. Typical "strain-stress curves" are shown in Figure 3.9.

A linear elastic range is exhibited by a number of materials. In this region, the strain is proportional to the stress (Hooke's Law) and upon the removal of the stress, the strain returns to zero. The constant of proportionality is Young's Modulus, E, for the material. If the stress exceeds a certain value, that is the yield stress, σ_y, the strain does not return to zero upon the removal of the load. Table 3.1 gives the typical values of E and σ_y for some common engineering materials.

A tensile specimen, with a circular cross-section, made from mild steel, which is a ductile material, breaks with a "cup and cone" mode of failure, see Figure 3.10, the cone angle is ~45°.

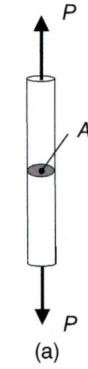

(a)

FIGURE 3.8   Uniaxial tensile loading.

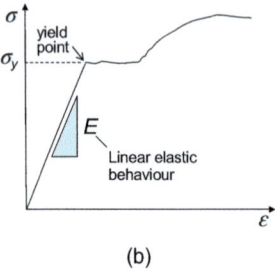

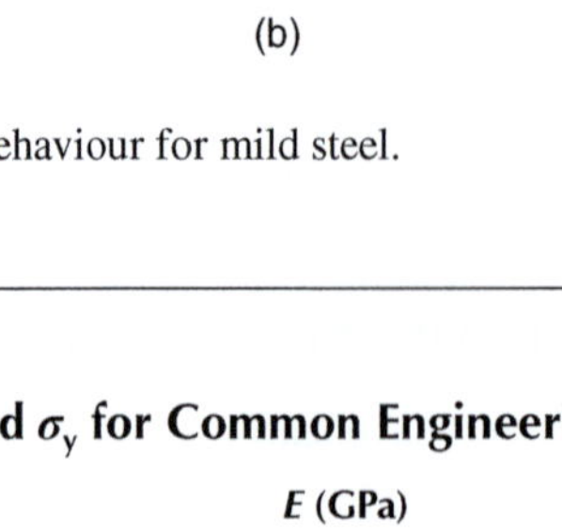

(b)

FIGURE 3.9   Uniaxial stress–strain behaviour for mild steel.

TABLE 3.1
Typical Values of E and σ_y for Common Engineering Materials at 20°C

Material	E (GPa)	σ_y(MPa)
Mild steel	200	350
Copper	120	310
Brass	100	390
Aluminium alloy	70	140—500

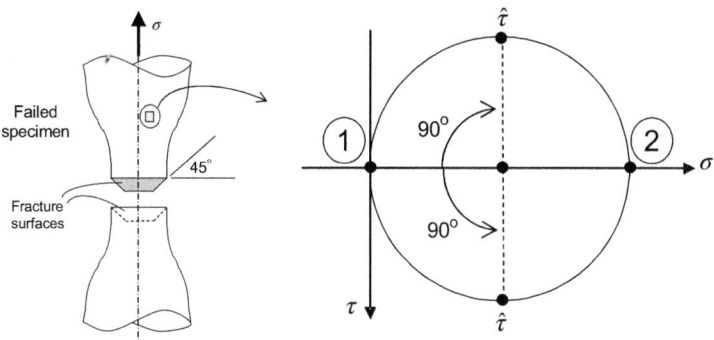

FIGURE 3.10   Cup and cone failure for mild steel and Mohr's circle for uniaxial tensile loading.

FIGURE 3.11   Pure torsional loading for a circular cross-sectioned bar.

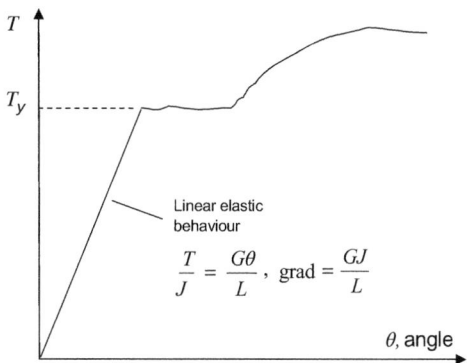

FIGURE 3.12   Torque–angle behaviour for mild steel.

A Mohr's circle for the loading condition (see Figure 3.10) indicates that the 45° plane is that on which the maximum shear stress, $\hat{\tau}$, occurs.

If a circular cross-sectioned, mild steel bar is subjected to torsional loading (Figure 3.11), the torque angle response is as shown in Figure 3.12. For $T \le T_y$, $T \propto \theta$ and the torque–angle behaviour is reversed upon the removal of the torque. If the torque is continuously increased until failure occurs, the fracture plane is transverse to the axis of the specimen (Figure 3.13). A Mohr's circle for the torsion loading case, Figure 3.13, indicates that failure occurs on the maximum shear stress plane.

Therefore, the results of both the tension and the torsion tests indicate that failure occurs on the planes which contain the maximum shear stresses. This behaviour is similar to that of many "ductile" materials.

Similar torsion tests on "brittle" materials (e.g. cast iron) give different behaviour. The torque–angle response is shown in Figure 3.14 and the fracture occurs on a 45° helix as shown in Figure 3.15.

A Mohr's circle for this loading condition, see Figure 3.14, shows that the point on the Mohr's circle associated with the maximum principal stress is 90° (ccw) from point 1. Therefore, this represents a plane at 45° from plane 1 on the element. This maximum principal stress, $\hat{\sigma}$, plane

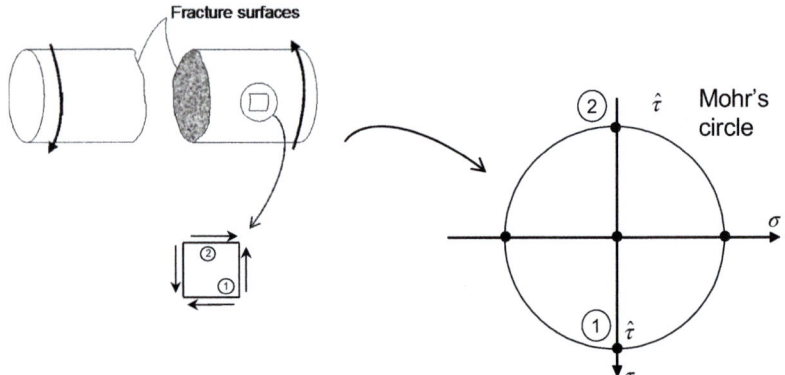

FIGURE 3.13   Transverse failure mode exhibited by mild steel and pure torsional loading and corresponding Mohr's circle.

Note: The shear stresses on faces 1 and 2 are the $\hat{\tau}$ values, i.e. the maximum

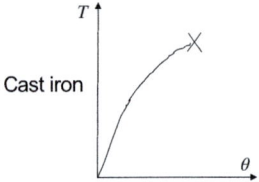

FIGURE 3.14   Torque–angle behaviour for cast iron.

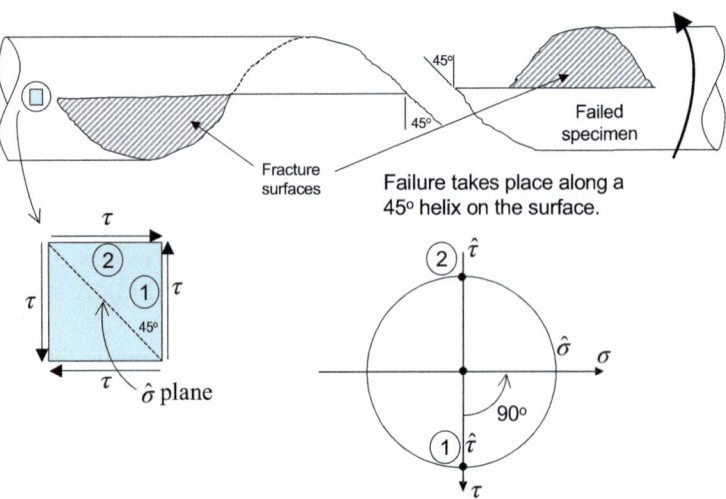

FIGURE 3.15   The 45° helix failure mode exhibited by cast iron under pure torsional loading and corresponding Mohr's circle.

corresponds to the 45° helix angle of the fracture on the surface. This indicates that failure in this material has occurred on a plane on which the maximum principal stress exists.

The tests and stress states used to come to the above conclusions for "ductile" and "brittle" failures are very simple and it would therefore be unwise to base failure criteria on this evidence alone.

3.3.2 YIELDING OF DUCTILE MATERIALS

The topic of "Yield Criteria" is limited to the prediction of the initiation of yielding in "ductile" materials. A yield criterion closely related to the maximum shear stress criterion (Tresca) is the maximum shear strain energy criterion (von Mises); these two criteria generally provide a good indication of yield and are widely used in elastic-plastic analysis.

3.3.3 THE MAXIMUM SHEAR STRESS (TRESCA) YIELD CRITERIA

The Tresca yield criterion states that the material will yield when the maximum shear stress in the material exceeds a limiting value.

If σ_1, σ_2 and σ_3 are the three principal stresses ($\sigma_1 > \sigma_2 > \sigma_3$) then, Figure 3.16 shows that

$$\tau_{max} = \frac{\sigma_1 - \sigma_3}{2}$$

The limiting value can be related to the uniaxial yield stress, σ_y, obtained from a uniaxial tensile test. In this case, $\sigma_1 = \sigma_y$ and $\sigma_2 = \sigma_3 = 0$

$$\tau_{max} = \frac{\sigma_y - 0}{2} = \frac{\sigma_y}{2}$$

The τ_{max} criterion therefore states that the material will yield if:

$$\sigma_1 - \sigma_3 \geq \sigma_y . (\sigma_1 > \sigma_2 > \sigma_3)$$

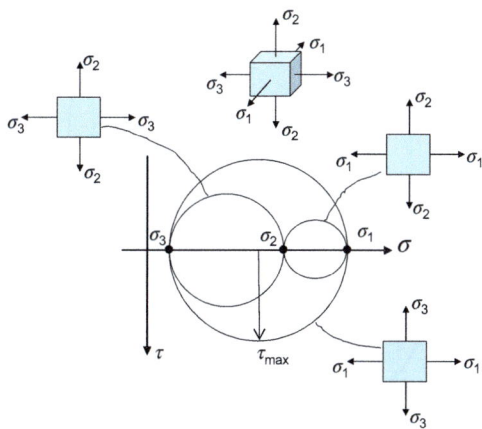

FIGURE 3.16 Mohr's circle representation of a three-dimensional stress state.

3.3.4 The Maximum Shear Strain Energy (von Mises) Yield Criterion

The von Mises yield criterion states that the material will yield when the maximum shear strain energy (per unit volume) exceeds a limiting value. If σ_1, σ_2 and σ_3 are the three principal stresses ($\sigma_1 > \sigma_2 > \sigma_3$), then:

$$\frac{\text{shear strain energy}}{\text{unit volume}} = \frac{1}{12G}\left\{\left(\sigma_1 - \sigma_2\right)^2 + \left(\sigma_2 - \sigma_3\right)^2 + \left(\sigma_3 - \sigma_1\right)^2\right\}$$

Again, the limiting value can be related to the uniaxial yield stress, σ_y, obtained from a uniaxial tensile test. Thus, at yield $\sigma_1 = \sigma_y$ and $\sigma_2 = \sigma_3 = 0$:

$$\frac{\text{shear strain energy}}{\text{unit volume}} = \frac{1}{12G}\left\{2\sigma_y^2\right\}$$

The von Mises yield criterion can thus be expressed as follows:

$$\frac{1}{12G}\left\{\left(\sigma_1 - \sigma_2\right)^2 + \left(\sigma_2 - \sigma_3\right)^2 + \left(\sigma_3 - \sigma_1\right)^2\right\} \geq \frac{1}{12G}\left\{2\sigma_y^2\right\}$$

which can be reduced to the following, more common expression for the onset of yield, according to the von Mises yield criterion:

$$\left(\sigma_1 - \sigma_2\right)^2 + \left(\sigma_2 - \sigma_3\right)^2 + \left(\sigma_3 - \sigma_1\right)^2 \geq 2\sigma_y^2$$

3.3.5 Two-Dimensional Stress Systems (i.e. $\sigma_3 = 0$)

For plotting purposes here, σ_1 and σ_2 can take on any values, that is σ_1 is not necessarily always greater than σ_2. The yield boundaries on the $\sigma_1 - \sigma_2$ plane are shown in Figure 3.17.

In general, the von Mises yield criterion is easier to handle analytically because it is continuous. This is particularly important for the calculation of incremental plastic strains since the plastic

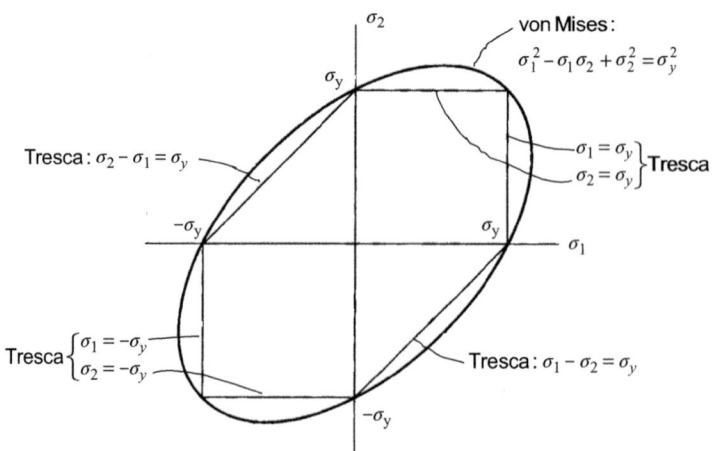

FIGURE 3.17   The yield boundaries, for a two-dimensional stress state, on the $\sigma_1 - \sigma_2$ plane ($\sigma_3 = 0$).

strains are related to the normals to the yield surface, and at the corners of the Tresca yield locus, there is ambiguity about the directions of the normals, whereas there is no such ambiguity about the von Mises yield locus.

3.3.6 Three-Dimensional Stress Systems

The Tresca yield criterion

$$\sigma_1 - \sigma_3 = \sigma_y \ (\sigma_1 > \sigma_2 > \sigma_3)$$

and the von Mises yield criterion

$$\left(\sigma_1 - \sigma_2\right)^2 + \left(\sigma_2 - \sigma_3\right)^2 + \left(\sigma_3 - \sigma_1\right)^2 = 2\sigma_y^{\,2}$$

are not altered if a constant stress component, say σ, is added to each stress component,

$$\text{i.e.} \left(\sigma_1 + \sigma\right) - \left(\sigma_3 + \sigma\right) = \sigma_1 - \sigma_3 = \sigma_y$$

and

$$\left(\left(\sigma_1 + \sigma\right) - \left(\sigma_2 + \sigma\right)\right)^2 + \left(\left(\sigma_2 + \sigma\right) - \left(\sigma_3 + \sigma\right)\right)^2 + \left(\left(\sigma_3 + \sigma\right) - \left(\sigma_1 + \sigma\right)\right)^2$$
$$= \left(\sigma_1 - \sigma_2\right)^2 + \left(\sigma_2 - \sigma_3\right)^2 + \left(\sigma_3 - \sigma_1\right)^2$$
$$= 2\sigma_y^{\,2}$$

This implies that the addition of a "hydrostatic stress state", that is $\sigma_1 = \sigma_2 = \sigma_3 = \sigma$ does not change the shapes of the yield surfaces shown in the section on two-dimensional stress systems.

The mean principal stress $\sigma_h = \dfrac{1}{3}(\sigma_1 + \sigma_2 + \sigma_3)$, which is known as the *hydrostatic stress* for a given stress state $(\sigma_1, \sigma_2, \sigma_3)$, is the stress which causes volume change. Now, the independence of the yield criteria with respect to hydrostatic stress means that the three-dimensional yield criteria are prismatic surfaces with the axes of the prisms in each case being the line $\sigma_1 = \sigma_2 = \sigma_3$. This is called the *hydrostatic line* in 3D stress space (Haigh–Westergaard stress space) and it has direction cosines $\left(\dfrac{1}{\sqrt{3}}, \dfrac{1}{\sqrt{3}}, \dfrac{1}{\sqrt{3}}\right)$. The yield boundaries can thus move any distance in the direction $\sigma_1 = \sigma_2 = \sigma_3$. The yield surfaces for both the von Mises and Tresca yield criteria; therefore, have a constant oblique section and hence a constant perpendicular cross-section, whose true shape can be seen in the view along the line $\sigma_1 = \sigma_2 = \sigma_3$. Any arbitrary stress 'vector' $(\sigma_1, \sigma_2, \sigma_3)$, for example, $\overrightarrow{OB}$ and $\overrightarrow{OD}$, (Figure 3.18), in the stress space can be decomposed into two components, one parallel to the hydrostatic line, for example, $\overrightarrow{OA}$ and $\overrightarrow{OC}$, and one perpendicular to the hydrostatic line, for example, $\overrightarrow{AB}$ and $\overrightarrow{CD}$. The oblique planes which are perpendicular to the hydrostatic line are called deviatoric planes and are given by equations of the form $\sigma_1 + \sigma_2 + \sigma_3 = \text{const}$, each representing a different level of hydrostatic stress. The deviatoric plane $\sigma_1 + \sigma_2 + \sigma_3 = 0$ is known as the π-plane. It can be shown that the component $(\sigma_1, \sigma_2, \sigma_3)$ parallel to the hydrostatic line is $(\sigma_h, \sigma_h, \sigma_h)$, for example, $\overrightarrow{OA}$ and $\overrightarrow{OC}$, while the component parallel to the deviatoric planes is $(\sigma_1 - \sigma_h, \sigma_2 - \sigma_h, \sigma_3 - \sigma_h)$, $\overrightarrow{AB}$ and $\overrightarrow{CD}$. Only the latter component of stress is important in determining yield according to the von Mises and Tresca criteria.

The view along the $\sigma_1 = \sigma_2 = \sigma_3$ line of the von Mises and Tresca yield criteria is an isometric view which shows the three axes included at 120° intervals. This is sometimes called a view on the π-plane, on which the Tresca yield surface is a hexagon and the von-Mises yield surface is a circle.

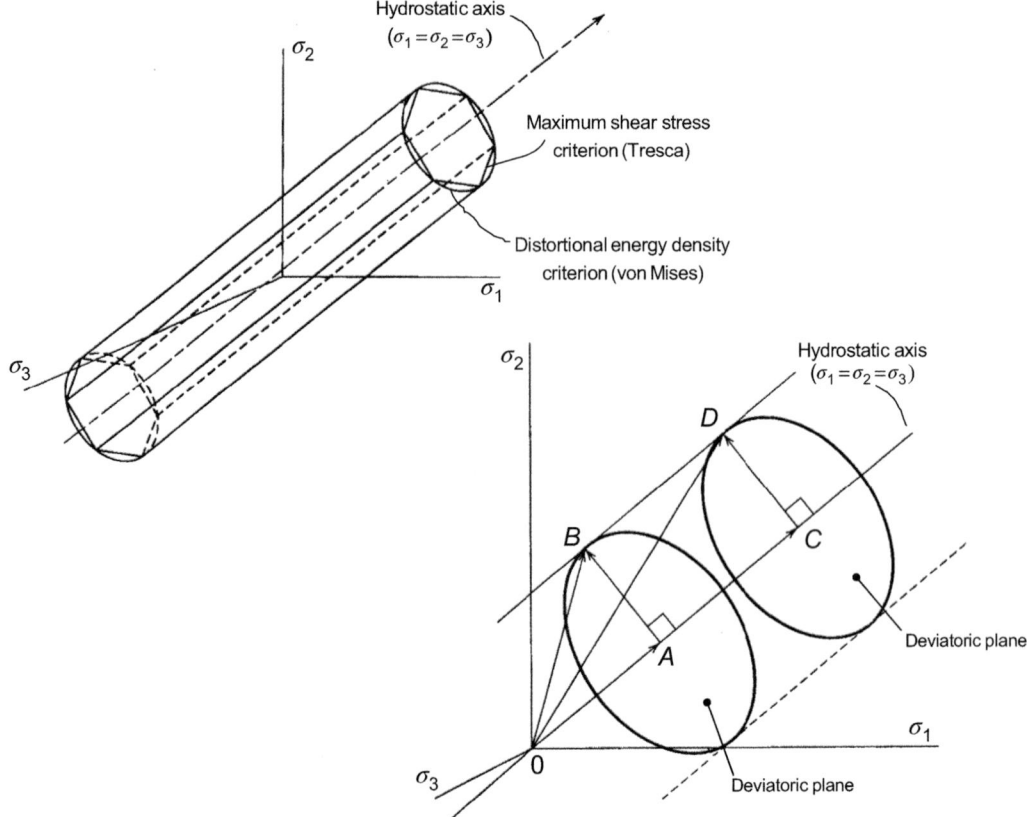

FIGURE 3.18 Representation of yield surfaces for a three-dimensional stress state and figure showing the decomposition of the stress into hydrostatic and deviatoric components.

Therefore, large principal stresses do not necessarily result in yield; it is the stress differences and the route to the final stress state that govern whether yielding will occur.

Figure 3.19 can be used, instead of the equations, to decide whether a certain stress state will be safe. Simply plot on the diagram each of the three principal stresses parallel to each of the three axes and see whether the final point lies inside the appropriate yield surface.

NB:

i. The yield condition can be examined by either using the appropriate equation, that is, Tresca:

$$\sigma_1 - \sigma_3 = \sigma_y \ (\sigma_1 > \sigma_2 > \sigma_3)$$

von-Mises:

$$\left(\sigma_1 - \sigma_2\right)^2 + \left(\sigma_2 - \sigma_3\right)^2 + \left(\sigma_3 - \sigma_1\right)^2 = 2\sigma_y^2$$

or by plotting principal stresses on the π-plane.

ii. All three principal stresses are important. At free surfaces, the normal stress is usually zero, but it may be important, particularly if the other two principal stresses are of the same sign.

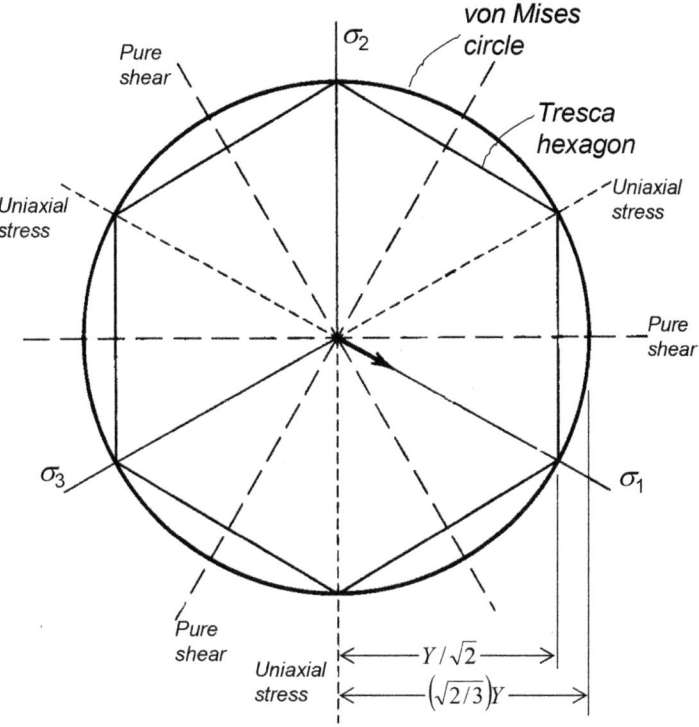

FIGURE 3.19   View of the yield surfaces on the π-plane.

LEARNING SUMMARY

By the end of this section, you should have learnt:

✓ The difference between ductile and brittle failure, illustrated by the behaviour of bars subjected to uniaxial tension and torsion;
✓ The meaning of yield stress and proof stress, in uniaxial tension, for a material;
✓ The Tresca (maximum shear stress) yield criterion and the 2D and 3D diagrammatic representations of it;
✓ The von Mises (maximum shear strain energy) yield criterion and the 2D and 3D diagrammatic representations of it.

3.4   DEFLECTION OF BEAMS

In this section, we address the important topic of beam deflections. Whereas the design of engineering structures and components is very often dictated by the strength of the materials used and consequently the stresses within the structure, often the limiting factor is the allowable deflection. This is particularly important for engineering artefacts made from materials of lower stiffness, for example, aluminium, plastics, composites etc. but may also be critical for high-stiffness steel structures comprising slender flexible members. It is, therefore, important as part of the design process to be able to calculate maximum deflections in a structure in addition to the position at which they occur.

This section focuses on deflections in beam structures. Following the derivation of the fundamental deflection equation for a beam, a flexible procedure is introduced, called Macaulay's Method, which allows deflections to be calculated at any position along a beam span. In particular, the method allows us to deal with different types of loading such as point loads, UDLs and point moments including discontinuities in these loads. Although not the only method for calculating deflections, as we will see later in the Strain Energy subsection, it is a particularly powerful and flexible method.

3.4.1   EQUATION OF THE ELASTIC LINE

The section of the span of a beam, shown in Figure 3.20, is under pure bending, that is, there is a constant bending moment along this section and no shear force. Under pure bending conditions, the neutral surface of the beam is a circular arc with a radius of curvature, R, as shown. The transverse deflection of the neutral surface is given by the coordinate y of any position along the surface [n.b. do not confuse this 'y' definition for deflection with the 'y' denoting distance from the neutral axis/surface in the beam bending equation – as detailed in 'An Introduction to Mechanical Engineering – Part 1']. The line denoting the neutral surface in Figure 3.20 is known as the **'elastic line'** or the **'deflection curve'** of the beam.

Referring to Figure 3.20, the angle $d\theta$ defines an element AB, ds in length, of the elastic line, and, because $d\theta$ is small,

$$ds = R\, d\theta$$

in other words, the **curvature**,

$$\frac{1}{R} = \frac{d\theta}{ds} \tag{4.1}$$

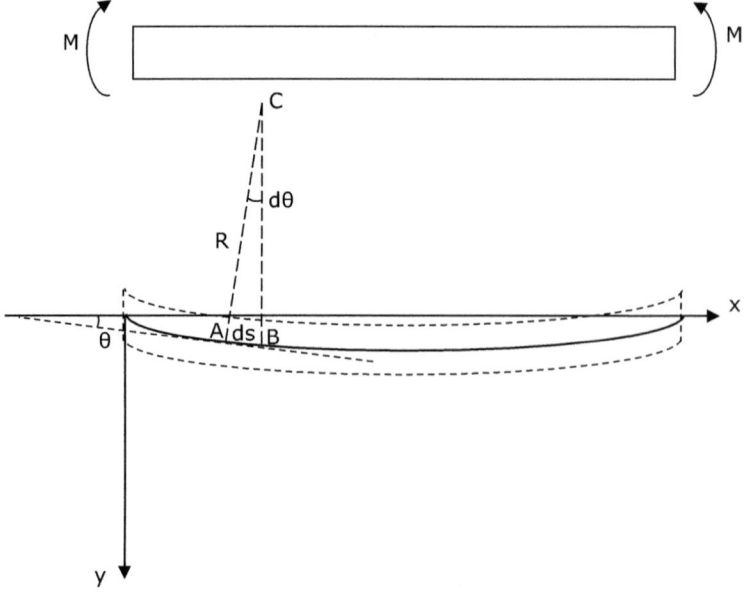

FIGURE 3.20   The elastic line (i.e. deflection curve) of a beam under pure bending.

For a beam bent in the elastic range, the elastic line is a very flat curve, that is, deflections are small. In this case,

$$ds \approx dx$$

where dx is the increment of x between A and B and Equation (4.1) becomes

$$\frac{1}{R} = -\frac{d\theta}{dx} \tag{4.2}$$

Note the introduction of the **minus** $(-)$ sign because θ decreases as x increases for positive bending, that is, sagging. This maintains a positive radius of curvature for positive bending.

Also, θ is the slope of the elastic line at position A:

$$\theta = \frac{dy}{dx} \tag{4.3}$$

Combining Equations (4.2) and (4.3) gives,

$$\frac{1}{R} = -\frac{d^2 y}{dx^2} \tag{4.4}$$

Finally, combining Equation (4.4) with the part of the beam bending equation relating the radius of curvature to the bending moment (M), Young's modulus (E) and second moment of area (I), that is,

$$\frac{M}{I} = \frac{E}{R} \text{ or } \frac{1}{R} = \frac{M}{EI} \tag{4.5}$$

gives,

$$-EI\frac{d^2 y}{dx^2} = M \tag{4.6}$$

Equation (4.6) is the **Differential Equation of the Elastic Line**, relating the deflection, y, to the applied bending moment, M, Young's modulus, E, and the second moment of area, I. The product of E and I, that is, EI, is termed the **Flexural Rigidity** of the beam.

The successive integration of Equation (4.6) will yield the slope, dy/dx, and the deflection, y, at all points along the beam.

Equation (4.6) has been derived for the case of pure bending, that is, constant bending moment along the section, and does not take into account deflections due to shear. For long slender beams, shear deflections can be neglected.

Equation (4.6) may also be integrated and solutions found if the bending moment, M, is a continuous function of x. This condition applies only to beams under specific simple loading conditions. A complication arises where **discontinuities** in M exist, such as where there are point loads and moments or where there is an abrupt change in distributed loading.

Various methods have been developed to solve such problems with discontinuities. In Subsection 3.4.2, we will introduce and develop the procedure called **Macaulay's Method**, a versatile solution procedure which can handle most discontinuities we are likely to come across.

3.4.2   Macaulay's Method (Also Termed the Method of Singularity Functions)

Named after the mathematician W H Macaulay, Macaulay's Method uses a mathematical technique to deal with discontinuous loading. The moment expression $M(x)$, that is, M as a function of x, is replaced with the step function $M<x - a>$, in which 'a' defines the points where discontinuities arise. The procedure will now be developed from first principles.

Figure 3.21 shows a simply-supported beam carrying three point loads W_A, W_B and W_C, along its span. Each of the three loads gives rise to a discontinuity in the bending moment. Knowing the applied loads, the end reactions, R_1 and R_2 can be calculated from equilibrium conditions for the whole beam. Now, considering each part of the span between the loads separately:

Span 1
Figure 3.22 (a) shows the free body diagram of the LH end of the beam, cut at a section in span 1. The unknown bending moment, M, and shear force, S, at this section are shown in the diagram. Considering the bending moment only and taking moments about the cut section,

$$M - R_1 x = 0$$

and

$$EI\frac{d^2y}{dx^2} = -M = -R_1 x \tag{4.7}$$

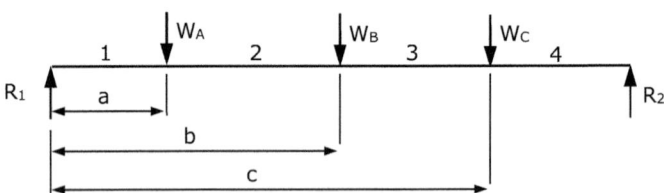

FIGURE 3.21   Simply-supported beam carrying point loads.

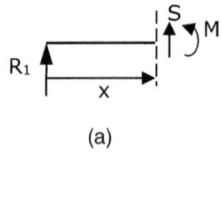

(a)

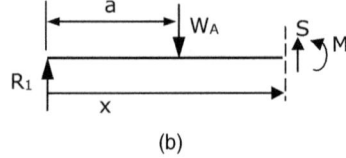

(b)

FIGURE 3.22   Freebody diagrams of beam cut at different sections.

This equation applies to span 1 only. Now considering span 2:

Span 2

Figure 3.22 (b) shows the free body diagram of the LH end of the beam, now cut at a section in span 2. As before, the unknown bending moment, M, and shear force, S, in this section are shown in the diagram. Again, considering the bending moment only and taking moments about the cut section,

$$M + W_A(x - a) - R_1 x = 0$$

and now,

$$EI\frac{d^2 y}{dx^2} = -M = -R_1 x + W_A(x - a) \tag{4.8}$$

This equation now applies to span 2 only.

Similar equations can be derived for spans 3 and 4 as follows:

Span 3

$$EI\frac{d^2 y}{dx^2} = -M = -R_1 x + W_A(x - a) + W_B(x - b) \tag{4.9}$$

and

Span 4

$$EI\frac{d^2 y}{dx^2} = -M = -R_1 x + W_A(x - a) + W_B(x - b) + W_C(x - c) \tag{4.10}$$

It is interesting to note that the forms of Equations (4.7)–(4.10) are very similar, in that extra terms are added to take account of the point loads as we move towards the right-hand (RH) end of the beam. In fact, due to this similarity, Equation (4.10) can be used to apply to all spans by rewriting it in a slightly different form as follows:

$$EI\frac{d^2 y}{dx^2} = -M = -R_1 x + W_A\langle x - a \rangle + W_B\langle x - b \rangle + W_C\langle x - c \rangle \tag{4.11}$$

Note the change of bracket shape in Equation (4.11). The <> brackets are termed 'Macaulay Brackets' and Equation (4.11) is now applicable to any point in the beam if we adopt what is called 'Macaulay's' convention.

> Macaulay's Convention: "Whenever a bracketed term becomes negative it must be ignored"

Adopting this convention, the general expression for M in Equation (4.11) can be integrated twice to give the slope, dy/dx, and the deflection, y, at any point along the beam. During the integration, boundary conditions for the beam are used to determine the integration constants. It should also be noted that the bracketed terms <u>must not</u> be multiplied out in any general expression for y, dy/dx or d^2y/dx^2; otherwise, the Macaulay convention will be lost.

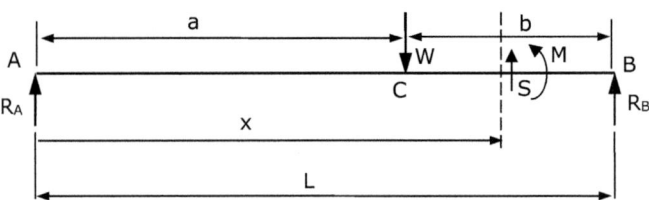

FIGURE 3.23 Simply-supported beam carrying a single-point load.

3.4.3 Macaulay's Method Example – Point Load on a Beam

Figure 3.23 shows a simply supported beam carrying a point load, distance a from the LH end and distance b from the RH end.

Question: Derive an equation for the deflection of this beam at <u>any</u> point along its length.

First, determine the reaction force, R_A, by taking moments for the whole beam about B,

$$R_A L - Wb = 0$$

$$\therefore\ R_A = \frac{Wb}{L}$$

Now, cutting the beam at a section in the extreme RH part of the span, distance x from the LH end, the LH part of the beam can be considered as a free body with an unknown bending moment, M, and shear force, S, on the cut section. Taking moments about this cut section for the free body, we have,

$$M + W\langle x - a\rangle - R_A x = 0 \tag{4.12}$$

Here we have used Macaulay brackets to indicate that the expression applies anywhere along the beam on the condition that the Macaulay Convention is adhered to. The second term in Equation (4.12), $W<x–a>$, is specific to a point load and is termed *a* 'singularity function'. Different forms of singularity function appear for different types of loading as we will see in Section 3.4.4.

From Equation (4.12) and the equation of the elastic line (i.e. (4.6)),

$$EI\frac{d^2 y}{dx^2} = -M = -R_A x + W\langle x - a\rangle = -\frac{Wb}{L}x + W\langle x - a\rangle$$

Integrating

$$EI\frac{dy}{dx} = -\frac{Wb}{2L}x^2 + W\frac{<x-a>^2}{2} + A \tag{4.13}$$

where A is an integration constant

Integrating again

$$EIy = -\frac{Wb}{6L}x^3 + W\frac{<x-a>^3}{6} + Ax + B \tag{4.14}$$

where B is also an integration constant

We use the boundary conditions to solve for the two integration constants, A and B:

$y = 0$ at $x = 0$

$\therefore$ from (4.14)

$$0 = 0 + 0 + 0 + B$$

$$\therefore \underline{B = 0}$$

[Note that the second term after the '=' sign is zero because the quantity in the Macaulay bracket, $<x–a>$, is negative at $x = 0$ – 'The Macaulay Convention']

$y = 0$ at $x = L$

$\therefore$ from (4.14)

$$0 = \frac{-WbL^2}{6} + \frac{W(L-a)^3}{6} + AL$$

$$\therefore A = \frac{-WbL}{6} - \frac{Wb^3}{6L} = \frac{Wb}{6}\left(L - \frac{b^2}{L}\right)$$

Substituting for A and B in Equation (4.14), a general expression for y can be obtained as follows:

$$y = \frac{1}{EI}\left[-\frac{Wbx^3}{6L} + \frac{W<x-a>^3}{6} + \frac{Wbx}{6}\left(L - \frac{b^2}{L}\right)\right] \qquad (4.15)$$

n.b. The Macaulay bracketed term in Equation (4.15) is ignored when $x < a$

The slope, dy/dx can be obtained from Equation (4.13) or by differentiating Equation (4.15).

A special case for this problem occurs when the load is applied at the centre of the beam span. The deflection at point C, the load point, in this case, is given by putting $b = a$ and $x = a$ in (4.15), giving,

$$y = \frac{WL^3}{48EI}$$

which is the well-known result for the central deflection of a simply-supported beam with the load applied at the centre span.

3.4.4   OTHER LOADING CONDITIONS

3.4.4.1   Uniformly Distributed Load

Consider a UDL, q Nm^{-1}, acting over a part of a beam's span as shown in Figure 3.24. The UDL runs from position C, distance a from the LH end of the beam all the way to the RH end of the beam. A discontinuity occurs where the UDL commences at position C.

Cutting a section to the right of the discontinuity and considering the part of the beam to the left of this section as a free body, moments can be taken about the section, giving the following equilibrium equation,

$$M + \frac{q<x-a>^2}{2} - R_1 x = 0 \qquad (4.16)$$

Combining Equation (4.16) with the equation of the elastic line,

$$EI\frac{d^2y}{dx^2} = -M = -R_1 x + \frac{q<x-a>^2}{2}$$

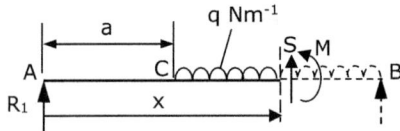

FIGURE 3.24 Simply-supported beam carrying a uniformly distributed load (UDL).

As for a point load, this equation can be integrated twice to give the deflection at any position along the beam. Integration constants are determined from the boundary conditions as before. The procedure is identical to the point load case except for the singularity function, $\dfrac{q<x-a>^2}{2}$, which is different. The Macaulay Convention applies equally in this case, with the Macaulay brackets in the singularity function indicating whether the term is included or not depending on the sign of the bracketed term. It is excluded if $<x–a>$ is negative, that is, if $x < a$.

How do we deal with a UDL which only acts over part of the beam span as shown in Figure 3.25(a)?

In this case, there are two discontinuities at distances a and b from the LH end of the beam. The UDL can be extended to the RH end of the beam and an additional negative counterbalance UDL superimposed over the newly extended part, also shown in Figure 3.25(b). This gives a statically equivalent system to the partially extended UDL.

Cutting a section in the newly extended part and looking at the equilibrium of moments on the LH part of the beam, we have

$$M + \frac{q<x-a>^2}{2} - \frac{q<x-b>^2}{2} - R_1 x = 0$$

which, combined with the equation of the elastic line, gives

$$EI \frac{d^2y}{dx^2} = -M = -R_1 x + \frac{q<x-a>^2}{2} - \frac{q<x-b>^2}{2}$$

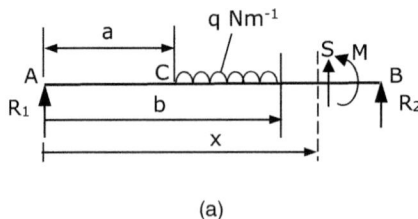

(a)

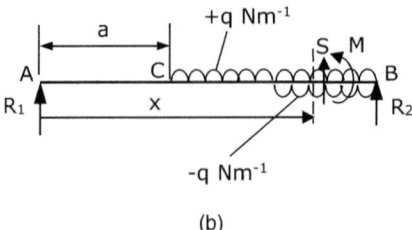

(b)

FIGURE 3.25 Dealing with a UDL acting over part of the beam's span – 'counterbalancing' the load.

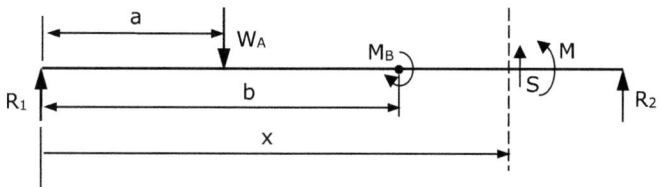

FIGURE 3.26   Simply-supported beam carrying a point bending moment and a point load.

As before, this equation can be integrated twice to give the deflection at any position along the beam. There are now two singularity functions corresponding to the discontinuities in UDL at positions a and b (n.b. the negative sign for the second function arises from the negative counterbalance UDL].

3.4.4.2   Point Bending Moment

Figure 3.26 shows both a point load, W_A, acting at position a, and a point bending moment, M_B, acting at position b on a simply supported beam. This problem has a combination of two loads (point load and point moment) acting on the beam and shows the versatility of Macaulay's Method.

Cutting a section to the right of the point bending moment discontinuity and considering the part of the beam to the left of this section as a free body, moments can be taken about the section, giving the following equilibrium equation,

$$M - R_1 x + W_A \langle x - a \rangle - M_B < x - b >^0 = 0 \qquad (4.17)$$

Note the form of the discontinuity function for the point bending moment, that is, $M_B < x - b >^0$. The zero exponent is a special form which allows subsequent integration to give higher-level exponents when deriving the expressions for slope and deflection as shown below.

Combining Equation (4.17) with the equation of the elastic line,

$$EI \frac{d^2 y}{dx^2} = -M = -R_1 x + W_A \langle x - a \rangle - M_B < x - b >^0$$

Integrating gives,

$$EI \frac{dy}{dx} = -\frac{R_1 x^2}{2} + W_A \frac{< x - a >^2}{2} - M_B \langle x - b \rangle + A$$

The point bending moment singularity function has integrated to have an exponent of order 1 and A is an integration constant.

Further integration gives,

$$EIy = -\frac{R_1 x^3}{6} + W_A \frac{< x - a >^3}{6} - M_B \frac{< x - b >^2}{2} + Ax + B \qquad (4.18)$$

The final form of the applied bending moment singularity has an exponent of 2. There are two integration constants, A and B, which can be solved by using the boundary conditions. As before, Equation (4.18) can be used to determine the deflection at any position along the beam on the condition that Macaulay's Convention is used when evaluating the bracketed terms.

3.4.5   Summary of Singularity Functions

We have seen that Macaulay's method can be used to find deflections at any position along a beam where point loads, UDLs and point moments produce discontinuities in the bending moment distribution. The method can also be used where there is a combination of these loads acting on a beam.

When developing the bending moment equation for the beam with load discontinuities, the following singularity functions are used for each different type of load:

Load	Singularity Function
Point Load (W)	$W<x-a>$
UDL (q) with single discontinuity	$\dfrac{q<x-a>^2}{2}$
UDL (q) with double discontinuity	$\dfrac{q<x-a>^2}{2}-\dfrac{q<x-b>^2}{2}$
Point Moment (M_B)	$M_B<x-a>^0$

It is interesting to note that in these singularity functions the exponent is 1 for a point load, 2 for a UDL, and 0 for a point moment.

3.4.6   Summary of Macaulay's Method for Beam Deflections

a. Take the origin at the LH end of the beam.
b. When necessary, extend and counterbalance UDL to the RH end of the beam.
c. Obtain the expression for M in the extreme RH part of the beam span, that is, beyond the final discontinuity.
d. Use Macaulay brackets in relevant singularity functions for discontinuities.
e. Use the equation of the elastic line, that is, $EI\ d^2y/dx^2 = -M$.
f. Integrate to give dy/dx and y. Do not multiply bracketed terms during these integrations.
g. Use boundary conditions to determine integration constants.
h. Evaluate slope (dy/dx) and deflection (y) at required positions. Bracketed terms are ignored when the value within brackets is negative.

3.4.7   Worked Example

Figure 3.27 (a) shows a simply supported beam carrying a point load, $P = 2$ kN, at the centre span and a UDL, $q = 4$ kN/m over part of the RH half of the span. The length of the beam is 2 m and the flexural rigidity of its cross-section is $EI = 10^5$ Nm^2.

Use Macaulay's method to determine the deflection of the beam at the centre span below the point load.

3.4.7.1   Reaction Forces

Considering the equilibrium of the full beam,

$$\text{Vertical Forces: } R_A + R_B = P + \frac{qL}{4} \tag{4.19}$$

$$\text{Moments about A } \frac{PL}{2} + \frac{qL}{4}\frac{5L}{8} - R_B L = 0$$

$$R_B = \frac{P}{2} + \frac{5qL}{32} = 2250\ N$$

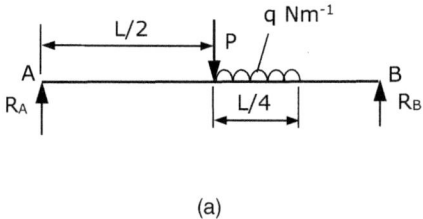

(a)

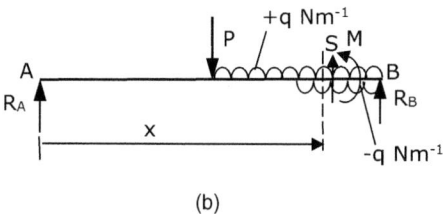

(b)

FIGURE 3.27 Worked example: Point load and UDL discontinuities acting on the same beam.

and from (4.18)

$$R_A = \frac{P}{2} + \frac{3qL}{32} = 1750 \ N$$

3.4.7.2 Moment Expression and Integration

The UDL is extended to the LH end of the beam and a counterbalancing negative UDL is added as shown in Figure 3.27(b). Taking moments about a section in the extreme RH end of the beam beyond the last discontinuity, we have,

$$M = R_A x - P \left\langle x - \frac{L}{2} \right\rangle - \frac{q < x - \frac{L}{2} >^2}{2} + \frac{q < x - \frac{3L}{4} >^2}{2}$$

and combining with the equation of the elastic line,

$$EI \frac{d^2 y}{dx^2} = -M = -R_A x + P \left\langle x - \frac{L}{2} \right\rangle + \frac{q < x - \frac{L}{2} >^2}{2} - \frac{q < x - \frac{3L}{4} >^2}{2}$$

Interating twice gives,

$$EI \frac{dy}{dx} = -\frac{R_A x^2}{2} + \frac{P < x - \frac{L}{2} >^2}{2} + \frac{q < x - \frac{L}{2} >^3}{6} - \frac{q < x - \frac{3L}{4} >^3}{6} + A$$

$$EIy = -\frac{R_A x^3}{6} + \frac{P < x - \frac{L}{2} >^3}{6} + \frac{q < x - \frac{L}{2} >^4}{24} - \frac{q < x - \frac{3L}{4} >^4}{24} + Ax + B \qquad (4.20)$$

where A and B are integration constants

3.4.7.3 Boundary Conditions

Setting the boundary conditions in Equation (4.20) we have,

When $x = 0$, $y = 0$

$\therefore B = 0$

[n.b. negative Macaulay bracketed terms are ignored, i.e. set to zero)

When $x = L$, $y = 0$

$$\therefore 0 = \frac{R_A L^3}{6} + \frac{PL^3}{48} + \frac{qL^4}{16.24} - \frac{qL^4}{256.24} + AL$$

Substituting values for R_A, P, q and L and solving for A gives,

A = 922 Nm²

3.4.7.4 Deflection at Centre Span

Equation (4.20) can now be used to evaluate the deflection at the centre span. As $x = L/2$, the three Macaulay bracketed terms are zero (using Macaulay Convention). Therefore,

$$EIy = -\frac{R_A x^3}{6} + Ax = -\frac{1750.1^3}{6} + 922.1 = 630$$

$$\therefore y = \frac{630}{10^5} = 6.3 \text{ mm}$$

3.4.8 STATICALLY INDETERMINATE BEAMS

Macaulay's method can also be used to solve for deflections and slopes of a <u>statically indeterminate</u> beam which is a beam where the reaction forces and moments cannot be determined by the equations of statics alone. An example is a clamped-clamped beam subjected to a point load, as shown in Figure 3.28(a).

Drawing the free body diagram for this beam, Figure 3.28(b), the end reactions comprise moments M_A and M_B which restrain rotation and the vertical reaction forces R_A and R_B. There are therefore

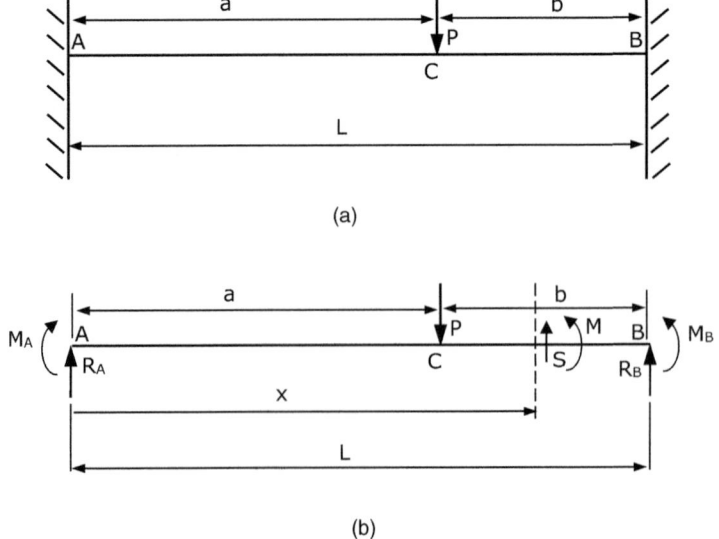

(a)

(b)

FIGURE 3.28 Statically indeterminate beam: Clamped-clamped beam subjected to a point load.

four unknowns which cannot be solved for by equilibrium conditions alone. They have to satisfy both equilibrium (two conditions) and the boundary conditions, that is, slope and deflection at both ends (four conditions). Equilibrium is incorporated in the Macaulay bending moment equation, and boundary conditions, as we have seen, are also an integral part of the method.

Taking a section, X-X, at distance x from the left-hand (LH) support, beyond the load discontinuity, and considering the LH part of the beam as a free body, the equation for the unknown moment, M, on the section is,

$$M = -EI\frac{d^2y}{dx^2} = M_A + R_A x - P\langle x - a \rangle$$

Integrating this expression twice gives

$$-EI\frac{dy}{dx} = M_A x + \frac{R_A x^2}{2} - \frac{P<x-a>^2}{2} + A$$

and

$$-EIy = \frac{M_A x^2}{2} + \frac{R_A x^3}{6} - \frac{P<x-a>^3}{6} + Ax + B \qquad (4.21)$$

This expression for the deflection, y, includes four unknowns, namely M_A and R_A and the integration constants A and B. We can use the boundary conditions to solve for these unknowns.

3.4.8.1   Boundary Conditions

When $x = 0$, $y = 0$

∴$\underline{B = \underline{0}}$

When $x = 0$, $dy/dx = 0$

∴$\underline{A = \underline{0}}$

[n.b. negative Macaulay bracketed terms are ignored i.e. set to zero]

When $x = $ L, $y = 0$

$$\therefore \quad 0 = \frac{M_A L^2}{2} + \frac{R_A L^3}{6} - \frac{P(L-a)^3}{6} \qquad (4.22)$$

When $x = $ L, $dy/dx = 0$

$$\therefore \quad 0 = M_A L + \frac{R_A L^2}{2} - \frac{P(L-a)^2}{2} \qquad (4.23)$$

[n.b. Macaulay bracketed terms are converted back to normal brackets at this stage]

Equations (4.22) and (4.23) can be solved simultaneously to give R_A and M_A as follows:

$$R_A = \frac{P(L-a)^2(L+2a)}{L^3} \qquad (4.24)$$

$$M_A = -\frac{Pa(L-a)^2}{L^2} \qquad (4.25)$$

Substituting R_A and M_A from Equations (4.24) and (4.25) into (4.21) gives the required general expression for deflection, y, at any position along the beam.

$$y = \frac{1}{EI}\left[\frac{Pa(L-a)^2 x^2}{2L^2} - \frac{P(L-a)^2(L+2a)x^3}{6L^3} + \frac{P<x-a>^3}{6}\right]$$

3.4.8.2 Special Case – Centrally Loaded Beam

For a beam loaded at the centre span, that is $a = L/2$, Equations (4.23) and (4.24) give,

$$R_A = \frac{P}{2}$$

$$M_A = -\frac{PL}{8}$$

The maximum deflection at the centre of the span, under the load, is,

$$y_{max} = -\frac{1}{EI}\left[\frac{M_A L^2}{8} + \frac{R_A L^3}{48} - \frac{P(0)^3}{6}\right]$$

$$= \frac{PL^3}{192EI}$$

which is the well-known expression given in many textbooks.

Comparing this result with a simply-supported beam, where the central deflection is given by $y_{max} = PL^3/48EI$, it can be seen that clamping the ends of the beam results in a deflection which is 25% of the deflection of a simply-supported beam.

LEARNING SUMMARY

By the end of this section, you should have learnt:

- ✓ To derive the differential equation of the elastic line (i.e. deflection curve) of a beam;
- ✓ To solve this equation by successive integration to yield the slope, dy/dx, and the deflection, y, of a beam at any position along its span;
- ✓ To use Macaulay's Method, also called the Method of Singularities, and how it can be used to solve for beam deflections where there are discontinuities in the bending moment distribution arising from discontinuous loading;
- ✓ To use different singularity functions in the bending moment expression for different loading conditions including point loads, UDLs and point bending moments;
- ✓ To use Macaulay's method for statically indeterminate beam problems.

3.5 ELASTIC PLASTIC DEFORMATIONS

3.5.1 ELASTIC PLASTIC MATERIAL BEHAVIOUR MODELS

3.5.1.1 Elastic Perfectly Plastic (EPP)

In this case, there is assumed to be no hardening, that is, the yield stress, σ_y, is assumed to remain constant at $\pm \sigma_y$, regardless of any previous plastic deformation (see Figures 3.29 and 3.30). Therefore, the yield surface doesn't change in either shape or position in the principal stress-space, see Figure 3.31.

This is a good model for mild steel with moderate plasticity, but is also used very generally for materials without well-defined yield, that is $\sigma_y \approx \sigma_{0.2\%}$, for moderate plasticity.

3.5.1.2 Isotropic Hardening

For an isotropically hardening material, any plastic deformation causes an increase in the yield stress and the yield surface continuously expands isotropically (i.e. by the same amount in all directions) with loading, reverse loading and re-loading involving plastic deformation, as indicated in Figures 3.32 and 3.33. Thus, cyclic tension-compression loading, that is, varying between equal and opposite fixed values of applied stress, which initially gives yield (in tension) will quickly 'shake down' to elastic behaviour, since the subsequent compressive stress will not be large enough to cause yield in compression.

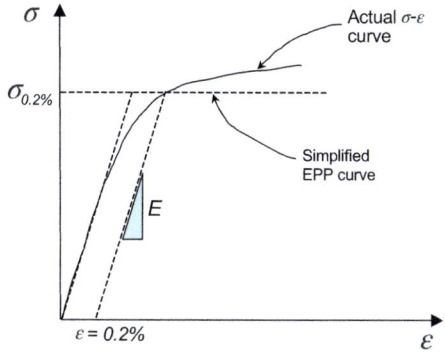

FIGURE 3.29 EPP stress-strain curve.

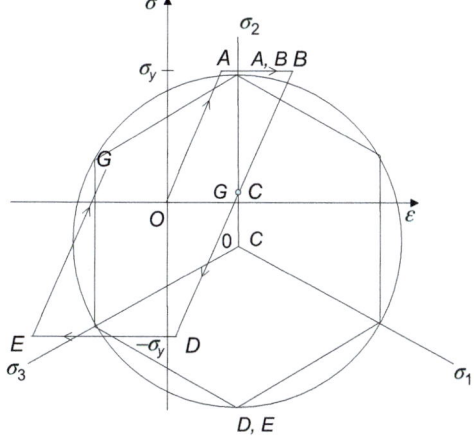

FIGURE 3.30 Uniaxial cyclic σ-ε behaviour for an EPP material model.

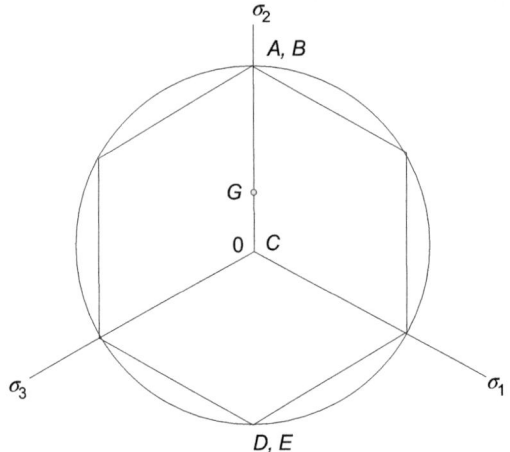

FIGURE 3.31 Representation of uniaxial stress–strain behaviour (Figure 3.30) in principal stress space.

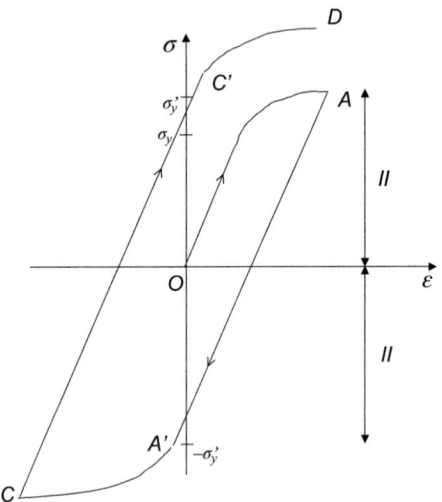

FIGURE 3.32 Uniaxial cyclic σ-ε behaviour for an isotropic material model.

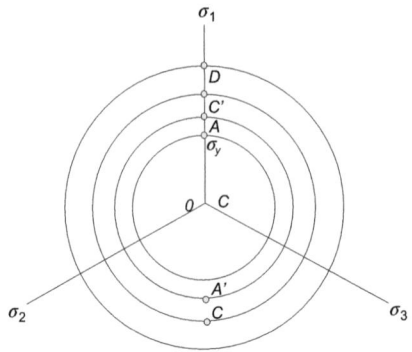

FIGURE 3.33 Representation of isotropic stress–strain behavior in principal stress space.

3.5.1.3   Kinematic Hardening (Assuming Linear Plasticity for Simplicity)

For a kinematically hardening material, the yield stress in compression following yield in tension is reduced due to the tensile yield, as indicated in Figures 3.34 and 3.35.

Kinematic hardening assumes that the yield range remains constant at a value of $2\sigma_y$, so that the subsequent yield in compression will be the highest tensile stress with plasticity, minus $2\sigma_y$ shown in Figure 3.34. This can be represented in principal stress space as a translation of the yield surface without changing shape or size, as shown in Figure 3.35. In this case, cycling between tensile and compressive stresses of equal and opposite values, which initially gives yield (in tension) will also give yield due to the subsequent compressive stress since the compressive yield stress will have been reduced. Thus, the kinematic hardening model predicts that constant alternating plastic strains will occur after the first loading cycle.

In the following analyses related to the elastic-plastic deformation of components (e.g. beams in bending and torsion of shafts), only EPP material behaviour models will be considered (Figure 3.36).

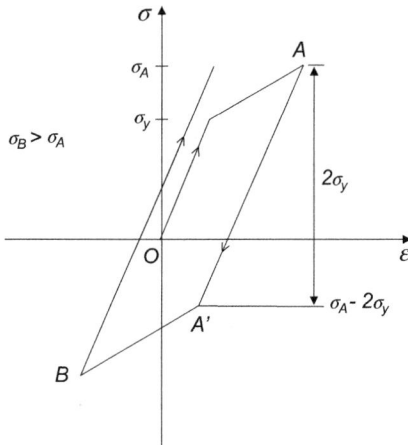

FIGURE 3.34   Uniaxial cyclic σ-ε behaviour for a kinematically hardening material behaviour.

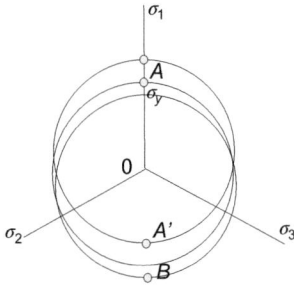

FIGURE 3.35   Representation of kinematic stress–strain in principal stress space.

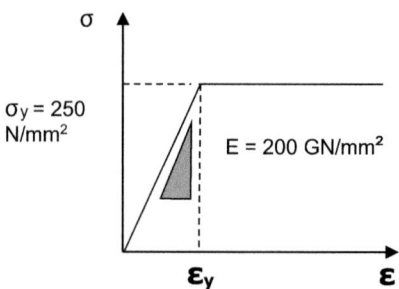

FIGURE 3.36 The EPP material behaviour model.

3.5.2 BENDING OF A BEAM WITH RECTANGULAR CROSS-SECTION, ASSUMING ELASTIC-PERFECTLY PLASTIC MATERIAL BEHAVIOUR MODEL

The following examples will illustrate the basic approach.

Example 3.5.1 A rectangular section beam (100 mm wide × 200 mm deep) is made from an EPP material with $E = 200$ GN/m² and $\sigma_y = 250$ N/mm². Calculate the radius of curvature and the bending stress distribution when a pure moment of 190 kNm is applied and after the moment is removed.

Check whether or not yield has occurred. The applied moment at first yield, M_y, is that which causes the maximum elastic bending stress to become equal to the yield stress. Since the maximum elastic stress occurs at the top and bottom (extreme) fibres of the cross-section, the first yield occurs when $\sigma = \mp\sigma_y$ at the position $y = \pm100$ mm. Thus, M_y is given by

$$M_y = \frac{\sigma_y I}{y} = \frac{250\,\text{N}/\text{mm}^2 \times \left(\dfrac{100 \times 200^3}{12}\right)\text{mm}^4}{100\,\text{mm}}$$

i.e. $M_y = 166.7\,\text{N}/\text{mm} = 166.7\,\text{kNm}$

The applied moment (190 kNm) exceeds M_y and therefore yielding will occur.

Assume yielding occurs at $y \geq a$ and $y \leq -a$.

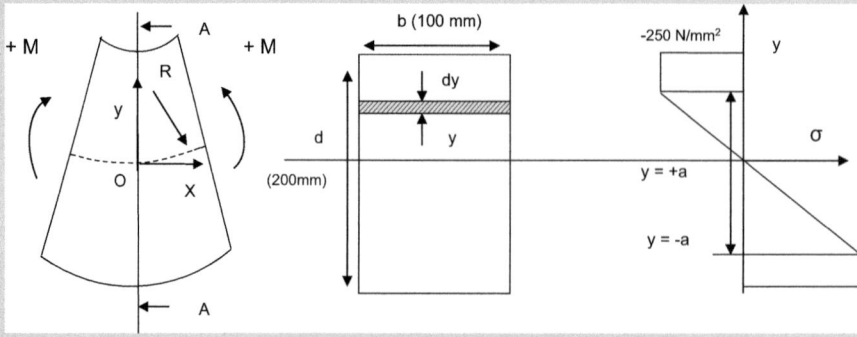

FIGURE 3.37 A section of the beam in bending, the beam cross-section and the variation of stress through the beam cross-section.

3.5.2.1   Variation of Stress with y

$$\text{For } -a < y < a, \quad \sigma = -250\frac{y}{a}\,\text{N}/\text{mm}^2$$

$$\text{For } a < y < \frac{d}{2}, \quad \sigma = -250\,\text{N}/\text{mm}^2$$

and

$$\text{for } -\frac{d}{2} < y < -a, \quad \sigma = 250\,\text{N}/\text{mm}^2$$

3.5.2.2   Moment Equilibrium

$$\therefore M = \int_A y(\sigma dA) = \int_{y=-\frac{d}{2}}^{y=+\frac{d}{2}} \sigma y(bdy)$$

$$\text{i.e. } M = 2\int_{y=0}^{y=-\frac{d}{2}} \sigma y b dy$$

$$\therefore M = 2\left\{\int_{y=0}^{y=-a}\left(-250\frac{y}{a}\right)ybdy + \int_{y=-a}^{y=-\frac{d}{2}}250\,ybdy\right\}$$

$$= 2\times 250b\left\{\left[-\frac{y^3}{3a}\right]_0^{-a} + \left[\frac{y^2}{2}\right]_{-a}^{-\frac{d}{2}}\right\}$$

$$= 500b\left\{\frac{a^2}{3} + \frac{d^2}{8} - \frac{a^2}{2}\right\}$$

$$\therefore M = 250b\left\{\frac{d^2}{4} - \frac{a^2}{3}\right\}$$

If $M = 190 \times 10^6$ Nmm

$$\text{then } 190\times 10^6 = 250\times 100\left(10^4 - \frac{a^2}{3}\right)$$

$$\therefore a = 84.85 \text{ mm}$$

3.5.2.3   Compatibility Requirement

$$\varepsilon = y/R$$

At $y = -84.85$ mm, $\sigma = 250\,\text{N/mm}^2$ and since this point is within the elastic range, the use of the linear elastic stress–strain equation gives

$$\varepsilon = \frac{\sigma}{E} = \frac{250}{200\times10^3} = \frac{84.85}{R}\left(= -\frac{y}{R}\right)$$

$\therefore R = 67.9 \times 10^3$ mm $= 67.9$ m

On unloading, assume that the stress change which occurs is elastic, but check the resulting solution to establish whether or not reverse yielding has occurred in order to validate this assumption.

Assuming elastic unloading, then the maximum stress change will occur at $y = -\dfrac{d}{2}$ and is given by

$$\Delta\sigma_{max}^{el} = -\frac{My}{I} = -\frac{-190\times10^6\,\text{Nmm}\times(-100\,\text{mm})}{\left(\dfrac{100\times200^3}{12}\right)\text{mm}^4}$$

$$= -285\,\text{N}/\text{mm}^2$$

$$\therefore \Delta\sigma_{max}^{el} = -285\,\text{N}/\text{mm}^2 \text{ at } y = -\frac{d}{2}$$

with a corresponding stress change of $+285$ N/mm² at $y = \dfrac{d}{2}$

By adding the stresses on the cross-section when loaded to the stress changes which occur on unloading, the residual stresses can be obtained, as shown in Figure 3.38.

It is clear that the residual stress is well below the yield stress of ± 250 N/mm². So, reverse yielding does not occur.

There will also be a residual curvature corresponding to the residual stresses. At $y = 84.85$ mm, there has been no plastic deformation. Therefore, the residual strain is given by

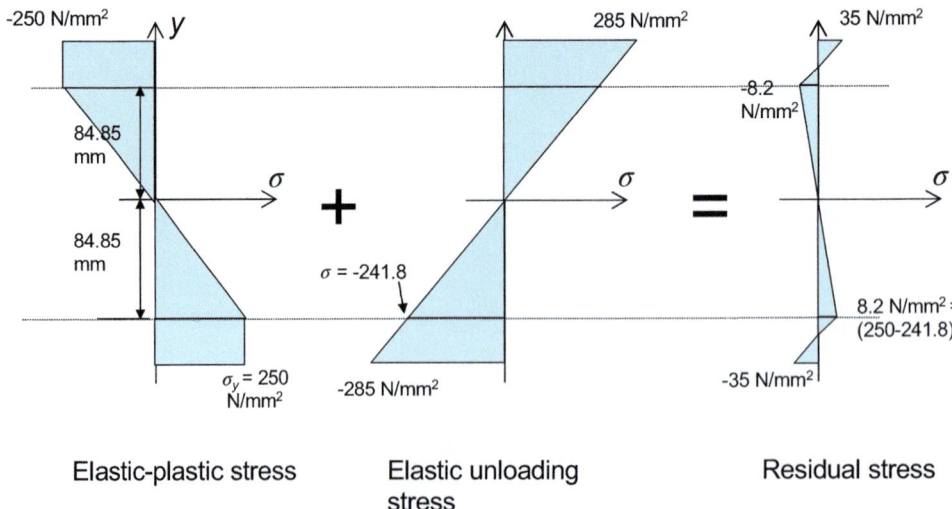

Elastic-plastic stress   Elastic unloading   Residual stress
                                        stress

FIGURE 3.38 The stress distribution on loading, the stress change on unloading and the residual stress distribution.

$$\varepsilon\left(\text{residual}\right) = \frac{\sigma}{E} = \frac{250}{200 \times 10^3} = \frac{84.85}{R} \left(= -\frac{y}{R}\right)\left(= -\frac{\gamma}{R}\right)$$

$$\therefore R = 2.070 \times 10^6 \text{ mm}$$

i.e. the residual radius of curvature is 2070 m

On releasing the moment, the radius of curvature changes from 67.9 m to 2070 m. This change of curvature is called 'spring back' and is particularly important when bending bars of metal to the specified radii of curvature.

3.5.3 Torsion of Circular Shafts, Assuming an Elastic-Perfectly-Plastic Material Behaviour Model

Figure 3.39 shows a circular shaft subjected to pure torque, T. This results in a twist, θ, and the rotation of an initially axial line by an angle γ. This causes the distortion of an element of material on the surface of the shaft as indicated in Figure 3.40.

3.5.3.1 Compatibility Requirement
Assuming plane sections remain plane and radii remain straight, then at $r = R$

$$l \times \left(\gamma_{r=R}\right) = R\theta \Rightarrow \gamma_{r=R} = \frac{R\theta}{l}$$

and hence at a general radial position, r,

$$\gamma = \frac{r\theta}{l} \tag{5.1}$$

i.e. the shear strain varies linearly with radius, r.

Since the highest strains occur at the maximum radius $r = R$, the yield will start at the outer surface and spread inwards as the torque is increased as indicated in Figure 3.41.

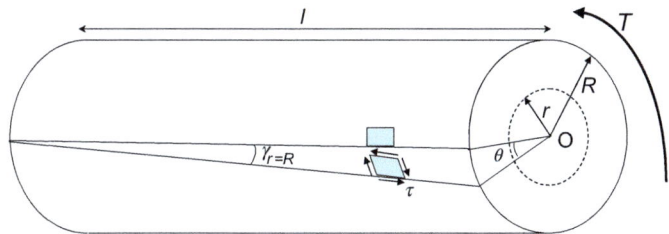

FIGURE 3.39 A circular shaft subjected to a pure torque, T.

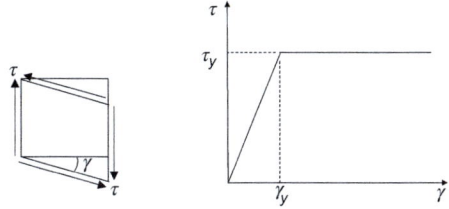

FIGURE 3.40 Distortion of the initially rectangular element and T-γ material behaviour.

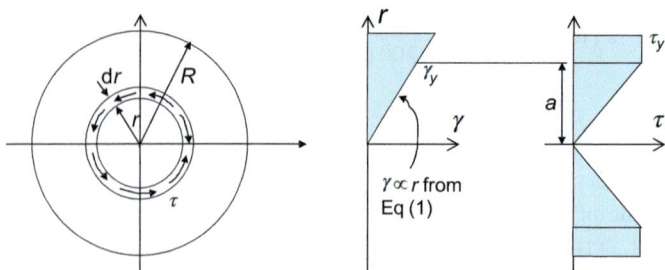

FIGURE 3.41   Shear stress on the shaft cross-section, variation of shear strain with radius and variation of shear stress with radius.

3.5.3.2   Equilibrium (Moments about the Axis)

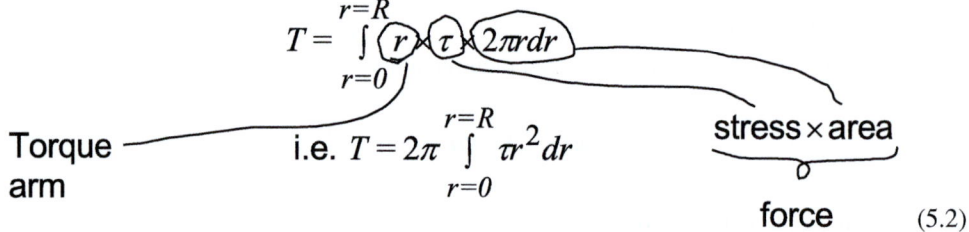

As with the case of beam bending, residual stresses can be obtained by determining (elastically) the change of stress which occurs when the load is removed. Remember that it is necessary to check that the residual stresses are not great enough to cause reverse yield. For <u>elastic unloading</u>, the following elastic torsion formulae can be used:-

$$\frac{T}{J} = \frac{\tau}{r} = \frac{G\theta}{l}$$

The relationship between T-γ and the more commonly available uniaxial $\sigma = \varepsilon$ data depends on the yield criterion. The Mohr's circle for this case, Figure 3.42, indicates that $\sigma_{,} = -\sigma_3 = \tau$ and $\sigma_2 = 0$. Hence, for a material obeying the von Mises' criterion.

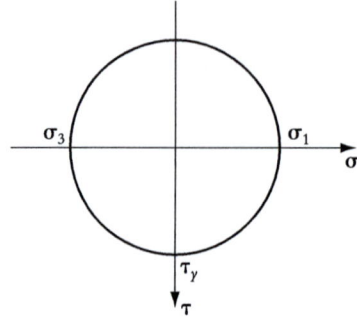

FIGURE 3.42   Mohr's circle.

For a material obeying the Tresca yield criterion,

$$\sigma_y = \sigma_1 - \sigma_3 = \tau_y - \left(-\tau_y\right) = 2\tau_y = \sigma_{vm}$$

$$\therefore \tau_y = \frac{\sigma_y}{2} = 0.5\sigma_y$$

Example 3.5.2 A solid circular shaft of diameter 50 mm and length 100 mm is subjected to a torque, T. The shaft is made from an EPP material with $\tau_y = 100$ N/mm^2 and $G = 70$ GN/m^2. Determine the magnitude of the torque required to cause yielding to occur at a radius of 15 mm (and greater) and the angle of twist.

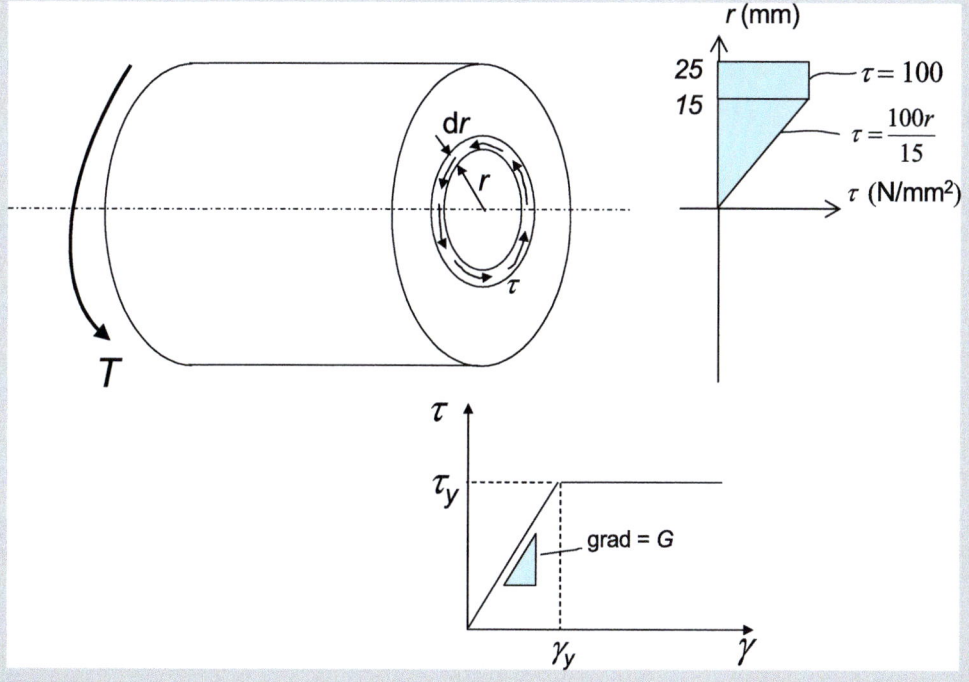

Equilibrium (moments about the axis)

$$T = \int_{r=0}^{r=25} r \times \tau \times 2\pi r \, dr$$

Torque arm × force

$$T = \int_0^{15} 2\pi \frac{100}{15} r^3 dr + \int_{15}^{25} 200\pi r^2 dr$$

$$= 200\pi \left\{ \left[\frac{r^4}{15 \times 4} \right]_0^{15} + \left[\frac{r^3}{3} \right]_{15}^{25} \right\} = 200\pi \left[\frac{15^4}{15 \times 4} + \frac{25^3}{3} - \frac{15^3}{3} \right]$$

$$\therefore T = 3.096 \times 10^6 \, \text{Nmm} = 3.096 \, \text{kNm}$$

Relationship between T_y and γ_y

At the outermost elastic point, $\tau = \tau_y$ and $\gamma = \gamma_y$ and the elastic relation $G = \dfrac{\tau_y}{\gamma_y}$ is applicable. It should be noted that outside this outermost elastic point, that is $r = 15$ mm, the strain, which will be larger, will be elastic-plastic and consequently will not be governed by $\tau = G\gamma$, which is the elastic relation. Nonetheless, at $r = 15$ mm, we have

$$\gamma_y = \frac{\tau_y}{G} = \frac{100}{70000} = 1.4286 \times 10^{-3}\,\text{rad}$$

by invoking the compatibility requirement, that is,

$$r_y\theta = \gamma_y l$$

and hence

$$\theta = \frac{\gamma_y l}{r_y}$$

$$\therefore \theta = \frac{1.4286 \times 10^{-3} \times 1000\,\text{mm}}{15\,\text{mm}} \times \left(\frac{360\,\text{deg}}{2\pi\,\text{rad}}\right)$$

$ie.\ \theta = 5.456\,\text{deg}.$

EXAMPLE 3.5.3

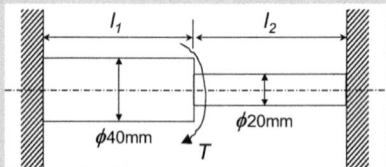

A stepped circular shaft is built in at both ends; the two shaft segments are of the same length, l. If the stepped shaft is made from an EPP material with $\tau_y = 100\text{N/mm}^2$, find the magnitude of the torque applied at the step to just cause yield in the smaller shaft.

3.5.3.3   Stress–Strain Behaviour

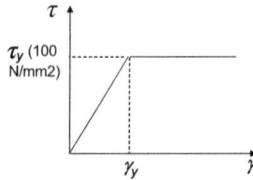

3.5.3.4 Equilibrium

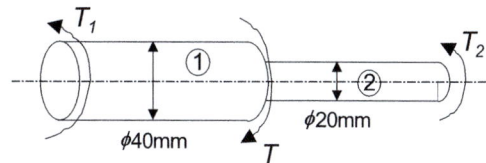

$$T = T_1 + T_2$$

The small shaft just reaches yield; therefore, the linear elastic torsion equation can be used for that shaft, that is,

$$\frac{T}{J} = \frac{\tau}{r} = \frac{G\theta}{l}$$

$$\Rightarrow \frac{T_2}{\left(\dfrac{\pi \times 20^4}{32}\right) \text{mm}^4} = \frac{100\,\text{N}/\text{mm}^2}{10\,\text{mm}}$$

$$\Rightarrow T_2 = 157.1 \times 10^3\,\text{Nmm} = 157.1\,\text{Nm}$$

For the large shaft, θ and l are the same as the small shaft, and since $\gamma = \dfrac{r\theta}{l}$ (elastic or plastic), then $\gamma_1\,|_{r=10\text{mm}} = \gamma_2\,|_{r=10\text{mm}}$.

Therefore, shaft 1 must also be elastic for $r < 10$ mm and will be plastic for $r > 10$ mm.

Thus, for the large shaft,

for $r < 10$ mm,

$$\tau = \tau_y \frac{r}{10} = 10r$$

for $r > 10$ mm,

$$\tau = \tau_y = 100\,\text{N}/\text{mm}^2$$

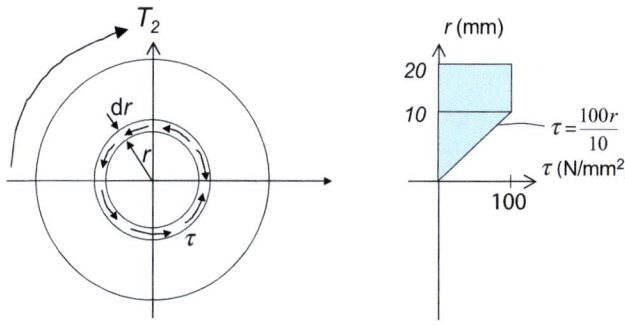

$$T_1 = \underbrace{\int_{r=0}^{r=10} r \times \underbrace{10r \times 2\pi r dr}_{\text{Stress} \times \text{area}}}_{\text{Torque arm} \times \text{force}} + \underbrace{\int_{r=10}^{r=20} r \times \underbrace{100 \times 2\pi r dr}_{\text{Stress} \times \text{area}}}_{\text{Torque arm} \times \text{force}}$$

$$T_1 = 20\pi \int_0^{10} r^3 dr + 200\pi \int_{10}^{20} r^2 dr$$

$$= 20\pi \left[\frac{r^4}{4} \right]_0^{10} + 200\pi \left[\frac{r^3}{3} \right]_{10}^{20}$$

$$= 20\pi \times \frac{10^4}{4} + 200\pi \left[\frac{20^3}{3} - \frac{10^3}{3} \right]$$

$$ie \; T_1 = 1623 \times 10^3 \, \text{Nmm} = 1623 \, \text{Nm}$$

$$ie \; \text{total torque} = T = T_1 + T_2$$

$$= (1623 + 157.1) \, \text{Nm}$$

$$\therefore T = 1780 \, \text{Nm}$$

LEARNING SUMMARY

By the end of this section, you should have learnt:

✓ The shapes of uniaxial stress–strain curves and the EPP approximation to uniaxial stress–strain curves;
✓ The kinematic and isotropic material behaviour models used to represent cyclic loading behaviour;
✓ About the elastic-plastic bending of beams and the need to use equilibrium, compatibility and σ-ε behaviour to solve these types of problems;
✓ About the elastic-plastic torsion of shafts and the need to use equilibrium, compatibility and $\tau - \gamma$ behaviour to solve these types of problems;
✓ How to determine residual deformations and residual stresses.

3.6 ELASTIC INSTABILITY

3.6.1 INTRODUCTION

For many structural problems, it is reasonable to assume that the system is in stable equilibrium. However, not all structural arrangements are stable. For example, consider a one-metre-long stick with the cross-sectional area of a pencil. If this stick stood on its end, the axial stress would be small, but the stick could easily topple over sideways. This simple example demonstrates that in some configurations, stability considerations can be primary.

This section is concerned with the stability of struts. Struts are compression members with cross-sectional dimensions which are small compared to the length, that is, they are slender. If a circular rod of, say, 5 mm diameter, which has its ends machined flat and perpendicular to the axis, were made 10 mm long to act as a column, there would not be a problem of instability and it could carry considerable force. However, if the same rod were made a metre long, the rod would become laterally unstable at a much smaller applied force and could collapse.

Buckling also occurs in many other situations with compressive forces. Examples include thin sheets which have no problem carrying tensile loads, narrow beams unbraced laterally, and vacuum tanks, as well as submarine hulls. Thin-walled tubes can wrinkle like paper when subjected to torque.

3.6.2 BUCKLING PHENOMENON

Consider the response of a marble when subjected to disturbances from an initial equilibrium position on different types of surfaces as shown in Figure 3.43. If the surface is concave, the marble will return to its original equilibrium position and the marble is said to be in a stable equilibrium position; if the surface is flat, the marble will move to another equilibrium position and the marble is said to be in a neutral equilibrium position. Finally, if the surface is convex, the marble will roll off uncontrollably in an unstable fashion and the marble is said to be in an unstable equilibrium position.

This analogy is useful for understanding the energy approach to buckling problems. Every deformed structure has a potential energy associated with it, which depends on the strain energy stored in the structure and the work done by the external loads. A concave potential energy function at equilibrium gives a stable equilibrium while a convex potential energy function gives unstable equilibrium.

Alternatively, buckling problems may be treated as bifurcation problems. Referring to Figure 3.44(a), it is clear that the tensile force will tend to restore the bar to equilibrium if there is a slight displacement to the right. However, the same bar under the action of a compressive force, Figure 3.44(b), will continue to fall when subjected to a slight displacement. This illustrates an unstable equilibrium.

Figures 3.45(a) and (b) illustrate a slightly more complicated example of the same phenomenon. The vertical bar is supported horizontally by two springs of stiffness, k. If the bar of length L is displaced a small amount, x, horizontally, there is a displacing moment of Px about O and a restoring moment of $2kxL$. Hence, we get $Px < 2kxL$ for stable equilibrium and $Px > 2kxL$ for unstable equilibrium.

The critical condition occurs when

$$Px = 2kxL$$

or

$$P_c = 2kL$$

where P_c is termed the critical load between stable and unstable equilibrium.

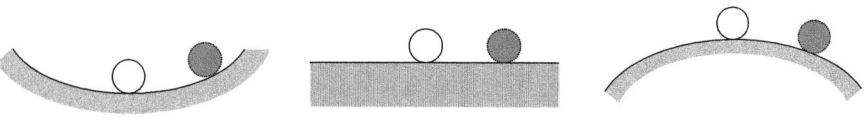

Stable equilibrium position   Neutral equilibrium position   Unstable equilibrium position

FIGURE 3.43 Equilibrium states for a marble on various surfaces.

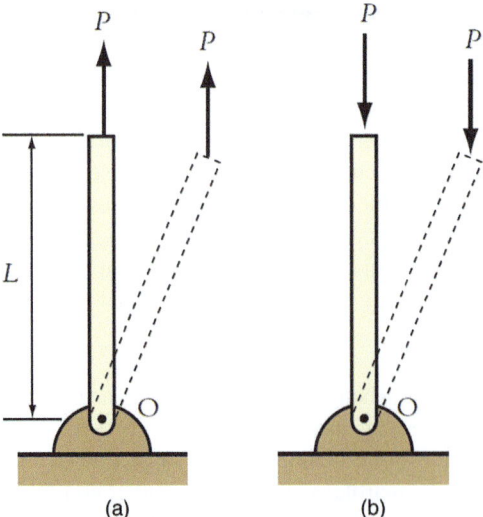

FIGURE 3.44   Examples of stable and unstable equilibrium.

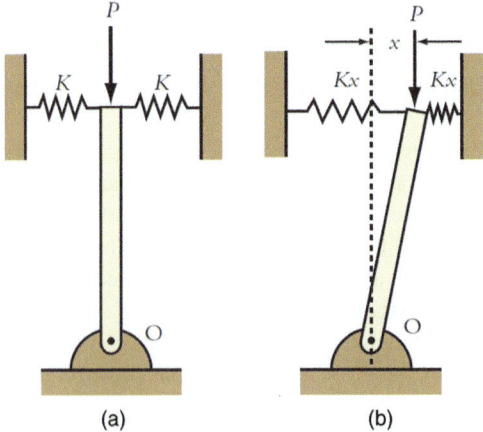

FIGURE 3.45   (a) Axially loaded rigid bar with transverse springs.

Figure 3.46 shows a rigid bar subjected to a compressive axial load with a torsional spring at its base; a free body diagram of the problem is also shown. Taking moments about point O gives

$$PL\sin\theta = K_\theta\theta$$

$$\frac{PL}{K_\theta} = \frac{\theta}{\sin\theta}$$

Figure 3.47 shows a graph of $\dfrac{PL}{K_\theta}$ versus θ. There is a stable region for low loads and an unstable region for high loads. Below point A, the bar will return to its equilibrium position if rotated slightly to either the right or left. Once the load exceeds the value at point A, then any disturbance will cause the bar to rotate along either the right branch or the left branch of the bifurcation curve. Point A is called the bifurcation point, at which there are three possible solutions. The associated load at point

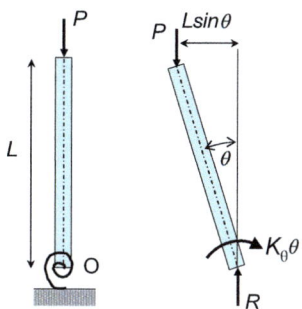

FIGURE 3.46 Rigid bar supported by a torsional spring.

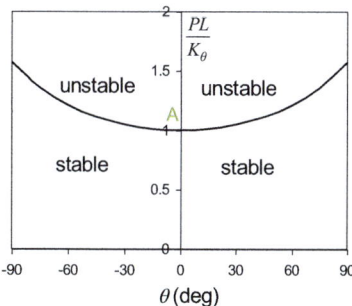

FIGURE 3.47 Variation of $PL/K\sigma$ with θ, indicating stable and unstable regions.

A is called the critical (buckling) load. $\theta = 0°$ is the trivial solution. Only non-trivial solutions are generally of interest, that is for $\theta \neq 0°$.

3.6.3 IDEAL STRUTS

Ideal struts are assumed to be initially perfectly straight and of uniform section properties and subjected to purely axial loading. Expressions will be developed by relating the critical buckling load to the applied load, the material properties and the member dimensions, for different support conditions of the struts. At a critical load, members which are circular or tubular in cross-section will buckle sideways in any direction. Often, compression members do not have equal flexural rigidity, EI, in all directions and there will be one axis about which the flexural rigidity is a minimum, depending on the dimensions. The member will therefore buckle about this axis and the I-value (second moment of area) referred to in this section is assumed to be the minimum value, based on the nominal dimensions of the member.

CASE 1   HINGED-HINGED

Consider an initially straight strut with its ends free to rotate around frictionless pins, as shown in Figure 3.48, which will be referred to as a hinged-hinged case. The dashed line represents the initially straight strut. The strut is now considered to be perturbed, from its initially straight position, as shown in Figure 3.48. This perturbation is equivalent to the movement of the marble as shown in Figure 3.43.

The bending moment, M, shown in Figure 3.49, depends on the deflection, y, and is hence a function of position, x.

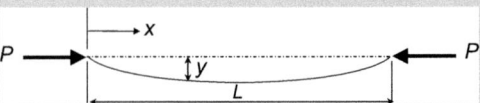

FIGURE 3.48   A hinged-hinged strut.

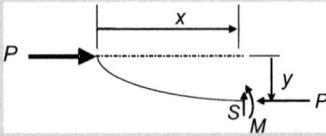

FIGURE 3.49   Free body diagram for the LH portion of the strut.

The deflection, y, is related to the moment, M, by the relationship

$$EI\frac{d^2y}{dx^2} = -M$$

and

$$M = Py$$

Therefore,

$$\frac{d^2y}{dx^2} + \frac{P}{EI}y = 0$$

or

$$\frac{d^2y}{dx^2} + \alpha^2 y = 0$$

where $\alpha^2 = \dfrac{P}{EI}$

Hence,

$$y = A\sin\alpha x + B\cos\alpha x$$

In order to determine A and B, we need two "boundary conditions", that is

at $x = 0$, $y = 0$

and at $x = l$, $y = 0$

Therefore, B + 0 and A $\sin(\alpha l) = 0$

The condition $a = 0$ results in a trivial solution, that is, $y = 0$, which is the case for an undeflected strut. Hence, the non-trivial solution is $\sin(\alpha l) = 0$, which gives $\alpha l = n\pi$, where $n = 0, 1, 2, \ldots$

$$\therefore \alpha^2 l^2 = n^2 \pi^2$$

$$\frac{P}{EI} l^2 = n^2 \pi^2$$

$$\text{or } P = \frac{n^2 \pi^2 EI}{l^2}$$

$n = 0$ gives another trivial solution, that is, $P = 0$,

$n = 1$ gives $P_c = \dfrac{\pi^2 EI}{l^2}$

This is called the **Euler buckling (or crippling) load**, $PB_{c,B}$; it is the lowest load at which buckling can occur (Euler solved this problem in 1757).

$n = 2$ gives $P = \dfrac{4\pi^2 EI}{l^2}$

and this corresponds to a different deflected (buckling) shape of the strut.

For $n = 1$, $y = y_{max}$ at $x = \dfrac{l}{2}$ and therefore $A = y_{max}$ and the deflected shape of the strut is given by the following expression:

$$y = y_{max} \sin(\alpha x) = y_{max} \sin\left(\frac{n\pi x}{l}\right)$$

The magnitude of y_{max} cannot be determined from the boundary conditions and it can become arbitrarily large, leading to elastic instability of the structure. The first three buckling mode shapes are shown in Figure 3.50. If buckling mode I is prevented from occurring by installing a restraint (support), then the column will buckle at the next higher mode at critical load values that are higher than for the lower modes. There is zero deflection at the inflexion point, I, for each deflection curve. Recalling that the curvature $\dfrac{d^2 y}{dx^2}$ at an inflexion point is zero indicates that the internal moment at these points is zero. If roller supports are put at any other point than Point I, the boundary value problem must be solved for new eigenvalues (buckling loads) and eigenvectors (mode shapes).

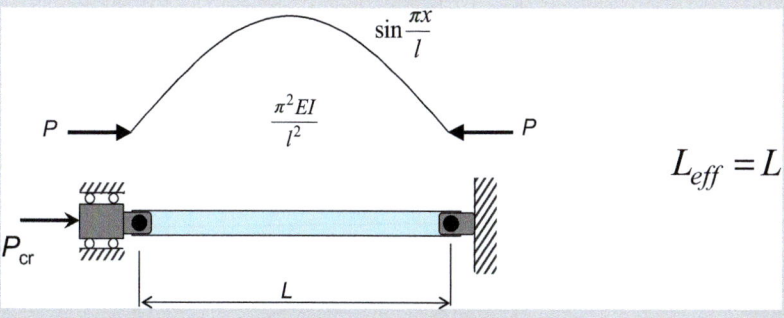

FIGURE 3.50 Buckling mode shapes for a hinged-hinged strut with $n = 1$, $n = 2$ and $n = 3$.

(*Continued*)

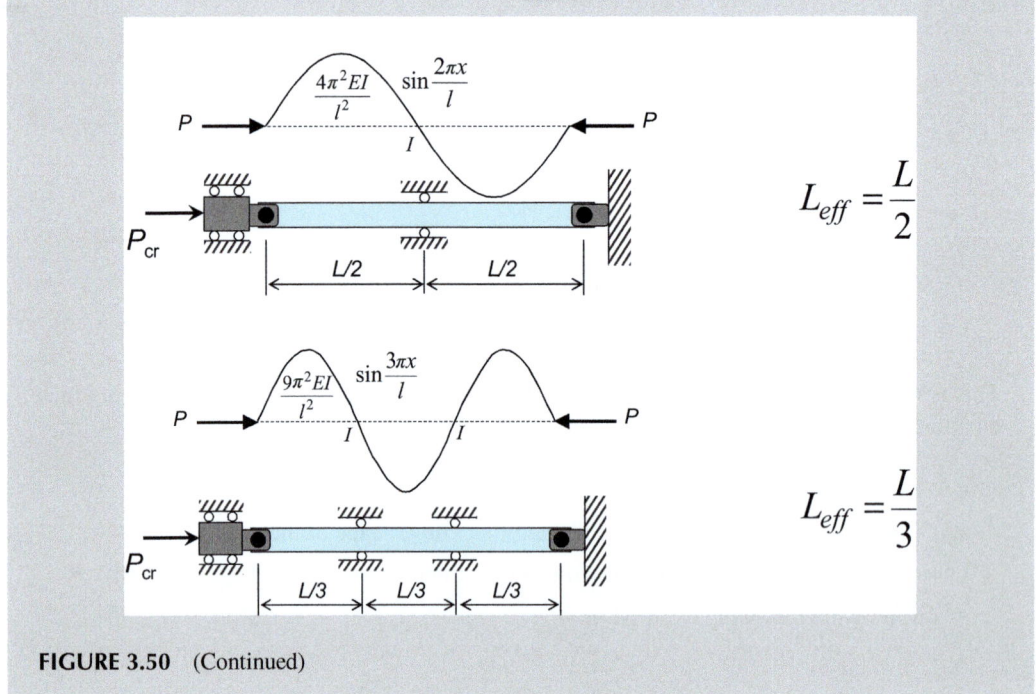

FIGURE 3.50   (Continued)

CASE 2   FREE-FIXED

Figure 3.51 shows the deflected shape and a free body diagram for a fixed-free strut.

$$EI\frac{d^2y}{dx^2} = -M$$

Where $M = Py$

$$\therefore EI\frac{d^2y}{dx^2} + Py = 0$$

or $\dfrac{d^2y}{dx^2} + \alpha^2 y = 0$

where $\alpha^2 = \dfrac{P}{EI}$

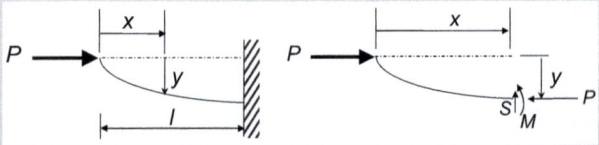

FIGURE 3.51   A fixed-free strut.

The solution to this differential equation is:

$$y = A\sin\alpha x + B\cos\alpha x$$

At $x = 0$, $y = 0$ and therefore $B = 0$ and at $x = l$, $\dfrac{dy}{dx} = 0$ and therefore $A\,\alpha\cos(\alpha l) = 0$

So far, the mathematical solution is identical to that of a free-free strut. However, the boundary conditions are different, that is in this case, $A = 0$ or $\alpha = 0$ or $\cos(\alpha l) = 0$, leading to trivial solutions (as before). The non-trivial solution results from taking $\cos(\alpha l) = 0$, which implies that $\alpha l = \dfrac{n\pi}{2}$.

$$\text{i.e. } \frac{P}{EI}l^2 = \frac{n^2\pi^2}{4} \text{ where } n = 1,3,\ldots$$

The smallest, non-trivial, value of P occurs with $n = 1$
   i.e.

$$P_c = \frac{\pi^2 EI}{4l^2}$$

In comparison with Case 1, that is, the hinged-hinged case, it can be seen that the solution is the same except that "l" is replaced by "$2l$", that is the fixed-free case can be treated as the hinged-hinged case for a strut with an equivalent length of $2l$.

CASE 3   FIXED-FIXED

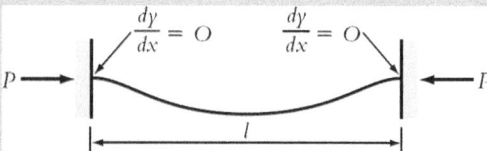

FIGURE 3.52   A fixed-fixed strut.

The solution procedure is the same as that for Cases 3.2.1 and 3.2.2, leading to:-

$$P_c = \frac{4\pi^2 EI}{l^2}$$

The fixed-fixed case, Figure 3.52, shows a significant increase in the buckling capacity relative to the hinged-hinged case.

CASE 4   FIXED-HINGED

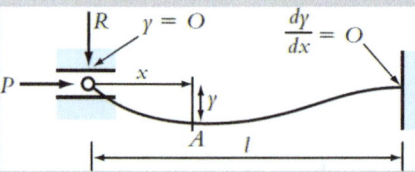

FIGURE 3.53   A fixed-hinged strut.

$$P_c = \frac{2.045\pi^2 EI}{l^2} \quad \left(\approx \frac{2\pi^2 EI}{l^2} \right)$$

This case differs from the previous examples in that a transverse force, R, shown in Figure 3.53, is necessary to create this mode of deformation.

$$EI \frac{d^2 y}{dx^2} = -M$$

$$\text{But } M + Rx = Py$$

$$\therefore \Rightarrow EI \frac{d^2 y}{dx^2} + Py = Rx \Rightarrow \frac{d^2 y}{dx^2} + \frac{P}{EI} y = \frac{R}{EI} x$$

$$\text{or} \Rightarrow EI \frac{d^2 y}{dx^2} + \alpha^2 y = \frac{R}{EI} x \text{ where } \alpha^2 = \frac{P}{EI}$$

$$\text{where } \alpha^2 = \frac{P}{EI}$$

The solution to this type of differential equation consists of two parts, a homogenous solution and a particular integral. The homogeneous solution is given by the following:

$$y = A \sin \alpha x + B \cos \alpha x$$

The particular integral for such second-order differential equations is generally obtained by taking

$$P.I. = y = C.f(x)$$

where

$$\frac{d^2 y}{dx^2} + \alpha^2 y = \text{const.} f(x)$$

Substituting for y in the differential equation gives the solution for C. In this particular case, $f(x) = x$, so that

$$\alpha^2 C.f(x) = \frac{R}{EI} f(x)$$

$$C = \frac{R}{\alpha^2 EI} = \frac{R}{P}$$

$$P.I. = y = \frac{R}{P}.f(x) = \frac{R}{P}x$$

Hence, the complete solution is:

$$y = A \sin \alpha x + B \cos \alpha x + \frac{R}{P}x$$

There are three unknowns this time, they are A, B and R. Therefore, we need three boundary conditions, that is:

at $x = 0$, $y = 0$ and so $B = 0$,

$$\text{at } x = I, y = 0, \text{ and so, } 0 = A \sin \alpha l + \frac{R}{P}l$$

$$\text{Therefore } A = -\frac{Rl}{P \sin \alpha l}$$

Also, at $x = l$, $\dfrac{dy}{dx} = 0$, $\quad -\dfrac{Rl\alpha}{P \sin \alpha l} \cos \alpha l + \dfrac{R}{P} = 0$

$$\text{Therefore } \tan \alpha l = \alpha l$$

The smallest non-trivial root to this equation is

$$\alpha l = 4.493 (\approx 1.43p)$$

$$\therefore \alpha^2 l^2 = \frac{P}{EI}l^2 = 1.43^2 \pi^2$$

$$\text{i.e. } P_c = \frac{2.045\pi^2 EI}{l^2}$$

CASE 5   HINGED-HINGED WITH INITIAL CURVATURE

Figure 3.54 shows a hinged-hinged strut with initial curvature, y_o, and the deflected shape, y. Suppose that $y_o = \varepsilon \sin\left(\dfrac{\pi x}{l}\right)$ describes the initial shape.

From the free-body diagram (Figure 3.55),

$$M = Py$$

$$EI \frac{d^2(y - y_o)}{dx^2} = -Py$$

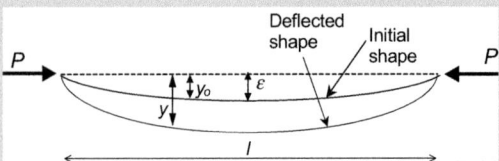

FIGURE 3.54   A hinged-hinged strut with initial curvature.

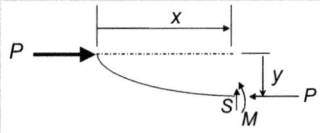

FIGURE 3.55   Free-body diagram for a hinged-hinged strut with initial curvature.

Where $\alpha^2 = \dfrac{P}{EI}$

$$\therefore \frac{d^2y}{dx^2} + \alpha^2 y = -\frac{\pi^2}{l^2}\varepsilon \sin\left(\frac{\pi x}{l}\right)$$

In this case, there is a particular integral as well as a homogeneous function. Assume the particular integral is obtained by taking $y = C\sin\left(\dfrac{\pi x}{l}\right)$, where C is an unknown constant, and substituting this into the differential equation, that is,

$$-C\frac{\pi^2}{l^2}\sin\left(\frac{\pi x}{l}\right) + \alpha^2 C\sin\left(\frac{\pi x}{l}\right) = -\frac{\pi^2}{l^2}\varepsilon \sin\left(\frac{\pi x}{l}\right)$$

$$\Rightarrow \qquad C = \frac{-\dfrac{\pi^2}{l^2}\varepsilon}{\left(\alpha^2 - \dfrac{\pi^2}{l^2}\right)}$$

Thus, the overall solution, that is the homogeneous function and the particular integral, is

$$y = A\sin(\alpha x) + B\cos(\alpha x) - \frac{\dfrac{\pi^2}{l^2}\varepsilon \sin\left(\dfrac{\pi x}{l}\right)}{\left(-\dfrac{\pi^2}{l^2} + \alpha^2\right)}$$

The A and B values are obtained by substituting the boundary conditions.
    At $x = 0$, $y = 0$

$$\therefore B = 0$$

At $x = l$, $y = 0$

$$\therefore 0 = A\left(\sin\left(\alpha l\right)\right), \text{i.e. } A = 0$$

$$\therefore \; y = -\frac{\dfrac{\pi^2}{l^2} \varepsilon \sin\left(\dfrac{\pi x}{l}\right)}{\left(-\dfrac{\pi^2}{l^2} + \alpha^2\right)}$$

When $\dfrac{\pi^2}{l^2} = \alpha^2$, $y = \infty$ (except at $x = 0$ or l)

$$\text{i.e. } \frac{\pi^2}{l^2} = \frac{P}{EI}$$

$$\therefore \; P = \frac{\pi^2 EI}{l^2}$$

that is, the initially curved strut buckles when

$$P\left(= P_c\right) = \frac{\pi^2 EI}{l^2}$$

$$\text{and} \quad y = -\frac{\varepsilon \sin\left(\pi x / l\right)}{\left(\dfrac{l^2 P}{\pi^2 EI} - 1\right)} = -\frac{\varepsilon \sin\left(\pi x / l\right)}{\left(\dfrac{P}{P_c} - 1\right)} = \frac{\varepsilon \sin\left(\pi x / l\right)}{\left(1 - \dfrac{P}{P_c}\right)}$$

Therefore, the assumed initial shape has little effect on the solution, that is,

$$P_c = \frac{\pi^2 EI}{l^2}$$

and the deflected shape, y, would be practically the same.

3.6.4 Eccentrically-Loaded Struts

Consider a long, slender member subjected to a slightly eccentric compressive load as shown in Figure 3.56; the axial load is P and the eccentricity of the load is e. This form of loading will cause the member to curve.

So that at a distance x from the LH end, the eccentricity of the load becomes $(e + y)$ see Figure 3.57. The deflection, y, is related to the moment, M, by the relationship

$$EI \frac{d^2 y}{dx^2} = -M$$

and, x from the LH-end,

$$\therefore M = P\left(e + y\right)$$

FIGURE 3.56 An eccentrically loaded strut.

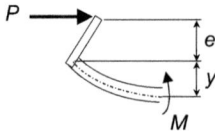

FIGURE 3.57 Free-body diagram for an eccentrically loaded strut.

$$EI\frac{d^2y}{dx^2} = -P(e+y)$$

or

$$\frac{d^2y}{dx^2} + \alpha^2 y = -\alpha^2 e$$

where

$$\alpha^2 = \frac{P}{EI}$$

The solution to this differential equation consists of two parts, a homogeneous solution and a particular integral. The homogeneous solution is given by

$$y = A\sin\alpha x + B\cos\alpha x$$

In general, the particular integral is obtained by taking

$$P.I. = y = C.f(x)$$

where

$$\frac{d^2y}{dx^2} + \alpha^2 y = \text{const.}f(x)$$

and then substituting into the differential equation for y and solving for C. In this case, the particular integral is taken to be C, since $f(x)$ is 1 and hence $C = -e$. Thus, the total solution to the differential equation is:

$$y = A\sin\alpha x + B\cos\alpha x - e$$

In order to determine A and B, we use two "boundary conditions".

At $x = 0$, $y = 0$ and at $x = \dfrac{l}{2}$, $\dfrac{dy}{dx} = 0$

Thus, $0 = B - e$, ie $B = e$, and $0 = A\alpha\cos\left(\dfrac{\alpha l}{2}\right) - B\alpha\sin\left(\dfrac{\alpha l}{2}\right)$

$$ie\,A = e\tan\left(\frac{\alpha l}{2}\right)$$

So,

$$y = e\tan\left(\frac{\alpha l}{2}\right)\sin\left(\alpha x\right) + e\cos\left(\alpha x\right) - e$$

$$\therefore\ y = e\left(\tan\left(\frac{\alpha l}{2}\right)\sin\left(\alpha x\right) + \cos\left(\alpha x\right) - 1\right)$$

The maximum deflection, $\hat{y}$, will occur at $x = \dfrac{l}{2}$, so that,

$$\hat{y} = e\left(\tan\left(\frac{\alpha l}{2}\right)\sin\left(\frac{\alpha l}{2}\right) + \cos\left(\frac{\alpha l}{2}\right) - 1\right)$$

or

$$\hat{y} = e\left(\frac{\sin\left(\dfrac{\alpha l}{2}\right)\sin\left(\dfrac{\alpha l}{2}\right)}{\cos\left(\dfrac{\alpha l}{2}\right)} + \frac{\cos^2\left(\dfrac{\alpha l}{2}\right)}{\cos\left(\dfrac{\alpha l}{2}\right)} - 1\right)$$

$$\hat{y} = e\left(\frac{\sin^2\left(\dfrac{\alpha l}{2}\right) + \cos^2\left(\dfrac{\alpha l}{2}\right)}{\cos\left(\dfrac{\alpha l}{2}\right)} - 1\right)$$

$$\hat{y} = e\left(\sec\left(\frac{\alpha l}{2}\right) - 1\right)$$

Therefore, when $\dfrac{\alpha l}{2} = \dfrac{\pi}{2}$, then $\hat{y} = \infty$

However, $\alpha^2 = \dfrac{P}{EI}$, so that when $\dfrac{P}{EI} = \dfrac{\pi^2}{l^2}$, $\hat{y} = \infty$

The value of P which causes $\hat{y} \to \infty$ is the buckling load (**or the Euler crippling load**), P_c, so $\hat{y} = \infty$, when

$$P = P_c = \frac{\pi^2\,EI}{l^2}$$

It is important to note that P_c is independent of e, so that the buckling load is the same no matter how large or small the eccentricity, that is even very small eccentricities will give the same buckling load.

$$\text{Then } \hat{y} = e\left(\sec\left(\frac{\alpha l}{2}\right) - 1\right) = e\left(\sec\left(\sqrt{\frac{P}{EI}}\,\frac{l}{2}\right) - 1\right)$$

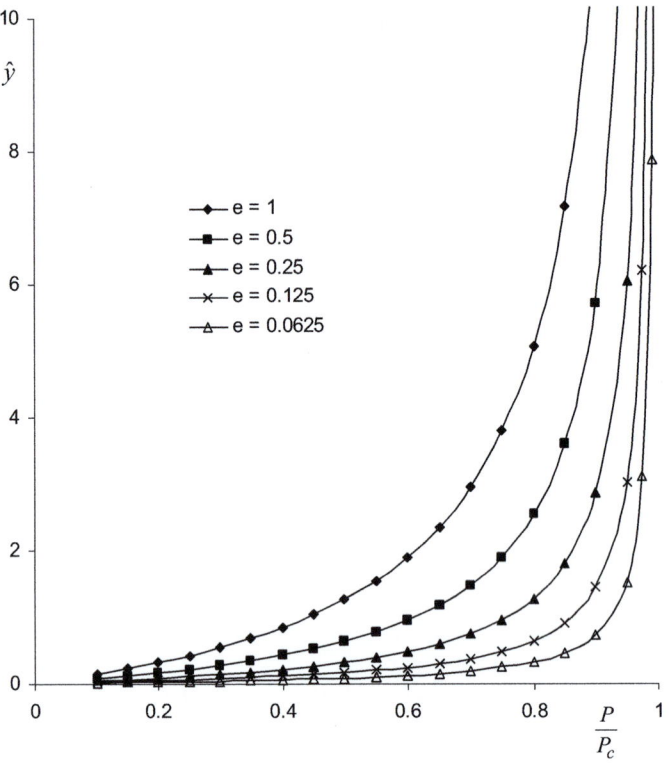

FIGURE 3.58   Variation of $\hat{y}$ with P/P_c for a range of eccentricity values.

$$\Rightarrow \ \hat{y} = e\left(\sec\left(\sqrt{\frac{P\pi^2}{P_c l^2}}\ \frac{l}{2} \right) - 1 \right)$$

$$\text{i.e.} \ \hat{y} = e\left(\sec\left(\frac{\pi}{2}\sqrt{\frac{P}{P_c}} \right) - 1 \right)$$

Figure 3.58 shows the relationship between transverse deflection $\hat{y}$ and the applied load ratio $\dfrac{P}{P_c}$ for a range of eccentricity values, e.

3.6.4.1   Summary of Euler Buckling Loads of Struts

General formula: $P_c = \dfrac{\pi^2 EI}{L_{eff}^2}$

Description	Schematic	Critical Buckling load, P_c	Effective Length, L_{eff}
Free—fixed	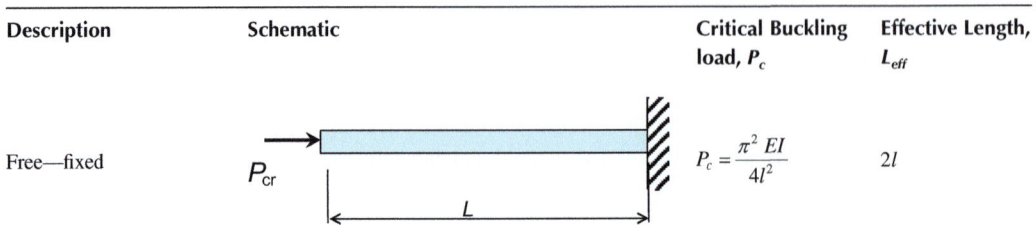	$P_c = \dfrac{\pi^2 EI}{4l^2}$	$2l$

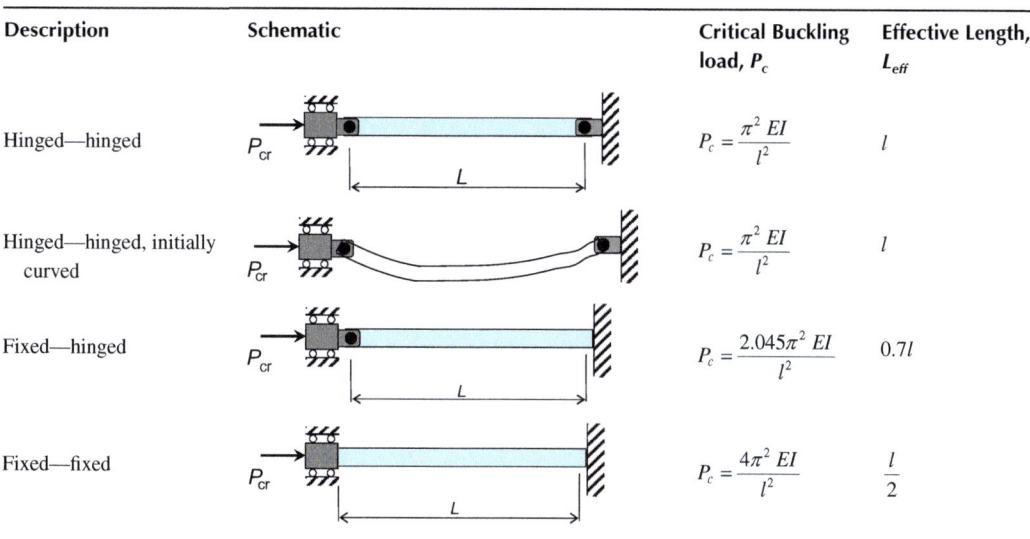

Description	Schematic	Critical Buckling load, P_c	Effective Length, L_{eff}
Hinged—hinged		$P_c = \dfrac{\pi^2 \, EI}{l^2}$	l
Hinged—hinged, initially curved		$P_c = \dfrac{\pi^2 \, EI}{l^2}$	l
Fixed—hinged		$P_c = \dfrac{2.045\pi^2 \, EI}{l^2}$	$0.7l$
Fixed—fixed		$P_c = \dfrac{4\pi^2 \, EI}{l^2}$	$\dfrac{l}{2}$

The effective length, L_{eff}, is a measure of how much longer (and thus more unstable) a given strut configuration appears to be in terms of critical buckling load, relative to the hinged-hinged case. Thus, the fixed-fixed case, for example, has a shorter effective length because it is more stable and thus appears to be shorter with respect to buckling.

3.6.5 Beams with Both Axial and Transverse Loads

A beam which is acted upon by a compressive axial force in addition to transversely applied loads is referred to as a beam column – see Figures 3.59 and 3.60. In this section, the buckling load for such a member, under the action of a point load at the mid-span, is derived.

$$\text{For } 0 \le x \le \frac{l}{2},$$

$$M = Py + \frac{W}{2}x \quad \text{and} \quad M = -EI\frac{d^2y}{dx^2}$$

$$\therefore \frac{d^2y}{dx^2} + \alpha^2 y = -\frac{W}{EI}\frac{x}{2} \quad \text{where} \quad \alpha^2 = \frac{P}{EI}$$

The solution to this differential equation, including using both the homogeneous function and the particular integral, is

$$y = A\sin(\alpha x) + B\cos(\alpha x) - \frac{Wx}{2P}$$

FIGURE 3.59 A beam column with a central transverse load.

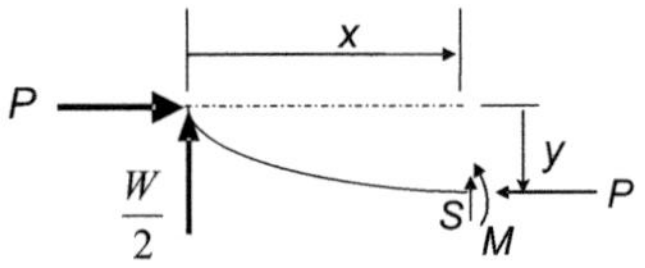

FIGURE 3.60   A free-body diagram for the left-hand section of the beam-column.

The constants A and B are obtained by using the boundary condition, that is,

   at $x = 0$, $y = 0$

   $\therefore B = 0$

   and at $x = \dfrac{l}{2}$, $\dfrac{dy}{dx} = 0$

$$\therefore 0 = A\,\alpha\cos\left(\frac{\alpha l}{2}\right) - \frac{W}{2P}$$

giving

$$A = \frac{W}{2P}\,\frac{1}{\alpha\cos\left(\dfrac{\alpha l}{2}\right)}$$

Therefore, the solution is $y = \dfrac{W}{2P}\left(\dfrac{\sin(\alpha x)}{\alpha\cos\left(\dfrac{\alpha l}{2}\right)} - x\right)$

The maximum deflection, y_{max}, occurs at $x = \dfrac{l}{2}$, that is,

$$y_{max} = \frac{W}{2P}\left(\frac{\sin\left(\dfrac{\alpha l}{2}\right)}{\alpha\cos\left(\dfrac{\alpha l}{2}\right)} - \frac{l}{2}\right)$$

$$ie\quad y_{max} = \frac{W}{2P\alpha}\left(\tan\left(\frac{\alpha l}{2}\right) - \frac{\alpha l}{2}\right)$$

The maximum absolute bending moment occurs at the mid-span, that is,

$$M_{max} = \left|-\frac{Wl}{4} - Py_{max}\right| = \frac{W}{2\alpha}\tan\frac{\alpha l}{2}$$

These expressions for y, y_{max} and M_{max} become infinite when $\dfrac{\alpha l}{2}$ is a multiple of $\dfrac{\pi}{2}$, that is, when $\dfrac{\alpha l}{2} = \sqrt{\dfrac{P}{EI}}\,\dfrac{l}{2} = \dfrac{n\pi}{2}$ where n is an integer.

   When $n = 1$,

$$P_c = \frac{\pi^2 EI}{l^2}$$

that is, the critical buckling load is unaffected by the presence of the transverse load, W.

The expression for y_{max} can be rewritten as

$$y_{max} = \frac{W}{2EI\alpha^3}\left(\tan\left(\frac{\alpha l}{2}\right) - \frac{\alpha l}{2}\right)$$

$$y_{max} = \left[\frac{Wl^3}{48EI}\right]\left\{\frac{3\left(\tan\left(\frac{\alpha l}{2}\right) - \frac{\alpha l}{2}\right)}{\left(\frac{\alpha l}{2}\right)^3}\right\}$$

Since $\dfrac{Wl^3}{48EI}$ would be the deflection without P, it is clear that the presence of the compressive force has magnified the deflection by a factor $3\left(\tan\left(\dfrac{\alpha l}{2}\right) - \left(\dfrac{\alpha l}{2}\right)\right)/\left(\dfrac{\alpha l}{2}\right)^3$ which is approximately equal to 1.33 for $\dfrac{\alpha l}{2} = \dfrac{\pi}{4}$, for example. In contrast, a tensile axial load would have the effect of reducing the transverse deflections.

It also follows that the bending moments in slender members can be substantially increased by the presence of compressive axial forces.

3.6.5.1   Some Important Notes

1. In contrast to the classical cases considered here, actual compression members are seldom truly pinned or completely fixed against rotation at the ends. Because of this uncertainty regarding the fixity of the ends, struts or columns are often assumed to be pin-ended. This procedure is conservative.
2. The above equations are not applicable in the inelastic range, that is for $\sigma > \sigma_y$, and must be modified.
3. The critical load formulae for struts or columns are remarkable in that they do not contain any strength properties of the material and yet they determine the load-carrying capacity of the member. The only material property required is the elastic modulus, E, which is a measure of the stiffness of the strut.

3.6.6   Compressive Loading of Rods

If we assume that the rod loading is perfectly axial, and the material can be represented by an EPP stress-strain curve, as in Figure 3.61, then: the plastic collapse failure would occur in compression if $\sigma\left(=\dfrac{-P}{A}\right)$ reaches $-\sigma_y$ before the buckling load is reached. Now, $P_c = \dfrac{\pi^2 EI}{l^2}$ and defining the second moment of area, I, as $I = Ak^2$ where k is the radius of gyration, gives $P_c = \dfrac{\pi^2 EAk^2}{l^2}$ and

$$\sigma = \frac{P_c}{A} = \frac{\pi^2 Ek^2}{l^2} = \frac{\pi^2 E}{(l/k)^2}$$ l/k is the slenderness ratio.

Therefore, buckling will occur if $\sigma = \dfrac{\pi^2 E}{(l/k)^2}$, whereas plastic collapse will occur if $\sigma = \sigma_y$.

This can be represented diagrammatically as shown in Figure 3.64.

If the load is now assumed to be applied eccentrically as shown in Figure 3.63,

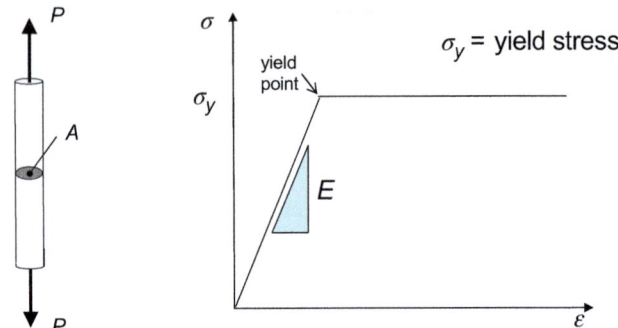

FIGURE 3.61   Tensile test specimen and elastic perfectly plastic stress-strain behaviour.

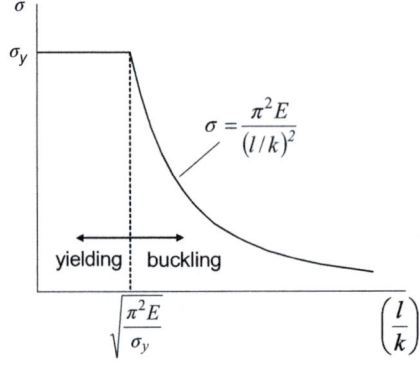

FIGURE 3.62   Plot of σ versus l/k indicating the buckling and plastic collapse regions.

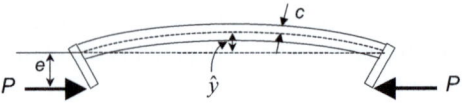

FIGURE 3.63   Eccentrically loaded bar.

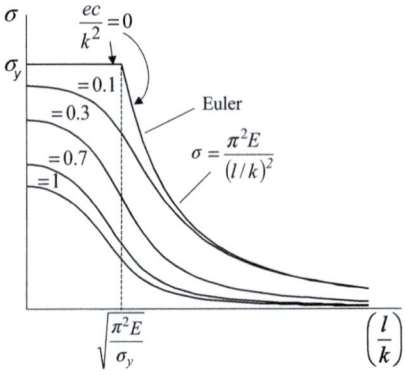

FIGURE 3.64   Effect of ec/k^2 on the buckling characteristics of an eccentrically loaded strut.

We find that:

$$\hat{\sigma}_c = -\left(\frac{P}{A} + \frac{P(e+\hat{y})c}{Ak^2}\right)$$

$$i.e.\,\hat{\sigma}_c = -\frac{P}{A}\left(1 + \frac{(e+\hat{y})c}{k^2}\right)$$

Recall $\hat{y} = e\left(\sec\left(\frac{\alpha l}{2}\right) - 1\right)$ for an eccentrically-loaded strut

$$\therefore \hat{\sigma}_c = -\frac{P}{A}\left(1 + \frac{ec}{k^2}\left(1 + \sec\left(\frac{\alpha l}{2}\right) - 1\right)\right)$$

$$ie\,\hat{\sigma}_c = -\frac{P}{A}\left(1 + \frac{ec}{k^2}\sec\left(\frac{\alpha l}{2}\right)\right)$$

$$giving -\frac{P}{A} = \frac{\hat{\sigma}_c}{1 + \frac{ec}{k^2}\sec\left(\frac{\alpha l}{2}\right)}$$

It should be noted that the latter equation is a non-linear equation in P since α also contains P. The plotted results were obtained by substituting in the material yield stress for $\hat{\sigma}_c$ and then solving for an allowable load P using known values of eccentricity e, thus:

$$\sigma_y = -\frac{P}{A}\left(1 + \frac{ec}{k^2}\sec\left(\sqrt{\frac{P}{EI}}\frac{l}{2}\right)\right)$$

It should also be noted that the buckling load for an eccentrically loaded strut is the same as that for a hinged-hinged ideal strut. Therefore, the plotted results show that yield failure will always occur first except for small eccentricities at high slenderness ratios where the yield curve approaches the Euler buckling curve.

EXAMPLE PROBLEM

A hoist is constructed using two wooden bars ($E = 12.4$ GPa) as shown in the figure. The allowable normal stress is 13.8 MPa. Determine the maximum permissible weight W that can be lifted using the hoist for the two cases: (a) $L = 1.2$ m; (b) $L = 1.5$ m.

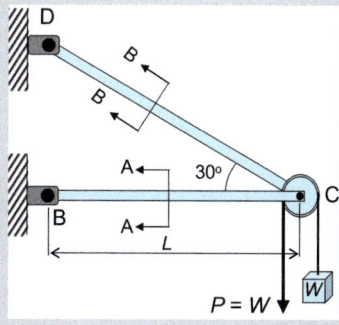

Solution

By inspection, we see that member BC will be in compression. Thus, to determine the maximum value of W, we need to consider the buckling failure of member BC and the strength failure of both members due to the axial stress exceeding the given allowable stress. The internal forces in members BC and CD can be found in terms of W. The axial stresses in the member found are then compared with the given allowable values to determine one set of limits on W. To determine the critical buckling load, the smaller second moment of area about the perpendicular axes through the section centroid should be used and the upper limit on W to prevent buckling failure can be found. The maximum value of W that satisfies the strength and buckling criteria can now be determined.

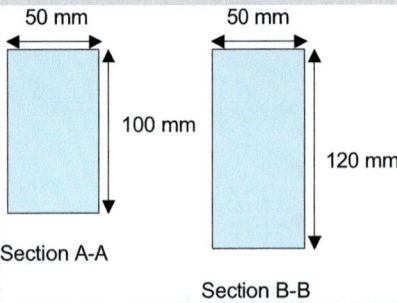

The free-body diagram for the pulley is shown below, with F_{BC} drawn as compressive and F_{CD} as tensile. The internal axial forces are found from

$$\sum F_y = 0 \Rightarrow F_{CD} \sin 30 = 2W \Rightarrow F_{CD} = 4W$$

$$\sum F_x = 0 \Rightarrow F_{BC} = F_{CD} \cos 30 \Rightarrow F_{BC} = 3.464W$$

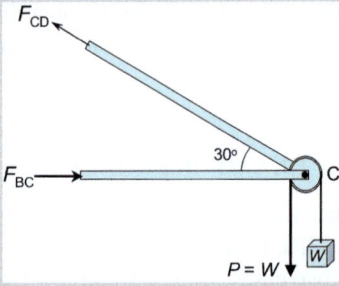

The cross-section areas for the two members are $A_{BC} = 5000$ mm^2 and $A_{CD} = 6000$ mm^2. Thus, the axial stresses in terms of W can be found and these must be less than 13.8 MPa, giving two limits on W:

$$\sigma_{CD} = \frac{F_{CD}}{A_{CD}} = \frac{4W}{6000} \leq 13.8 \text{ MPa or } W \leq 20.7 \text{ kN}$$

$$\sigma_{BC} = \frac{F_{BC}}{A_{BC}} = \frac{3.464W}{5000} \leq 13.8 \text{MPa or } W \leq 17.9 \text{kN}$$

For cross-section A-A, we note that $I_{\min} = \frac{1}{12} \times 100 \times 50^3 = 1041666.7 \text{mm}^4$

Thus, for length (a),

$$P_{crit} = \frac{\pi^2 EI}{L^2} = \frac{\pi^2 \times 12.4 \times 10^3 \times 1041667}{1200^2} = 88529.4 \text{N}$$

Thus, a second limit on F_{BC} is $F_{BC} < 88529$ N, thus $3.464W < 88529$ N or $W < 25.6$ kN. Since this is greater than the value that causes yielding, the yielding value will occur first, so that the maximum permissible value of W for $L = 1.2$ m is 17.9 kN.

Then, for length (b),

$$P_{crit} = \frac{\pi^2 EI}{L^2} = \frac{\pi^2 \times 12.4 \times 10^3 \times 1041667}{1500^2} = 56658.8 \text{N}$$

Thus, a second limit on F_{BC} is $F_{BC} < 56658.8$ N, thus $3.464W < 56658.8$ N or $W < 16.36$ kN. Since this is less than the value that causes yielding, the buckling value becomes limiting, so that the maximum permissible value of W for $L = 1.5$ m is 16.4 kN.

Note: In case (a), the design was governed by material strength whereas in case (b) buckling governed the design. If there were several bars of different lengths and cross-sectional dimensions, it would save significant time to calculate the slenderness ratio that separates long columns from short columns, that is buckling failure regime from the yielding failure regime, respectively. Using $\sigma_{cr} = 13.8$ MPa in $\sigma = \dfrac{\pi^2 E}{(L/k)^2}$ gives $L/k = 94.2$ as the threshold ratio separating long columns from short columns. For case (a), the slenderness ratio is 83.1, so that it is classified as a short column and material strength governs W_{max}. For case (b), the slenderness ratio is 103.9 so that buckling governs W_{max}.

LEARNING SUMMARY

By the end of this section, you will have learnt:

- ✓ How to use Macaulay's Method for determining beam deflection in situations with axial loading;
- ✓ The meanings of and the differences between stable, unstable and neutral equilibrium;
- ✓ How to determine the buckling loads for ideal struts;
- ✓ The effects of eccentric loading, initial curvature and transverse loading on the buckling loads;
- ✓ How to include the interaction of yield behaviour with buckling and how to represent this interaction graphically.

3.7 SHEAR STRESSES IN BEAMS

Whereas bending stresses in beams arising from transverse loading are important, **transverse** (i.e. through-thickness) **shear stresses** due to these same loads also exist. For long slender beams, the shear stresses can generally be neglected, and it is only necessary to do a bending calculation for the beam. However, as the beam span-to-depth ratio reduces, that is if the beam is shorter and thicker, shear stresses become more important and should be calculated in any design evaluation. This can be particularly important for laminated beams, for example, plywood or composite beams, where the transverse shear can cause failure between individual layers (plies) making up the beam. In this section, we will derive a general formula for calculating the shear stress distribution through the thickness of a beam. We will then introduce the concept of **shear centre** which is the point through which the resultant of the shear stresses always act. The shear centre becomes important for beam sections which have low torsional rigidity, that is can twist easily, such as thin-walled sections. For such beams, if the resultant of the applied transverse loads do not act through the shear centre, they can cause twisting of the beam, that is there is a bending–twisting interaction in the system. The designer should avoid this situation if possible or, at least, evaluate the degree of twisting which might take place.

3.7.1 Transverse Shear Stress Distribution

The through-thickness shear force in a beam is the integral of the shear stresses over the cross-section. In this section, we will determine an expression for the shear stress distribution (transverse i.e. through-thickness) at a section as a function of the shear force at that position. Consider an element of beam length, δx, as shown in Figure 3.65. The bending moment at x, section AC, is M and at $x + \delta x$, section BD, is $M + \delta M$. The direct bending stresses on AC are,

$$\sigma_{AC} = \frac{My}{I}$$

Where

y = distance from the neutral surface
I = second moment of area of the section

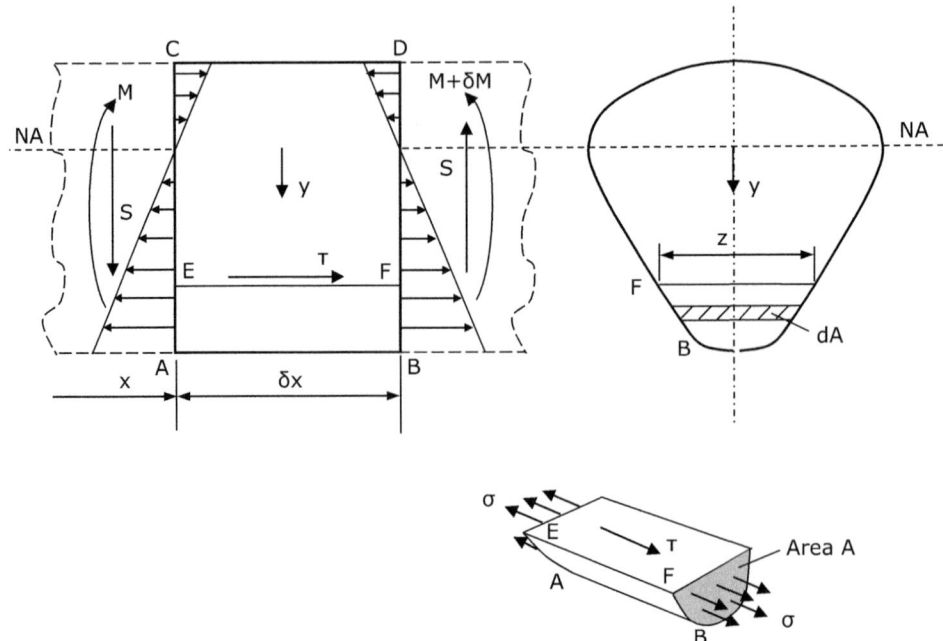

FIGURE 3.65 Element of beam length subjected to a varying bending moment.

and on BD, the bending stresses are,

$$\sigma_{BD} = \frac{(M+\delta M)y}{I}$$

Thus, when the bending moment varies along the length of the beam on an element such as ABEF, also shown in Figure 3.65, there is a net axial force due to a change in the bending stresses. The force on the face FB is the integral of the bending stresses over the area FB,

$$F_{FB} = \int_A \frac{(M+\delta M)}{I} y dA$$

Similarly, the force on the face of EA is,

$$F_{EA} = \int_A \frac{M}{I} y dA$$

The net force to the right acting on the element ABEF is the difference in these, that is,

$$\text{Net Force}\left(\text{bending}\right) = \int_A \frac{\delta M}{I} y dA \tag{7.1}$$

In order to maintain the equilibrium of ABEF, shear stresses must act on the plane EF, of average value τ, as shown in Figure 3.66. These shear stresses are complementary to the <u>transverse</u> shear stresses. For positive transverse shear stresses, as shown, the complementary shear stresses act in the positive x direction. The net force to the right due to these complementary shear stresses is,

$$\text{Net Force}\left(\text{shear}\right) = \tau.z.\delta x$$

where z is the width of the section at that depth

Now, the equilibrium of ABEF requires the net force due to bending to balance the net force due to the complementary shear. Thus,

$$\tau\, z\, \delta x + \int_A \frac{\delta M}{I} y dA = 0$$

$$\therefore \tau = -\frac{1}{Iz}\frac{\delta M}{\delta x}\int_A y dA$$

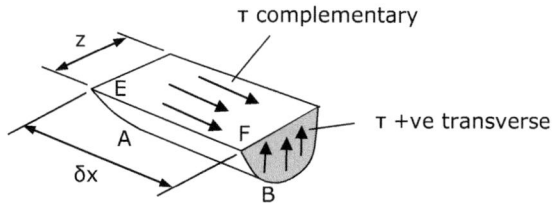

FIGURE 3.66 Equilibrium of stresses on ABEF.

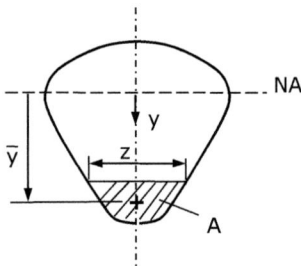

FIGURE 3.67 Definition of shear stress parameters in a cross-section.

But, in the limit, $\dfrac{\lim}{\delta x \to 0} \ \dfrac{\delta M}{\delta x} = \dfrac{dM}{dx} = -S$

where S = shear force at the section

$$\therefore \tau = \frac{S}{Iz}\int_A y\,dA \tag{7.2}$$

This is the general expression for transverse shear stress at any position y through the thickness. The integral can also be written in a discrete form as follows:

$$\tau = \frac{S}{Iz}A\,\bar{y} \tag{7.3}$$

where A is the area of the part of the cross-section outside the position at which τ is determined and $\bar{y}$ is the distance of the centroid of this area from the neutral axis as shown in Figure 3.67.

3.7.2 DETERMINATION OF SHEAR STRESS DISTRIBUTION FOR DIFFERENT CROSS-SECTIONAL SHAPES

3.7.2.1 Rectangular Section

Referring to Figure 3.68 and using the discrete form for shear stress distribution, that is Equation (7.3), we have,

$$A=\left(\frac{d}{2}-y\right)b \text{ and } \bar{y}=\left(\frac{d}{2}+y\right)\frac{1}{2}$$

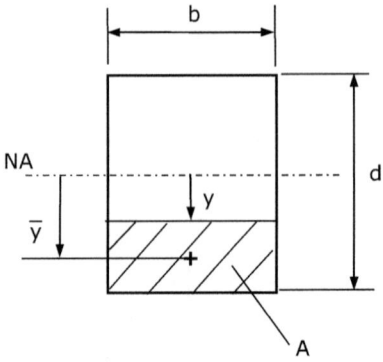

FIGURE 3.68 Rectangular cross-section.

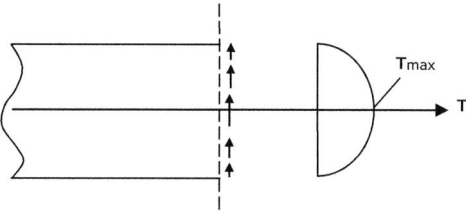

FIGURE 3.69   Shear stress distribution in a rectangular section.

$$\tau = \frac{S}{\left(bd^3/12\right).b}\left(\frac{d}{2}-y\right)b\left(\frac{d}{2}+y\right)\frac{1}{2}$$

$$\therefore \tau = \frac{6S}{bd^3}\left[\left(\frac{d}{2}\right)^2 - y^2\right] \tag{7.4}$$

Note the parabolic distribution of shear stress (i.e. τ varies with y^2), illustrated in Figure 3.69. Also, at the top and bottom of the section, where $y = \pm d/2$, Equation (7.4) gives $\tau = 0$. As expected, there is no complementary shear stress on the top and bottom free surfaces; therefore, the transverse shear stress is also zero at these positions.

At the neutral axis, that is where $y = 0$, Equation (7.4) gives

$$\tau = \frac{6S}{bd^3}\frac{d^2}{4} = 1.5\frac{S}{bd}$$

This is the position of maximum shear stress whose magnitude is 1.5× the average shear stress S/bd.

Note also that, in this analysis, the shear stress does not vary across the width of the section.

3.7.2.2   Circular Section

Figure 3.70 shows a solid circular cross-section of a beam. To calculate the transverse shear stress distribution in this section, we use the integral form of the shear equation, that is Equation (7.2).

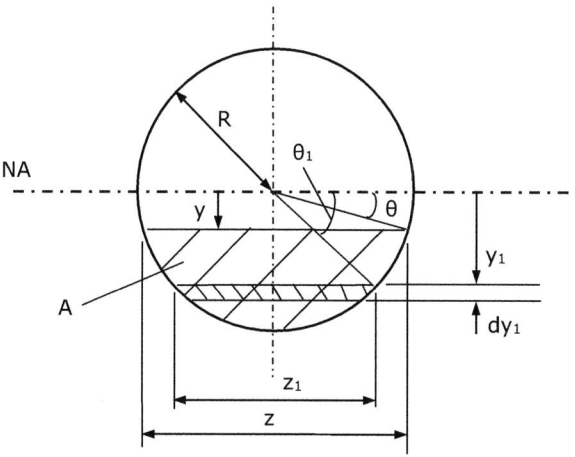

FIGURE 3.70   Solid circular cross-section.

However, because of the circular shape, it is convenient to change the variables y and z in this equation to polar variables, R and θ. Referring to Figure 3.70 we have,

$$y_1 = R \sin \theta_1$$

$$dy_1 = R \cos \theta_1 d\theta_1$$

$$z_1 = 2R \cos \theta_1$$

$$z = 2R \cos \theta$$

and the second moment of area, $I = \pi D^4/64 = \pi R^4/4$

The shear equation now becomes

$$\tau = \frac{S}{Iz} \int y dA = \frac{S.4}{\pi R^4.2R\cos\theta} \int_y^R y_1 z_1 dy_1$$

$$= \frac{S.4}{\pi R^4.2R\cos\theta} \int_0^{\pi/2} R\sin\theta_1.2R\cos\theta_1.R\cos\theta_1 d\theta_1$$

$$= \frac{4SR^3}{\pi R^5 \cos\theta} \int_0^{\pi/2} \cos^2\theta_1.\sin\theta_1 d\theta_1$$

$$= \frac{4S}{\pi R^2 \cos\theta} \left[\frac{-\cos^3\theta_1}{3} \right]_\theta^{\pi/2}$$

$$\therefore \tau = \frac{4S}{3\pi R^2} \cos^2\theta$$

But $\cos^2\theta = 1 - \sin^2\theta = 1 - \left(\dfrac{y}{R}\right)^2$

$$\therefore \tau = \frac{4S}{3\pi R^2}\left[1 - \left(\frac{y}{R}\right)^2\right] \tag{7.5}$$

Again, a parabolic distribution and the maximum value of τ, at the neutral axis, when $y = 0$ is,

$$\therefore \tau_{max}\left(at\ y = 0\right) = \frac{4S}{3\pi R^2} = \frac{4}{3}\tau_{average}$$

In this case, τ must vary across the width of the section. As can be seen in Figure 3.71, at the free surface the shear stress must be zero. Therefore, the complementary shear on the cross-section, normal to the boundary, is also zero. Thus, shear must be tangential to the boundary as drawn.

3.7.2.3 I-Section

To determine the transverse shear stress distribution in an I-section, we need to consider the web and flange areas separately.

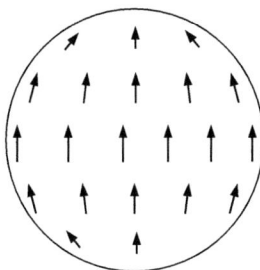

FIGURE 3.71 Shear stress distribution in a solid circular section.

3.7.2.3.1 Transverse Shear in the Web

Figure 3.72(a) shows an I-section and the position y where we wish to determine the shear stress. Using the discrete form of the shear stress equation we have

$$\tau = \frac{S}{Iz} A\bar{y} = \frac{S}{Iz}\left[A_1\bar{y}_1 + A_2\bar{y}_2\right]$$

$$= \frac{S}{Ib}\left[\left(\frac{d}{2} - y\right)\cdot b \cdot \frac{1}{2} \cdot \left(\frac{d}{2} + y\right) + B\left(\frac{D}{2} - \frac{d}{2}\right) \cdot \frac{1}{2} \cdot \left(\frac{D}{2} + \frac{d}{2}\right)\right]$$

$$= \frac{S}{Ib}\left[\frac{b}{2} \cdot \left(\frac{d^2}{4} - y^2\right) + \frac{B}{2} \cdot \left(\frac{D^2}{4} - \frac{d^2}{4}\right)\right]$$

and $I = \dfrac{BD^3}{12} - \dfrac{(B-b)d^3}{12}$

The maximum τ at $y = 0$: $\tau_{max} = \dfrac{S}{Ib}\left[\dfrac{BD^2}{8} - \dfrac{(B-b)d^2}{8}\right]$

At the bottom and top of the web, where $y = \pm d/2 : \tau = \dfrac{S}{Ib} \cdot \dfrac{B}{8} \cdot \left(D^2 - d^2\right)$ \hfill (7.6)

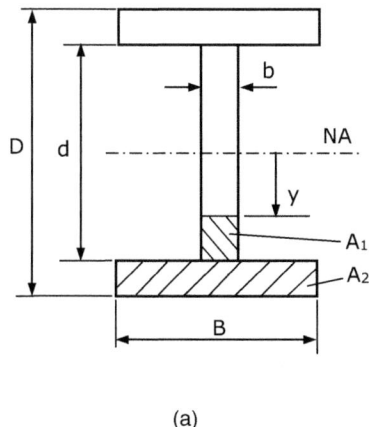

(a)

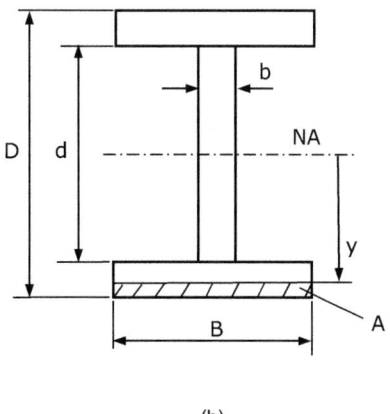

(b)

FIGURE 3.72 I-section.

3.7.2.3.2   Transverse Shear in the Flange

Figure 3.72(b) shows the position y where we wish to determine the shear stress in the flange. Again, using the discrete form of the shear stress equation, we have

$$\tau = \frac{S}{Iz}A\bar{y}$$

$$= \frac{S}{Ib}\left[B\cdot\left(\frac{D}{2}-y\right)\cdot\frac{1}{2}\cdot\left(\frac{D}{2}+y\right)\right]$$

$$= \frac{S}{2I}\left(\frac{D^2}{4}-y^2\right)$$

At $y = D/2 \tau = 0$, as expected that is zero shear complementary to the free surface

At $y = d/2 \; \tau = \dfrac{S}{8I}\left(D^2 - d^2\right)$. Comparing this expression with Equation (7.6), there is a step change in τ from the web to the flange of magnitude B/b, that is the ratio of the flange width to the web width.

Figure 3.73 shows the transverse shear stress distribution down the centre line of the section and illustrates the step change discussed above. The shear in the flanges is small compared to the web, and the shear stress in the web is approximately uniform with vertical position. Because of the small shear in the flanges, the average shear stress in the web is $\approx S/bd$, that is, the shear force divided by the area of the web.

The above distribution only applies down the centre line of the web. The shear stresses in the flanges are small and non-uniform across the width. This must be the case as they must be zero at the top and bottom surfaces (i.e. free surfaces) of the flanges. There are, however, more significant shear stresses in the flanges which act parallel to the flanges, that is, horizontally. These can be determined by a similar analysis as follows:

3.7.2.3.3   Horizontal Shear in the Flange

Figure 3.74 shows a small length, δx, of the I beam over which the bending moment changes from M to $M + \delta M$. To determine the hidden horizontal shear stress, τ, at distance a from the edge of the flange, the equilibrium of an element of the flange is considered. The equilibrium of stresses acting on the element gives,

$$\int_A \frac{(M+\delta M)}{I}y\,dA - \int_A \frac{M}{I}y\,dA + \tau z\delta x = 0$$

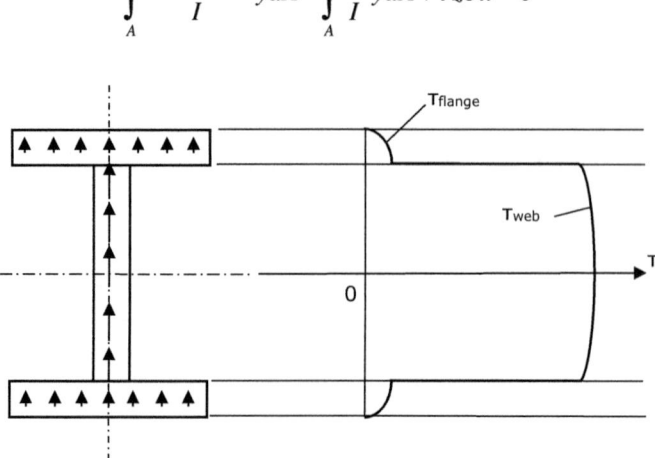

FIGURE 3.73   Transverse shear stress distribution in an I-section.

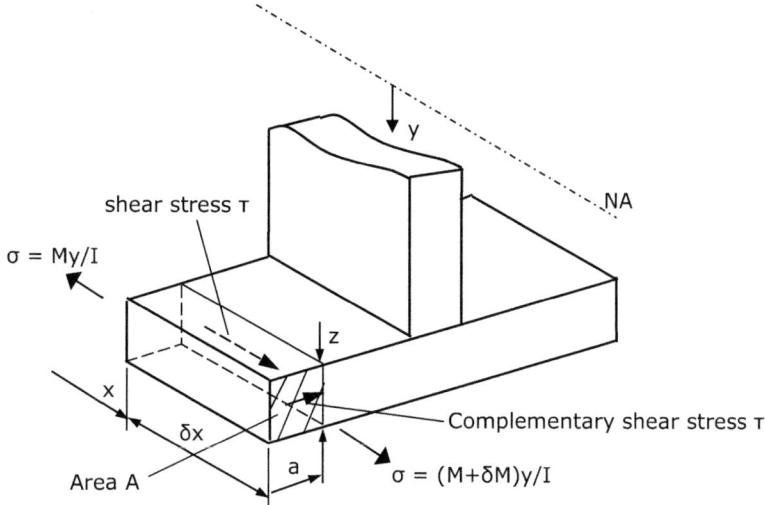

FIGURE 3.74   Determining horizontal shear in the flange of an I-section.

$$\therefore \tau = -\frac{1}{Iz}\frac{\delta M}{\delta x}\int_A ydA$$

In the limit as $\delta x \to 0$, $\dfrac{\delta M}{\delta x} = \dfrac{dM}{dx} = -S$

$$\therefore \tau = -\frac{S}{Iz}\int_A ydA = \frac{S}{Iz}A\overline{y}$$

This is the same shear equation as before except the interpretation of the quantities A, $\overline{y}$ and z is different as shown in Figure 3.74. At a distance a from the edge of the flange, the horizontal shear stress is given by

$$\therefore \tau = \frac{S}{Iz}\cdot(az)\cdot\frac{1}{2}\left(\frac{D}{2}+\frac{d}{2}\right)$$

$$= \frac{Sa}{4I}(D+d)$$

τ therefore varies linearly with a from zero at the flange edge to a maximum value at the flange centre ($a = B/2$),

$$\tau_{max} = \frac{SB}{8I}(D+d)$$

T is also parallel to the flange, that is, horizontal.

We can now draw the dominant shear stresses in both the flange and the web. Figure 3.75 shows the distribution of these horizontal and vertical (transverse) shear stresses. The critical stress position is likely to be at the joint of the web and flange where both the shear and bending stresses are high.

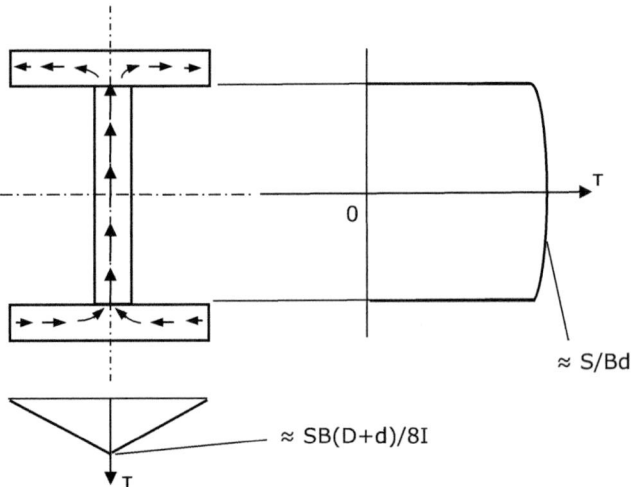

FIGURE 3.75 Shear stress distribution in the flange (horizontal) and the web (vertical-transverse) of an I-section.

3.7.3 SHEAR CENTRE

Consider the shear stress distribution in a symmetric, thin-walled channel section bending in the plane of the web as shown in Figure 3.76.

For the flange at distance a from the edge, the horizontal shear stresses are,

$$\tau = \frac{S}{Iz} A\bar{y} = \frac{S}{It}(at) \cdot \left(\frac{d}{2}\right) = \frac{S.d.a}{2I}$$

[analysed as above for the flange in an I-section]

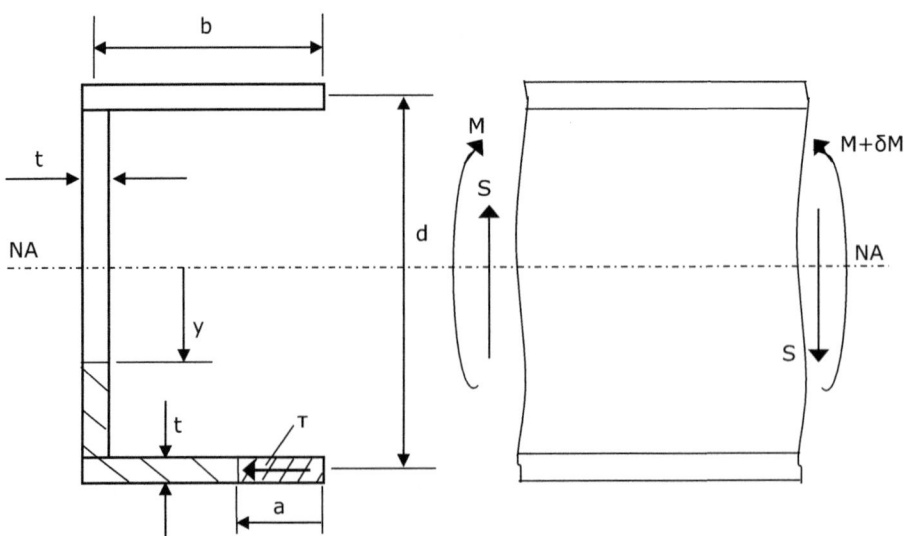

FIGURE 3.76 Determining shear stress distribution in a thin-walled asymmetric channel section.

For the web at distance y from the neutral axis, the transverse shear stresses are

$$\tau = \frac{S}{Iz}A\bar{y} = \frac{S}{It}\left[bt\frac{d}{2}+\left(\frac{d}{2}-y\right)t\left(\frac{d}{2}+y\right)\frac{1}{2}\right]$$

$$= \frac{S}{2I}\left(bd+\left(\frac{d}{2}\right)^2-y^2\right)$$

We can now draw the shear stress distribution in both the flanges and the web, as shown in Figure 3.77(a). For this shear stress distribution, note that the shear stress in the upper flange is in the opposite sense to that in the lower flange that is there is no horizontal resultant. Also, as there are no shear stresses on the free surfaces; the shear stresses act along the walls, that is horizontal in the flanges and vertical in the web.

We can now look at the resultant forces arising from this shear stress distribution as shown in Figure 3.77(b).

The total shear force in the lower flange, S_1, is the integral of the shear stresses in this flange as follows:

$$S_1 = \int_0^b \tau t\,da = \int_0^b \frac{Sda}{2I}t\,da$$

$$= \frac{Sdtb^2}{4I}$$

An equal and opposite shear force acts in the upper flange.

The shear force in the web is approximately S, that is, the total vertical shear load [assuming thin flanges carry negligible vertical shear load].

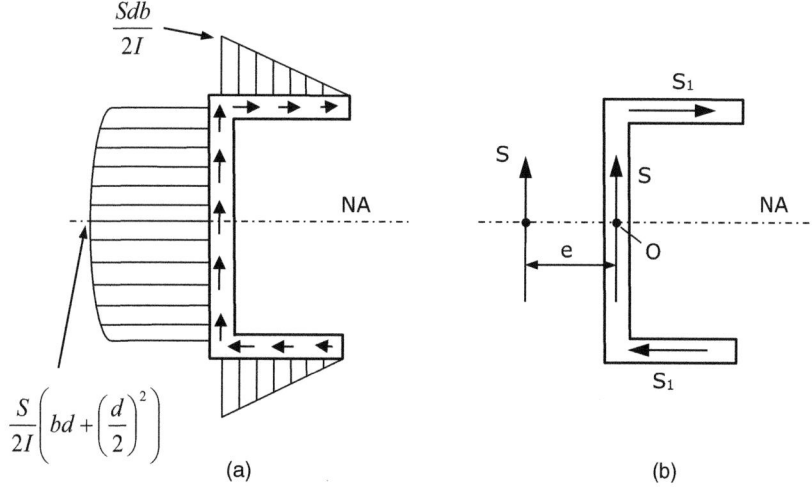

(a) (b)

FIGURE 3.77 Shear stress distribution in the flanges (horizontal) and the web (vertical-transverse) of an asymmetric channel section.

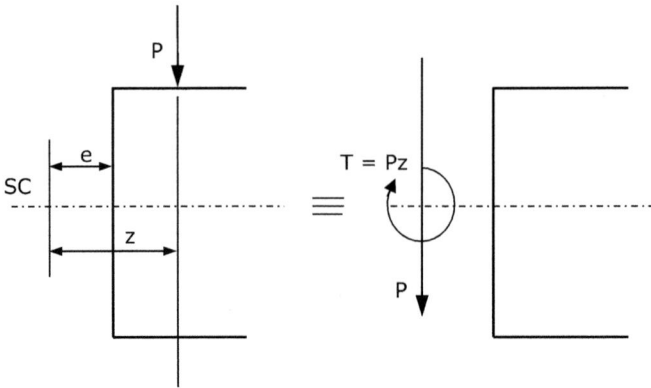

FIGURE 3.78 Location of the shear centre in an asymmetric channel section.

The resultant of all the shear stresses must be the vertical shear force S, and its line of action is distance e outside the web. Now taking moments about O in the web,

$$S.e = 2S_1 \frac{d}{2}$$

$$\therefore e = \frac{S_1 d}{S} = \frac{d^2 t b^2}{4I}$$

It can be shown that the resultant of the shear stresses for a section, for bending in any plane, always acts through one point, the **shear centre**. The shear centre always lies on an axis of symmetry. For sections with two axes of symmetry, the shear centre is at the centroid.

If the applied vertical loads do not act on the plane of the resultant of the shear stresses, that is through the shear centre, then there is a torsional load on the section as shown in Figure 3.78. For arbitrary solid sections, the location of the shear centre is a complicated problem. However, it is <u>not</u> usually important to determine the shear centre for solid sections because such sections usually have considerable torsional rigidity and twist very little due to bending loads. However, for thin-walled open sections, which have low torsional rigidity, the position of the shear centre may be very important.

3.7.4 WORKED EXAMPLES

3.7.4.1 Shear Stresses in a Beam

The section shown in Figure 3.79 is subjected to a vertical shear force, $S = 50$ kN, acting down the vertical centre line, that is the y-axis. The second moment of area of the section, about the x-axis, which passes through the centroid of area, G, is $I_{xx} = 2.31 \times 10^6$ mm^4. G is positioned 14 mm below the flange.

a. Determine the magnitude of the transverse (i.e. vertical) shear stress at positions A, B, G and C on the vertical centre line.
b. Sketch the variation of the transverse shear stress down the vertical centre line

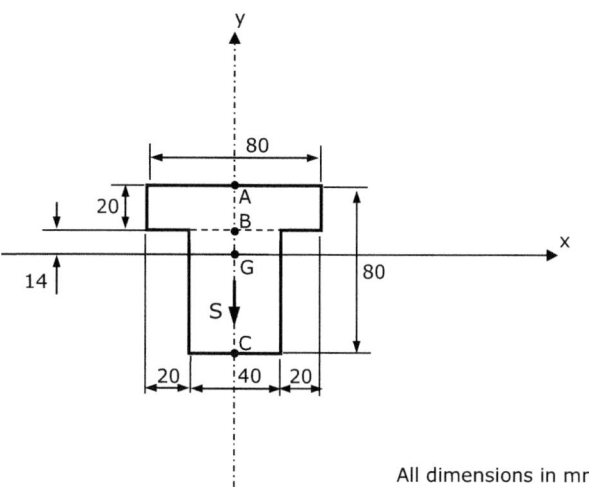

FIGURE 3.79   Worked example: a T-section.

3.7.4.1.1   In the Top Flange

Consider a position in the top flange, vertical distance y from the centroid, G, as shown in Figure 3.80(a). Using the discrete form of the shear formula,

$$\tau = \frac{S}{Iz}A\bar{y} = \frac{S}{2.31x10^6.80}\cdot\left(80.(34-y)\right).\frac{(34+y)}{2}$$

$$= \frac{S}{4.62x10^6}\left(1156-y^2\right) = 0.0108\left(1156-y^2\right)$$

At position A, $y = 34 \therefore \tau = 0$
  At position B, $y = 14 \therefore \tau = 10.4 \text{ Nmm}^{-2}(\text{MPa})$

3.7.4.1.2   In the Lower Section

Consider a position in the lower section, again vertical distance y from the centroid, G, as shown in Figure 3.80(b).

  At position B, there is a step change in the shear stress given by the ratio of the section widths at this point. Thus,

At position B $\tau = \dfrac{10.4.80}{40} = 20.8 \text{ Nmm}^{-2}(\text{MPa})$

At position G, that is the neutral axis, we can use the discrete formula for the shear stress. In this case, to simplify the calculation, the relevant area can be regarded as the area below the neutral axis. Thus,

At position G $\tau = \dfrac{S}{Iz}A\bar{y} = \dfrac{S}{2.31x10^6.40}\cdot(46.40).23$

$$= 22.91 \text{ Nmm}^{-2}(\text{MPa})$$
At position C $\tau = 0$, that is, a free surface

A sketch of the variation of the shear stress down the vertical centre line is given in Figure 3.81.

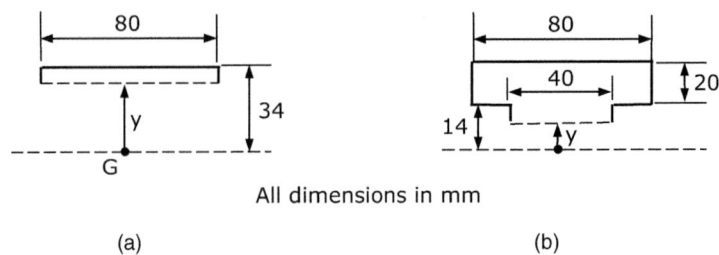

All dimensions in mm

(a)   (b)

FIGURE 3.80   Determining the shear stress distribution in a T-section.

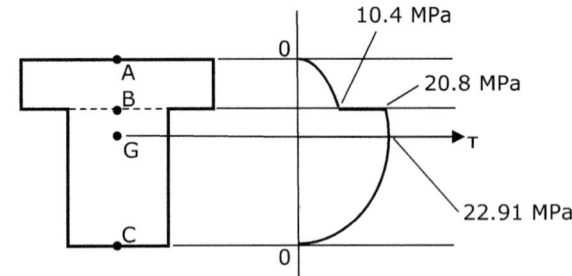

FIGURE 3.81   Transverse (vertical) shear stress distribution in a T-section.

3.7.4.2   Shear Centre

For the thin-walled semicircular cross-section shown in Figure 3.82, determine the position of the shear centre (assume bending about the axis of symmetry X-X).

3.7.4.2.1   Shear Stress Distribution

To solve this problem it is necessary to change from a rectangular coordinate system (x-y) to a polar coordinate system (r-θ). Referring to Figure 3.82 and using the integral form of the shear stress formula, we obtain a general expression for the shear stress distribution parallel to the wall of the section. Thus,

$$\tau = \frac{S}{I z} \int_A y \, dA$$

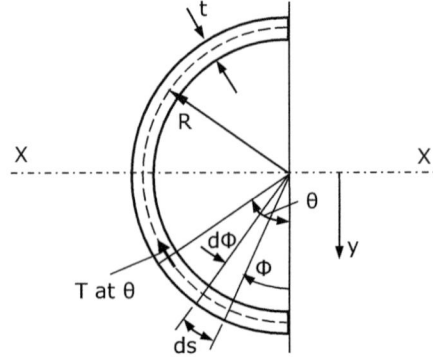

FIGURE 3.82   A thin-walled semicircular cross-section.

with

$$y = R\cos\varphi$$
$$dA = ds.t = Rtd\varphi$$
$$z = t$$

giving

$$\tau = \frac{S}{Iz}\int_0^\theta R\cos\varphi.Rtd\varphi$$

$$= \frac{SR^2}{I}\int_0^\theta \cos\varphi d\varphi$$

$$\therefore \tau = \frac{SR^2 \sin\theta}{I} \tag{7.7}$$

Now,

$$I = \int_A y^2 dA = \int_0^\pi (R\cos\theta)^2.Rd\theta.t$$

$$= \int_0^\pi R^3 t\cos^2\theta d\theta = \int_0^\pi \frac{R^3 t}{2}(1+\cos 2\theta)d\theta$$

$$= \frac{R^3 t}{2}\left[\theta + \frac{\sin 2\theta}{2}\right]_0^\pi$$

$$\therefore I = \frac{\pi R^3 t}{2} \tag{7.8}$$

From Equations (7.7) and (7.8),

$$\tau = \frac{SR^2 \sin\theta.2}{\pi R^3 t}$$

$$\therefore \tau = \frac{2S\sin\theta}{\pi Rt}$$

3.7.4.3   Shear Centre

The twisting moment (torque) associated with the above shear stress distribution for the whole cross-section is found by taking moments about O,

$$\text{Torque} = \int_0^\pi \tau.(Rd\theta)t.R = \frac{R^2 t.2S}{\pi Rt}\int_0^\pi \sin\theta d\theta$$

$$= \frac{2SR}{\pi}\left[-\cos\theta\right]_0^\pi = \frac{4SR}{\pi}$$

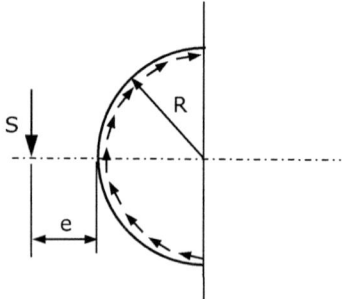

FIGURE 3.83   Location of -walled semi-circular section.

To counteract this twisting moment, as shown in Figure 3.83, the shear force, S, must be applied at the shear centre, a distance e, given by,

$$S(e+R) = \frac{4SR}{\pi}$$

$$\therefore e = \frac{4R}{\pi} - R = \underline{0.273R}$$

LEARNING SUMMARY

By the end of this section, you should have learnt,

✓ That, in addition to longitudinal bending stresses, beams also carry transverse shear stresses arising from the vertical shear loads acting within the beam;
✓ How to derive a general formula, in both integral and discrete form, for evaluating the distribution of shear stresses through a cross-section;
✓ How to determine the distribution of the shear stresses through the thickness in a rectangular, circular and I-section beam;
✓ That, in an I-section, in addition to transverse vertical shear stresses in the flange and web, more dominant horizontal shear stresses also occur in the flange;
✓ That, the resultant of the shear stresses for a section always act through one point, the shear centre;
✓ How to calculate the position of the shear centre;
✓ If the applied loads do not act through the shear centre, then there is a resultant torsional load which can result in twisting of the section if the torsional rigidity is low, for example, in thin-walled sections

---

3.8   THICK CYLINDERS

3.8.1   INTRODUCTION

Thick cylinders differ from thin cylinders in that the variation of stress through the wall thickness is significant in thick cylinders when subjected to internal and/or external pressure whereas for thin cylinders, the variation of stress is negligible. Figure 3.84 includes drawings of thick cylinders with

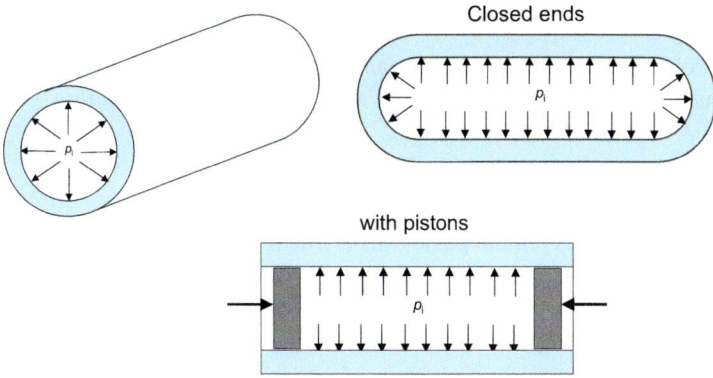

FIGURE 3.84 Thick cylinders subjected to internal pressure.

closed ends and with pistons. For closed-ended, internally pressured cylinders, the axial force on the inside of the end closures produces a distribution of axial stress in the cylinder while for cylinders with pistons the resultant axial force in the cylinder and hence the axial stress also are zero.

3.8.2 ANALYSIS OF THIN CYLINDERS

For an internally pressurised thin cylinder situation, it is reasonable to assume that the variations of the stresses through the wall thickness are negligible. This results in the problem being *statically determinate*, that is expressions for the stresses can be obtained by considering equilibrium alone, as described below (See Figure 3.85).

$$2\sigma_\theta t l = pdl$$

$$\therefore \sigma_\theta = \frac{pd}{2t} = \frac{pR}{t}$$

$$p\frac{\pi d^2}{4} = \sigma_a \pi dt$$

$$\therefore \sigma_a = \frac{pd}{4t} = \frac{pR}{2t}$$

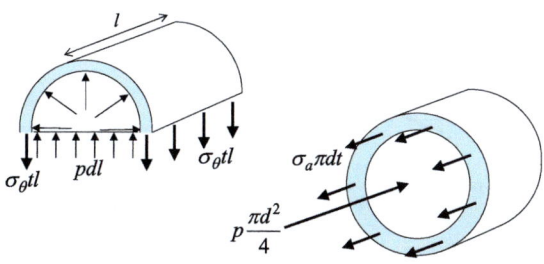

FIGURE 3.85 Free body diagrams used to obtain hoop and axial stresses in thin cylinders.

3.8.3 ANALYSIS OF THICK CYLINDER

Thick cylinder problems are *statically indeterminate*. Therefore, in order to obtain a solution, it is necessary to consider equilibrium, compatibility and material behaviour (stress–strain relationship) (See Figure 3.86).

Assumptions

    i. Plane transverse sections remain plane (this is true remote from the ends).

    ii. Deformations are small.

    iii. The material is linear elastic, homogenous and isotropic.

       a. **Equilibrium**

$$\left(\sigma_r + \frac{d\sigma_r}{dr}\delta r \right)(r + \delta r)\delta\theta\delta z = \sigma_r \left(r\delta\theta\delta z \right) + 2\sigma_\theta \left(\delta r\delta z \right)\sin\left(\frac{\delta\theta}{2} \right)$$

For small $\delta\theta$, $\sin\left(\dfrac{\delta\theta}{2} \right) \approx \dfrac{\delta\theta}{2}$

Therefore,

$$\sigma_r \left(r + \delta r \right)\delta\theta + \frac{d\sigma_r}{dr}\delta r \left(r + \delta r \right)\delta\theta = \sigma_r r\delta\theta + \sigma_\theta \delta r\delta\theta$$

$$r\sigma_r + \sigma_r \delta r + r\frac{d\sigma_r}{dr}\delta r + \frac{d\sigma_r}{dr}\delta r^2 = \sigma_r r + \sigma_\theta \delta r$$

As $\delta r \to 0, \dfrac{d\sigma_r}{dr}\delta r^2 \to 0$

Therefore,

$$\sigma_\theta - \sigma_r = r\frac{d\sigma_r}{dr} \tag{8.1}$$

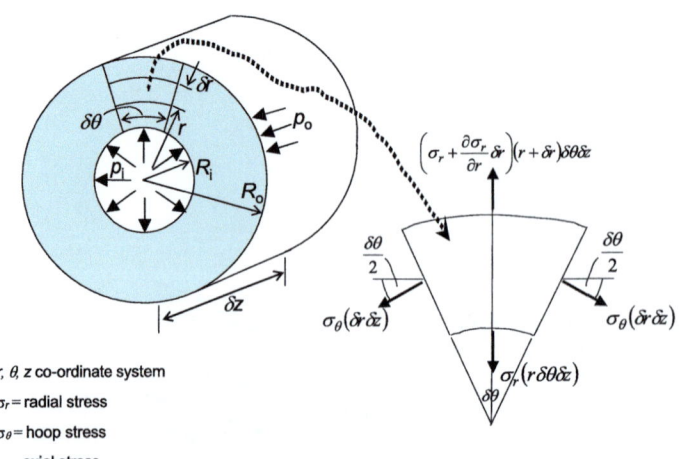

r, θ, z co-ordinate system

σ_r = radial stress

σ_θ = hoop stress

σ_z = axial stress

FIGURE 3.86 A thick cylinder and a free body diagram of an element of material within the cylinder.

b. **Compatibility**

$$\varepsilon = \frac{\text{extension}}{\text{original length}}$$

$$\text{Hoop strain,}\, \varepsilon_\theta = \frac{(r+u)\delta\theta - r\delta\theta}{r\delta\theta} = \frac{u}{r} \tag{8.2}$$

$$\text{Radial strain,}\quad \varepsilon_r = \frac{\left(u + \dfrac{du}{dr}\delta r\right) - u}{\delta r} = \frac{du}{dr} \tag{8.3}$$

$$\text{Axial strain,}\, \varepsilon_z = \text{constant} \tag{8.4}$$

c. **Material behaviour** (stress–strain relationships)
Generalised Hooke's Law (linear elastic and isotropic)

$$\varepsilon_\theta = \frac{1}{E}\left(\sigma_\theta - v\left(\sigma_r + \sigma_z\right)\right) \tag{8.5}$$

$$\varepsilon_r = \frac{1}{E}\left(\sigma_r - v\left(\sigma_\theta + \sigma_z\right)\right) \tag{8.6}$$

$$\varepsilon_z = \frac{1}{E}\left(\sigma_z - v\left(\sigma_r + \sigma_\theta\right)\right) \tag{8.7}$$

Equations (8.1)–(8.7) have seven unknowns, that is u, σ_θ, σ_r, σ_z, ε_θ, ε_r and ε_z which are all functions of r, p_o, p_i, R_o, R_i, v and E.

Substituting $u = r\varepsilon_\theta$ from Eq. (8.2) into (8.3) gives

$$\varepsilon_r = \frac{d}{dr}\left(r\varepsilon_\theta\right)$$

i.e.

$$\varepsilon_r = \varepsilon_\theta + r\frac{d\varepsilon_\theta}{dr} \tag{8.a}$$

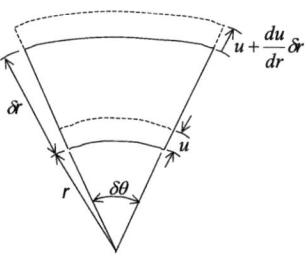

FIGURE 3.87 Initial and deformed shape of an element of material

Using Eq. (8.5) and (8.6) in Eq. (8.a) gives

$$\frac{1}{E}\left(\sigma_r - v\left(\sigma_\theta + \sigma_z\right)\right) = \frac{1}{E}\left(\sigma_\theta - v\left(\sigma_r + \sigma_z\right)\right) + \frac{r}{E}\left(\frac{d\sigma_\theta}{dr} - v\frac{d\sigma_r}{dr} - v\frac{d\sigma_z}{dr}\right)$$

$$ie\ (1+v)\sigma_r = (1+v)\sigma_\theta + r\frac{d\sigma_\theta}{dr} - rv\frac{d\sigma_r}{dr} - rv\frac{d\sigma_z}{dr} \tag{8.b}$$

Using Eq. (8.4), $\dfrac{d\varepsilon_z}{dr} = 0$, then Eq. (8.7) gives:

$$0 = \frac{d\sigma_z}{dr} - v\frac{d\sigma_r}{dr} - v\frac{d\sigma_\theta}{dr}$$

$$\therefore\ \frac{d\sigma_z}{dr} = v\frac{d\sigma_r}{dr} + v\frac{d\sigma_\theta}{dr} \tag{8.c}$$

Substituting (8.c) in (8.b)

$$(1+v)\sigma_r = (1+v)\sigma_\theta + r\frac{d\sigma_\theta}{dr} - rv\frac{d\sigma_r}{dr} - rv^2\frac{d\sigma_r}{dr} - rv^2\frac{d\sigma_\theta}{dr}$$

$$ie\ (1+v)\sigma_r = (1+v)\sigma_\theta + r\left(1-v^2\right)\frac{d\sigma_\theta}{dr} - rv\left(1+v\right)\frac{d\sigma_r}{dr}$$

$$\therefore\ \sigma_\theta - \sigma_r = rv\frac{d\sigma_r}{dr} - r\left(1-v\right)\frac{d\sigma_\theta}{dr} \tag{8.d}$$

Subtracting into Eq. (8.1) from Eq. (8.d) gives

$$rv\frac{d\sigma_r}{dr} - r\left(1-v\right)\frac{d\sigma_\theta}{dr} = r\frac{d\sigma_r}{dr}$$

$$ie\ r\left(1-v\right)\left[\frac{d\sigma_r}{dr} + \frac{d\sigma_\theta}{dr}\right] = 0$$

$$\therefore\ \frac{d}{dr}\left(\sigma_r + \sigma_\theta\right) = 0$$

i.e.

$$\sigma_r + \sigma_\theta = \text{constant} = 2A, \text{say} \tag{8.e}$$

$$\sigma_\theta - \sigma_r = r\frac{d\sigma_r}{dr}$$

$$ie\ 2\sigma_r = 2A - r\frac{d\sigma_r}{dr}$$

$$ie\ r\frac{d\sigma_r}{dr} + 2\sigma_r = 2A$$

$$\therefore \frac{1}{r}\frac{d}{dr}\left(r^2\sigma_r\right) = 2A$$

Hence: $r^2\sigma_r = \dfrac{2Ar^2}{2} - B$

i.e. $\sigma_r = A - \dfrac{B}{r^2}$

and using Eq. (e) leads to $\sigma_\theta = A + \dfrac{B}{r^2}$

Note that, since $\varepsilon_z = $ const and $\sigma_r + \sigma_\theta = $ const, then Eq. (8.7) shows that $\sigma_z = $ const, that is it is independent of r. The value of σ_z can therefore be obtained by considering axial equilibrium.

$$\sigma_r = A - \frac{B}{r^2}$$

and

$$\sigma_\theta = A + \frac{B}{r^2}$$

The constants, A and B, are the so-called Lame's constants, which are the constants of integration, can be obtained from the boundary conditions, that is,

$$\text{at } r = R_i, \sigma_r = -p_i$$

$$\text{at } r = R_o, \sigma_r = -p_o$$

$$\therefore \ -p_i = A - \frac{B}{R_i^2}$$

$$\text{and} -p_o = A - \frac{B}{R_o^2}$$

Hence, A and B can be determined.

For closed-ended cylinders,

$$\pi\left(R_o^2 - R_i^2\right)\sigma_z + \pi R_o^2 p_o = \pi R_i^2 p_i$$

$$ie\,\sigma_z = \frac{R_i^2 p_i - R_o^2 p_o}{\left(R_o^2 - R_i^2\right)}$$

For a solid cylinder, that is $R_i = 0$

$$\sigma_{r(r=R_i=0)} = A - \frac{B}{0^2} = \infty, \text{unless } B = 0$$

Therefore, B must be zero, since the stresses cannot be infinite, and so, for a solid cylinder, the radial and hoop stresses are equal to each other and they are constant,

that is $\sigma_r = \sigma_\theta = A$

Also, since a solid cylinder can only have external pressure, the constant A must be equal to the external pressure.

Displacements are most conveniently obtained by using Eqs. (8.5) and (8.7) together with Eqs. (8.2) and (8.4), that is,

$$\varepsilon_\theta = \frac{u}{r} = \frac{1}{E}\left(\sigma_\theta - v\left(\sigma_r + \sigma_z\right)\right)$$

$$\varepsilon_z = \frac{\Delta l}{l} = \frac{1}{E}\left(\sigma_z - v\left(\sigma_r + \sigma_\theta\right)\right) = \text{constant}$$

where l is the cylinder length, Δl is the increase in cylinder length and u is the radial displacement at radius r.

3.8.4 ANALYSIS OF ROTATING DISCS

Rotating components such as flywheels and turbine discs can be regarded as **thick cylinders with body forces**, as well as possible pressure loads and as such represent an extension of the thick cylinder theory discussed in Section 3.8.3.

a. Equilibrium

At the radius r, $v = \omega r$ and $a_r = \dfrac{v^2}{r} = \omega^2 r$, directed towards the centre of rotation.

$$\sigma_r r \delta\theta \delta z + 2\sigma_\theta \delta r \delta z \left(\frac{\delta\theta}{2}\right) - \left(\sigma_r + \frac{d\sigma_r}{dr}\delta r\right)(r + \delta r)\delta\theta\delta z$$

$$= \left[\rho\left(r + \frac{\delta r}{2}\right)\delta\theta\delta z\delta r\right]\left(r + \frac{\delta r}{2}\right)\omega^2 \sigma_r r + \sigma_\theta \delta r - r\sigma_r$$

$$-\sigma_r \delta r - r\frac{d\sigma_r}{dr}\delta r - \frac{d\sigma_r}{dr}(\delta r)^2 = \rho\left(r + \frac{\delta r}{2}\right)^2 \delta r\omega^2$$

$$\therefore \sigma_\theta - \sigma_r - r\frac{d\sigma_r}{dr} - \frac{d\sigma_r}{dr}\delta r = \rho r^2 \omega^2 + \rho\left(\frac{\delta r}{2}\right)^2 \omega^2 + \rho r \delta r\omega^2$$

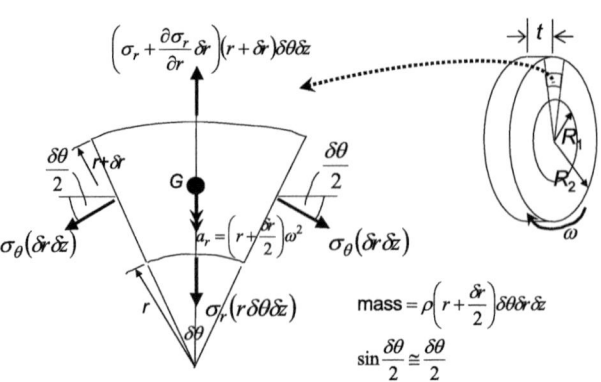

FIGURE 3.88   Free body diagram of an element of material within a disc.

Neglecting small terms, that is, those containing δr and $(\delta r)^2$

$$\sigma_\theta - \sigma_r = r\frac{d\sigma_r}{dr} + \rho r^2 \omega^2 \qquad (8.8)$$

also

$$\sigma_z = 0$$

since it is a disc with no applied axial forces and is not constrained axially along its faces.

3.8.5   COMPATIBILITY AND STRESS–STRAIN RELATIONSHIPS (LINEAR ELASTIC)

$$\varepsilon_\theta = \frac{u}{r} = \frac{1}{E}\left(\sigma_\theta - v\sigma_r\right) \qquad (8.9)$$

$$\varepsilon_r = \frac{du}{dr} = \frac{1}{E}\left(\sigma_r - v\sigma_\theta\right) \qquad (8.10)$$

Substituting for u from Eq. (8.9) into Eq. (8.10) gives:

$$\frac{d}{dr}\left(\frac{r}{E}\left(\sigma_\theta - v\sigma_r\right)\right) = \frac{1}{E}\left(\sigma_r - v\sigma_\theta\right)$$

$$ie,\ \sigma_\theta - v\sigma_r + r\left(\frac{d\sigma_\theta}{dr} - v\frac{d\sigma_r}{dr}\right) = \sigma_r - v\sigma_\theta$$

$$\therefore \left(\sigma_\theta - \sigma_r\right)\left(1+v\right) + r\left(\frac{d\sigma_\theta}{dr} - v\frac{d\sigma_r}{dr}\right) = 0 \qquad (8.f)$$

Substituting for $\sigma_\theta - \sigma_r$ from Eq. (8.8) into Eq. (8.f) gives:

$$\left(r\frac{d\sigma_r}{dr} + \rho r^2 \omega^2\right)\left(1+v\right) + r\frac{d\sigma_\theta}{dr} - rv\frac{d\sigma_r}{dr} = 0$$

$$\therefore r\frac{d\sigma_r}{dr} + rv\frac{d\sigma_r}{dr} + \left(1+v\right)\rho r^2 \omega^2 + r\frac{d\sigma_\theta}{dr} - rv\frac{d\sigma_r}{dr} = 0$$

$$ie, \frac{d}{dr}\left(\sigma_\theta + \sigma_r\right) = -\left(1+v\right)\rho\omega^2 r$$

$$\therefore\ \ \sigma_\theta + \sigma_r = -\left(1+v\right)\rho\omega^2 \frac{r^2}{2} + 2A \qquad (8.g)$$

Subtracting Eq. (8.8) from Eq. (8.g) gives:

$$2\sigma_r + r\frac{d\sigma_r}{dr} = -\left(1+v\right)\rho\omega^2 \frac{r^2}{2} - \rho\omega^2 r^2 + 2A$$

$$\frac{1}{r}\frac{d}{dr}\left(r^2\sigma_r\right) = -\frac{\rho\omega^2 r^2\left(3+\upsilon\right)}{2} + 2A$$

$$r^2\sigma_r = -\rho\omega^2\left(3+v\right)\frac{r^4}{8} + Ar^2 - B$$

where B is a constant of integration. Therefore,

$$\sigma_r = A - \frac{B}{r^2} - \frac{\rho\omega^2\left(3+v\right)}{8}r^2$$

and from Eq. (8.g):

$$\sigma_\theta = A + \frac{B}{r^2} + \frac{\rho\omega^2\left(3+v\right)}{8}r^2 - \frac{\rho\omega^2\left(1+v\right)}{2}r^2$$

$$\sigma_\theta = A + \frac{B}{r^2} - \frac{\rho\omega^2\left(1+3v\right)}{8}r^2$$

EXAMPLE 3.8.1   THICK CYLINDER WITH PISTONS

A cylinder with a 50 m bore and 100 mm OD is subjected to an internal pressure of 400 bar. The end loads are supported by pistons which seal without restraint. Determine the distributions of stress across the cylinder wall.

Solution

$$P = 400 \text{ bar} = 400\times100 \text{ KPa}$$
$$= 40\times1000 \text{ kPa}$$
$$= 40 \text{ N/mm}^2$$

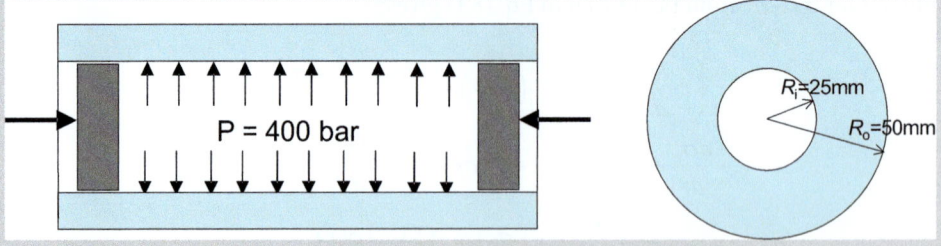

Since there is no axial load on the cylinder, then => $\sigma_z = 0$.
For a thick cylinder,

$$\sigma_r = A - \frac{B}{r^2}$$

and $\sigma_\theta = A + \dfrac{B}{r^2}$

At $r = 25$mm, $\sigma_r = -40$ N/mm^2

$$\therefore -40 = A - \frac{B}{625} \tag{8.10}$$

At $r = 50\,\text{mm}$, $\sigma_r = 0$

$$0 = A - \frac{B}{2500} \tag{8.11}$$

Eliminating A from Equation (8.h) gives:

$$40 = B\left(\frac{1}{625} - \frac{1}{2500}\right)$$

$$= B\left(\frac{4-1}{2500}\right)$$

$$\therefore B = \frac{40 \times 2500}{3}$$

Substituting for B into Equation (8.i) gives:

$$-40 = A - \frac{40 \times 2500}{3 \times 625}$$

$$\therefore A = \frac{40}{3}$$

Hence,

$$\sigma_\theta = \frac{40}{3} + \frac{40 \times 2500}{3r^2} = \frac{40}{3}\left(1 + \frac{2500}{r^2}\right)$$

and $\sigma_r = \dfrac{40}{3} - \dfrac{40 \times 2500}{3r^2} = \dfrac{40}{3}\left(1 - \dfrac{2500}{r^2}\right)$

At $r = 25\,\text{mm}$, $\sigma_\theta = \dfrac{40}{3} \times 5\,\text{N}/\text{mm}^2 = 66.7\,\text{N}/\text{mm}^2$ and $\sigma_r = \dfrac{40}{3} \times (-3)\,\text{N}/\text{mm}^2 = -40\,\text{N}/\text{mm}^2$

At $r = 50\,\text{mm}$, $\sigma_\theta = \dfrac{40}{3} \times 2\,\text{N}/\text{mm}^2 = 26.7\,\text{N}/\text{mm}^2$ and $\sigma_r = 0$

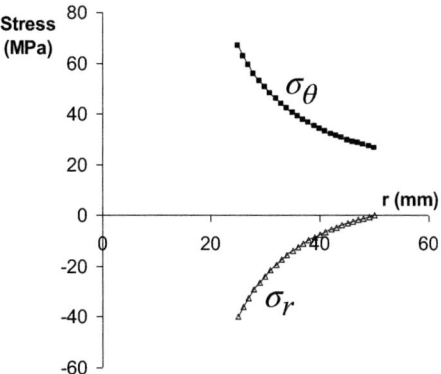

EXAMPLE 3.8.2   SHRINK/INTERFERENCE FIT

A pair of mild steel cylinders ($E = 200$ GPa) of equal length have the following dimensions:

1. 40 mm bore and 80.06 mm outside diameter
2. 80 mm bore and 120 mm outside diameter

that is, there is a diametral interference of 0.06 mm. The larger cylinder is heated, placed around and allowed to shrink onto the smaller cylinder. Calculate the stresses after assembly.

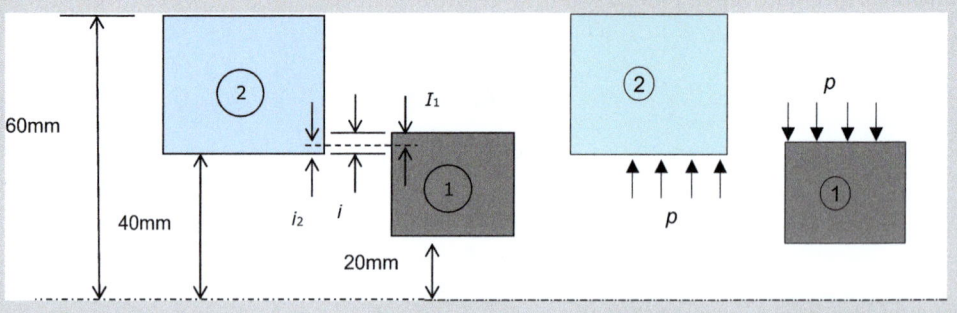

Conditions

    i. After assembly, the radial interference pressure, p, will be the same on both cylinders, that is, Cylinder 1 will have an external pressure, p, and Cylinder 2 will have an internal pressure, p, as indicated in the above figure.

   ii. The decrease in the outside radius of Cylinder 1, i_1, plus the increase in the inside radius of Cylinder 2, i_2, will be equal to the radial interference, that is $i = i_1 + i_2$.

  iii. axial stresses are assumed to be zero (or negligible).

For Cylinder (1):

$$\sigma_r = A_1 - \frac{B_1}{r^2}$$

and $\sigma_\theta = A_1 + \dfrac{B_1}{r^2}$

at $r = 20$ mm, $\sigma_r = 0$,

$$\therefore B_1 = 400 A_1$$

at $r = 40$ mm (no significant difference with 40.03 mm), $\sigma_r = -p$

$$\therefore -p = A_1 - \frac{20^2}{40^2} A_1 = A_1 - \frac{400}{1600} A_1$$

$$ie\ A_1 = -\frac{4}{3} p$$

and

$$B_1 = -\frac{1600}{3} p$$

Thus,

$$\left(\sigma_r\right)_1 = -\frac{4p}{3}\left(1-\frac{400}{r^2}\right)$$

and

$$\left(\sigma_\theta\right)_2 = -\frac{4p}{5}\left(1+\frac{3600}{r^2}\right)$$

$$\varepsilon_\theta = \frac{u}{r} = \frac{1}{E}\left(\sigma_\theta - v\left(\sigma_r + \sigma_z\right)\right) = \frac{1}{E}\left(\sigma_\theta - v\sigma_r\right)$$

At the outside of Cylinder (1), $r = 40$ mm,

$$\frac{-i_1}{40} = \frac{1}{200,000}\left(\sigma_\theta - v\sigma_r\right)$$

$$ie\; \frac{-i_1}{40} = \frac{1}{200,000}\left(-\frac{4p}{3}\right)\left(1+\frac{400}{1600} - v\left(1-\frac{400}{1600}\right)\right)$$

$$i_1 = \frac{8p}{30000}\left(\frac{5}{4}-\frac{3v}{4}\right)$$

$$\therefore i_1 = \frac{2p}{30000}\left(5-3v\right)$$

For Cylinder (2):

$$\sigma_r = A_2 - \frac{B_2}{r^2}$$

and

$$\sigma_\theta = A_2 + \frac{B_2}{r^2}$$

At $r = 60$ mm, $\sigma_r = 0$

$$\therefore B_2 = 3600 A_2$$

At $r = 40$ mm, $\sigma_r = -p$

$$\therefore -p = A_2 - \frac{60^2}{40^2}A_2 = A_2 - \frac{3600}{1600}A_2$$

$$ie\; A_2 = \frac{4}{5}p$$

and

$$B_2 = 3600 \times \frac{4}{5}p$$

Thus,

$$\left(\sigma_r\right)_2 = \frac{4p}{5}\left(1 - \frac{3600}{r^2}\right)$$

and

$$\left(\sigma_\theta\right)_1 = -\frac{4p}{3}\left(1 + \frac{400}{r^2}\right)$$

$$\varepsilon_\theta = \frac{u}{r} = \frac{1}{E}\left(\sigma_\theta - v\left(\sigma_r + \sigma_z\right)\right) = \frac{1}{E}\left(\sigma_\theta - v\sigma_r\right)$$

At the inside of Cylinder (2), $r = 40$ mm,

$$\frac{+i_2}{40} = \frac{1}{200,000}\left(\frac{4p}{5}\right)\left(1 + \frac{3600}{1600} - v\left(1 - \frac{3600}{1600}\right)\right)$$

$$ie\, i_2 = \frac{8p}{50000}\left(\frac{13}{4} + \frac{5v}{4}\right)$$

$$\therefore i_2 = \frac{2p}{50000}\left(13 + 5v\right)$$

But $i_1 + i_2 = i = 0.03$ mm

$$\therefore \frac{2p}{30000}\left(5 - 3v\right) + \frac{2p}{50000}\left(13 + 5v\right) = 0.03$$

$$\frac{10p}{30000} - \frac{2vp}{10000} + \frac{26p}{50000} + \frac{2vp}{10000} = 0.03$$

$$\frac{50p + 78p}{150,000} = 0.03$$

$$ie\, p = \frac{4500}{128}\,\mathrm{N/mm^2} = 35.2\,\mathrm{N/mm^2}$$

For Cylinder (1),

$$\left(\sigma_r\right)_1 = -46.9\left(1 - \frac{400}{r^2}\right)$$

$$\left(\sigma_\theta\right)_1 = -46.9\left(1 + \frac{400}{r^2}\right)$$

and for Cylinder (2),

$$\left(\sigma_r\right)_2 = 28.2\left(1 - \frac{3600}{r^2}\right)$$

$$\left(\sigma_\theta\right)_2 = 28.2\left(1 + \frac{3600}{r^2}\right)$$

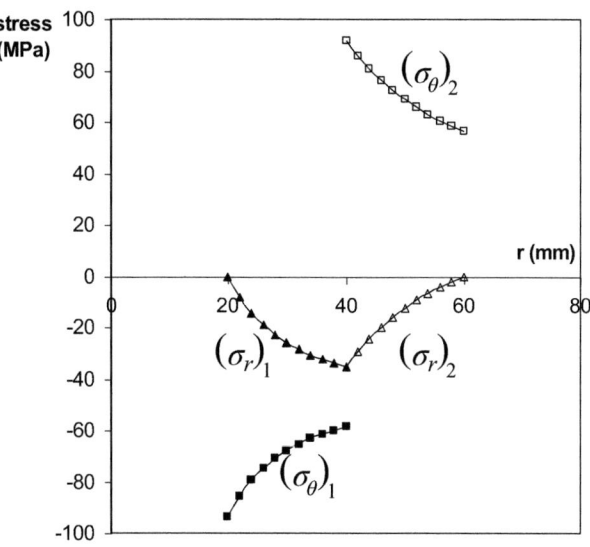

EXAMPLE 3.8.3

A turbine rotor disc with an angular velocity of 4000rpm has an external diameter of 1.2m and has a 0.1m diameter hole bored along its axis. Determine the stress distributions in the disc.
  Take:

$$\rho = 7850\text{kg} / \text{m}^3 \text{ and } \nu = 0.3$$

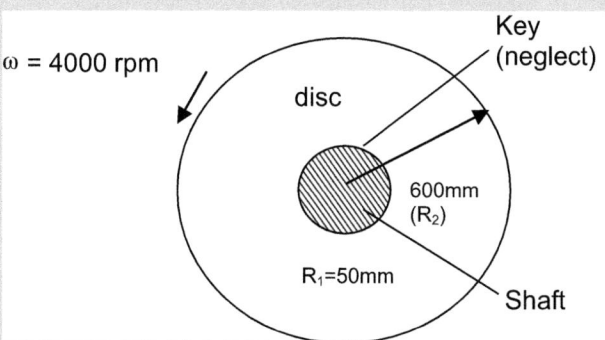

$$\sigma_r = A - \frac{B}{r^2} - \frac{\rho w^2 (3+\nu) r^2}{8}$$

$$\sigma_\theta = A + \frac{B}{r^2} - \frac{\rho w^2 (1+3\nu) r^2}{8}$$

at

$$r = 50 \text{ mm}, \sigma_r = 0$$

$$\therefore 0 = A - \frac{B}{50^2} - \frac{7850 \times 10^{-9}}{8} \times \left(4000 \times \frac{2\pi}{60}\right)^2 (3 + 0.3) \times 50^2 \times 10^{-3} \frac{N}{mm^2}$$

i.e.

$$O = A - \frac{B}{2500} - 1.4204 \tag{8.12}$$

at

$$r = 600 \text{ mm}, \sigma_r = O$$

$$\therefore O = A - \frac{B}{600^2} - \frac{7850 \times 10^{-9}}{8} \times \left(4000 \times \frac{2\pi}{60}\right)^2 \times (3 + 03) \times 600^2 \times 10^{-3} \frac{N}{mm^2}$$

i.e.

$$O = A - \frac{B}{3.6 \times 10^5} - 204.54 \tag{8.13}$$

Subtracting Equation (8.12) from Equation (8.13),

$$\frac{B}{2500} - \frac{B}{600^2} = 204.54 - 1.42$$

$$\therefore B = 511350.0$$

and

$$A = 205.95$$

Hence,

$$\sigma_r = 205.95 - \frac{511350}{r^2} - 0.000568r^2$$

and

$$\sigma_\theta = 205.95 + \frac{511350}{r^2} - 0.000327r^2$$

at

$$r = 50 \text{ mm}, \sigma_r = O \text{ and } \sigma_\theta = 409.7 N / mm^2$$

at

$$r = 600 \text{ mm}, \sigma_r = O \text{ and } \sigma_\theta = 89.6 \text{ N/mm}^2$$

The maximum σ_r value could be obtained by

$$\frac{d\sigma_r}{dr} = O$$

to find the r-value and the substituting this value of r into expression for σ_r.

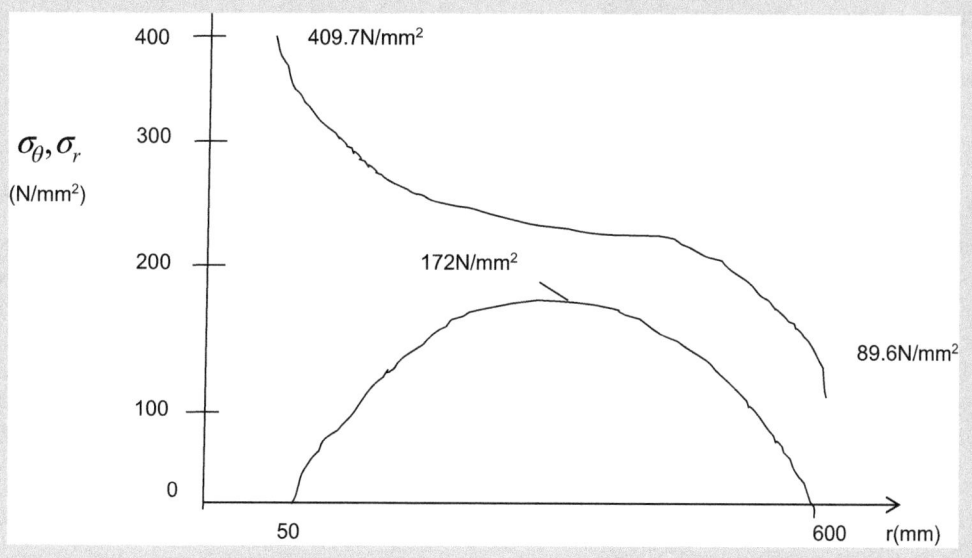

LEARNING SUMMARY

By the end of this section, you should have learnt:

- ✓ The essential differences between the stress analysis of thin and thick cylinders, leading to an understanding of statistically determinate and statically indeterminate situations;
- ✓ How to derive the equilibrium equations for an element of material in a solid body (e.g. a thick cylinder);
- ✓ The derivation of Lame's equations;
- ✓ How to determine stresses caused by shrink fitting of one cylinder onto another;
- ✓ How to include "inertia" effects into the thick cylinder equations in order to calculate the stresses in a rotating disc.

3.9   ASYMMETRICAL BENDING

The beam bending equation, $\dfrac{M}{I} = \dfrac{\sigma}{y} = \dfrac{E}{R}$, has been derived and is generally used to determine stresses in a beam with a <u>symmetrical</u> cross-section. The symmetry is usually about an axis perpendicular to the neutral axis of the section. For a section where this symmetry does not apply, that is asymmetric sections, a complication arises, making bending analysis more difficult. In these cases, applying a bending moment about the neutral axis will, in general, result not only in bending about that axis but also in simultaneous bending about the perpendicular axis, that is there is an interaction effect. To analyse such sections, we introduce a new geometric quantity called the product moment of area and this leads to the concept of principal second moments of area and principal axes for the section. These are axes for which the product moment of area is zero and the above interaction effect during bending does not occur. Thus, it is convenient to analyse the bending of asymmetric sections about these axes. In this section, we will look at the theory behind this effect and develop a general procedure for dealing with asymmetrical bending situations.

3.9.1   Second Moments of Area of a Complex-Shaped Cross-Section

3.9.1.1   Second Moments of Area about Parallel Axes

Consider an arbitrary shaped cross-section, as shown in Figure 3.89. The centroid of area, G, is at the origin, O, of the O-x-y axes set. A parallel axes set, O'-x'-y', also exists, distance a and b from O-x-y, as shown in the figure. The centroid of area, G, is positioned at coordinates $(x',y') = (a,b)$ in this parallel axes set.

We know that the second moments of area, I_x and I_y, of the section with respect to the x and y axes are given by

$$I_x = \int_A y^2 dA$$

and

$$I_y = \int_A x^2 dA$$

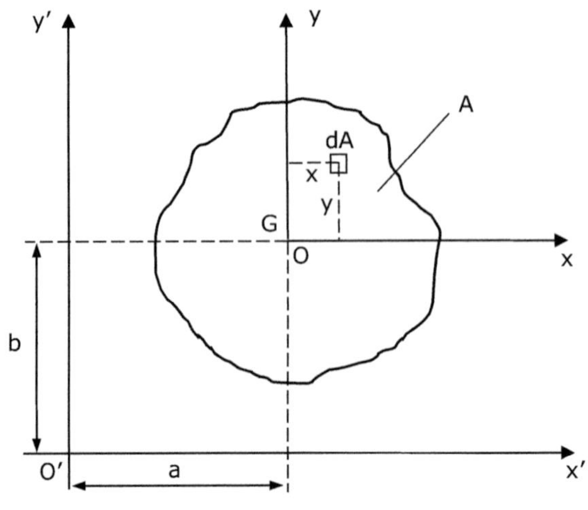

FIGURE 3.89   Parallel axis theorem.

that is the product of an element of area, dA, and its distance squared from the particular axis (x or y), integrated over the full cross-sectional area, A.

The **Parallel Axis Theorem** allows the calculation of the second moments of area, $I_{x'}$ and $I_{y'}$, with respect to the x' and y' axes as follows:

$$I_{x'} = I_x + Ab^2 \qquad (9.1)$$

and

$$I_{y'} = I_y + Aa^2 \qquad (9.2)$$

I_x and I_y are the second moments of area about a set of axes through the centroid and are always the minimum second moments. $I_{x'}$ and $I_{y'}$ will always be greater because the second terms in Equations (9.1) and (9.2) are always positive as the distances between the axes, a and b, are squared.

3.9.1.2   The Product Moment of Area

We now introduce a new quantity, the **product moment of area**, I_{xy}, which is defined as,

$$I_{xy} = \int_A xy\,dA$$

I_{xy} is the summation of the elements of area x the product of their coordinates. We can now develop the parallel axis theorem for the product moments of the area as follows:

$$I_{x'y'} = \int_A x'y'\,dA = \int_A (x+a)(y+b)\,dA$$
$$= \int_A xy\,dA + a\int_A y\,dA + b\int_A x\,dA + ab\int_A dA$$

but, $\int_A y\,dA$ and $\int_A x\,dA$ are both zero because the origin of axes O_{xy} is at the centroid of area, G. Thus,

$$I_{x'y'} = I_{xy} + abA \qquad (9.3)$$

This is the **Product Parallel Axis Theorem**. Again, I_{xy} is the product moment of area about a set of axes through the centroid. In this case, $I_{x'y'}$, can be either positive or negative, depending on the signs of a and b.

3.9.1.3   Symmetric Sections

Figure 3.90 shows a section where one axis (the y-axis in this case) is an axis of symmetry. The sum of the contributions to the product moment of the area from elements of area, dA, on the opposite sides of the axis of symmetry will cancel out because of the change of sign of the x coordinate. Thus, in general, if a section has an axis of symmetry, then I_{xy} is zero.

3.9.1.4   Rotation of Axes

Referring to Figure 3.91, the question arising is that, given I_x, I_y and I_{xy} for a particular section, what are the second moments of area I_u, I_v and I_{uv} about a set of axes, O_{uv}, through the same origin but rotated through an angle, θ?

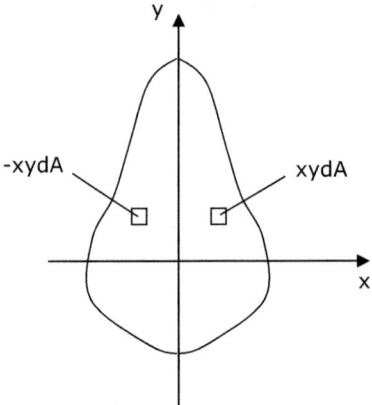

FIGURE 3.90 Symmetric section.

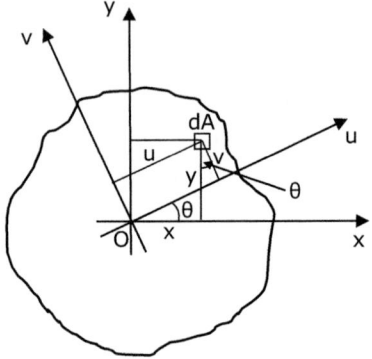

FIGURE 3.91 Rotation of axes.

The coordinate transformation equations for rotation through an angle, θ, are,

$$u = x \cos \theta + y \sin \theta$$

$$v = -x \sin \theta + y \cos \theta$$

Using these equations, we can now derive the second moment of area about the u-axis, I_u, as follows:

$$I_u = \int_A v^2 dA = \int_A \left(x^2 \sin^2 \theta + y^2 \cos^2 \theta - 2xy \sin \theta \cos \theta \right) dA$$

$$= I_y \sin^2 \theta + I_x \cos^2 \theta - I_{xy} \sin 2\theta$$

but,

$$\cos^2 \theta = \frac{1}{2}\left(1 + \cos 2\theta \right)$$

and

$$\sin^2\theta = \frac{1}{2}(1-\cos 2\theta)$$

$$\therefore I_u = \frac{1}{2}(I_x+I_y)+\frac{1}{2}(I_x-I_y)\cos 2\theta - I_{xy}\sin 2\theta \qquad (9.4)$$

Similarly, we can show that,

$$I_v = \frac{1}{2}(I_x+I_y)+\frac{1}{2}(I_x-I_y)\cos 2\theta + I_{xy}\sin 2\theta \qquad (9.5)$$

Equations (9.4) and (9.5) differ only by the sign of the last term. Thus, adding Equations (9.4) and (9.5) gives

$$I_u + I_v = I_x + I_y = \int_A (x^2+y^2)dA = \int_A r^2 dA = J$$

This is a statement of the **perpendicular axis theorem**, that is the sum of the second moments of area about two perpendicular axes is equal to the second moment of area about the third perpendicular axis. In this case, the latter second moment of area is the polar second moment of area, J.

Now considering the product moment of area, I_{uv}, we have,

$$I_{uv} = \int_A uv\, dA = \int_A (x\cos\theta + y\sin\theta)(-x\sin\theta + y\cos\theta)dA$$

$$= \int_A \left[-x^2\cos\theta\sin\theta + y^2\sin\theta\cos\theta + xy(\cos^2\theta - \sin^2\theta)\right]dA$$

$$\therefore I_{uv} = \frac{1}{2}(I_x-I_y)\sin 2\theta + I_{xy}\cos 2\theta \qquad (9.6)$$

Equations (9.4)–(9.6) enable us to calculate I_u, I_v and I_{uv}, knowing I_x, I_y and I_{xy} and the angle of rotation, θ.

3.9.1.5 Principal Second Moments of Area

Now, looking more closely at Equation (9.6). If θ changes by 90°, 2θ changes by 180° and I_{uv} changes sign but not its magnitude. Thus, since it is a continuous function of θ, I_{uv} must have zero value.

The axes which give rise to a zero product moment are called the **principal axes**.

For simplicity, let us assume that x and y are the principal axes. Then,

$$I_x = I_p$$

$$I_y = I_q$$

and

$$I_{xy} = I_{pq} = 0$$

where I_p and I_q are called the principal second moments of area

Substituting I_p and I_q into Equations (9.4)–(9.6), we have

$$I_u = \frac{1}{2}\left(I_p + I_q\right) + \frac{1}{2}\left(I_p - I_q\right)\cos 2\theta$$

$$I_v = \frac{1}{2}\left(I_p + I_q\right) - \frac{1}{2}\left(I_p - I_q\right)\cos 2\theta$$

$$I_{uv} = \frac{1}{2}\left(I_p - I_q\right)\sin 2\theta$$

These equations can be represented by a Mohr's circle as shown in Figure 3.92. Second moments are plotted on the *x*-axis and the product moments are plotted on the *y*-axis [*note that the y-axis for the circle is positive upwards, unlike Mohr's circle for stress which is positive downwards for shear*].

Point A on the circle has coordinates which correspond to the second moment and product moment, that is, (I_u, I_{uv}), relevant to the *u*-axis. Point B on the circle has coordinates which correspond to the second moment and product moment, that is $(I_v, I_{vu} = -I_{uv})$, relevant to the *v*-axis. These two points enable the circle to be drawn.

The centre of the circle, *C*, and radius, *R*, are given by

$$\text{Centre } C = \frac{I_u + I_v}{2} \tag{9.7}$$

$$\text{Radius } R = \sqrt{\left(\frac{I_u - I_v}{2}\right)^2 + I_{uv}^{\,2}} \tag{9.8}$$

The points P and Q on the circle correspond to the **principal planes** for which the product moment of areas are zero and the second moments are the **principal second moments of area**, I_p and I_q. Their magnitudes are given by,

$$I_p = \text{Centre} + \text{Radius}$$

and

$$I_q = \text{Centre} - \text{Radius}$$

where the centre and radius are given by Equations (9.7) and (9.8).

Thus, by knowing I_u, I_v and I_{uv}, the principal second moments of area, I_p and I_q, can be determined.

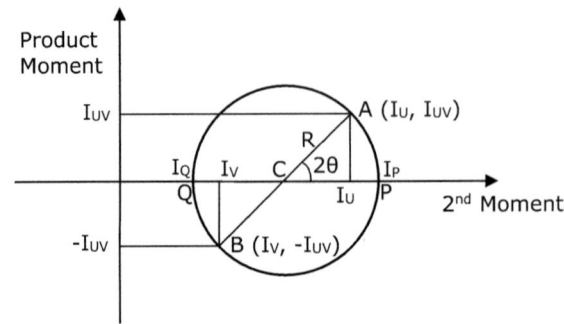

FIGURE 3.92 Mohr's circle.

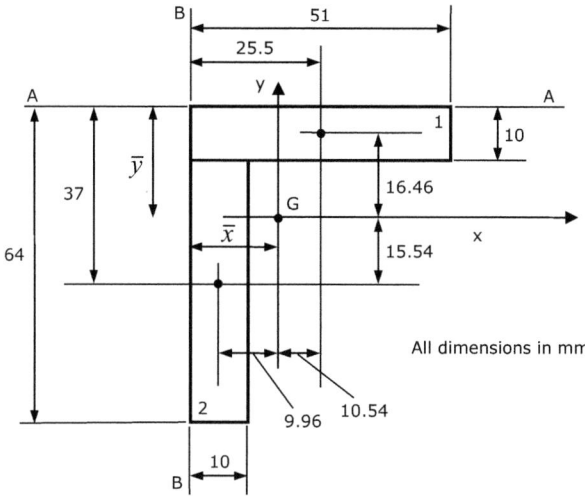

FIGURE 3.93   Second Moments of area of an angle section.

The angle of the principal axes with respect to the U-V axes is the angle θ, where 2θ is shown in Figure 3.92 and is given by

$$\sin 2\theta = \frac{I_{uv}}{R}$$

or alternatively

$$\tan 2\theta = \frac{I_{uv}}{\left(\dfrac{I_u - I_v}{2}\right)}$$

Key points about the Mohr's circle for second moments of area

1. the +ve upward direction for the product moment ensures that rotation in Mohr's circle has the same sense as the rotation of the axes in space.
2. I_p and I_q are both +ve.
3. If $I_p = I_q$, all product moments are zero and all axes in all directions are principal axes. For example, this is the case for a circular section.
4. The sign of the product moment is important. $I_{uv} = \int_A uv\,dA$ is associated with the u-axis and can be +ve or −ve. The product moment associated with the v-axis is $I_{vu} = -I_{uv}$

Summary of procedure to calculate the principal second moments of area and the directions of the principal axes

1. Divide the cross-section into subsections for which the centroid of areas and second moments of area about their own axes can be determined.
2. Choose a convenient set of orthogonal axes with its origin at the centroid of the full cross-section.
3. Use the parallel axis theorem to determine the second moment of area for the full cross-section.
4. Use a Mohr's circle construction to determine the principal second moments of area and the directions of the principal axes.

WORKED EXAMPLE – PRINCIPAL SECOND MOMENTS OF AREA

Figure 3.93 shows an asymmetric angle cross-section. Determine:

  a. The principal second moments of area
  b. The directions of the principal axes

The section is divided into rectangular subsections 1 and 2.
    Position of the Centroid:

$$\text{Total Area} = 51 \times 10 + 54 \times 10 = 1050 \, \text{mm}^2$$

Taking moments of the areas about the datum AA,

$$1050 \times \bar{y} = \left(51 \times 10\right) \times 5 + \left(54 \times 10\right) \times 37$$
$$\therefore \bar{y} = 21.46 \text{mm}$$

Taking moments of the areas about the datum BB,

$$1050 \times \bar{x} = \left(51 \times 10\right) \times 25.5 + \left(54 \times 10\right) \times 5$$
$$\therefore \bar{x} = 14.96 \text{mm}$$

Second moments of area about a convenient set of axes:
    The x and y axes are drawn as a convenient set of axes through the centroid. Using the parallel axis theorem,

$$I_x = \left(\frac{51 \times 10^3}{12} + 51 \times 10 \times 16.46^2\right) + \left(\frac{10 \times 54^3}{12} + 10 \times 54 \times 15.54^2\right)$$
$$= \underline{404,051 \ \text{mm}^4}$$

$$I_y = \left(\frac{10 \times 51^3}{12} + 10 \times 51 \times 10.54^2\right) + \left(\frac{54 \times 10^3}{12} + 54 \times 10 \times 9.96^2\right)$$
$$= \underline{225,268 \text{mm}^4}$$

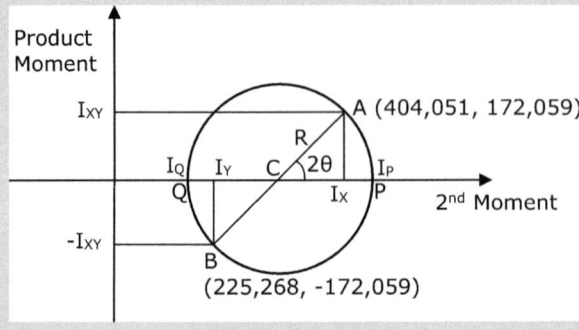

FIGURE 3.94 Mohr's circle.

And the product parallel axis theorem,

$$I_{xy} = \left(0 + 51 \times 10 \times 10.54 \times 16.46\right) + \left(0 + 54 \times 10 \times \left(-9.96\right) \times \left(-15.54\right)\right)$$
$$\underline{= 172,059 \text{mm}^4}$$

Note that, in the product moment of area calculation above, the product moment of each subsection about its own axis is zero due to the symmetry of each subsection. It is also important that the correct sign for the coordinates of each subsection centroid with respect to the full cross-section centroid is taken. Thus, for subsection 3.9.1.1, the coordinates are both positive (10.54 and 16.46), while for subsection 3.9.1.2, they are both negative (−9.96 and −15.54).

Mohr's Circle:

A Mohr's circle can now be drawn to represent the axes about which I_x, I_y and I_{xy} act, as shown in Figure 3.94. The centre and radius are calculated as follows:

$$\text{Centre } C = \frac{I_x + I_y}{2} = 314,659 \text{ mm}^4$$

$$\text{Radius } R = \sqrt{\left(\frac{I_x - I_y}{2}\right)^2 + I_{xy}^2} = 193,895 \text{ mm}^4$$

Principal second moments of area:

The principal second moments of area can now be determined from the circle as follows:

$$I_p = C + R = 508,554 \text{ mm}^4$$

$$I_q = C - R = 120,764 \text{ mm}^4$$

and the angle, θ, of the principal axes with respect to the x-axis is given by

$$\sin 2\theta = \frac{I_{xy}}{R} = \frac{172,059}{193,895} = 0.887$$
$$\therefore \underline{\theta = 31.27°}$$

From Mohr's circle it can be seen that the principal axis 1, that is, the p-axis is 31.27° clockwise from the x-axis. The principal axes can now be drawn on a sketch of the element as shown in Figure 3.95.

3.9.2 BENDING OF BEAMS WITH ASYMMETRIC SECTIONS

Figure 3.96 shows an arbitrary cross-section of a beam subjected to a bending moment, M, acting at an angle θ to the x-axis. The origin of the x-y axes coincides with the centroid of the section. The bending moment has two components, M_x and M_y, as shown, acting about the x-axis and y-axis, respectively. [*note that the bending moment and its components are drawn in vector form with a double arrow head. The RH screw rule defines the sense of each bending moment component as shown in the figure inset*].

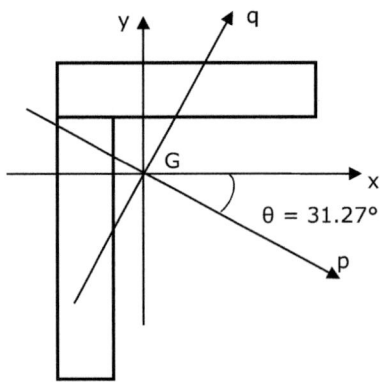

FIGURE 3.95 Angle of principal axes.

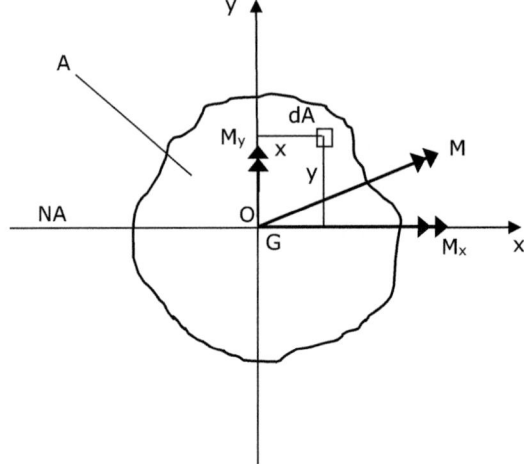

FIGURE 3.96 Bending about an arbitrary cross-section.

Assume that bending takes place about the x-axis, that is, O-x is the neutral axis. Then, the bending stress, σ, is proportional to the distance, y, from the neutral axis, or alternatively,

$$\sigma = c.y$$

where c is an arbitrary constant.

The resultant moment about the x-axis is given by the sum moments of the forces acting on each elemental area in the cross-section. In the limit, this sum can be written as an integral as follows:

$$M_x = \int_A \sigma.y dA$$

$$= \int_A c.y.y dA$$

$$= c.I_x$$

where I_x = second moment of area about the x-axis

But $c = \sigma/y$

$$\therefore \sigma = \frac{M_x}{I_x} y$$

which is the beam bending equation as expected.

However, there is also a resultant moment about the y-axis, which is as follows:

$$M_y = -\int_A \sigma.x dA$$

$$= -\int_A c.y.x dA$$

$$= -c.I_{xy}$$

where $I_{xy} = \int xy dA$ is the Product Moment of Area

[note the −ve sign arising because a positive stress results in a −ve moment about the y-axis]

Thus, in general, a moment has to be applied about the y-axis as well as the x-axis to produce bending about the x-axis only. A +ve moment is required about the y-axis to counterbalance the −ve moment set up by the stresses arising from M_x. This is not the case if I_{xy} is zero, that is, for sections which are symmetric about the y-axis.

To ensure bending about the x-axis only, a resultant moment $M = \sqrt{M_x^2 + M_y^2}$ must be applied at angle, θ, given by

$$\theta = \tan^{-1}\left(\frac{M_y}{M_x}\right) = \tan^{-1}\left(\frac{-I_{xy}}{I_x}\right)$$

The resultant moment is only applied to the x-axis when $I_{xy} = 0$.

Figure 3.97 illustrates the effect for the z-section. If a bending moment is applied about the x-axis only, then the stresses in the flanges will create a resulting moment about the y-axis. Consequently, bending will take place about both the x- and y-axes. This is a consequence of I_{xy} not being zero for this asymmetric section.

To avoid this moment coupling effect, it is usually convenient to solve bending problems by considering bending about the **principal axes** of a section for which the **product moment of area** is zero.

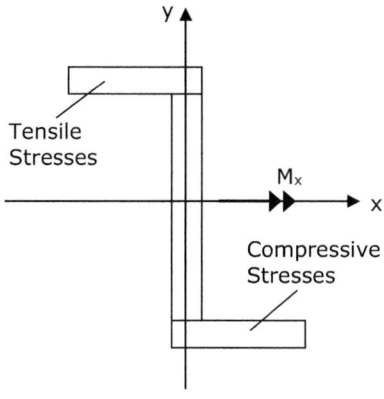

FIGURE 3.97   Z-section.

3.9.2.1 Solving Asymmetrical Bending Problems

Consider the arbitrary asymmetric section shown in Figure 3.98(a). O is the centroid and O-p and O-q are the **principal axes** of the section. The principal axes are inclined at angle θ to the x-y axes. The components of an applied moment M, that is M_x and M_y, act about the O-x and O-y axes, respectively. First, M_x and M_y are resolved onto the principal directions, as illustrated in Figure 3.98(b), giving

$$M_p = M_x \cos\theta + M_y \sin\theta$$

and

$$M_q = -M_x \sin\theta + M_y \cos\theta$$

We can now calculate the total bending stress at any position, (p, q), which arises from these two bending moment components and is given by,

$$\sigma_b = \frac{M_p.q}{I_p} - \frac{M_q.p}{I_q} \tag{9.9}$$

[note that when p and q are both +ve, i.e. in the first quadrant of the p-q axes set, a +ve M_p gives rise to a +ve bending stress while a +ve M_q gives rise to a –ve bending stress]

The maximum stress in the section will occur at an extreme distance from the **neutral axis**. We therefore need to determine the position/orientation of the neutral axis which can be found by setting the bending stress, that is Equation (9.1), to zero. Thus, the neutral axis occurs where,

$$\sigma_b = \frac{M_p.q}{I_p} - \frac{M_q.p}{I_q} = 0$$

$$\therefore \frac{M_p.q}{I_p} = \frac{M_q.p}{I_q}$$

$$\therefore \frac{q}{p} = \frac{M_q}{M_p} \cdot \frac{I_p}{I_q}$$

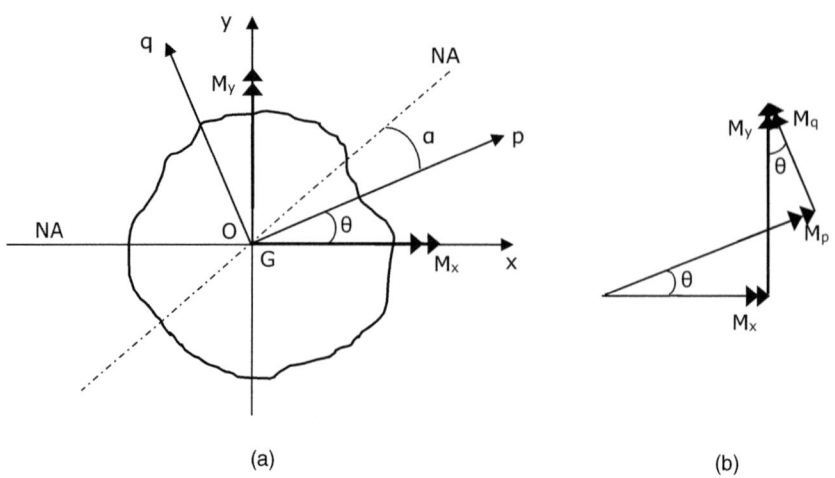

(a)   (b)

FIGURE 3.98 Asymmetrical section.

This value for q/p defines the angle, α, of the neutral axis, with respect to the p-axis, shown in Figure 3.98(a), as follows:

$$\alpha = \tan^{-1}\left(\frac{q}{p}\right) = \tan^{-1}\left(\frac{M_q.I_p}{M_p.I_q}\right) \qquad (9.10)$$

Equation (9.9) can therefore be used to determine the magnitude of the stress at any position (p, q) and Equation (9.10) can be used to determine the orientation of the neutral axis and hence the position of the maximum stress which is at the extreme distance from the neutral axis.

Summary of the procedure for solving asymmetrical bending problems

1. Determine the principal axes of the section, p and q, about which $I_{xy} = 0$.
2. Consider bending about the principal axes, that is, resolve bending moments onto these axes.
3. Knowing M_p, M_q, I_p and I_q, determine the general expression for the bending stress at position (p, q) as follows:

$$\sigma_b = \frac{M_p.q}{I_p} - \frac{M_q.p}{I_q}$$

4. Determine the angle of the neutral axis with respect to the p-axis as follows:

$$\alpha = \tan^{-1}\left(\frac{q}{p}\right) = \tan^{-1}\left(\frac{M_q.I_p}{M_p.I_q}\right)$$

5. Evaluate the bending stress at any position in the section including the extreme positions from the neutral axis which give the maximum bending stresses.

WORKED EXAMPLE – ASYMMETRICAL BENDING

The angle section, shown in Figure 3.99, with principal axes and principal second moments of area indicated, is subjected to a bending moment of 300 Nm about the x-axis.

Determine

i. The position/orientation of the neutral axis.

ii. The bending stresses at positions A, B and C.

Resolving the applied moment:

Referring to Figure 3.100(a), the components of the applied bending in the p and q directions are,

$$M_p = M \cos(31.27) = 0.855\,M$$

$$M_q = M \sin(31.27) = 0.519\,M$$

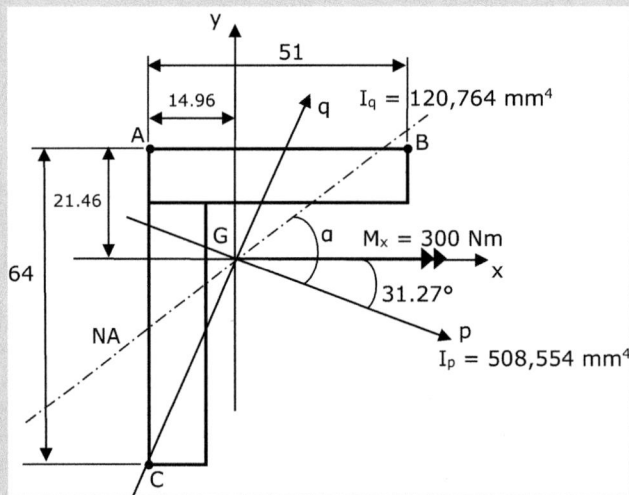

FIGURE 3.99   Bending about an asymmetrical angle section.

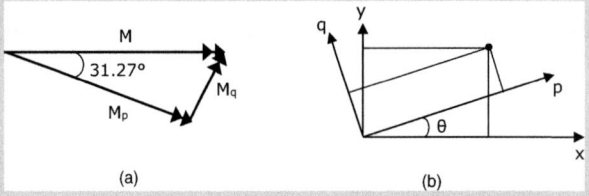

FIGURE 3.100   Transformation onto principal axes.

Bending stress equation:

The general expression for bending stress at position (p, q) is,

$$\sigma_b = \frac{M_p.q}{I_p} - \frac{M_q.p}{I_q}$$

$$= \frac{300.10^3 \, 0.855q}{508,554} - \frac{300.10^3 \, 0.519p}{120,764}$$

$$= 0.5042q - 1.2894p$$

Note that for p and q in mm, this expression gives bending stress in N/mm², that is, MPa.

Orientation of the neutral axis, α, with respect to the p-axis:

$$\alpha = \tan^{-1}\left(\frac{q}{p}\right) = \tan^{-1}\left(\frac{M_q.I_p}{M_p.I_q}\right)$$

$$= \tan^{-1}\left(\frac{1.2894}{0.5042}\right) = \tan^{-1}(2.558)$$

$$= 68.64°$$

with respect to the p-axis

Orientation of the neutral axis with respect to the x-axis = 68.64 − 31.27 = 37.37°. These orientations are illustrated in Figure 3.99.

Bending stresses:
To determine the bending stresses at A, B and C, we need the p and q coordinates of these points. Referring to Figure 3.100(b), the general coordinate transformation equations for a set of axes, p-q, that are inclined θ <u>anticlockwise</u> from another set, x-y, are,

$$p = x\cos\theta + y\sin\theta$$

and

$$q = -x\sin\theta + y\cos\theta$$

For this problem, the p-axis is inclined at 31.27° clockwise to the x-axis. Thus, $\theta = -31.27°$ and the above transformation equations become,

$$p = 0.8547x - 0.5191y$$

and

$$q = 0.5191x + 0.8547y$$

We can now draw a table for calculating the coordinates of A, B and C as follows:

Position	X	Y	p	q
A	−14.96	21.46	−23.92	10.88
B	36.04	21.46	19.66	37.05
C	−14.96	−42.54	9.3	−44.12

and the stresses follow from the general equation $\sigma_b = 0.5042q - 1.2894p$ as follows:

at A σ_A = 36.33 MPa

B σ_B = −6.67 MPa

C σ_C = −34.24 MPa

LEARNING SUMMARY

By the end of this section, you should have learnt:

✓ that an asymmetric cross-section, in addition to its second moments of area about the x and y axes, I_x and I_y, possesses a geometric quantity called the product moment of area, I_{xy}, with respect to these axes;

✓ how to calculate the second moments of area and the product moment of area about a convenient set of axes;

✓ that an asymmetric section will have a set of axes at some orientation for which the product moment of area is zero and that these axes are called the principal axes;

✓ that the second moments of area about the principal axes are called the principal second moments of area;

✓ that a Mohr's circle construction can be used to determine the second moments of area and the product moment of area about any oriented set of axes, including the principal axes;

✓ that it is convenient to analyse the bending of a beam with an asymmetric section by resolving bending moments onto the principal axes of the section;

✓ a basic procedure for analysing the bending of a beam with an asymmetric cross-section.

3.10   STRAIN ENERGY

We saw in Section 3.4 how deflections of a beam can be determined by solving the differential equation of the elastic line, that is Macaulay's method. This method is not appropriate for more complex-shaped structures or bodies or where combined loading is applied. In this section, we introduce the concept of **strain energy** which will enable us to calculate deflections of complex structures. When an elastic body is subject to loading, strain energy is stored within the body. An Italian railway engineer, called Castigliano, derived a theorem and procedure for using the strain energy to determine deflections in structures or bodies. Castigliano's theorem is a powerful and flexible method for solving deflection problems.

3.10.1   STRAIN ENERGY EXPRESSIONS

The strain energy in an elastic body is equal to the work done on the body by the applied loads. Thus, when an elastic body is subjected to a single applied load P, causing it to displace by a distance u at the position of load application, as illustrated by Figure 3.101, the strain energy (denoted by the symbol capital U) is given by,

$$\text{Strain Energy, } U = \text{Work Done} = \int_{0}^{u} P \, du$$

For a linear elastic body,

$$U = \frac{1}{2} P.u \tag{10.1}$$

where

P = final load
u = final displacement

Similar expressions to this can be derived for moments, torques and couples. Thus, when a moment is applied to an elastic body, causing it to rotate through an angle θ, the strain energy is given by,

$$U = \frac{1}{2} M.\theta \tag{10.2}$$

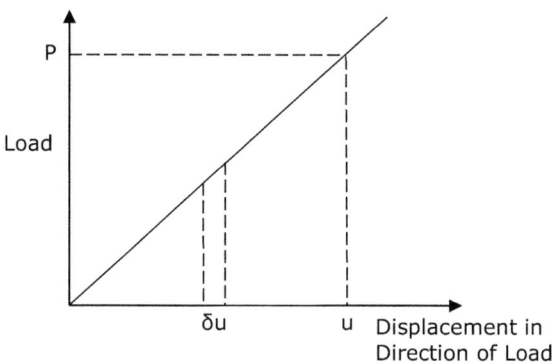

FIGURE 3.101   Strain energy is given by the area under the load–displacement curve.

where

   M = final moment
   θ = final angle of rotation

We will now use Equations (10.1) and (10.2) to derive expressions for the strain energy in a bar under tension, torsion and bending.

3.10.1.1   Tension in a Bar

Figure 3.102 shows an element of the length of a bar, length δs, cross-sectional area, A, which is extended, δu, due to a tensile load P.

   The strain energy for the element, δU, is given by

$$\delta U = \text{Work Done} = \frac{1}{2} P \delta u \qquad (10.3)$$

[n.b. there are transverse strains/displacements due to Poisson's effects but <u>no</u> transverse stresses/loads. Thus, there is no work done in the transverse direction]
   Now,

$$\text{Strain } \varepsilon = \text{extension / original length}$$

$$= \frac{\delta u}{\delta s}$$

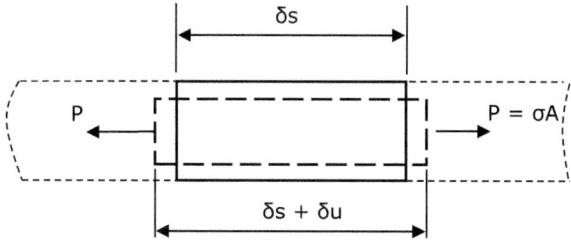

FIGURE 3.102   Strain energy in a bar under tension.

Also Strain $\varepsilon = \dfrac{\sigma}{E} = \dfrac{P}{EA}$ i.e. Hooke's law

$$\therefore \frac{\delta u}{\delta s} = \frac{P}{EA}$$

$$\delta u = \frac{P}{EA}\delta s \tag{10.4}$$

Substituting Equation (10.4) into (10.3) gives

$$\delta U = \frac{P^2}{2AE}\delta s \tag{10.5}$$

This is the expression for strain energy in an element of bar. Thus, for a bar of length L, with possibly varying A and E, integrating Equation (10.5) gives

$$U = \int_0^L \frac{P^2}{2AE}ds \tag{10.6}$$

that is the general expression for strain energy in a bar under tension. This expression can also be used for a bar under compression.

For constants A and E,

$$U = \frac{P^2 L}{2AE} \tag{10.7}$$

So, by knowing the load, geometry and material stiffness, we can use Equation (10.6) or (10.7) to determine the strain energy in the bar.

3.10.1.2   Strain Energy per Unit Volume

It is interesting to note that, from Equation (10.3), the strain energy per unit volume is given by

$$\frac{\delta U}{\delta V} = \frac{1}{2}\frac{P\delta u}{A\delta s} = \frac{1}{2}\sigma.\varepsilon$$

i.e. ½ × stress × strain (for uniaxial stress)

3.10.1.3   Torsion in a Shaft

Figure 3.103 shows an element of the length of a shaft, length δs, which is twisted through angle $\delta\theta$ due to a torque T.

The strain energy for the element, δU, is given by,

$$\delta U = \text{Work Done} = \frac{1}{2}T\delta\theta \tag{10.8}$$

Now for a circular shaft of polar second moment of area, J, and shear modulus, G, the torsion equation gives

$$\delta\theta = \frac{T}{GJ}\delta s \tag{10.9}$$

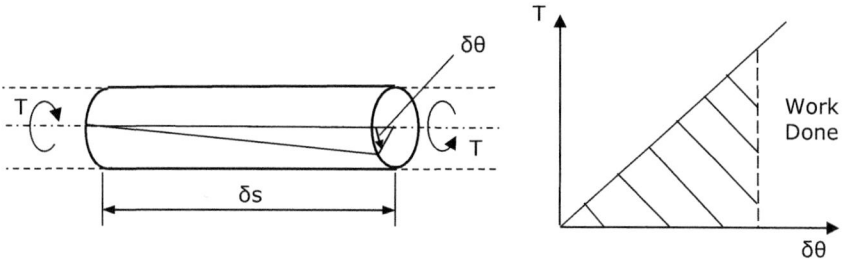

FIGURE 3.103 Strain energy in a shaft under torsion.

[i.e. from $\dfrac{T}{J} = \dfrac{G\theta}{L}$]

Substituting Equation (10.9) into (10.8) gives

$$\delta U = \frac{T^2}{2GJ}\delta s \tag{10.10}$$

This is the expression for strain energy in an element of the shaft. Thus, for a shaft of length L, integrating Equation (10.10) gives

$$U = \int_0^L \frac{T^2}{2GJ}\,ds \tag{10.11}$$

that is the general expression for strain energy in a shaft under torsion *[note the similarity in form with Equation (10.6) for a bar under tension]*

3.10.1.4 Bending of a Bar

Figure 3.104 shows an element of a straight beam, length δs, which bends to curvature R, due to an applied bending moment M. The angle subtended by the element of beam is $\delta\varphi$, which is also equal to the change in the slope of the beam over δs.

The strain energy for the element, δU, is given by

$$\delta U = \text{Work Done} = \frac{1}{2}M\delta\varphi \tag{10.12}$$

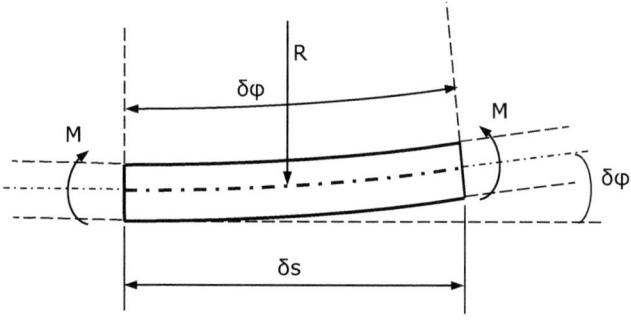

FIGURE 3.104 Strain energy in a beam under bending.

Now,

$$R.\delta\Phi = \delta s \text{ and } \frac{M}{I} = \frac{E}{R}, \text{ the beam bending equation.}$$

$$\therefore \delta\varphi = \frac{\delta s}{R} = \frac{M}{EI}\delta s \tag{10.13}$$

Substituting Equation (10.13) into (10.12) gives

$$\delta U = \frac{M^2}{2EI}\delta s \tag{10.14}$$

Thus, for a beam of length L, integrating Equation (10.14) gives

$$U = \int_0^L \frac{M^2}{2EI}ds \tag{10.15}$$

that is the general expression for strain energy in a beam under bending [*note the similarity in form with Equations (10.6) and (10.11)*]

The three expressions for calculating strain energy can now be summarised as follows:

$$U_{\text{tension/compression}} = \int_0^L \frac{P^2}{2AE}ds$$

$$U_{\text{torsion}} = \int_0^L \frac{T^2}{2GJ}ds$$

$$U_{\text{bending}} = \int_0^L \frac{M^2}{2EI}ds \tag{10.16}$$

In practical engineering structures, where members are relatively long and slender, strain energy due to tension and compression can usually be neglected. Bending is usually dominant [*strain energy due to shear deflections can also exist, but again can normally be neglected*].

3.10.2 CASTIGLIANO'S THEOREM

Consider an **elastic** body loaded by forces, P_i, as shown in Figure 3.105(a). The corresponding displacements of the load points in the direction of the loads are u_i. For the general non-linear elastic case, the load–displacement curve for the ith load is shown in Figure 3.105(b).

If all forces are applied simultaneously and proportionately, the total stored strain energy, U, in the body, is equal to the work done by the forces as follows:

$$U = \sum_i U_i = \sum_i \int P_i du_i$$

where U_i is the area under the load–displacement curve for one loading.

We can also define the complementary energy, U^*, as follows:

$$U^* = \sum_i U_i^* = \sum_i \int u_i dP_i$$

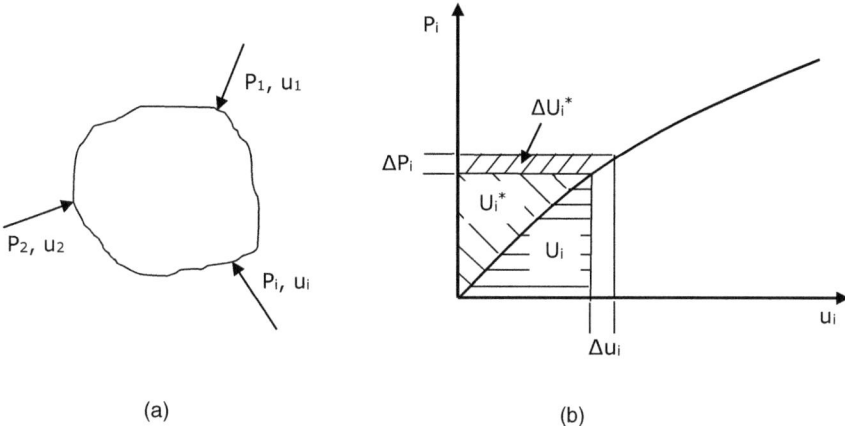

FIGURE 3.105 (a) Elastic body subject to several loads and (b) the load–displacement curve for the ith load acting on the body; U_i is the strain energy and U_i^* is the complementary energy.

where U_i^* is the area above the load–displacement curve for one loading.

Now, consider a small increment of the load P_i, ΔP_i, while the other loads remain constant. This causes an increment of the complementary strain energy as shown in Figure 3.105(b) which is given by

$$\Delta U_i^* = \int u_i \, dP_i \approx u_i \, \Delta P_i$$

For the other loads, P_j, $j \neq i$, P_j is constant and $\Delta U_j^* = \int u_j dP_j = 0$

The total change in complementary strain energy is, therefore,

$$\Delta U^* = \sum_i \Delta U_i^* = \Delta U_i^* = u_i \Delta P_i$$

In the limit, $\Delta P_i \geq 0$ and,

$$u_i = \frac{\lim}{\Delta P_i = 0} \frac{\Delta U^*}{\Delta P_i} = \frac{\partial U^*}{\partial P_i} \qquad (10.17)$$

Note that this is a partial derivative because all other loads except P_i were kept constant.

Equation (10.17) is the general form of **Castigliano's Theorem**, which, as stated in the introduction, is named after the Italian railway engineer, Carlo Alberto Castigliano (1847–1884), who developed the method. His theorem states that deflection, u_i, at a given load point i, may be obtained by differentiating the complementary strain energy, U^*, with respect to the load, P_i, acting at the point.

Note that u_i is the deflection at the point of application of the load, P_i, in the direction of P_i and P_i is independent of other loads.

Equation 10.17 applies to any non-linear elastic body. For linear elastic bodies, strain energy is equal to the complementary strain energy as shown in Figure 3.106. Thus,

$$U = U^* \quad \text{for a linear elastic body}$$

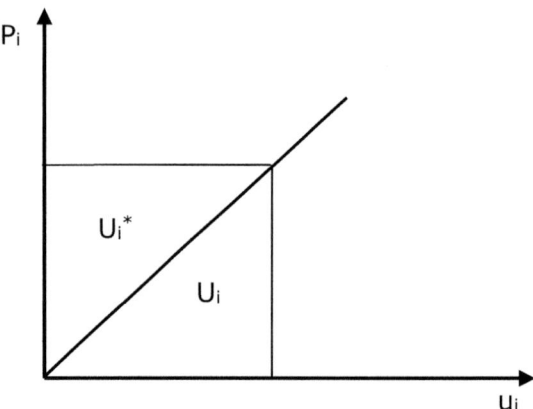

FIGURE 3.106 The load–displacement line for a linear elastic body.

and

$$u_i = \frac{\partial U}{\partial P_i} \tag{10.18}$$

Equation (10.18) is the well-known form of **Castigliano's Theorem** and states that **the partial derivative of the strain energy, U, of a linear elastic system with respect to any independent force is equal to the displacement of the structure at the point of application of the force in the direction of the force**.

Castigliano's theorem may be extended to include rotations due to moments and torques as follows:

$$\theta_i = \frac{\partial U}{\partial M_i} \tag{10.19}$$

where θ_i is the rotation at the point of application of the moment M_i about the axis of M_i.

The above forms of Castigliano's theorem (Equations 10.18 and 10.19) can be used to determine deflections/displacements and rotations at a point in a body or structure where a load or moment is applied. The procedure is as follows:

a. Obtain expressions for the total strain energy within a structure or body in terms of the applied loads or moments using Equation (10.16).
b. Differentiate the strain energy expressions with respect to the applied loads or moments, as in Equations (10.18) and (10.19), to determine the expressions for deflections or rotations, respectively.
c. Evaluate numerically the deflections and/or rotations.

Notes on further use of Castigliano's theorem

1. To determine the deflection or rotation at a point where a load or moment is not applied, include a dummy load or moment at the point. Obtain and differentiate a strain energy expression incorporating the dummy load/moment and set the latter to zero when numerically evaluating the deflections/rotations. Thus, to determine the horizontal tip deflection

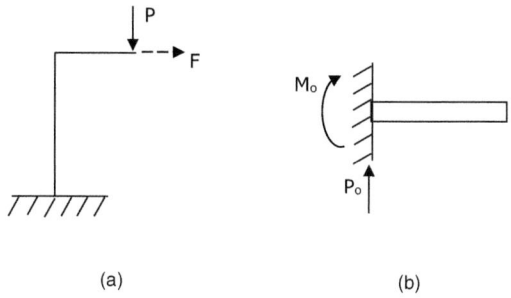

(a)   (b)

FIGURE 3.107   Castigliano's theorem: (a) Dummy load, F, added to determine the horizontal deflection and (b) the use of the zero displacement condition to solve for unknown reaction loads and moments.

of the structure shown in Figure 3.107(a), with an applied vertical load, P, it is necessary to add a dummy load F at the tip. Then,

$$u_F = \left(\frac{\partial U}{\partial F} \right)_{(F=0)}$$

2. If the displacement or rotation at a point is restrained, for example, at a support as shown in Figure 3.107(b), and the reactions are R_o and M_o, then,

$$\frac{\partial U}{\partial R_o} = 0$$

and

$$\frac{\partial U}{\partial M_o} = 0$$

These conditions can be used to determine the reactions between R_o and M_o.

3.10.3   WORKED EXAMPLES

3.10.3.1   Beam Deflection Example

Figure 3.108 shows a simply supported beam, ABC, with a point load at the centre of its span.
   Use strain energy to derive an expression for the central deflection of the beam
   [assume bending strain energy only, no shear].

Reaction forces

From symmetry $R_A = R_C = W/2$
Bending moment distribution
   Take a section, X-X, in the part of the span, AB, and consider the LH part of the beam as a free body. Moments about X-X give

$$M - \frac{Wx}{2} = 0$$

$$\therefore M = \frac{Wx}{2}$$

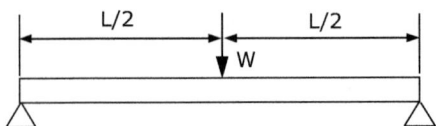

FIGURE 3.108   Simply-supported beam with a point load at the centre span.

Strain energy

The strain energy stored in AB is given by

$$U_{AB} = \int \frac{M^2}{2EI}\, ds = \int_{0}^{L/2} \frac{W^2 x^2}{4.2EI}$$

$$= \frac{W^2}{8EI}\left[\frac{x^3}{3}\right]_{0}^{L/2}$$

$$= \frac{W^2 L^3}{192EI}$$

From symmetry, $U_{BC} = U_{AB}$

The total strain energy $U_{tot} = 2U_{AB} = \dfrac{W^2 L^3}{96EI}$

Deflection

The central deflection, that is deflection at the load, is therefore given by

$$y_B = \frac{\partial U}{\partial W} = \frac{WL^3}{48EI}$$

3.10.3.2   Dummy Load Example

A curved beam, whose geometric form is a quadrant in the vertical plane, as shown in Figure 3.109(a), has small cross-sectional dimensions compared with the quadrant radius.

Use strain energy to derive an expression for the horizontal and vertical deflections of its tip due to vertical load P.

Because of the slender nature of the curved beam, direct stress and transverse shear strain energies may be neglected compared to bending strain energy.

To determine the horizontal deflection, a horizontal dummy load, Q, is added at the tip of the beam as shown in Figure 3.109(b).

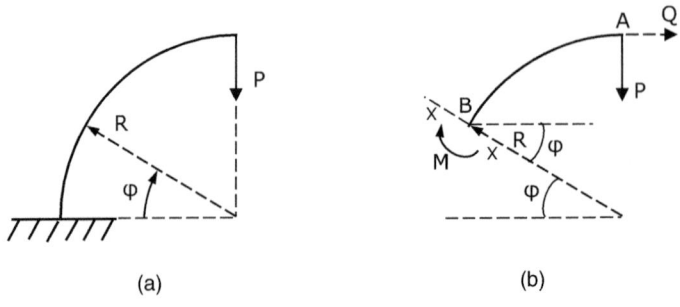

(a)   (b)

FIGURE 3.109   Curved cantilever beam with point load applied at the free end.

Bending moment expression:

As shown in Figure 3.109(b), to determine the bending moment distribution, a section X-X is taken at position B, angle φ from the horizontal, and equilibrium of a section AB of the beam is considered. Taking moments about X-X,

$$M + PR\cos\varphi + QR(1 - \sin\varphi) = 0$$
$$\therefore M = -PR\cos\varphi - QR(1 - \sin\varphi) \tag{10.20}$$

Note that the bending moment is a function of both the applied load, P, and the dummy load, Q. Also, the sign of the bending moment is unimportant as it is squared in the strain energy integral.

Strain energy:

The strain energy expression can now be written as follows:

$$U = \int \frac{M^2}{2EI} ds = \int_0^{\pi/2} \frac{\left[-PR\cos\varphi - QR(1 - \sin\varphi)\right]^2}{2EI} R d\varphi \tag{10.21}$$

Note that $ds = R d\varphi$ and that the limits of integration are 0 to $\pi/2$.

Vertical displacement:

The vertical displacement is now given by differentiating Equation (10.21) with respect to the applied load, P, as follows:

$$u_v = \frac{\partial U}{\partial P} = \frac{1}{EI} \int_0^{\pi/2} \left[-PR\cos\varphi - QR(1 - \sin\varphi)\right].(-R\cos\varphi).R d\varphi$$

To simplify matters, the dummy load, Q, can be set to zero before the integration is performed. We then have,

$$u_v = \frac{PR^3}{EI} \int_0^{\pi/2} \cos^2\varphi d\varphi = \frac{PR^3}{EI} \int_0^{\pi/2} \frac{1}{2}(1 + \cos 2\varphi) d\varphi$$

$$= \frac{PR^3}{2EI} \left[\varphi + \frac{1}{2}\sin 2\varphi \right]_0^{\pi/2}$$

$$\therefore u_v = \frac{\pi PR^3}{4EI}$$

Horizontal displacement:

The horizontal displacement is now given by differentiating Equation (10.21) with respect to the dummy load, Q, as follows:

$$u_h = \frac{\partial U}{\partial Q} = \frac{1}{2EI} \int_0^{\pi/2} 2\left[-PR\cos\varphi - QR(1 - \sin\varphi)\right].\left[-R(1 - \sin\varphi)\right].R d\varphi$$

Again, by setting the dummy load, Q, equal to zero before integrating, we have

$$u_h = \frac{PR^3}{EI} \int_0^{\pi/2} \cos\varphi(1-\sin\varphi)d\varphi = \frac{PR^3}{EI} \int_0^{\pi/2} \cos\varphi - \frac{1}{2}\sin 2\varphi d\varphi$$

$$= \frac{PR^3}{EI}\left[\sin\varphi + \frac{1}{4}\cos 2\varphi\right]_0^{\pi/2}$$

$$\therefore u_h = \frac{PR^3}{2EI}$$

Note that the vertical deflection is $\pi/2 \times$ the horizontal deflection.

3.10.3.3 Combined Strain Energy Example

A circular cross-section offset cantilever, ABC, as shown in Figure 3.110(a), is loaded at its free end by a load, P.

Neglecting strain energy due to bending shear, derive an expression for the vertical deflection at the load point.

In this example, section AB of the cantilever stores bending strain energy only, while section BC stores both bending and torsional strain energy.

Strain energy in AB:

Referring to Figure 3.110(b), which shows a section, X-X, taken distance x from A, the bending moment is obtained by considering equilibrium as follows:

$$M + Px = 0 \qquad \therefore M = -Px$$

The strain energy, U_{AB}, is given by,

$$U_{AB} = \int \frac{M^2}{2EI}ds = \int_0^{L_1} \frac{P^2x^2}{2EI}dx = \frac{P^2L_1^3}{6EI}$$

[note the limits 0 to L_1 for the length]

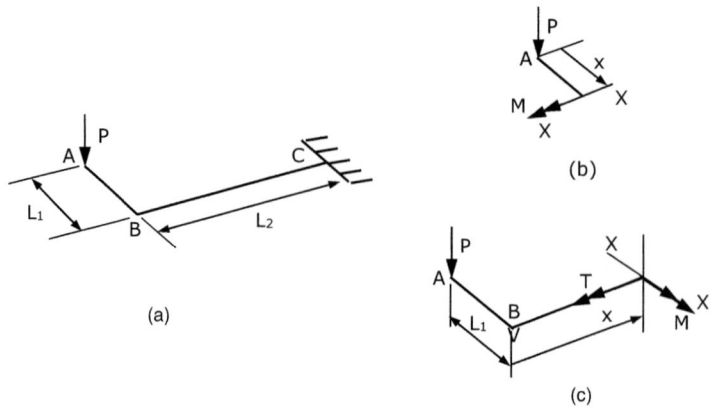

(a)

(b)

(c)

FIGURE 3.110   Offset cantilever beam with a point load applied at the free end.

Strain energy in BC:

Referring to Figure 3.110(c), which shows a section, X-X, in BC, taken distance x from B, the bending moment and torque are obtained by considering equilibrium as follows:

$$M + Px = 0 \qquad \therefore M = -Px$$

$$T + PL_1 = 0 \qquad \therefore T = -PL_1$$

The strain energy, U_{BC}, is now given by

$$U_{BC} = \int \frac{M^2}{2EI}\,ds + \int \frac{T^2}{2GJ}\,ds$$

$$= \int_0^{L_2} \frac{P^2 x^2}{2EI}\,dx + \int_0^{L_2} \frac{P^2 L_1^{\ 2}}{2GJ}\,dx$$

$$= \frac{P^2 L_2^{\ 3}}{6EI} + \frac{P^2 L_1^{\ 2} L_2}{2GJ}$$

So, the total strain energy in the cantilever is

$$U = U_{AB} + U_{BC} = P^2 \left[\frac{L_1^{\ 3} + L_2^{\ 3}}{6EI} + \frac{L_1^{\ 2} L_2}{2GJ} \right] \qquad (10.22)$$

Deflection of the tip:

The deflection of the tip is now given by differentiating Equation (10.22) with respect to the applied load, P, as follows:

$$u_v = \frac{\partial U}{\partial P} = 2P \left[\frac{L_1^{\ 3} + L_2^{\ 3}}{6EI} + \frac{L_1^{\ 2} L_2}{2GJ} \right]$$

LEARNING SUMMARY

By the end of this section, you should have learnt:

✓ The basic concept of strain energy stored in an elastic body under loading;
✓ How to calculate strain energy in a body/structure arising from various types of loading, including tension/compression, bending and torsion;
✓ Castigliano's theorem for linear elastic bodies which enables the deflection or rotation of a body at a point to be calculated from strain energy expressions

3.11 FATIGUE

3.11.1 INTRODUCTION

Fatigue failure of components and structures results from cyclic or repeated loading and the associated cyclic stresses and strains as opposed to failure due to monotonic or static stresses or strains, such as buckling or plastic collapse due to excessive plastic deformation. The topic of fatigue is extremely important in mechanical engineering since machines have moving parts, which in turn give rise to stresses and strains which may vary with time, typically in a repetitive fashion. For example, the axle of a car will transmit a time-varying torque, that changes from zero to some finite value when the car is put into gear and driven (and back to zero again when the car is taken out of gear).

An important design consideration, with respect to fatigue, is the fact that fatigue failure can occur at stresses which are well below the ultimate tensile strength of the material and often below the yield strength (or 0.2% proof stress).

3.11.2 BASIC PHENOMENA

The failure mechanism for an initially un-cracked component with a smooth (polished) surface can be split into three parts, namely crack initiation, crack propagation and final fracture, as follows:

   i. Stage I Crack growth: The micro-structural phenomenon which causes the initiation of a fatigue crack is the development of persistent slip bands at the surfaces of the specimen (Figure 3.111). These persistent slip bands are the result of dislocations moving along crystallographic planes leading to both slip band intrusions and extrusions on the surface. These act as excellent stress concentrations and can thus lead to crack initiation. A crystallographic slip is primarily controlled by shear stresses rather than normal stresses so that fatigue cracks initially tend to grow in a plane of maximum shear stress range. This stage leads to short cracks, usually only of the order of a few grains (Figure 3.112).
  ii. Stage II crack growth: As cycling continues, the fatigue cracks tend to coalesce and grow along the planes of the maximum tensile stress range.
 iii. Final fracture: This occurs when the crack reaches a critical length and results in either ductile tearing (plastic collapse) at one extreme or cleavage (brittle fracture) at the other extreme.

3.11.3 FATIGUE LIFE ANALYSIS

In order to allow for fatigue in the analysis and design of components, a number of different approaches are adopted; two of these approaches are described here. The more traditional approach is what is now referred to as the *total life approach*, based on laboratory tests, which are carried

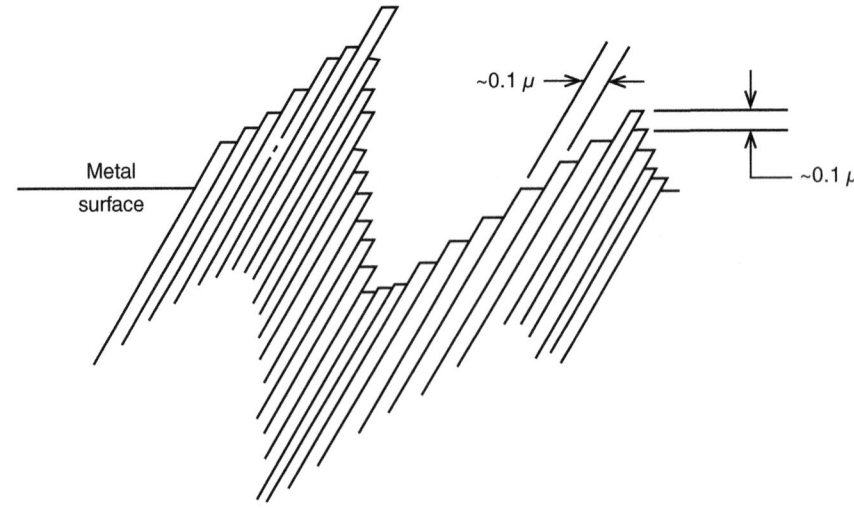

FIGURE 3.111   Persistent slip bands in ductile metals subjected to cyclic stress.

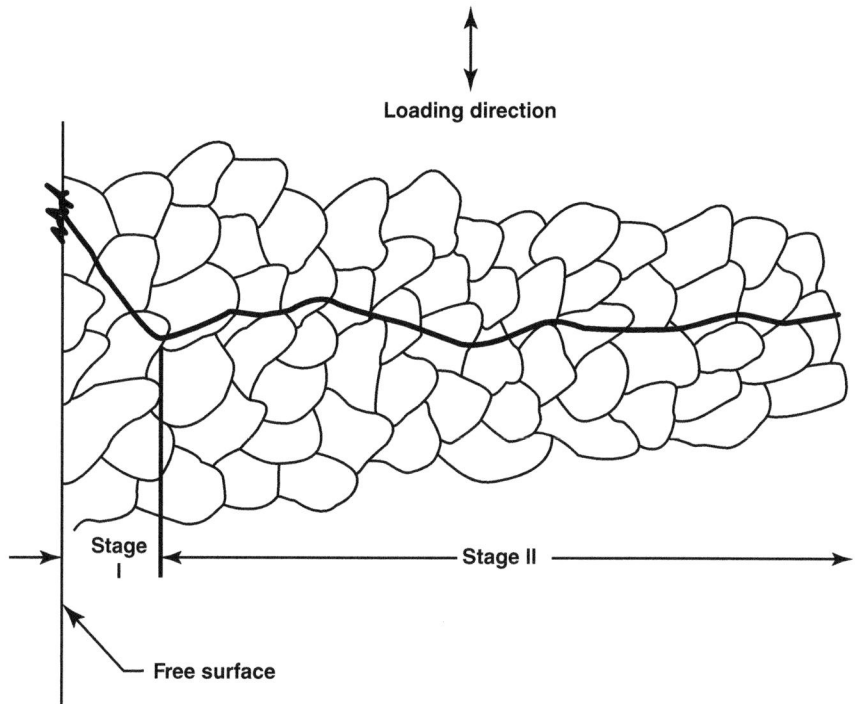

FIGURE 3.112   Schematic of Stages I and II transcrystalline microscopic fatigue crack growth.

out under either stress- or strain-controlled loading conditions on idealised specimens. These tests furnish the number of loading cycles to the initiation of a 'measurable' crack as a function of applied stress or strain parameters. The 'measurability' is dictated by the resolution accuracy of the crack detection method employed. A typical 'measurable' crack is about 0.75 mm to 1 mm. The challenge of fatigue design is then to relate these test results to the actual component that lives under real loading conditions. The second approach is known as the *damage-tolerant approach*. This approach is based on the inclusion of fatigue as a crack growth process, taking into account the fact that all components have inherent flaws or cracks. The development of fracture mechanics techniques to predict crack growth has facilitated this approach as a competing technique to the total life approach. Both of the approaches have advantages and disadvantages; the former has more appeal to design engineers while the latter is more often used by material scientists and researchers. Nonetheless, even in routine design, the damage-tolerant approach is gaining popularity.

3.11.4   TOTAL LIFE APPROACH

The total life approach is based on the results of stress- and strain-controlled cyclic testing of laboratory test specimens of material in order to obtain the number of cycles to failure as a function of the applied alternating stress, for example. Figure 3.113 shows a rotating bending test machine set-up. This is a constant load amplitude machine since the load doesn't change even with crack growth. The specimens usually have finely polished surfaces to minimise surface roughness effects, which would particularly affect Stage I growth. In this approach, no distinction is made between crack initiation and propagation. Stress concentration effects can be studied by machining in grooves, notches or holes.

Traditionally, most fatigue testing was based on fully-reversible (i.e. zero mean stress, $S_m = 0$), stress-controlled conditions and the fatigue design data were presented in the form of *S–N* curves,

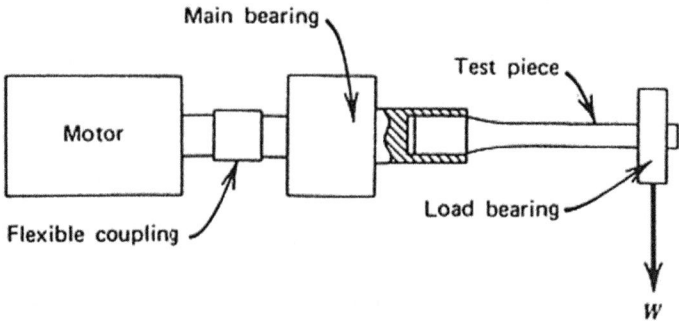

FIGURE 3.113   Rotating bending moment test apparatus for fully-reversed fatigue loading.

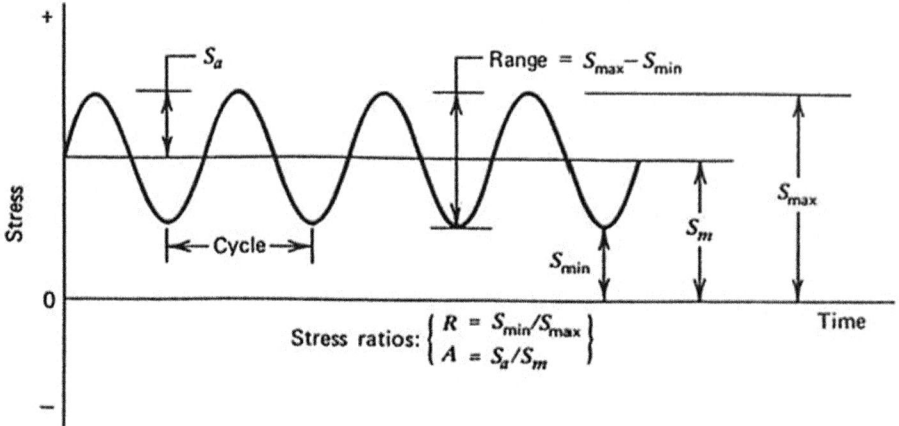

FIGURE 3.114   The notation used to describe constant load fatigue test cycles.

which are either semi-log or log-log plots of alternating stress, S_a, against the measured number of cycles to failure, N, where failure is defined as fracture. Some of the important stress parameters for cyclic loading are shown in Figure 3.114.

Figure 3.115 contains the schematic representations of two typical S–N curves obtained under load or stress-controlled tests on smooth specimens. Figure 3.115(a) shows a continuously sloping curve while Figure 3.115(b) shows a discontinuity or "knee" in the curve. A "knee" is only found in a few materials (notably low-strength steels) between 10^6 and 10^7 cycles under non-corrosive conditions. The curves are normally drawn through the median life value (of the scatter in N) and thus represent 50 percent expected failure. The *fatigue life*, N, is the number of cycles of stress or strain range of a specified character that a given specimen sustains before failure of a specified nature occurs. *Fatigue strength* is a hypothetical value of stress range at failure for exactly N cycles as obtained from an S–N curve. The *fatigue limit* (sometimes called the *endurance limit*) is the limiting value of the median fatigue strength as N becomes very large, for example, $>10^8$ cycles.

3.11.5   EFFECT OF MEAN STRESS

The alternating stress, S_a, and the mean stress, S_m, are defined in Figure 3.114. Early investigators of fatigue assumed that only the alternating stress affected the fatigue life of a cyclically-loaded component. However, it has since been established that the mean stress has a significant effect on fatigue behaviour, as shown in Figure 3.116. It can be seen that tensile mean stresses are detrimental while compressive mean stresses are beneficial.

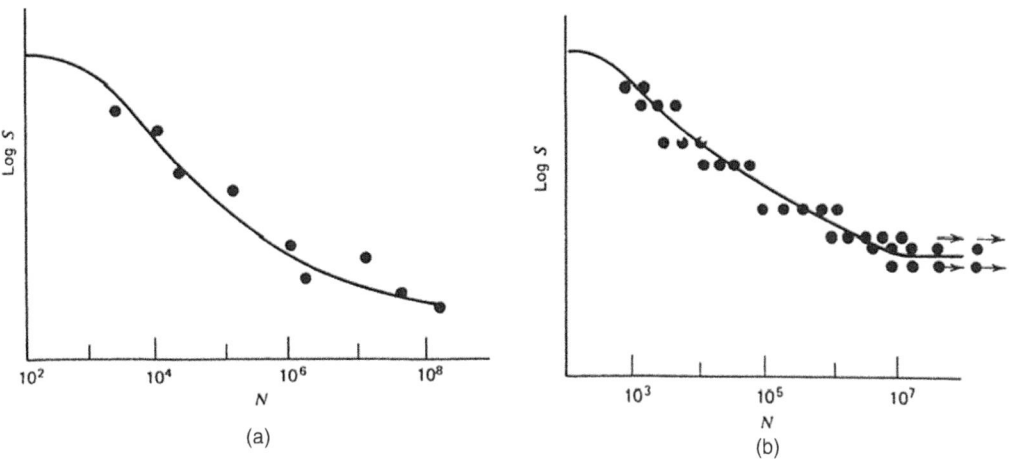

FIGURE 3.115   Typical S–N diagrams.

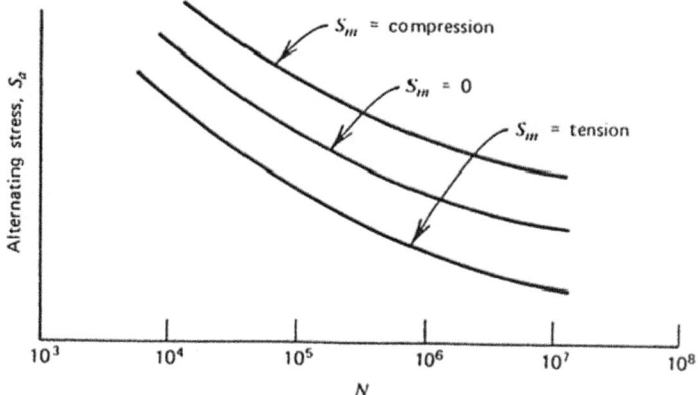

FIGURE 3.116   The effect of mean stress on fatigue life.

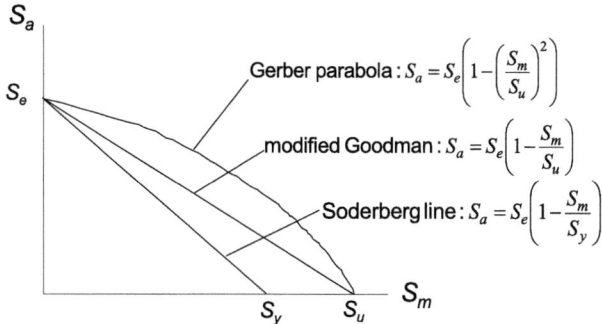

FIGURE 3.117   The effect of mean stress on fatigue life.

The effect of mean stress is commonly represented as a plot of S_a versus S_m for a given fatigue life. Attempts have been made to develop this relationship into general relations. Three of these common relations between allowable alternating stress for a given life as a function of mean stress are shown in Figure 3.117. The modified Goodman line assumes a linear relationship between the allowable S_a and the corresponding mean stress S_m, where the slope and intercepts are defined by the fatigue life, S_e, and the material UTS, S_u, respectively. The Gerber parabola employs the same end-points but, in this case, the relation is defined by a parabola. Finally, the Soderberg line again assumes a linear relation, but this time the mean stress axis end-point is taken as the yield stress, S_y. The modified Goodman line, for example, can be extended into the compressive mean stress region to give increasing allowable alternating stress with increasing compressive mean stress, but this is normally taken to be horizontal for design purposes and conservatism.

3.11.6 EFFECT OF STRESS CONCENTRATIONS

Ever since the first occurrences of fatigue failure, it has been recognised that such failures are most commonly associated with notch-type features in components. It is impossible to avoid notches in engineering structures, although the effects of such notches can be reduced through appropriate design. The stress concentration associated with notch-type features leads typically to local plastic strain which eventually leads to fatigue cracking. Consequently, the estimation of stress concentration factors associated with various types of notches and geometrical discontinuities has received a lot of attention. This is typically expressed in terms of an elastic stress concentration factor (SCF), K_t, which is simply the relationship between the maximum local stress and an appropriate nominal stress, as follows:

$$K_t = \frac{\sigma_{max}^{el}}{\sigma_{nom}}$$

It was once thought that the fatigue strength of a notched component could be predicted as the strength of a smooth component divided by the SCF. However, this is not the case. The reduction is, in fact, often less than K_t and is defined by the *fatigue notch factor, K_f*, which is defined as the ratio of the smooth fatigue strength to the notched fatigue strength as follows:

$$K_f = \frac{S_{a,smooth}}{S_{a,notch}}$$

However, this fatigue notch factor is also found to vary with both alternating and mean stress levels and thus with the number of cycles to failure. Figure 3.118 shows the effect of a notch, with an SCF of 3.4, on the fatigue behaviour of a wrought aluminium alloy, where the smooth lines are for the smooth specimen and the dotted lines are for the notched specimen. The table shows how the fatigue notch factor changes with mean stress level and fatigue life. Clearly, the fatigue notch factor increases from 3.2 to 5.7 from 10^4 cycles to 10^7 cycles at 172 MPa mean stress but remains unchanged between these lives at 2.3 for zero mean stress.

Mean stress	10^4 cycles	10^7 cycles
0 MPa	$51/22 = 2.3$	$22/9 = 2.3$
172 MPa	$42/13 = 3.2$	$17/3 = 5.7$

3.11.6.1 *S–N* Design Procedure for Fatigue

Constant life diagrams plotted as S_a versus S_m, also called Goodman diagrams, can be used in design to give safe estimates of fatigue life and loads.

  i. The Goodman line connects the endurance limit, S_e (or long life fatigue strength), to the U.T.S., S_u.

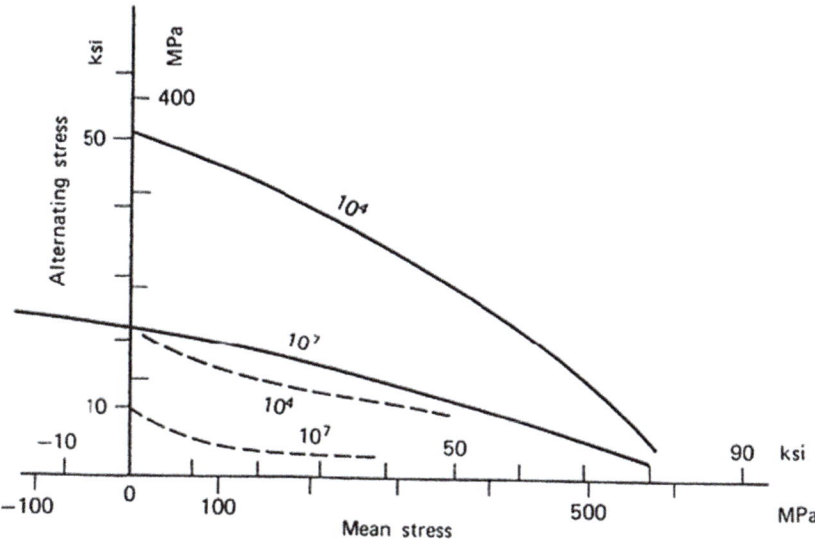

FIGURE 3.118 Constant life diagrams for a wrought aluminium alloy for both smooth and notched specimens.

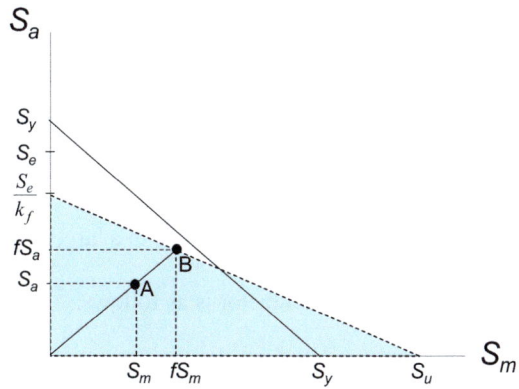

ii. The fatigue strength for zero mean stress is reduced by the fatigue notch factor, K_f. The stress concentration factor, K_t, is used if K_f is not known.

iii. For static loading of a ductile component with a stress concentration, failure still occurs when the mean stress is equal to the U.T.S. Failure at intermediate values of mean stress is assumed to be given by the dotted line.

iv. In order to avoid yield of the whole cross-section of the component, the maximum nominal stress must be less than the yield stress, S_y *i.e.* $S_m + S_a < S_y$

    a.   This relationship gives the yield line joining S_y to S_y.

v. The region of the diagram nearest to the origin is the 'safe' region (can also be extended to include compressive yield).

vi. A component is assessed by plotting the point corresponding to the nominal alternating stress, S_a, and the nominal mean stress, S_m, i.e., not the maximum values associated with the notch. The factor of safety is determined from the position of the point relative to the safe/fail boundary.

that is the factor of safety F = OB/OA

from similar triangles $\dfrac{S_a}{\left(\dfrac{S_u}{F} - S_m\right)} = \dfrac{S_e}{k_f S_u}$

$$\frac{1}{F} = \frac{S_a k_f}{S_e} + \frac{S_m}{S_u}$$

A procedure similar to that described above for long life can also be used to design for a specified number of cycles. In this case, the endurance limit and the fatigue notch factor are replaced by the fatigue strength and the fatigue notch factor for the specified number of cycles.

LEARNING SUMMARY

By the end of this section, you should have learnt:

✓ The various stages leading to fatigue failure;
✓ The basis of total life and the damage-tolerant approaches to estimating the number of cycles to failure;
✓ How to include the effects of mean and alternating stress on cycles to failure using the Gerber and modified Goodman and Soderberg methods;
✓ How to include the effect of a stress concentration on fatigue life;
✓ The S–N design procedure for fatigue life.

3.12 FRACTURE MECHANICS

3.12.1 LINEAR ELASTIC FRACTURE MECHANICS (LEFM)

Consider the stress concentration factor for an elliptical hole in a large, linear-elastic plate subjected to a remote, uniaxial stress depicted in Figure 3.119.

It can be shown that the stress concentration factor is as follows:

$$K_t = \frac{\hat{\sigma}}{\sigma} = 1 + \frac{2a}{b}$$

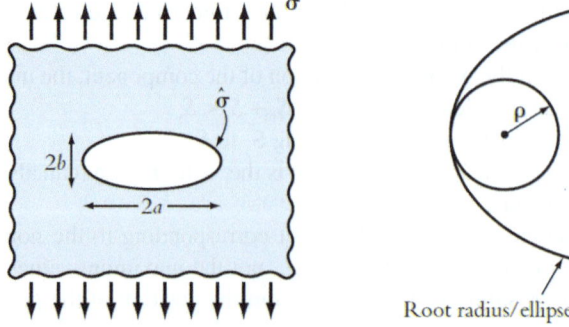

FIGURE 3.119 Elliptical hole in an infinite plate. Root radius ellipse.

Thus, as $b \rightarrow 0$, the elliptical hole degenerates to a crack, and $\dfrac{a}{b} \rightarrow \infty$, so that the notch stress also goes to infinity (i.e. becomes singular), $\dfrac{\hat{\sigma}}{\sigma} \rightarrow \infty$, provided the material behaviour remains linear elastic.

The root radius for an ellipse is given by

$$\rho = \frac{b^2}{a} \therefore b = \sqrt{a\rho}$$

so that

$$\frac{\hat{\sigma}}{\sigma} = 1 + 2\sqrt{\frac{a}{\rho}}$$

and again, as the notch tip radius goes to zero, that is, $\rho \rightarrow 0$, the notch tip stress again goes to infinity,

$$\frac{\hat{\sigma}}{\sigma} \rightarrow 2\sqrt{\frac{a}{\rho}} \rightarrow \infty$$

The singular (infinite) state of stress at a crack tip is one of the fundamental and most important aspects of fracture mechanics.

3.12.2 BASIS OF THE ENERGY APPROACH TO FRACTURE MECHANICS

Griffith (1921) studied the brittle fracture of glass and adopted an energy approach to solve the problem. He reasoned that unstable crack propagation occurs only if an increment of crack growth, da, results in more strain energy being released than is absorbed by the creation of the new crack surfaces. This can be re-expressed as the change in strain energy U, due to crack extension, being greater than the energy absorbed by the creation of the new crack surfaces. Thus, if we designate the surface energy per unit area of the crack γ_s, then the surface energy associated with a crack of length $2a$ in a body of thickness B (as shown in Figure 3.120) is given by:

$$W_s = 4aB\gamma_s$$

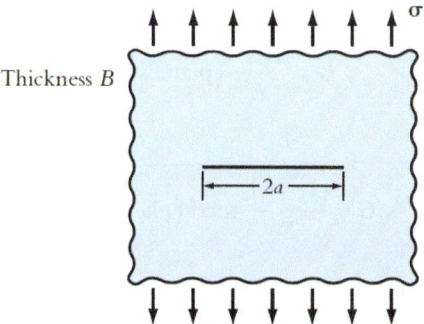

FIGURE 3.120 Crack in an infinite plate.

Detailed stress analysis of an elliptical hole in an infinite elastic plate has established that the strain energy in such a body is

$$W_p = -\frac{\pi a^2 \sigma^2 B}{E'}$$

where σ is the remote stress (away from the hole) and where, for plane strain and plane stress, respectively,

$$E' = \frac{E}{1-v^2} \text{ and } E' = E$$

The total system energy is thus

$$W = -\frac{\pi a^2 \sigma^2 B}{E'} + 4aB\gamma_s$$

According to Griffith, the critical condition for the onset of crack growth is

$$\frac{dU}{dA} = -\frac{\pi a \sigma^2}{E'} + 2\gamma_s = 0$$

Therefore,

$$\frac{\pi a \sigma^2}{E'} = 2\gamma_s$$

where $A = 2aB$ is the crack area and dA denotes an incremental increase in the crack area. The total surface area of the two crack surfaces is $2A$. This relationship is conventionally re-expressed as

$$G = G_c$$

where G is called the *strain energy release rate*, the *crack tip driving force* or the *crack extension force*. G_c is a material property, which is known as the *critical strain energy release rate*, the *toughness* or the *critical crack extension force*. A high value of G_c means that it is difficult to cause unstable crack growth in the material whereas a low value means it is easy to make a crack grow unstably. Thus, copper, for example, has a value of $G_c \approx 10^6$ Jm^{-2}, whereas glass has a value of $G_c \approx 10$ Jm^{-2}. The following relationships for plane stress and plane strain, respectively, follow from Equation (12.9):

$$G = \frac{\pi a \sigma^2}{E} \left(\text{plane stress}\right)$$

$$G = \frac{\left(1-v^2\right)}{E} \pi a \sigma^2 \left(\text{plane strain}\right)$$

Note that plane stress and plane strain are two contrasting two-dimensional assumptions which permit the simplification of three-dimensional problems to two-dimensional ones. Plane stress corresponds physically to thin-plate-type situations while plane strain corresponds to thick-plate-type situations. Plane strain testing of fracture leads to lower values of G_c, so the material property value of G_c for design purposes is taken as the plane strain value and is designated as G_{Ic}.

The critical stress, which causes a crack to propagate in an unstable fashion, giving fracture, is governed by the following relationships:

$$\sigma\sqrt{\pi a} = \sqrt{EG_c} \ (\text{plane stress})$$

$$\sigma\sqrt{\pi a} = \sqrt{\frac{EG_c}{1-v^2}} \ (\text{plane strain})$$

Since the term on the RH side of these equations is a material constant and since the term on the LH side is so common, it is usually abbreviated to the symbol, K, which is referred to as the *stress intensity factor* and the equations can be re-expressed as:

$$K = K_c$$

where K_c is called the *critical stress intensity factor* or the *fracture toughness*. Thus, $K_c = \sqrt{EG_c}$.
In summary,

$K = \sigma\sqrt{\pi a}$ is called the *stress intensity factor*
K_c is called the fracture *toughness* of *critical stress intensity factor*
G_c is called the *toughness* or the *critical strain energy release rate*.

Note Most materials are not linear elastic up to failure. However, the energy approach can still be used if the plastic strains are restricted to a region very close to the crack tip; this is referred to as *small-scale yielding*. Under these conditions, the energy release rate can still be reasonably accurate based on a linear elastic analysis. Also, G_c or G_{Ic} now includes a component associated with plastic deformation of the crack tip as well as the creation of the surfaces. So far, we have only considered the so-called Mode I loading case. There are actually three different loading modes considered in fracture mechanics – see Figure 3.121 and 3.122.

In general, the energy release rate under mixed-mode loading is given by

$$G_{\text{total}} = G_I + G_{II} + G_{III}$$

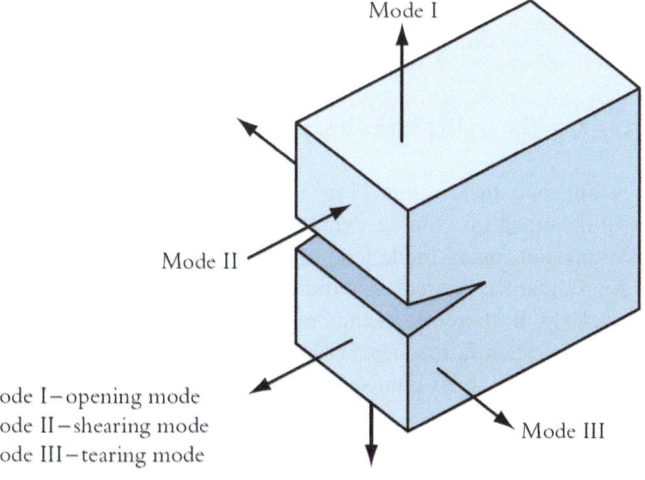

Mode I – opening mode
Mode II – shearing mode
Mode III – tearing mode

FIGURE 3.121 Crack tip loading modes.

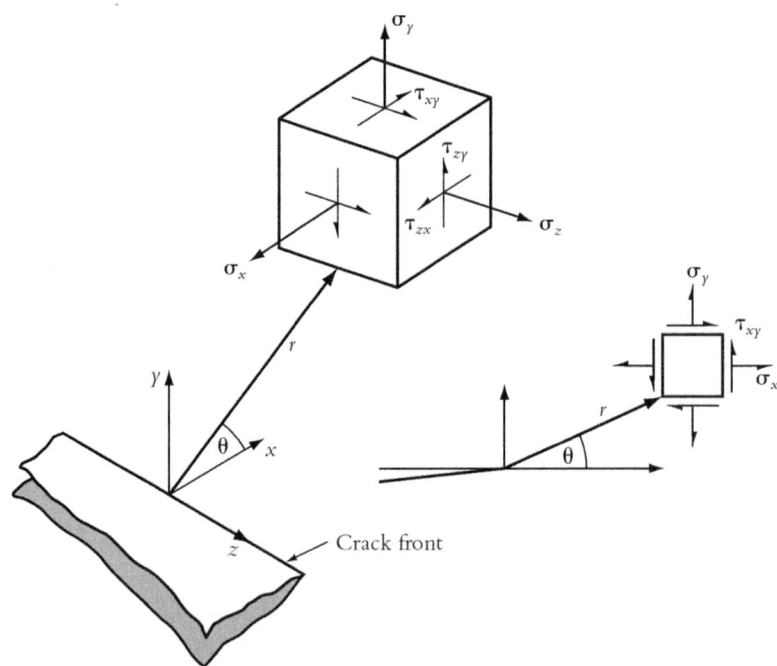

FIGURE 3.122   Crack tip stress fields.

3.12.3   ELASTIC CRACK TIP STRESS FIELDS

Westergaard (1939) established the following equations for the elastic stress field in the vicinity of a crack tip:

$$\sigma_x = \frac{K_I}{\sqrt{2\pi r}} \cos\left(\frac{\theta}{2}\right)\left[1 - \sin\left(\frac{\theta}{2}\right)\sin\left(\frac{3\theta}{2}\right)\right] + \text{non} - \text{singular terms}$$

$$\sigma_y = \frac{K_I}{\sqrt{2\pi r}} \cos\left(\frac{\theta}{2}\right)\left[1 + \sin\left(\frac{\theta}{2}\right)\sin\left(\frac{3\theta}{2}\right)\right] + \text{non} - \text{singular terms}$$

$$\tau_{xy} = \frac{K_I}{\sqrt{2\pi r}} \sin\left(\frac{\theta}{2}\right)\cos\left(\frac{\theta}{2}\right)\cos\left(\frac{3\theta}{2}\right) + \text{non} - \text{singular terms}$$

$$\tau_{zx} = \tau_{zy} = 0; \sigma_z = 0 \left(\text{plane stress}\right); \sigma_z = \nu\left(\sigma_x + \sigma_y\right)\left(\text{plane strain}\right)$$

K_I is the Mode I stress-intensity factor (units $N/m^{3/2}$) which defines the magnitude of the elastic stress field in the vicinity of the crack tip. Similar expressions exist, in terms of K_{II} and K_{III}, for Modes II and III loading situations. For mixed-mode loading, the stress fields can be added together directly. It can be seen that K_I, K_{II} and K_{III} characterise the entire stress field (and hence the strain fields) in the vicinity of the crack tip. It, therefore, seems reasonable to assume that for Mode I loading, for example, failure will occur when K_I reaches a critical value K_c (K_{Ic} under plane strain conditions).

The energy approach and the stress intensity approach are equivalent. Generally, for plane stress

$$G_{\text{total}} = G_I + G_{II} + G_{III} = \frac{1}{E}\left(K_I^2 + K_{II}^2 + K_{III}^2\right)$$

and for plane strain, $\dfrac{1}{E}$ is replaced by $\dfrac{\left(1-v^2\right)}{E}$.

Generally, for geometries with finite boundaries, the following expression is employed for the stress intensity factor

$$K_{\mathrm{I}} = Y\sigma\sqrt{\pi a}$$

and similarly for K_{II} and K_{III}, where Y is a function of the crack and component dimensions.

3.12.4 TYPICAL FRACTURE TOUGHNESS VALUES

Material	$\sigma_y \left(\mathrm{MN/m^2}\right)$	$K_{Ic}\left(\mathrm{MN/m^{3/2}}\right)$
Mild Steel	220	140 to 200
Pressure Vessel Steel (HY130)	1700	170
Aluminium Alloys	100 to 600	45 to 23
Cast Iron	200 to 1000	20 to 6

Solutions for Y can be found in the literature for a wide range of geometries and loadings.

The effect of finite boundaries on expressions for stress intensity factors

$$K_{\mathrm{I}} = \sigma\sqrt{\pi a}\left(\sec\frac{\pi a}{W}\right)^{1\backslash 2}$$

$$K_{\mathrm{II}} = \tau\sqrt{\pi a}\left(\text{small}\,\frac{a}{W}\right)$$

$$K_{\mathrm{I}} = 1.12\sigma\sqrt{\pi a}\left(\text{small}\,\frac{a}{W}\right)$$

or $\quad K_{\mathrm{I}} = Y\sigma\sqrt{a}$

with $\quad Y = 1.99 - 0.41\frac{a}{W} + 18.7\left(\frac{a}{W}\right)^2 - 38.48\left(\frac{a}{W}\right)^3$

$$+ 53.85\left(\frac{a}{W}\right)^4$$

$(1.99 = 1.12\sqrt{\pi})$

$$K_{\mathrm{I}} = 1.12\sigma\sqrt{\pi a}\left(\text{small}\,\frac{a}{W}\right)$$

or $\quad K_{\mathrm{I}} = Y\sigma\sqrt{a}$

with $\quad Y = 1.99 - 0.76\frac{a}{W} - 8.48\left(\frac{a}{W}\right)^2 + 27.36\left(\frac{a}{W}\right)^3$

$(1.99 = 1.12\sqrt{\pi})$

Thickness B

$$K_{\mathrm{I}} = \frac{PS}{BW^{3/2}}\left[2.9\left(\frac{a}{W}\right)^{1/2} - 4.6\left(\frac{a}{W}\right)^{3/2} + 21.8\left(\frac{a}{W}\right)^{5/2} - \right.$$

$$\left. - 37.6\left(\frac{a}{W}\right)^{7/2} + 38.7\left(\frac{a}{W}\right)^{9/2}\right]$$

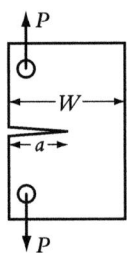

Thickness B

$$K_I = \frac{P}{BW^{1/2}}\left[29.6\left(\frac{a}{W}\right)^{1/2} - 185.5\left(\frac{a}{W}\right)^{3/2}\right.$$
$$\left. + 655.7\left(\frac{a}{W}\right)^{5/2} - 1017\left(\frac{a}{W}\right)^{7/2} + 63.9\left(\frac{a}{W}\right)^{9/2}\right]$$

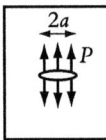

p per unit thickness

$$K_I = p\sqrt{\pi a}$$

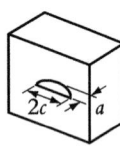

$$K_{Imax} = 1.12\frac{\sigma}{\Phi}\sqrt{\pi a}$$

$$K_{Imin} = 1.12\frac{\sigma}{\Phi}\sqrt{\pi a^2/c}$$

$$\Phi = \int_0^{\pi/2}\left[1 - \frac{c^2 - a^2}{c^2}\sin^2\varphi\right]d\varphi$$

$$\Phi \approx \frac{3\pi}{8} + \frac{\pi}{8}\frac{a^2}{c^2}$$

EXAMPLE

A large high-carbon steel plate with a thumbnail crack which $K\sqrt{\pi a}_{max}$ has a fracture toughness of 72 MN/m$^{3/2}$ and $\sigma_y = 1450$ MN/m^2.

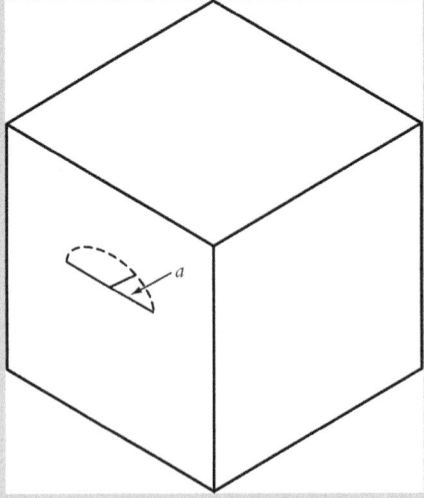

If $\sigma = \frac{2}{3}\sigma y$, determine the critical initial crack size assuming linear elastic material.

Solution

At fracture, with

$$\sigma = \frac{2}{3}\sigma y = \frac{2 \times 1450}{3} \frac{\text{MN}}{\text{m}^2}$$

and

$$K_{\text{IC}} = 72 \frac{\text{MN}}{m^{\frac{3}{2}}}$$

Then,

$$72 \frac{\text{MN}}{m^{\frac{3}{2}}} = 1.2 \times \left(\frac{(2 \times 1450)}{3}\right) \sqrt{\pi a_{\text{crit}}}$$

Therefore,

$$a_{\text{crit}} = \frac{1}{\pi}\left(\frac{72 \times 3}{1.2 \times 2 \times 1450}\right)^2 \text{m}$$
$$= 1.226 \times 10^{-3} \text{m}$$
$$= 1.226 \text{mm}$$

If the material was mild steel, with

$$\sigma_y = 210 \frac{\text{MN}}{m^2} \text{ and } K_{IC} = 200 \frac{\text{MN}}{m^{\frac{3}{2}}}$$

Then $a_{\text{crit}} = 0.451\text{m} = 451\text{mm}$.
ie, it is much more likely to be detected during inspection!

3.12.5 FATIGUE CRACK GROWTH

It has been shown by Paris and co-workers (1961) that, for a wide range of conditions, there is a logarithmic linear relationship between crack growth rate and the stress intensity factor range during cyclic loading of cracked components. Although this proposition had difficulty being accepted initially, it has become the basis of the damage-tolerant approach to fatigue life estimation and is widely used both in industry and in research. Essentially, it means that crack growth can be modelled and estimated based on the knowledge of crack and component geometry, loading conditions and using experimentally-measured crack growth data to furnish material constants. This section describes the basics of this approach.

Consider a load cycle as shown in Figure 3.123 which gives rise to a load range

$$\Delta P = P \min_{\max} \tag{12.1}$$

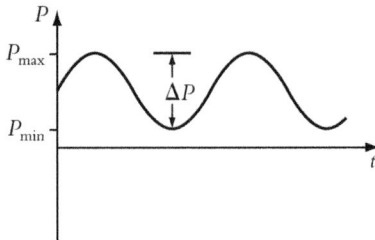

FIGURE 3.123 Variation of P(load) with t (time).

acting on a cracked body. The load range and crack geometry give rise to a cyclic variation in the stress intensity factor, which is given by

$$\Delta K = K \min_{\max} \tag{12.2}$$

Even though the stress intensity factor may be less than the critical stress intensity factor for unstable crack growth, stable crack growth may occur if the stress intensity range, ΔK, is greater than an empirically determined material property called the *threshold stress intensity factor range*, designated ΔK_{th}. In addition, Paris showed that the subsequent crack growth can be represented by an empirical relationship as follows:

$$\frac{da}{dN} = C\left(\Delta K\right)^m \tag{12.3}$$

where C and m are empirically-determined material constants. This relationship is known as the Paris equation. Fatigue crack growth data is often plotted as the logarithm of crack growth per load cycle, da/dN, and the logarithm of the stress intensity factor range. There are three stages. Below ΔK_{th}, no observable crack growth occurs; Region II shows an essentially linear relationship between $\log(da/dN)$ and $\log(\Delta K)$, where m is the slope of the curve and C is the vertical axis intercept; in Region III, rapid crack growth occurs and little life is involved. Region III is primarily controlled by K_c or K_{Ic} (Figures 3.124 and 3.125).

The linear regime (Region II) is the region in which engineering components that fail by fatigue propagation occupy most of their life. Knowing the stress intensity factor expression for a given component and loading, the fatigue crack growth life of the component can be obtained by integrating the Paris Equation between the limits of initial crack size and final crack size.

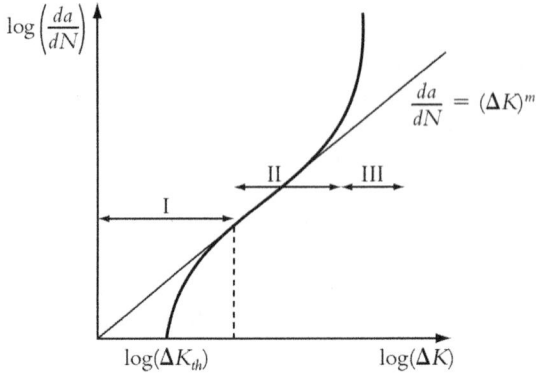

FIGURE 3.124 Typical (schematic) variation of log (da/dN) with log (ΔK).

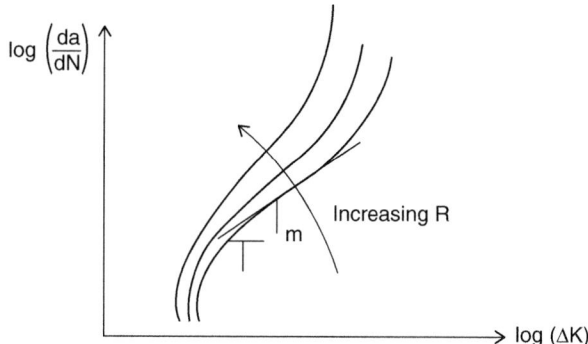

FIGURE 3.125   Effect of R on fatigue crack growth

For most materials, the constant C is found to be dependent on R where R is a measure of the mean stress which is defined as

$$R = \frac{K_{min}}{K_{max}}$$

Typical values of Δk_{th} m and Δk...:

Material	ΔK_{th} (MN/m$^{3/2}$)	m	ΔK (MN/m$^{3/2}$) or $da/dN = 10^{-6}$ mm/cycle
Mild Steel	4 to 7	3.3	6.2
316 Stainless Steel	4 to 6	3.1	6.3
Aluminium	1 to 2	2.9	2.9
Copper	1 to 3	3.9	4.3
Brass	2 to 4	4.0	4.3 to 66.3
Nickel	4 to 8	4.0	8.8

Source: From L.P. Pook, *J. Strain Analysis*, 1978, pp 114–135.

The ΔK_{th} and ΔK (for da/dN = 10^{-6} mm/cycle) values depend on the R-value.

LEARNING SUMMARY

By the end of this section, you should have learnt:

✓ The meaning of LEFM (linear elastic fracture mechanics);
✓ The three crack tip loading modes;
✓ The energy and stress intensity factor (Westergaard crack tip stress field) approaches to LEFM;
✓ The meaning of small-scale yielding and fracture toughness;
✓ About the Paris equation for fatigue crack growth and the effects of the mean and alternating components of the stress intensity factor.

3.13   THERMAL STRESSES

3.13.1   INTRODUCTION

Changes in the temperature of a body cause expansion/contraction. This phenomenon is quantified by the coefficient of thermal expansion α. In any direction, the change of length due to a temperature change, T, is $l\alpha T$, where l is the original length. Therefore, direct thermal strain is given by:

$$\varepsilon_{thermal} = \frac{l\alpha T}{l} = \alpha T$$

For isotropic materials, α is the same for all directions and there are no thermal shear strains.

Uniform temperature change throughout an unrestrained body produces uniform strain but no stress, that is free expansion/contraction in all directions.

3.13.2   GENERALISED HOOKE'S LAW

$$\varepsilon_x = \frac{1}{E}\left(\sigma_x - v\left(\sigma_y + \sigma_z\right)\right) + \alpha T$$

$$\varepsilon_y = \frac{1}{E}\left(\sigma_y - v\left(\sigma_x + \sigma_z\right)\right) + \alpha T$$

$$\varepsilon_z = \frac{1}{E}\left(\sigma_z - v\left(\sigma_x + \sigma_y\right)\right) + \alpha T$$

$$Y_{xy} = \tau_{xy} / G$$

$$Y_{yz} = \tau_{yz} / G$$

$$Y_{zx} = \tau_{zx} / G$$

where T is the temperature at a point relative to some datum.

By introducing these thermal strains into the generalised Hooke's Law, we can obtain solutions to thermal stress problems which are often very important in, for example, power and chemical plants, aeroengines and internal combustion engines (e.g. pistons and cylinder walls) etc.

CASE 6   AN INITIALLY STRAIGHT UNIFORM BEAM

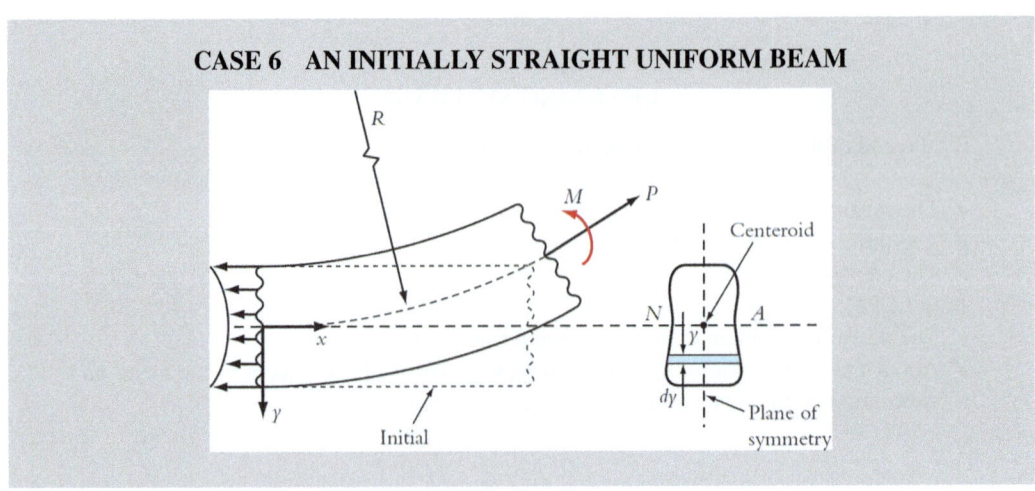

- Determine deformations and stresses (small deformations)

The temperature is (assumed) purely a function of y, ie $T = T(y)$.

The coefficient of thermal expansion $= \alpha$. Axial force P, and pure bending, about the z–z axis, M, are also applied.

$\sigma_z, \sigma_y, \tau_{xz}$ and $\tau_{yz} = 0$, because the cross-sectional dimensions are small compared with the length.

Also

$$\tau_{xy} = 0, \text{because M does not vary with } x, \therefore S = \frac{dM}{dx} = C$$

Compatibility

Remote from the ends, strain varies linearly with y,

$$\text{ie } \varepsilon_x = \bar{\varepsilon} + \frac{y}{R}$$

where $\bar{\varepsilon}$ is the mean strain (at $y = 0$) and R is the radius of curvature.

Stress-strain

$$\varepsilon_x = \frac{\sigma_x}{E} + \alpha T \left\{ = \bar{\varepsilon} + \frac{y}{R} \right\}$$

$$\therefore \sigma_x = E \left\{ \bar{\varepsilon} + \frac{y}{R} - \alpha T \right\} \tag{13.1}$$

Axial Equilibrium

$$P = \int_A \sigma_x dA = E \int_A \left\{ \bar{\varepsilon} + \frac{y}{R} - \alpha T \right\} dA = E\bar{\varepsilon}A + \frac{E}{R} \int_A y dA - E\varepsilon \int_A T dA$$

$\int_A y dA = 0$ because y is measured from an axis passing through the centroid.

$$\therefore P = E\bar{\varepsilon}A - E\alpha \int_A T dA \tag{13.2}$$

Moment Equilibrium

$$M = \int_A y\sigma_x dA = E \int_A \left\{ \bar{\varepsilon} + \frac{y}{R} - \alpha T \right\} y dA = E\bar{\varepsilon} \int_A y dA + \frac{E}{R} \int_A y^2 dA - E\alpha \int_A TydA$$

Now

$$\int_A y^2 dA = I \ (\text{second moment of area, by definition})$$

$$\therefore M = \frac{EI}{R} - E\alpha \int_A TydA \qquad\qquad (13.3)$$

EXAMPLE 3.13.1   A RECTANGULAR BEAM OF WIDTH b AND DEPTH d HAS A TEMPERATURE VARIATION GIVEN BY:

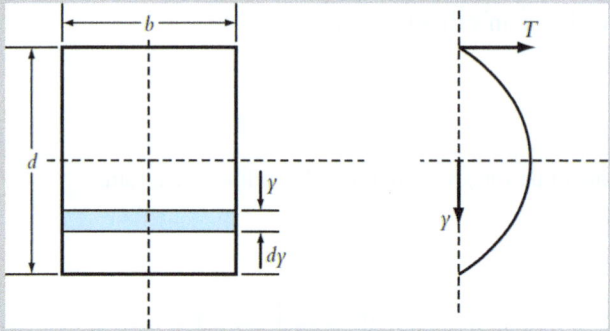

$$T = T_o\left\{1 - \frac{4y^2}{d^2}\right\}$$

There is no restraint or applied loading (ie $P = M = O$). Obtain the stress distribution.

Axial Force Equilibrium (Equation 13.2)

$$P = E\bar{\varepsilon} A - E\alpha \int_A TdA$$

$$0 = E\bar{\varepsilon}.bd - E\alpha \int_{-\frac{d}{2}}^{\frac{d}{2}} T_o\left\{1 - \frac{4y^2}{d^2}\right\}(bdy)$$

$$\therefore \bar{\varepsilon} = \frac{\alpha}{d} T_o \int_{\frac{-d}{2}}^{\frac{d}{2}}\left\{1 - \frac{4y^2}{d^2}\right\}dy$$

$$ie\ \bar{\varepsilon} = \frac{\alpha}{d} T_o \left[y - \frac{4y^3}{\left(\frac{d^2}{2}\right)3} \right]_{\frac{-d}{2}}^{\frac{d}{2}}$$

$$\therefore \bar{\varepsilon} = \frac{2}{3}\alpha.T_o$$

With $M = 0$ we can obtain $1/R$ from the moment equilibrium (Equation 13.3) but from symmetry we can see that $(1/R) = 0$.

Using Equation (13.1)

$$\sigma_x = E\left(\bar{\varepsilon} + \frac{y}{R} - \alpha T\right)$$

$$\sigma_x = E\left(\bar{\varepsilon} + \frac{y}{R} - \alpha T\right)$$

$$= E\left(\frac{2}{3}\alpha T_o + 0 - \alpha T_o\left(1 - \frac{4y^2}{d^2}\right)\right)$$

$$= E\alpha T_o\left(\frac{4y^2}{d^2} - \frac{1}{3}\right)$$

$$\text{At } y = 0, \sigma_x = \frac{-E\alpha T_o}{3}$$

$$\text{At } = \frac{\pm d}{2}, \sigma_x = E\alpha T_o\left(1 - \frac{1}{3}\right) = \frac{2E\alpha T_o}{3}$$

$$\sigma_x = 0 \text{ when } \frac{4y^2}{d^2} = \frac{1}{3} \text{ ie at } y = +/-0.287\,d$$

This is the stress distribution away from the ends. At the ends, $\sigma_x = 0$ and there is a gradual transition between the two.

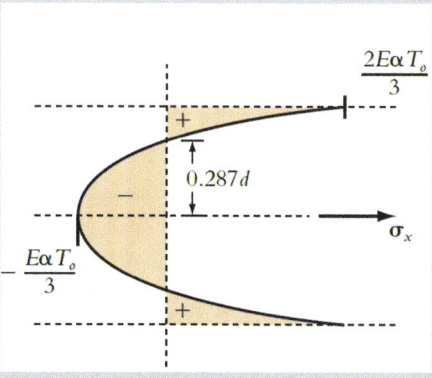

EXAMPLE 3.13.2   A RECTANGULAR BEAM (AGAIN $b \times d$), BUT WITH:

$$T = T_o \times \frac{2y}{d}$$

And the constrained so that $\varepsilon = 0$ and $1/R = 0$.
  Determine the stresses and restraints.

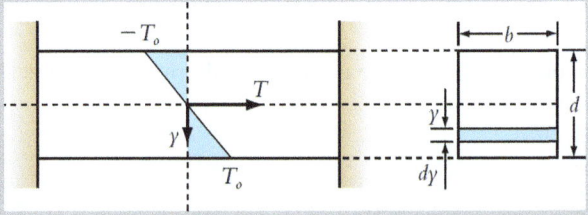

Axial Force Equilibrium (Equation 13.2)

$$= \bar{\varepsilon}A - E\alpha \int_A T dA$$

$$\int_A T dA = \frac{2T_o b}{d} \int_{\frac{-d}{2}}^{\frac{d}{2}} y \, dy = \frac{2T_o b}{d}\left[\frac{y^2}{2}\right]_{\frac{-d}{2}}^{\frac{d}{2}} = 0$$

Also,

$$\bar{\varepsilon} = 0 \therefore \underline{P = 0}$$

Moment Equilibrium (Equation 13.3)

$$M = \frac{EI}{R} - E\alpha \int_A T \, y dA$$

$$\int_A T y dA = \frac{2T_o b}{d} \int_{\frac{-d}{2}}^{\frac{d}{2}} y^2 dA = \frac{2Tob}{d}\left[\frac{y^3}{3}\right]_{\frac{-d}{2}}^{\frac{d}{2}}$$

$$= \frac{2T_o b}{d}\left[\left(\frac{d^3}{24}\right) - \left(\frac{-d^3}{24}\right)\right] = \frac{T_o b d^2}{6}$$

Also,

$$1/R = 0, \therefore M = \frac{-E\alpha b d^2 T_o}{6}$$

Using Equation (13.1) (with $\varepsilon = (1/R = 0)$

$$\sigma_x = \left(\bar{\varepsilon} + \frac{y}{R} - \alpha T \right) = -E\alpha.T$$

$$\therefore \sigma_x = -E\alpha T_o \frac{2y}{d}$$

i.e.

$$\sigma_X = \frac{-2E\alpha T_o}{d} y$$

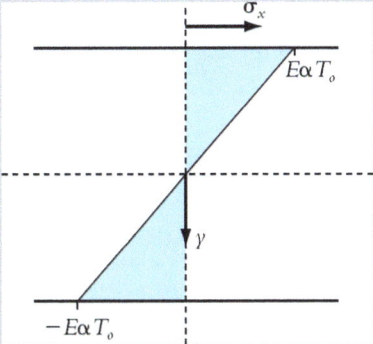

CASE 7   A THIN DISC OF UNIFORM THICKNESS

The equilibrium and compatibility equations are as those for rotating discs, with $\omega = 0$, that is

Equilibrium (radial)

$$\sigma_r + r\frac{d\sigma_r}{dr} - \sigma_\theta = 0 \tag{13.4}$$

Compatibility

$$\varepsilon_\theta = \frac{u}{r} \tag{13.5a}$$

$$\varepsilon_r = \frac{du}{dr} \tag{13.5b}$$

$$\text{ie } \varepsilon_r = \frac{d}{dr}\left(r\varepsilon_\theta\right) = \varepsilon_\theta + r\frac{d\varepsilon_\theta}{dr} \tag{13.6}$$

However, the stress–strain equations now contain thermal terms.

$$ie \ \varepsilon_\theta = \frac{1}{E}\left(\sigma_\theta - v\sigma_r\right) + \alpha T \tag{13.7a}$$

$$\varepsilon_r = \frac{1}{E}\left(\sigma_r - v\sigma_\theta\right) + \alpha T \tag{13.7b}$$

[Note: Because the disc is thin, $\sigma_z = 0$].
Substituting Equations (13.7a) and (13.7b) into Equation (13.6) gives

$$\left(\frac{1}{E}\left(\sigma_r - v\sigma_\theta\right) + \alpha T\right) = \left(\frac{1}{E}\left(\sigma_\theta - v\sigma_r\right) + \alpha T\right) + \frac{r}{E}\left(\frac{d\sigma_\theta}{dr} - v\frac{d\sigma_r}{dr}\right) + r\alpha\frac{dT}{dr}$$

$$\therefore \left(1+v\right)\left(\sigma_\theta - \sigma_r\right) + r\frac{d\sigma_\theta}{dr} - vr\frac{d\sigma_r}{dr} + E\alpha r\frac{dT}{dr} = 0$$

Substitute $\sigma_\theta - \sigma_r = r\dfrac{d\sigma_r}{dr}$ from (13.4)

$$\therefore \left(1+v\right)r\frac{d\sigma_r}{dr} + r\frac{d\sigma_\theta}{dr} - vr\frac{d\sigma_r}{dr} + E\alpha.r\frac{dT}{dr} = 0$$

$$ie, r\frac{d}{dr}\left(\sigma_\theta + \sigma_r\right) = -E\alpha.r\frac{dT}{dr} = 0$$

Also from (13.4):

$$\sigma_\theta = \sigma_r + r\frac{d\sigma_r}{dr}$$

$$\therefore r\frac{d}{dr}\left(2\sigma_r + r\frac{d\sigma_r}{dr}\right) = -E\alpha r\frac{dT}{dr}$$

$$r\left(2\frac{d\sigma_r}{dr} + r\frac{d^2\sigma_r}{dr} + \frac{d\sigma_r}{dr}\right) = -E\alpha r\frac{dT}{dr}$$

$$ie, r^2\frac{d^2\sigma_r}{dr^2} + 3r\frac{d\sigma_r}{dr} = -E\alpha r\frac{dT}{dr}$$

$$\therefore \sigma_r = A - \frac{B}{r^2} + P.I.$$

From (13.4):

$$\sigma_\theta = \sigma_r + r\frac{d\sigma_r}{dr}$$

Example: A steel disc with a 300 mm outside diameter, 40 mm bore diameter and 5 mm thickness is subjected to a temperature distribution of the following form:

$$T = \left(\frac{200r^2}{150^2}\right)^o C$$

where r is the radius in mm. Find the hoop stress at the bore.

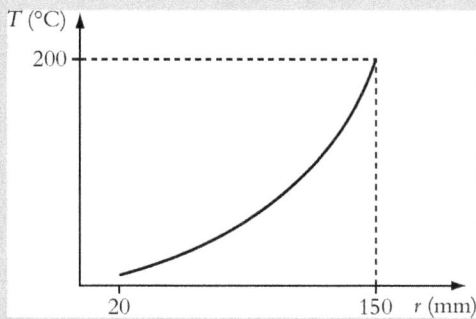

The governing differential equation is:

$$r^2 \frac{d^2\sigma_r}{dr^2} + 3r\frac{d\sigma_r}{dr} = -E\alpha r\frac{dT}{dr} = -E\alpha r\frac{400r}{150^2} = 400E\alpha\frac{r^2}{150^2}$$

$$P.I. = Cr^2 \text{ (say)}$$

$$\therefore r^2.2C + 3r.2Cr = -400E\alpha\frac{r^2}{150^2}$$

$$\text{ie } C = \frac{-50E\alpha}{150^2}$$

$$\therefore \sigma_r = A - \frac{B}{r^2} - 50E\alpha\frac{r^2}{150^2}$$

Boundary conditions:
At $r = 20$ mm, $\sigma_r = 0$

$$\therefore 0 = A - \frac{B}{20^2} - 50E\alpha\frac{20^2}{150^2} \qquad (13.a)$$

At $r = 150$ mm, $\sigma_r = 0$

$$\therefore 0 = A - \frac{B}{150^2} - 50E\alpha\frac{150^2}{150^2} \qquad (13.b)$$

Subtracting Equation (13.b) from Equation (13.a) gives:

$$0 = B\left(\frac{1}{22500} - \frac{1}{400}\right) + 50E\alpha\left(1 - \frac{400}{150^2}\right)$$

Substituting Equation b in Equation (13.b) gives:

$$\text{ie } \underline{B = 20000E\alpha}$$

$$A = \frac{20000E\alpha}{22500} + 50E\alpha = \underline{50.88E\alpha}$$

$$\text{So } \sigma_r + r\frac{d\sigma_r}{dr} = A - \frac{B}{r^2} - 50E\alpha\frac{r^2}{150^2} + r\left(\frac{2B}{r^3} - \frac{100E\alpha r}{150^2}\right)$$

$$\text{ie } \sigma_\theta = A + \frac{B}{r^2} - 150E\alpha\frac{r^2}{150^2} = \left(50.88 + \frac{20000}{r^2} - \frac{r^2}{150}\right)E\alpha$$

At $r = 20$:

$$\sigma_\theta = \left(50.88 + \frac{20000}{400} - \frac{400}{150}\right)E\alpha = 98.22E\alpha$$

$$\text{ie } \sigma_\theta = 223.9\text{N}/\text{mm}^2 \ (\text{at } r = 20\,\text{mm})$$

$$\text{Also, } \sigma_\theta = A - \frac{B}{r^2} - 50\frac{E\alpha r^2}{150^2}$$

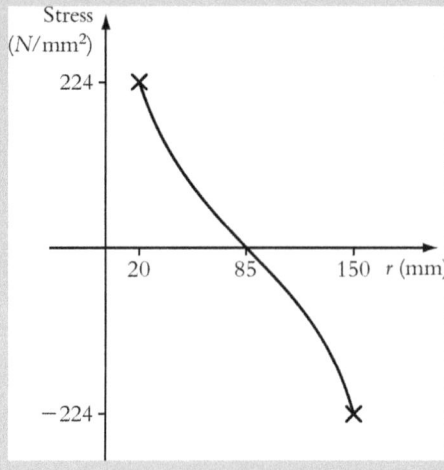

CASE 8   THIN CYLINDERS

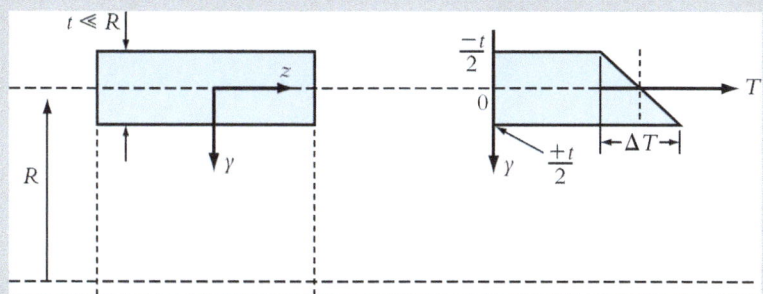

Thin cylinders are in common use in power and chemical plants, e.g. pipes, pressure vessels, etc. Often, temperature variations are approximately linear through the thickness, considering positions remote from ends, flanges, etc. It is convenient to consider the effect of the uniform temperature change and the temperature gradient separately. If the cylinder is not restrained, then the uniform temperature change causes overall dimensional changes, but no stress. The stresses due to axial restraint are easily calculated.

For the temperature gradient, we have:

$$T = \Delta T \frac{y}{t}$$

where ΔT is the temperature difference across the wall.

For a thin cylinder:

$$\sigma_r \cong 0$$

$$\therefore \varepsilon_{,\theta} = \frac{1}{E}\left(\sigma_\theta - v\sigma_z\right) + \alpha T$$

and

$$\varepsilon_z = \frac{1}{E}\left(\sigma_z - v\sigma_\theta\right) + \alpha T$$

Remote from the ends of the cylinder sections remain plane and circular. Therefore, from compatibility considerations (with zero mean temperature change), the hoop and axial strains must both be zero. Therefore,

$$0 = \frac{1}{E}\left(\sigma_{,\theta} - v\sigma_z\right) + \alpha\left(\Delta T\right)\frac{y}{t}$$

and

$$0 = \frac{1}{E}\left(\sigma_z - v\sigma_\theta\right) + \alpha\left(\Delta T\right)\frac{y}{t}$$

Solving, gives

$$\sigma_\theta = \sigma_z = \frac{E\alpha(\Delta T)}{(1-v)}\left(\frac{y}{t}\right)$$

LEARNING SUMMARY

By the end of this section, you should have learnt:

✓ The definition of thermal strain and how "thermal stresses" are caused by thermal strains;
✓ How to include thermal strains in the generalised Hooke's Law equations;
✓ How to include the temperature distribution within a solid component (e.g. a beam, a disc or a tube) in the solution procedure for the stress distribution;
✓ The stress/strain equations include thermal strain terms but the equilibrium and compatibility equations are the same whether the component is subjected to thermal loading or not.

4 Electromechanical Drive Systems

Arthur Jones and Alan Howe
University of Nottingham, Nottingham, United Kingdom

4.1 INTRODUCTION TO ELECTROMECHANICAL DRIVE SYSTEMS

In many engineering situations, it is necessary to design a system that involves a source of mechanical power to drive (directly or indirectly) a mechanical load of some kind, with one or more rotating shafts often being used within the means of transmitting the power from source to load. Such systems can range from very small (for instance, focusing mechanisms within cameras) to extremely large (e.g. driving the propellers of a transatlantic liner) in size. In all cases, it is essential to understand the characteristics of the load we are trying to drive, the characteristics of the source of mechanical power that we are using to drive it and the need for (and function of) any transmission mechanism to match the characteristics of the power source and load. This unit will therefore examine in depth the steady-state and transient characteristics of loads and typical sources of the mechanical power used to drive them, including electric motors and other forms of motors and engines. It is also useful to be able to alter the characteristics of the power source in order to provide control over the way that the overall system behaves, so the techniques by which electric motors can be controlled will be explored in some depth. The manner in which the power source and load interact is examined, with particular reference to how apparently incompatible characteristics can be matched in order to achieve stable operation in the desired manner. Although this unit draws together topics from the disciplines of dynamics, drive systems, electrical engineering, fluid power and mechanical design, various texts are available that follow this kind of approach in greater depth, for example, the text by Fraser and Milne (1994).

4.2 CHARACTERISTICS OF LOADS

As hinted in the introduction, in order to arrive at an optimal decision on how to drive a machine, it is necessary to have a full understanding of how that machine behaves as a mechanical load. Specifically, the following questions need to be addressed.

- How hard is it to get the machine moving from rest? (friction: 'stiction' and Coulomb friction)
- How hard is it to speed it up? (inertia)
- How much harder does it get to drive as we make it go faster? (torque–speed characteristics including viscous friction and windage)
- How hard is it to make it stop when we want to? (inertia again, with some friction considerations)
- How well do the characteristics of the machine match with those of the device we are using to drive it, and how can we overcome any mismatch?

Some of these issues can be illustrated with reference to issues familiar to automobile drivers.

DOI: 10.1201/9780429319495-4

- A car filled to its maximum design capacity with heavy people is much less responsive (e.g. when pulling away from traffic lights) than when carrying only the driver. **The inertia of the load has increased and is no longer very well matched to the power source.**
- A car driven at a steady 50 mph uses considerably less energy (e.g. in the form of petrol/gasoline or stored electrical energy) to travel a given distance than one driven at the current (2024) legal UK motorway limit of 70 mph. The **steady-state torque required from the car engine is determined (in a large part) by wind drag which is roughly proportional to the square of the speed**. (The drag force falls to something like $(50/70)^2 = 0.51$ of its original value, though there are various other losses that show less of a reduction).
- It is possible to tow quite a heavy trailer with a car as long as the driver allows for poorer acceleration. However, such a trailer must include its own braking system (various mechanisms exist) because it would otherwise be difficult to stop the car/trailer combination. **The inertia of the system is no longer well matched to the braking system (nor, for that matter to the power source).**
- When a steep hill is reached, the car will tend to slow down, and if it is driven by an internal combustion engine (ICE), the driver will need to change down a gear (and probably settle for driving at a lower speed if currently travelling fast). **The power of the motor or engine is limited, and in top gear, an ICE is badly matched to the increased load. So, to obtain the required torque it is necessary to gear the engine down to obtain a high torque at low speed. An electric vehicle will also have less torque at higher speeds and hence may also tend to slow down.**

4.3   LINEAR AND ROTARY INERTIA

The concept of inertia is based on Newton's second law:

$$F = ma \tag{4.1}$$

where F is the force required to cause a mass m to undergo an acceleration of a. In this context, mass is a measure of how difficult it is to accelerate something – in other words, mass is a measure of **linear inertia**.

In engineering, we are often concerned with rotational movements, and so we need to introduce the concept of **rotational inertia**; in other words, how difficult it is to cause rotational acceleration in a body that is capable of rotation about an axis. This is called the **moment of inertia**, a concept which was covered in Section 6.7 of Part 1 and is used within Unit 6 of the present volume in the context of torsional vibration. It is briefly revised here for continuity.

4.3.1   MOMENT OF INERTIA

To avoid confusion with electric current I, the symbol J will be used here for the moment of inertia of a body about its axis of rotation. Moment of inertia is analogous to mass in the rotational version of Newton's second law:

$$\text{Torque} = \text{moment of inertia} \times \text{angular acceleration}$$
$$\text{i.e.} \quad L = J\alpha \tag{4.2}$$

For a body B rotating about a given axis OZ (Figure 4.1), J_{OZ} is calculated as follows:

$$J_{OZ} = \int_B r_z^2 \, dm \tag{4.3}$$

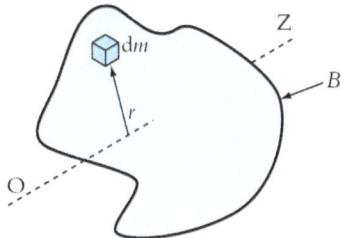

FIGURE 4.1   A body with the axis of rotation OZ.

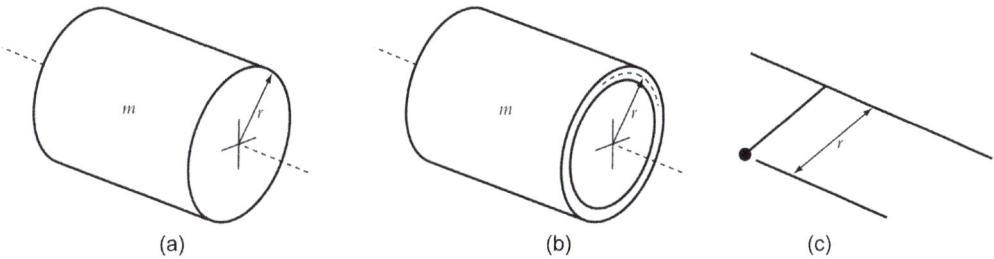

FIGURE 4.2   (a) Solid cylinder, (b) thin-walled tube (c) concentrated mass at radius r from the axis of rotation.

By way of a specific example, a **solid cylinder** or a disc of mass m and radius r (Figure 4.2(a)) has a moment of inertia

$$J = \frac{1}{2}mr^2 \tag{4.4}$$

about its axis of symmetry.

By contrast, a **thin-walled tube** or a thin circular ring of mass m and radius r (Figure 4.2(b)) has a moment of inertia

$$J = mr^2 \tag{4.5}$$

about its axis of symmetry.

Finally, a **concentrated mass** m being swung at radius r about a given axis (Figure 4.2(c)) has a moment of inertia

$$J = mr^2 \tag{4.6}$$

about that axis. An example of such a system would be a golf club with a heavy head and a light handle, pivoting around the golfer's wrist.

Inertias on the same axis are simply added together; objects with portions removed are treated by subtracting the inertias of the 'missing' portions from the inertia of the overall object. For example, the moment of inertia of a hollow cylinder of outer radius r_o and inner radius r_i about its axis

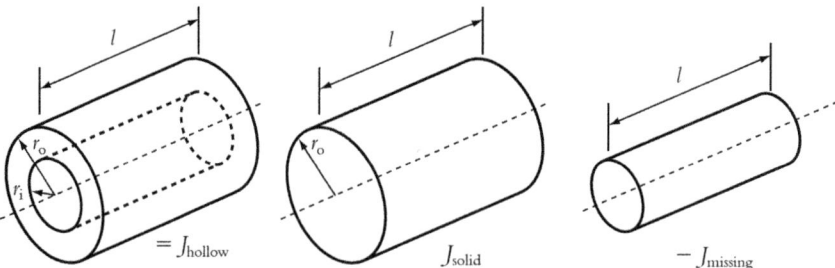

FIGURE 4.3 Combining inertias.

(Figure 4.3) is found by subtracting the moment of inertia of the 'missing' inner cylinder (of mass m_i) from the moment of inertia of the solid cylinder (of mass m_o):

$$
\begin{aligned}
J_{\text{hollow}} &= J_{\text{solid}} - J_{\text{missing}} \\
&= \frac{1}{2} m_o r_o^2 - \frac{1}{2} m_i r_i^2 \\
&= \frac{1}{2} \left(\pi r_o^2 \rho l \right) r_o^2 - \frac{1}{2} \left(\pi r_i^2 \rho l \right) r_i^2 \\
&= \frac{\pi}{2} \rho l \left(r_o^4 - r_i^4 \right)
\end{aligned}
\tag{4.7}
$$

which can be expressed as

$$
\begin{aligned}
J_{\text{hollow}} &= \frac{\pi}{2} \rho l \left(r_o^4 - r_i^4 \right) \\
&= \frac{\pi}{2} \rho l \left(r_o^2 - r_i^2 \right) \left(r_o^2 + r_i^2 \right) \\
&= \frac{1}{2} m_{\text{hollow}} \left(r_o^2 + r_i^2 \right)
\end{aligned}
\tag{4.8}
$$

Superficially, this appears to give a different answer from the formula relating to thin-walled cylinders (Figure 4.2(b)), but for such a cylinder r_o and r_i tend towards being equal as the wall thickness tends to zero, and J_{hollow} approaches the familiar value of mr^2.

LEARNING SUMMARY

At the end of this section, you should have learnt:

- ✓ The similarities and differences between linear and rotational inertias and how they are analysed.
- ✓ The concept of moment of inertia.
- ✓ To calculate the moment of inertia for simple components made up of cylinders and tubes.

4.4   GEARED SYSTEMS

In order to extend the above concepts to geared systems, the concept of a **gear ratio** must also be defined (Figure 4.4):

$$\text{Gear ratio } n = \frac{N_2}{N_1} = \frac{\omega_1}{\omega_2} = \frac{\alpha_1}{\alpha_2} = \frac{L_2}{L_1} \tag{4.9}$$

where N_1 and N_2 are the number of teeth on the gears mounted on the input and output shafts, respectively, ω_1 is the angular velocity (in rad/s), α_1 is the angular acceleration (in rad/s^2) and L_1 is the torque (in Nm) for the input shaft, with corresponding terms ω_2 etc. referring to the output shaft. Note that Equation (4.9) is also valid if the angular velocities are expressed consistently in other units, for example, rev/min or rev/s; this validity does *not* extend to equations that involve calculations of actual values of power. A similar argument applies to accelerations: calculations involving inertias and accelerations must be performed using rad/s^2 as the unit of angular acceleration because Equation (4.2) is not valid if units such as rev/s^2 are used.

The angular accelerations experienced by the different shafts are related in just the same way as the speeds (angular velocities). The last term in Equation (4.9) comes from the law of conservation of energy for a perfectly efficient geared system:

$$\text{power out} = \text{power in}$$

i.e.

$$L_2\omega_2 = L_1\omega_1 \tag{4.10}$$

so that input torque L_1 can be calculated from L_2

$$L_1 = \frac{L_2\omega_2}{\omega_1} = \frac{L_2}{n} \tag{4.11}$$

The same theory relates to other toothed drive systems, specifically sprocket/chain systems (such as those used in bicycles, motorcycles and some industrial situations) and toothed pulley/belt systems,

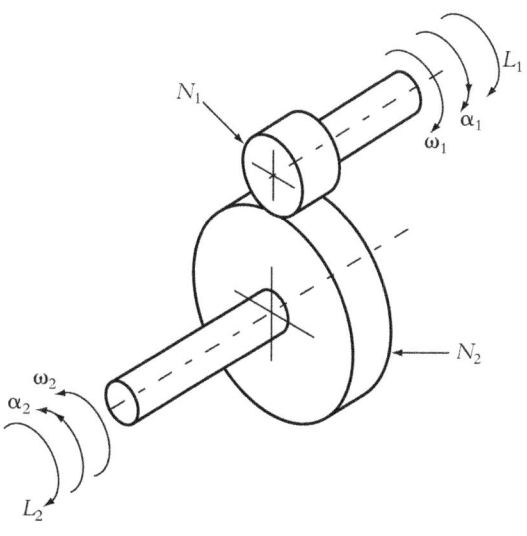

FIGURE 4.4   A pair of gears showing torques, speeds etc.

used for the camshaft drive in cars and within numerous motion control and positioning systems in industry and within more light-duty systems such as traversing the heads of inkjet printers and moving the scanning sensor on photocopiers. All of these may be regarded as synchronous drives in that the drive ratios are exactly determined by the number of teeth on the driving and driven components. One difference between gears and other drive systems such as chain drives and toothed belt drives is that gears reverse the direction of drive while belts and chains normally do not.

WORKED EXAMPLE

A machine is driven at 35 rad/s with a torque of 30 Nm. The drive is provided via a transmission with a 4:1 reduction ratio drive. What is the combination of torque and speed required at the input to the transmission system? Assume that the transmission is perfectly efficient.

For speed,

$$n = \frac{\omega_1}{\omega_2}$$

where $n = 4$, $\omega_2 = 35 \,\text{rad}/\text{s}$

$$\therefore \omega_1 = 4\omega_2 = 4 \times 35 = 140 \,\text{rad}/\text{s}$$

For torque,

$$n = 4 = \frac{L_1}{L_2}$$

where $L_2 = 30 \,\text{Nm}$

$$\therefore L_1 = \frac{L_2}{4} = \frac{30}{4} = 7.5 \,\text{Nm}$$

A broadly similar analysis can be performed for friction-based drive systems, though these are not synchronous because the components no longer possess teeth, and some degree of slippage can take place between driving and driven components. The nominal drive ratio is now determined by the ratio of the effective radii (or diameters) of the components (Figure 4.5):

$$\text{Drive ratio } n = \frac{r_2}{r_1} = \frac{d_2}{d_1} = \frac{\omega_1}{\omega_2} = \frac{\alpha_1}{\alpha_2} \qquad (4.12)$$

The most commonly encountered friction-based transmissions are vee belt/pulley systems (used extremely widely, an everyday example being the drive from a car engine to alternator), flat belt systems as shown in Figure 4.6(a) (less widely used today; examples include belt-drive turntables used for playing vinyl LP records) and systems based upon friction rollers (rarely used today, but still found within fairground and amusement rides, for example for 'gearing' rotating carriages to the rotation of the frame which carries them or to provide the impetus to swinging 'boats' from a motorised wheel/tyre running on the bottom of the boat). Toothed belts (Figure 4.6(b) are widely used in a variety of applications where a positive drive is needed and are sometimes known as 'timing belts' due to their use in driving the camshaft etc. in ICEs as they control the timing of the valve events. In the cases of vee belts, flat belts etc., care must be taken to understand the concept of the effective radius or pitch circle radius, since this may be significantly different from any of the measurable radii of the pulley (Figure 4.6).

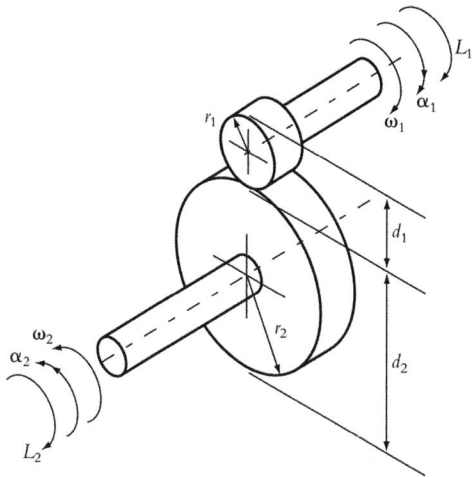

FIGURE 4.5 A pair of wheels coupled by friction.

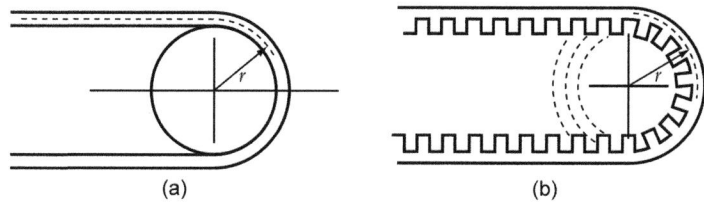

FIGURE 4.6 (a) Flat belt and (b) toothed belt showing effective radius.

WORKED EXAMPLE

A centrifugal blower needs to be driven at 3000 rev/min from a motor that runs at 1460 rev/min. If the effective diameter of the pulley on the blower is 100 mm, what should the effective diameter of the pulley on the motor be?

Answer

$$\text{Drive ratio } n = \frac{d_2}{d_1} = \frac{\omega_1}{\omega_2}$$

where $d_2 = 100\,\text{mm}, \omega_1 = 1460\,\text{rev}/\text{min}, \omega_2 = 3000\,\text{rev}/\text{min}$

$$\therefore d_1 = \frac{\omega_2}{\omega_1} d_2 = \frac{3000}{1460} \times 100 = 205.5\,\text{mm}$$

An interesting variant of the friction-based drive is the variable-ratio drive. Various mechanisms are available, including systems based on vee pulleys of variable width (and hence of variable effective radius) and tilting spheres contacting conical rollers (Figure 4.7) so that the spheres behave as rollers with varying effective radii. Hydraulic variable-ratio drives also exist – see Section 4.13.8. These mechanical approaches to variable speed drive are appropriate if electricity is unavailable or cannot be used for safety reasons. By contrast, an electrical approach to variable speed drives is described in Section 4.11, involving varying the motor speed rather than the transmission ratio.

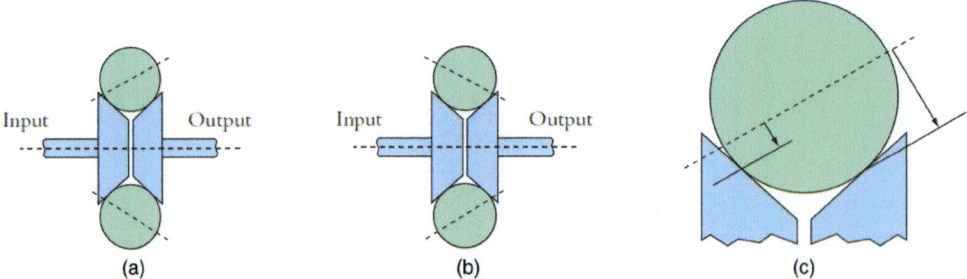

FIGURE 4.7 Sphere-based variable speed drive (a) set to increase speed, (b) set to decrease speed, (c) effective radii of contact of spheres against the two cones.

4.4.1 COMPOUND DRIVE TRAINS

In many engineering situations, it is not practical to obtain the required drive ratio via a single stage of transmission. In this case, multiple stages are used, with the overall drive ratio being the product of the drive ratios of each stage in the transmission system.

For example, the drive ratio of the gear train shown in Figure 4.8 is:

$$\text{Drive ratio } n = \frac{N_2}{N_1} \times \frac{N_4}{N_3} \times \frac{N_6}{N_5} = \frac{\omega_1}{\omega_6} = \frac{\alpha_1}{\alpha_6} \tag{4.13}$$

4.4.2 EFFICIENCY OF GEARED SYSTEMS

While real gears, toothed belts, chains etc. scale kinematic quantities (rotational angle, angular velocity, angular acceleration) according to the drive ratio, frictional forces and torques result in inefficiencies, so the ratios of torques do not in fact precisely follow the relationships given in Equations (4.10) and (4.11). These equations may therefore be rewritten to take into account the efficiency η of the drive train:

$$\text{power out} = \text{efficiency} \times \text{power in}$$
$$\text{i.e. } L_2\omega_2 = \eta L_1\omega_1 \tag{4.14}$$

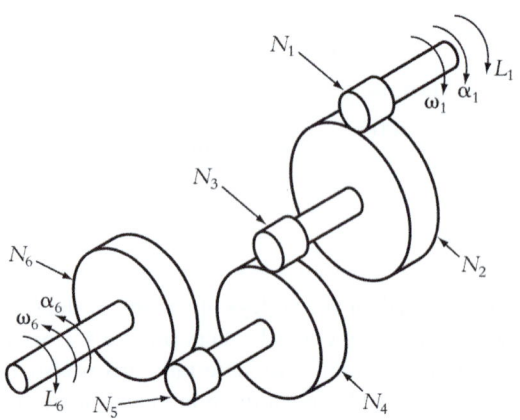

FIGURE 4.8 A compound drive train.

so that input torque L_1 can be calculated from L_2:

$$L_1 = \frac{L_2 \omega_2}{\eta \omega_1} = \frac{L_2 N_1}{\eta N_2} = \frac{L_2}{\eta n} \tag{4.15}$$

Efficiency is always less than unity, so input power is always greater than output power, with the difference being lost as heat. Typical efficiencies for geared systems may range from values close to unity (e.g. 0.95) for well-designed single-stage gear systems, down to much lower values (less than 0.5) for very high-ratio systems such as worm gear drives, which give a very compact reduction drive at the expense of involving significant sliding action and frictional losses. It must be emphasised, however, that kinematic quantities are unaffected by inefficiencies, so the following relationship still applies:

$$\text{Gear ratio } n = \frac{N_2}{N_1} = \frac{\omega_1}{\omega_2} = \frac{\alpha_1}{\alpha_2} \tag{4.16}$$

In a compound gear system, the overall efficiency of the gear train is the product of the efficiencies of each stage.

WORKED EXAMPLE

A vehicle is driven from an electric motor via the compound gear system shown in Figure 4.9, where $N_1 = 20$, $N_2 = 100$, $N_3 = 25$ and $N_4 = 75$. The efficiency of the first stage is 94 per cent and of the second stage is 96 per cent. If the final drive requires a torque of 100 Nm at 200 rev/min, what torque and speed are required from the motor?

Adapting Equation (4.13) to the present situation:

$$\text{Drive ratio } n = \frac{N_2}{N_1}\frac{N_4}{N_3} = \frac{\omega_1}{\omega_4}$$

where $\omega_4 = 200 \text{ rev / min}$

$$\therefore \omega_1 = \frac{N_2}{N_1} \times \frac{N_4}{N_3} \times \omega_4 = \frac{100}{20} \times \frac{75}{25} \times 200 = 300 \text{ rev / min}$$

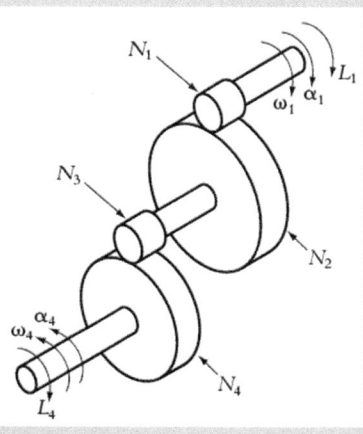

FIGURE 4.9 Compound gear system.

Overall efficiency is the product of the efficiencies of the two stages; so,

$$\eta = \eta_1 \times \eta_2 = 0.94 \times 0.96 = 0.902 \text{ or } 90.2\%$$

Adapting Equation (4.15) to the present situation gives

$$L_1 = \frac{L_4}{\eta n}$$

where $L_4 = 100 \text{ Nm}$

and $n = \dfrac{N_2}{N_1} \times \dfrac{N_4}{N_3} = \dfrac{100}{20} \times \dfrac{75}{25} = 15$

so that

$$L_1 = \frac{100}{0.902 \times 15} = 7.39 \text{ Nm}$$

4.4.3 INERTIA OF A GEARED SYSTEM

Consider two shafts 1 and 2 (Figure 4.10) geared together with a perfectly efficient transmission of gear ratio n, with shaft 2 having mounted on it a moment of inertia J_2. Shaft 1 is driven by a motor, which causes the shafts to accelerate with angular accelerations α_1 and α_2, respectively.

The motor is assumed to provide a torque L_1 to shaft 1. This torque is scaled up by the gear ratio n and transmitted to shaft 2 causing inertia J_2 to accelerate with angular acceleration α_2. Because the shafts are geared together, the motor observes shaft 1 as accelerating with angular acceleration α_1. Overall, therefore, the motor is providing a torque L_1 which is causing an angular acceleration α_1 to the machine being driven. As far as the motor is concerned, this is simply an inertia load J_1' defined as:

$$J'_1 = \frac{L_1}{\alpha_1} \tag{4.17}$$

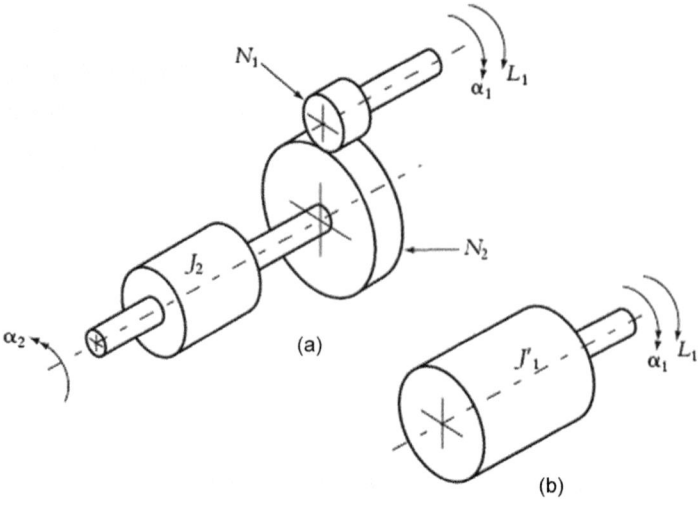

FIGURE 4.10   (a) An inertia driven via gears and (b) its inertia referred to the input shaft.

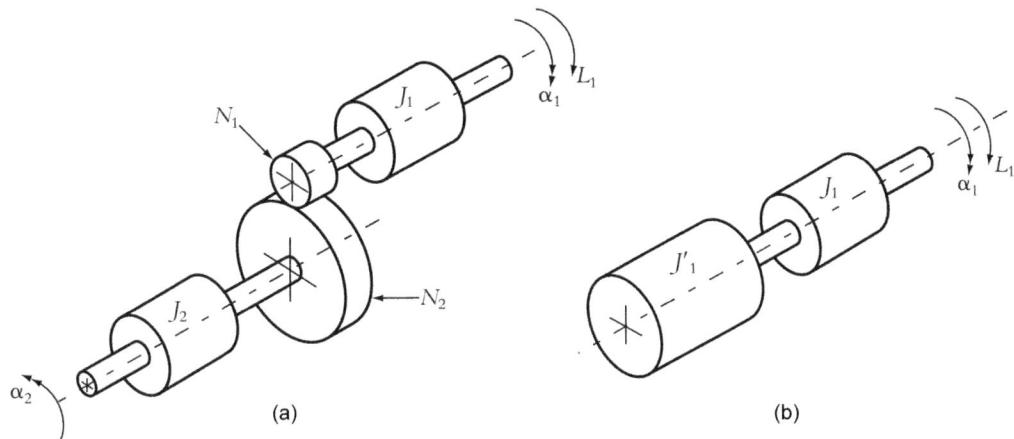

FIGURE 4.11   (a) A more complex system involving geared inertia and (b) summation of inertias referred to the input shaft.

where (in general) the notation J' will be used for the referred or apparent inertia typically involving an indirectly driven inertia load. In the present case, the load actually consists of inertia J_2 driven via the gear train of ratio n, experiencing a torque $L_1 n$. This causes J_2 to experience an angular acceleration of $\alpha_2 = (L_1 n)/J_2$ which is observed at shaft 1 as $\alpha_1 = \alpha_2 n = (L_1 n^2)/J_2$, It can, therefore, be shown that inertia J'_1 experienced by the motor is given by

$$J'_1 = \frac{J_2}{n^2} \tag{4.18}$$

Of particular note is the n^2 term and the fact that **inertias are scaled by the square of the gear ratio**, rather than the gear ratio itself, as for kinematic quantities as shown in Equation (4.9).

   By extension, a geared system with inertia J_1 on shaft 1 geared to shaft 2 carrying another inertia J_2 (Figure 4.11) will involve the sum of J_1 and the referred inertia due to J_2; in other words,

$$J'_1 = J_1 + \frac{J_2}{n^2} \tag{4.19}$$

WORKED EXAMPLE

A machine (Figure 4.12) consists of two shafts A and B, each fitted with a flywheel of inertia $J_A = J_B = 1$ kg.m². Shaft B is driven from shaft A by a gear train of 3:1 reduction ratio. To a blindfolded observer, what does the system feel like if a handle is attached to shaft A (i.e. what is the inertia of the system referred to shaft A)?

$$J_{total} = J_A + \frac{J_B}{n^2}$$

$$= 1 + \frac{1}{3^2} = 1.11 \, \text{kgm}^2$$

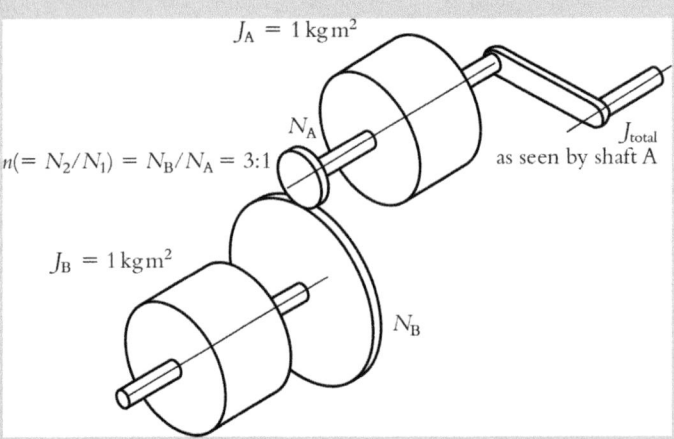

FIGURE 4.12   A geared system driven from shaft A.

WORKED EXAMPLE

The handle is instead attached to shaft B (Figure 4.13). What does the system feel like now (i.e. what is the inertia of the system referred to shaft B)?

Note that since the first shaft is now B and the second is A, the gear ratio $n = N_2/N_1$ is now 1:3 which is used in the calculation as 1/3:

$$J_{total} = J_B + \frac{J_A}{n^2} = 1 + \frac{1}{(1/3)^2}$$

$$= 1 + 3^2 = 10 \text{ kg m}^2$$

All the above concepts are applicable to drivetrains consisting of pulleys and belts, or sprockets and chains.

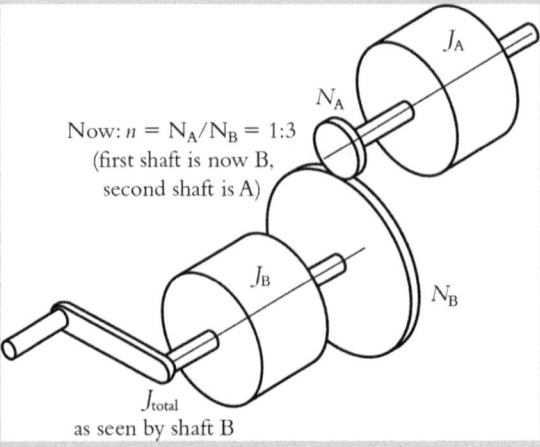

FIGURE 4.13   A geared system driven from shaft B.

4.4.4   Influence of Efficiency on the Apparent Inertia of Geared Systems

It has already been shown that practical geared systems (and similar systems) are not perfectly effi-cient, and this has important implications for considering the practical effects of inertia in such sys-tems. For an imperfectly efficient version of the example shown earlier, torque L_1 results in inertia J_2 experiencing a reduced torque $\eta L_1 n$ and hence undergoing a smaller value of angular acceleration $(L_1 R)/J_2$. Thus, the ratio of torque to angular acceleration on shaft 1 can be shown to be:

$$\frac{L_1}{\alpha_1} = \frac{L_1 J_2}{\eta L_1 n^2} = \frac{J_2}{\eta n^2} = \frac{J'_1}{\eta} \tag{4.20}$$

Note that the true value of referred inertia J'_1 is not affected by the efficiency term – one cannot, for instance, store more rotational energy in a geared flywheel by connecting it via an inefficient gear train! However, it is useful to realize that a heavy system being accelerated via an inefficient trans-mission system will *appear* to have a higher value of inertia than if the transmission system were efficient. Conversely, it will appear to have a lower value of inertia when it is decelerating.

LEARNING SUMMARY

At the end of this section, you should have learnt:

✓ The concept of a gear ratio.
✓ To relate angular velocities, angular accelerations and torques between input and output shafts of a geared system (or other system involving belts, friction drives etc.).
✓ The concept of 'referred inertia'.
✓ The effect of inefficiency on the transmission of torque and the apparent value of referred inertia.

4.5   TANGENTIALLY DRIVEN LOADS

There are numerous situations where rotational motion is converted into linear motion or vice versa, such as:

* Tangentially driven systems involving belts, chains and ropes.
* Vehicles, which are also a form of tangentially driven load.
* Screw-driven systems (e.g. machine tools, testing machines and presses).
* Mechanisms for converting between rotational and reciprocating motion such as cranks and cams.

Within the present analysis, we will concern ourselves only with systems that give a constant rela-tionship between rotational and linear motion, namely tangentially driven and screw-driven systems. Tangentially driven systems will be covered initially, and screw-driven systems will also be covered briefly. Cranks, cams etc. involve a non-uniform relationship between rotation and displacement, and a rigorous treatment is beyond the scope of the present discussion.

There are numerous situations in engineering where a rotating shaft causes continuous linear motion. Such situations are generally referred to as **tangential drives**, and examples include:

* Conveyer belts.
* Toothed belt linear drives, e.g. the positioning of print heads on inkjet printers.

- Cranes, winches and hoists.
- Vehicles.

Indeed, any belt or chain drive involves the tangentially driven mass of the belt or chain itself.

4.5.1 TANGENTIAL DRIVES VIA BELTS, CHAINS ETC.

Consider the simplest tangential load situation where a mass m is driven by a light pulley of effective radius r via a light belt or cord (Figure 4.14). The inertia J' of this system referred to the axis of the pulley is:

$$J' = mr^2 \tag{4.21}$$

Practical tangential drive situations involve pulleys or sprockets, together with belts, conveyors or chains, of non-negligible mass, and each of these contributes to the total inertia. For instance, consider a conveyor belt of mass m carrying a load of mass M, mounted on a pair of rollers, one of which is connected to the drive input shaft and has a moment of inertia J_1 and radius r_1, with the second being an idler of inertia J_2 and radius r_2 (Figure 4.15).

Each of these contributes to the inertia J' of the system referred to the input shaft:

- The roller connected to the input shaft obviously contributes a moment of inertia J_1.
- The load of mass M is driven tangentially by the first roller of radius r_1 and so contributes Mr_1^2 to the total inertia

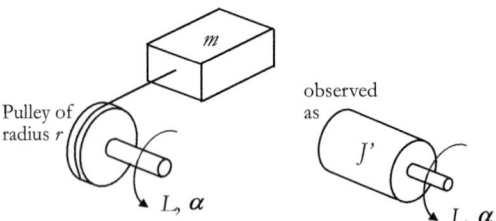

FIGURE 4.14 Tangentially driven mass.

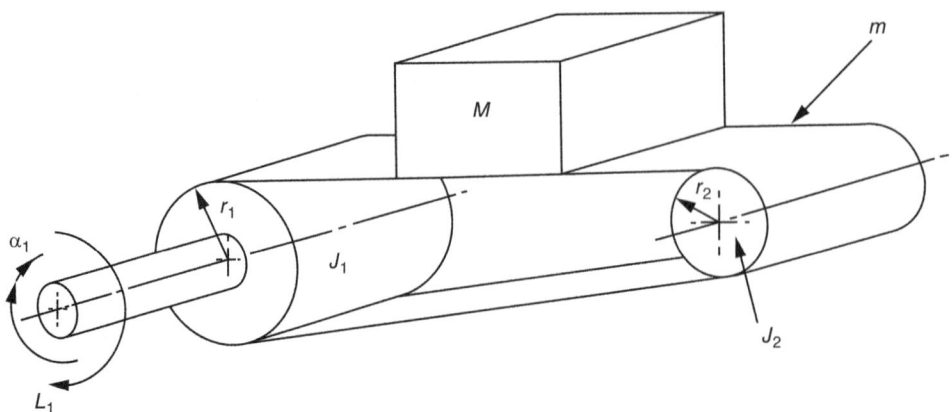

FIGURE 4.15 Conveyor belt system.

- The conveyor belt of mass m is also driven tangentially by the first roller and so contributes mr_1^2 to the total inertia.
- The idler roller is driven (via the conveyor belt) from the first roller with a pulley ratio of $r_2{:}r_1$. So, its inertia referred to the drive input shaft is $J_2(r_1/r_2)^2$.

Therefore, the total moment of inertia referred to the input drive shaft is:

$$J' = J_1 + (M + m)r_1^2 + J_2(r_1/r_2)^2 \tag{4.22}$$

WORKED EXAMPLE

A probe assembly of mass 100 g on this item of laboratory equipment is to move 0.2 m in 1 second. It is driven by a toothed belt that weighs 30 g and has an effective radius of 20 mm where it passes over the pulleys (Figure 4.16(a)). A uniform acceleration/deceleration (triangular) velocity profile is to be used (Figure 4.16(b)). Each pulley has a moment of inertia of 2×10^{-6} kg.m^2

a. Calculate the inertia of the system referred to the axis of the drive pulley.
b. Calculate the maximum acceleration of the probe and hence the maximum angular acceleration of the pulley.
c. Calculate the torque required to cause the desired acceleration.

Solution

(a)

$$J' = 2 \times J_{\text{pulley}} + (M + m)r^2$$
$$= 2 \times 2 \times 10^{-6} + (100 \times 10^{-3} + 30 \times 10^{-3}) \times 0.02^2 = 5.6 \times 10^{-5}\,\text{kgm}^2$$

(b) The probe covers 0.2 m in 1 s divided equally between acceleration and deceleration; therefore, the acceleration phase covers 0.1 m in 0.5 s from rest.

Applying $s = ut + \dfrac{1}{2}at^2$ gives

$$a = \frac{2s}{t^2} = \frac{2 \times 0.1}{0.5^2} = 0.8\,\text{ms}^{-2}$$

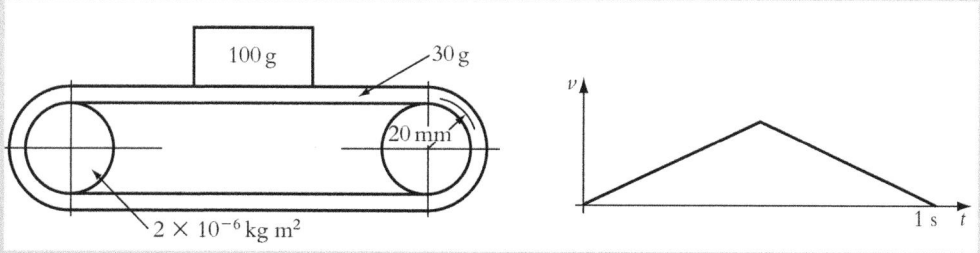

FIGURE 4.16 (a) Tangentially driven probe assembly showing belt and pulleys and (b) uniform acceleration/deceleration profile.

Angular acceleration is therefore given by

$$\alpha = \frac{a}{r} = \frac{0.8}{0.02} = 40 \, \text{rads}^{-2}$$

(c)   Torque is given by

$$L = J'\alpha = 5.6 \times 10^{-5} \times 40 = 0.00224 \, \text{Nm}$$

4.5.1.1   Vehicles

A very important type of tangentially driven load is a vehicle running on a road (or rail or other surface), and it can be useful to consider the effective inertia of a vehicle referred to (for example) the vehicle's axle or the engine's output shaft. There are two contributions to inertia J' referred to the axle of a vehicle (Figure 4.17), ignoring the inefficiencies in wheel/road contact, such as hysteresis in a rubber tyre:

- mr^2 where m is the overall mass of the vehicle (**including the wheels**) and r is the effective radius of the wheels where they contact the road.
- NJ_W where N is the number of wheels and J_W is the inertia of each wheel (whether the wheels are directly mounted on the axle or not – coupling between the wheels is provided via their contact with the road or rail).

This gives

$$J' = mr^2 + NJ_W \tag{4.23}$$

In practice, the contribution due to the wheels is likely to be small compared with that of the overall mass of the vehicle.

4.5.1.2   Screw Drives

These provide an alternative to tangential drives where only a small movement is desired for each revolution of the input shaft. They are widely used in machine tools for accurate positioning of slideways (where they are known as leadscrews) and within the aerospace industry for the actuation of landing gear.

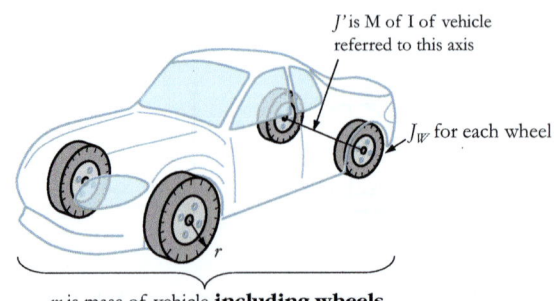

FIGURE 4.17   Simplified diagram of a vehicle showing contributions to inertia referred to axle.

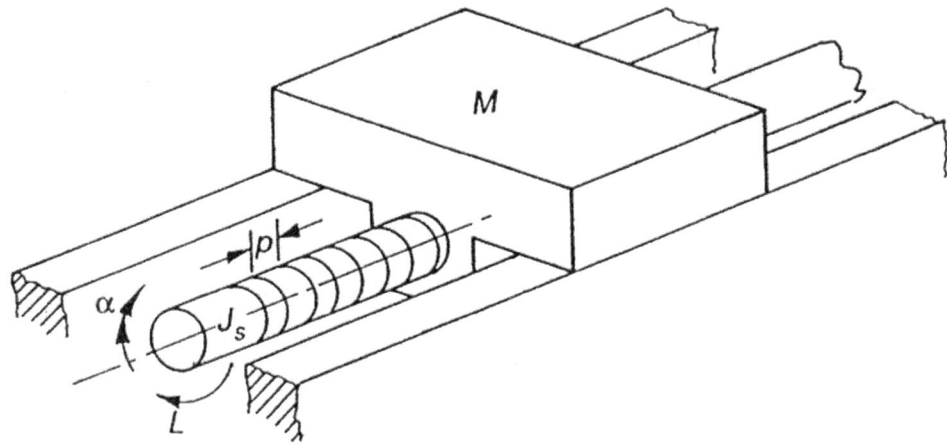

FIGURE 4.18   System driven by a leadscrew.

The apparent moment of inertia of a screw-driven system (Figure 4.18), referred to the leadscrew axis, depends on the leadscrew pitch p, the mass M being driven and the efficiency η of the leadscrew as well as the moment of inertia J_s of the screw itself. The torque for a given angular acceleration α of the screw is therefore:

$$L = \left[J_s + \frac{Mp^2}{4\pi^2\eta} \right] \alpha \qquad (4.24)$$

Unless the load is very heavy or the pitch is large, the biggest contribution to the total inertia is often from the inertia J_s of the screw itself.

LEARNING SUMMARY

By the end of this section, you should have learnt:

✓ To calculate the referred inertia of a tangentially-driven system.
✓ The inertia behaviour of screw-driven systems.

4.6   STEADY-STATE CHARACTERISTICS OF LOAD

So far, the analysis has only considered the effects of inertia. With this theory alone, it could be assumed that once a machine has been accelerated to its running speed, it will carry on running forever with no further input of energy. Clearly, this is not true, and it is now necessary to consider the various contributions of frictional losses that tend to oppose movement when the machine is running.

4.6.1   FRICTIONAL EFFECTS AND LOSSES

Three types of frictional losses are typically encountered within machines:

4.6.1.1 Coulomb Friction

This is based on Coulomb's law for contacting bodies (Figure 4.19) that are able to move linearly with respect to each other, stating that the limiting frictional force F for two bodies experiencing contact force N is given by

$$F = \mu N \tag{4.25}$$

where μ is the coefficient of friction. In fact, two distinct regimes can be considered:

- Static friction ('stiction') is the force required to get two bodies moving with respect to each other; in effect, this is the force required to break any intermolecular or interatomic bonds that have built up between the bodies over time.
- Sliding friction or dynamic friction is a frictional force between two objects that are already sliding over each other; this remains (in theory) constant irrespective of the relative speed of the two objects.

In practice, the coefficients of static friction, μ_S, and dynamic friction, μ_D, tend to be noticeably different (it takes a slightly smaller force to keep two components moving with respect to each other, compared to the force required to get them to move). Typical values might be $\mu_S = 1.5\,\mu_D$. However, as stated above, μ_D may be assumed to be independent of relative speed, so the force–speed relationship may be represented graphically as shown in Figure 4.20:

Mathematically, this can be represented for the static situation as:

$$-F_S \le F \le F_S \text{ for } v = 0 \tag{4.26}$$

where

$$F_S = \mu_S N \tag{4.27}$$

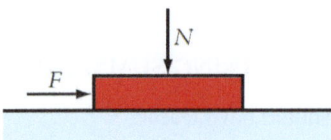

FIGURE 4.19   Block in contact with a surface.

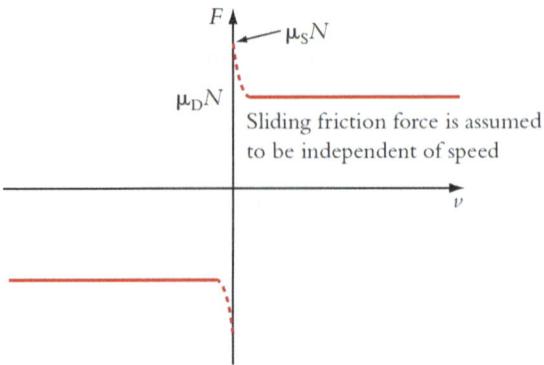

FIGURE 4.20   'Stiction' and dynamic friction for linear movement.

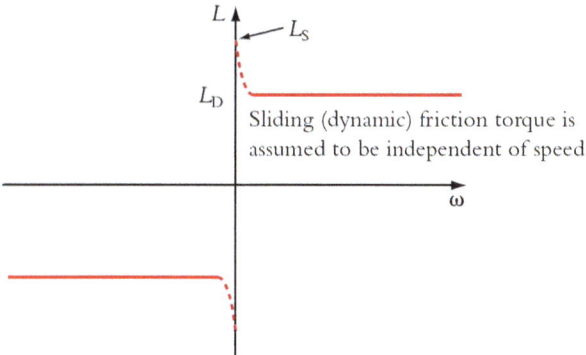

FIGURE 4.21   'Stiction' and dynamic friction for rotational movement.

and for the dynamic situation:

$$F = F_D \text{ for } v > 0 \tag{4.28}$$

and

$$F = \left(-F_D\right) \text{ for } v < 0 \tag{4.29}$$

where

$$F_D = \mu_D N \tag{4.30}$$

While components undergoing linear sliding certainly are encountered in engineering (for example, machine tool saddles moving with respect to the stationary bed, pistons sliding in cylinders), the above concepts are equally applicable to rotational movement, and the characteristics shown earlier may be represented graphically in Figure 4.21 and mathematically as

$$-L_S \leq L \leq L_S \text{ for } \omega = 0 \tag{4.31}$$

$$L = L_D \text{ for } \omega > 0 \tag{4.32}$$

and

$$L = \left(-L_D\right) \text{ for } \omega < 0 \tag{4.33}$$

4.6.1.2 Viscous Friction

Viscous friction is caused by the shearing of a Newtonian fluid, for example, a lubricating oil that separates the two bodies that are 'sliding'. Newton's model of viscous friction states that the shear stress on a fluid is proportional to the shear strain rate of the fluid. For a linear bearing properly lubricated with a Newtonian fluid, this implies that the viscous friction force is proportional to the relative velocity of the two components. For the linear motion of two plates of area A separated by distance d, the force F required to maintain a relative velocity v (sometimes known as the 'drag force') is:

$$F = \frac{Av\mu}{d} \tag{4.34}$$

where μ is used in this context to represent the viscosity (strictly the dynamic viscosity) of the fluid.

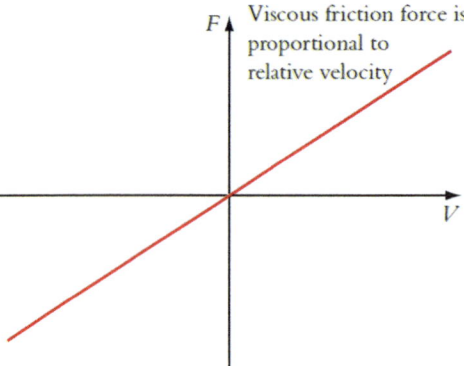

FIGURE 4.22   Viscous friction for linear movement.

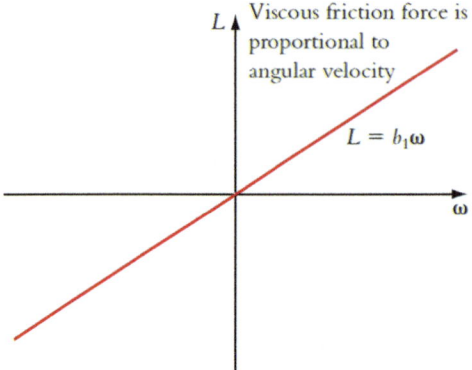

FIGURE 4.23   Viscous friction for rotational movement.

This gives a simple linear relationship between force and velocity which can represented as

$$F = a_1 v \tag{4.35}$$

This can be shown graphically (Figure 4.22).

Again, it is useful to represent this in rotational form (Figure 4.23):

$$L \propto \omega \tag{4.36}$$

or

$$L = b_1 \omega \tag{4.37}$$

4.6.1.3   Windage

Windage is a force caused by aerodynamic effects such as turbulence, for example, drag on a car, windage drag on a motor due to its cooling fan and the 'churning up' of air in its casing. The magnitude of the windage force is proportional to the **square** of velocity, noting that the direction of the force opposes the direction of movement.

$$|F| \propto v^2 \tag{4.38}$$

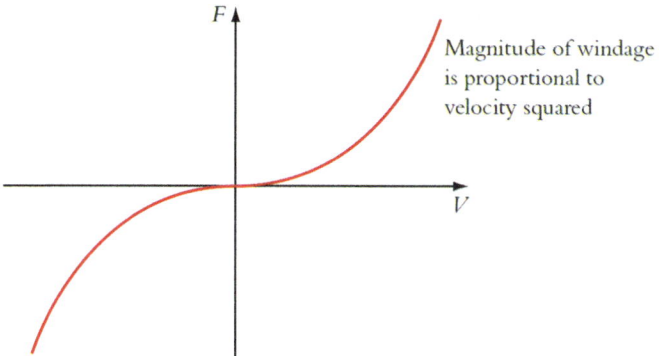

FIGURE 4.24   Windage for linear movement.

This is represented graphically in Figure 4.24.

For a body of frontal area A moving through a fluid of density ρ with velocity v, the windage force it experiences is:

$$|F| = \frac{1}{2}\rho v^2 A C_\mathrm{d} \tag{4.39}$$

where C_d is an approximately constant value called the **coefficient of drag**, which depends upon the shape of the body and is in the order of 1 or slightly larger for 'bluff' bodies (such as cubes, plates etc.) going down to much lower values for aerodynamically designed, streamlined shapes such as aircraft. Such coefficients are widely tabulated in the fluid mechanics literature and are presented in Unit 1 of the present book. In a simplified form, this kind of relationship can be represented as

$$|F| = a_2 v^2 \tag{4.40}$$

In a similar manner to rotation, torque is proportional to the square of angular velocity:

$$|L| \propto \omega^2 \tag{4.41}$$

or

$$|L| = b_2 \omega^2 \tag{4.42}$$

This can be represented graphically in Figure 4.25.

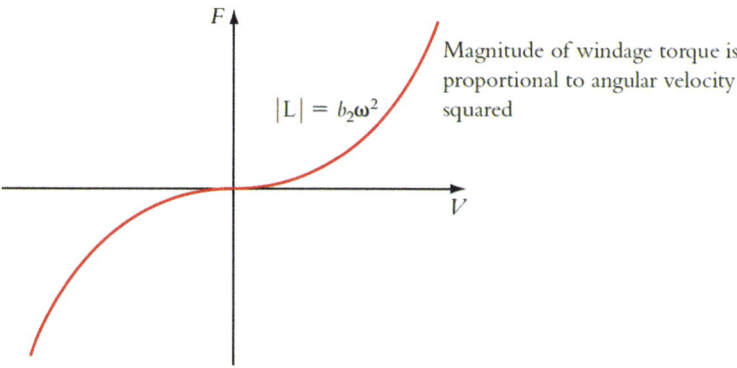

FIGURE 4.25   Windage for rotational movement.

4.6.1.4 Summary of Frictional Effects

For linear motion, assuming the motion is in a positive direction,

- The at rest/threshold of movement: $F = F_S$ = static friction force.
- At velocity v: $F = F_D + a_1 v + a_2 v^2$.
  i.e. total frictional force = Coulomb friction force + viscous friction force + windage.

For rotation in a positive direction,

- At rest/threshold of movement: $L = L_S$ = static friction torque.
- At velocity v: $L = L_D + b_1 \omega + b_2 \omega^2$.
  i.e. total frictional torque = Coulomb friction torque + viscous friction torque + windage.

4.6.1.5 Characteristics of Fans, Blowers, Pumps etc.

Not surprisingly, windage-type effects are the predominant contribution to the torque–speed characteristics of rotodynamic fans and blowers. Broadly similar characteristics are also encountered for similar devices such as centrifugal pumps, such as those encountered in many chemical plants. It should be noted that, in approximate terms, the pressure that can be supplied by rotodynamic pumps, blowers, fans etc. is also proportional to the square of the speed. However, an important issue is that if the torque required to drive such a device is proportional to the square of the speed, then the power (which is given by torque × angular velocity) will increase with the cube of the speed. This has very important implications for energy efficiency, in that considerable savings can be made by reducing the speed of such devices to the minimum which will deliver the desired performance, rather than throttling their flow rates or allowing them to supply higher pressures and flow rates than are required.

Note that the issues surrounding the torque–speed characteristics of positive-displacement pumps, compressors etc. are rather different. The torque required to drive them is dominated by the pressure or head against which they are pumping: if this is constant (e.g. if supplying a header tank), the torque required (when averaged over a cycle) will not change greatly with speed; if the pressure changes with flow rate, then the torque will also change.

4.6.1.6 Other Contributions to Steady-State Running Characteristics

While it is clearly impossible to list all possible contributions to running characteristics, some common examples include gravitational load; for example, due to items being carried upwards on a sloping conveyor, winched upward using a hoist or where a vehicle is climbing a slope. Such a contribution to load torque will be independent of speed, and can be found using simple engineering mechanics (refer to Unit 6 of *An Introduction to Mechanical Engineering: Part 1*).

WORKED EXAMPLE: TORQUE ON A HIGH-SPEED ROTATING ANTENNA

Figure 4.26 illustrates the characteristics of a system that exhibits all of the above contributions. It is interesting to note that the windage element varies with the ambient wind speed.

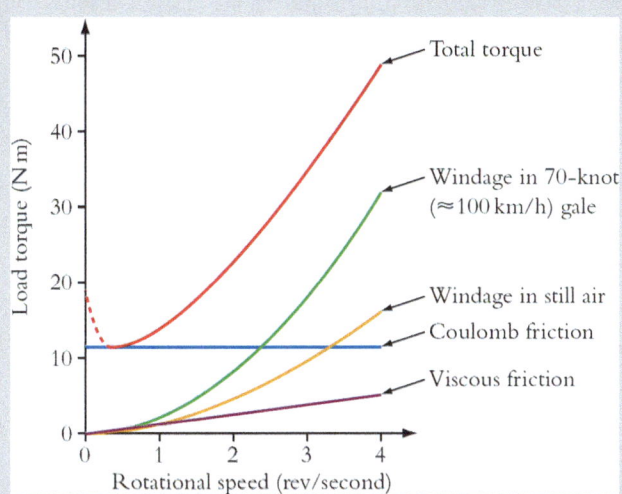

FIGURE 4.26 Contributions to torque–speed characteristics of a high-speed rotating antenna. (Adapted from Blackburn et al., 1960.)

4.6.1.7 Some Positive Aspects of Friction – It Wastes Energy, But Is It Always a Bad Thing?

Coulomb friction, viscous friction and windage all involve the conversion of mechanical power into heat. Within mechanical power transmission situations, this is normally undesirable as it involves wasted energy, yet friction is usefully harnessed as an essential feature of several engineering devices.

- Brakes, which apply a frictional torque to oppose motion, are widely used to convert unwanted kinetic energy (for instance, in a vehicle) or potential energy (in the load of a crane or elevator) into heat, which clearly needs to be dissipated.
- Clutches are used to provide a means of temporarily matching differing speeds while transmitting a given value of torque. This is achieved by frictional coupling of the input shaft to the output shaft via two contacting plates. The function of a clutch will be examined further in Section 4.14.
- Worm gears (Figure 4.27) are widely used to provide a very large speed reduction ratio (and torque increase) between a rapidly rotating input shaft and a much more slowly rotating output shaft. This is achieved in much less space than would be possible via conventional (spur or helical) gears, which would require the use of compound gearing to achieve the required gear ratio. Unlike conventional gears, which involve primarily rolling friction between gear teeth, worm gears involve a primarily sliding action as the screw-like worm rotates against the flanks of the teeth of the worm wheel, driving it around. This friction results in the undesirable feature of very low efficiency (values of 30 per cent are typical for worm gear assemblies). However, friction also accounts for an important feature of most worm gear systems which contrasts with conventional gear systems: provided the helix angle of the worm is sufficiently shallow (typically corresponding to the use of a single-start or possibly two-start worm), they can normally only transmit drive in one direction through the power train, with the drive going only from the input shaft to the output shaft. In other words, they are self-locking if an attempt is made to provide the drive from the output shaft to the input shaft. This makes them a particularly useful feature of devices such as hoists, cranes, torsional testing machines etc. For example, if the drive to a worm-driven hoist fails, the load will not descend as the worm gear assembly will lock up. Similarly, a worm-driven torsional testing machine will retain the torsional loading on the specimen if

FIGURE 4.27   Picture of worm mechanism. (AndreyProekt/Shutterstock.com.)

no input torque is applied. In both of these cases, it is said that the drive system cannot be 'over-hauled'. There are exceptions to this: in some cases, a worm with a very steep helix angle (often a multi-start worm) is used to provide a compact speed increase gear system. These are relatively rare in engineering – perhaps, the classic example of such a system was within the speed-limiting governor used in mechanical telephone dials up to the 1970s, where a rapidly rotating centrifugal brake was operated from the slowly turning dial via a reverse-driven worm gear mechanism. Reverse-driven, steep helix angle worms are similarly used to drive the speed governor fans of musical box movements.

LEARNING SUMMARY

By the end of this section, you should have learnt:

✓ The different contributions to steady-state running characteristics of loads.

✓ To express these different contributions to the load characteristics in the form of a mathematical expression, both for linear and rotational motion.

✓ That friction can have a beneficial role in some situations as well as having a detrimental effect on efficiency in other situations.

4.7   MODIFYING STEADY-STATE CHARACTERISTICS OF A LOAD USING A TRANSMISSION

In order to match the torque–speed characteristics of a load to the characteristics of the mechanical power source that will drive it, it is often necessary to drive the load via a transmission of some kind, which may take the form of gears, belts, chain drive etc. as described in Section 4.4. It is therefore very useful to find the characteristics of the load referred to the input shaft of the transmission. This is straightforwardly achieved by substituting the relations for torque and speed given in Equation (4.9) (or, if inefficiencies are taken into account, the torque relations given in Equation 4.15) into the equation giving the load's torque–speed relationship. For example, consider a load with the torque–speed (or, strictly, torque–angular velocity) characteristic as follows:

$$L = L_D + b_1\omega + b_2\omega^2 \tag{4.43}$$

Such a characteristic is shown diagrammatically in Figure 4.28(a). The load is driven via a transmission with a ratio of $n{:}1$ and an efficiency of η, and it is desired to find the characteristic of the system expressed as the relationship between torque L' and speed ω' at the input shaft of the transmission. Making use of Equation (4.9), ω is expressed in terms of ω':

$$\omega = \frac{\omega'}{n} \tag{4.44}$$

Similarly, making use of Equation (4.15), L' is expressed in terms of L as:

$$L' = \frac{L}{n\eta} \tag{4.45}$$

Inserting these into Equation (4.45) gives

$$
\begin{aligned}
L' &= \frac{1}{n\eta}\left(L_D + b_1\omega + b_2\omega^2\right) \\
&= \frac{1}{n\eta}\left(L_D + b_1\frac{\omega'}{n} + b_2\frac{(\omega')^2}{n^2}\right) \\
&= \frac{L_D}{n\eta} + b_1\frac{\omega'}{n^2\eta} + b_2\frac{(\omega')^2}{n^3\eta}
\end{aligned}
\tag{4.46}
$$

It is instructive to observe that a family of torque–speed curves (Figure 4.28(b)) can be constructed as the ratio n is varied: a low value of n will cause the load to require high values of torque to be provided at a relatively low range of speeds, while a large value of n will require lower values of torque to be provided, but with the drive being at higher speeds.

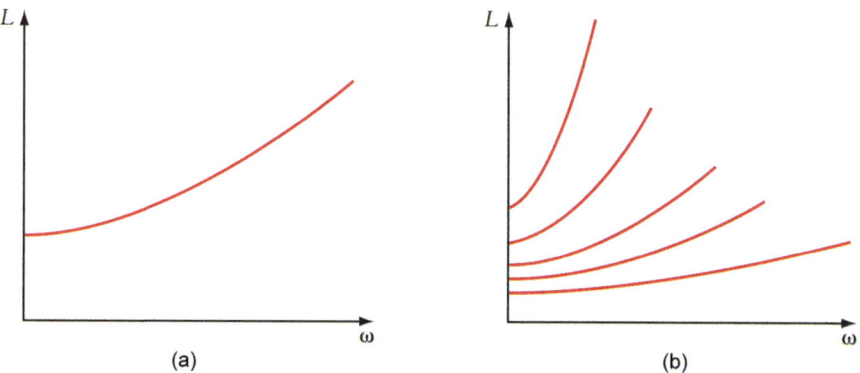

(a)   (b)

FIGURE 4.28 (a) Typical torque–speed curve for a directly driven machine involving dynamic Coulomb friction, viscous friction and windage (b) family of torque–speed curves when driven with varying transmission ratios.

It is crucial to note that changing the transmission ratio does *not* alter the power requirements of the load – it merely allows the characteristics of the load to be moved to provide a better match to those of the mechanical power source. The issue of matching of load and power source will be explored further in Section 4.14.

LEARNING SUMMARY

By the end of this section, you should have learnt:

✓ To refer the torque–speed characteristics of a load to the input shaft of a transmission system in order to obtain the characteristics observed by the mechanical power source driving it.

✓ That a transmission will affect the combination of torque and speed required to drive a load but will not help to overcome a shortfall in the power available for providing the drive.

4.8 SOURCES OF MECHANICAL POWER AND THEIR CHARACTERISTICS

In some cases, a machine will be driven manually via a handle, pedals etc., but in most cases, some other source of mechanical power will be needed. There are various categories of such systems and it can be useful to distinguish them. Exact terminology can vary (see Chambers (2007) *Science and Technology Dictionary* for one set of formal definitions) but as a guideline:

1. The term 'motor' is usually used to describe a machine that causes motion or generates mechanical power, often drawing its energy from some other easily managed form (electrical, hydraulic, pneumatic, possibly chemical). Within the present unit, we will consider a variety of electric, pneumatic and hydraulic motors. However, the term 'motor' historically also covered ICEs, leading to the formal term 'motor vehicle' and related terms such as 'motorist', 'motorcycle' etc.
2. Prime movers are devices that convert a natural source of energy into mechanical power; these can include engines which make use of burning fuel, but could also include wind turbines, water turbines etc.
3. The term 'engine' tends to be reserved for a machine that creates mechanical power from heat energy; examples are ICEs, jet engines (a form of gas turbine), steam engines etc. Within the present discussion, we will pay some attention to the characteristics of ICEs.

In order to be able to design a drive system, it is important to understand how the torque available from a source of mechanical power varies with speed. With this information, it is then possible to analyse the way in which the mechanical power source interacts with the load it is driving.

4.8.1 CHARACTERISTICS OF THE INTERNAL COMBUSTION ENGINE

Although there is no immediate derivation of the torque–speed characteristics, the ICE has for many years been one of the most important prime movers. It remains of major interest for industrial, marine, rail and heavy goods vehicle applications even though it is starting to decrease in importance for automobiles etc. Its torque–speed characteristics vary depending upon the position of the throttle (on a petrol engine) or the setting of the fuel injection pumps (diesel engine). It is usual to give only the maximum torque–speed characteristics (for a petrol engine, this is known as the wide

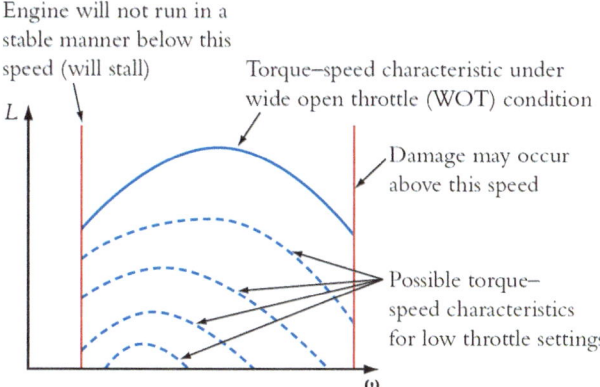

FIGURE 4.29 Hypothetical set of torque–speed characteristics for an internal combustion engine at different throttle settings.

open throttle (WOT) condition); it is not usual to give the torque–speed characteristics for different settings of the throttle etc., but for illustrative purposes, the possible curves of this type are included in Figure 4.29.

(Note that it is also usual to present ICE characteristics such as those in Figure 4.29 in terms of a quantity known as 'brake mean effective pressure', a quantity related to torque and the engine capacity. For simplicity, however, the graphs of torque and speed are plotted directly here). A particular feature of ICEs is that they cannot run in a sustained manner at low speeds: instead, they *stall*, a phenomenon very familiar to learner drivers of such vehicles! Rather than represent the infinite set of characteristics, it is usual to represent instead the maximum possible torque (under WOT conditions etc.) for each speed and fill the area under the curve with a 'map' showing the *specific fuel consumption* (SFC) of the engine for each combination of torque and speed (Figure 4.30a and b). This gives the amount of fuel consumed by the engine per unit of mechanical energy produced (usually expressed in g/kWh). It may be noted that there is an optimal combination of torque and speed (close to the maximum torque value, but well below the maximum safe running speed) at which the engine runs at maximum efficiency, consuming the least amount of fuel for a given output. This combination is sometimes known as the 'sweet spot'.

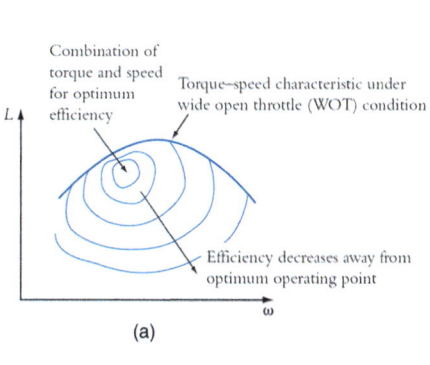

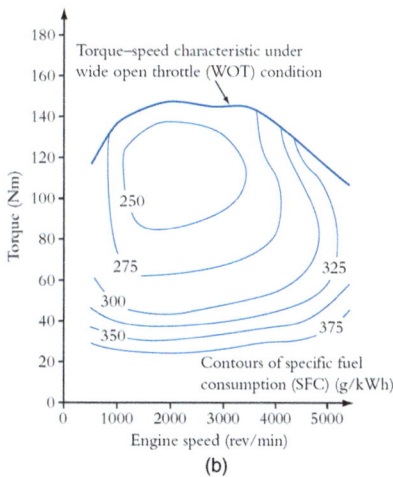

FIGURE 4.30 (a) Features of a torque–speed–SFC map for an internal combustion engine, (b) Torque–speed–SFC map for a 1.9-litre internal combustion engine. (adapted from Shayler et al., 1999.)

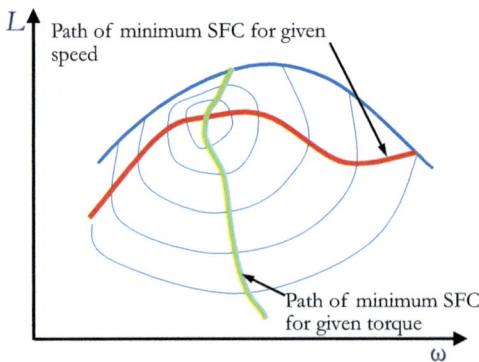

FIGURE 4.31 Two possible trajectories of torque–speed combinations to optimise fuel efficiency with varying torque (green) or speed (red).

It is not possible always to run at this 'sweet spot', but hybrid ICE–electric vehicles exploit the map by choosing the optimal values of torque and/or speed to run the engine in order to maximise efficiency, often using the battery to store excess energy or to make up for a shortfall in energy available from the engine. Different strategies may be used depending on the exact type of hybrid vehicle; see (for example) Nino-Baron et al. (2011). Trajectories of optimal torque–speed combinations for two different engine control strategies, which give the optimal torque for a given speed or vice versa, are illustrated in Figure 4.31.

The great advantage of the ICE is its ready availability and independence from other sources of power; its most obvious applications are in motor vehicles, ships and railway locomotives. In addition, diesel engines, in particular, are often used for powering generators for medium-sized communities and standby pumping or power generation applications. Conventionally, they have, of course, been used with fossil fuels, though some non-fossil alternative fuels are available. At the time of writing, there are plans to phase out ICEs in automobiles within the UK over approximately the next 10 years. Prime movers such as ICEs are essential in situations where externally supplied power cannot be imported. However, rather than using a prime mover such as an engine, it is often far more appropriate to use some kind of motor which is supplied with power in the form of electricity or pressurized fluid and delivers mechanical power locally in a controllable way. This approach may relate to a fixed location with a source of electrical and/or fluid power available on-site or (for example) to vehicles where electrical power is stored in a battery.

LEARNING SUMMARY

By the end of this section, you should have learnt:

✓ The different types of mechanical power sources used within drive systems.

✓ The meaning of a torque–speed–SFC diagram for an ICE, and in particular, understand the implications of its main features for applications such as hybrid vehicles.

4.9   PERMANENT MAGNET DIRECT CURRENT MOTORS

Permanent magnet direct current (PMDC) motors are widely used in domestic and automotive applications such as:

- Battery-operated appliances like food mixers, hand-held vacuum cleaners, toothbrushes, electric shavers and cooling fans.
- Portable tools such as electric drills and hedge trimmers.
- Car window winders, windscreen wipers, washers, fans and blowers.
- Battery-operated and track-powered toys.

They are also used in various forms as direct current (DC) servo motors in control applications. The usual construction is a DC motor with high-quality permanent magnets (PMs), a wound armature and a commutator as described below but there are variants such as:

- Ironless motors (in which the armature windings form a rigid lightweight structure of their own rather than being wound onto iron laminations).
- Pancake and 'printed armature' motors in which the ironless armatures are flat in shape.
- Brushless DC motors in which the design is turned inside-out so that the armature windings are stationary, the field magnets form a rotor and the switching action of the brushes and commutator is carried out instead by solid-state switches.

Brushless DC motors are used in some electric vehicles; these and other types of motors are discussed in Section 4.12.

4.9.1   CONSTRUCTION OF PMDC MOTOR WITH PERMANENT MAGNETS, WOUND ARMATURE AND COMMUTATOR

A PMDC motor has two constituent parts; a stator and a rotating armature. The stator comprises a steel cylinder with PMs mounted on the inner circumference of the cylinder. The magnets are usually made from hard rare earth magnetic materials. Figure 4.32(a) shows the stator of a four-pole PMDC motor, with two N-poles and two S-poles (indicating north-seeking and south-seeking, respectively).

The armature has circular varnish-insulated steel laminations fixed to a central shaft. Slots are cut in or just below its outer circumference as shown in Figures 4.32(b) and (c). Also fixed to the shaft, adjacent to the steel core, is a commutator constructed of copper segments that are separated by thin sheets of insulation. Coils are laid in the slots in the steel laminations and their ends brazed to individual commutator segments. Carbon brushes connect the commutator to an external direct voltage supply.

Magnetic flux circulates from the N-poles, across the air gap, through the armature and the second air gap into the S-poles and returns through the low-reluctance steel cylinder as illustrated in Figure 4.32(a). This is called the main field.

4.9.2   OPERATION OF PMDC MOTORS

The simplest PMDC motor has two poles, an armature with only two conductors, 1 and 2 forming a single coil and a commutator with two segments, as shown in Figure 4.33.

A direct voltage supply feeds current, I_a, to the armature winding via the carbon brushes. This current produces subsidiary magnetic fields that circulate around the armature conductors. Figure 4.34(a) shows a cross-section of the conductors encircled by the subsidiary fields and the main field passing from the N-pole to the S-pole.

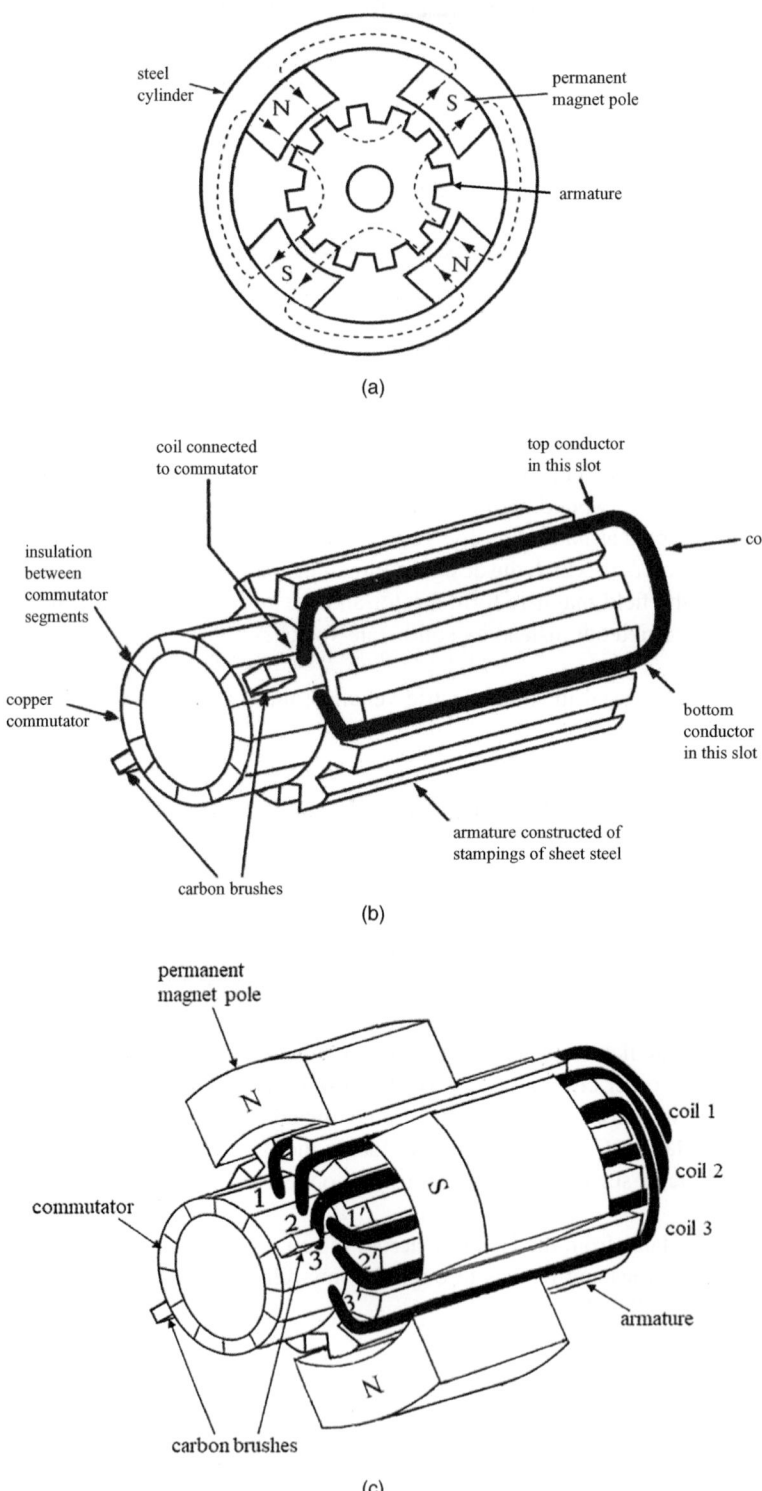

FIGURE 4.32 Construction of a permanent magnet DC motor with wound armature and commutator. (a) cross-section through a four-pole motor, (b) simplified diagram of an armature winding, (c) partially completed motor showing the relationship between the armature and PM poles.

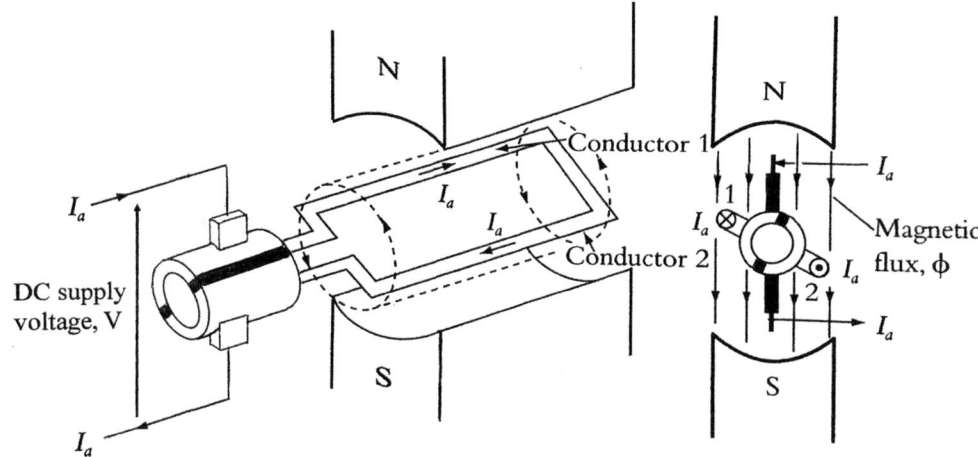

FIGURE 4.33   The simplest form of a PMDC motor showing an armature with only one coil.

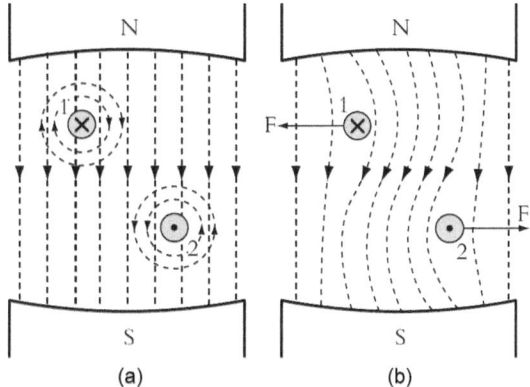

FIGURE 4.34   Generation of forces on a pair of conductors carrying current into the page ⊗ and out of the page ⊙.

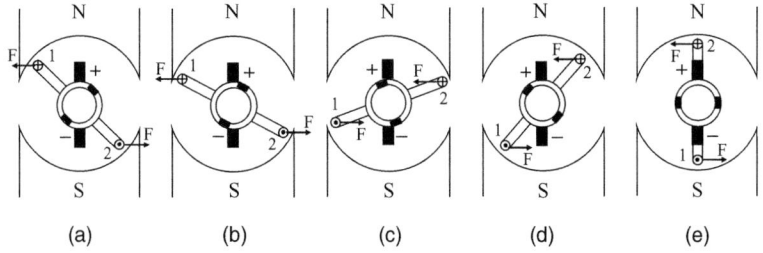

FIGURE 4.35   Forces on armature windings at different armature orientations.

The subsidiary fields interact with the main field to produce the resultant field shown in Figure 4.34(b). This produces forces on the conductors, which create a torque on the armature making it rotate.

By supplying current to the armature via carbon brushes and commutators, the currents flowing through the individual armature conductors reverse every time the coil is perpendicular to the main field, maintaining the torque in the same direction. Figure 4.35(a) to (e) shows the change in current direction as the armature rotates. In Figure 4.35(a) and (b), the current in conductor 1 flows 'into the page' and the current in conductor 2 flows 'out of the page' making the coil rotate anticlockwise. When the coil passes the horizontal (Figure 4.35(c)), the currents in the two conductors are reversed by the switching action of the commutator to maintain the torque and hence the rotation in the same direction.

4.9.3 ELECTROMOTIVE FORCE INDUCED IN THE ARMATURE WINDING

As the armature coil rotates, an electromotive force (emf) is induced in it which opposes the externally supplied armature current, I_a. The external supply voltage must overcome this emf if the machine is to motor and deliver mechanical power through the shaft.

Figure 4.36 shows the variation in flux linkage between the main field and the coil as the latter rotates, where the flux linkage is defined as the flux linking with (i.e. passing through) a coil multiplied by the number of turns in the coil. In Figure 4.36(a), when the coil is perpendicular to the field there is maximum flux linkage, ψ_m. With only one coil this equals the flux between the poles, ϕ

When the coil is parallel to the field, as shown in Figure 4.36(b), the flux linkage is zero. We will take this position to be the reference point and set the time $t = 0$ and the angle between the coil and main field $\theta = 0$.

When the coil has reached an angle θ to the field (see Figure 4.36(c)), the flux linkage is given by

$$\text{Flux linkage, } \psi = \psi_m \sin \theta = \phi \sin \theta \tag{4.47}$$

Assuming that the speed of rotation is ω rad s^{-1}, this occurs when the time, t, is:

$$t = \frac{\theta}{\omega} \text{ seconds}$$

Substituting for θ in Equation (4.47),

$$\psi = \phi \sin \omega t \tag{4.48}$$

Michael Faraday showed that the magnitude of the emf, e, induced in a coil was proportional to the rate of change of flux linkages (Details in Part 1). This may be expressed mathematically as

$$e = \frac{d\psi}{dt} \tag{4.49}$$

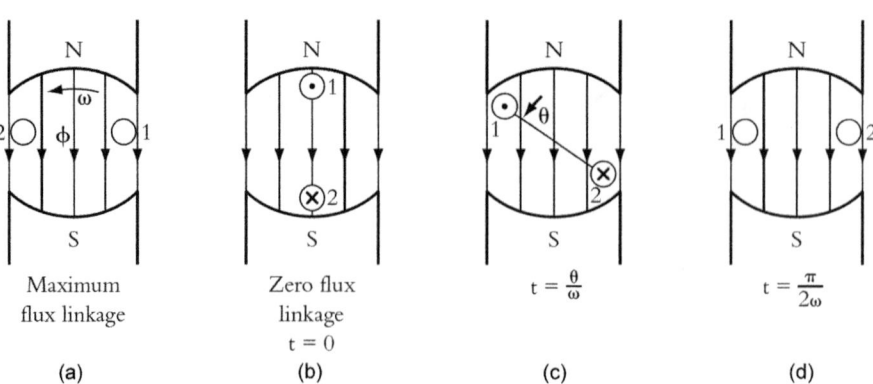

| Maximum flux linkage | Zero flux linkage $t = 0$ | $t = \frac{\theta}{\omega}$ | $t = \frac{\pi}{2\omega}$ |
| (a) | (b) | (c) | (d) |

FIGURE 4.36 Variation in flux linkage between the main field and armature coil. The crosses indicate that the direction of the induced emf is into the page and the dots show it coming out of the page.

$$e = \frac{d(\phi \sin \omega t)}{dt}$$

$$e = \phi \omega \cos \omega t \tag{4.50}$$

The coil continues to rotate to $(\pi/2)$ (Figure 4.36(d)) arriving at time, $t = \pi/(2\omega)$

If ε is the average emf induced in the coil as it rotates from being parallel to the field at time $t = 0$, (Figure 4.36(b)) to being perpendicular to the field at time $t = \frac{\pi}{2\omega}$ (Figure 4.36(d)), then:

$$\varepsilon = \frac{1}{\left[\frac{\pi}{2\omega}\right]} \int_{0}^{\frac{\pi}{2\omega}} \phi \omega \cos \omega t \tag{4.51}$$

$$\varepsilon = \frac{2\omega}{\pi} \phi \left[\sin \omega t\right]_{0}^{\frac{\pi}{2\omega}}$$

$$\varepsilon = \frac{2\omega}{\pi} \phi \text{ volts} \tag{4.52}$$

Many machines have several pairs of poles. The motor shown in Figure 4.32 has two pairs of poles; two N- and two S-poles. For a given speed of armature rotation, additional poles reduce the time the coil links with flux from a particular pole. This affects the induced emf. For a machine with p pairs of poles, the coil only links with the magnetic flux from each pole for $\left(\frac{\pi}{p\omega}\right)$ seconds. If the flux per pole is ϕ, then the average emf induced in the coil, ε, is given by:

$$\varepsilon = \frac{2\omega p}{\pi} \phi \text{ volts} \tag{4.53}$$

In practice, an armature winding will have many coils connected in series, and the total emf induced in the armature winding is given by:

total emf induced in armature winding = average emf induced in one armature coil

× number of coils in series

$$E = \frac{2\omega p}{\pi} \phi \times A_s \tag{4.54}$$

The magnetic flux, (ϕ), the number of poles $(2p)$ and the number of armature coils in series (A_s) are all constant for a particular machine; so, it is convenient to combine these three parameters into a single design constant, k, where:

$$k = \frac{2p\phi A_s}{\pi} \tag{4.55}$$

and express the total emf induced in the armature winding, E, as:

$$E = k\omega \text{ volts} \tag{4.56}$$

4.9.4 TORQUE

The electrical power delivered to the armature is converted to mechanical power, creating a torque which makes the armature rotate. Assuming that there are no losses in the system:

$$\text{electrical power delivered to armature} = \text{mechanical power at the shaft}$$
$$P_{\mathrm{a}} = P_{\mathrm{m}} \tag{4.57}$$

$$\text{electrical power} = \text{armature emf} \times \text{armature current}$$
$$P_{\mathrm{a}} = E \times I_{\mathrm{a}} \tag{4.58}$$

$$\text{mechanical power} = \text{torque} \times \text{angular velocity}$$
$$P_{\mathrm{m}} = L \times \omega \tag{4.59}$$

Combining Equations (4.57) and (4.59)

$$L = \frac{P_{\mathrm{a}}}{\omega}$$
$$\text{Torque} = \frac{\text{Electrical power delivered to the armature}}{\text{Angular velocity}} \tag{4.60}$$

Substituting Equations (4.56) and (4.58) into (4.60),

$$\text{Torque} = \frac{k\omega I_{\mathrm{a}}}{\omega}$$

giving:

$$\text{Torque} = kI_{\mathrm{a}} \text{ newton metres} \tag{4.61}$$

In practice, there will be some losses in the system due to friction at the bearings, windage and iron losses in the steel core. These reduce the mechanical torque produced at the shaft. However, as these losses are usually small, they will be ignored for the rest of this section and it will be assumed that the torque calculated using Equation (4.61) equals the mechanical torque delivered at the shaft.

Figure 4.37 shows the equivalent circuit for the armature winding. It has resistance R_{a}; so, the voltage applied to the armature V is given by:

$$V = E + I_{\mathrm{a}}R_{\mathrm{a}} \tag{4.62}$$

$$V = k\omega + I_{\mathrm{a}}R_{\mathrm{a}} \tag{4.63}$$

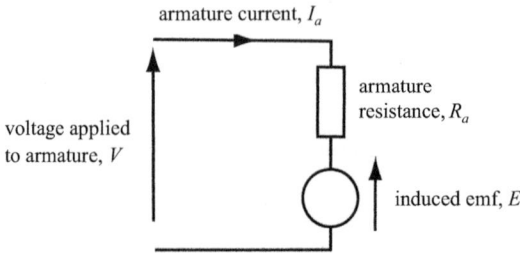

FIGURE 4.37   Equivalent circuit of armature.

$$\omega = \frac{\left(V - I_a R_a\right)}{k} \qquad (4.64)$$

where k is known as both the speed–voltage constant with the unit Vs $(\text{rad})^{-1}$ and the torque constant with the unit Nm A^{-1}. Although this seems confusing, dimensional analysis shows that Vs$(\text{rad})^{-1}$ and Nm A^{-1} are equivalent and therefore identical in meaning. Hence, we end up having two different names and apparently two conflicting sets of units for the same constant.

4.9.5 TORQUE AND SPEED CONTROL OF PMDC MOTORS

Equation (4.61) shows that the torque produced by a PMDC motor is proportional to armature current, I_a. One simple method to control the torque is to connect a variable resistor in series with the armature winding. More complex methods are based on the rapid switching or chopping of the supply to match the measured current to a desired value.

Motor speed depends on both the armature voltage, V, and the armature current and may be controlled by either varying the armature voltage or using a DC chopper circuit to convert a fixed DC supply voltage to a variable DC voltage and supplying the latter to the armature winding.

WORKED EXAMPLE

A 24V PMDC motor produces a torque of 0.1 Nm. The armature current is 1.2 A and the armature resistance is 1.5 Ω. Assuming that there are no losses, calculate the speed in rev s^{-1} and the output power.

The DC supply voltage is reduced to 16 V. The load torque is unchanged. Calculate the new speed and output power.

$$L = k I_a$$

$$k = \frac{L}{I_a} = \frac{0.1}{1.2} = 0.0833 \, \text{Vs} \left(\text{rad}\right)^{-1}$$

$$V = E + I_a R_a$$

$$V = k\omega + I_a R_a$$

$$24 = 0.0833\omega + 1.2 \times 1.5$$

$$24 = 0.0833\omega + 1.8$$

$$\omega = \frac{\left(24 - 1.8\right)}{0.0833} = \frac{22.2}{0.0833}$$

$$\omega = 266.5 \, \text{rad s}^{-1}$$

$$\text{speed} = \frac{\omega}{2\pi} = \frac{266.5}{2\pi} = 42.4 \, \text{rev s}^{-1}$$

mechanical output power, $P_m = L\omega$

$$P_m = 0.1 \times 266.5 = 26.6 \, \text{W}$$

Check:

$$\text{With no losses } P_a = P_m$$

Electrical power delivered to the armature, $P_a = E \times I_a$

$$P_a = 22.2 \times 1.2 = 26.6 \, \text{W}$$

The supply voltage is reduced to 16V.

$$L = k\,I_a$$

If the load torque is unchanged, then the armature current will remain at 1.2 A.

$$V = k\omega + I_a R_a$$

$$16 = 0.0833\omega + 1.2 \times 1.5$$

$$\omega = \frac{(16 - 1.8)}{0.0833} = \frac{14.2}{0.0833}$$

$$\omega = 170.5 \, \text{rad s}^{-1}$$

$$\text{speed} = \frac{\omega}{2\pi} = \frac{170.5}{2\pi} = 27.1 \, \text{rev s}^{-1}$$

mechanical output power, $P_m = L\omega$

$$P_m = 0.1 \times 170.5 = 17.1 \, \text{W}$$

LEARNING SUMMARY

By the end of this section, you should have learnt:

✓ The construction and uses of PMDC motors.

✓ How these motors work.

✓ To analyse their operation.

4.10   STEPPER MOTORS

Stepper motors move in incremental steps so they are used in applications where precise positioning is necessary. Generally, they produce torques of up to 15 newton metres with output powers ranging from milliwatts to hundreds of watts. Applications include:

- Numerical control of machine tools, routers and 3D printers.
- Computer peripherals and robotics.
- Integrated circuit manufacture.
- Digitally controlled (quartz crystal based) watches and clocks with analogue faces.
- Textile production.

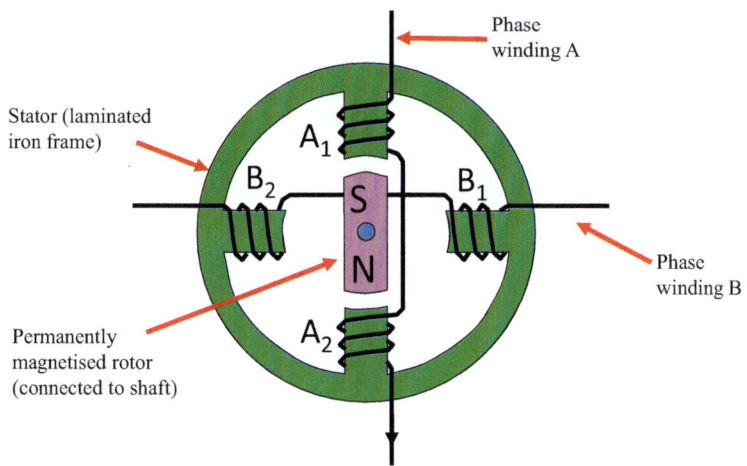

FIGURE 4.38   Construction of a stepper motor.

4.10.1   CONSTRUCTION OF STEPPER MOTORS

Figure 4.38 shows the basic construction of a four-pole stepper motor. The motor has two parts, a rotor and a stator. The rotor is a PM that is free to rotate. The stator comprises a stack of insulated iron or steel laminations clamped together in a frame. The laminations are shaped to create four poles on the inner circumference. For purposes of explanation, the poles have been labelled A_1, B_1, A_2 and B_2. A coil is wound around each pole and diametrically opposite coils are connected in series to form a phase winding.

We will define the positive direction for current in the A-phase as flowing downwards, entering at the top of coil A_1 and leaving from the bottom of coil A_2 and the positive direction for current in the B-phase winding as flowing from right to left, entering through coil B_1 and leaving through coil B_2.

4.10.2   PRINCIPLES OF OPERATION

We will take as our starting point the situation shown in Figure 4.39(a) when the A-phase current is flowing in the positive direction. We will call this I_{A+}. The current creates a magnetic field in the vertical plane that attracts the S-pole (south pole) of the PM rotor towards pole A_1 and the rotor N-pole (north pole) to pole A_2.

The A-phase current is then switched off and the B-phase current is turned on which flows in the positive direction from right to left, I_{B+}. This creates a magnetic field in the horizontal plane and the rotor moves 90° clockwise to align with the new field as shown in Figure 4.39(b).

The next step is to turn off the B-phase current and switch on the A-phase current but this time it flows in the negative direction, I_{A-} This makes the rotor turn another 90° clockwise to the position shown in Figure 4.39(c) with the *N*-pole of the rotor directly under pole A_1.

When the A-phase current is turned off and the negative B-phase current, I_{B-} is switched on, the rotor rotates a further 90° clockwise to the position shown in Figure 4.39(d).

Finally, the B-phase current is switched off and the positive A-phase current, I_{A+}, is turned on again attracting the rotor back to its original position, from where the process can repeat.

4.10.3   TRANSISTOR H-BRIDGES

To turn the phase currents on and off in the correct sequence, two transistor H-bridges are used as shown in Figure 4.40. Transistors make good switches and are ideal for this type of application. They only conduct when a small base current flows, which establishes a far larger collector current to circulate in the main circuit. So, by controlling the base current it is possible to turn the main

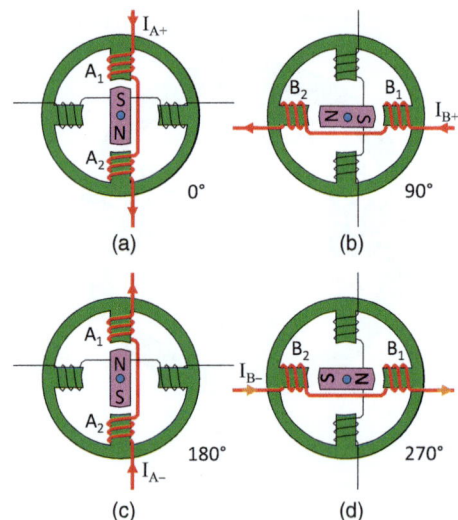

FIGURE 4.39 Rotation of a stepper motor.

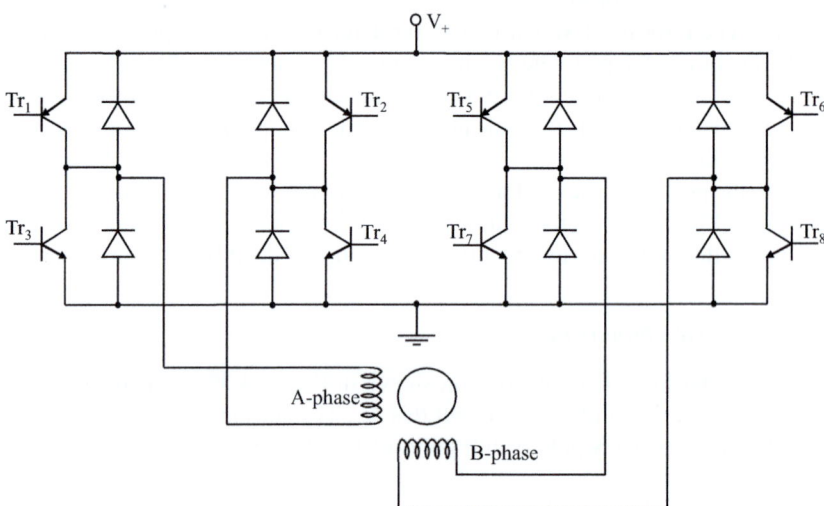

FIGURE 4.40 Transistor H-bridges.

current on and off as desired. Full details on how to operate transistors as switches may be found in Part 1. The 'free-wheeling' diodes connected across the transistors prevent the latter from being destroyed by the high voltages induced in the coils when currents are turned off.

The operation of the stepper motor relies on the transistors being turned on and off in the correct sequence. To demonstrate this sequence, we will replace the transistors with simple on–off switches as illustrated in Figure 4.41.

For I_{A+} to flow, switches Tr_1 and Tr_4 must be closed; see Figure 4.41(a). Tr_5 and Tr_8 have to be closed for I_{B+} to flow (Figure 4.41(b)). It is Tr_2 and Tr_3 for I_{A-} (Figure 4.41(c)) and Tr_6 and Tr_7 for I_{B-} (Figure 4.41(d)).

We now have the order that the transistors must be switched on and off. For the motor to run smoothly, the base currents must be switched on and off in the correct order and at precisely the right times. To achieve this, a ring counter is used.

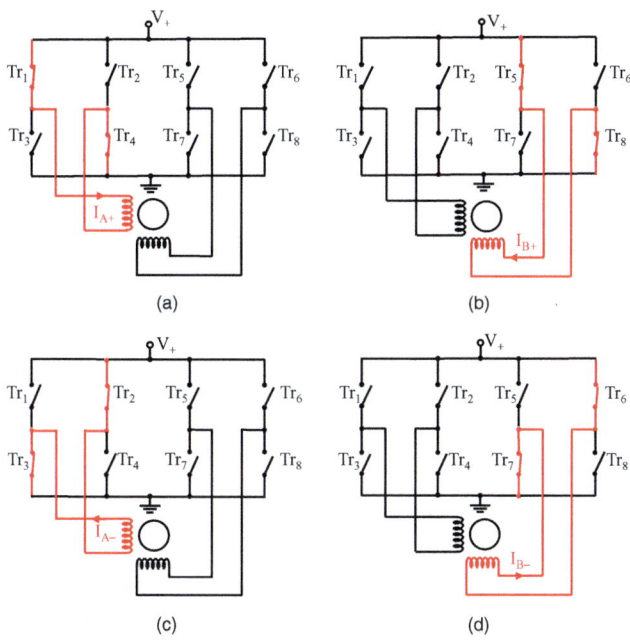

FIGURE 4.41 Transistor switching sequence.

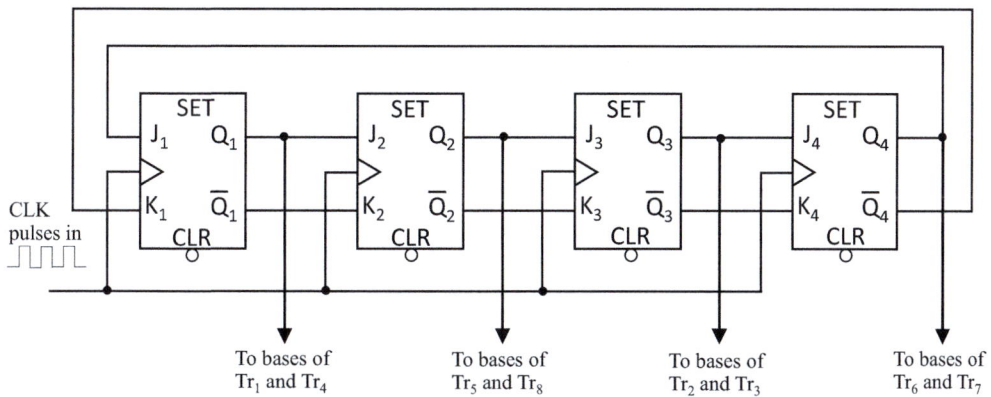

FIGURE 4.42 Ring counter.

4.10.4 RING COUNTER

Figure 4.42 shows a four-stage ring counter with a JK bistable at each stage. Bistables are sequential logic devices. Their operation is explained fully in Part 1 Unit 5. In the present application, the clock (CLK) signal has the effect of incrementing the motor position to the next step.

The bistable has two inputs J and K and complementary outputs Q and $\bar{Q}$. The logic level or state of the outputs remains constant until one of the following is received.

- A set pulse that immediately makes $Q = 1$ and $\bar{Q} = 0$.
- A clear (CLR) pulse, which instantly puts $Q = 0$ and $\bar{Q} = 1$.
- A clock pulse, when the outputs are set at the logic levels defined in the JK bistable state table, see Table 4.1.

TABLE 4.1
JK Bistable State Table

J	K	Q_{n+1}	
0	0	Q_n	No change at Q after the clock pulse
0	1	0	
1	0	1	
1	1	$\overline{Q_n}$	Output changes after the clock pulse

Key:

Q_{n+1} is the logic level at output Q after $(n + 1)$ clock pulses.
Q_n is the logic level at output Q after n clock pulses.

When the ring counter is switched on, bistable 1 receives a set pulse to make

$$Q_1 = 1 \text{ and } \overline{Q_1} = 0.$$

The other bistables all receive clear (CLR) pulses to put

$$Q_2 = Q_3 = Q_4 = 0 \text{ and } \overline{Q_2} = \overline{Q_3} = \overline{Q_4} = 1.$$

There are direct connections from Q_1 to J_2 and $\overline{Q_1}$ to K_2; so, J_2 also becomes 1 and K_2 equals 0.
With $Q_2 = Q_3 = Q_4 = 0$, the other J inputs J_3, J_4 and J_1 are all at 0.

Similarly, with $\overline{Q_2} = \overline{Q_3} = \overline{Q_4} = 1$, the K inputs K_3, K_4 and K_1 are all 1.

This is all laid out in the first and second rows of the ring-counter state table (Table 4.2).

Figure 4.42 shows that Q_1 is connected to the bases of transistors Tr_1 and Tr_4. With Q_1 at logic 1, currents flow into the bases of these devices enabling the two transistors to conduct and current I_{A+} to flow in the stator A-phase winding.

After the first clock pulse, the outputs change. To determine how they alter, one needs to use the JK bistable state table (Table 4.1) in conjunction with the procedure on how to analyse sequential

TABLE 4.2
Ring Counter Output Changes after Each Clock Pulse

No. of Clock Pulses		$J_1 = Q_4$	$K_1 = \overline{Q_4}$	Q_1	$J_2 = Q_1$	$K_2 = \overline{Q_1}$	Q_2	$J_3 = Q_2$	$K_3 = \overline{Q_2}$	Q_3	$J_4 = Q_3$	$K_4 = \overline{Q_3}$	Q_4	
0	Outputs			1			0			0			0	
	Inputs	0	1		1	0		0	1		0	1		
1	Outputs			0			1			0			0	
	Inputs	0	1		0	1		1	0		0	1		
2	Outputs			0			0			1			0	
	Inputs	0	1		0	1		0	1		1	0		
3	Outputs			0			0			0			1	
	Inputs	1	0		0	1		0	1		0	1		
4	Outputs			1			0			0			0	cycle
	Inputs	0	1		1	0		0	1		0	1		repeats
5	Outputs			0			1			0			0	
	Inputs	0	1		0	1		1	0		0	1		

TABLE 4.3

Operational Sequence

Number of Clock Pulses Received	Q Output at Logic 1	Transistors Conducting	Current Flowing in the Stator Windings	Rotation Clockwise	
0	Q_1	Tr_1 and Tr_4	I_{A+}	$0°$	
1	Q_2	Tr_5 and Tr_8	I_{B+}	$90°$	
2	Q_3	Tr_2 and Tr_3	I_{A-}	$180°$	
3	Q_4	Tr_6 and Tr_7	I_{B-}	$270°$	
4	Q_1	Tr_1 and Tr_4	I_{A+}	$360° = 0°$	Rotor completes one revolution and the sequence repeats
5	Q_2	Tr_5 and Tr_8	I_{B+}	$90°$	

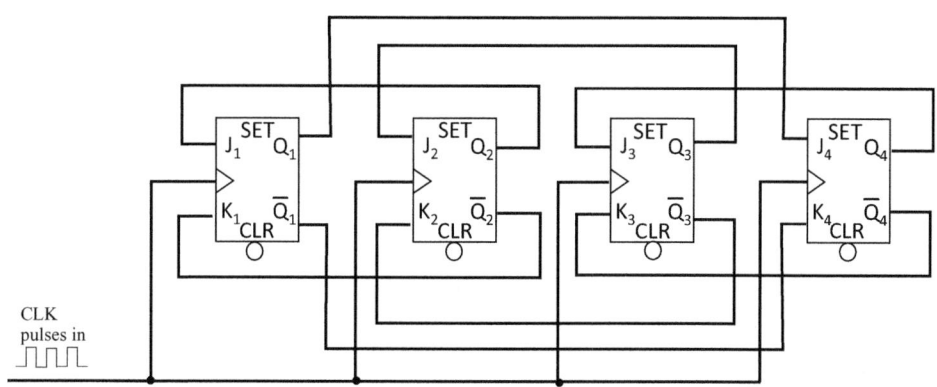

FIGURE 4.43 Ring counter connections for operation in reverse direction.

logic circuits described in Part 1 Unit 5. The results are shown in the third and fourth rows of Table 4.2 where the number of clock pulses received is 1. Q_1 falls to 0, turning off transistors Tr_1 and Tr_4 and stopping current I_{A+} from flowing in the stator A-phase winding. Simultaneously, Q_2 rises to 1 switching on transistors Tr_5 and Tr_8, establishing current I_{B+} in the B-phase winding and making the rotor move 90° clockwise.

The complete sequence is laid out in Table 4.3.

The angle of rotation is proportional to the number of clock pulses received, and hence the speed of rotation is proportional to the frequency. However, in order to use a stepper motor for positioning, it is essential to be able to change the direction of rotation on the command of a signal. At present, the permanent connection of the Q and $\bar{Q}$ outputs to J and K inputs, respectively, fixes the direction of rotation. To reverse the direction of the rotor, it is necessary to reconfigure the wiring as shown in Figure 4.43. For example, the connections from Q_1 to J_2 and $\bar{Q}_1$ to K_2 must be replaced by links from Q_3 to J_2 and $\bar{Q}_3$ to K_2, respectively. These reverse the direction of rotation but the motor remains unidirectional.

4.10.5 Bi-Directional Operation

For the motor to have a two-way or bi-directional operation, eight switches are required. Figure 4.44 shows the one that connects either Q_1 or Q_3 to J_2.

For this to work properly, all eight switches must operate simultaneously and on the command of an external logic signal and this is impossible using mechanical switches. Instead, four identical

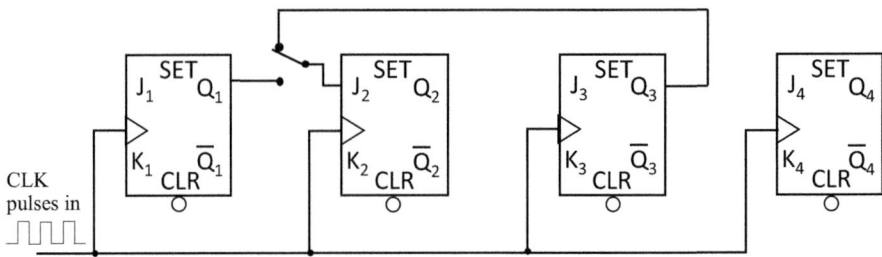

FIGURE 4.44   Switch to reverse motor direction.

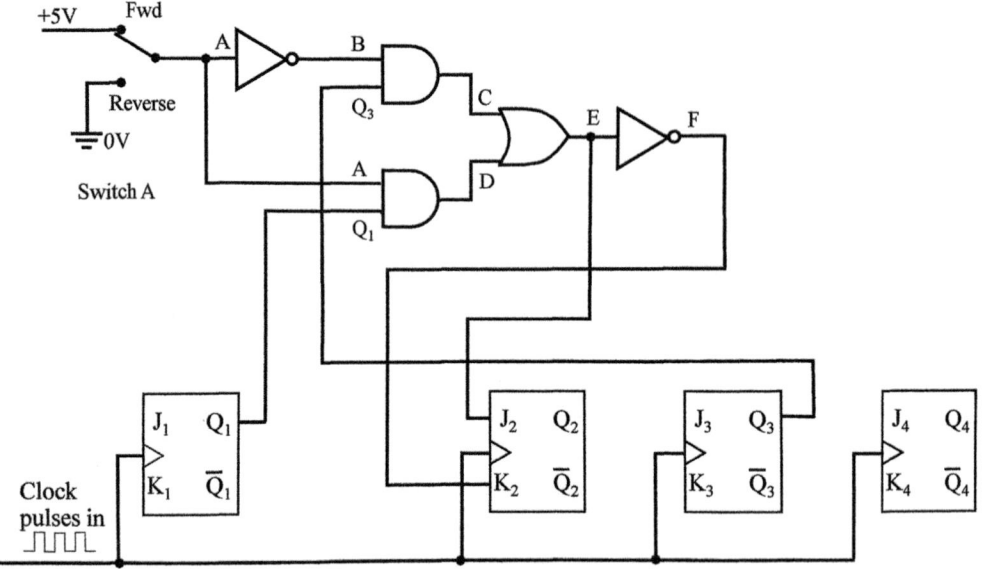

FIGURE 4.45   Using combinational logic to control motor direction.

combinational logic circuits are used, one for each ring counter stage. Figure 4.45 shows the second-stage circuit in place.

Combinational logic is different from sequential logic. Here, the outputs respond instantly to changes in the inputs. The reversing circuit employs three combinational logic gates NOT, AND and OR. The truth tables, Boolean expressions and logic symbols for all these devices are shown in Table 4.4. Boolean algebra uses a unique set of mathematical symbols: a . (dot) represents AND and + (plus sign) represents OR.

To analyse a combinational logic circuit, all the gate inputs and outputs must be given a unique identifier. A truth table is then constructed to show the logic levels at every point in the circuit with a column for every device input and output and a row for every combination of circuit input states. See Table 4.5. With only two directions of rotation to consider, just two rows are required. The process of analysis is fully described in Part 1.

Table 4.5 shows that for forward rotation, J_2 and K_2 are effectively connected to Q_1 and $\overline{Q_1}$, respectively, and for reverse operation, J_2 and K_2 are effectively joined to Q_3 and $\overline{Q_3}$, respectively. So, with a click of the switch or, in practice, a change in the state of the logic signal for direction, the direction of rotation is reversed.

Figure 4.46 shows the complete network for bi-directional operation. The ring counter and reversing circuits are known collectively as the translator and the transistor H-bridges are called the driver.

TABLE 4.4
Combinational Logic Gates

Type of Gate	Truth Table	Boolean Expression	Logic Symbol
NOT	A B 0 1 1 0	$Q = \bar{A}$	
AND	A B Q 0 0 0 0 1 0 1 0 0 1 1 1	$Q = A.B$	
OR	A B Q 0 0 0 0 1 1 1 0 1 1 1 1	$Q = A + B$	

TABLE 4.5
Logic Levels for the Two Directions of Rotation

Direction of Rotation	A	$B = \bar{A}$	Q_3	$C = B.Q_3$	Q_1	$D = A.Q_1$	$E = C + D$	$F = \bar{E}$	$J_2 = E$	$K_2 = F$
Forward	1	0	Q_3	0	Q_1	Q_1	Q_1	$\bar{Q_1}$	Q_1	$\bar{Q_1}$
Reverse	0	1	Q_3	Q_3	Q_1	0	Q_3	$\bar{Q_3}$	Q_3	$\bar{Q_3}$

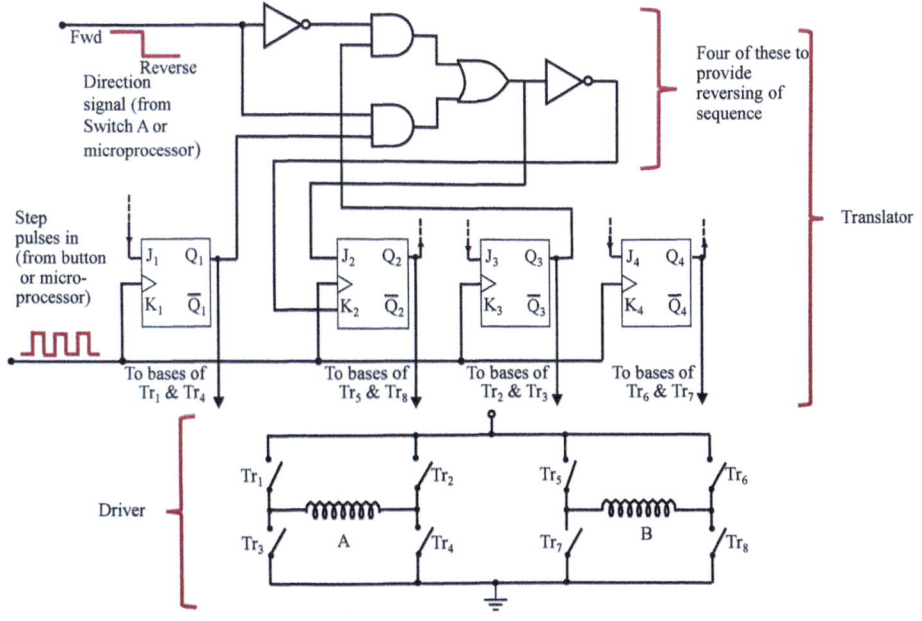

FIGURE 4.46 Complete driving system.

As noted earlier, the CLK signal has the effect of stepping the system through the four states of energising the windings, thus causing the motor to step from position to position. To use clock pulses from an oscillator to turn the transistors on and off and to have a switch to control the direction of rotation is impractical. In reality, both signals would be derived from a microprocessor. Self-contained units containing the translator and drive circuits are readily available. These often incorporate microstepping (as described below) and the control of currents. They range from heavy-duty industrial units down to low-cost units for use in hobby and student project applications.

4.10.6 STEPPER MOTORS AND DRIVERS IN PRACTICE

A common application for stepper motors is to provide positioning of machine parts under the control of a computer or microprocessor, which (via a suitable digital output interface) generates the step and direction logic signals to command the motor to make the movements required. The conversion of the step and direction signals into stator phase currents I_{A+}, I_{B+}, I_{A-} and I_{B-} is handled by the translator and driver circuits as illustrated in Figure 4.47.

The computer or microprocessor provides the motion commands to the translator/driver hardware. With the basic system of operation described above, there is a fixed sequence for supplying current to each coil once every rotor revolution. This mode of operation is called single-coil excitation. The current sequence is shown in Figure 4.48(a).

In the full-step drive mode, illustrated in Figure 4.48(b), current is supplied to two coils simultaneously to produce a larger torque output but it still takes four steps to complete one revolution.

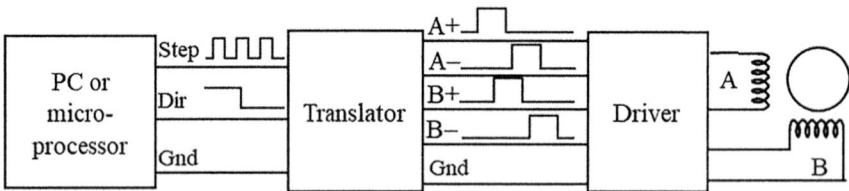

FIGURE 4.47 Computer-controlled stepper motor system.

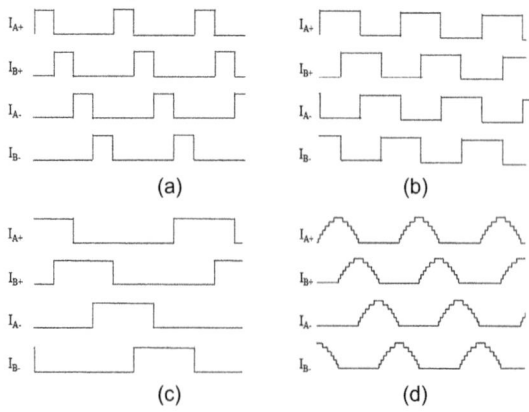

FIGURE 4.48 Driving current waveforms, (a) Single-coil excitation, (b) full-step drive, (c) half-step drive and (d) microstepping.

To improve the resolution, the half-step drive mode has to be used. Here, the sequence is that one coil receives current, followed by two coils being supplied simultaneously. This is then continuously repeated as shown in Figure 4.48(c). Eight steps are required for one revolution.

The resolution can be improved further using micro-stepping. Like half-stepping, it involves energising successive coils simultaneously, but this time in varying proportions. Currents resembling sine waves are supplied to the coils, see Figure 4.48(d), to produce a smooth rotation and provide greater accuracy and resolution. Micro-stepping is most commonly used today and even the lowest-cost driver units can typically be configured to provide micro-stepping to various resolutions.

Another way to improve the resolution of a stepper motor is to increase the ratio of rotor poles relative to stator poles. To examine this concept, compare the motor with four stator poles and two rotor poles shown in Figure 4.39 with the one illustrated diagrammatically in Figure 4.49 which has four stator poles and six rotor poles. As the stator phase currents march through the sequence, starting at I_{A+} through I_{B+} and I_{A-} and finishing at I_{B-}, the machine with two rotor poles rotates 270° whilst the motor with six rotor poles only moves 90°. This demonstrates clearly that an increase in the number of rotor poles relative to stator poles improves the resolution. Many practical stepper motors use a more complex construction involving a toothed rotor and stator and typically have a step angle of 1.8°.

Stepper motors are widely used in printers, scanners etc., but also form a convenient approach for motion control in student projects and form the basis of many 3D printing machines and light-duty CNC (computer numerical control) routers such as those available as low-cost, self-assembly kits. These are typically controlled using a microcontroller to generate the step and direction signals and low-cost microstepping translator/driver units to control the motors.

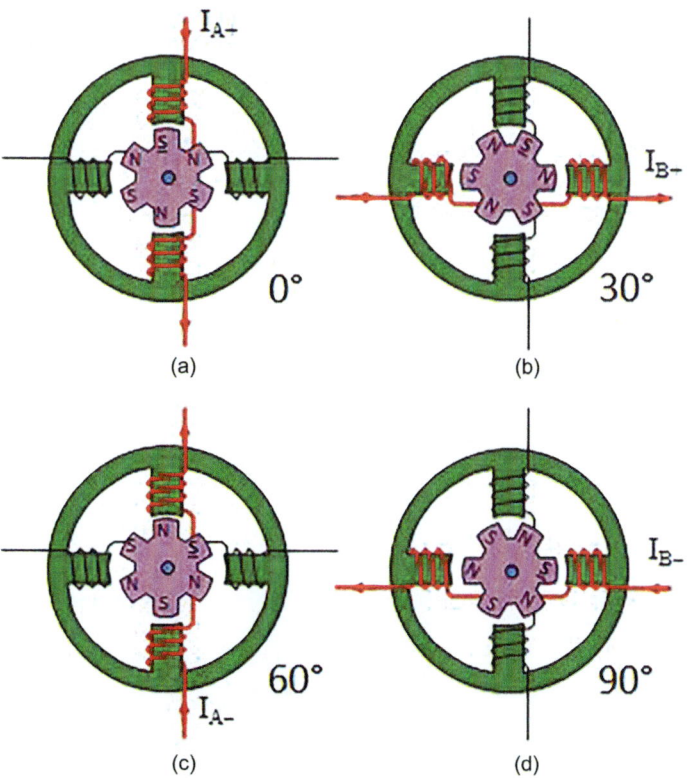

FIGURE 4.49   Stepper motor with six rotor poles and four stator poles (diagrammatic representation).

LEARNING SUMMARY

By the end of this section, you should have learnt:

- ✓ The principles of operation of stepper motors.
- ✓ The use of transistor H-bridges to switch the stator winding currents on and off.
- ✓ The role that ring counters play in the operation of stepper motors.
- ✓ The differences between combinational logic and sequential logic.
- ✓ How the direction of rotation can be reversed.
- ✓ How variations in driving current waveforms affect motor resolution.
- ✓ How fine resolution of stepper motor systems can be achieved via microstepping and the use of toothed rotors and stators.

4.11 INVERTER-FED INDUCTION MOTORS AND THEIR CHARACTERISTICS

The construction and operation of a three-phase alternating current (AC) induction motor connected to a fixed voltage and fixed frequency supply is explained in detail in Part 1 Unit 5. It will suffice to repeat the key points here.

Induction motors are the most common motors encountered in industry and are the most rugged type of motors. They have two parts; an outer stationary frame called the stator and an inner rotating part called the rotor. These are shown in Figure 4.50.

The simplest form of induction motor, the two-pole motor, has three separate coils wound in the slots on the inner circumference of the stator. The coils are mutually displaced by 120°. The rotor

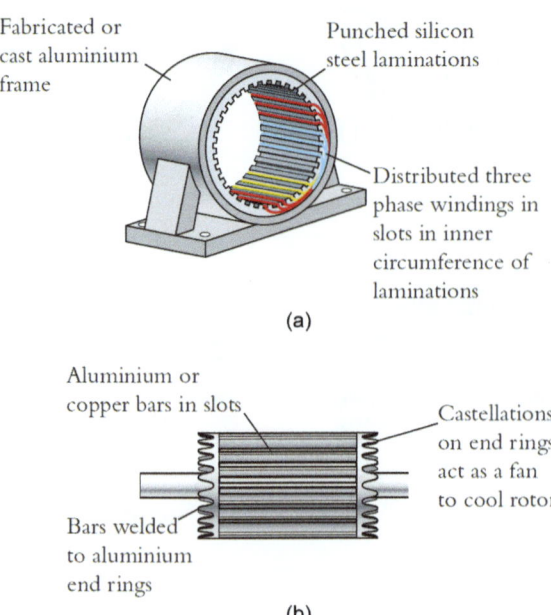

Fabricated or cast aluminium frame

Punched silicon steel laminations

Distributed three phase windings in slots in inner circumference of laminations

(a)

Aluminium or copper bars in slots

Castellations on end rings act as a fan to cool rotor

Bars welded to aluminium end rings

(b)

FIGURE 4.50   (a) Stator and (b) rotor of a simple (squirrel-cage) induction motor.

has silicon steel laminations keyed to a central shaft. Slots are cut in the laminations into which aluminium or copper conductors are fitted. The ends of the conductors are welded to aluminium end rings. There are no electrical connections to the rotor.

When a balanced three-phase fixed frequency supply is connected to the stator windings, currents flow in the coils which produce a uniform magnetic field that rotates at constant speed in the air gap between the stator and rotor. In the time it takes the currents to complete one cycle, the magnetic field makes one revolution of the air gap. This is illustrated in the sequence of diagrams in Figure 4.51.

The speed at which the field rotates is called the synchronous speed, n_s. Measured in revolutions per minute, it is related to the frequency of the stator currents, f, in hertz (Hz) as follows:

$$n_s = 60\,f \tag{4.65}$$

The rotating magnetic field interacts with the rotor, inducing EMFs in the rotor conductors. The latter are short circuited by the end rings and hence currents flow in the conductors. These currents create subsidiary magnetic fields around the conductors. The subsidiary fields interact with the rotating stator field to produce forces on the rotor conductors to create a torque on the rotor and make

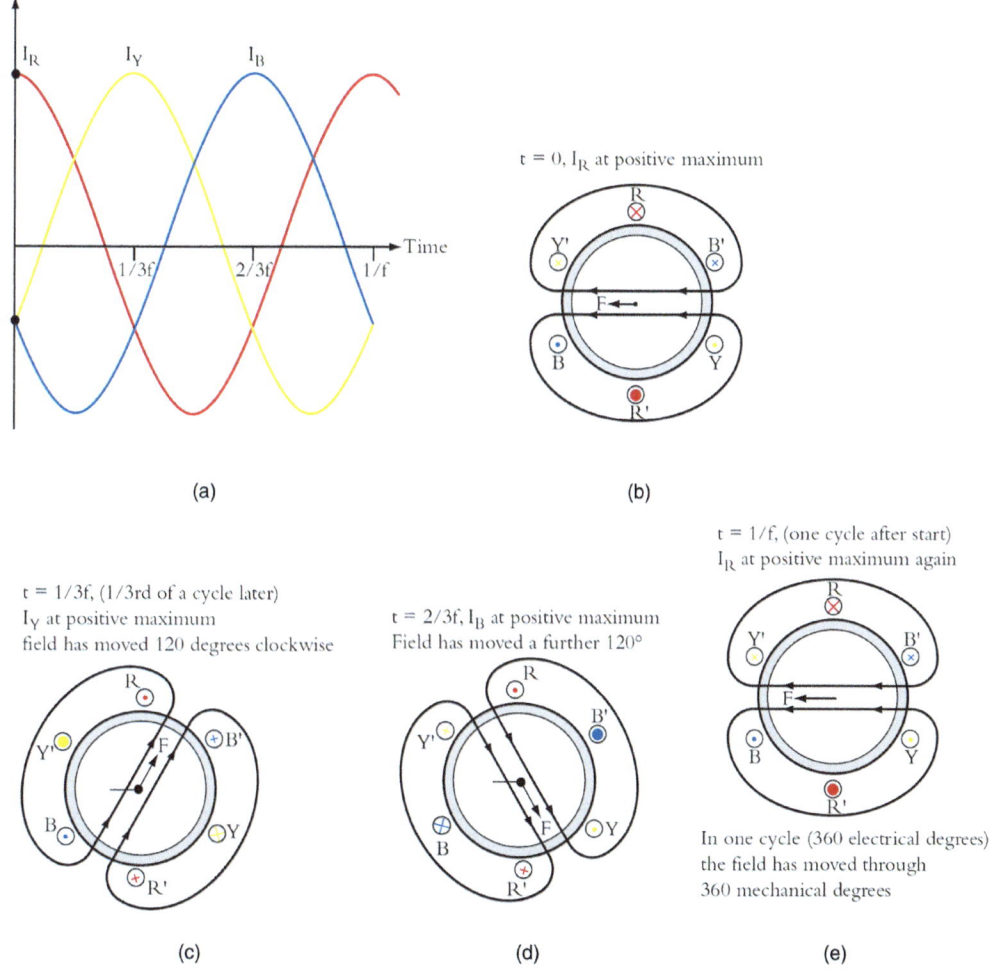

FIGURE 4.51   (a) Three-phase AC waveforms; (b)–(e) sequence of energising the windings of a two-pole induction motor.

it turn. This torque is called the electrical or driving torque and its magnitude is proportional to the magnitude of the rotating magnetic field.

The rotor accelerates until the electrical torque produced by the currents exactly equals the mechanical load torque on the shaft. At this point, the rotor is running at a speed, n, slightly slower than the synchronous speed. This difference in speed is expressed as a ratio and is known as the 'slip' or 'per unit slip', s:

$$s = \frac{n_s - n}{n_s} = 1 - \frac{n}{n_s} \tag{4.66}$$

The small difference between the speed of the rotating field and that of the rotor is fundamental to the operation of the induction motor.

Figure 4.52 shows a typical torque *versus* slip characteristic for a two-pole motor.

When it is operating in steady-state conditions, an increase in the load torque causes the rotor to slow down and the slip to increase. This raises the emfs induced in the rotor conductors and the rotor currents and creates more field distortion and more driving torque. The slip increases until the driving torque equals the load torque and steady conditions are re-established. For most motors, the steady-state slip varies between around 0.01 on no load and 0.10 when the motor is driving its full rated load. This means that a two-pole motor connected to a 50 Hz supply will run at a steady-state speed somewhere between 45 rev s^{-1} (2700 rev min^{-1}) when $s = 0.1$ and 49.5 rev s^{-1} (2970 rev min^{-1}) when $s = 0.01$.

For many applications, these speeds are too high. Manufacturers can reduce the synchronous speed, and hence the rotor speed by winding the stator coils in a way that effectively increases the number of poles in the rotating magnetic field. This was demonstrated pictorially in Part 1. For example, a four-pole motor, with two N-poles and two S-poles, has a synchronous speed equal to half the supply frequency. Expressing the synchronous speed in rev min^{-1}:

$$n_s = 60f / 2 = 30f \tag{4.67}$$

Connected to a 50 Hz supply, the synchronous speed is 1,500 rev min^{-1} and the rotor speed varies between around 1,350 rev min^{-1} when $s = 0.1$ and 1,485 rev min^{-1} when $s = 0.01$.

Further increases in the number of poles reduce even more the synchronous speed and with it the rotor speed as demonstrated in Table 4.6. Industrial induction motors are typically rated to run at a slip value in the order of 0.04 (4%).

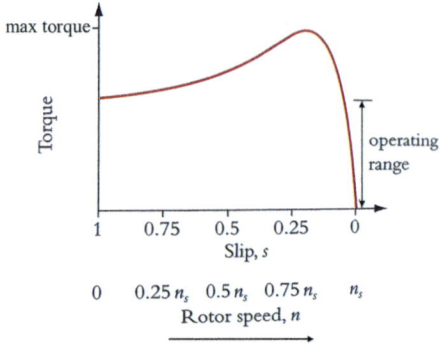

FIGURE 4.52   Typical torque vs. slip characteristic for a two-pole motor.

TABLE 4.6

Effect on the Speed of Increasing the Number of Poles

No of Poles	Pairs of Poles	Synchronous Speed in rev min^{-1} when f = 50 Hz	Rotor Speed in rev min^{-1} when s = 0.1	Rotor Speed in rev min^{-1} when s = 0.04	Rotor Speed in rev min^{-1} when s = 0.01
2	1	3000	2700	2880	2970
4	2	1500	1350	1440	1485
6	3	1000	900	960	990
8	4	750	675	720	742.5
2p	p	3000/p	2700/p	2880/p	2970/p

4.11.1 CREATING AN AC SUPPLY OF VARIABLE FREQUENCY AND VOLTAGE USING AN INVERTER

With a fixed number of poles and fixed supply frequency, the variation in rotor speed is very limited. To operate an induction motor over a wide speed range, it must be connected to a variable frequency supply.

Inverters produce variable frequency supplies. They convert DC to alternating current of a chosen frequency. To understand their operation, you should consider the simple single-phase bridge inverter shown in Figure 4.53. Diagonally opposite pairs of transistors conduct in turn. When transistors Tr$_1$ and Tr$_4$ are switched on and Tr$_2$ and Tr$_3$ are off, the current flows from the positive DC rail through Tr$_1$, the load and Tr$_4$ to the negative rail, travelling through the load from left to right. When Tr$_2$ and Tr$_3$ are turned on and Tr$_1$ and Tr$_4$ are switched off, the current flows through the load in the reverse direction as it travels from the positive DC rail to the negative. 'Freewheeling' diodes are connected in parallel with the transistors to prevent the latter from being destroyed by the large back emfs generated when the transistors are turned off and the current through the inductive load (i.e. stator field windings) is reversed.

By turning each diagonal pair of transistors on for slightly less than half of a cycle, the load voltage waveform shown in Figure 4.54(a) is created. Having a short period with no current at the beginning and end of each half cycle reduces the risk of two transistors in the same arm conducting simultaneously and short circuiting the DC supply. The output frequency is varied by altering the conduction times of the transistors as illustrated in Figure 4.54(b). A more comprehensive description of inverter operation can be found in Hughes and Drury (2013).

To control the magnitude of the load current, pulse width modulation (PWM) is often used. This involves the removal of a series of notches in each half cycle to produce a voltage waveform like that in Figure 4.55(a). One way this is achieved is to switch Tr$_4$ on and off when Tr$_1$ is conducting

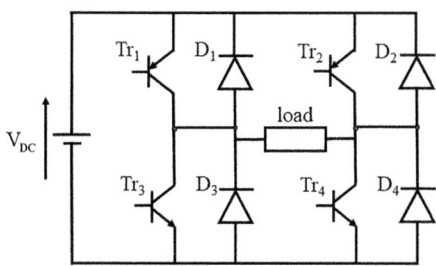

FIGURE 4.53 Single-phase bridge inverter circuit.

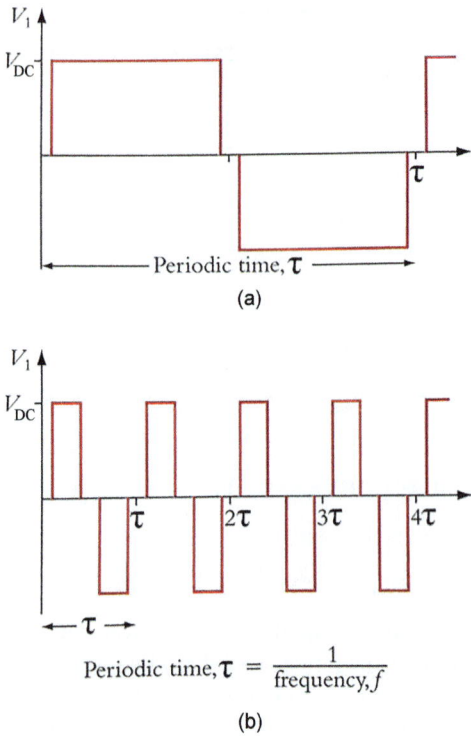

FIGURE 4.54   Output waveforms for simple bridge inverter circuit (a) giving a low frequency and (b) giving a higher frequency output.

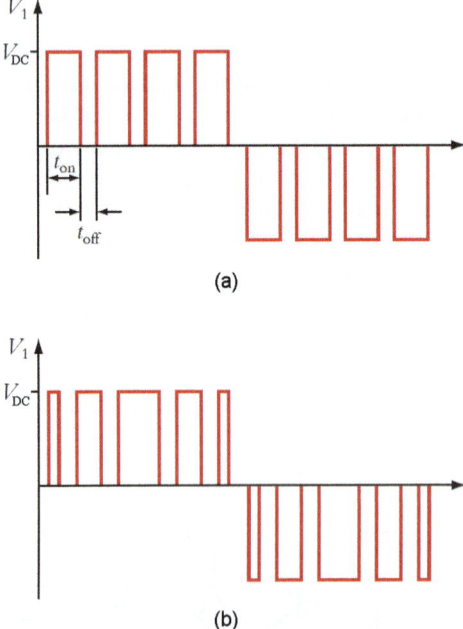

FIGURE 4.55   PWM output waveforms with (a) constant pulse cycle over each half-wave and (b) variable pulse cycle.

and turn Tr_3 on and off when Tr_2 is on, taking precautions to avoid two transistors on the same arm being switched on simultaneously. The mean output voltage is then equal to:

$$\text{mean output voltage}, V_1 = V_{\text{DC}} \frac{t_{\text{on}}}{\left(t_{\text{on}} + t_{\text{off}}\right)} \tag{4.68}$$

where t_{on} = duration of the pulses and t_{off} = time between pulses.

To alter the mean output voltage, the times at which Tr_3 and Tr_4 are switched on and off are adjusted to change the ratio of $\dfrac{t_{\text{on}}}{\left(t_{\text{on}} + t_{\text{off}}\right)}$.

In more sophisticated systems, microprocessors are used to vary the duration of the pulses and the spaces between pulses. This is shown in Figure 4.55(b). In this case, the average output voltage is given by

$$\text{Mean output voltage} = V_{\text{DC}} \times \frac{\text{total duration of pulses in a half cycle}}{\text{duration of half cycle}} \tag{4.69}$$

$$= V_{\text{DC}} \times \frac{\Sigma t_{\text{on}}}{\Sigma\left(t_{\text{on}} + t_{\text{off}}\right)} \tag{4.70}$$

For an induction motor supplied by an inverter to operate properly, the number, width and spacing of the pulses must be chosen with extreme care and controlled by a microprocessor to make the load current waveform resemble as closely as possible a sine wave. Looking at the voltage waveforms in Figure 4.55, this might seem impossible. But, all repetitive waveforms with a fixed periodic time comprise an infinite series of sine and cosine waves of different magnitudes and frequencies. These are known as a Fourier Series (see Bird, 2004). For the example of the square wave illustrated in Figure 4.56(a), the Fourier Series begins:

$$V\left(t\right) = \frac{4V_{\text{DC}}}{\pi}\left(\sin \omega t + \frac{1}{3}\sin 3\omega t + \frac{1}{5}\sin 5\omega t + \ldots\ldots\right) \tag{4.71}$$

where

$$\omega = 2\pi f$$

The first term in the bracket of Equation (4.71) is a sine wave of the same frequency as the square wave. This is called the fundamental. The other sine waves are called the harmonics. They have smaller magnitudes and their frequencies are multiples of the fundamental. Figure 4.55(b) shows the result of adding the first three terms in the series shown in Equation (4.71). As more terms are added, the waveform becomes closer to a square wave.

To explain what happens when such voltages are applied to motor coils, we need to go back to basic principles. Motor coils have both inductance and resistance, although the latter is usually small and is often ignored. In Part 1 Unit 5, it was shown that when a voltage, v is applied to an inductor, L the instantaneous current through the inductor, i is given by the equation:

$$v = L\frac{di}{dt} \qquad \text{(Part 1 Equation 5.24)}$$

where L is the inductance in henries (H).

[The reader should note that L is used by electrical and electronic engineers to represent inductance, and it takes this meaning in the two equations reproduced here from Part 1 Unit 5 and in Equation (4.73) below. All other uses of L within the present unit symbolise torque]

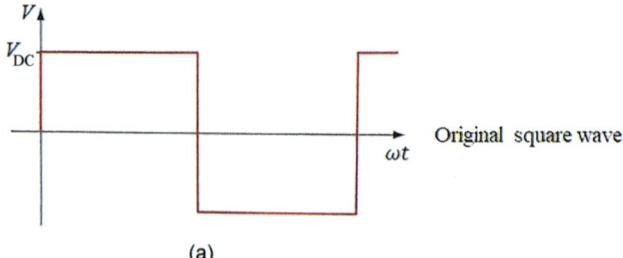

(a)

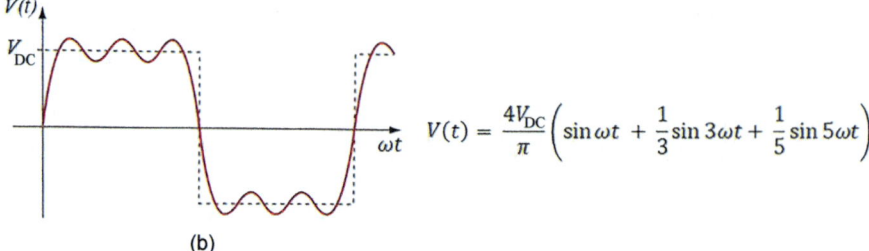

$$V(t) = \frac{4V_{DC}}{\pi}\left(\sin\omega t + \frac{1}{3}\sin 3\omega t + \frac{1}{5}\sin 5\omega t\right)$$

(b)

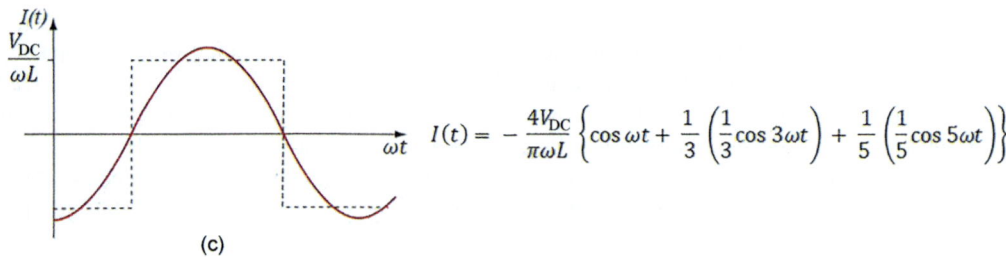

$$I(t) = -\frac{4V_{DC}}{\pi\omega L}\left\{\cos\omega t + \frac{1}{3}\left(\frac{1}{3}\cos 3\omega t\right) + \frac{1}{5}\left(\frac{1}{5}\cos 5\omega t\right)\right\}$$

(c)

FIGURE 4.56 (a) Voltage waveform in the form of a square wave, (b) square wave voltage waveform approximated by the summation of fundamental and first two harmonics and (c) the resulting current waveform through an inductive load.

The instantaneous current, i, is found by integrating the above equation

$$i = 1/L\int v\,dt \qquad (4.72)$$

So, when the voltage $V(t)$ defined in the first three terms of the series in Equation (4.71) is applied to an inductor, L the resultant current $I(t)$ is found to be

$$I(t) = -\frac{4V_{DC}}{\pi\omega L}\left\{\cos\omega t + \frac{1}{3}\left(\frac{1}{3}\cos 3\omega t\right) + \frac{1}{5}\left(\frac{1}{5}\cos 5\omega t\right)\right\} \qquad (4.73)$$

Inductive reactance, X_L was defined in Part 1 as

$$X_L = 2\pi f L \qquad \text{(Part 1 Equation 5.52)}$$

and so it is directly proportional to frequency. The result is that the reactance of an inductor (and hence a motor winding) is larger at the harmonic frequencies than at the fundamental. So, when a non-sinusoidal voltage, as shown in Figure 4.56(b) and represented mathematically by Equation (4.71), is applied to a motor coil, the harmonic components in the current waveform are attenuated

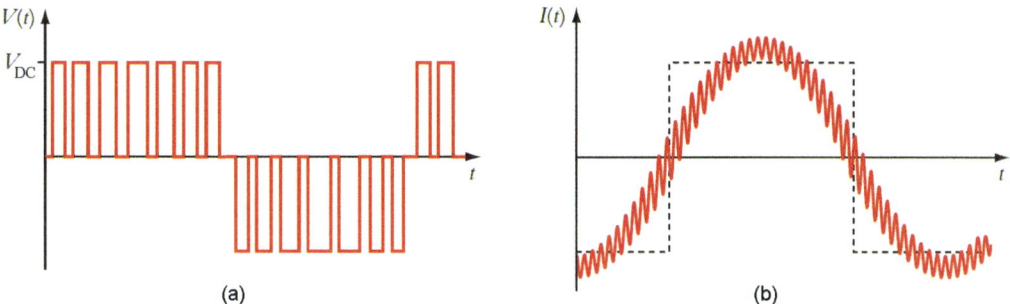

FIGURE 4.57 (a) Voltage waveform synthesised via pulse width modulation and (b) the resulting current waveform through an inductive load.

more than the fundamental, producing a current waveform that more closely resembles a sine wave, illustrated in Figure 4.56(c) and expressed mathematically by Equation (4.73).

Even when a pulse-modulated voltage waveform as shown in Figure 4.57(a) is applied, the current waveform is approximately sinusoidal (see Figure 4.57(b)). The reader needing more details is again directed to Hughes and Drury (2013).

Three-phase variable frequency inverters are used to supply induction motors, as shown in Figure 4.58. Three-phase inverters are similar to single-phase bridge inverters. The main difference is that the former has three arms to the latter's two. The inverter supplies the currents to the motor stator windings to produce a rotating magnetic field in the air gap between the stator and rotor. Adjusting the frequency of the currents controls the rotational speed of the field (i.e. synchronous speed), and hence the rotor speed.

Figure 4.51 shows the stator currents and magnetic field distribution in a two-pole induction motor at four points in a cycle. The currents were balanced and the field had constant magnitude and rotated at constant speed. Analysing the current distribution at every instant in the cycle, which is beyond the scope of this book, it can be shown that the currents combine to produce a magnetomotive force (mmf) of constant magnitude equal to:

$$\text{mmf} = 1.5\,IN \tag{4.74}$$

where I = stator phase current and N = turns in each phase of the stator winding.

The magnitude of the mmf is independent of frequency and equals the combined mmf drops in the air gap between the stator and rotor, $H_{air}\,l_{air}$, and in the steel cores, $H_{steel}\,l_{steel}$.

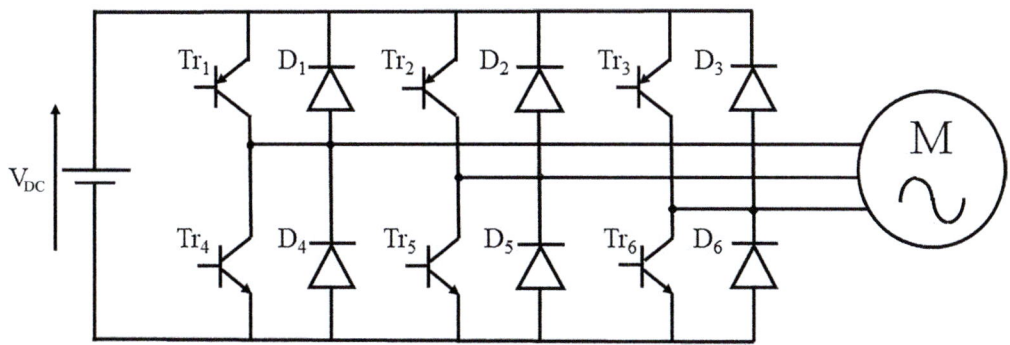

FIGURE 4.58 Three-phase inverter circuit.

where

H_{air} = magnetic field strength in the air gap.
l_{air} = total distance flux travels through air.
l_{air} = length of the air gap between the stator and rotor multiplied by 2, as the magnetic field
    has to cross the air gap twice.
H_{steel} = magnetic field strength in the steel cores.
l_{steel} = total length of the magnetic path in the steel cores.

The reluctance of the air gaps is greater than that of the steel cores, and so, it will be assumed that the latter can be ignored, and the mmf drop in the air gaps ($H_{air}\, l_{air}$) is directly proportional to the mmf produced by the stator currents.

$$H_{air}\, l_{air} \propto IN \tag{4.75}$$

The magnetic flux density, B, in the air is given by

$$B = \mu_0\, H_{air} \tag{4.76}$$

where μ_0 = constant, $4\pi \times 10^{-7}$ Hm^{-1} (for all practical purposes[1])
    So, if the flux density is proportional to the mmf produced by the stator currents, then

$$B \propto IN \tag{4.77}$$

As the air gap between the stator and rotor is uniform, it may be concluded that the air gap flux, ϕ, is also proportional to the mmf produced by the sinusoidal stator currents and will have constant magnitude.
    When the magnetic field rotates, it links with the stator windings. Figure 4.51(b) shows the situation of a two-pole motor at time, $t = 0$, when maximum flux links with the red-phase coil, $R\text{-}R'$.
    The field is rotating at the synchronous speed, n_s rev min^{-1}, which is related to the frequency, f

$$n_s = 60f$$

The angular velocity of the field, ω, equals

$$\omega = 2\pi f$$

and the general expression for flux linkage with the red phase is given by

$$\psi = \psi_m \cos \omega t \tag{4.78}$$

but

flux linkage = flux linking with a coil × number of turns in the coil

$$\psi = \phi N \text{ and } \psi_m = \phi_m N$$

So,

$$\phi = \phi_m \cos \omega t \tag{4.79}$$

According to Faraday's Law, the emf induced in a coil is given by

$$e = N \frac{d\phi}{dt} = \frac{d\psi}{dt} \tag{4.80}$$

So, the emf induced in the red coil will equal

$$e_R = -N \omega \phi_m \sin \omega t \tag{4.81}$$

The three-phase windings are mutually displaced by $120°\left(\dfrac{2\pi}{3} \text{ radians}\right)$. With the magnitude of the magnetic flux remaining constant as it rotates around the air gap, the emfs induced in the yellow and blue phases will be equal in magnitude to that induced in the red phase but will lag behind it by $\dfrac{2\pi}{3}$ and $\dfrac{4\pi}{3}$ radians, respectively. Thus,

$$e_Y = -N \omega \phi_m \sin\left(\omega t - \frac{2\pi}{3}\right) \tag{4.82}$$

$$e_B = -N \omega \phi_m \sin\left(\omega t - \frac{4\pi}{3}\right) \tag{4.83}$$

and the root mean square (rms) value for induced emf in each phase winding is given by

$$E = \frac{N \omega \phi_m}{\sqrt{2}} \tag{4.84}$$

with $\omega = 2\pi f$

$$E = \frac{2\pi f N \phi_m}{\sqrt{2}} \tag{4.85}$$

The induced emf in each phase of the stator in an AC induction motor opposes the applied phase voltage, V. If the stator winding impedance is considered to be negligible, there will be no volt drop in the stator winding and

$$|E| = |V_p|.$$

Rearranging Equation (4.85) and substituting for E,

$$\phi_m = \frac{\sqrt{2} V_p}{2\pi f N} \tag{4.86}$$

Let

$$k = \text{constant} = \frac{1}{\sqrt{2}\pi N} \tag{4.87}$$

$$\phi_m = k \frac{V_p}{f} \tag{4.88}$$

The magnitude of the rotating magnetic flux is directly proportional to the applied voltage, V_p and inversely proportional to the frequency, f. To keep the magnitude of the flux constant, the frequency and voltage must be adjusted simultaneously (Hughes and Drury, 2013).

As explained earlier, the motor torque is proportional to the magnitude of the rotating magnetic flux. So, with $\phi_m = k\dfrac{V_p}{f}$ (Equation 4.88), we can say that the torque is also proportional to the applied voltage divided by the frequency (V/f).

When a variable frequency inverter is used to drive an induction motor, the output frequency of the inverter is set to produce the required speed of flux rotation in the motor, that is synchronous speed, and the inverter output voltage is adjusted to produce full load torque. To alter the motor speed and maintain full load torque, the inverter frequency and output voltage must be adjusted simultaneously to keep the V/f ratio and hence the magnetic flux constant.

Figure 4.59(a) shows the effect on the motor torque *versus* speed characteristics of keeping (V/f) constant at different frequencies.

The simultaneous increase in inverter frequency and voltage is only possible until the voltage reaches the rated value for the inverter and motor, specified by the manufacturers. The frequency at this point is called the 'base speed' and is usually set by manufacturers at 50 Hz or 60 Hz.

To prevent overload or damage to the insulation, it is unsafe to operate electrical equipment above its rated voltage. If the inverter frequency is increased above the base speed, the voltage must be held constant at its rated value and the magnitude of the flux allowed to fall. The drop in flux reduces the emfs induced in the rotor conductors, the currents flowing through the conductors, the forces on the conductors and hence the torque on the rotor. As a result, there is a fall in torque at frequencies above the base speed as shown in Figure 4.59(b). The reader needing more details is again directed to Hughes and Drury (2013).

In the constant voltage region (i.e. when running at frequencies above the rated frequency), the torque–speed curve is less steep. Above the base speed, the maximum allowable power of the motor is constant and the allowable torque reduces inversely with the speed. Hughes and Drury (2013) present an interesting discussion on the limits to the torque–speed envelope for induction motors. They suggest that the constant power region extends to around twice the base speed (limited by the peak torque available at a given frequency) and within this region, the motor may be operated at higher slips than permitted below the base speed.

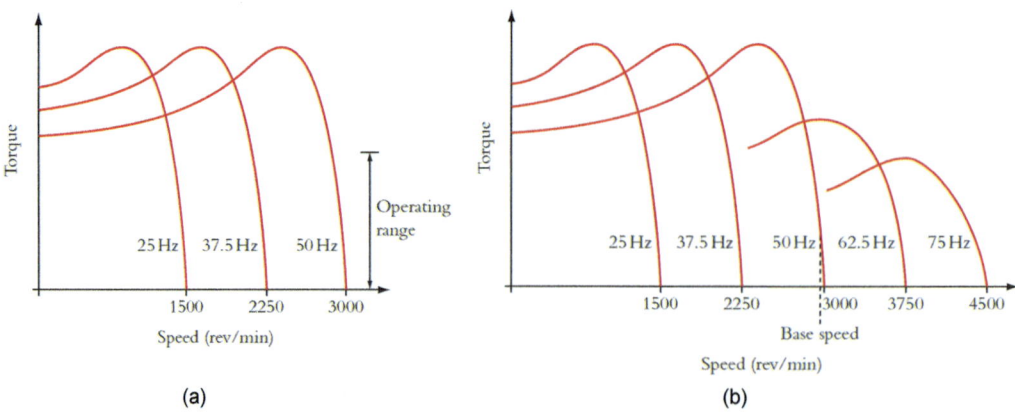

FIGURE 4.59   (a) Torque–speed characteristics of an induction motor operated at varying frequencies with a constant V/f ratio; (b) reduced torque above 'base speed' where the voltage is held constant as frequency increases.

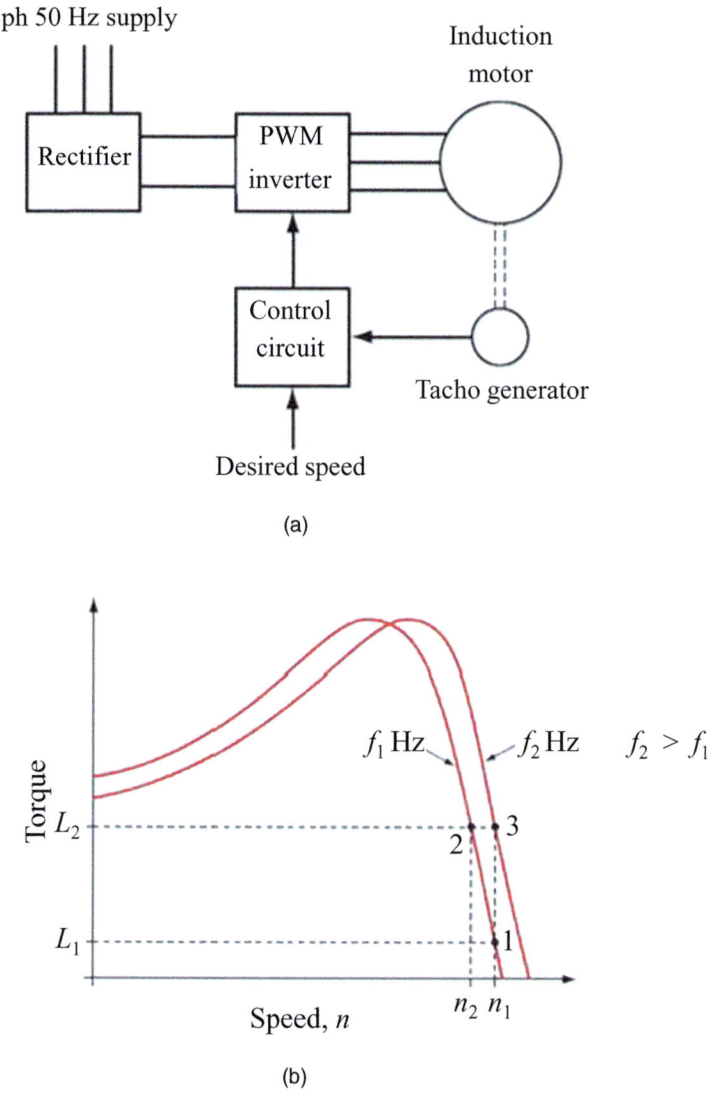

FIGURE 4.60 (a) Closed-loop control of inverter drive and (b) frequency changed by control loop to restore motor running speed to desired value n_1 despite torque increase from L_1 to L_2.

All the curves in Figure 4.59(a) and (b) show a small drop in motor speed as the load torque is increased in the normal operating range. In situations where this fall in speed is unacceptable, a closed-loop speed control system, like that shown in Figure 4.60(a), must be used.

To compensate for this fall in motor running speed as the load torque increases, the control loop automatically raises the inverter output frequency to restore the speed to the desired value. Figure 4.60(b) illustrates this. Assume that the motor is running on no load, the torque is L_1 and the speed is n_1 (at operating point 1). Without closed-loop speed control, an increase in load torque to L_2 will cause the speed to fall to n_2 (operating point 2) but with closed-loop control, the frequency will be increased incrementally as the load torque rises to L_2 to maintain the speed constant at n_1 (operating point 3).

4.11.2 APPROXIMATE CHARACTERISTICS OF INDUCTION MOTORS FED FROM INVERTERS

The analysis of an induction motor was covered in Part 1 Unit 5 and can be used to draw a torque–speed curve as shown in Figure 4.52. The nonlinear nature of the torque–speed equation underlying this curve makes it difficult to use this equation within drive system calculations, so it can be convenient to derive approximate torque–speed relationships relating to the operating region of the motor as shown in Figure 4.59. Within this range, the torque–speed characteristic of the motor is approximately linear. It was shown in Equation (5.117) in Part 1 that the torque–slip characteristics of a simple three-phase AC induction motor are given by the following equation:

$$L = \frac{3p}{\omega} \frac{E_2^2 a s}{X_2 \left(a^2 + s^2\right)} \tag{4.89}$$

where

L = motor torque (N m).
p = number of pairs of poles on the stator.
$\omega = 2\pi f$ = circular frequency of mains supply (rad/s).
E_2 is the emf induced in each phase of the rotor winding when the rotor is stationary.
X_2 is the rotor reactance per phase when the rotor is stationary.
$a = \dfrac{R_2}{X_2}$ = ratio of resistance to reactance of rotor winding.
s = per unit slip = $\dfrac{n_s - n}{n_s}$.

where n_s and n are the synchronous speed and the running speed of the motor, respectively.

Using Equations (5.86), (5.87) and (5.91) from Part 1, the stator phase voltage V_1 can be expressed as

$$V_1 = E_1 = \frac{N_1}{N_2} E_2 \tag{4.90}$$

and the rotor standstill reactance per phase, referred to the stator, is

$$X_1 = \frac{N_1^2}{N_2^2} X_2 \tag{4.91}$$

So, Equation (4.89) becomes

$$L = \frac{3p}{\omega} \frac{V_1^2 a s}{X_1 \left(a^2 + s^2\right)} \tag{4.92}$$

Note also that the rotor resistance per phase, referred to the stator, is

$$R_1 = \frac{N_1^{\,2}}{N_2^{\,2}} R_2 \tag{4.93}$$

So, a can be re-expressed as

$$a = \frac{R_1}{X_1} \tag{4.94}$$

As all quantities are henceforward only referred to the stator, the subscripts for V_1, X_1 and R_1 can be dropped from Equations (4.92) to (4.94).

The operating region of the motor typically involves slip values in the range of 1–10%, with the most common value being around 4%. This is around a fifth of the typical value of a (which is likely to be around 0.2 or 20%), so that $(a^2 + s^2)$ can be reasonably approximated as a^2. Noting the definitions of a, ω and s, and that

$$n_s = \frac{60f}{p} \left(\text{in rev} / \text{min} \right)$$

Equation (4.92) can be approximated as

$$L \approx \frac{3p}{\omega} \frac{V^2 s}{R} \tag{4.95}$$

and hence,

$$L \approx \frac{1}{60} \frac{3p^2}{2\pi R} \left(\frac{V}{f} \right)^2 \left(\frac{60f}{p} - n \right) \tag{4.96}$$

or relating the torque instead of the angular velocity of the rotor, ω_r, measured in rad/s,

$$L \approx \frac{3p^2}{4\pi^2 R} \left(\frac{V}{f} \right)^2 \left(\frac{2\pi f}{p} - \omega_r \right) \tag{4.97}$$

In practice, nearly all induction motors encountered today are squirrel-cage motors; so, R will be constant but unknown. By adapting an approach described by Wildi (1991) for the constant frequency situation, it is often easier to infer the approximate torque–speed characteristics from the design data ('nameplate data') for the motor involving its rated power P_{rated}, rated speed n_{rated}, its rated frequency f_{rated}, rated voltage V_{rated}, and to write an expression equivalent to Equation (4.96) as

$$L \approx K \left(\frac{f_{rated}}{V_{rated}} \right)^2 \left(\frac{V}{f} \right)^2 \left(\frac{60f}{p} - n \right) \tag{4.98}$$

or equivalent to Equation (4.97) as

$$L \approx \frac{60K}{2\pi} \left(\frac{f_{rated}}{V_{rated}} \right)^2 \left(\frac{V}{f} \right)^2 \left(\frac{2\pi f}{p} - \omega_r \right) \tag{4.99}$$

where

$$K = \frac{60P_{rated}}{2\pi n_{rated} \left(\dfrac{60f_{rated}}{p} - n_{rated} \right)}$$

Strictly speaking, Equations (4.96) and (4.97) relate to drawing a straight-line tangent (at $s = 0$) to the torque–speed curve, while the approach of Wildi (1991) corresponds to drawing a straight line or chord between the points on the curve corresponding to zero slip and rated slip. In practice, the difference between these lines is small.

Using these equations, it is now possible to generate the approximately linear regions of the families of the torque–speed curves given in Figures 4.52 and 4.59.

- For the single curve shown in Figure 4.52 relating to a motor driven directly from its rated power supply $f = f_{\text{rated}}$, the approximate torque–speed characteristic is simply given by inserting the rated frequency and voltage into Equation (4.98), which reduces to

$$L = K\left(\frac{60 f_{\text{rated}}}{p} - n \right) \tag{4.100}$$

- For the curves relating to $f < f_{\text{rated}}$ as shown in Figure 4.59(a) and the left-hand part of Figure 4.59(b), f and V are varied in proportion as described within the present section so that their ratio remains constant and Equation (4.98) now reduces to

$$L = K\left(\frac{60 f}{p} - n \right) \tag{4.101}$$

- For the curves relating to $f > f_{\text{rated}}$, as shown in the right-hand part of Figure 4.59(b), the voltage remains at the rated value V_{rated} while the frequency increases and Equation (4.98) now reduces to

$$L = K\left(\frac{f_{\text{rated}}}{f} \right)^2 \left(\frac{60 f}{p} - n \right) \tag{4.102}$$

- For completeness, it can be stated that for the situation where a motor is driven at its rated frequency and at reduced voltage (e.g. via a variable transformer or a so-called 'soft starter'), Equation (4.98) reduces to

$$L = K\left(\frac{V}{V_{\text{rated}}} \right)^2 \left(\frac{60 f}{p} - n \right) \tag{4.103}$$

All of these approximations are valid within the approximately linear region of the motor's torque–speed curve, which applies when the actual value of torque lies within the following limit (extended from Wildi (1991)) to consider varying frequency:

$$L \leq \left(\text{Rated torque} \right) \times \left(\frac{f_{\text{rated}}}{V_{\text{rated}}} \right)^2 \left(\frac{V}{f} \right)^2 \tag{4.104}$$

where

$$\text{Rated torque} = \frac{60 P_{\text{rated}}}{2\pi n_{\text{rated}}} \tag{4.105}$$

It should be emphasized that this limitation on the applicability of the linearly approximated torque–speed characteristic does not necessarily correspond to the limitation on allowable torque for the motor. Once again, the reader is referred to Hughes and Drury (2013) for a discussion of the allowable operating envelope for an inverter-fed motor.

WORKED EXAMPLE

A 25 kW, 415 V (line-to-line voltage), 50 Hz, 1440 rev/min squirrel-cage induction motor is fed from a variable frequency inverter. The voltage and frequency are varied in proportion up to the rated frequency; the voltage is held constant above this frequency. This arrangement is used to drive a blower that has a torque–speed characteristic given by $L_{load} = 5.556 \times 10^{-5} n^2$ where torque L_{load} is in Nm and n is in rev/min. Determine the inverter frequency and line-to-line voltage at blower speeds of 720 and 1560 rev/min.

For this motor, the rated power is 25000 W and the rated speed is $n_{rated} = 1440$ rev/min.

It can be seen from Table 4.6 that for a rated speed of 1440 rev/min, the synchronous speed is almost certainly 1500 rev/min and hence the motor is a four-pole motor with $p = 2$. Therefore,

$$K = \frac{60 P_{rated}}{2\pi n_{rated}\left(\dfrac{60 f_{rated}}{p} - n_{rated}\right)}$$

$$= \frac{60 \times 25000}{2\pi \times 1440 \times \left(\dfrac{60 \times 50}{2} - 1440\right)}$$

$$= 2.763 \text{ Nm min/ rev}$$

For the blower at $n = 720$ rev/min,

$$\text{Load torque} = L_{load} = 5.556 \times 10^{-5} \times 720^2 = 28.8 \text{ Nm}$$

720 rev/min is below the rated speed, so $V/f = V_{rated}/f_{rated}$, and Equation (4.101) applies. Setting the motor torque equal to L_{load} and rearranging gives

$$f = \frac{p}{60}\left(\frac{L_{load}}{K} + n\right)$$

$$= \frac{2}{60}\left(\frac{28.8}{2.763} + 720\right)$$

$$= 24.35 \text{ Hz}$$

The rated line voltage is 415 V and hence the rated phase voltage is $415/\sqrt{3} = 239.6$ V.

The voltage is obtained by reducing the rated voltage in proportion to the frequency.

$$V = V_{rated} \times \frac{f}{f_{rated}} = 239.6 \times \frac{24.35}{50} = 116.7 \text{ V}$$

so the line-to-line voltage is $116.7 \times \sqrt{3} = 202.1$ V.

It is important to check that the approximately linear operating region, and hence the applicability of the approximate equation, have not been exceeded. Noting that the rated phase voltage is 239.6 V and the actual phase voltage is 116.7 V, then using Equation (4.104),

$$\text{Maximum permissible torque} = (\text{Rated torque}) \left(\frac{f_{\text{rated}}}{V_{\text{rated}}} \right)^2 \left(\frac{V}{f} \right)^2$$

$$= \frac{25000}{2\pi \times 24} \left(\frac{50}{239.6} \right)^2 \left(\frac{116.7}{24.24} \right)^2$$

$$= 165.8 \text{ Nm}$$

which is very far from being exceeded by the actual torque of 28.8 Nm.
   For the blower at $n = 1560$ rev/min,

$$\text{Load torque} = L_{\text{load}} = 5.556 \times 10^{-5} \times 1560^2 = 135.2 \text{ Nm}$$

1560 rev/min is above the rated speed, so $V = V_{\text{rated}}$, and Equation (4.98) applies. This can be rearranged to give

$$\text{Motor torque} = L_{\text{motor}} = 60K \frac{f_{\text{rated}}^2}{pf} - K \left(\frac{f_{\text{rated}}}{f} \right)^2 n$$

Equating L_{motor} to L_{load}, multiplying by f^2 and rearranging gives

$$L_{\text{load}} f^2 - 60K \frac{f_{\text{rated}}^2}{p} f + K f_{\text{rated}}^2 n = 0$$

Inserting the values for L_{load}, K, n and f_{rated} for the present problem gives

$$135.2 f^2 - 60 \times 2.763 \times \frac{50^2}{2} f + 2.763 \times 50^2 \times 1560 = 135.2 f^2 - 207250 f + 10775700 = 0$$

This is a quadratic in f which has the following solutions:

$$f = \frac{207250 \pm \sqrt{207250^2 - 4 \times 135.2 \times 10775700}}{2 \times 135.2}$$

$$= 53.9 \text{ Hz or } 1479 \text{ Hz}$$

As the new running speed is only slightly above the rated speed of 1440 rev/min, we would expect the frequency to be slightly above the rated frequency of 50 Hz. So, the first of the alternative solutions is clearly the correct one. Recall also that $V = V_{\text{rated}} = 239.6$V.
   Again it is important to check using Equation (4.104) that the applicability of the approximate equation has not been exceeded:

$$\text{Maximum permissible torque} = (\text{Rated torque}) \left(\frac{f_{\text{rated}}}{V_{\text{rated}}} \right)^2 \left(\frac{V}{f} \right)^2$$

$$= \frac{25000}{2\pi \times 24} \left(\frac{50}{239.6} \right)^2 \left(\frac{239.6}{53.9} \right)^2$$

$$= 142.7 \text{ Nm}$$

which is slightly greater than the actual motor (and load) torque of 135.2 Nm. So, the motor is (by a small margin) within its linear region and the calculations are valid.

LEARNING SUMMARY

By the end of this section, you should have learnt:

✓ The principles of operation of an induction motor.

✓ The operation of a simple inverter.

✓ The principles of pulse width modulation (PWM).

✓ The induction motor torque is proportional to the square of the ratio of applied voltage to frequency, that is $(V_p/f)^2$.

✓ At frequencies above the 'base speed', the torque falls as the frequency increases.

✓ Approximately the linear torque–speed characteristics of induction motors can be obtained for relatively low values of slip.

4.12 MOTORS FOR ELECTRIC VEHICLES

Environmental protection and energy conservation are major concerns for our planet. To reduce the emissions from motor vehicles, car manufacturers have hugely increased their production of zero and low-emission electric vehicles. In the United Kingdom, for example, at the time of writing, it is planned that the sales of new internal combustion engine (ICE)-driven cars (including hybrid vehicles) will be prohibited within a fairly short timescale, possibly as early as 2035 both for fully ICE cars and hybrids.

There are two main forms of electric vehicles: fully electric vehicles and hybrid electric vehicles, where the drive is derived (in one of several possible configurations) from a combination of an electric motor and ICE. The present discussion will focus mainly on fully electric vehicles but is applicable also to many hybrid vehicles.

Figure 4.61 shows a schematic of the drive system for a fully electric vehicle. The power is obtained from a battery which is charged from an external electricity supply. In most hybrid vehicles, it is instead charged from a generator driven by an ICE. The drive to the wheels is provided by one or more electric motors, which may be connected to the wheels either directly or via a fixed-ratio transmission system. The motors are controlled by the semiconductor drive or power electronics unit (often a three-phase inverter) supplying currents to the individual motor windings. There are also various other ancillary electronic devices including power steering (not shown) and an energy management system.

Li (2019) summarises the main requirements for an electric vehicle drive to include 'high instant power, high initial torque for accelerating, high power for cruising, fast torque response, high power density, wide constant torque, constant power regions, high efficiency over wide speed range, high reliability and reasonable cost'. In order to achieve these aims, the electric motors employed in electric vehicles have to be

- Reasonably compact and light in weight for their power, that is, must have good power density in terms of power-to-weight and power-to-space ratios.
- Reasonably low cost in order for the vehicle cost to be competitive.
- Efficient in order to keep running costs low and range high.
- Capable of being driven in a controlled manner from a DC battery supply.
- Capable of providing high torque at standstill and low speeds in order to provide good acceleration from rest, at the expense of lower torque at higher speeds, without the use of the variable-ratio gearbox and clutch needed in conventional ICE-powered vehicles as discussed in Sections 4.4 and 4.14.

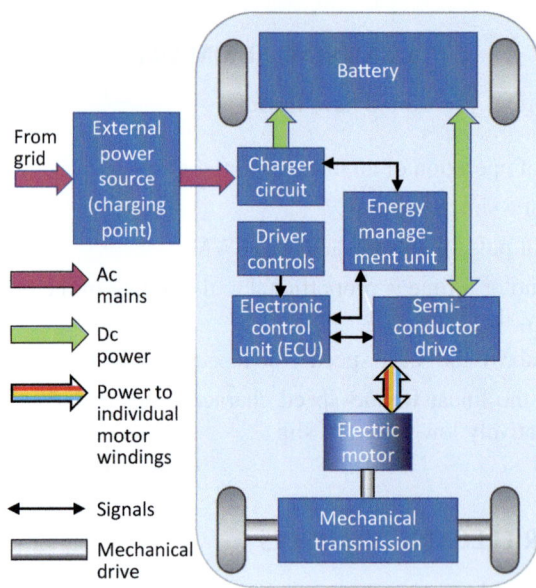

FIGURE 4.61   A simplified schematic diagram of a typical fully electric vehicle, drawing upon information from Chan (2002).

There is no single motor type which optimally satisfies these requirements and car manufacturers employ a variety of electric motors in their vehicles. Chau and Li (2014), Yıldırım et al. (2014), Riba et al. (2016), Li (2017) and Karthik (2019) all present overviews of the area, with various emphases. From the above reviews, it can be seen that the most popular motors used in modern electric vehicles are as follows:

1. *DC permanent magnet (PM) motors*, both the conventional (brushed) type discussed in Section 4.9 and the brushless variety which use semiconductor switches (with rotational position sensors) in lieu of the mechanical commutator for switching of the armature windings. Most brushless motors place the PMs on the rotor and relocate the windings on the stator (Figure 4.62), effectively turning the motor inside out. Brushed motors suffer from

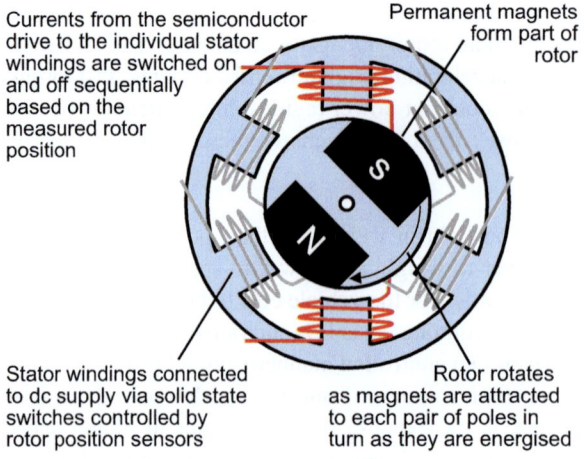

FIGURE 4.62   DC brushless motor.

wear issues which the brushless motors overcome, but both types require powerful, compact PMs which typically depend on rare earth elements of which there is only a limited supply. DC brushed PM motors tend to be used widely in small electric vehicles such as electric wheelchairs, mobility scooters, golf trolleys etc.

2. *Permanent magnet (PM) synchronous motors.* These superficially resemble DC brushless motors but here the semiconductor drive is an inverter supplying a variable frequency three-phase supply similar to those discussed in Section 4.11 to energise the windings on the stator and hence provide a rotating magnetic field. PM synchronous motors use a PM on the rotor to lock onto the rotating magnetic field. To start the vehicle, the frequency must be varied from zero to the frequency corresponding to the desired driving speed. The more commonly used configuration is the 'inrunner' motor (Figure 4.63(a)), where the stator typically resembles that of an AC induction motor, though an alternative configuration (known as the 'outrunner' motor, Figure 4.63(b)) turns the motor inside out and has a set of magnets linked to the wheel hub and rotating around a fixed set of stator coils. PM synchronous motors are efficient and compact but like the DC PM motors they rely on the availability of powerful, compact PMs and the supply of rare earth elements used to make them. Electrically magnetised (wound rotor) synchronous motors have occasionally been used to overcome this problem.

3. *Induction motors, as described in Section 4.11.* Again, the semiconductor drive is an inverter providing a three-phase AC supply. Unlike PM DC and synchronous motors, they do not require PMs and their attendant rare earth elements, but they inherently have inefficiencies directly related to their slip associated with all induction motors and their self-starting nature confers no advantage.

4. *Switched reluctance (SR) motors.* These motors have a wound stator (typically with several phases of windings) and a toothed rotor superficially resembling that of the PM stepper motor illustrated in Figure 4.49, though the rotor here is not permanently magnetised. The motors operate by energising opposing pairs of stator turns in sequence to make the rotor turn (Figure 4.64) to a position which minimises the magnetic reluctance of the path. In practice, this means moving the rotor to enable a pair of rotor poles to align with the energised pair of stator poles. In this sense, their principle of operation is essentially the same as that of the electromagnetic actuators described in Part 1 Section 5.3. To control

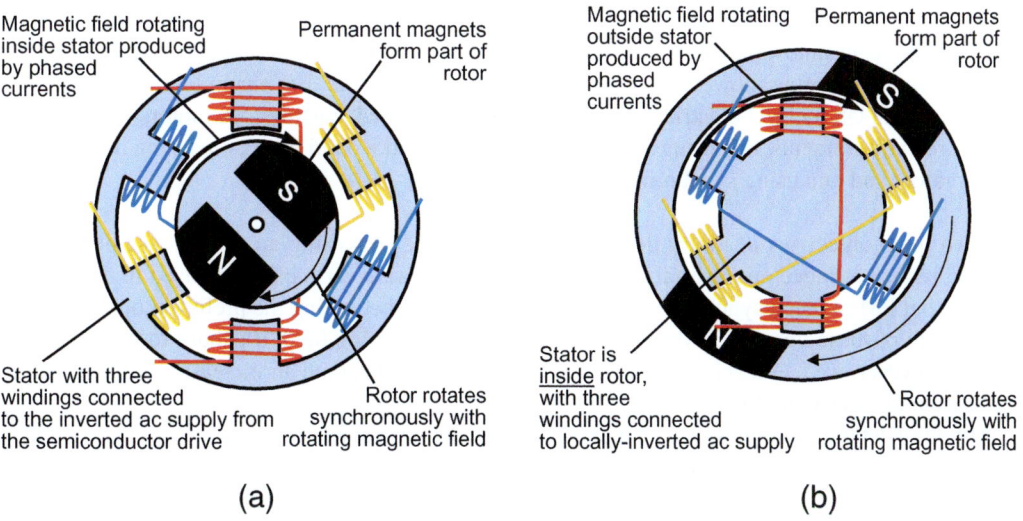

FIGURE 4.63   Permanent magnet synchronous motors (a) inrunner and (b) outrunner.

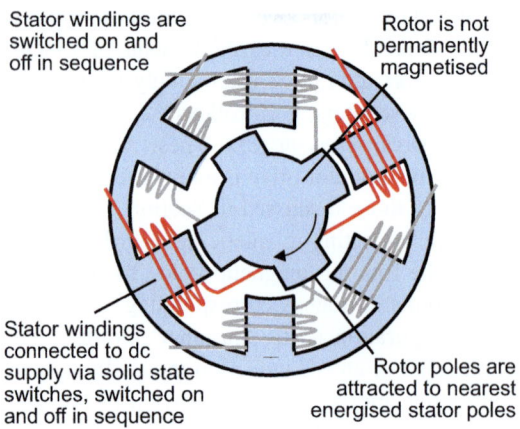

Stator windings are switched on and off in sequence

Rotor is not permanently magnetised

Stator windings connected to dc supply via solid state switches, switched on and off in sequence

Rotor poles are attracted to nearest energised stator poles

FIGURE 4.64 Switched reluctance motor.

the operation of SR motors, a special form of switching circuit must be used, which is too complex to describe here. These motors have long been considered to have excellent characteristics for electric vehicles but they can be noisy (Omekanda, 2013) and their commercial uptake in vehicles appears to have been limited.

5. *Synchronous reluctance motors.* These motors combine the variable-reluctance principle of the SR motor with the configuration of a synchronous or induction motor and consist of a conventional (typically three-phase) wound stator surrounding a ferromagnetic (laminated iron) rotor. A simple two-pole motor is shown in Figure 4.65(a). This is shaped to give low-reluctance magnetic paths only in certain orientations (Figure 4.65(b)). Unlike the induction motor rotor, it contains no conductors, and unlike the synchronous PM motor rotor, it contains no magnets. A more typical, four-pole synchronous reluctance motor is shown again in a simplified form in Figure 4.65(c). However, a variant of this kind of motor (PM-assisted synchronous reluctance motor) incorporates some PMs in the rotor (see Figure 4.65(d)) which provides various improvements, so the overall method of operation is a hybrid of the PM and variable reluctance principles. The PM-assisted synchronous reluctance motor has the attractions of requiring only limited use of PM materials while avoiding the slip-related inefficiencies of induction motors, and hence, has been the subject of much recent innovation. It is now being used in several production models of vehicles. At the time of writing, however, its adoption is not yet complete. For example, Credo (2019) lists nine US and EU electric vehicles, of which seven had synchronous motors (six using rare earth elements, one using an electrically magnetised wound rotor), one using induction motors and one using synchronous reluctance motors.

The two main objectives of motor development at the time of writing are to increase power density and reduce reliance on unsustainable materials such as rare earth elements used in magnets. It appears that PM-assisted synchronous reluctance motors are becoming the preferred option as they satisfy both of these objectives, though, in this rapidly developing field, it is hard to predict future trends. However, we can be certain that electric vehicles will become an increasingly large area of application of electric motors.

Although detailed torque–speed characteristics for electric vehicle motors are not presented here for reasons of space, it is noteworthy that the envelopes of allowable torque–speed combinations are typically as shown in Figure 4.66, with higher torques being permitted for short-term or transient conditions such as brief periods of acceleration and with the torque limits being lower for steady-state operation. At higher speeds, the maximum power rating of the motor typically limits

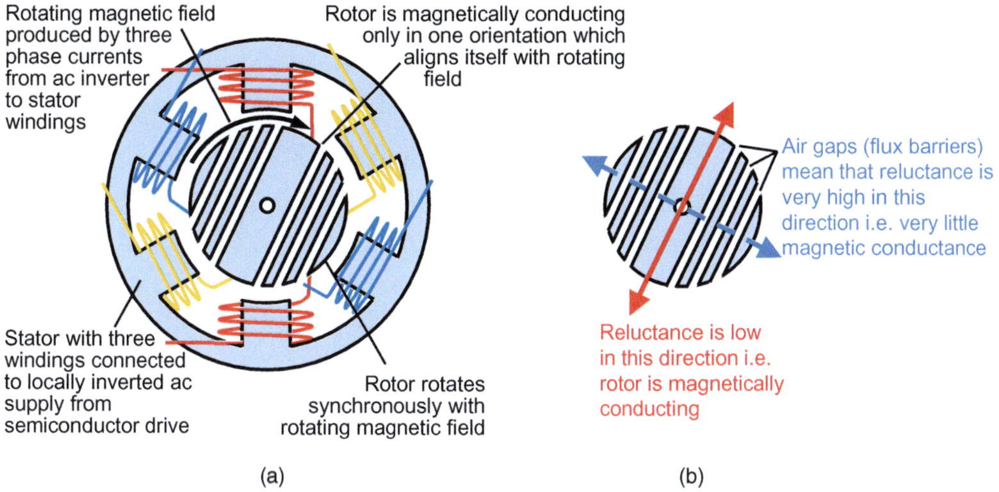

Rotating magnetic field
produced by three
phase currents
from ac inverter
to stator
windings

Rotor is magnetically conducting
only in one orientation which
aligns itself with rotating
field

Stator with three
windings connected
to locally inverted ac
supply from
semiconductor drive

Rotor rotates
synchronously with
rotating magnetic field

(a)

Air gaps (flux barriers)
mean that reluctance is
very high in this
direction i.e. very little
magnetic conductance

Reluctance is low
in this direction i.e.
rotor is magnetically
conducting

(b)

High
reluctance

Low reluctance

(c)

Air gaps
(flux barriers)

Permanent
magnets in air
gaps

(d)

FIGURE 4.65 Synchronous reluctance motor. (a) Simplified diagram of two-pole motor. (b) The rotor showing reluctance in different directions within the rotor. (c) More typical four-pole motor. (d) The rotor of a permanent magnet–assisted synchronous reluctance motor. Note that all these diagrams omit the narrow webs of rotor material required to hold the rotor together and prevent it from falling apart.

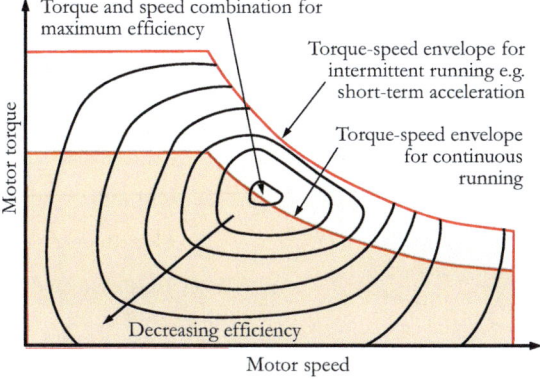

Torque and speed combination for
maximum efficiency

Torque-speed envelope for
intermittent running e.g.
short-term acceleration

Torque-speed envelope
for continuous
running

Motor torque

Decreasing efficiency

Motor speed

FIGURE 4.66 Typical torque–speed envelopes for an electric vehicle motor, showing efficiency contours.

the available torque. The contours of motor efficiency (analogous to the inverse of SFC for an IC engine) can be drawn on the torque–speed envelopes.

LEARNING SUMMARY

By the end of this section, you should have learnt:

✓ The range of motors used in electric vehicles.
✓ That the characteristics of electric vehicle motors allow them to be used with a fixed transmission ratio.

4.13   OTHER SOURCES OF POWER: PNEUMATICS AND HYDRAULICS

Electric motors (particularly DC PM motors for light duties and AC induction motors for heavier duties) are by far the most common means of providing drive to mechanical systems. However, under certain circumstances, it may be appropriate to use other means of providing drive, notably the use of pneumatic and hydraulic drives. Electric motors are cheap, efficient, reliable and controllable, but they can be heavy in relation to the power needed, they can overheat or burn out if stalled and they are not ideal for providing either very large forces or long linear movements. Pneumatic and hydraulic actuators and transmission systems can in some situations overcome these problems.

4.13.1   PNEUMATICS

Pneumatic actuators can be used in preference to electrical actuators for the following reasons:

- Better power-to-weight ratio (i.e. less weight for a given power output).
- Better power-to-space ratio (i.e. smaller for a given power output).
- High torque of pneumatic motor at stall, controlled via pressure and with no damage to the motor being caused by the stall.
- Long stroke of pneumatic cylinder.

A widely used form of pneumatic motor, used to provide continuous drive, is the vane motor, which operates as shown in Figure 4.67.Other types of pneumatic motors are based on pistons.

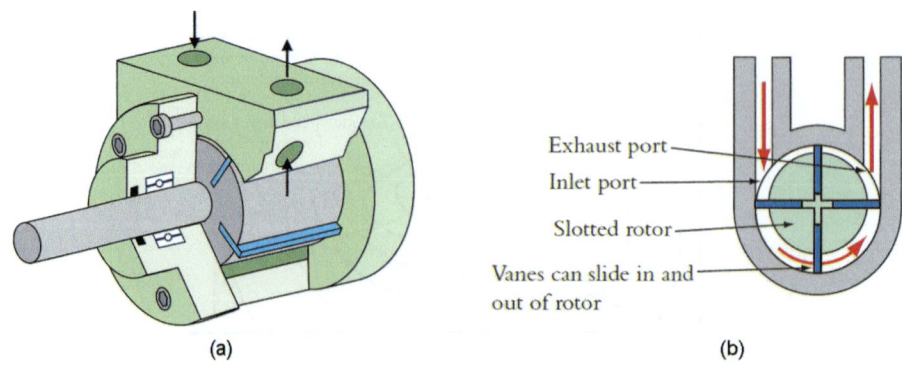

(a)   (b)

FIGURE 4.67   Pneumatic vane motor. (a) Cutaway view and (b) simplified diagram.

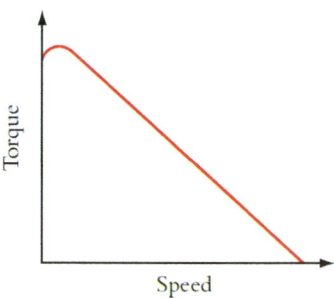

FIGURE 4.68   Characteristic of a pneumatic vane motor.

The characteristics of pneumatic vane motors can be approximated as a straight-line graph join-ing the high stall torque to the high no-load speed (Figure 4.68), typically with a small elbow repre-senting reduced torque on start-up:
A particular advantage of pneumatic motors is that their characteristics can easily be altered by

- Varying the pressure (which changes both the stall torque and the no-load speed) as shown in Figure 4.69(a).
- Throttling the inlet or the outlet (which primarily changes the no-load speed) as shown in Figure 4.69(b).

Typical applications of rotary pneumatic motors include:

- Pneumatic wrenches and similar devices for tightening bolts in assembly lines, tyre service stations etc. – these exploit both the good power-to-weight ratio and the stall-tolerant char-acteristics of pneumatic motors.
- Bottling plants, where bottle tops need to be tightened, again exploiting the ability of pneu-matic motors to provide a pressure-limited value of stall torque without damage.

Pneumatics are also particularly useful for providing linear motion, primarily via pneumatic cylinders (Figure 4.70). These provide moderately large forces over a stroke or displacement ranging from a few millimetres up to the order of a metre. When used in conjunction with solenoid-operated valves, they provide a means of obtaining rapid linear motion under electrical or computer control and are widely used in automated production environments. The force from a pneumatic cylinder is given by

$$F = pA \qquad\qquad (4.106)$$

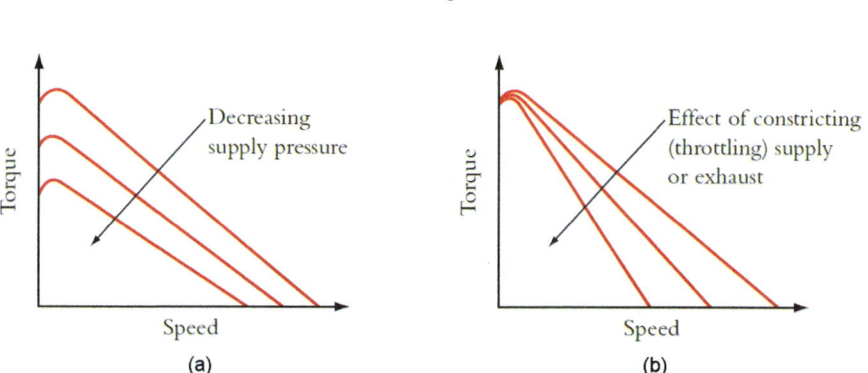

FIGURE 4.69   Effect of (a) varying the supply pressure and (b) throttling a pneumatic vane motor [Based loosely on technical literature from Atlas-Copco and Ingersoll-Rand].

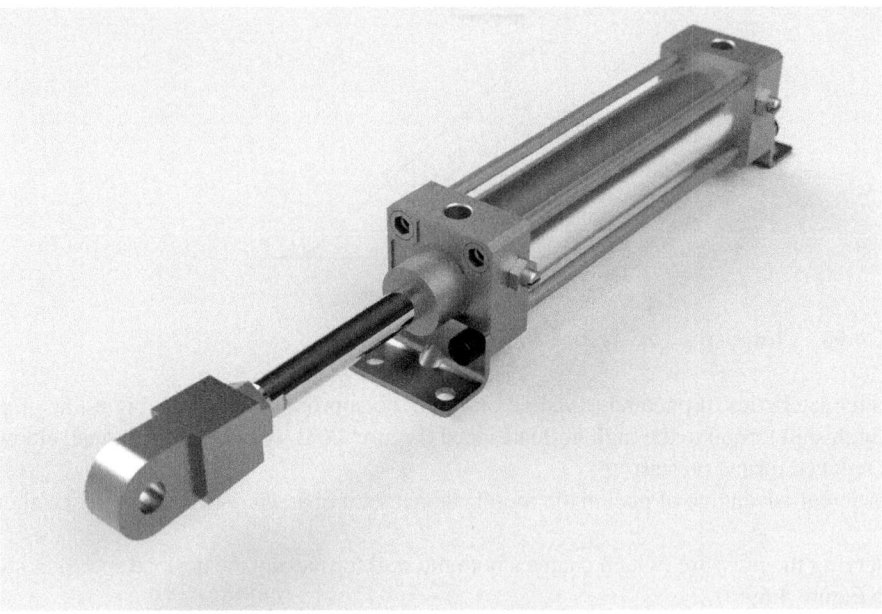

FIGURE 4.70   Typical pneumatic cylinder actuator. (Simils/Shutterstock.com.)

where F is the force obtained, p is the pressure of the compressed air supply and A is the area of the piston. In practice, the available force may be a little lower than this, due to friction between the piston seals and the cylinder.

4.13.2   Pneumatic Systems

In order to be able to use a pneumatic motor or actuator, it is of course necessary to have a source of compressed air. There is no pneumatic equivalent of the electricity supply grid, although most factories, engineering laboratories etc. are often equipped with a compressed air supply, which may be supplied centrally within the factory etc. or via local compressor units. While details of different systems vary depending on size, cost etc., a compressed air system typically includes the following components (Figure 4.71):

- Electric motor (or other mechanical power source) to provide the mechanical power to the system.
- Several filters, including an inlet filter, to remove airborne particles, which may cause damage to the compressor or actuators.
- Compressor – typically a one- or two-stage reciprocating compressor, although other types (e.g. screw compressors) may be used.
- A control system to ensure that the correct pressure is maintained without excessive wastage of energy – this may be a simple pressure switch for switching the motor on and off as required, or may involve, for instance, the lifting of the compressor's inlet valves so that it idles without compressing air or may involve varying the speed of the motor.
- 'Receiver' or pressure vessel for storing the compressed air.
- Pressure relief valve, or safety valve, to prevent excessive pressure building up in the event of the failure of the control system.
- Network of pipes to carry the compressed air around the factory.
- Oil-mist lubricators to add lubricant to the air for use within motors.

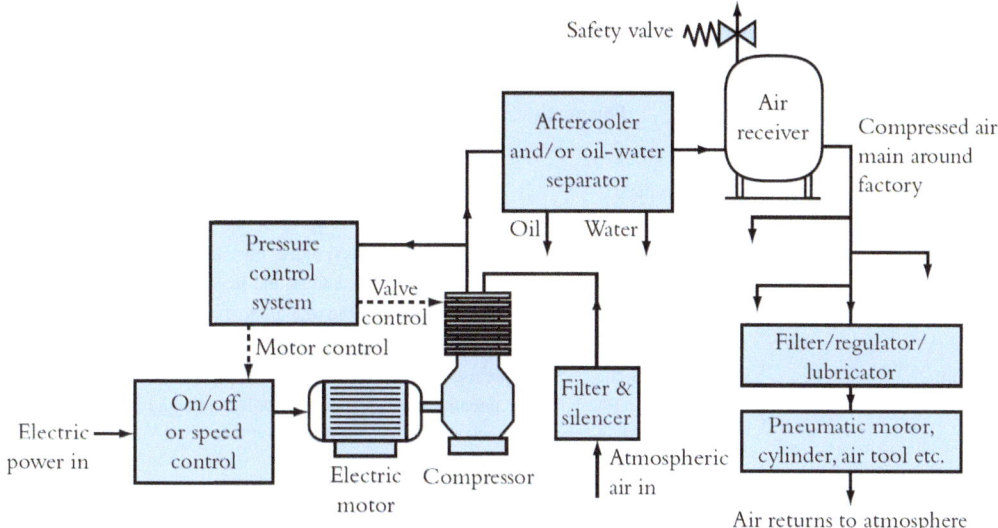

FIGURE 4.71 Typical pneumatic system for a factory or similar installation.

Larger systems will also include a cooler and/or condenser to allow the dissipation of heat generated by the adiabatic compression of air and the condensation of moisture from the air as it cools after compression. An oil/water separator may also be included to avoid contaminating the drains with lubricating oil from the compressor.

4.13.3 EFFICIENCY AND ENERGY UTILISATION ISSUES OF PNEUMATIC SYSTEMS

Pneumatic systems and actuators give a more intense, robust and lightweight source of power than is achievable with electricity alone and provide an excellent route to high-speed, low-cost automation. However, poor design and working practices can easily lead to greater inefficiencies than those inherent in purely electrical systems. The Carbon Trust (2012) provided a detailed guide to avoiding waste due to poor practice in the application of compressed air, and much of this advice remains relevant. At that time, they stated 'a 3mm hole could cost over £1,000/year in wasted energy' and stated that leak rates on compressed air systems could be 40–50% or higher, compared with a desirable target of 5–10%. They also explained that much of the energy supplied to a compressor is wasted via issues such as leakage, poor maintenance, inappropriate use, pressure drops in filters and poor control. Careful searches for leaks, elimination of wasteful applications of compressed air and the use of filters with low pressure drops can all be used to reduce the wastage of compressed air and unnecessary use of energy. In the right applications, compressed air provides a very valuable approach for powering and controlling suitable items of equipment, but it is important for engineers to be aware of the scope for inefficiencies, and to be aware that the energy needs and associated cost and environmental issues involved in air compression can be significant.

4.13.4 HYDRAULIC SYSTEMS

Hydraulic systems involve the use of liquids under very high pressure and achieve even greater power-to-weight and power-to-space ratios than are achievable via pneumatic systems. There are obvious similarities with pneumatics (e.g. both involve fluids under pressure and both can be used with cylinder actuators and rotational motors) but there are major differences which can be summarised in Table 4.7:

TABLE 4.7

Comparison and Contrast between Pneumatic and Hydraulic Systems

Characteristic	Pneumatics	Hydraulics
Working fluid	Air	Liquid (usually oil or ester)
Pressure	6–8 bar typical	50–200 bar typical, can go up to 250–700 bar or higher
Construction of actuators	Lightweight, typically from aluminium extrusions and die castings	Much more substantial, typically ferrous castings and steel tubing
Forces and torques available	Medium (tens of N up to around 10 kN from linear actuators)	Large (e.g. up to 10,000 kN from linear actuators)
Precision	Low – not usually used for precise positioning (though some pneumatic control systems provide precise control of flow etc. via the use of feedback) – more usually used for simple two-position operation	Used for precise positioning e.g. within machine tools, and under the control of servo valves
Stiffness	Not stiff because air is compressible	Very stiff as hydraulic fluid is virtually incompressible – the actuator becomes rigid if inlets/outlets are blocked off, as in the case of cranes, jacks etc.
Cost	Relatively low cost	Relatively high cost due to robust construction
Power-to-weight and power-to-space ratios	Very good – makes pneumatics suitable for power-intensive handheld tools, for example, wrenches and grinders.	Excellent – makes hydraulics suitable for providing power locally within aerospace applications, for example for operating landing gear and control surfaces in aircraft

One of the major features of hydraulic systems is the ability to obtain a very large mechanical advantage. The classic example is a hydraulic jack (Figure 4.72), where a reciprocating pump with a small bore requires a relatively small force (but many operating strokes) to provide the fluid to operate a cylinder of a large bore which is used to raise a heavy load by a small distance.

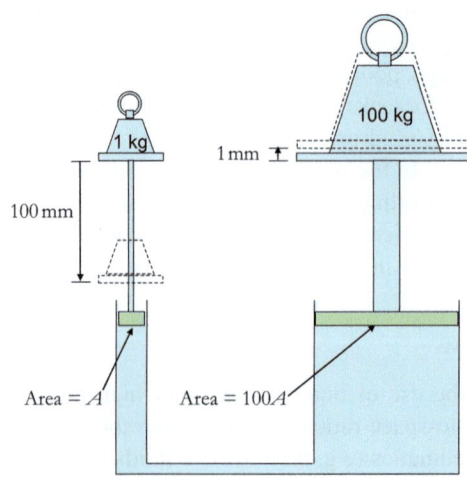

FIGURE 4.72 A simple hydraulic jack system.

In the context of mechanical drive systems, a pump with a small displacement (which pumps a small amount of fluid per revolution) can be used to generate a very large force or torque when connected to a linear or rotational actuator with a large total piston area.

4.13.5 HYDRAULIC CYLINDERS

As indicated above, these are similar in concept to pneumatic cylinder actuators but operate at much higher pressures and are consequently much more solidly built (Figure 4.73). They may be single-acting or double-acting and provide a large force proportional to the pressure supplied to them. They are widely used where a large force is required, for example in raising and lowering the buckets of excavators and the jibs of cranes.

4.13.6 HYDRAULIC MOTORS

Hydraulic motors are used in a variety of situations where a very large torque is required in a given space. Examples include providing drive to the tracks of track-laying excavators, driving rollers and other large-torque applications in industry and within variable-speed hydraulic drives. Hydraulic motors are typically based on the use of pistons and are generally radial (e.g. Figure 4.74, with the pistons typically acting on some form of radial cam) or axial (e.g. Figure 4.75, with the pistons acting on some form of inclined plate or swashplate).

Because pressure drops due to dynamic head losses are very small in comparison with the pressures at which hydraulic systems operate, the torque–speed characteristic of a hydraulic motor is very nearly a horizontal line corresponding to a constant torque proportional to pressure (Figure 4.76).

FIGURE 4.73 Cutaway view of a typical hydraulic cylinder actuator. (Hotdogcartman.)

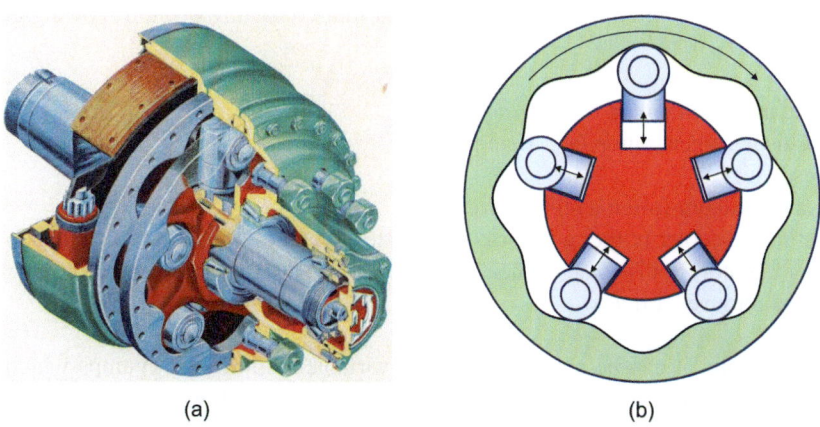

(a)   (b)

FIGURE 4.74 (a) Construction and (b) operation of a radial piston hydraulic motor. ((a): Oy Suomen Autoteollisuus AB, now Sisu Auto AB.)

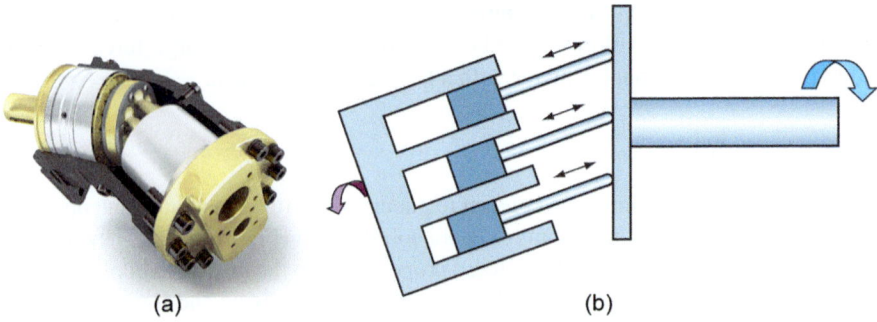

(a)   (b)

FIGURE 4.75   (a) Construction and (b) operation of an axial piston hydraulic motor. ((a): NosorogUA/Shutterstock.com.)

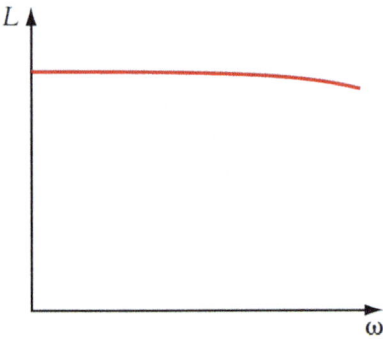

FIGURE 4.76   Torque–speed characteristic of a hydraulic motor.

4.13.7   HYDRAULIC POWER SUPPLIES

In order to power hydraulic actuators, a pump must be used to provide a source of hydraulic fluid at high pressure. This clearly requires a mechanical power source, which is typically an electric motor (normally an induction motor) but can be a diesel engine. In a factory or laboratory environment, a 'power pack' is often used, a self-contained unit consisting of the following components (Figure 4.77):

- Electric induction motor
- Fixed-displacement pump (e.g. vane pump)
- Oil reservoir
- Pressure regulator (often a relief valve)
- Sometimes a cooler (either a water-cooled heat exchanger or an air-cooled device similar to a car radiator) to dissipate the heat generated in the pump and regulator
- Filters to remove particles

In other applications, it can be convenient to use variable-displacement pumps which allow the flow rate from the pump to be adjusted directly, for example, to allow a piece of machinery (such as the wheels or tracks of a piece of construction machinery) to be driven at variable speed and direction.

FIGURE 4.77 Photograph of a hydraulic power pack showing a 75 kW motor and pump, and below them the oil storage tank.

4.13.8 HYDRAULIC VARIABLE-RATIO DRIVES

One convenient application of hydraulics is within variable-ratio drives, which behave as a 'gear-box' with a ratio that is infinitely variable between given positive and negative values. The mechanism (Figure 4.78) consists simply of a variable-displacement pump driving a fixed displacement motor. With the swashplate on the pump in its neutral position, no fluid is pumped and the motor and output shaft do not rotate. However, if the swashplate is moved from neutral towards its positive or negative limit, the motor and output shaft will rotate at speeds up to the maximum rate in the required direction.

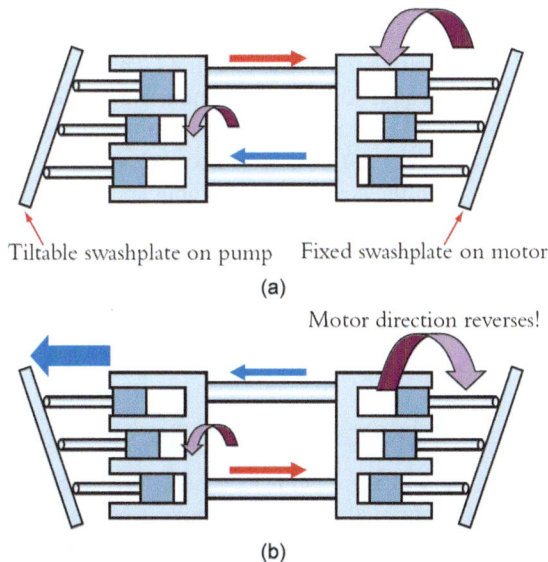

FIGURE 4.78 Diagram of a variable-ratio hydraulic drive providing an output rotation (a) in the forward direction and (b) in the reverse direction.

LEARNING SUMMARY

By the end of this section, you should have learnt:

✓ The similarities and differences between pneumatic and hydraulic systems.

✓ Why pneumatic or hydraulic actuators might be used in preference to electric motors.

✓ The ancillary equipment needed to power pneumatic and hydraulic systems.

✓ How hydraulics can be used to provide a variable-ratio drive.

✓ The cost and energy-efficiency issues associated with pneumatic power and compressed air.

4.14   STEADY-STATE OPERATING POINTS AND MATCHING OF LOADS TO POWER SOURCES

Consideration of torque–speed characteristics of loads and engines etc. provides an opportunity to introduce the concept of the steady-state operating point at which the motor or engine will drive the load without acceleration or deceleration. This concept occurs widely in all branches of engineering (mechanical, electrical, pneumatic etc.) and is closely related to the concept of 'impedance matching'. In practice, it is often necessary to use some kind of transmission to match the characteristics of the power source to the load, so that the power source and load interact in an optimal manner.

4.14.1   STEADY-STATE OPERATING POINTS

The concept of an operating point is straightforward: if the torque–speed characteristics of the load and the motor or engine are both known, then the combination of torque and speed at which the motor will drive the load corresponds to the point at which the characteristics cross. This is straightforward (and obvious) to determine graphically if the characteristics are plotted; it can also be found if the equations of the characteristics are known, simply by expressing torque as a function of speed for both characteristics and then equating the two expressions. The solution of the resulting equation may or may not be possible analytically; so, an iterative numerical solution may be required.

An issue related to the concept of operating points concerns the scenario if, at any instant, the system is running at a speed that does *not* correspond to the operating point, for example, if the system is running more slowly than the steady-state running speed. Assuming (for the purpose of the argument) that the characteristics are valid for transient (non-steady state) conditions, it will be observed if the torque available from the engine or motor exceeds that required by the load. In that case, the surplus torque is available for accelerating the load, which will speed up until the steady-state running speed is approached asymptotically.

EXAMPLE OF AN INTERNAL COMBUSTION ENGINE PROBLEM

a. A naturally aspirated diesel engine with the characteristic given in Figure 4.79 is driving a pump at 2000 rev/min; the pump consumes a power of 30 kW at this speed. How many litres of fuel are required per hour? Assume the specific gravity of the fuel (effectively, the density in kg/litre) is 0.84.

b. A sudden flood occurs, and the engine is set to maximum injection by the control system. If the inertia of the pump is 4 kg.m^2, at what rate (rev/sec^2) does the system

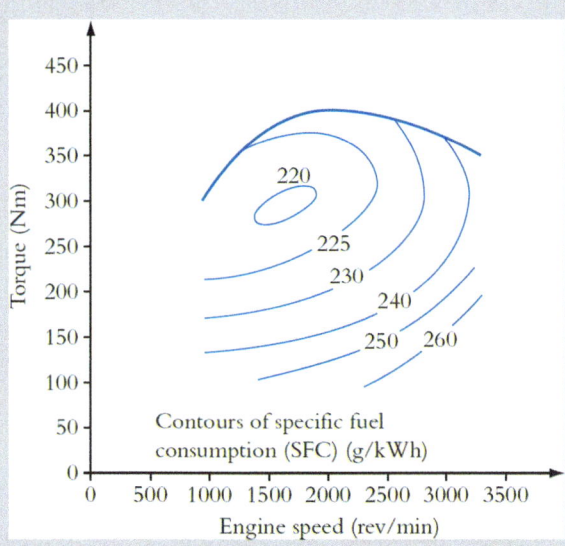

FIGURE 4.79 Representative torque–speed–SFC map for a diesel engine. (Adapted from Heywood, 1988.)

accelerate at the instant the throttle is opened fully? You may assume that the engine and pump characteristics both remain valid for this sudden change.

c. If the pump has a square-law torque–speed characteristic (i.e. torque proportional to speed squared), what is the maximum speed at which the engine can drive it?

Outline solution

a. Torque = $30000/(2\pi \times 2000/60) = 143.2$ Nm
   Specific fuel consumption = 245 g/kWh
   Mass of fuel consumed per hour = $30 \times 245 = 7350$ g = 7.35 kg
   Volume of fuel per hour 7.35/0.84 = 8.75 litres

b. Maximum engine torque at 2000 rpm = 400 Nm
   Load torque = 143.2 Nm
   Excess torque for acceleration = $400 - 143.2 = 256.8$ Nm
   Acceleration = net torque/moment of inertia = $256.8/4 = 64.2$ rad/s^2 = 10.2 rev/s^2

c. Assume the equation of form $L_{pump} = c\,N^2$ where L_{pump} is torque in Nm, c is a constant and N is the speed in rev/min
   Constant c = initial torque/(initial speed)^2 = $143.3/2000^2 = 0.0000358075$
   Characteristic of pump: $L_{pump} = 0.0000358075\,N^2$ (Nm)
   $N = 3150$ rpm is where this characteristic crosses the maximum torque line

4.14.2 Matching of Loads to Power Sources

It is very unusual for the characteristics of a power source to match closely the requirements of a load without the need for any kind of transmission. One example is an electric vehicle with direct drive of the electric motors to the wheels (Figure 4.80(a)), an approach not widely used at the time of writing, but with potential advantages for torque control of each wheel (torque vectoring). Another

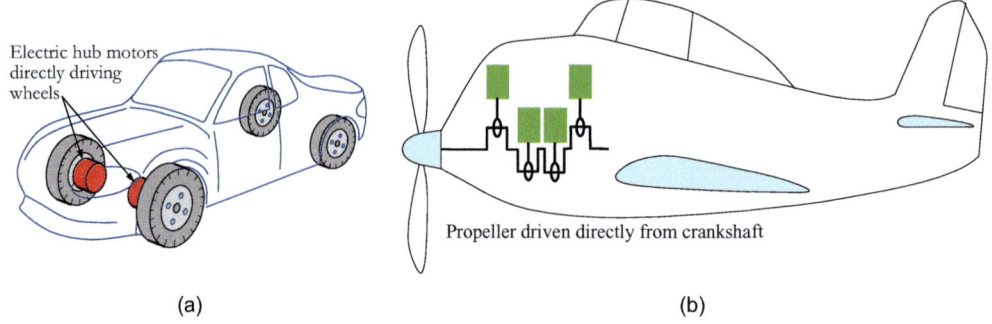

FIGURE 4.80 Two examples of direct drive from engine to load. (a) An electric vehicle with direct-drive hub motors and (b) a piston-engine aircraft.

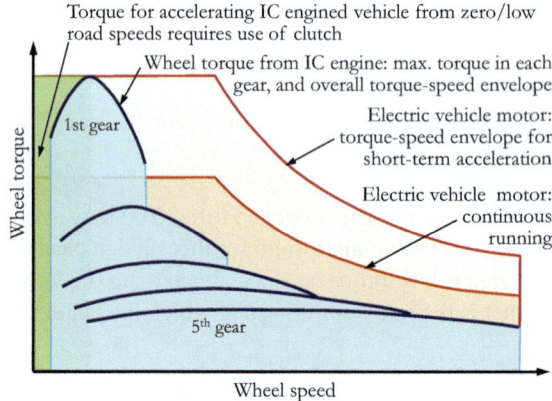

FIGURE 4.81 Torque–speed envelopes for a typical ICE vehicle with multi-ratio gearing and clutch and an electric motor with only fixed ratio drive.

is a simple piston-engine aircraft (Figure 4.80(b)), where the propeller is mounted directly on the end of the engine's crankshaft.

Contrast the first of these examples with the majority of currently manufactured electric vehicles, which have a mechanical transmission (including reduction gearing and differential) between the motor and the wheels. A conventional ICE vehicle needs both a clutch and a multi-ratio gearbox to match the engine characteristics to the load; otherwise, it would be impractical either to start the vehicle moving from rest (as the engine cannot produce torque at zero and low speeds) or drive the vehicle at a wide range of speeds. Figure 4.81 shows how the multi-ratio gearing and clutch are used to overcome these issues for an ICE, whereas electric motors (whether directly connected or with fixed gearing) need neither multiple ratios nor a clutch to obtain the required characteristics over the full range from standstill to full speed. Similarly, in a 'turboprop' aircraft, such as a military transport plane, the high-speed gas turbine engine is linked to the relatively low-speed propeller via a reduction gear.

4.14.2.1 Matching of Steady-State Characteristics

Probably, the most common challenge in matching load to a power source is where the steady-state running characteristics of the load do not cross the characteristics of the power source in a useful manner, for instance where the power source runs too fast and does not provide enough torque (see Figure 4.82).

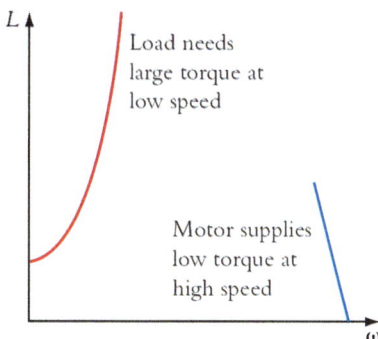

FIGURE 4.82   Mismatched torque–speed characteristics.

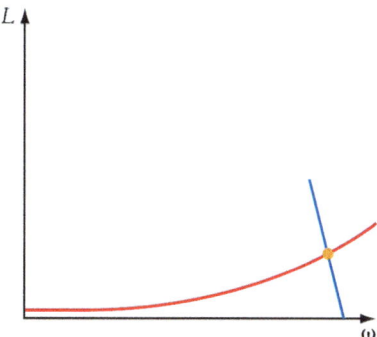

FIGURE 4.83   Torque–speed characteristics which cross at a useful operating point.

In such a case, it is straightforward to shift the torque–speed characteristic via the use of a transmission ratio, as described in Section 4.7, so that a useful operating point can be chosen as shown in Figure 4.83.

WORKED EXAMPLE

An example of such a problem might be to consider the matching of a load such as a blower to a motor (see Figure 4.84).
   Assume that the characteristics of a load are

$$L_{\text{load}} = 20 + 0.2\omega + 0.025\omega^2$$

where L_{load} is the torque in Nm and ω is its angular velocity in rad/s (both measured at the shaft directly driving the load). The load is to be driven via a 3:1 reduction ratio from a 5 kW induction motor whose characteristics are approximated as

$$L'_{\text{motor}} = 5\left(50\pi - \omega'\right)$$

using the prime (′) to denote that torque and angular velocity relate to the motor axis. The raw characteristics of the motor and load do not cross in a useful manner (Figure 4.85)

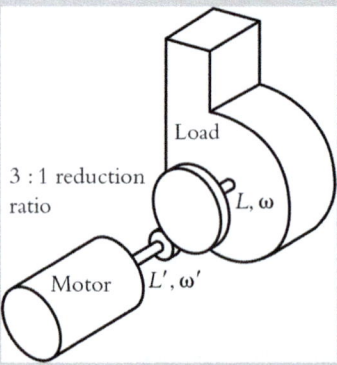

FIGURE 4.84 Blower driven from a motor via a reduction gear.

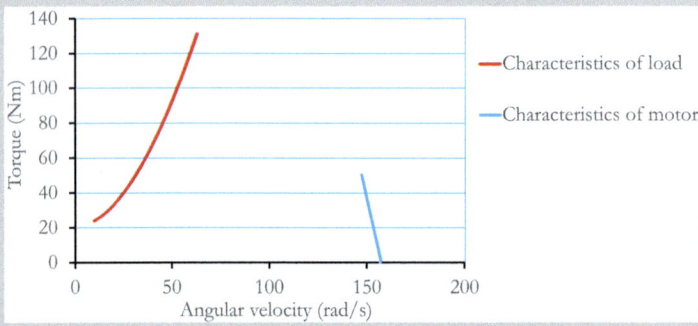

FIGURE 4.85 Mismatched characteristics of a motor and blower load.

The problem is, of course, tackled by connecting the motor and the load via the transmission system illustrated in Figure 4.79. The torque–speed characteristics of the load can now be referred to the axis of the motor:

$$L'_{\text{load}} = L_{\text{load}}/3$$
$$\omega' = 3\omega \Rightarrow \omega = \omega'/3$$
$$L_{\text{load}} = 20 + 0.2\omega + 0.025\omega^2$$
$$L'_{\text{load}} = \frac{1}{3}\left(20 + 0.2\omega + 0.025\omega^2\right)$$
$$= \frac{1}{3}\left(20 + 0.2\frac{\omega'}{3} + 0.025\left(\frac{\omega'}{3}\right)^2\right)$$
$$= 6.667 + 0.02222\omega' + 0.0009259\omega'^2$$

This referred characteristic now crosses the characteristic of the motor to give a stable operating point as shown in Figure 4.86.

This operating point can be found graphically as the point where the characteristics of the power source and load (referred, for instance, to the power source axis) cross. However, if (as in the present case) the torque–speed characteristic can be expressed in the form of an equation, the operating point can be found by equating the torques for the motor and load and solving to find the angular velocity. In the present case, this solution involves constructing and solving a quadratic equation using the standard formula:

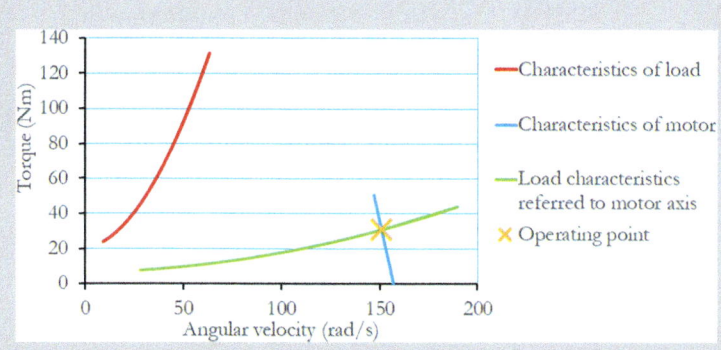

FIGURE 4.86 Matching of load and motor characteristics.

$$L'_{\text{load}} = L'_{\text{motor}}$$
$$6.667 + 0.02222\omega' + 0.0009259\omega'^2 = 5(50\pi - \omega')$$
$$(6.667 - 250\pi) + (0.02222 + 5)\omega' + (.0009259)\omega'^2 = 0$$
$$.0009259\omega'^2 + 5.02222\omega' - 778.7 = 0$$
$$\omega' = \frac{-5.02222 \pm \sqrt{5.02222^2 + 4 \times .0009259 \times 778.7}}{2 \times .0009259}$$
$$= 150.86 \text{ rad}/\text{s} \ (\text{or} \ -5575 \text{ rad}/\text{s, which is clearly incorrect})$$
$$L'_{\text{load}} = L'_{\text{motor}} = 5(50\pi - 150.86) = 31.1 \text{ Nm}$$

4.14.2.2 Matching of Starting Characteristics

Another common situation is where it is desired to start a load from rest. If the load does not have a large inertia or if the mechanical power source has a large torque at zero speed (and is not damaged by stalling or running at low speed), this may not be a problem. However, in many practical situations, starting can be a problem. For example, ICEs will not run in a stable manner at low speed (they stall easily, as anyone who has learned to drive on a manual-transmission ICE-powered vehicle will know!), while large electric motors will burn out if they are required to provide a large torque at low speed for a significant length of time. Several strategies can be adopted:

- Alteration of the characteristics of the mechanical power source, for example, by driving an induction motor from an inverter so that it gives a good torque as the frequency is raised from rest (see, for example, Figure 4.87). (Note: this diagram assumes that the frequency is increased slowly so that the steady-state operating points are reached at each value of speed. For rapid ramping of frequency, the torque required to cause acceleration will, of course, be significant and the locus of combinations of torque and speed will not coincide with the torque–speed curve of the load).
- Clutch used to allow a mismatch in speed between the mechanical power source and load, whilst allowing a given value of torque (equal to the slippage torque of the clutch) to be transmitted. The power source can now run at a stable speed while the load accelerates from rest under the influence of that torque. Examples include:
  - The process of getting a car moving from rest, allowing the clutch to slip gently as the speed of the car picks up.
  - The engagement of a clutch on a lathe to start the spindle (which may carry a heavy workpiece with a large value of inertia), having earlier started the lathe's induction motor under no-load conditions.

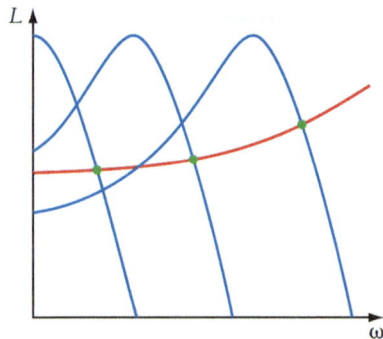

FIGURE 4.87 Locus of operating points as a load consisting of both friction and windage is accelerated slowly from standstill using an inverter-driven induction motor.

- The use of a transmission system which allows the mechanical power source to run at speeds within its design range while allowing the load to run up from rest. This may be as simple as the clutch described above, and in the case of an ICE car, also a transmission ratio with a series of gear ratios that enable the vehicle speed to be increased progressively (see Figure 4.81). Alternatively, it may involve the use of variable-ratio mechanical or hydraulic drives as briefly mentioned earlier in this Unit or may involve using the mechanical power source to drive a generator which in turn (via a controller) drives the load. This last approach is widely used in diesel-electric railway locomotives and trains where an electrified line is not available, with the diesel engine driving the generator and electric motors driving the wheels. The concept of using a generator and electric motor also forms the basis of hybrid ICE/electric vehicles, though these also usually involve battery storage.
- Choosing a mechanical power source that has a torque–speed characteristic which is well suited to the differing demands of starting and running. This involves a high torque at low speed (without damage to the power source or any ancillary equipment, e.g. power supply, occurring due to running under stall conditions) combined with lower torques at high speeds. Good examples of motors with these characteristics (see Figure 4.88) include:
  - Pneumatic motors.
  - Hydraulic motors.
  - Electric motors with suitable driver circuitry to allow them to produce significant torque at low speed without damage, as are used within electric vehicles and many other applications.

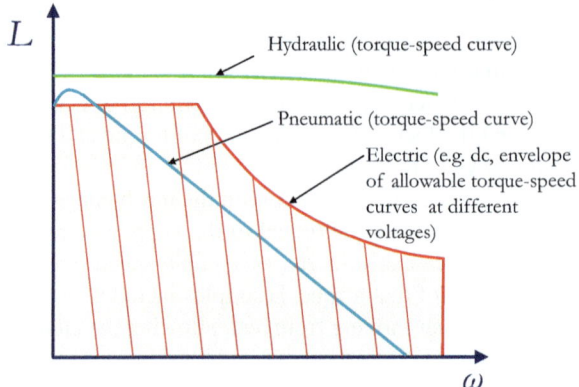

FIGURE 4.88 Torque–speed characteristics of motors that are well suited to starting.

LEARNING SUMMARY

By the end of this section, you should have learnt:

✓ How mechanical power sources and loads interact.

✓ How a transmission system may be used to match the power source to the load.

✓ To calculate the combination of torque and speed at which a load and power source will operate under steady-state conditions.

✓ The function of a clutch with particular reference to starting a mechanical load from rest.

✓ Why some sources of mechanical power are better suited to starting loads than others.

NOTE

1   Strictly, this mathematically exact definition of μ_0 was superseded in 2019 by a definition derived from experimentally determined quantities, though the difference in values is far too small to be of any practical significance and the definition given here is still used in practice.

REFERENCES

Bird, J., 2004, *Electrical Circuit Theory and Technology*, Newnes.

Blackburn, J.F., Reethof, G. and Shearer, J.L. (eds), 1960, *Fluid Power Control*, MIT Press.

Carbon Trust, 2012. Compressed Air - Opportunities for Business. Carbon Trust guide CT-V050. London, UK, January 2012.

Chambers Dictionary of Science and Technology, 2007, (general editor, John Lackie). Chambers, Edinburgh.

Chan, C.C., 2002. The State of the Art of Electric and Hybrid Vehicles. *Proceedings of the IEEE*, 90(2), pp. 247–275.

Chau, K.T. and Li, W., 2014. Overview of Electric Machines for Electric and Hybrid Vehicles by K.T. Chau; Wenlong Li. *International Journal of Vehicle Design (IJVD)*, 64(1). DOI: 10.1504/IJVD.2014.057775

Credo, A., Fabri, I., Villani, M. and Popescu, M., 2019. High Speed Synchronous Reluctance Motors for Electric Vehicles: A Focus on Rotor Mechanical Design. *2019 IEEE International Electric Machines & Drives Conference (IEMDC)*, pp. 165–171, available at: https://api.semanticscholar.org/CorpusID:199490333

Fraser, C. and Milne, J., 1994. *Integrated Electrical and Electronic Engineering for Mechanical Engineers*, McGraw-Hill.

Heywood, John B., 1988. *Internal Combustion Engine Fundamentals*. McGraw-Hill.

Hughes, A. and Drury, W., 2013. *Electric Motors and Drives: Fundamentals, Types and Applications*, fourth edition, Newnes.

Karthik, S. H., 2019. Types of Motors used in Electric Vehicle Electronics May 03, 2019, available at https://circuitdigest.com/article/different-types-of-motors-used-in-electric-vehicles-ev Accessed 12 November 2020.

Li, S., 2017. *A Review of Electric Motor Drives for Applications in Electric and Hybrid Vehicles*. Published online in March 2017 at https://www.researchgate.net/publication/330533021_A_Review_of_Electric_Motor_Drives_for_Applications_in_Electric_and_Hybrid_Vehicles accessed 12 November 2020.

Nino-Baron, C.E., Tariq, A.R., Zhu, G., and Strangas, E.G., July 2011. Trajectory Optimization for the Engine–Generator Operation of a Series Hybrid Electric Vehicle. *IEEE Transactions on Vehicular Technology*, 60(6), 2438–2447.

Omekanda, A. M., 2013. Switched Reluctance Machines for EV and HEV Propulsion: State-of-the-Art. *2013 IEEE Workshop on Electrical Machines Design, Control and Diagnosis (WEMDCD)*, Paris, pp. 70–74. DOI: 10.1109/WEMDCD.2013.6525166

Riba, J.-R., López-Torres, C., Luís Romeral. L. and Garcia, 2016. A.Rare-earth-free Propulsion Motors for Electric Vehicles: A Technology Review. *Renewable and Sustainable Energy Reviews*, 57, pp. 367–379.

Shayler, P. J., Chick, J., Darnton, N. J. and Eade, D., 1999. Generic Functions for Fuel Consumption and Engine-Out Emissions of HC, CO and NOx of Spark-Ignition Engines. *Proc. Instn. Mech. Engrs* Vol. 213 Part D, pp. 365–378.

Wildi, T., 1991, *Electrical Machines, Drives and Power Systems*, 2nd ed., Prentice-Hall, pp. 292–293.

Yıldırım, M., Polat, M. and Kurum, H., 2014. A Survey on Comparison of Electric Motor Types and Drives Used for Electric Vehicles. *16th International Power Electronics and Motion Control Conference and Exposition*, PEMC 2014. pp. 218–223.

5 Feedback and Control Theory

Arthur Jones

5.1 OVERVIEW

There are numerous situations in engineering where it is desirable to provide power to a machine in order to make it run at a chosen (and variable) speed, flow rate, pressure etc. In many of these cases, there is a need to maintain the speed or other attributes of a system at a fixed value despite the changes to the environment in which it operates. Examples including maintaining the speed of a car at a fixed value regardless of road gradient and wind, maintaining the course of an aircraft regardless of the motion of the air in which the aircraft flies and adjusting the rotational speed of a Blu-Ray disc within its player to maintain the required data rate regardless of the changing radius at which the laser beam reads the disc.

This unit initially concentrates on the concept of feedback and control and on the mathematical concepts used in modelling a control system. The basic concepts are outlined with a minimum of mathematical overhead and then the mathematical tools required to analyse realistic situations (such as those involving dynamics) are introduced one by one until systems can be modelled, analysed and characterised in terms of their performance and stability.

The concept of automatic control is exemplified by systems such as the cruise control on a vehicle, which maintains a set speed regardless of gradient, wind speed etc., and the autopilot on an aircraft, which maintains the course, altitude and speed without human intervention. However, although control systems are now present in numerous everyday situations, they are frequently so invisible or transparent in their operation that the user may be totally unaware of them. For instance, modern car engines continuously take data from sensors to adjust the engine to optimum performance regardless of ambient temperature and cameras automatically focus themselves and compensate for varying levels of illumination. By contrast, 40 to 50 years ago, cars were equipped with a manually operated choke to compensate for a low engine temperature and cameras needed manual settings of focus, shutter and aperture. The present unit aims to introduce some of the concepts involved in automatic control, beginning with a simplified approach involving a minimum of mathematics in order to examine:

- Feedback
- Block diagrams
- Error
- The control algorithm.

The unit then progresses to examining some of the mathematical tools of classical control theory that are required to analyse and understand real control systems. These tools include:

- The concept of Laplace transforms to allow the analysis of systems within the 's domain'
- The concept of the transfer function
- The analysis of initial and final behaviour and frequency response
- The techniques for determining the stability of a control system
- A technique for visualising the behaviour of a system.

DOI: 10.1201/9780429319495-5

This unit can only give an overview of some of the simpler topics in control engineering, and the approach taken here is to present a variety of worked examples rather than going into excessive depth on the theory or techniques. Numerous texts are available that cover control engineering in more detail and from a more rigorous mathematical perspective including the current edition of the classic text by Dorf and Bishop (2021). A slightly earlier but excellent text is that by Van de Vegte (1993). Bolton (1998) provides an introduction to control in a manner that does not rely heavily on mathematics.

5.2 FEEDBACK AND THE CONCEPT OF CONTROL ENGINEERING

5.2.1 CONCEPT, HISTORY AND SIMPLE EXAMPLES

Feedback may broadly be defined as **the comparison of a measured physical variable with the desired value of that variable with the aim of taking action to minimise the difference between the measured and desired values**.

The original application of feedback in the control of engineering plants appears to have been the centrifugal governor originally applied to steam power by James Watt in 1788. As the governor rotates and the weights fly outwards, a steam valve is closed so that the engine speed is maintained at an approximately constant value. In fact, Watt's development of the steam engine governor was based closely on the use of a similar device invented a year earlier by the millwright Thomas Mead for the control of the settings (but not necessarily the speed) of grinding wheels in windmills (Bennett, 1979). Figure 5.1(a) shows one of the governors on what is thought to be the last set of engines built by James Watt and Co. in 1882–4, still in working order in the preserved Papplewick Pumping Station near Nottingham, UK. At the time of writing, there is a similar engine (also with a centrifugal governor) in working order in Wollaton Park immediately to the north of the University of Nottingham, within the Nottingham Industrial Museum.

The process by which the governor operates (Figure 5.1b) is:

*   If the engine speeds up beyond its desired speed, the masses on the governor move outwards under centrifugal action against the retaining spring or gravity
*   The movement of the masses is transferred, via a collar, to a control rod that slightly closes a control valve supplying steam to the engine

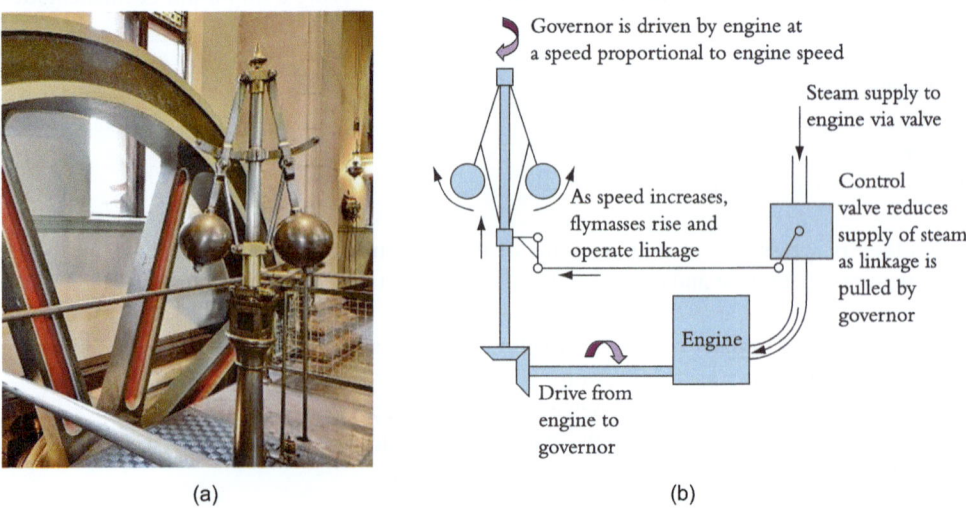

(a)   (b)

FIGURE 5.1   (a) Governor at Papplewick Pumping Station (b) schematic operation of a typical centrifugal governor.

- This causes the engine to slow down
- By means of this process, known as negative feedback (e.g. where an increase in speed results in an action which causes the speed to be reduced), the engine's speed is automatically maintained at a value close to the desired value.

Conversely, if the engine speed falls unexpectedly, the governor will cause the control valve to open slightly above its current setting, causing the engine to speed up.

Since the development of the centrifugal governor, control engineering has evolved into the discipline of classical control which forms the basis of the present Unit. The history of control is described by Bennett in detail in two books (Bennett, 1979 and 1993) and summarised succinctly in an article (Bennett, 1993). A detailed history of aircraft control is also given by McRuer (1974). A selection of the key challenges overcome over the years, and the resulting building blocks of classical control theory, can be itemised as follows, based on the above historical reviews amongst other sources:

- The need to assemble models of the behaviour of complete systems in order to analyse their stability, leading to the development of block diagrams by Brown (1946), the concept with which we start our exploration of control in the present section;
- The need to analyse the time-dependent behaviour of systems, leading to the pragmatic application by Heaviside of a method equivalent to Laplace transforms (Lützen, 1979; Heaviside, 1893), to provide a rational framework for the analysis of such systems and their later use within transfer functions (relationships between outputs and inputs of a sub-system, expressed using Laplace transforms) as introduced in Sections 5.4 to 5.6;
- The need to obtain the overall equations describing the behaviour of a system, leading to the use by Hall (1943) of Laplace transform-based transfer functions within block diagrams to provide the equations which can then be analysed for stability as described in Sections 5.5 and 5.13;
- The need encountered in practice to consider the value, rate of change and accumulation of control errors, leading to the development by Minorsky (1922) of what is now termed proportional, integral and derivative (PID or three-term) control, introduced in Section 5.11.
- The need to determine the stability of control systems, leading to the various stability analysis techniques described in Sections 5.13 and 5.14 for analysing the stability of control systems, for example, that of Nyquist (1932), the root locus approach of Evans (1948) and the stability criterion of Hurwitz (1895) who independently re-discovered an existing solution by Routh (1877).

Major parts of this work took place in the years leading up to, during and shortly after the Second World War and can be traced to a variety of industrial and military contexts including telephony, power generation, ship stability and the targeting and guidance of weapons. Developments since then tend to be considered as 'modern' control theory and form the subject of advanced control courses which build on the material covered by the present Unit but are not covered here. The interested reader should refer to (for example) the text by Dorf and Bishop (2021).

5.2.1.1 Everyday Example of Closed-Loop Feedback Control: Driving a Vehicle at Constant Speed

The task of maintaining the constant speed of a vehicle provides an excellent example of closed-loop control. In the case of a car without cruise control (or with the cruise control not engaged), one aspect of the driving process is:

- A speed is chosen (for example, the driver might choose to drive as fast as possible to reach an appointment in time but without breaking the law) – this is taken to be the 'demand' for the system.

- The current speed is measured via the speedometer, which is observed by the driver.
- The driver compares the two and assesses the difference (known technically as the **error**, which is the desired value minus the measured value).
- Equipped with the skill of how to drive a car, the driver takes a decision on how to correct the speed based on the value of error and also based on an understanding of how well the speed is converging to the desired value – this decision-making process is very important even though it is probably taken for granted within a human control situation.
- The driver attempts to correct the speed by raising or lowering their foot on the accelerator (gas) pedal.
- Depending on issues such as the mass of the car, the power of the engine etc., the car will respond by gradually increasing or decreasing its speed; if the driver is experienced, the car's speed will rapidly converge on the desired value, but if the driver is a learner, the speed is very likely to oscillate about the desired value until the driver hits upon the correct position of the accelerator.

This process is illustrated via a simple block diagram in Figure 5.2.

An obvious but essential point to note is that feedback is **negative**: for example, if the car is travelling too fast, the driver partially releases the accelerator to allow it to slow down. (Positive feedback is explored later in this section but would not be helpful in this situation – it would involve the driver pressing the accelerator even harder as the car's speed was seen to increase!).

This process can of course be automated, and a system for doing so (known as the cruise control) is available in many modern cars. The control loop (Figure 5.3) remains essentially the same as before, except that the decision-making process requiring the skill of the driver is now replaced with the automated decision-making process (known formally as the **control algorithm**) forming

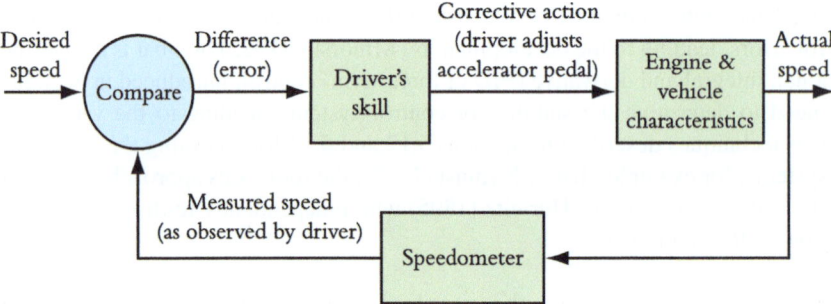

FIGURE 5.2   Block diagram of the process of driving a car.

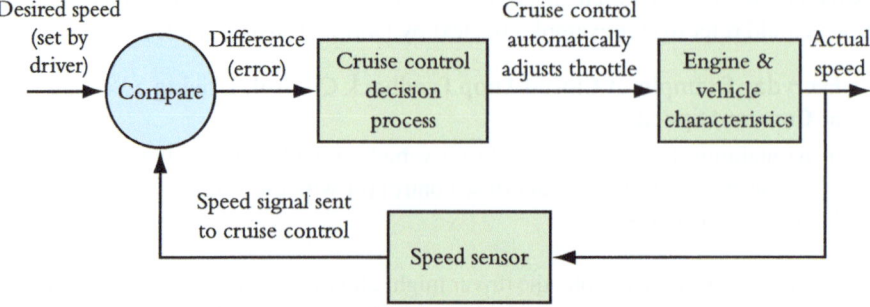

FIGURE 5.3   Block diagram of a cruise control system (greatly simplified).

the heart of the cruise control system and the actuation of the engine's throttle takes place under the command of the cruise control rather than via the driver's operation of the pedal.

The control algorithm is carefully designed to take into account the way in which the car's speed is changing in order to avoid over- or undershooting the desired speed or causing the speed to fluctuate about its desired value.

However, it is worth noting that, even without an automated control system (and indeed under open-loop conditions, i.e. with no feedback at all), the speed of a car is essentially stable for a given throttle setting. In other words, keeping one's foot on the accelerator in a fixed position may not result in the desired speed, but the speed is unlikely to go wildly out of control, at least on a road with only moderate gradients. This contrasts with the next example, which uses a control system to stabilise an inherently highly unstable system.

5.2.1.2 A More Challenging Set of Control Problems: Self-Balancing Vehicles

As noted above, cruise control represents a straightforward example of closed-loop control in a situation where the underlying system does not show significant instability when used under open-loop conditions. A rather extreme counter-example, consisting of a system which is totally unstable in open-loop conditions, is the unicycle and the rider (Figure 5.4), where only the rider's skill prevents them from falling over either along the line of travel or sideways; there is no mechanical restraint whatsoever against this tendency to fall over.

FIGURE 5.4   Unicycle and rider. (Aedonis/Shutterstock.com.)

The unicyclist is continually taking information from their eyes and their sense of balance (i.e. making use of feedback) and uses their skills (a human implementation of a highly tuned control algorithm) to adjust continually the torque to the pedals and their overall body movements to correct the tendency of the unicycle to fall over. Until this skill is acquired and fully developed (i.e. until the control algorithm embodied in their skill is fully functional), they will have numerous falls. (Similar arguments apply to the stability of a bicycle, though there is essentially only a single mode of instability, i.e. sideways falling, so the control algorithm required is less complex and the skill embodying it is easier to acquire. Presumably, that is why there are vastly more bicycle riders than unicyclists!).

A slightly more manageable example of a vehicle stabilised only by closed-loop control is a two-wheeled self-balancing vehicle (Figure 5.5). It consists of a platform, seat or board on which the rider is carried, supported only by two wheels, one on either side of the vehicle. It exhibits one of the modes of inherent instability of the unicycle (a tendency to fall forwards or backwards in its direction of travel) and similarly would tend to pivot around the axles and fall over unless active efforts are made to stabilise it.

Unless the centre of gravity of the vehicle and rider is precisely in a vertical line with the axis of the wheels (the position of unstable equilibrium) (Figure 5.6(a)), there will be a moment of the gravitational force about the wheel axis which causes rotation and hence makes the moment larger, rapidly resulting in a fall! (Figure 5.6(b)). However, stability of the two-wheeled vehicle can be achieved using a combination of sensors (typically rate gyros for measuring the rate of tilt of the body and rotation sensors (see https://bestelectrichoverboard.com/how-do-self-balancing-scooters-work/) to

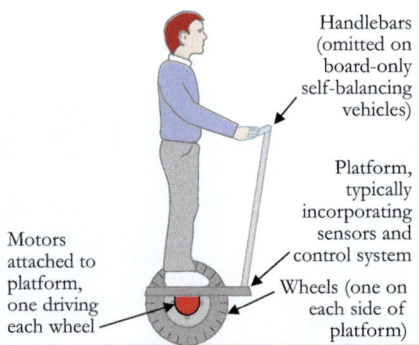

FIGURE 5.5   Features of a two-wheeled self-balancing vehicle.

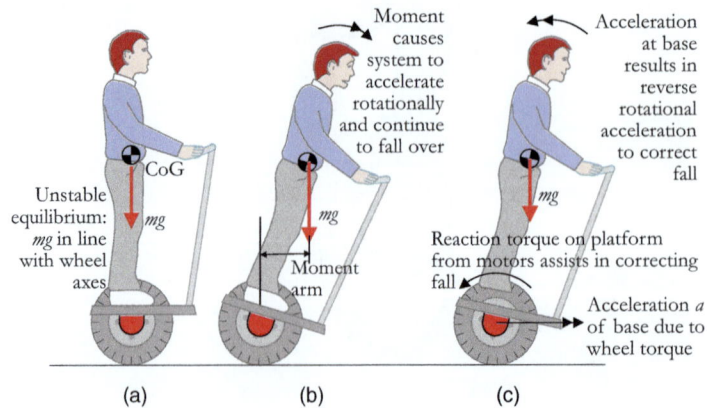

FIGURE 5.6   Behaviour of a two-wheeled vehicle showing (a) unstable equilibrium, (b) mode of instability and (c) method by which instability is counteracted.

measure the angle and speed of rotation of the wheels relative to the body etc.) as the inputs to a control algorithm which can calculate the torques which must be supplied by the motors to the wheels in order to overcome the tendency to fall over. If (for example) the rider starts to tilt forward from vertical and hence starts to fall forward, the motors will begin to accelerate the vehicle forward (Figure 5.6(c)) so that the vehicle starts to catch up with the falling rider; there will also be reaction moment from the wheel motors which additionally helps to tilt the rider back to vertical. This clearly requires a clever set of decision-making processes embodied in the control algorithm. If the algorithm is correct, excellent stability is achieved and the vehicle is easy to ride. However, if the algorithm is incorrect, or the control is inactive (i.e. the system is switched off), then the rider will fall over.

A highly simplified version of the control loop for a self-balancing vehicle is shown in Figure 5.7. This approach has been applied to human transport devices, two-wheeled wheelchairs and so-called 'hoverboards' from a variety of manufacturers; it also forms the basis of a variety of student and hobby projects involving small, microcontroller-based self-balancing two-wheeled robots. Several models of single-wheeled, self-balancing vehicles are also commercially available, automating the more complex balancing task achieved by the unicyclist described earlier.

Although the self-balancing vehicle is unlikely to be encountered within an undergraduate laboratory exercise, some students may encounter the inverted pendulum experiment (Figure 5.8) which is a classic demonstration of control concepts and is effectively a much-simplified version of the self-balancing vehicle problem. Like the self-balancing vehicle, it involves a mass whose centre of gravity lies over a pivot (Figure 5.8(a)) and is therefore inherently unstable in all practical situations

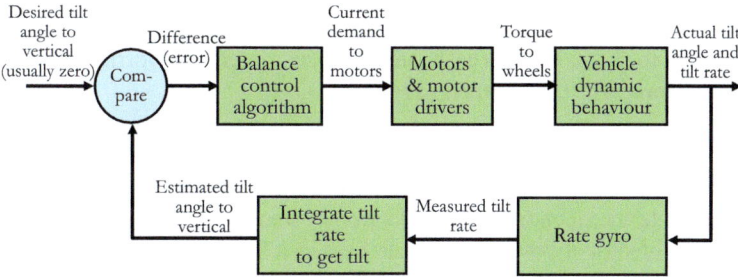

FIGURE 5.7 A simplified block diagram of the control system for a self-balancing vehicle, showing only control of tilt angle.

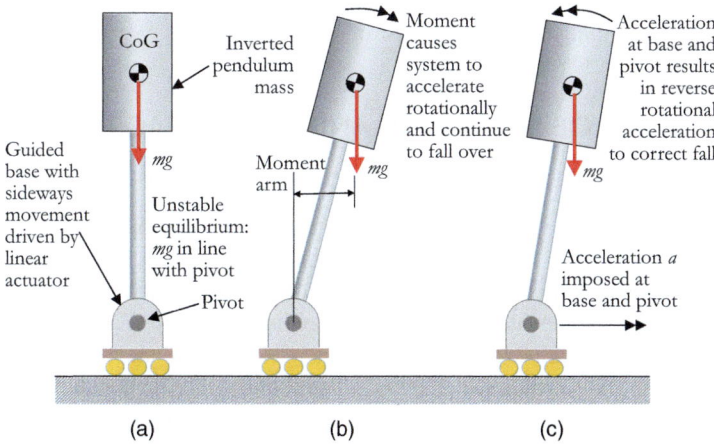

FIGURE 5.8 Inverted pendulum demonstration showing (a) unstable equilibrium, (b) mode of instability and (c) method by which instability is counteracted.

(Figure 5.7(b). It is slightly simpler than the self-balancing vehicle in that the corrective action is provided by directly causing the base to accelerate sideways using a linear actuator (Figure 5.8(c)) rather than by applying torque to the wheels. If the control loop is operating correctly, the pendulum mass remains vertically above the pivot, and the system automatically restores the pendulum to vertical even if it is externally disturbed. Obviously, if the control system is not working correctly, the pendulum will simply fall over.

5.2.1.3   Less Constructive Examples of Feedback

The fact that control systems generally employ negative feedback has already been noted – for instance, if a car is going too fast, the corrective action involves sending a signal to slow the car down. Although **positive feedback** is occasionally used to beneficial effect, it is generally an undesirable phenomenon, as any discrepancy between desired and actual values of system behaviour results in a signal that makes that discrepancy greater, giving a form of **instability**. A simple example of positive feedback is the inherent instability of an (uncontrolled) inverted pendulum as described in the previous example, with increasing departure from vertical leading to more tendency for the pendulum to rotate from vertical (Figure 5.8b). A powerful example of the destructive nature of positive feedback is the situation (a 'bank run') which occurs when a bank starts to encounter 'liquidity problems', that is when it starts to run out of money:

- Customers begin to hear that the bank is short of cash and become worried about the security of their savings
- They therefore withdraw their savings from the bank so that they can keep them safely elsewhere
- This means that the bank has less cash, and more customers get worried about its problems
- Even more customers withdraw their savings

… and so on until either the bank fails completely (as happened numerous times in the 1930s) or the process is halted by state intervention (as in the case of Northern Rock in the UK in 2007). Other examples include the phenomenon in electronics known as thermal runaway, which relates to a device that turns electrical power into waste heat as a by-product of its normal operation and tends to do this more vigorously as its temperature rises. Unless precautions are taken, the amount of power it converts to heat will keep increasing as its temperature rises, in turn causing it to get hotter still until the component burns out. An even more catastrophic example was the Chernobyl nuclear disaster, which can be regarded as a complex form of thermal runaway. The more the cooling water in the nuclear reactor boiled to form steam, the more heat the nuclear reaction generated. This created yet more steam, eventually causing a disastrous explosion.

5.3   ILLUSTRATIONS OF MODELLING AND BLOCK DIAGRAM CONCEPTS

5.3.1   Simple Example of Feedback and Control: Modelling of a Heated System

Real control systems usually include some form of time-dependent behaviour, but it is useful to illustrate the modelling process via a very simple control system in which time-related terms are neglected. Consider a heated enclosure for a small set of trackside railway equipment, designed to maintain the equipment at an elevated temperature to avoid it icing up (Figure 5.9).

The railway equipment has an exposed surface area of $A = 0.5$ m^2 and a convective heat transfer coefficient of $h = 5$ W/m^2.°C (a value which takes account of the thermal insulation of the equipment box and more correctly represented as $h = 5$ W/m^2.K). A heater provides a power p of heat input into the railway equipment. It will be assumed for simplicity that the temperature of the system's surroundings is 0°C. It is easy to show that if (for example) $p = 10$ W and the system is allowed to

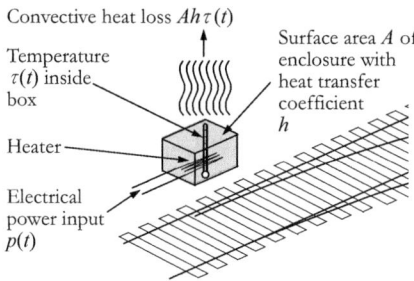

FIGURE 5.9 Heated enclosure for trackside equipment.

settle down to a steady temperature τ, the temperature achieved (which we may regard as the output of the system) is obtained by equating the power input p to the rate of heat loss:

$$p(t) = Ah\tau(t) \tag{5.1}$$

This is solved to obtain

$$\tau(t) = \frac{p(t)}{Ah} \tag{5.2}$$

which, for the above values, gives a temperature of

$$\tau(t) = \frac{10}{0.5 \times 5} = 4°C \tag{5.3}$$

Note that the values that are properties of the system (A and h) are constant, that is invariant with time, while all variable quantities p and τ are expressed in the form of functions of time t though in this example we will assume that such quantities have had time to settle down to their steady-state values or that that the thermal capacity of the system is negligibly small. (The importance of this apparently obvious statement will become clear later as a slightly different convention is normally followed in control engineering: variable quantities will be expressed as a function of a complex variable s, rather than time; see Section 5.4.). This very simple relationship between power input and temperature may be (trivially) represented using a block diagram as shown in Figure 5.10.

In the block diagram, the mathematical relationship between an input (in this case, power p) and an output (actual temperature τ) is shown as a 'block' which acts upon 'flows' of values or data. Specifically, a block represents the act of multiplying the input by the value of the constant or expression contained in the block in order to obtain the output. A separate convention is used for addition and subtraction and will be considered shortly.

Note: In Figure 5.10, the fact that the variables p and τ are functions of time is deliberately not shown. In fact, we would never normally draw the block diagram directly with variables as a function of time; they would instead be expressed as a function of the complex variable s which will be introduced in Section 5.4.

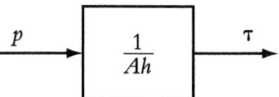

FIGURE 5.10 Relationship between power input and temperature.

Suppose that we now wish to maintain the temperature of the equipment at a given value $\tau_D(t)$. In order to automate this process, a very simple control system is introduced (Figure 5.11) which drives the heater with a power that is proportional to the difference between the desired temperature and the actual temperature, so that for each degree Celsius of temperature difference, an extra power of 10 W will be supplied where this constant of proportionality will be termed the *gain K* of the controller. As will be explained later, this is a very simple form of the control algorithm (decision-making process) and is known as proportional control. It may be represented mathematically as

$$p(t) = K\left(\tau_D(t) - \tau(t)\right) \tag{5.4}$$

and within a block diagram as shown in Figure 5.12.

The block diagram in Figure 5.12 now involves not only the input setting τ_D (demand temperature) and the output τ (actual temperature) but also an intermediate quantity $\tau_D - \tau$, known as the 'error'. Also visible is a 'differencing junction' where τ is subtracted from τ_D; this is the convention used to represent the subtraction of different signals or variables (rather than the multiplication by an expression within the rectangular blocks), and 'adding junctions' are similarly used to represent the addition of variables. The block diagram for the complete control system can now be assembled by 'joining the loose ends' in the block diagrams for the different aspects of the system as shown in Figure 5.13.

Note that, because we wish to treat τ as an output from the system as well as feeding it back, a branch is introduced into the flow; no additional symbols are involved in this. The block diagram can be simplified slightly by combining successive blocks as shown in Figure 5.14.

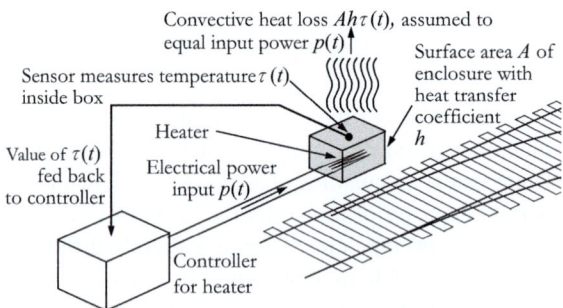

FIGURE 5.11   Trackside equipment with automated control system.

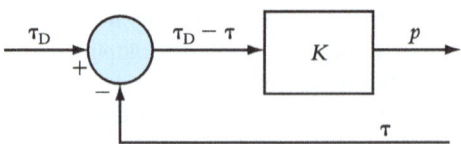

FIGURE 5.12   Temperature controller with gain K.

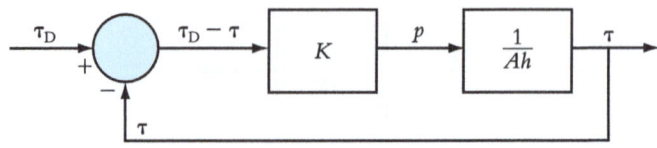

FIGURE 5.13   Assembly of control system model.

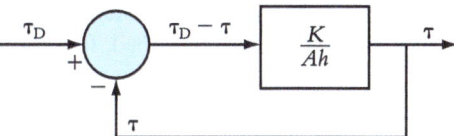

FIGURE 5.14 Combining blocks in temperature control system model.

Assembling the block diagram corresponds to combining Equations (5.2) and (5.4) to eliminate $p(t)$:

$$\tau(t) = \frac{1}{Ah} K\left(\tau_D(t) - \tau(t)\right)$$
$$= \frac{K\left(\tau_D(t) - \tau(t)\right)}{Ah} = \frac{K\tau_D(t)}{Ah} - \frac{K\tau(t)}{Ah} \qquad (5.5)$$

It would appear that we have a circular relationship in which $\tau(t)$ appears twice. However, Equation (5.5) can be rearranged to group the $\tau(t)$ terms together:

$$\tau(t) + \frac{K\tau(t)}{Ah} = \tau(t)\left(1 + \frac{K}{Ah}\right) = \frac{K\tau_D(t)}{Ah} \qquad (5.6)$$

It is now straightforward to rearrange this to give the output $\tau(t)$ in terms of the input (demand) $\tau_D(t)$:

$$\tau(t) = \frac{K\tau_D(t)}{Ah\left(1 + \frac{K}{Ah}\right)} = \frac{K}{(Ah + K)}\tau_D(t) \qquad (5.7)$$

so that the error is:

$$\tau_D(t) - \tau(t) = \tau_D(t) - \frac{K}{Ah + K}\tau_D(t) = \frac{Ah}{Ah + K}\tau_D(t) \qquad (5.8)$$

The block diagram for the whole system can now be expressed in a simplified form in Figure 5.15, where the contents of the 'box' are the ratio of the output to the input. This concept will be extended within the more complex analysis given in Sections 5.4 and 5.5 and will be referred to as the *transfer function*.

The issues of manipulating and simplifying block diagrams will be explored in detail in Sections 5.8 and 5.9. It is important to note that, with this simple (proportional) control algorithm, the desired temperature is never actually reached. For example, with the above choice of gain $K = 10$ W/°C, the $\tau(t)$ settles down to the following value:

$$\tau(t) = \frac{10}{(0.5 \times 5 + 10)}\tau_D(t) = 0.8\tau_D(t) \qquad (5.9)$$

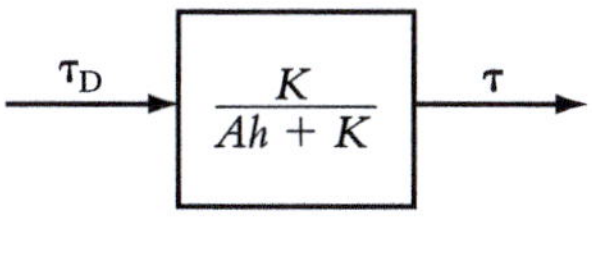

FIGURE 5.15 Overall block diagram containing transfer function for temperature control system.

so that a demand temperature of 5°C will result in an actual temperature of 4°C. Increasing the gain will result in the temperature being more closely approached; for example, a gain of $K = 100$ W/°C will result in a temperature of

$$\tau(t) = \frac{100}{(0.5 \times 5 + 100)} \tau_D(t) = 0.976\tau_D(t) \tag{5.10}$$

being achieved; for example, 4.88°C for a demand of 5°C. It will be seen later, however, that there can be a penalty (in terms of system stability) associated with making the gain of a system larger than necessary.

5.3.2   EXAMPLE OF FEEDBACK AND CONTROL: MODELLING OF A FIBRE TENSIONING SYSTEM

Another introductory example involves a slightly more complex control system, again with time-related terms such as inertia being neglected and involving only mechanical components. Consider a purely mechanical tension control system for a filament winding machine (used for producing fibre-reinforced composite containers) in which dry fibres are held on a cardboard spool of radius r, braked via a drum of radius R and coefficient of friction μ with a force related to the fibre tension f_T and a pre-load setting f_P applied to the braking system (Figure 5.16). The concept is that the brake provides a frictional force on the drum in order to tension the fibre, the pre-load contributes to the frictional force (and can be adjusted to vary the tension) and the tension feedback pulley provides a mechanism for reducing the braking force if the fibre tension becomes too large. In this system, the pre-load setting f_P is the input to the system and the fibre tension f_T is the output.

The block diagram, and hence the relationship (transfer function) between input f_P and output f_T can be derived by considering each part of the system in turn, noting that each block in the diagram multiplies the input by the contents of the block to give the output.

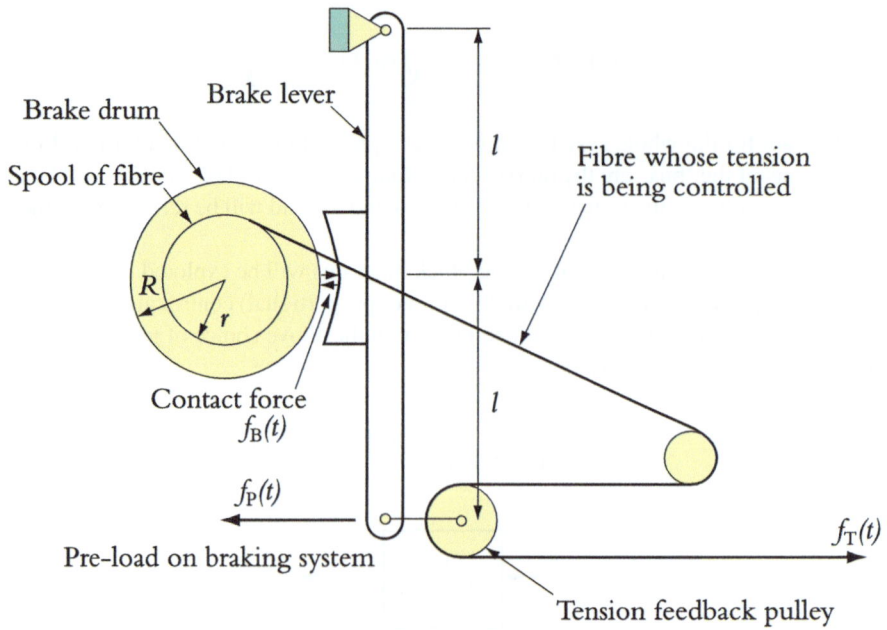

FIGURE 5.16   Diagram of mechanical fibre tensioning system (frictional forces omitted for clarity).

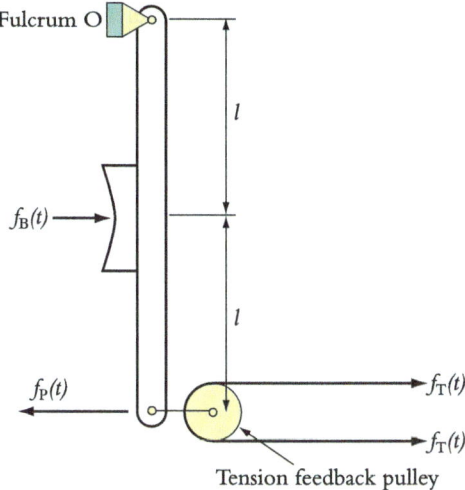

FIGURE 5.17 Lever and brake assembly (braking friction is assumed to have negligible moment and is omitted for clarity; reaction force at O also omitted).

By taking moments about the fulcrum O in a free-body diagram of the brake lever and pulley (Figure 5.17), the force f_B on the brake shoe is found from:

$$2l\left(2f_T\left(t\right)-f_P\left(t\right)\right)+lf_B\left(t\right)=0$$
$$\therefore f_B\left(t\right)=2f_P\left(t\right)-4f_T\left(t\right)$$
$$\text{or}: f_B\left(t\right)=2\left(f_P\left(t\right)-2f_T\left(t\right)\right) \tag{5.11}$$

Again we note that the length l, and for that matter, quantities such as R, r and μ, are all constant, that is invariant with time, while all variable quantities f_P, f_T and f_B are expressed in the form of functions of time t. In the present version of the example, we will assume that such quantities only change very slowly and/or that the inertia of the rotating components is negligibly small. The behaviour of the brake lever and pulley is represented as a block diagram in Figure 5.18.

In the block diagram, the mathematical relationships between quantities such as the input setting f_P and the output f_T (fibre tension) and intermediate quantities such as brake force f_B are shown as 'blocks' that act on 'flows' of values or data. In this particular block diagram, blocks involving multiplication by constant factors (gains) of values 2 and 4 are evident. Also visible is a 'differencing junction' where $4f_T$ is subtracted from $2f_P$ to give f_B. Again, the fact that f_P, f_T and f_B are functions of time is deliberately not shown; the correct representation of time-dependent quantities within block diagrams via the complex variable s will be introduced in Section 5.4.

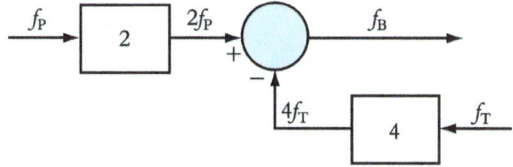

FIGURE 5.18 Block diagram for a lever and pulley.

Provided that inertia effects are ignored, or that no acceleration or deceleration is occurring, the brake drum (Figure 5.19) is similarly analysed as follows:

$$f_T(t)r - \mu f_B(t)R = 0$$

$$\therefore f_T(t) = \frac{\mu R}{r} f_B(t) \tag{5.12}$$

This analysis can be represented via a block diagram (Figure 5.20).

By virtue of the feedback inherent in the system, there is a circular relationship involving f_B and f_T since the application of the braking force f_B results in an increase in the fibre tension f_T, but the lever system results in f_T acting to reduce the braking force f_B. This simple system could be solved algebraically without further ado, but it is instructive first to draw the complete block diagram for the system, in which the feeding-back of f_T for eventual subtraction from f_P is now clearly visible in Figure 5.21.

Figure 5.21 may be rearranged to bring the leftmost block inside the loop as shown in Figure 5.22 and may be further simplified into the form shown in Figure 5.23 by combining the transfer functions of successive blocks by multiplication.

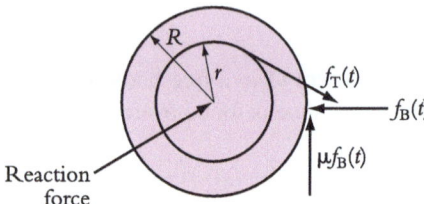

FIGURE 5.19   Brake drum.

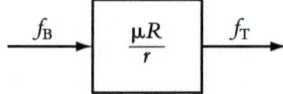

FIGURE 5.20   Block diagram for brake drum.

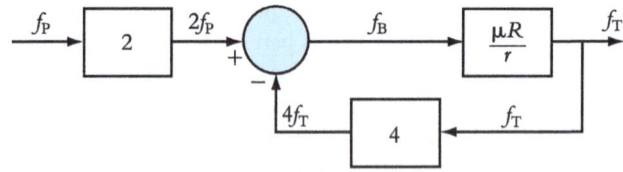

FIGURE 5.21   Block diagram for fibre tensioning system.

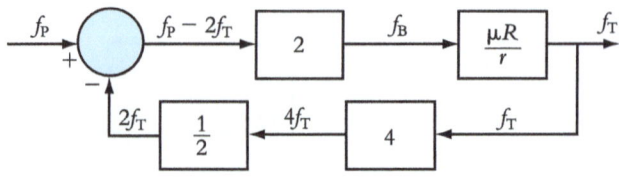

FIGURE 5.22   Block diagram of a tensioning system after rearrangement.

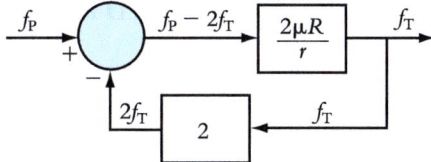

FIGURE 5.23 Block diagram of a tensioning system after multiplying successive blocks.

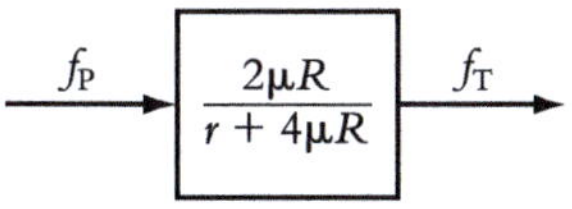

$$\xrightarrow{\;f_P\;}\boxed{\dfrac{2\mu R}{r+4\mu R}}\xrightarrow{\;f_T\;}$$

FIGURE 5.24 Block diagram for a tensioning system represented using a single transfer function.

The system may now be solved as follows:

$$f_T(t) = 2\frac{\mu R}{r}\big(f_P(t) - 2f_T(t)\big) = 2\frac{\mu R}{r}f_P(t) - 4\frac{\mu R}{r}f_T(t)$$

$$\therefore f_T(t)\left(1 + 4\frac{\mu R}{r}\right) = 2\frac{\mu R}{r}f_P(t)$$

$$\therefore f_T(t) = \frac{2\dfrac{\mu R}{r}}{\left(1 + 4\dfrac{\mu R}{r}\right)}f_P(t) = \left(\frac{2\mu R}{r + 4\mu R}\right)f_P(t) \tag{5.13}$$

In this case, the ratio $f_T(t)/f_P(t)$ of fibre tension to pre-load force (effectively the ratio of output force to input force) is simply $2\mu R/(r + 4\mu R)$, and a block diagram which represents this is shown in Figure 5.24.

The effectiveness of the control provided by this system can now be observed for the following parameters: $r = 50$ mm, $R = 100$ mm, $\mu = 0.4$ and $F_P = 2.5$ N. A 20% reduction in r (to 40 mm) as the spool of fibre unwinds results only in a 5% increase in tension from 0.95 N to 1 N – without the feedback system, the change in tension would be around 20%.

If we wish to consider the inertia of the spool and brake assembly (which would be significant and would need to be included in the analysis in order to understand the tension in the fibre), then the simple (time domain) mathematics described above is insufficient to express the situation and an additional tool is required to take into account the rates of change within a system. This will be described in the next section.

LEARNING SUMMARY

By the end of this section, you should have learnt:

✓ The concepts of closed-loop control and feedback and their typical applications
✓ The difference between positive and negative feedback
✓ The concept of a block diagram representing the characteristics of a control system and the manner in which feedback is provided.

5.4   THE *s* DOMAIN: A NOTATION BORROWED FROM MATHEMATICS

It is very difficult to analyse anything except the most elementary systems (such as the ones described above) in the time domain. This is because practical systems involve features such as inertia and damping, which require derivatives such as acceleration and velocity to be incorporated into the model. In such a case, the block diagram effectively becomes a graphical representation of a differential equation rather than a simple algebraic one. In order to tackle such systems, analysis within classical control engineering is generally undertaken using an adaptation of the methods of solving differential equations, using a notation based on one of two complex variables:

- The variable known as *s*, involving analysis in the *s* domain, used for modelling continuous control situations such as the ones covered in this unit. This notation is closely related to the mathematical approach involving **Laplace transforms**, which are just a useful way of solving differential equations by converting them to polynomials in *s*
- A different complex variable, known as *z*, involving analysis in the *z* domain, used for modelling discrete and digital control situations. This will not be considered further in the present unit but will be of interest in the case of more advanced control situations (see, for example, Dorf and Bishop (2021)).

Within the present unit, the behaviour of control systems and their components, including modes of behaviour that involve derivatives and integrals with respect to time, will henceforth be expressed in terms of the variable *s*. The concept of the *s* domain, the meaning of *s* as a variable and the relationship to Laplace transforms can all be discouraging or confusing to engineers whose main concern is to solve practical control problems. It is therefore convenient here to view the variable *s* and the *s* domain as a language or notation for processes involving integration and differentiation, rather than becoming too concerned about either the physical meaning of *s* or its use as a tool for solving differential equations; there are numerous introductions to the technique in the literature, such as chapter 6 of the engineering mathematics text of Kreyszig (2011). For completeness, however, it can be stated that the Laplace transform $F(s) = \mathscr{L}\{f(t)\}$ of a time-domain function $f(t)$ is defined as:

$$F(s) = \mathscr{L}\{f(t)\} = \int_0^\infty e^{-st} f(t)\,dt \tag{5.14}$$

where *s* may be interpreted as complex angular frequency or radians per unit time. By convention, as shown in Equation (5.14), variable quantities expressed in the time domain (i.e. as a function of time) are denoted by lowercase letters, for example, $f(t)$, while the quantities expressed in the *s* domain are denoted by upper-case letters, for example $F(s)$.

Crucial to the application of Laplace transforms to control systems are the rules (explored in more detail later in the present section) that multiplication by *s* (and subtraction of the initial condition, which in a control system is typically zero) represents differentiation, while division by *s* represents integration. This 'shorthand' for these calculus operations is encountered often in dealing with control systems. For readers familiar with the 'operator D' method of solving differential equations (Kreyszig, 2011, pp. 60–61; Spencer et al., 1977, pp. 17–18 and 41, originated by Heaviside, 1893), there are significant similarities between the operator D (where multiplication by D represents differentiation and division by D represents integration) and the way in which differentiation and integration in the time domain are represented respectively as multiplication and division by *s*.

The procedure for using Laplace transforms to solve ordinary differential equations is:

1. with knowledge of the initial conditions, the differential equation is transformed from the time domain to the *s* domain

2. by use of partial fractions or otherwise, the transformed equation is simplified to give the solution in the s domain
3. the simplified equation is transformed back into the time domain to give the solution (in control engineering, this step is usually not necessary and the behaviour of the system is generally left as an expression in s). If a time domain solution really is required, this is achieved by applying an inverse transform or (more commonly) with the aid of tables of Laplace transforms.

A straightforward example illustrates the power of Laplace transforms in solving differential equations before we proceed to look in more detail at the transforms of specific functions and the rules relating to their use. A system (Figure 5.25) consisting of a spring of stiffness k in parallel with a viscous damper of coefficient c (relating viscous force to linear velocity) is subjected to a constant force $p(t) = q$ beginning at time $t = 0$. For traceability throughout the following derivation, the term for the applied external excitation input (known as the forcing term) is highlighted in green, the spring term in red and the damping term in blue, up to the point where the equations are rearranged and the individual terms are no longer identifiable.

It will be assumed that the link joining the spring to the damper has no significant mass. Applying equilibrium to the link,

$$p(t) = kx(t) + c\frac{dx(t)}{dt}$$

$$t = 0, x = 0$$
$$p(t) = 0 \text{ for } t < 0$$
$$p(t) = q \text{ for } t > 0 \tag{5.15}$$

Mathematically, this load can be expressed in terms of the unit step function or Heaviside step function $H(t)$ which is defined as a function that takes the value 0 for $t < 0$ and 1 for $t > 0$:

$$p(t) = qH(t) \tag{5.16}$$

so that the equation of motion now becomes

$$qH(t) = kx(t) + c\frac{dx(t)}{dt} \tag{5.17}$$

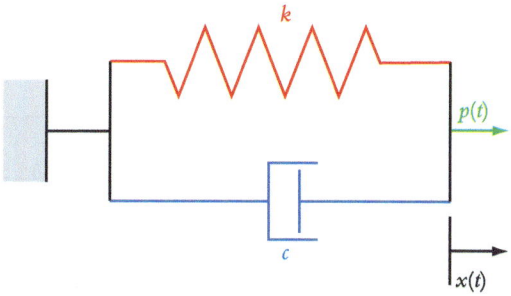

FIGURE 5.25 Spring–damper system.

TABLE 5.1
Laplace Transforms of Some Commonly Encountered Functions

Function $f(t)$ (Valid only for $t \geq 0$; Assumed to Be Zero for $t < 0$)	Graph	Laplace Transform $F(s)$
1 Unit step function (Heaviside function) $H(t)$: $f(t) = 0$ for $t < 0$ $f(t) = 1$ for $t > 0$ (Note: definitions vary for the value of $f(t)$ for $t = 0$)	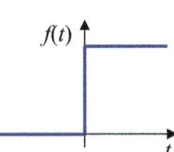	$\dfrac{1}{s}$
2 Unit ramp function $tH(t)$: $f(t) = 0$ for $t \leq 0$ $f(t) = t$ for $t > 0$	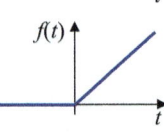	$\dfrac{1}{s^2}$
3 Unit impulse function (Dirac delta function) $\delta(t)$: $f(t) = 0$ for $t < 0$ $f(t) = \infty$ for $t = 0$ $f(t) = 0$ for $t > 0$ (area enclosed is unity)	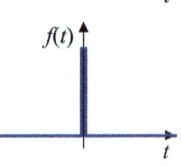	1
4 e^{-at}	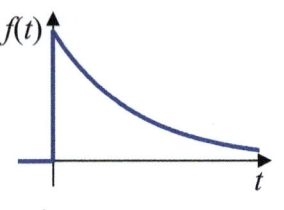	$\dfrac{1}{s+a}$
5 $1 - e^{-at}$	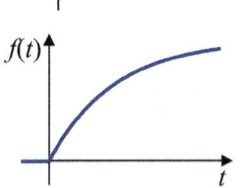	$\dfrac{a}{s(s+a)}$
6 $t - \dfrac{1}{a}\left(1 - e^{-at}\right)$	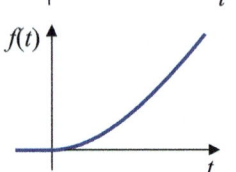	$\dfrac{a}{s^2(s+a)}$
7 $\sin(\omega t)$	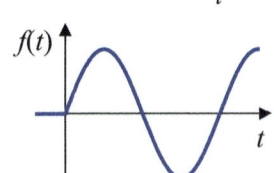	$\dfrac{\omega}{s^2 + \omega^2}$
8 $\cos(\omega t)$	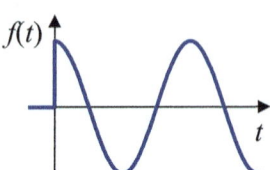	$\dfrac{s}{s^2 + \omega^2}$

(Continued)

TABLE 5.1   (CONTINUED)

Function $f(t)$ (Valid only for $t \geq 0$; Assumed to Be Zero for $t < 0$)	Graph	Laplace Transform $F(s)$
9 $\dfrac{\omega}{\sqrt{1-\zeta^2}} e^{-\zeta\omega t} \sin\left(\omega t\sqrt{1-\zeta^2}\right)$	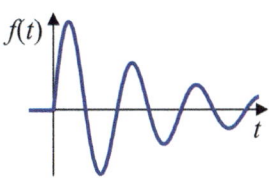	$\dfrac{\omega^2}{s^2 + 2\zeta\omega s + \omega^2}$
10 $1 - \dfrac{e^{-\zeta\omega t}}{\sqrt{1-\zeta^2}} \sin\left(\omega t\sqrt{1-\zeta^2} + \varphi\right)$ where $\cos\varphi = \zeta$	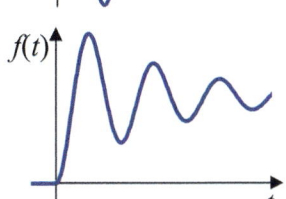	$\dfrac{\omega^2}{s\left(s^2 + 2\zeta\omega s + \omega^2\right)}$

It can be seen from the mathematical literature (and from item 1 in Table 5.1) that the Laplace transform of a Heaviside step function is given by $\mathscr{L}\{h(t)\} = 1/s$. Therefore, taking Laplace transforms of Equation (5.17) gives

$$\frac{q}{s} = kX(s) + csX(s) - cx(0)$$

$$= (k + cs)X(s) - cx(0) \tag{5.18}$$

where multiplication by s within the term $csX(s)$ represents differentiation and the term involving $x(0)$ is needed to take into account the initial conditions when $t = 0$ (as will be explained in Equation (5.24) and the text that follows it). Rearrangement gives

$$X(s) = \frac{q}{s(k+cs)} + \frac{cx(0)}{k+cs} \tag{5.19}$$

Using partial fractions, this can be expressed as

$$X(s) = \frac{q}{sk} - \frac{qc}{k(k+cs)} + \frac{cx(0)}{k+cs}$$

Inserting the initial condition that $x(0) = 0$, and rearranging, gives

$$X(s) = \frac{q}{k}\frac{1}{s} - \frac{q}{k}\frac{1}{(s+k/c)} \tag{5.20}$$

From the mathematical literature (and tabulated as item 1 in Table 5.1), it is seen that the inverse transform of $1/s$ is 1 for $t > 0$, and similarly, the inverse transform of $1/(s + a)$ is e^{-at} for $t > 0$; these items are of the same form as the terms in Equation (5.20). Assuming that $a = k/c$ and inserting these inverse transforms into Equation (5.20) gives the well-known solution to the problem of a spring–damper system under a step load:

$$x(t) = \frac{q}{k}.1 - \frac{q}{k}e^{-\frac{kt}{c}} = \frac{q}{k}\left(1 - e^{-\frac{kt}{c}}\right) \tag{5.21}$$

Instead of using partial fractions, it is often possible to make use of more complicated inverse transforms to get straight to the answer. For example, Equation (5.20) can be expressed as:

$$X(s) = \frac{q}{k}\left(\frac{k/c}{s\left(s + k/c\right)}\right) \tag{5.22}$$

Again, from the mathematical literature (and from item 5 in Table 5.1), it can be seen that the inverse transform of

$$\frac{a}{s(s+a)}$$

is

$$1 - e^{-at}$$

Assuming as before that $a = k/c$ and substituting this inverse transform into Equation (5.22) once again recovers the solution Equation (5.21), this time without any further working being required.

Having introduced the concept of Laplace transforms and illustrated their use via a simple example, it is now appropriate to state more formally the rules which they obey, which include the following:

1. Laplace transforms are **linear**, that is they obey the principle of superposition. For example:

$$\mathcal{L}\{af(t) + bg(t)\} = \mathcal{L}\{af(t)\} + \mathcal{L}\{bg(t)\}$$
$$= a\mathcal{L}\{f(t)\} + b\mathcal{L}\{g(t)\} = aF(s) + bG(s) \tag{5.23}$$

   In particular, it can be seen from this example that

   - The original functions $f(t)$ and $g(t)$ are multiplied by the constants a and b, respectively, and so are their transforms $F(s)$ and $G(s)$
   - The transform of the sum of $af(t)$ and $bg(t)$ is the sum of their transforms.

   Similarly, the transform of a difference, for example $af(t) - bg(t)$ would be $aF(s) - bG(s)$. However, the unwary student should note that this concept does not extend to the situation where functions of t are multiplied together: the transform of $f(t)g(t)$ is *not* the product of the transforms, as this does not involve superposition or linear combination.

2. The Laplace transform of a derivative is

$$\mathcal{L}\left\{\frac{df(t)}{dt}\right\} = s\mathcal{L}\{f(t)\} - f(0) \tag{5.24}$$

where $f(0)$ is the value of $f(t)$ at $t = 0$ in order to take into account the initial conditions. In turn, the transform of a second derivative is

$$\mathscr{L}\left\{\frac{d^2 f(t)}{dt^2}\right\} = s\mathscr{L}\left\{\frac{df(t)}{dt^2}\right\} - f'(0) = s^2\mathscr{L}\left\{f(t)\right\} - f'(0) - sf(0) \qquad (5.25)$$

where $f'(0)$ is the value of $\dfrac{df(t)}{dt}$ at $t = 0$.

3. The Laplace transform of an integral is

$$\mathscr{L}\left\{\int_0^t f(\tau)d\tau\right\} = \frac{1}{s}\mathscr{L}\left\{f(t)\right\} \qquad (5.26)$$

4. The Laplace transform of a time-shifted function (where the time axis has been changed by a constant value τ is

$$\text{If } \mathscr{L}\left\{f(t)\right\} = F(s) \text{ then } \mathscr{L}\left\{f(t - \tau)\right\} = e^{-s\tau}F(s) \qquad (5.27)$$

Items 2 and 3 illustrate more formally the all-important points that (in the Laplace domain and a block diagram) multiplication by s represents differentiation with respect to time and division by s represents integration.

In the worked example, use was made without proof of the two of the Laplace transforms available from the literature. In fact, the Laplace transforms of a large variety of functions are tabulated in the mathematical literature (e.g. Kreyszig, 2011, pp. 248–251; Healey, 1967) as well as the mainstream control engineering texts (e.g. Dorf and Bishop (2021), Bolton (1998)). For the purpose of the present unit, those tabulated in Table 5.1 may be considered to be the most useful.

LEARNING SUMMARY

At the end of this section, you should have learnt:

✓ The concept of the Laplace transform and its role in the solution of differential equations

✓ The basic rules relating to the application of Laplace transforms to problems involving integration, differentiation and time-shifting

✓ How to transform a range of expressions from the time domain into the s domain.

5.5 BLOCK DIAGRAMS AND THE s NOTATION: THE HEATER CONTROLLER AND TENSIONING SYSTEM EXAMPLES REVISITED

5.5.1 HEATER CONTROLLER

The real strength of the s notation, when applied to control engineering, comes when it is used in conjunction with block diagrams to model systems whose behaviour involves differentiation or integration with respect to time. Revisiting the simple example of the heated system modelled earlier, in Section 5.3.1, consider the heated equipment to have a mass of $m = 50$ kg and a specific heat capacity of $C_p = 500$ J/kg K. Now, it is assumed that the heating power $p(t)$ supplied and the rate of heat loss

$Ah\tau(t)$ are changing significantly with time and are no longer equal (so the system is no longer in thermal equilibrium) and the net rate of heat gain causes the temperature to increase:

$$p(t) - Ah\tau(t) = mC_p \frac{d\tau(t)}{dt} \qquad (5.28)$$

where the term mC_p may be regarded as the thermal capacity or thermal inertia of the equipment (Figure 5.26).

Assuming the initial conditions (i.e. assuming that $\tau(t)$ is initially zero, meaning it is at the ambient temperature of 0°C), and making use of Equation (5.24), the Laplace transform of this expression is

$$P(s) - AhT(s) = mC_p sT(s) \qquad (5.29)$$

This can be rearranged as

$$T(s) = \frac{P(s) - AhT(s)}{mC_p s} \qquad (5.30)$$

The block diagram (Figure 5.27) for this aspect of the system is now slightly more complex than before and shows the circular relationship inherent in Equation (5.29), in which $T(s)$ is defined via a function involving $T(s)$ itself.

This circular relationship can be eliminated from the mathematics by straightforward manipulation of Equation (5.30):

$$T(s) + \frac{AhT(s)}{mC_p s} = \frac{mC_p sT(s) + AhT(s)}{mC_p s} = \frac{mC_p s + Ah}{mC_p s} T(s) = \frac{P(s)}{mC_p s}$$

$$\Rightarrow T(s) = \frac{1}{mC_p s + Ah} P(s) \qquad (5.31)$$

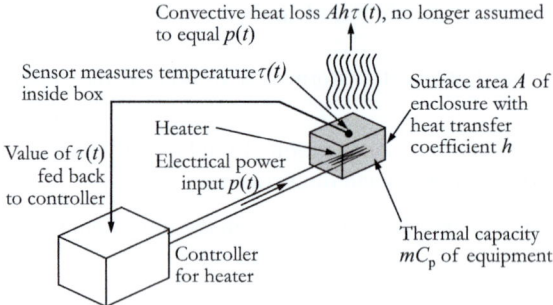

FIGURE 5.26 Heating system with thermal capacity and time-dependent temperature.

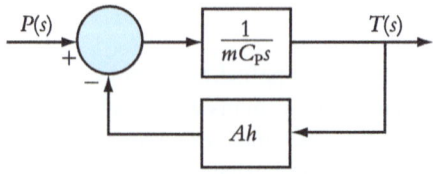

FIGURE 5.27 Block diagram relating to the heat balance of the heating system.

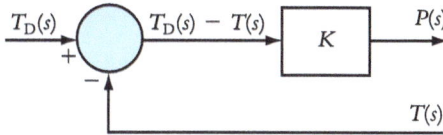

FIGURE 5.28 Block diagram relating to the controller.

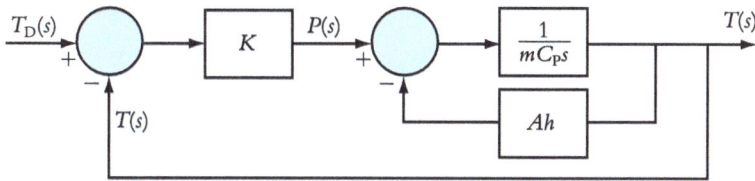

FIGURE 5.29 Complete block diagram of heater controller system.

We could use this to simplify the block diagram relating to this aspect of the problem. But for illustrative purposes, the loop in the block diagram will be left in place and instead be tackled within Section 5.9.

The controller still follows the same approach as in the simpler version of the problem, that is the power supplied to heat the equipment is proportional to the error, which is the difference between desired and achieved temperatures:

$$p(t) = K\left(\tau_D(t) - \tau(t)\right) \tag{5.32}$$

This expression is very straightforward to transform to the s domain:

$$P(s) = K\left(T_D(s) - T(s)\right) \tag{5.33}$$

and is represented by substantially the same block diagram as before (Figure 5.28). Compare this with Figure 5.12; it is now (correctly) expressed in terms of the complex variable s.

Assembly of the complete block diagram now results in a diagram which involves two nested loops (Figure 5.29).

This block diagram will be simplified to a straightforward transfer function (ratio of output to input in the s domain) in Section 5.9.

5.5.2 Fibre Tensioning System

Revisiting also the fibre tensioning system example, now consider the spool/brake drum assembly to have inertia J and the tension, velocity etc. now to have significant variations with time (Figure 5.30).

Also, consider that the fibre is being drawn through the system at a velocity $u(t)$ (which is equal to $r\dot{\theta}(t)$) and acceleration $du(t)/dt$, so that the angular acceleration $\ddot{\theta}(t)$ of the drum is

$$\ddot{\theta}(t) = \frac{1}{r}\frac{du(t)}{dt} \tag{5.34}$$

The analysis starts in almost the same manner as the static analysis of the system undertaken earlier. Equation (5.11) is still applicable:

$$f_B(t) = 2\left(f_P(t) - 2f_T(t)\right) \tag{5.35}$$

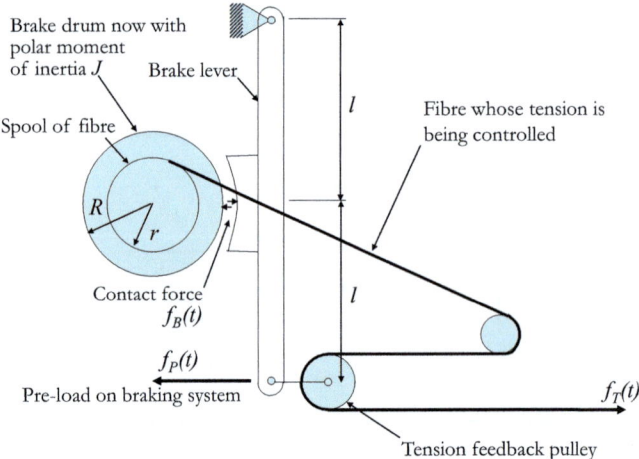

FIGURE 5.30   Fibre tensioning system, now with rotational inertia, still omitting frictional forces and reaction force at the pivot.

Equation (5.35) straightforwardly transforms to the s domain to give

$$F_B(s) = 2F_P(s) - 4F_T(s)$$
$$\text{or} : F_B(s) = 2\big(F_P(s) - 2F_T(s)\big) \qquad (5.36)$$

The block diagram for the lever is now unchanged from the version in Figure 5.18 other than being expressed in the s domain (Figure 5.31).

However, when we consider the drum, the moments of the tension force and the braking force are no longer in equilibrium and Newton's second law for rotation $L = J\ddot{\theta}$ must be applied. This is illustrated in Figure 5.32.

Applying Newton's second law for rotation introduces an additional term into Equation (5.12):

$$f_T(t)r - \mu f_B(t)R = J\ddot{\theta} = \frac{J}{r}\frac{du(t)}{dt} \qquad (5.37)$$

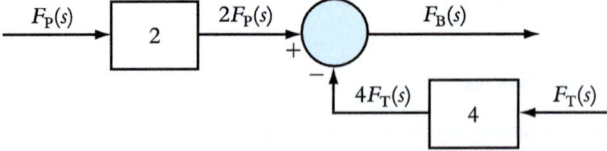

FIGURE 5.31   Block diagram for lever system.

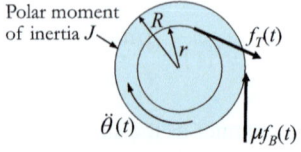

FIGURE 5.32   Brake drum showing angular acceleration $\ddot{\theta}$ due to an imbalance of torques (only forces with non-zero moments are shown).

Rearranging gives

$$f_T(t) = \frac{\mu R}{r} f_B(t) + \frac{J}{r}\frac{du(t)}{dt} \tag{5.38}$$

Making use of Equation (5.24), the Laplace transform of this expression (assuming that $u(t)$ and $du(t)/dt$ are both zero initially, i.e. at $t = 0$) is now

$$F_T(s) = \frac{\mu R}{r} F_B(s) + s\frac{J}{r}U(s) \tag{5.39}$$

The block diagram (Figure 5.33) for the brake/drum assembly is therefore somewhat more complex than before and now involves a summing junction.

By combining the block diagrams for the lever and for the brake and joining the arrows relating to $F_B(t)$ and $F_T(t)$, the overall block diagram for the system can now be assembled in Figure 5.34.

It will no longer be possible to express the transfer function purely as a ratio of output to input, as effectively there are now two inputs to the system (pre-load $f_P(t)$ and filament speed $u(t)$).

By way of introduction to the manipulation and simplification of block diagrams, note that proceeding along the main flow of the block diagram leads to the following (circular) expression:

$$\begin{aligned}F_T(s) &= \left(2F_P(s) - 4F_T(s)\right)\frac{\mu R}{r} + U(s)\frac{sJ}{r}\\ &= 2F_P(s)\frac{\mu R}{r} - 4F_T(s)\frac{\mu R}{r} + U(s)\frac{sJ}{r}\end{aligned} \tag{5.40}$$

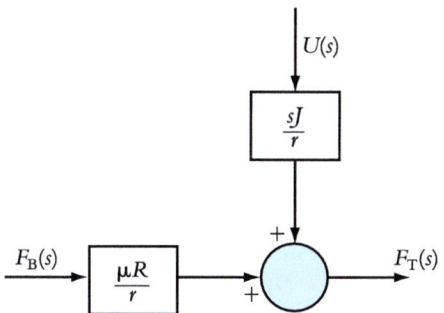

FIGURE 5.33 Block diagram of brake/drum assembly including inertia.

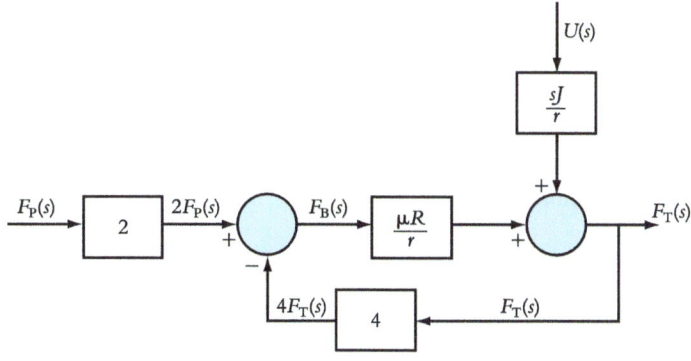

FIGURE 5.34 Block diagram of the tensioning system.

Rearranging to take $F_T(s)$ to the left-hand side of the equation gives

$$F_T(s) + 4F_T(s)\frac{\mu R}{r} = 2F_P(s)\frac{\mu R}{r} + U(s)\frac{sJ}{r} \qquad (5.41)$$

Multiplying throughout by r gives

$$F_T(s)(r + 4\mu R) = 2F_P(s)\mu R + U(s)sJ \qquad (5.42)$$

Finally, dividing throughout by $(r + 4\mu R)$ gives

$$F_T(s) = \frac{2\mu R}{(r + 4\mu R)}F_P(s) + \frac{Js}{(r + 4\mu R)}U(s) \qquad (5.43)$$

LEARNING SUMMARY

By the end of this section, you should have learnt:

✓ The use of block diagrams involving the s domain
✓ The construction of the block diagram of a system involving dynamic behaviour such as mechanical or thermal inertia

5.6 WORKING WITH TRANSFER FUNCTIONS AND THE s DOMAIN

5.6.1 OPEN- AND CLOSED-LOOP TRANSFER FUNCTIONS

It has already been mentioned in passing that a transfer function gives the relationship between the output and input. In general, the transfer function of a simple system (e.g. that given in Figure 5.35) is expressed as a function of s giving the Laplace transform $Y(s)$ of the output $y(t)$ of a system divided by the Laplace transform $X(s)$ of the input $x(t)$, for example:

$$G(s) = \frac{Y(s)}{X(s)} \qquad (5.44)$$

While this is straightforward for a simple system having a block diagram consisting of a single box, practical control systems are more complex and, like the example of the tensioning system, nearly always involve some form of (usually negative) feedback. Such systems typically have a block diagram of the form shown in Figure 5.36.

In such a system:

- $G(s)$ typically represents the transfer function of the item of the equipment being controlled, along with the transfer function of the controller. $G(s)$ is sometimes known as the forward path transfer function.

FIGURE 5.35 A block diagram representing a transfer function $G(s)$.

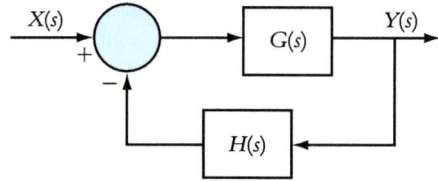

FIGURE 5.36 A typical feedback loop.

- $H(s)$ typically represents the transfer function of the sensor or transducer being used to measure the output of the system and compare it with the input or demand signal. $H(s)$ is sometimes known as the feedback path transfer function.

Two versions of the transfer function for this system can be defined:

- Open-loop transfer function. There are two conventions for this, which can be confusing for the unwary student. For the typical system shown in Figure 5.36, the open-loop transfer function is sometimes defined simply as $G(s)$, the relationship between output $Y(s)$ and input $X(s)$ *for the condition that the feedback does not actually take place* (Figure 5.37). However, the open-loop transfer function is also sometimes defined as the relationship between the input and the value which would be subtracted from the input, again for the condition that no feedback actually takes place. In other words, it is sometimes defined as $G(s)H(s)$, the product of the transfer functions of the two boxes through which the signal passes on its way back to the differencing junction:

    The latter version of the open-loop transfer function is of particular interest in predicting the stability of a control system from experimental measurements without actually causing the control system to go unstable, and it is sometimes used for deriving the 'loop gain' of the system.
- Closed-loop transfer function. This gives the relationship between the output of the closed-loop control system (Figure 5.36) and its input. The closed-loop transfer function can be derived by noting that:

$$Y(s) = G(s)\big[X(s) - H(s)Y(s)\big]$$ (5.45)

Rearranging gives

$$Y(s) + G(s)H(s)Y(s) = Y(s)\big[1 + G(s)H(s)\big] = G(s)X(s)$$ (5.46)

and hence

$$\frac{Y(s)}{X(s)} = \frac{G(s)}{1 + G(s)H(s)}$$ (5.47)

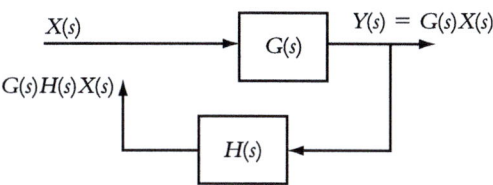

FIGURE 5.37 A control system operating in open loop mode.

which is the closed-loop transfer function of the system, the ratio of output $Y(s)$ to input $X(s)$ in the s domain. The closed-loop transfer function is of particular interest in terms of understanding the error of the system (how the actual output from the system compares with the desired output).

In order to determine the behaviour of the output of the system, two further mathematical tools are needed: the **initial-value theorem** and the **final-value theorem**.

5.6.2   INITIAL-VALUE THEOREM

It is convenient to assume within classical control modelling that the initial state of a given control system is known. In other words, the starting point for future behaviour is unambiguously defined. It can usually be assumed that everything is initially at rest and in its 'neutral' position (i.e. all values defining the state are zero). It can be useful to check that the transfer function, in conjunction with the inputs to the system, correctly calculates this state. This theorem may be stated as:

$$\lim_{t \to 0^+} f(t) = \lim_{s \to \infty} sF(s) \tag{5.48}$$

Note that this finds the value of $f(t)$ *immediately after* time $t = 0$ – in other words, an infinitesimal amount of time is allowed for the analysis to begin and for inputs to be 'switched on' (via the unit step and unit impulse functions given in items 1 and 3 of Table 5.1, for example).

As an example, consider the spring–damper system described earlier, consisting of a spring of stiffness k and a damper of constant c in parallel, subjected to a load which is zero for $t < 0$ and of value q for $t > 0$. Mathematically, this load can be expressed in terms of the Heaviside step function $H(t)$:

$$p(t) = qH(t) \tag{5.49}$$

The Laplace transform of the Heaviside step function is given in row 1 of Table 5.1, so that Equation (5.49) transforms to

$$P(s) = \frac{q}{s} \tag{5.50}$$

From earlier work, it was shown (Equation (5.25)) that the displacement of the spring–damper system under this load of q applied at $t = 0$ can be expressed as

$$X(s) = \frac{q}{sk} - \frac{q}{k\left(\dfrac{k}{c} + s\right)} \tag{5.51}$$

Inserting this expression into the initial-value theorem (Equation (5.48)) gives

$$\lim_{t \to 0^+} x(t) = \lim_{s \to \infty} sX(s) = \lim_{s \to \infty} s\left(\frac{q}{sk} - \frac{q}{k\left(\frac{k}{c} + s\right)}\right) = \lim_{s \to \infty}\left(\frac{sq}{sk} - \frac{sq}{k\left(\frac{k}{c} + s\right)}\right)$$

$$= \lim_{s \to \infty}\left(\frac{q}{k} - \frac{q}{\dfrac{k^2}{sc} + k}\right) = \frac{q}{k} - \frac{q}{\dfrac{k^2}{\infty \times c} + k} = \frac{q}{k} - \frac{q}{k} = 0 \tag{5.52}$$

Note that the very large terms involving s swamp the smaller terms such as k/c, so that the latter may be neglected; the zero initial displacement of the system is correctly recovered.

Now consider a simplified version of this system, which omits the damper. In this case, the equation describing the spring is very simple:

$$kx(t) = p(t); x(0) = 0 \tag{5.53}$$

Dividing by k, taking Laplace transforms and noting the zero initial conditions gives

$$X(s) = \frac{1}{k} P(s) \tag{5.54}$$

Once again the load is represented as a Heaviside step function (item 1 in Table 5.1):

$$p(t) = qH(t) \tag{5.55}$$

so that

$$P(s) = \frac{q}{s} \tag{5.56}$$

Combining these expressions gives the expression for displacement under the applied load:

$$X(s) = \frac{1}{k}\frac{q}{s} = \frac{q}{sk} \tag{5.57}$$

Inserting this into the initial-value theorem correctly recovers the displacement of the spring immediately after the application of the load:

$$\lim_{t \to 0^+} x(t) = \lim_{s \to \infty} sX(s) = \lim_{s \to \infty} s\left(\frac{q}{sk}\right) = \frac{q}{k} \tag{5.58}$$

It could usefully be argued that this example is unrealistic as the displacement has occurred instantaneously on the application of the load; in practice, of course, there would be some inertia, but that is not modelled here.

5.6.3 FINAL-VALUE THEOREM

Although the initial-value theorem provides a useful check that initial conditions are satisfied, a potentially more useful tool is the **final-value theorem**, which is a method of calculating the steady-state value of the response of a system. The theorem may be stated as

$$\lim_{t \to \infty} f(t) = \lim_{s \to 0} sF(s) \tag{5.59}$$

Note that this is only applicable where the final value is constant and finite.

A straightforward example is the calculation of the displacement of the spring–damper system after it has had a chance to settle down. Recall that

$$X(s) = \frac{q}{sk} - \frac{q}{k\left(\dfrac{k}{c} + s\right)} \tag{5.60}$$

Inserting this expression into the final-value theorem correctly recovers the steady-state displacement of the system:

$$\lim_{t \to \infty} x(t) = \lim_{s \to 0} sX(s) = \lim_{s \to 0} s\left(\frac{q}{sk} - \frac{q}{k\left(\frac{k}{c}+s\right)} \right) = \frac{q}{k} - 0 \times \frac{q}{k\left(\frac{k}{c}+0\right)} = \frac{q}{k} \tag{5.61}$$

A more sophisticated example will be given in Section 5.13.

LEARNING SUMMARY

At the end of this section, you should have learnt:

✓ To simplify a block diagram containing a feedback loop to give a simplified block diagram involving a single transfer function

✓ To find the values of an expression in the s domain at times $t = 0$ and $t = \infty$.

5.7   BUILDING A BLOCK DIAGRAM: PART 1

The strength of the techniques of control modelling lies in the fact that models of complete systems can be assembled from the building blocks consisting of the models of individual components. For each component, it is necessary to understand how the output and input are related (in the time domain). This relationship, often in the form of a differential equation, is then transformed into the s domain and then rearranged to give the transfer function of that component. Some typical examples of components are as follows:

5.7.1   MASS

The transfer function giving the displacement of a mass in terms of a force applied to it is obtained by transforming Newton's second law:

$$f(t) = ma(t) = m\frac{d^2x(t)}{dt^2} \tag{5.62}$$

Using Equation (5.25) and taking Laplace transforms gives

$$F(s) = ms^2X(s) - mx'(0) - msx(0) \tag{5.63}$$

where $x'(0)$ is the value of dx/dt at $t = 0$. If zero initial conditions are assumed (i.e. displacement and velocity are zero up to the instant when $t = 0$), this simplifies to

$$F(s) = ms^2X(s) \tag{5.64}$$

This may be represented in block diagram form as shown in Figure 5.38.

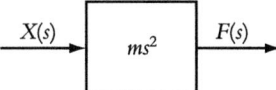

FIGURE 5.38   Block diagram for a mass.

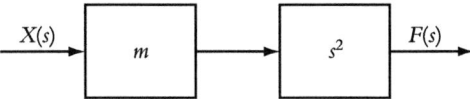

FIGURE 5.39   Alternative block diagram for a mass.

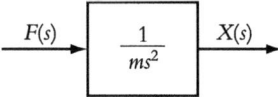

FIGURE 5.40   Another block diagram for a mass.

In this context (i.e. for this form of the block diagram), the transfer function of the mass is therefore the force divided by the displacement (where both are expressed as a function of s):

$$\frac{F(s)}{X(s)} = ms^2 \tag{5.65}$$

Sometimes, it is more convenient to consider the conversion to acceleration and the effect of the mass as separate 'boxes' in the block diagram as shown in Figure 5.39.

Alternatively, Equation (5.65) may be rearranged to give the displacement caused by a particular force:

$$X(s) = \frac{1}{ms^2} F(s) \tag{5.66}$$

The resulting block diagram is shown in Figure 5.40.

In this context, the transfer function is now:

$$\frac{X(s)}{F(s)} = \frac{1}{ms^2} \tag{5.67}$$

5.7.2   SPRING

The mathematical model of a spring is a very straightforward statement of proportionality (Hooke's law):

$$f(t) = kx(t) \tag{5.68}$$

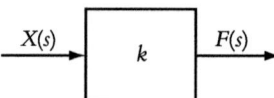

FIGURE 5.41   Block diagram for a spring.

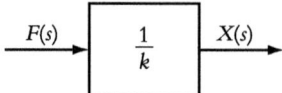

FIGURE 5.42   Alternative block diagram for a spring.

Taking Laplace transforms gives an equally simple result:

$$F(s) = kX(s) \tag{5.69}$$

so that the block diagram of a spring giving the displacement in terms of the force can be drawn (Figure 5.41).

The transfer function is

$$\frac{F(s)}{X(s)} = k \tag{5.70}$$

Conversely, the displacement may be expressed in terms of the force by simple rearrangement:

$$X(s) = \frac{1}{k}F(s) \tag{5.71}$$

so that the transfer function is now

$$\frac{X(s)}{F(s)} = \frac{1}{k} \tag{5.72}$$

The resulting block diagram is shown in Figure 5.42.

5.7.3   DASHPOT OR DAMPER

A dashpot or damper is a component that gives a reaction force proportional to the velocity of one end relative to the other. It is described mathematically as follows:

$$f(t) = cv(t) = c\frac{dx(t)}{dt} \tag{5.73}$$

where c is known as the viscous damping coefficient. Taking Laplace transforms,

$$F(s) = csX(s) - cx(0) \tag{5.74}$$

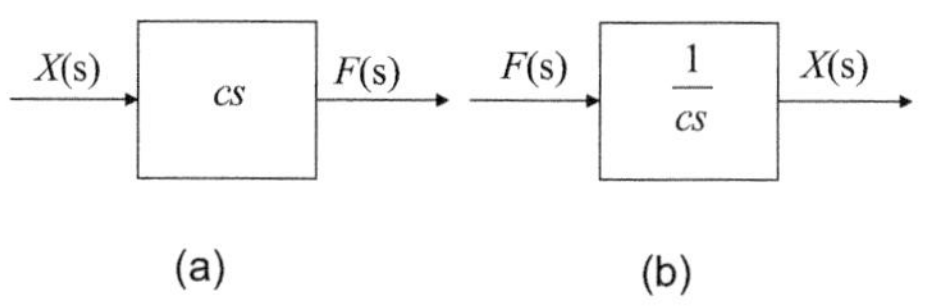

(a)   (b)

FIGURE 5.43 Alternative block diagrams for a damper.

Assuming zero initial conditions, and rearranging, gives the transfer function

$$\frac{F(s)}{X(s)} = cs \tag{5.75}$$

or alternatively

$$\frac{X(s)}{F(s)} = \frac{1}{cs} \tag{5.76}$$

The block diagrams for these two alternative representations are shown in Figure 5.43(a) and (b):

5.7.4 FLYWHEEL OR ROTATIONAL INERTIA

Just as Newton's law for linear motion relates force to acceleration via the equation $F = ma$, the torque l applied to a flywheel is related to that flywheel's angular acceleration $\alpha \left(= \ddot{\theta} = \dot{\omega} \right)$ by the following equation, which also involves the moment of inertia J, the rotational equivalent of mass:

$$l(t) = J\alpha(t) = J\ddot{\theta} = J\frac{d^2\theta(t)}{dt^2} \tag{5.77}$$

(Note that $l(t)$ is used for torque rather than $L(t)$ to be consistent with our notation that lowercase italic letters represent variables in the time domain, while uppercase italic letters represent variables transformed to the s domain.)

Taking Laplace transforms by applying Equation (5.16) to Equation (5.75) gives the following equation in the s domain:

$$L(s) = Js^2\Theta(s) - J\theta'(0) - Js\theta(0) \tag{5.78}$$

where $\Theta(s) = \mathcal{L}(\theta(t))$, that is the Laplace transform of the angular position $\theta(t)$. Setting initial conditions (angular position $\theta(0)$ and angular velocity $\theta'(0)$) to zero, and dividing by $\Theta(s)$ to rearrange the equation, gives the transfer function:

$$\frac{L(s)}{\Theta(s)} = Js^2 \tag{5.79}$$

This can be represented via the block diagram shown in Figure 5.44.

Alternatively, the same concept may be expressed in terms of angular velocity $\omega(t)$:

$$l(t) = J\alpha(t) = J\dot{\omega} = J\frac{d\omega(t)}{dt} \tag{5.80}$$

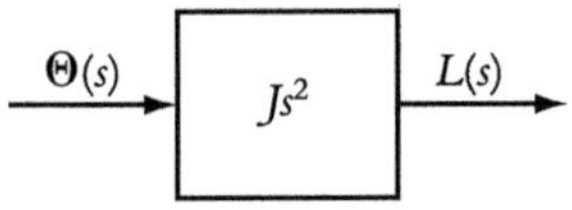

FIGURE 5.44   Block diagram for a rotational inertia load.

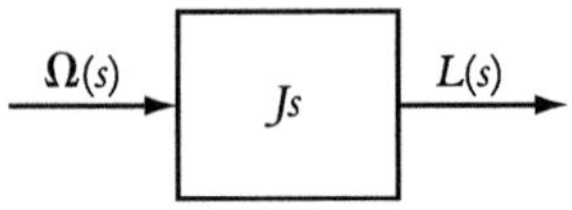

FIGURE 5.45   Alternative block diagram for a rotational inertia load.

Applying Equation (5.24) gives:

$$L(s) = Js\Omega(s) - J\omega(0) \tag{5.81}$$

Again assuming initial conditions, this gives:

$$\frac{L(s)}{\Omega(s)} = Js \tag{5.82}$$

where $\Omega(s)$ is the Laplace transform of angular velocity $\omega(t)$. The block diagram is shown in Figure 5.45.

5.7.5   ROTATIONAL LOAD WITH VISCOUS CHARACTERISTICS

This is the rotational equivalent of a dashpot, and again may be represented in terms of the angular position or angular velocity:

$$l(t) = b\omega(t) = b\frac{d\theta(t)}{dt} \tag{5.83}$$

where b is a rotational viscous damping coefficient relating viscous torque to angular velocity, analogous to the viscous damping coefficient c in Equations (5.15) and (5.73) for the linear motion situation. This gives the transfer function

$$\frac{L(s)}{\Omega(s)} = b \tag{5.84}$$

or

$$\frac{L(s)}{\Theta(s)} = bs \tag{5.85}$$

5.7.6   DC PERMANENT MAGNET MOTOR

The theory underlying DC permanent magnet motors is given in detail in Unit 4 Section 4.9. For steady-state purposes, the characteristics of a motor are very simple: torque $l(t)$ is proportional to current $i(t)$ supplied:

$$l(t) = K_{\mathrm{m}}i(t) \tag{5.86}$$

giving the transfer function

$$\frac{L(s)}{I(s)} = K_{\mathrm{m}} \tag{5.87}$$

It should be noted that a constant magnetic field is assumed within the motor (appropriate here, as the motor is of the permanent magnet type), and hence the constant K_{m} is identical in meaning to the constant k used within Unit 4, though the notation K_{m} is used here to avoid confusion with the spring constant k used elsewhere in the present unit. As a first approximation (and especially under no-load conditions), the angular velocity $\omega(t)$ is proportional to the motor supply voltage $v(t)$:

$$\omega(t) = \frac{v(t)}{K_{\mathrm{m}}} \tag{5.88}$$

Note that, for an ideal motor, the same constant K_{m} appears in both Equations (5.85) and (5.88). Equation (5.88) leads to the transfer function

$$\frac{\Omega(s)}{V(s)} = \frac{1}{K_{\mathrm{m}}} \tag{5.89}$$

A more sophisticated model of the motor takes account of the fact that the speed under load conditions depends on both the supply voltage $v(t)$ and the torque $l(t)$ which the motor is providing:

$$\omega(t) = \frac{v(t)}{K_{\mathrm{m}}} - \frac{l(t)R}{K_{\mathrm{m}}^2} \tag{5.90}$$

where R is the armature resistance of the motor. Note that, strictly, the torque in this expression includes any component of torque used for the acceleration of the motor armature as well as any provided as the useful mechanical output from the motor.

It is usual to plot torque vs speed rather than vice versa, so Equation (5.90) may be represented as a set of torque vs speed characteristics for a range of supply voltages (Figure 5.46).

The Laplace transform of Equation (5.90) (for zero initial conditions) is simply

$$\Omega(s) = \frac{V(s)}{K_{\mathrm{m}}} - \frac{L(s)R}{K_{\mathrm{m}}^2} \tag{5.91}$$

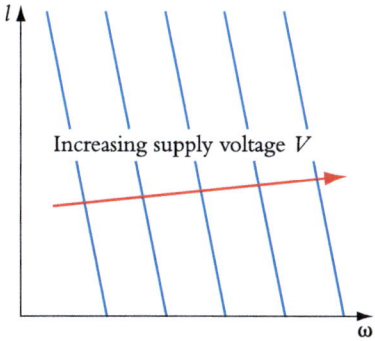

FIGURE 5.46   Typical set of torque-speed characteristics for a DC motor.

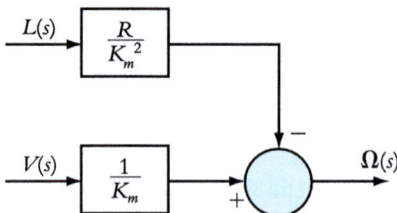

FIGURE 5.47   Block diagram for a DC motor.

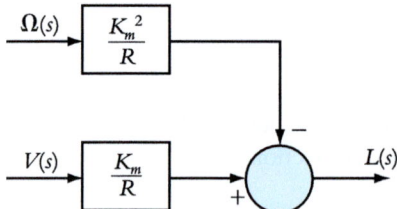

FIGURE 5.48   Alternative block diagram for a DC motor.

This situation can be represented via a block diagram (Figure 5.47).

Alternatively, the torque available for driving and/or accelerating the load (and the armature) can be represented in terms of the speed:

$$L(s) = V(s)\frac{K_\mathrm{m}}{R} - \Omega(s)\frac{K_\mathrm{m}^2}{R} \qquad\qquad (5.92)$$

so that the block diagram can now be drawn as in Figure 5.48.

5.7.7   AMPLIFIERS AND OTHER COMPONENTS WITH GAIN

The situation is frequently encountered in control engineering where a signal is amplified (made larger) using some form of amplifier. A familiar example is the amplifier used for an electric guitar, which takes a very small signal from the electromagnetic pickups adjacent to the strings and makes it much larger in order to drive the loudspeakers so that the music will be properly audible. In simple terms, an amplifier multiplies a signal (e.g. a voltage) by a given constant K and therefore involves a transfer function which can be expressed as

$$\frac{V_{\mathrm{out}}(s)}{V_{\mathrm{in}}(s)} = K \qquad\qquad (5.93)$$

The concept of a gain or constant of proportionality can be extended to numerous other situations, including sensors with calibration constants. A good example is a tachogenerator, which is a speed-sensing device (less commonly used today) that produces a DC voltage output signal proportional to the angular velocity. Such a device will have a certain gain in volts per unit of angular velocity. Other examples are simple components such as resistors (which allow a current to flow in proportion to voltage). A simple proportional gain (where the error signal is multiplied by a constant to give the corrective action) will form one possible component of the control algorithm to be examined in Section 5.11.

5.7.8 Pumps and Tanks

These components are widely encountered within courses on control theory as they relate to quantities (flow rate, liquid level) that are easily visible. The behaviour of a positive-displacement pump is very straightforward as a first approximation: the volume flow rate of fluid $q(t)$ delivered is proportional to the angular velocity $\omega(t)$ with a constant of proportionality K_p, giving the transfer function

$$\frac{Q(s)}{\Omega(s)} = K_p \tag{5.94}$$

The behaviour of a tank requires an understanding of its differential equation:

$$q_{in}(t) - q_{out}(t) = A\frac{dh(t)}{dt} \tag{5.95}$$

where $h(t)$ is the liquid level, A is the cross-sectional area of the tank, and $q_{in}(t)$ and $q_{out}(t)$ are, respectively, the rates of inflow to and outflow from the tank (Figure 5.49).

This results in the following expression in the s domain:

$$Q_{in}(s) - Q_{out}(s) = AsH(s) - Ah(0) \tag{5.96}$$

which, for zero initial conditions, can be represented as

$$H(s) = \frac{1}{As}\left(Q_{in}(s) - Q_{out}(s)\right) \tag{5.97}$$

This can be represented in the form of a block diagram (Figure 5.50).

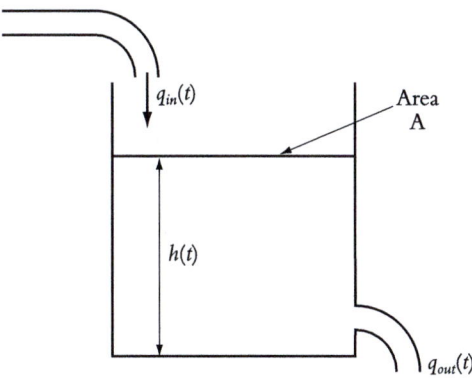

FIGURE 5.49 Water tank showing inflow $q_{in}(t)$ and outflow $q_{out}(t)$.

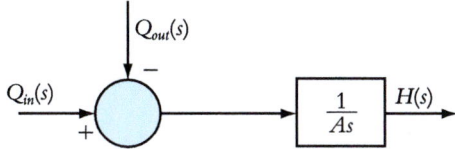

FIGURE 5.50 Block diagram of a tank.

5.7.9 FLOW RESTRICTIONS AND LINEARISATION

Components such as pipe bends, orifices, valves etc. are often encountered in control systems that involve fluid flow. It would be tempting to treat such flow restrictions as being analogous to resistors, until it is realised that, strictly speaking, they are not linear devices. For example, an orifice plate or valve has a pressure drop $\Delta p(t)$ across it that is approximately proportional to the *square* of the volume flow rate $q(t)$:

$$\Delta p(t) = C\big(q(t)\big)^2 \tag{5.98}$$

where C is a constant depending upon the nature of the flow restriction and the density of the fluid. It would appear superficially that it is impossible to represent such a component within the present (linear) framework. However, in practice, many control systems operate over a fairly narrow range of speeds, depths, flow rates etc., and it is possible to simplify the system's behaviour by considering *small perturbations* about a given operating point. This results in behaviour that can be assumed to be linear within a small range around the operating point. For the present (simplified) purposes, it is sufficient to assume (over-simplistically) that the flow rate $q(t)$ from a tank via a flow restriction is proportional to the height ('head') $h(t)$ of the liquid above the restriction:

$$q(t) = C_q h(t) \tag{5.99}$$

where C_q is the linearised flow constant of the restriction referred to the head of liquid. Taking Laplace transforms results in the following expression:

$$Q(s) = C_q H(s) \tag{5.100}$$

where $H(s) = \mathscr{L}(h(t))$

In general, linearisation is used to cope with all manner of situations where the true behaviour of a device is non-linear, but where small perturbations around an operating point are to be considered. Examples might include

- The relationship between liquid depth h (head) and flow rate q, described above (Figure 5.51(a)), leading to a more rigorous definition of the linearised flow constant C_q, though this would continue to be used in the simple linearised models given in Equations (5.99) and (5.100).
- The relationship between torque and speed of a centrifugal blower (Figure 5.51(b)).

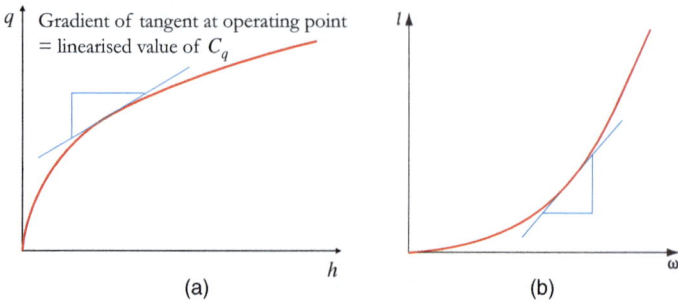

(a)　　　　　　　　　(b)

FIGURE 5.51 Linearisation of (a) flowrate–head characteristics for a tank and (b) torque–speed characteristics for a blower.

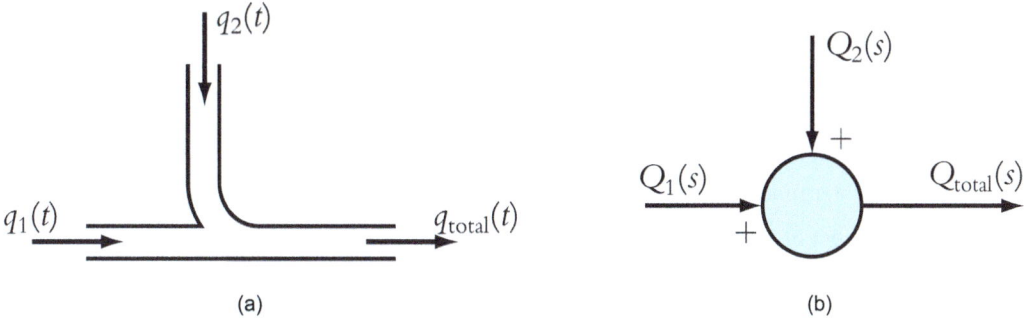

FIGURE 5.52 (a) Summation of two flows at a pipe junction and (b) a block diagram of a pipe junction represented as a summing junction.

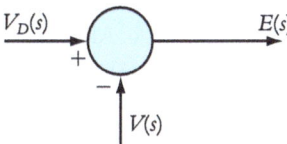

FIGURE 5.53 Typical differencing junction.

5.7.10 SUMMING AND DIFFERENCING JUNCTIONS

Within control systems, the effects of inputs and other influences are often combined via summing and differencing junctions; such junctions have already been used freely within the earlier examples. As another example, two flows may sum to form a larger flow (Figure 5.52(a)). The summing of the flows is represented via a summing junction in a block diagram (Figure 5.52(b)).

A very frequently encountered situation in control loops is the concept of negative feedback where the measured value of a signal is compared with the desired value, and the difference between them is known as the error. This has already been encountered in the examples given above. In the introductory example of the cruise control, the actual speed $v(t)$ may be compared with the desired or demand speed $v_\mathrm{D}(t)$, so that, in the s domain, the feedback quantity $V(s)$ is subtracted from the demand $V_\mathrm{D}(s)$ to obtain the error signal $E(s)$ (Figure 5.53).

5.7.11 TRANSFER FUNCTION OF A CONTROLLER

In order to correct the error consisting of the difference between the desired and actual values of the system response, some form of controller is used. The detailed behaviour (and transfer function) of typical systems will be explored in Section 5.11, but for the present time, it will be represented simply by the symbol $G_\mathrm{c}(s)$ (Figure 5.54).

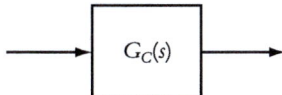

FIGURE 5.54 Block diagram of a controller.

5.8   BUILDING A BLOCK DIAGRAM: PART 2

By this stage, the reader should have sufficient information to be able to construct the transfer functions (and hence draw the block diagrams) for each of the components. The remaining challenge lies in correctly assembling the diagram for the whole system from its components. In constructing block diagrams, it is conventional to assume that all variable quantities or signals (such as displacement, speed etc.) are functions of s rather than time. Although each component may itself be trivial to model, the construction of the block diagram of a system may require careful thought, and it is not always obvious which way around a particular relationship (such as the relationship between torque and speed of a motor) should be applied.

Two examples of a block diagram construction have already been presented within Sections 5.3 and 5.5, but it is instructive to explore the assembly of the block diagram components from Section 5.7 via two further examples of varying complexities. A simple (and flawed) example is a positioning system (Figure 5.55) consisting of a motor with constant K_m driving a load of inertia J, with the motor being driven via a current amplifier that provides the motor with a current of K_a amperes per volt of input. The angular position is assumed to be sensed via a device that gives an output of K_s volts per (radian per second) of rotation of the motor. In practice, such a rotational position control system would typically be used to operate a linear motion control system, e.g. to position the table of a machine tool via a leadscrew, though for simplicity we will assume that we only require an accurate control of the angular position covering multiple rotations of the system. The input to the system is a demand voltage $v_D(t)$ which is proportional to the desired position. The voltage, v_P, corresponding to the actual position is subtracted from v_D to give an error voltage, which is amplified by the servo amplifier to give the input current i_m to the motor which will create a torque to rotate the load with the aim of correcting the positioning error:

$$i_m\left(t\right) = K_a\left(v_D\left(t\right) - v_P\left(t\right)\right)$$

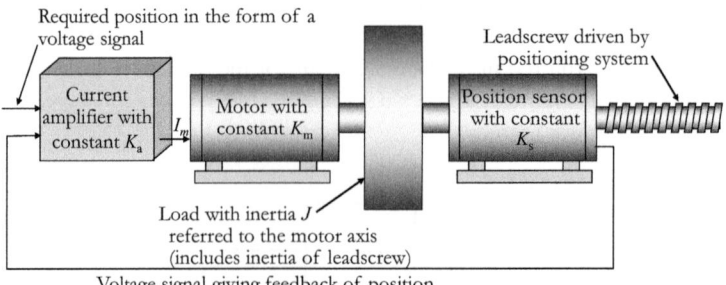

FIGURE 5.55   Oversimplistic positioning system.

Taking Laplace transforms and assuming that all voltages are initially zero:

$$I_m(s) = K_a \left(V_D(s) - V_P(s) \right)$$

This may be represented within a block diagram (Figure 5.56).

The motor will be assumed to be acting as a source of torque $l(t)$ dependent upon the current $i(t)$ by adapting Equations (5.86) and (5.87) to the present situation and rearranging so that

$$L(s) = K_m I_m(s) \tag{5.101}$$

The block diagram corresponding to this is straightforward and is shown in Figure 5.57.

The relationship between torque $l(t)$ and angular position $\theta(t)$ (and hence their transforms $L(s)$ and $\Theta(s)$) illustrates one of the pitfalls of constructing a block diagram from its components. Although the block diagram for an inertia load J is straightforward and was given earlier in Figure 5.44, it is necessary here to reverse the flow and invert the transfer function in order to obtain $\Theta(s)$ in terms of $L(s)$ (Figure 5.58).

The relationship between angular position $\theta(t)$ and the position-related voltage $v_p(t)$ is a straightforward linear one involving the sensor constant K_S (Figure 5.59).

For brevity, we will proceed straight to the complete block diagram for this system (Figure 5.60).

Interestingly, this would not form a very useful control system as its response would be both undamped and oscillatory (the control system is merely equivalent to an undamped rotational spring acting upon the inertia load, though in practice there would be some friction in the system which would introduce a certain degree of damping). This further illustrates the need for an effective

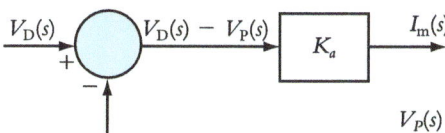

FIGURE 5.56   Block diagram of differencing junctions and amplifier for positioning system.

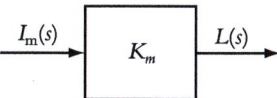

FIGURE 5.57   Block diagram of motor.

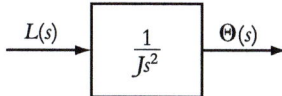

FIGURE 5.58   Rearranged block diagram of inertia.

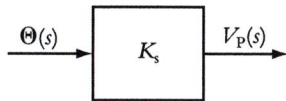

FIGURE 5.59   Block diagram of the sensor.

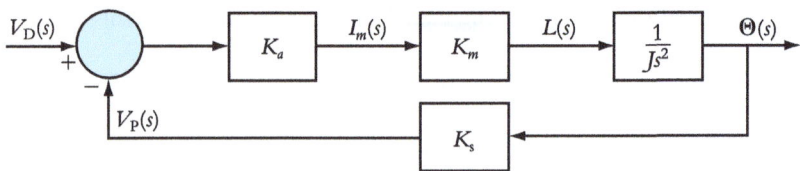

FIGURE 5.60 Block diagram for (over-simplistic) position control system.

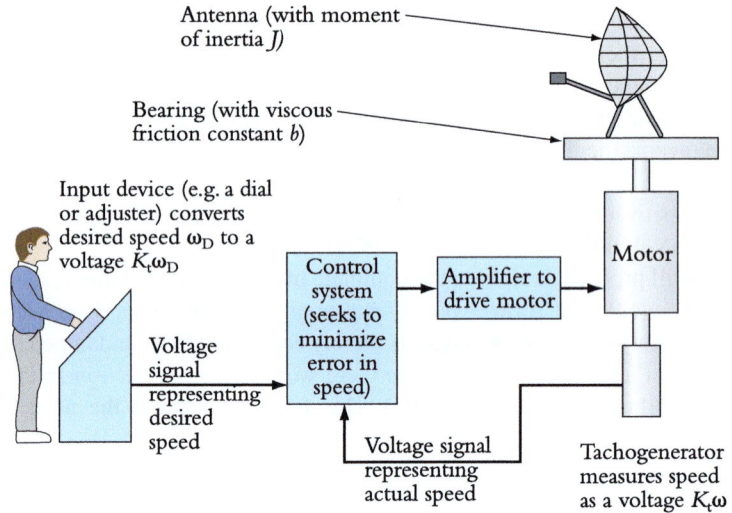

FIGURE 5.61 Radar antenna on viscous bearing, with tachometer control.

control algorithm and will be explored further in Sections 5.11 and 5.13 via the evaluation of more complex control algorithms applied to this example. A more realistic position control system will be described and modelled in Section 5.12.

A more complex example is shown in Figure 5.61. A radar antenna (or aerial) may be represented as viscous load with torque–speed constant b, that is in order to drive it at constant angular velocity ω, a constant torque l is required:

$$l = b\omega \tag{5.102}$$

It also has a moment of inertia J. It is driven by an electric motor with motor constant K_m and armature resistance R at a speed that must be closely controlled. The desired angular velocity is ω_D (which is shown for illustrative purposes in Figure 5.61 as being manually entered via an adjuster which creates a variable voltage proportional to ω_D), and the actual angular velocity ω is measured via a tachogenerator with a constant of K_t Vs/rad. The system is controlled via negative feedback with a controller with transfer function $G_\mathrm{c}(s)$ and via a voltage amplifier with gain K_a.

Consider first the antenna itself. In order to drive it at a given value of angular velocity ω, a constant torque l is required, as expressed in Equation (5.102). So, it is straightforward to treat the relationship between antenna torque and antenna speed in the time domain as

$$l(t) = b\omega(t) \tag{5.103}$$

After taking Laplace transforms this results in

$$L(s) = b\Omega(s) \tag{5.104}$$

which leads to the block diagram shown in Figure 5.62.

However, if torque l_m available from the motor for driving it exceeds torque l required to overcome viscous friction and maintain a steady speed, it will accelerate at the rate $(l_m - l)/J$, so that the equation of motion of the antenna is

$$J\ddot{\theta}(t) = J\frac{d\omega(t)}{dt} = l_m(t) - l(t) \tag{5.105}$$

This can be represented in the s domain as

$$Js\Omega(s) = L_m(s) - L(s) \tag{5.106}$$

which can be rearranged to give

$$\Omega(s) = \frac{1}{Js}\left(L_m(s) - L(s)\right) \tag{5.107}$$

The corresponding block diagram is therefore as shown in Figure 5.63

Inserting Equation (5.104) into Equation (5.107) gives

$$\Omega(s) = \frac{1}{Js}\left(L_m(s) - b\Omega(s)\right) \tag{5.108}$$

The corresponding block diagram (Figure 5.64) for the antenna, treated as a combined frictional and inertia load, can be created either directly from Equation (5.108) or by joining together the relevant flows from Figure 5.62 and Figure 5.63.

FIGURE 5.62 Block diagram of pure viscous load.

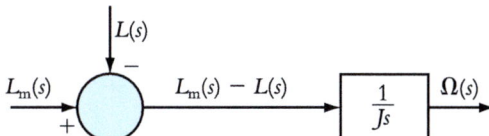

FIGURE 5.63 Block diagram of inertia load with external drag $L(s)$.

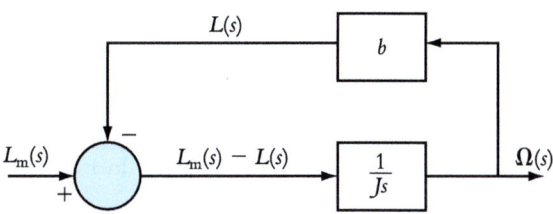

FIGURE 5.64 Block diagram of antenna treated as combined frictional and inertia load.

However, the antenna is driven by a motor that will provide a torque l_m that depends on the speed ω at which it is rotating:

$$l_m(t) = v(t)\frac{K_m}{R} - \omega(t)\frac{K_m^2}{R} \tag{5.109}$$

or in the s domain:

$$L_m(s) = V(s)\frac{K_m}{R} - \Omega(s)\frac{K_m^2}{R} \tag{5.110}$$

This is shown in block diagram form in Figure 5.65.

The block diagram for the antenna–motor assembly is therefore as shown in Figure 5.66

But recall that this system is driven via a voltage amplifier from the output of a controller, which acts upon the error signal. This error signal is in turn the difference between the required speed $\Omega_D(s)$ (converted into a voltage) and the voltage signal $K_t\Omega(s)$ from the tachogenerator. This control system driving the antenna and its motor may be represented as shown in Figure 5.67.

Inserting the block diagram for the antenna and motor results in the block diagram for the whole system (Figure 5.68). The behaviour of this system will be explored in Sections 5.13 and 5.14, though it is first necessary to convert the rather complex block diagram to a single transfer function.

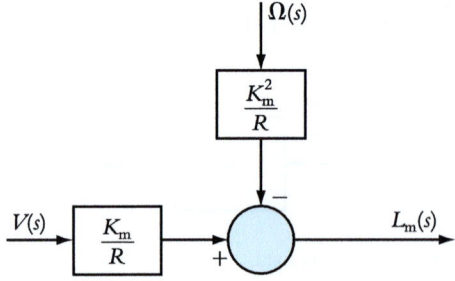

FIGURE 5.65   Block diagram for a motor.

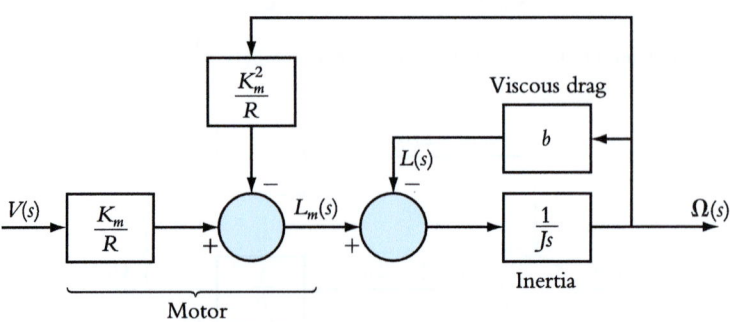

FIGURE 5.66   Block diagram for antenna–motor assembly.

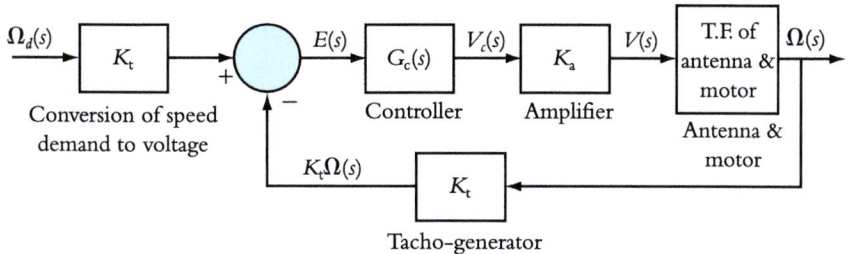

FIGURE 5.67   Block diagram for the control system.

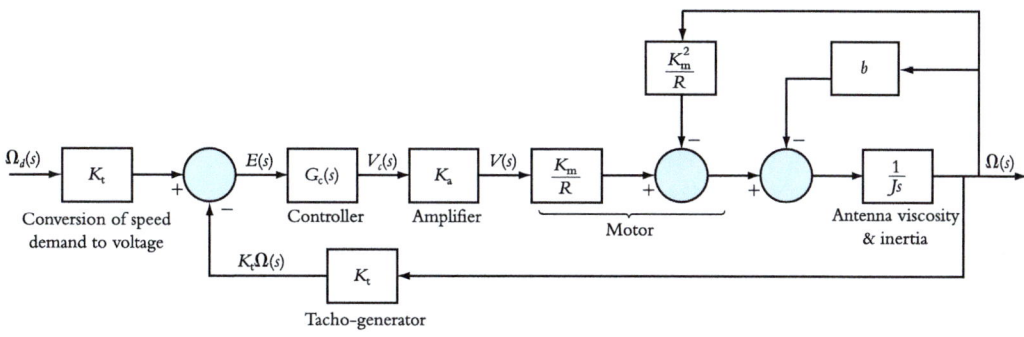

FIGURE 5.68   Block diagram for a complete antenna system.

LEARNING SUMMARY

By the end of this section, you should have learnt:

✓ To construct a block diagram for systems or subsystems based on the transfer functions of each of its components or features

✓ To assemble the block diagram for a complex system from the diagrams for its various subsystems

✓ To appreciate that distinguishing the input and output of particular processes is not always straightforward

---

5.9   CONVERSION OF THE BLOCK DIAGRAM TO THE TRANSFER FUNCTION OF THE SYSTEM

This process is broadly based on two rules:

1. The multiplication of the transfer functions of successive blocks in a sequence
2. The simplification of the block diagram shown in Figure 5.69(a) into a single transfer function shown in Figure 5.69(b), as demonstrated in Equations (5.45) to (5.47).

Essentially, this pair of actions is applied repeatedly to the feedback loops within a complex control system, starting with the innermost loop and working outwards.

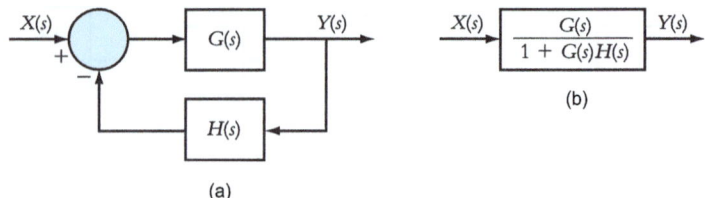

FIGURE 5.69 Block diagram of a typical control loop: (a) before and (b) after reduction to single transfer function.

Taking first the example of the block diagram for the controlled heat transfer system (Figure 5.29), the innermost loop can be tackled by replacing $G(s)$ with the transfer function $1/mC_\mathrm{p}s$ of the main flow and $H(s)$ with the transfer function Ah of the feedback flow. Substituting into Equation (5.47) gives

$$\frac{G(s)}{1+G(s)H(s)} = \frac{\dfrac{1}{mC_\mathrm{p}s}}{1+Ah \times \dfrac{1}{mC_\mathrm{p}s}} = \frac{1}{mC_\mathrm{p}s+Ah} \tag{5.111}$$

giving the same result as the more lengthy manipulation within Equation (5.31).

The overall block diagram then simplifies to Figure 5.70(a) and hence to Figure 5.70(b).

Now, the process of simplifying the structure of the block diagram is repeated. $G(s)$ is replaced with $K/(Ah+mC_\mathrm{p}s)$ and $H(s)$ with 1 since there is no transfer function associated with the feedback flow, which simply feeds the output directly back to the differencing junction. This manipulation results in the transfer function for the system:

$$\frac{T(s)}{T_\mathrm{D}(s)} = \frac{G(s)}{1+G(s)H(s)} = \frac{K/(mC_\mathrm{p}s+Ah)}{1+K/(mC_\mathrm{p}s+Ah)} = \frac{K}{mC_\mathrm{p}s+Ah+K} \tag{5.112}$$

The error (i.e. the difference between the desired and actual values of output) is

$$T_\mathrm{D}(s)-T(s)=T_\mathrm{D}(s)-T_\mathrm{D}(s)\frac{K}{mC_\mathrm{p}s+Ah+K}=T_\mathrm{D}(s)\frac{mC_\mathrm{p}s+Ah}{mC_\mathrm{p}s+Ah+K} \tag{5.113}$$

This expression will be used further in examining steady-state error in Section 5.13.

Dealing now with the radar antenna example, the innermost loop in Figure 5.68 can be considered by replacing $G(s)$ with the transfer function of the main flow $1/(Js)$ and $H(s)$ with the transfer function of the feedback flow b. The identity given in Equation (5.47) is then used to enable the innermost loop to be replaced with a single block containing the transfer function:

$$\frac{1/(Js)}{1+b/(Js)} = \frac{1}{Js+b} \tag{5.114}$$

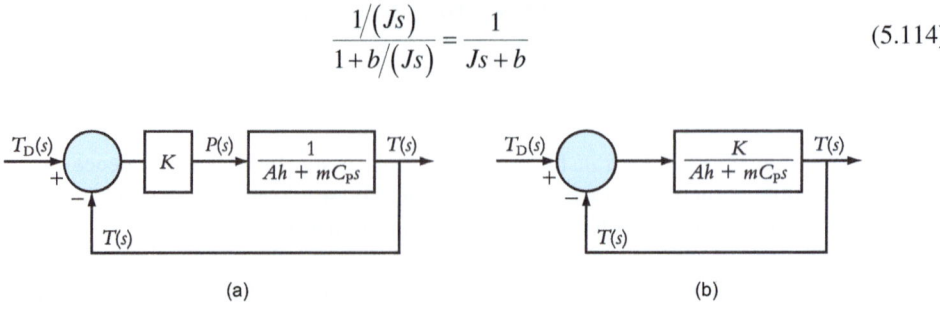

FIGURE 5.70 Simplified versions of the block diagram for a heating system (a) before and (b) after combining blocks by multiplication.

The resulting block diagram for the antenna and motor is now as shown in Figure 5.71.

Applying the same principle to the next loop involves replacing $G(s)$ with $1/(Js+b)$ and $H(s)$ with K_m^2/R, again giving a single block with the transfer function

$$\frac{1/(Js+b)}{1+K_m^2/\left[R(Js+b)\right]} = \frac{1}{Js+b+K_m^2/R} = \frac{R}{RJs+Rb+K_m^2} \quad (5.115)$$

giving a further simplified block diagram for the combination of antenna and motor (Figure 5.72).

The resulting block diagram for the whole system is now as shown in Figure 5.73.

The transfer functions of the four innermost blocks may be multiplied to give the a much simpler block diagram for the system (Figure 5.74).

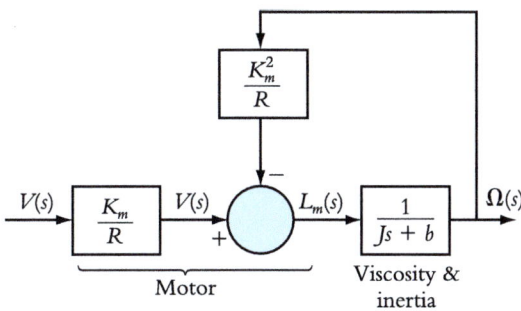

FIGURE 5.71 Block diagram for antenna and motor after simplification of the inner loop.

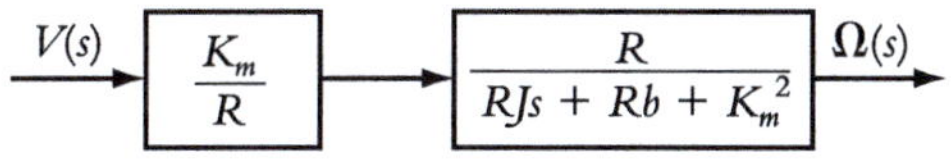

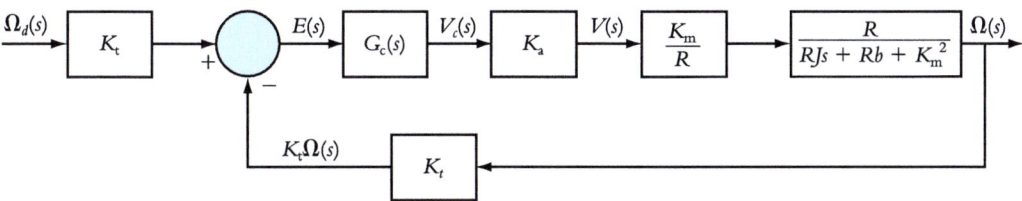

FIGURE 5.72 Block diagram for antenna and motor after simplification of the outer loop.

FIGURE 5.73 Block diagram for complete antenna system before combining blocks.

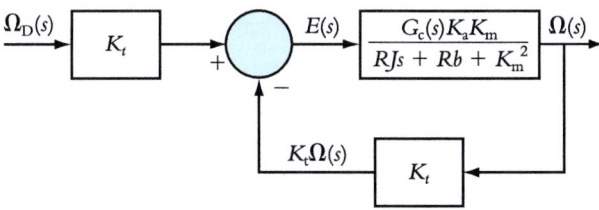

FIGURE 5.74 Simplified block diagram for a complete antenna system.

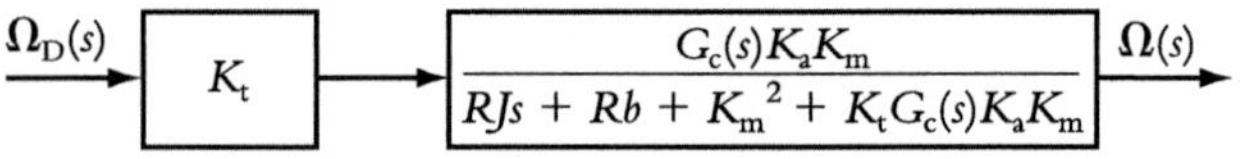

FIGURE 5.75 Elimination of loop for the complete antenna system.

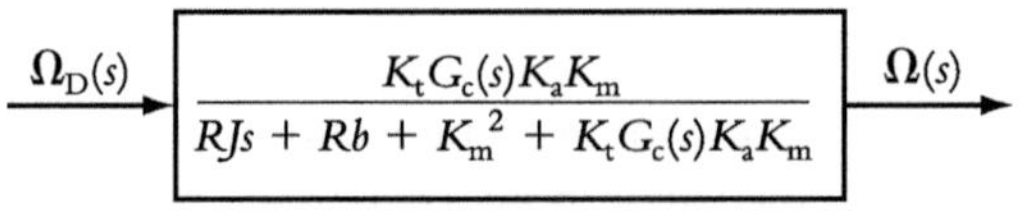

FIGURE 5.76 Final block diagram for the complete antenna system.

Substituting yet again for $G(s)$ and $H(s)$ within Equation (5.47) for the remaining loop allows us to replace that loop with the following transfer function (Figure 5.75):

$$\frac{G_c(s)K_aK_m}{RJs+Rb+K_m^2} \Bigg/ \left(1+\frac{K_tG_c(s)K_aK_m}{RJs+Rb+K_m^2}\right)=\frac{G_c(s)K_aK_m}{RJs+Rb+K_m^2+K_tG_c(s)K_aK_m} \tag{5.116}$$

Finally, multiplying by the transfer function of the leftmost block results in the transfer function for the whole system:

$$\frac{\Omega(s)}{\Omega_D(s)}=\frac{K_tG_c(s)K_aK_m}{RJs+Rb+K_m^2+K_tG_c(s)K_aK_m} \tag{5.117}$$

enabling the block diagram to be reduced to a single box (Figure 5.76).

An expression can therefore be written for the error, which is the difference between the desired value of angular velocity and the achieved value:

$$\Omega_D(s)-\Omega(s)=\left(1-\frac{K_tG_c(s)K_aK_m}{RJs+Rb+K_m^2+K_tG_c(s)K_aK_m}\right)\Omega_D(s)$$

$$=\left(\frac{RJs+Rb+K_m^2+K_tG_c(s)K_aK_m-K_tG_c(s)K_aK_m}{RJs+Rb+K_m^2+K_tG_c(s)K_aK_m}\right)\Omega_D(s)$$

$$=\left(\frac{RJs+Rb+K_m^2}{RJs+Rb+K_m^2+K_tG_c(s)K_aK_m}\right)\Omega_D(s) \tag{5.118}$$

This expression will be used in Section 5.13 when examining the steady-state response of the system.

LEARNING SUMMARY

By the end of this section, you should have learnt:

✓ The relationship between the transfer function of a closed-loop system and the transfer functions of its main and feedback paths

✓ The procedure for simplifying a block diagram with one or more loops in order to reduce it to a block diagram involving a single transfer function

5.10 HANDLING BLOCK DIAGRAMS WITH OVERLAPPING CONTROL LOOPS

A problem appears to arise when dealing with a block diagram including loops that are not simply nested one inside the other, but overlap in the manner shown in Figure 5.77.

This is handled via a simple trick. The point at which one of the feedback loops branches away from the main flow of the diagram is moved and the appropriate blocks are introduced into the feedback loop to compensate for the change in the topology of the diagram. For example, if a branch leading to a feedback loop is moved 'upstream' from the original branching point, the blocks that now lie outside the loop are now replicated within the loop itself so that the open-loop transfer function for that loop stays the same. In the example given above, the point at which the upper loop leaves the main flow is moved to coincide with the point at which the lower loop leaves. However, this would involve omitting the block containing $G_4(s)$ from that control path. That block is therefore reintroduced into the loop so that the overall transfer function for that loop stays the same (Figure 5.78).

An alternative approach to eliminating overlapping loops, involving altering the sequence of the differencing (or any summing) junctions is not described here, though is covered by the rules in standard texts such as Dorf and Bishop (2021).

It is now straightforward to reduce this block diagram to a single transfer function in the usual way. Replacing $G(s)$ in Equation (5.47) with $G_3(s)$ and $H(s)$ with $H_2(s)H_4(s)$, the inner loop involving G_3, H_2 and G_4 can be expressed as a single transfer function, so that the block diagram can be drawn as shown in Figure 5.79.

By a similar process, the remaining loop is eliminated (Figure 5.80).

The block diagram is now in the form of a straightforward sequence of blocks whose transfer functions can be multiplied together to give the transfer function of the overall system (Figure 5.81).

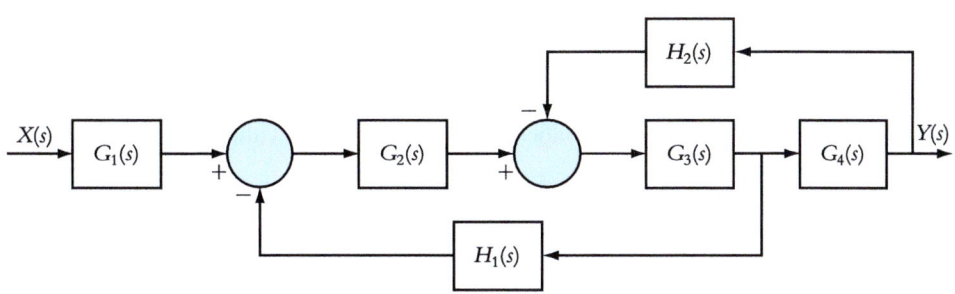

FIGURE 5.77   Block diagram with overlapping control loops.

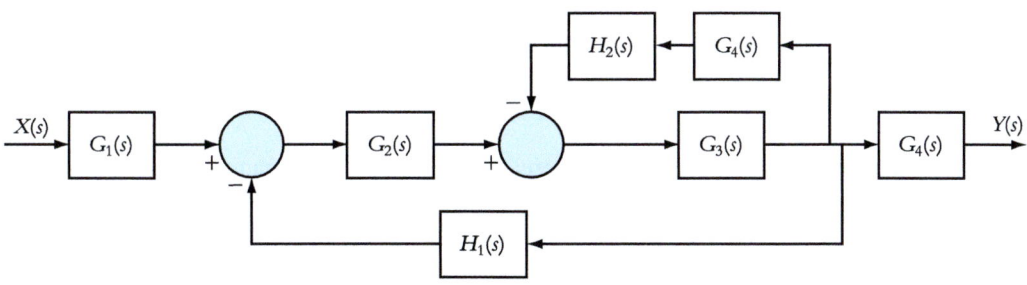

FIGURE 5.78   Block diagram rearranged to avoid overlapping of loops.

X(s)
G1(s)
+
−
G2(s)
G3(s) / 1 + G3(s)H2(s)G4(s)
G4(s)
Y(s)
H1(s)

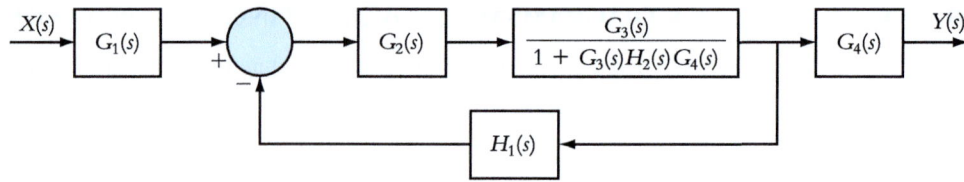

FIGURE 5.79 Block diagram redrawn to eliminate the inner loop.

X(s)
G1(s)
G2(s)G3(s) / 1 + G3(s)H2(s)G4(s) + G2(s)G3(s)H1(s)
G4(s)
Y(s)

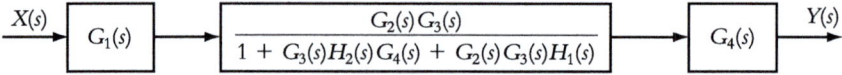

FIGURE 5.80 Block diagram redrawn to eliminate the outer loop.

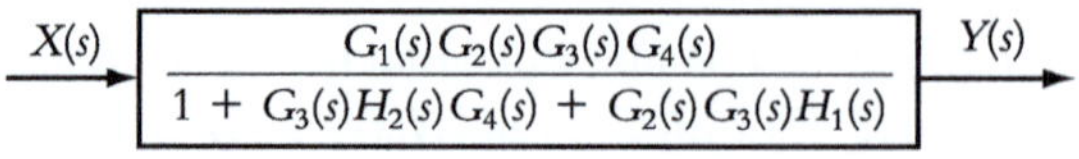

X(s)
G1(s)G2(s)G3(s)G4(s) / 1 + G3(s)H2(s)G4(s) + G2(s)G3(s)H1(s)
Y(s)

FIGURE 5.81 Block diagram redrawn in the form of a single transfer function.

5.10.1 Summary of Simplified Rules for Manipulating Block Diagrams

At this point, it is instructive to summarise the rules for manipulating simple block diagrams (specifically, those involving a single external input and no parallel forward paths) and converting them into transfer functions:

- Loops which overlap can be rearranged at the right-hand side and converted to nested (non-overlapping) loops by moving the outer branching point of the feedback loop upstream and copying into its feedback loop any blocks no longer contained within the loop, as shown in Figures 5.77 and 5.78.
- Loops which overlap can similarly be rearranged at the left-hand side by changing the sequence of any summing or differencing junctions, as described in (for example) Dorf and Bishop (2021).
- Successive blocks in series not incorporating any branching, summing or differencing are replaced with a single block containing the product of these transfer functions.
- Once a feedback loop has been reduced to a single forward transfer function $G(s)$, a single feedback transfer function $H(s)$ and a differencing junction, as shown in Figure 5.69(a), the loop is replaced with a single block containing the transfer function $G(s)/(1+G(s)H(s))$ as shown in Figure 5.69(b).
- The above rules are applied iteratively until the control system has been reduced to a single transfer function, which can then be characterised via the approaches described later in the present unit (Sections 5.13 and 5.14).

Block diagrams involving more than one external input require more manipulation and care and cannot of course be reduced to a transfer function involving a single multiplying term. For this situation, and the handling of forward blocks in parallel, reference should once again be made to more advanced texts, for example Dorf and Bishop (2021).

In Section 5.11, you will consolidate the knowledge of block diagrams by incorporating a typical control algorithm.

5.11   THE CONTROL ALGORITHM AND PROPORTIONAL-INTEGRAL-DERIVATIVE (PID) CONTROL

So far, only the simplest assumptions have been made regarding how the *error* (which may be colloquially expressed as 'the difference between what you want and what you get') is processed. The rule by which this error is processed in order to obtain the required corrective action is known as the *control algorithm* and is central to the behaviour of feedback and control systems. Ideally, the control system should:

- Take more corrective action if the error is larger
- Take more corrective action if the error has persisted for a long time
- Take more corrective action if the error is getting worse (and less if it is getting better)

A very widely used control algorithm is the **three-term** or **PID** controller, which formalises the three kinds of decisions defined verbally above. Specifically, it sets the corrective action to a value that depends on the sum of

- A multiple of the current value of the error
- A multiple of the integral of the error, that is the value of the error integrated or accumulated over time since the beginning of the process
- A multiple of the derivative of the error, that is its rate of change with time.

Mathematically, the control algorithm is defined as:

$$G_c(s) = K_c\left(1 + T_D s + \frac{1}{T_I s}\right)$$

$$= K_c + K_c T_D s + \frac{K_c}{T_I s} \tag{5.119}$$

where:

K_c is the gain of the controller

T_D is the derivative time constant (a larger time constant leads to more derivative action)

T_I is the integral time constant (a larger time constant leads to *less* integral action).

This is sometimes written alternatively as

$$G_c(s) = K_P + K_D s + \frac{K_I}{s} \tag{5.120}$$

where K_P, K_D and K_I are respectively the proportional, derivative and integral gains. In general, the terms within PID control have the following effects:

- Proportional control alone tends to result in **steady-state error**, which means that the system output never reaches the desired value, typically tending asymptotically to it
- Integral action in conjunction with proportional control tends to eliminate steady-state error, at the expense of tending to make the system's response oscillatory or even unstable
- Derivative action can in some circumstances damp down the tendency towards oscillatory behaviour without reintroducing steady-state error.

As an example, consider the positioning system described earlier. It is straightforward to show, from Figure 5.60, that the open-loop transfer of the position controller system is

$$\frac{V_P(s)}{V_D(s)} = \frac{K_a K_m K_s}{Js^2} \tag{5.121}$$

For illustrative purposes, the values $K_c = 10$, $K_m = 1$ A/V and $K_s = 1$ Vs/rad will be assumed. It can be demonstrated, either analytically by the application of inverse Laplace transforms to find the time domain solution or by numerical simulation (for instance, using the Simulink simulation package (http://www.mathworks.com/products/simulink), that the response of this system to a demand of 10 V (representing a request to move by 10 radians) results in an oscillatory response which continues forever at constant amplitude (Figure 5.82), which is clearly not a useful response. We have not yet considered what happens when there is some external torque trying to move the load away from its desired position.

Now let us replace the simple servo amplifier driving the motor with a controller/amplifier of gain K_c and integral and derivative time constants T_I and T_D, and additionally introduce a torque load $L_L(s)$ which is applied externally to the shaft carrying the flywheel etc. The resulting block diagram is shown in Figure 5.83.

The open loop transfer function can no longer be expressed as a simple ratio of output to input, but for completeness, the following expression can be derived:

$$V_P(s) = V_D(s)\frac{K_c K_m K_s T_D\left(s^2 + \frac{1}{T_D}s + \frac{1}{T_I T_D}\right) + sL_L(s)}{Js^3} \tag{5.122}$$

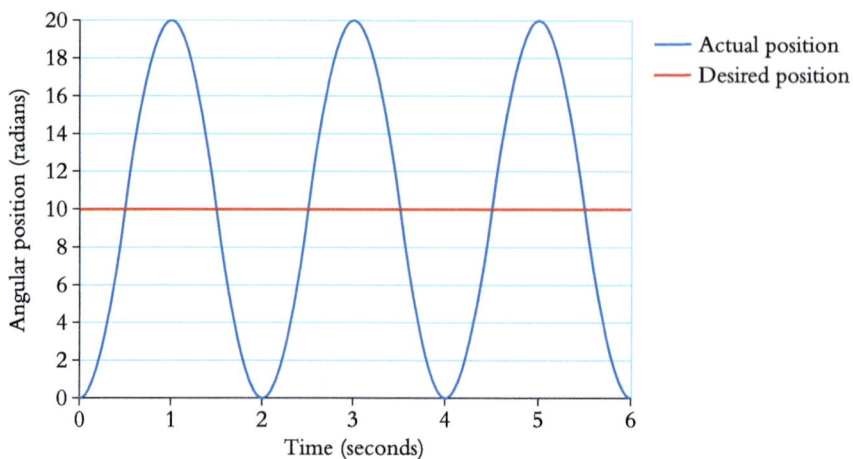

FIGURE 5.82　Response of positioning system under proportional control only.

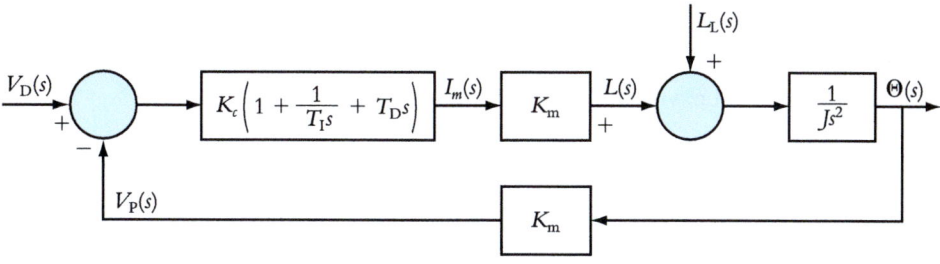

FIGURE 5.83 Block diagram of the positioning system with PID control and an external load.

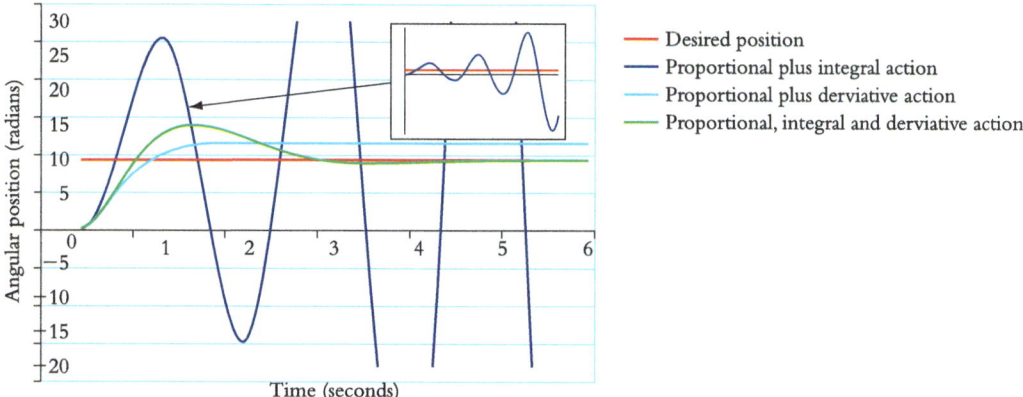

FIGURE 5.84 Response of positioning system with proportional plus integral control alone (blue line and inset), proportional plus derivative control (cyan line) and PID control (green line).

Along with the values of system constants assumed earlier, the values $T_D = 0.5$ s and $T_I = 1$ s will be assumed for illustrative purposes, and a demand of 10 V and an external load of 20 Nm will be imposed at $t = 0$. Proportional control only (omitting both integral and derivative terms) again results in a sinusoidal response of constant amplitude similar to that shown in Figure 5.82, but this time oscillates about a mean value of 12 radians. Proportional plus integral control only (i.e. omitting the derivative term) results in instability in the form of an exponentially growing sinusoidal response (Figure 5.84). Proportional plus derivative control gives a well-damped response rapidly settling to a steady value of 12 radians, indicating a steady-state error of 2 radians. None of these responses is really satisfactory; however, using all three terms (proportional, integral and derivative or PID control) results in a response that overshoots a little but rapidly settles to the desired value of 10 radians without any steady-state error. (For the sake of completeness, it should be noted that the derivative action in these simulations deliberately omits the impulse caused by the sudden change in demand at $t = 0$, as does the simulation of a more realistic position control system in Section 5.12).

The application of different control algorithms to the radar antenna problem is considered in a broadly similar manner in Section 5.12 with reference to steady-state error and dynamic response and the response of both of these systems is revisited yet again in Section 5.14 on the root locus method.

LEARNING SUMMARY

By the end of this section, you should have learnt:

✓ The importance of the control algorithm
✓ The purpose of the three terms in a PID controller
✓ Typical effects of the three contributions to PID control.

5.12 PRACTICAL DEMONSTRATIONS OF CONTROL SYSTEMS

Before proceeding further with the analysis techniques for control systems such as those described above, it is instructive to consider two practical demonstrations from within typical undergraduate laboratories. These will be examined both to illustrate the detailed implementation of practical control systems and to give more background to experiments which students might well encounter within their studies.

5.12.1 PUMPED TANK EXPERIMENT

The first practical demonstration is the pumped tank experiment (Figure 5.85) in which a tank is fed with water at a variable volume flow rate via a variable-speed positive displacement pump. The water can leave the tank via a valve which may be fully open or may be closed to varying degrees. The rate of outflow from the tank is related to the depth of water in the tank via linearised flow constant C_q as shown in Equations (5.99) and (5.100). The depth of water in the tank is measured using an electrical level probe, which gives an electrical signal output representing the water depth and typically measures the overall resistance of a pair of conductive tracks joined electrically by the water in the tank, giving a voltage related to the water depth by a calibration constant K_h. The electrical signal from the probe is taken into an analogue-to-digital converter or ADC (see Part 1 Unit 5 Section 5.11) which converts the varying voltage from the probe into the form of a digital quantity for use as a program variable within a personal computer (PC) which acts as the system controller. The desired water depth h_r is entered from the keyboard of the PC and converted into a form which can be compared (within the computer program) with the digitised depth measurement voltage v_h

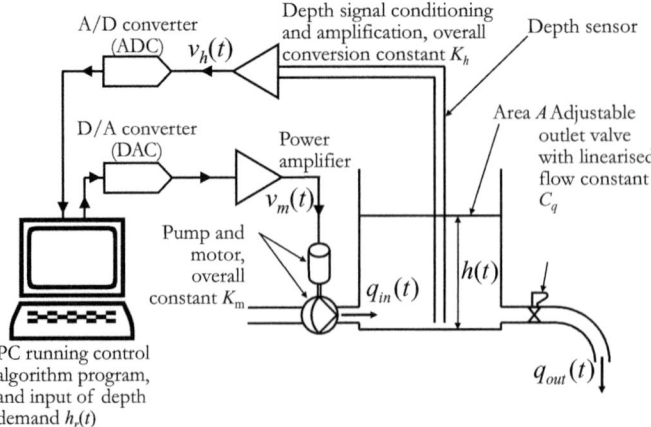

FIGURE 5.85 Pumped tank experiment.

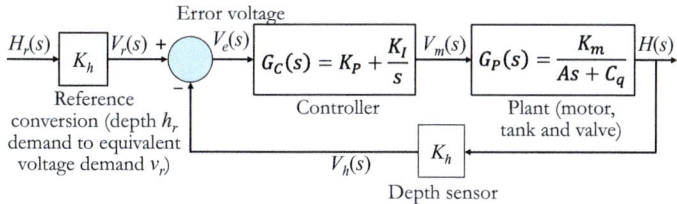

FIGURE 5.86   Block diagram for pumped tank experiment.

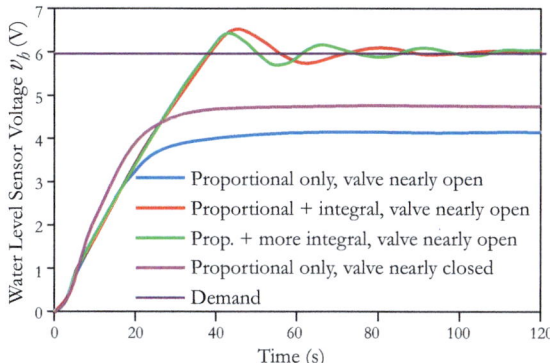

FIGURE 5.87   Typical experimental results for the pumped tank experiment.

from the probe and ADC. The difference is a value represented within the computer as error v_e. The control algorithm (specifically a PID) then calculates the corrective action in the form of a pump speed demand v_m, which is outputted as an electrical signal voltage via a digital-to-analogue con-verter (DAC). The DAC would not provide sufficient current to drive the pump motor so the signal is amplified so that the pump can then produce a flow input q_{in} proportional to that signal. This inflow, minus the outflow q_{out}, results in a change in the volume of water in the tank and hence a change in the depth h, as shown by Equations (5.95)–(5.97). This takes the system full circle as the depth is measured via the probe, and once again compared with the desired depth h_r giving a closed-loop control of the depth.

A typical block diagram for this apparatus is given in Figure 5.86.

With simple proportional control, steady-state errors are significant and are heavily influenced by the setting of the outlet valve (Figure 5.87). However, the introduction of an integral term enables the steady-state error to be eliminated without undue overshoot. The features of dynamic responses of this kind will be explored further in Section 5.13.

5.12.2   SERVOMOTOR-ENCODER EXPERIMENT

Another student experiment that provides an excellent illustration of control principles involves an apparatus based on the position control system described in Section 5.8. The specific version of the experiment described here is intended primarily as an exercise in mechatronics rather than as a quantitative experiment in classical control, and as such provides an opportunity to explore how a position control system can be constructed using commonly encountered sensors, actuators and electromechanical hardware.

As with the simplified and idealised position control system described earlier, the present exam-ple aims to provide closed-loop control of position using a DC motor and a position sensor though there are differences in both the detailed implementation and in the behaviour of the system. Ideally,

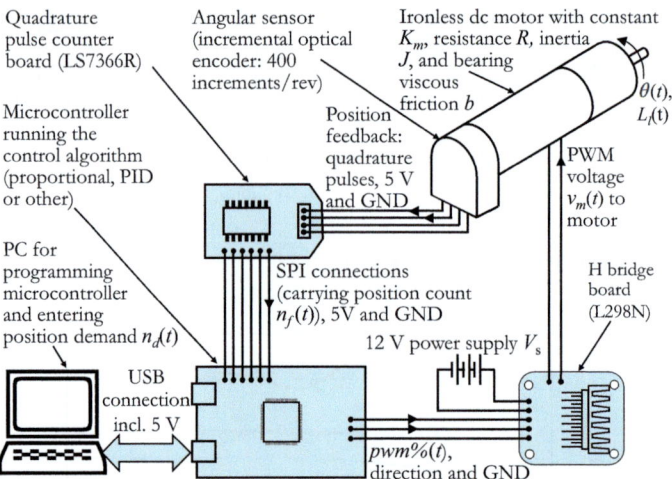

FIGURE 5.88   Mechatronics experiment involving rotational position control.

the motor is a high-quality ironless servomotor which does not exhibit large variations in torque (for a given motor current) as it rotates, though the experiment can still be performed with lower-cost motors. The control algorithm is implemented on a low-cost open-source microcontroller board. The system is illustrated diagrammatically in Figure 5.88.

The position control system in Section 5.8 might have used a digital-to-analogue converter to generate an analogue voltage which was amplified via a current amplifier to drive the motor. Instead, the present implementation uses techniques described in Unit 4 Section 4.11, specifically an H-bridge system (specifically an L298N integrated circuit) used in conjunction with pulse width modulation (PWM). This is much more straightforward to implement using a microcontroller, as it incorporates ready-made hardware for generating PWM signals under the command of the program running on it. The PWM effectively supplies the motor with a variable voltage rather than a variable current as in Section 5.8. So, the effect of this varying voltage is no longer a pure demand for torque but may be viewed as a more general demand for the motor to move in a given direction. More specifically, the details of the motor's response to voltage and load torque are assumed to be the same as given in Equations 5.109–5.110. The practical effects of this are that, if the motor is running only very slowly, the back-emf effects (see Unit 4, Section 4.9.3) are small and an applied voltage of v_m will lead to a current of v_m/R_a and hence to a torque of $K_m v_m/R_a$, so that the motor acts as a torque source as in the simplified model. Conversely, a steady voltage will (at steady state and if load torque is small) lead to an angular velocity close to v_m/K_m. So, the motor acts as an angular velocity source. Unlike the frictionless situation assumed earlier, there are various frictional effects which are idealised (rather imperfectly) as viscous bearing friction and a load torque which are assumed to be added to any external loading to give a total load torque L_1. The angular rotation is measured using an optical encoder (specifically of the type known as a quadrature encoder) rather than an analogue position sensor, giving an output in the form of pulses technically known as a quadrature signal. These can in principle be taken directly into the digital input interfaces of the microcontroller and counted using software, though for robustness these are instead counted using a dedicated counter circuit known as a bidirectional quadrature pulse counter, one such example being the LS7366R integrated circuit. This counter-circuit outputs the position count in the form of a digital signal which is transferred to the microcontroller via a Serial Peripheral Interface (SPI) communication connection. A PC is used for programming the microcontroller, entering the position demand and monitoring the system variables such as measured position, error and PWM duty cycle. A typical laboratory exercise with this apparatus involves coding up the control software required to implement the control loop, rather than detailed modelling or analysis. Simple proportional control can be coded from scratch, libraries

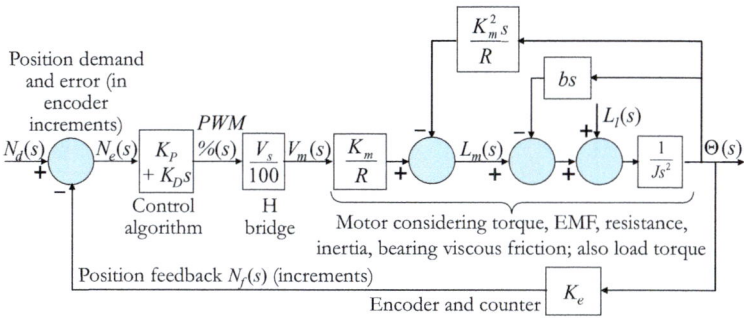

FIGURE 5.89   Block diagram of position control mechatronics experiment.

can be used to allow simple use of PID control (though the integral term gives no benefit within this experiment and is not explored), and more sophisticated variants of PID (e.g. adaptive control and changing the system parameters depending on the size of the positioning error) can be explored if required. As a result of these implementation differences, the block diagram representation of the system is shown in Figure 5.89 and is a little more complex than that assumed in Section 5.8; these differences relate specifically to the motor, driver and load (especially, the presence of back emf and frictional effects), noting that the different implementation of the position sensor makes no difference to the system model.

In terms of practical effects on system behaviour, the most important one is that the back-emf within the motor introduces some damping, as does friction in the system. The combination of these effects means that even pure proportional control can give reasonably stable behaviour in this situation at the expense of some steady-state error, provided the gain is not made too high. There is also an important point, beyond the scope of classical control theory, which is that the motor <u>cannot</u> be driven with a larger voltage than that which is available from the power supply. In other words, the motor/driver system will reach saturation (in colloquial terms, will 'max out') at 100% duty cycle and this cannot be exceeded even if the control algorithm requires larger corrective action. However, these complexities are of little practical importance within the context of the experiment. This aims to be a framework to enable a practical control loop to be coded from first principles and allows phenomena such as steady-state error, oscillation and instability to be observed in a position control system. Although numerical results are not gathered in the practical laboratory experiment and the detailed characteristics of the components are not accurately known, the system can be simulated numerically, based on the block diagram shown in Figure 5.89 (enhanced by taking into account the saturation and more realistic frictional behaviour and quantising the position measurements and error to integers, though these enhancements only slightly affect the results). Typical values of parameters for motor/encoder combinations of this type were used, specifically $J = 5 \times 10^{-7}$ kg m^2, $R = 5\ \Omega$, $K_{\mathrm{m}} = 0.015$ V s rad^{-1}, and $K_{\mathrm{e}} = 400$ increments rev$^{-1} = 63.33$ increments rad^{-1}. Friction was represented by a combination of $b = 5 \times 10^{-7}$ Nm s rad^{-1} and (instead of a constant value of L_{l}) a frictional torque of up to 0.0003 Nm opposing motion. V_{S} and H bridge gain took their laboratory values of 12 V and 0.01 respectively giving an overall transfer function for the H bridge of $V_{\mathrm{S}}/100 = 0.12$ V per PWM percentage point. Figure 5.90 shows the results of this simulation with four values of controller proportional gain K_{P} (0.01, 0.05, 0.1 and 1) and optional derivative gain $K_{\mathrm{D}} = 0.01$ s.

The results in Figure 5.90 usefully illustrate the following qualitative observations seen in the laboratory:

- A small value of proportional gain alone leads to a slow response and a large steady-state error;
- Increasing the value of proportional gain reduces the steady-state error but at the expense of overshoot and oscillation;

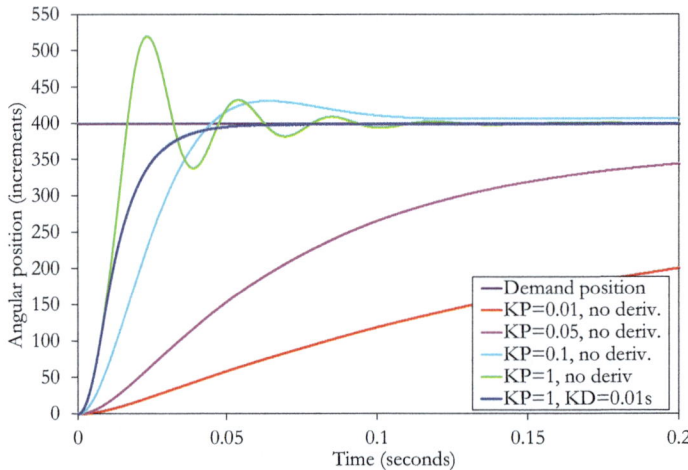

FIGURE 5.90   Simulated results for the position control system shown in Figure 5.88.

- Introduction of derivative action reduces or eliminates the oscillation arising from the large proportional gain.

It is instructive to note why the direct application of integral action is not likely to be beneficial in a simple laboratory implementation of this position control system. As explained above, saturation limits the control action that can be taken, meaning that corrective action on a large error will tend to take place more slowly than a linear model of the control system would assume and the error will be larger and more long-lasting than expected. This error will continue to be integrated by any integral term in the controller. In practice, this means that integral action will tend to accumulate, with the corrective action being unable (due to saturation) to keep up with the action increasingly required by the integral term, leading to overshoot and potentially to oscillation or even instability. This undesirable overshoot arising from integral action is known as 'integral windup'. More advanced control algorithms can be constructed which switch off the integral action until it is likely to be beneficial, but further details of anti-windup algorithms are far beyond the scope of the present unit.

LEARNING SUMMARY

By the end of this section, you should have learnt about the practical implementation of the control loops in two typical undergraduate laboratory experiments.

5.13   RESPONSE AND STABILITY OF CONTROL SYSTEMS

In order to understand the actual behaviour of a given control system, the chosen control algorithm must be substituted for $G_c(s)$. Once this has been performed, a number of issues can be examined:

- Is the system stable?
- Is its response oscillatory, that is, is it overdamped or underdamped?
- Is there a steady-state error, that is does the actual output from the system converge to the desired outcome or is there a difference that persists indefinitely?

These issues will be examined by continuing to examine the radar antenna problem.

5.13.1 STEADY-STATE ERROR

Steady-state error is the difference between the demand and achieved output which persists as time becomes infinite and is typically evaluated for one of two situations:

- A step input at $t = 0$ corresponds to a demand issued at $t = 0$ for the system to attain a steady value.
- A ramp input beginning at $t = 0$ corresponds to a demand for the output to increase linearly from zero with time, that is for a constant rate of motion or rate of change of output.

As a link with earlier work, it will be recalled that the error for the heated system under closed-loop control was given in Equation (5.113). If a step input is assumed, consisting of a demand for a temperature τ_1 imposed at $t = 0$, the demand can be expressed in the s domain (by using line 1 of Table 5.1).

$$T_D(s) = \tau_1 \mathscr{L}\{H(t)\} = \frac{\tau_1}{s} \tag{5.123}$$

Inserting this into Equation (5.113) and applying the final value theorem (Equation (5.59)) gives the error as time tends to infinity:

$$\text{Steady-state error} = \lim_{s \to 0} s \frac{\tau_1}{s} \frac{mC_p s + Ah}{mC_p s + Ah + K} = \frac{Ah}{Ah + K}\tau_1 \tag{5.124}$$

which recovers an expression for error equivalent to Equation (5.8) which was derived via the simple, time-independent analysis, confirming that this simple proportional control algorithm does not achieve a degree of control which eliminates steady-state error.

If, by contrast, a linearly increasing demand of $\tau_2\,°C/s$ is imposed, beginning at $t = 0$, the demand is modelled in the s space by making use of line 2 of Table 5.1:

$$T_D(s) = \tau_2 \mathscr{L}\{tH(t)\} = \frac{\tau_2}{s^2} \tag{5.125}$$

Attempting to calculate the steady-state errors gives

$$\text{Steady-state error} = \lim_{s \to 0} s \frac{\tau_2}{s^2} \frac{mC_p s + Ah}{mC_p s + Ah + K} = \infty$$

from which we can conclude that this simple control system will not only fail to achieve the desired (constantly increasing) temperature but indeed will fail even to provide a steady offset from the desired temperature and the error will become infinitely large with time.

Considering now the antenna problem, if proportional action K_c alone is assumed within Equation (5.119) so that $G_c(s) = K_c$, $T_D = 0$ and $T_1 = \infty$, and that the expression for $G_c(s)$ is further substituted into Equation (5.118) which gives the error in the antenna angular velocity, the following expression is obtained for the error as a function of s:

$$\Omega_D(s) - \Omega(s) = \Omega_D(s)\left(\frac{RJs + Rb + K_m^2}{RJs + Rb + K_m^2 + K_t K_c K_a K_m}\right) \tag{5.126}$$

Applying the final-value theorem enables the steady-state error to be calculated for the situation where the demand $\Omega_D(s)$ takes the form of a Heaviside step function at $t = 0$ and of magnitude Ω_1:

$$\lim_{s \to 0} s\left(\Omega_D(s) - \Omega(s)\right) = \lim_{s \to 0} s\,\frac{\Omega_1}{s}\left(\frac{RJs + Rb + K_m^2}{RJs + Rb + K_m^2 + K_t K_c K_a K_m}\right)$$

$$= \Omega_1\left(\frac{Rb + K_m^2}{Rb + K_m^2 + K_t K_c K_a K_m}\right) \tag{5.127}$$

which is a non-zero value.

The effect of proportional action alone is illustrated in Figure 5.91. For illustrative purposes, the value of R will be taken as $1\ \Omega$, J as $1\ kgm^2$, b as $1\ Ns/rad$, K_m and K_t both as $1\ Vs/rad$, K_a as 1 and K_c as 10. It is seen that the response to a demand beginning at $t = 0$ for a speed of $\Omega_1 = 10$ rad/s is an actual response tending asymptotically to 8.333 rad/s – a steady-state error of 1.667 rad/s correctly predicted by Equation (5.127).

In contrast, substituting the following expression (corresponding to proportional plus integral action, but no derivative action) into Equation (5.118):

$$G_c(s) = K_c\left(1 + \frac{1}{T_I s}\right) \tag{5.128}$$

results in the following expression for error:

$$\Omega_D(s) - \Omega(s) = \Omega_D(s)\left(\frac{RJs + Rb + K_m^2}{RJs + Rb + K_m^2 + K_t K_c\left(1 + 1/T_I s\right) K_a K_m}\right)$$

$$= \Omega_D(s)\left(\frac{T_I s\left(RJs + Rb + K_m^2\right)}{T_I s\left(RJs + Rb + K_m^2\right) + K_t K_c\left(T_I s + 1\right) K_a K_m}\right)$$

$$= \Omega_D(s)\left(\frac{T_I RJs^2 + T_I\left(Rb + K_m^2\right)s}{T_I RJs^2 + T_I\left(K_t K_c K_a K_m + Rb + K_m^2\right)s + K_t K_c K_a K_m}\right) \tag{5.129}$$

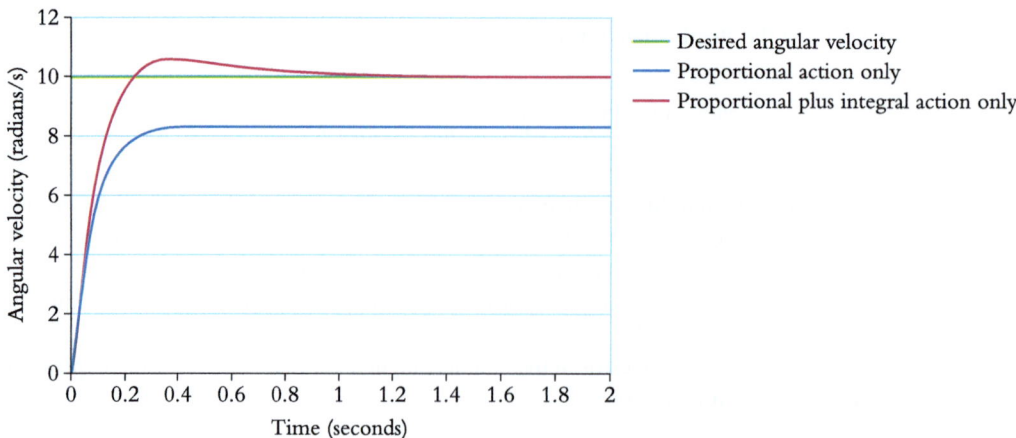

FIGURE 5.91 Response of radar antenna under proportional control alone and under proportional plus integral action.

Therefore, substituting for $\Omega_D(s)$ as before and applying the final-value theorem, gives the following expression for steady-state error:

$$\lim_{s \to 0} s\left(\Omega_D(s) - \Omega(s)\right) = \lim_{s \to 0} s \frac{\Omega_1}{s}\left(\frac{T_I R J s^2 + T_I\left(Rb + K_m^2\right)s}{T_I R J s^2 + T_I\left(K_t K_c K_a K_m + Rb + K_m^2\right)s + K_t K_c K_a K_m}\right) = 0 \quad (5.130)$$

In other words, the inclusion of the integral term has eliminated steady-state error. This is also shown in Figure 5.91 for an integral time constant T_I of 2.8 s, chosen to achieve a rapid settling time without oscillation (corresponding to critical damping) in response to a step input. A more detailed consideration of the form of the response (oscillatory or otherwise) and the effect of including derivative action are included in the discussion on the dynamic response of second-order systems contained in the next subsection.

5.13.2 DYNAMIC RESPONSE

So far, we have concentrated on what happens immediately after our system experiences its input and in the (infinitely) long term. No attempt has yet been made to quantify how the control system behaves in the short and medium term in response to an input. We are likely to be particularly interested in the following questions (refer to Figure 5.92 for typical responses):

- Does the system respond monotonically to an input or does its value overshoot and oscillate – in other words, is its response overdamped, underdamped or critically damped?
- If the response is underdamped, at what frequency does the output oscillate?
- Does the system settle down to a steady value or does the output grow increasingly large (possibly in an oscillatory manner) with time? In other words, is its response stable?

These issues are covered in significant depth in Unit 6, and an understanding of these types of damping will be assumed here.

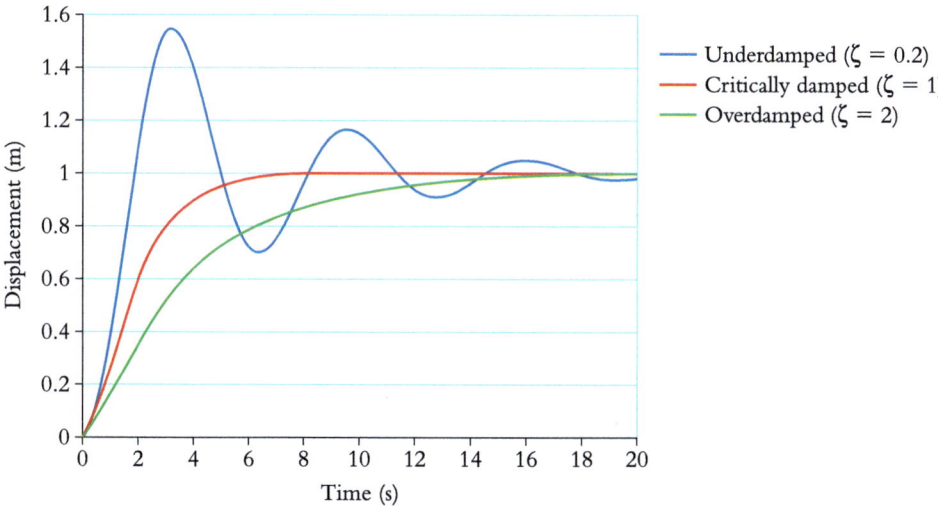

FIGURE 5.92 Graphs of underdamped, critically damped and overdamped responses of a second-order system.

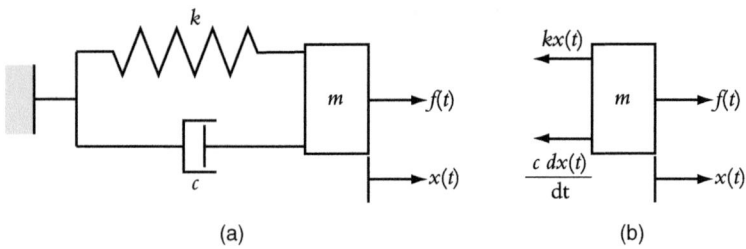

FIGURE 5.93 (a) Mass–spring–damper system and (b) free-body diagram of mass.

In practice, it is found that many systems typically exhibit behaviour analogous to a second-order (mass–spring–damper) system and it is instructive to answer the above questions by relating the system's response to that of a second-order system.

The key to answering the above questions lies in the closed-loop transfer function, and in particular, the **characteristic equation** which is obtained by setting the denominator of the transfer function (the **characteristic polynomial**) to zero. However, before examining a true control system, it is instructive to examine the simple mass–spring–damper system (Figure 5.93(a)) against which we will compare a control system's behaviour.

Consider the mass to be subjected to an externally applied force $f(t)$. From the free-body diagram of the mass under displacement $x(t)$ (Figure 5.93(b)) and noting that Newton's second law states that net force is equal to mass × acceleration, the equation of motion of the mass can be written down as

$$f(t) - kx(t) - c\frac{dx(t)}{dt} = m\frac{d^2x(t)}{dt^2} \tag{5.131}$$

This can be rearranged as

$$f(t) = m\frac{d^2x(t)}{dt^2} + c\frac{dx(t)}{dt} + kx(t) \tag{5.132}$$

Taking Laplace transforms, and assuming zero initial conditions, gives

$$F(s) = s^2mX(s) + scX(s) + kX(s) \tag{5.133}$$

This can be rearranged to give the transfer function of the system:

$$\frac{X(s)}{F(s)} = \frac{1}{ms^2 + cs + k} \tag{5.134}$$

However, it is instructive to rearrange it again into as

$$\frac{X(s)}{F(s)} = \frac{1/m}{s^2 + \dfrac{c}{m}s + \dfrac{k}{m}} \tag{5.135}$$

The characteristic equation can now be obtained by setting the denominator equal to zero:

$$s^2 + \frac{c}{m}s + \frac{k}{m} = 0 \tag{5.136}$$

However, it is well known that, for a simple mass–spring damper system, the following quantities can be defined:

- The natural frequency ω_n where

$$\omega_n = \sqrt{\frac{k}{m}}$$
(5.137)

- The damping ratio or damping factor ζ (pronounced 'zeta') where

$$\zeta = \frac{c}{\sqrt{4mk}}$$
(5.138)

The value of ζ gives an indication of the response of the system, giving a more precise way to characterise the behaviours shown in Figure 5.92:

- If $\zeta = 0$, the system is undamped and will continue to oscillate indefinitely if it is perturbed.
- If $0 < \zeta < 1$, the system is said to be **underdamped** and will exhibit overshoot and a decaying oscillatory response when perturbed.
- If $\zeta = 1$, the system is said to be **critically damped** and will exhibit a response that tends towards its steady-state response rapidly but without oscillation (or with only a small amount of overshoot).
- If $\zeta > 1$, the system is said to be **overdamped** and will tend asymptotically to its steady-state value and may take a long time to reach a value that can be regarded as close enough to the value desired.

An everyday example of a second-order system that could fall into any of these categories would be a door that can open in either direction (such as a door between sections of a factory or between a restaurant and a kitchen) and spring-loaded to return it to its closed position while allowing people or vehicles to push it open in either direction without operating a latch (Figure 5.94).

If the sprung door is underdamped, it will swing backwards and forwards after someone has passed through it; if it is overdamped, it will take a long time to close, and if it is critically damped

FIGURE 5.94   An everyday example of a second-order system.

it will close rapidly without swinging backwards and forwards. Common sense indicates that in this case, as in many practical situations, a critically damped response gives the best compromise between obtaining a quick response and avoiding prolonged oscillations and this reasoning is generally applicable also to control systems.

Note that the natural frequency is *not* necessarily quite the same as the frequency at which oscillations will occur if the system is perturbed and allowed to undergo free vibration – it is only equal to the free oscillation frequency for the undamped case ($\zeta = 0$). If the system is underdamped, the actual frequency will be $\omega_n \sqrt{(1 - \zeta^2)}$, whereas if it is critically damped or overdamped, no oscillation will occur at all.

Returning to the analysis of the second-order (mass–spring–damper) system, the characteristic equation can now be expressed in terms of the natural frequency and damping ratio as

$$s^2 + 2\zeta\omega_n s + \omega_n^2 = 0 \tag{5.139}$$

Second-order characteristic equations are frequently encountered in control systems, and it is often useful to determine by inspection of the characteristic equation what its response will be.

As an example, consider again the motor-driven, viscous-damped antenna described earlier. If proportional plus integral action is used, the closed-loop transfer function becomes

$$
\begin{aligned}
\frac{\Omega(s)}{\Omega_\mathrm{D}(s)} &= \frac{K_t K_c \left(1 + 1/(T_I s)\right) K_a K_m}{RJs + Rb + K_m^2 + K_t K_c \left(1 + 1/(T_I s)\right) K_a K_m} \\[2mm]
&= \frac{K_t K_c \left(T_I s + 1\right) K_a K_m}{T_I RJs^2 + RbT_I s + K_m^2 T_I s + K_t K_c \left(T_I s + 1\right) K_a K_m} \\[2mm]
&= \frac{K_t K_c K_a K_m \left(T_I s + 1\right)}{T_I RJs^2 + \left(Rb + K_m^2 + K_t K_c K_a K_m\right) T_I s + K_t K_c K_a K_m}
\end{aligned} \tag{5.140}
$$

The characteristic function (the denominator) is clearly that of a second-order system. The transfer function can then be rearranged to get the characteristic function (the denominator) into the form in which it appears in Equation (5.134):

$$\frac{\Omega(s)}{\Omega_\mathrm{D}(s)} = \frac{K_t K_c K_a K_m \left(T_I s + 1\right)/(T_I RJ)}{s^2 + \left(Rb + K_m^2 + K_t K_c K_a K_m\right)/(RJ)s + K_t K_c K_a K_m/(T_I RJ)} \tag{5.141}$$

The characteristic equation can therefore be written down by setting the denominator to zero:

$$s^2 + \left(Rb + K_m^2 + K_t K_c K_a K_m\right)/(RJ)s + K_t K_c K_a K_m/(T_I RJ) = 0 \tag{5.142}$$

From inspection of this characteristic equation, and by comparison with Equation (5.134), it is seen that

$$\omega_n = \sqrt{\frac{K_t K_c K_a K_m}{T_I RJ}} \tag{5.143}$$

and

$$\zeta = \frac{Rb + K_m^2 + K_t K_c K_a K_m}{2RJ\omega_n} = \frac{Rb + K_m^2 + K_t K_c K_a K_m}{2}\sqrt{\frac{T_I}{RJK_t K_c K_a K_m}} \tag{5.144}$$

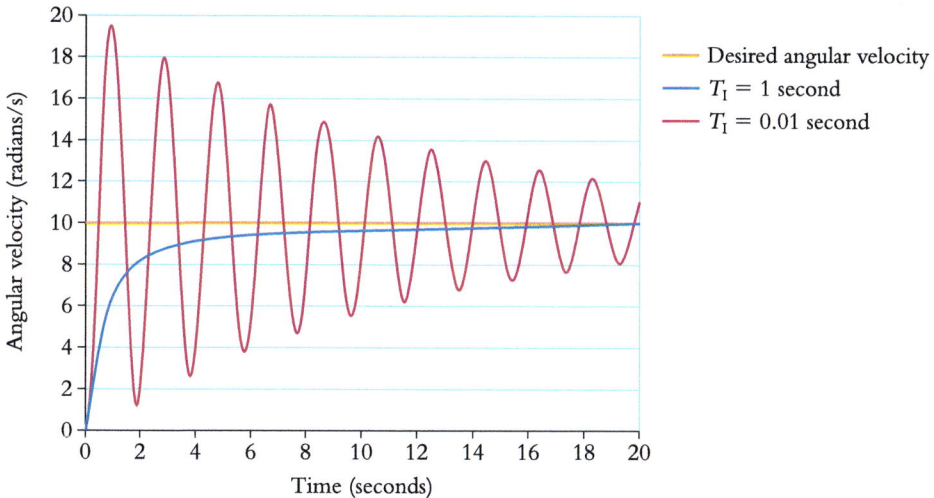

FIGURE 5.95 Response of a radar antenna system for two values of integral time constant.

It can be seen therefore that introducing integral action has the effect of turning the system from a first-order to a second-order one. Also, increasing the amount of integral action (specifically, reducing T_I) has the effect of reducing the damping ratio and making the system more oscillatory. For example, with the parameters used earlier but with $T_I = 1$, the system becomes overdamped ($\zeta = 1.9$) and approaches the demand speed slowly (Figure 5.95), whereas with $T_I = 0.01$, the system is noticeably underdamped ($\zeta = 0.19$) and the speed will fluctuate before settling down to its desired value.

If PID action is used, the corresponding analysis becomes rather messy but eventually results in the following expressions for natural frequency and damping ratio:

$$\omega_n = \sqrt{\frac{K_t K_c K_a K_m}{T_I R J + T_I T_D K_t K_c K_a K_m}} \tag{5.145}$$

and

$$\zeta = \frac{Rb + K_m^2 + K_t K_c K_a K_m}{2\left(RJ + T_D K_t K_c K_a K_m\right)\omega_n} = \frac{Rb + K_m^2 + K_t K_c K_a K_m}{2}\sqrt{\frac{T_I}{\left(RJ + T_D K_t K_c K_a K_m\right) K_t K_c K_a K_m}} \tag{5.146}$$

In fact, as this is a first-order system when in open loop, it can be seen that derivative action actually has no benefit in terms of increasing damping and can actually make stability worse by reducing the value of ζ; this is not the case for systems with a second-order open-loop transfer function, such as the positioning system analysed earlier. However, there can be pitfalls in differentiating real signals as this process can tend to accentuate noise. It is preferable to measure a derivative directly, for example to measure an angular velocity rather than obtaining it by numerical differentiation of rotational displacement.

5.13.3 FREQUENCY RESPONSE

This will only be covered briefly, though it forms a more significant part of texts on control theory, for example Dorf and Bishop (2021) and Bolton (1998). The frequency response of a control system can be determined by inserting the complex quantity $j\omega$ into the system's transfer function $G(s)$, where j is the name used by engineers for the imaginary unit (the square root of -1, often termed i by mathematicians). The result is a complex quantity whose real and imaginary parts $\text{Re}(G(j\omega))$

and Im($G(j\omega)$) will generally vary with the value of circular frequency ω. The result can either be left directly in the complex form or can be converted into magnitude and phase (or modulus and argument) as follows:

$$\left| G(j\omega) \right| = \sqrt{\left(\text{Re}\left(G(j\omega) \right) \right)^2 + \left(\text{Im}\left(G(j\omega) \right) \right)^2} \tag{5.147}$$

and

$$\arg\left(G(j\omega) \right) = \tan^{-1}\left(\frac{\text{Im}\left(G(j\omega) \right)}{\text{Re}\left(G(j\omega) \right)} \right) \tag{5.148}$$

noting that care must be taken to place $\arg\left(G(j\omega) \right)$ in the correct quadrant. This calculation process may be performed for both the open- and closed-loop situations. The results may be plotted in two different forms:

5.13.3.1 Bode Plot

This is actually a pair of graphs, presenting the modulus and argument of $G(j\omega)$ plotted against $j\omega$. Logarithmic axes are used for both variables of the modulus plot and the frequency axis of the argument plot. This is useful in order to identify resonances. Figure 5.96 gives a Bode plot for a second-order system, showing its single resonant frequency. Bode plots are also commonly used for characterising the open-loop transfer function $G(s)H(s)$ in order to determine the stability margins (gain margin and phase margin, i.e. values by which gain and phase lag could be increased before instability occurred).

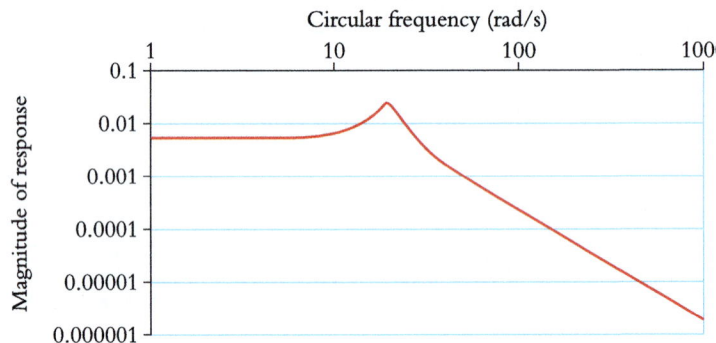

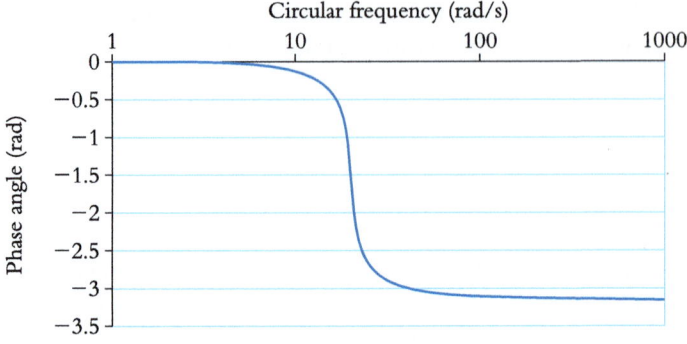

FIGURE 5.96 Example of Bode plot for a second-order system.

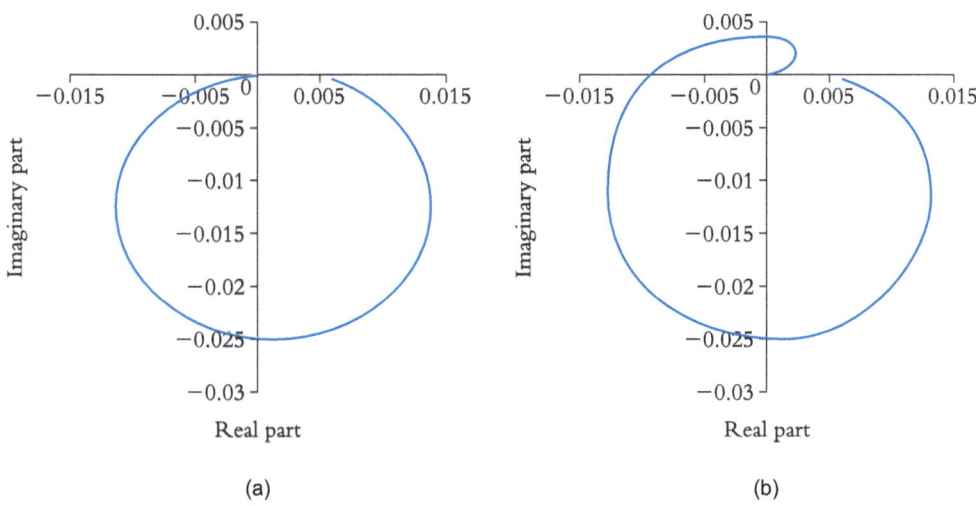

FIGURE 5.97 Nyquist plot for (a) a typical second-order system and (b) a typical higher-order system.

5.13.3.2 Nyquist Plot

This is a single graph plotted on an Argand diagram (i.e. a representation of the complex plane) that presents the locus of the points $(\text{Re}(G(j\omega)), \text{Im}(G(j\omega)))$ as ω is varied from zero to infinity. The result is typically a spiral-like plot that approaches the origin as ω tends to infinity. Although it contains less data than the Bode plot (the dependency upon frequency is lost, unless specific points are marked with values of ω), it is particularly helpful in understanding stability issues. Figure 5.97 shows the plots for a typical second-order system and a higher-order system.

5.13.4 STABILITY

Another very important issue is the stability of a control system. It has already been shown that a system may respond in an underdamped, critically damped or overdamped manner. However, it is also very important to determine whether or not a system exhibits **stability**, that is whether, for a bounded input (such as a unit impulse or step function), the output is also bounded, rather than (for example) undergoing oscillations which grow in amplitude with time rather than decaying. Such behaviour can occur owing to the use of excessive gain in a controller or the time delays or phase lags in the controller or elsewhere in the loop. In practice, the behaviour of real systems tends to be bounded, even under unstable conditions, due to the components of the control system 'maxing out' (saturating), for example feeding a motor with the maximum voltage available from the power supply or due to safety features deliberately shutting down or tripping out the system and returning it to a safe state when unexpectedly large amplitudes are encountered. Undesirably, bounding of the system behaviour may even result due to the failure of components occurring as a result of instability and large signal amplitudes.

A good example of an unstable behaviour of a feedback system is the unpleasant 'howling' that occurs with a public address (PA) system at a rock concert or in a lecture theatre, when the microphone is placed too close to the loudspeaker or if the volume control (gain) on the amplifier is turned up too far. The result is an oscillatory response that builds up until the maximum amplitude of the amplifier is reached.

There are three approaches to determining the stability of a control system, two of which will be considered here (the root locus method is dealt with separately in Section 5.14).

5.13.4.1 Routh–Hurwitz Criterion

This is perhaps the easier criterion to apply theoretically, though its physical meaning is not obvious. The closed-loop transfer function is determined as shown in Section 5.9, and the characteristic polynomial is found. It is then necessary to discover whether or not the characteristic function (obtained by equating the characteristic polynomial to zero) has any roots with positive real parts, which would result in a response that would become larger (rather than smaller) with time. The Routh–Hurwitz criterion is a rapid way to find this information without going to the trouble of actually finding the roots. The coefficients of the characteristic polynomial are used as the basis of figures that are inserted into a table known as the Routh array. Let the characteristic function take the form:

$$a_n s^n + a_{n-1} s^{n-1} + \ldots + a_2 s^2 + a_1 s + a_0 \tag{5.149}$$

The following table is completed:

s^n	a_n	a_{n-2}	a_{n-4}	a_{n-6}	$\ldots$
s^{n-1}	a_{n-1}	a_{n-3}	a_{n-5}	a_{n-7}	$\ldots$
s^{n-2}	b_1	b_2	b_3	$\ldots$	
s^{n-3}	c_1	c_2	c_3	$\ldots$	
s^{n-4}	d_1	d_2	$\ldots$		
$\ldots$	$\ldots$	$\ldots$	$\ldots$		

where

$$b_1 = \frac{a_{n-1} a_{n-2} - a_n a_{n-3}}{a_{n-1}}, \quad b_2 = \frac{a_{n-1} a_{n-4} - a_n a_{n-5}}{a_{n-1}} \quad \text{etc.}$$

and

$$c_1 = \frac{b_1 a_{n-3} - a_{n-1} b_2}{b_1}, \quad c_2 = \frac{b_1 a_{n-5} - a_{n-1} b_3}{b_1} \quad \text{etc.}$$

or more generally:

$$b_i = \frac{a_{n-1} a_{n-2i} - a_n a_{n-2i-1}}{a_{n-1}}$$

$$c_i = \frac{b_1 a_{n-2i-1} - a_{n-1} b_{i+1}}{b_1}$$

Diagrammatically, the terms are obtained as shown in Figure 5.98.

The number of changes in the signs of the left-hand column is equal to the number of roots of the characteristic polynomial having positive real parts. In the context of control system stability, if *any* of the roots have positive real parts, the system will be unstable. We can therefore conclude that if there are *any* changes in the signs of the left-hand column of the Routh array, the system will be unstable.

This can be illustrated via the following example.

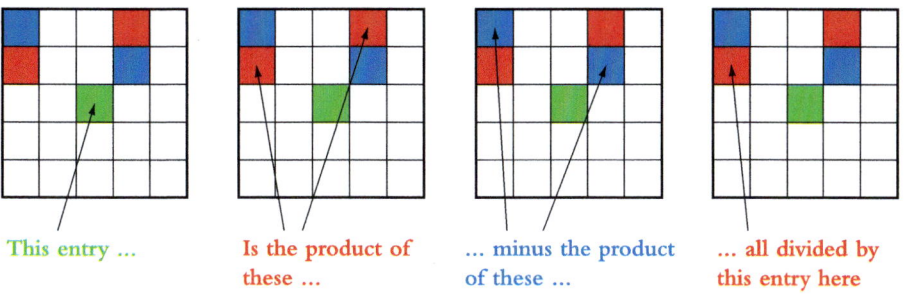

This entry ... Is the product of these ... ... minus the product of these ... ... all divided by this entry here

FIGURE 5.98 *Sequence of calculating a new term in the Routh array.*

WORKED EXAMPLE

Find how many positive roots the following characteristic equation has:

$$s^6 - 2s^5 - 12s^4 + 4s^3 - 149s^2 + 198s + 360 = 0 \qquad (5.150)$$

The Routh array is as follows:

s^6	1	−7	−214	400
s^5	−4	44	140	0
s^4	$\dfrac{(-7)\times(-4)-1\times44}{(-4)}=4$	$\dfrac{(-214)\times(-4)-1\times140}{(-4)}=-179$	$\dfrac{400\times(-4)-1\times0}{(-4)}=400$	0
s^3	$\dfrac{44\times4-(-4)\times(-179)}{4}=-135$	$\dfrac{140\times4-(-4)\times400}{4}=540$	0	0
s^2	$\dfrac{(-179)\times(-135)-4\times540}{(-135)}=-163$	$\dfrac{400\times(-135)-4\times0}{(-135)}=400$	0	0
s	$\dfrac{540\times163-(-135)\times400}{(-163)}=208.7$	0	0	0

There are four changes of sign in the left-hand column indicating four roots with positive real parts. In fact, this example of the characteristic function was generated by expanding the left-hand side of the following equation:

$$(s-5)(s+4)(s-1+3j)(s-1-3j)(s-2)(s+3) = 0 \qquad (5.151)$$

so the roots are −1, 2, −4, 5, 1 − 3j and 1 + 3j, with the four roots 2, 5, 1 − 3j and 1 + 3j all having positive real parts. A system with this characteristic equation would clearly be extremely unstable!

5.13.4.2 Nyquist Stability Criterion

This approach is more intuitive though it involves more numerical calculations. Moreover, it can be applied experimentally without actually causing the control system to become unstable. The Nyquist plot for the **open-loop** (not closed-loop) transfer function (including the contribution from the feedback loop, i.e. $G(s)H(s)$ within Figure 5.37), is plotted as shown in Figure 5.99(a) for a stable

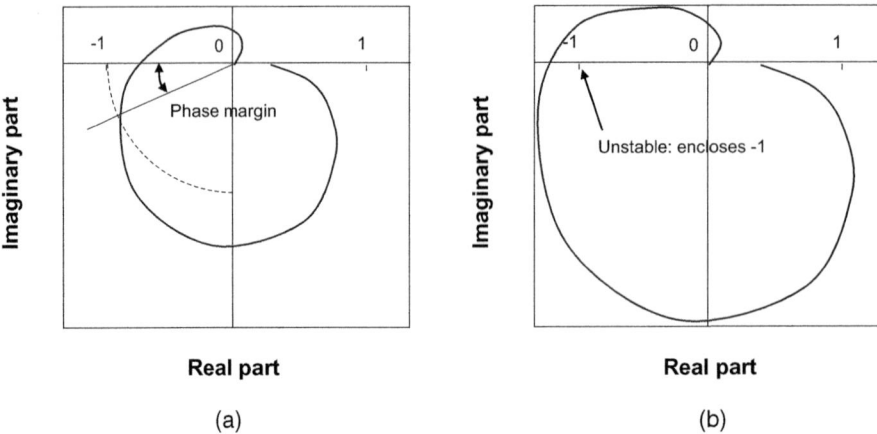

FIGURE 5.99 Nyquist plots of (a) the open-loop transfer function of a stable high-order system, showing the phase margin and (b) the open-loop transfer function of an unstable system showing how the plot encloses the point −1.

system. If the curve encloses the point in the complex plane corresponding to the real number −1 (i.e. the point one unit to the left of the origin, on the negative real axis) as shown in Figure 5.99(b), the system will be unstable. The closer the curve comes to enclosing this point, the less stable the system will be (meaning in practice that disturbances will die away more slowly).

The physical interpretation to this is as follows: if the feedback signal to the differencing junction is of the opposite sign to the demand signal (or, more accurately, contains a component which is of the opposite sign), the result from the differencing junction is actually an amplification of the input signal, and the net effect is equivalent to positive rather than negative feedback. If the amplitude of this 'positive feedback' signal is less than the input, the effect is not catastrophic as the feedback signal is still attenuated with respect to the input and each pass through the system attenuates it further. However, if the opposite sign (180° out-of-phase) component of the feedback signal equals or exceeds the amplitude of the input signal, the feedback signal will be further amplified by the system and a larger signal will be fed back. This implies that the signal will continue to grow and that instability will result.

The practical applicability of the Nyquist stability criterion is that the open-loop Nyquist plot can be constructed for a real (but theoretically uncharacterised) piece of control equipment, based on the experimental open-loop response (magnitude and phase) to a sinusoidal input. The Nyquist stability criterion provides a very useful tool for predicting the closed-loop behaviour of a piece of apparatus without taking any risk of allowing it to become unstable. The amount by which the feedback signal would have to be scaled to enclose −1 on the plot gives a measure of the amount by which the gain of the controller could be increased without instability occurring. This is known as the *gain margin* and is usually expressed in dB (decibels, defined as $20 \log_{10}$ of the factor by which the gain must be increased to cause instability) rather than as the factor itself.

Another useful quantity is the *phase margin*. This is the angle between the negative real axis of the Nyquist plot and the point at which the curve intersects the unit circle (Figure 5.99(a)). It represents the amount of the additional phase lag needed in the control system to cause instability. As noted earlier, the gain margin and phase margin can also be determined from Bode plots (see, for example, Dorf and Bishop (2021) and Bolton (1998)).

5.14   A FRAMEWORK FOR MAPPING THE RESPONSE OF CONTROL SYSTEMS: THE ROOT LOCUS METHOD

The response of second-order systems is straightforward to analyse by drawing on the analogy with a mass–spring–damper system and inferring the natural frequency and damping ratio, but for third-order (and higher-order) systems, a more sophisticated method of conceptualising the response is required. A useful tool is the *root locus plot* which is used for understanding the different aspects of a control system's behaviour, as represented by the various roots of the characteristic equation (which was introduced earlier in the present section as being the equation obtained by setting the denominator of the transfer function to zero).

The root locus method involves plotting the locations of the roots in the complex plane (i.e. plotting their imaginary parts versus their real parts), as the gain of the control system is varied and inferring the stability and nature of the response of the system from the position of the roots on the resulting diagram. Before looking in detail at the generation (and interpretation) of root locus plots, it is instructive to explore the physical meaning of different types of root, which can be represented in the complex plane shown as (a) to (d) in Figure 5.100:

(a) Positive real roots correspond to an exponentially growing (but non-oscillatory) transient response to a disturbance (which is clearly an inherently unstable response since it does not converge to a steady value)
(b) Negative real roots correspond to an exponentially decaying, non-oscillatory response that dies away to zero with time and can form part of a stable response
(c) Complex roots, which always occur in 'conjugate pairs' with equal positive and negative imaginary parts, may have positive real parts and be of the form $s = \alpha \pm j\beta$ giving an oscillatory response of exponentially growing amplitude – clearly another example of an unstable response
(d) Complex roots may have negative real parts and be of the form $s = -\alpha \pm j\beta$ giving an oscillatory response of exponentially decaying amplitude – this dies away with time and can again contribute to a stable response.

A simple example of a root locus plot, relating to a non-control example of a second-order system, namely the familiar example of a mass–spring–damper system, is discussed earlier in the present

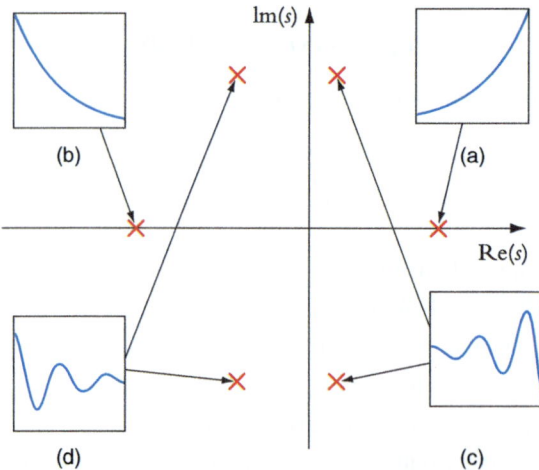

FIGURE 5.100 Typical responses corresponding to positions of roots in the complex plane.

section. It can be seen from the denominator in Equation (5.135) that this system has a character-istic function $ms^2 + cs + k$, which leads to the characteristic Equation (5.136) given again here for convenience:

$$s^2 + \frac{c}{m}s + \frac{k}{m} = 0 \qquad\qquad (5.152)$$

For illustrative purposes, let us choose mass $m = 1$ kg, damping constant $c = 10$ Ns/m and allow spring stiffness k to vary from zero to infinity so that the characteristic equation becomes

$$s^2 + 10s + k = 0 \qquad\qquad (5.153)$$

With very small k, the roots are real (for $k = 0$, the roots are $s = -10$ and $s = 0$) and the response to an initial input (such as a step or impulse) is heavily damped, decaying slowly but without oscillation. As the spring constant is made larger (in this case 25 N/m), the system becomes critically damped and the roots become equal but still real with the value -5. A slightly larger value of the spring con-stant causes the roots to break away from the real axis to form a conjugate pair of complex roots and the response becomes oscillatory. Further increases to the spring constant make the imaginary parts of the roots larger and the oscillations are faster but they do not die away any more slowly with time. The locus of the roots as the value of k varies is shown in Figure 5.101.

The same concepts can be applied to control systems, whose behaviour may or may not be oscil-latory. In practice, it is found that the response of a system can vary between any or all of the kinds of behaviour shown in Figure 5.100 as the gain of the controller in a closed-loop control system is varied.

In order to understand the relevance of the root locus method to control systems, recall that a typical control system has an open-loop transfer function $G(s)H(s)$ as shown earlier in Figure 5.37. Furthermore, the effect on the transfer function of closing the loop has also been demonstrated (Figure 5.36 and Equations (5.45) to (5.47)). Now let us assume that the transfer function of the main (forward) path involves a variable gain term K, arising for example from varying the gain of a controller lying within that forward path and two polynomials $Z_G(s)$ and $P_G(s)$ so that

$$G(s) = K\frac{Z_G(s)}{P_G(s)} \qquad\qquad (5.154)$$

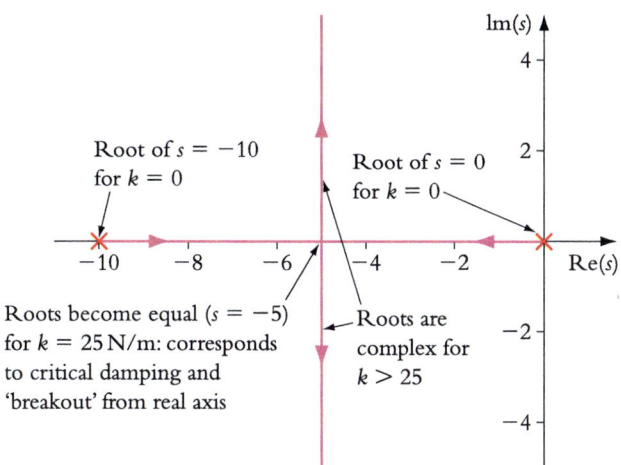

FIGURE 5.101   Root locus plot for a mass–spring–damper system with varying spring constant k.

Similarly, the transfer function of the feedback path can be expressed in terms of two more polynomials $Z_H(s)$ and $P_H(s)$:

$$H(s) = K \frac{Z_H(s)}{P_H(s)} \tag{5.155}$$

so that the open-loop transfer function is

$$G(s)H(s) = K \frac{Z_G(s)Z_H(s)}{P_G(s)P_H(s)} = K \frac{Z(s)}{P(s)} = K \frac{(s-z_1)(s-z_2)\ldots(s-z_{nZ})}{(s-p_1)(s-p_2)\ldots(s-p_{nP})} \tag{5.156}$$

where $P(s) = P_G(s)P_H(s)$ and $Z(s) = Z_G(s)Z_H(s)$. Also, $s = z_1$, $s = z_2 \ldots s = z_{nZ}$ are the roots of equation $Z(s) = 0$ and are known as the **zeros** of the open-loop transfer function, while $s = p_1, s = p_2, \ldots s = p_{nP}$ are the roots of equation $P(s) = 0$ and are known as the **poles** of the open-loop transfer function.

By inserting Equations (5.154) and (5.155) into Equation (5.47), giving the closed-loop transfer function in terms of $G(s)$ and $H(s)$, the following is obtained:

$$\text{Closed loop transfer function} = \frac{G(s)}{1 + G(s)H(s)} = \frac{\dfrac{Z_G(s)}{P_G(s)}}{1 + K \dfrac{Z_G(s)Z_H(s)}{P_G(s)P_H(s)}}$$

$$= \frac{Z_G(s)P_H(s)}{P_G(s)P_H(s) + KZ_G(s)Z_H(s)} = \frac{Z_G(s)P_H(s)}{P(s) + KZ(s)} \tag{5.157}$$

The closed-loop response of the system is governed by the roots of the characteristic equation, which once again is obtained by setting the denominator of the closed-loop transfer function to zero:

$$P(s) + KZ(s) = (s-p_1)(s-p_2)\ldots(s-p_{nP}) + K(s-z_1)(s-z_2)\ldots(s-z_{nZ}) = 0 \tag{5.158}$$

Now consider three possible situations:

1. When K is very small (i.e. K tends to zero), $P(s)$ is much larger than $KZ(s)$ so the characteristic equation can be approximated as

$$(s - p_1)(s - p_2)...(s - p_{np}) = 0 \tag{5.159}$$

whose roots are simply the open-loop poles of the system.

2. When K is very large (i.e. K tends to infinity) and s is not large, $P(s)$ is much smaller than $KZ(s)$ so the characteristic equation can be approximated as

$$(s - z_1)(s - z_2)...(s - z_{nZ}) = 0 \tag{5.160}$$

whose roots are simply the open-loop zeros of the system. However, there will also be some very large complex roots to the characteristic equation which (for very large complex s) can also be approximated by expanding it out and truncating the expansions of the two sets of brackets after the terms involving s^{np-1} and s^{nz-1}, respectively:

$$s^{np} - s^{np-1}p_1 - s^{np-1}p_2... - s^{np-1}p_{np} + K\left(s^{nz} - s^{nz-1}z_1 - s^{nz-1}z_2... - s^{nz-1}z_{n_z}\right) = 0 \tag{5.161}$$

or

$$s^{np-1}\left(s - p_1 - p_2... - p_{np}\right) + Ks^{nz-1}\left(s - z_1 - z_2... - z_{n_z}\right) = 0 \tag{5.162}$$

Noting that s is assumed to be large, Equation (5.156) can be further approximated (via a binomial expansion of the form $(1-a)^{-1} = 1 + a + a^2.... \approx 1 + a$ $(1-a)^{-1} = 1 + a + a^2.... \approx 1 + a$ for small a) and rearranged to give

$$s^{np-nz}\left(1 - \frac{p_1}{s} - \frac{p_2}{s}... - \frac{p_{np}}{s} + \frac{z_1}{s} + \frac{z_2}{s}... + \frac{z_{n_z}}{s}\right) = -K \tag{5.163}$$

Taking the $(n_p - n_z)$th root:

$$s\left(1 - \frac{p_1}{s} - \frac{p_2}{s}... - \frac{p_{np}}{s} + \frac{z_1}{s} + \frac{z_2}{s}... + \frac{z_{n_z}}{s}\right)^{\frac{1}{np-nz}} \approx (-K)^{\frac{1}{np-nz}} \tag{5.164}$$

Noting that $(1+a)^{1/n} = 1 + a/n + a^2/n^2 +.... \approx 1 + a/n$ for small a, and re-expressing the $(n_p - n_z)$th roots of $(-K)$ in terms of the $(n_p - n_z)$th roots of (-1):

$$s\left(1 - \frac{1}{np - nz}\left(\frac{p_1}{s} + \frac{p_2}{s}... + \frac{p_{np}}{s} - \frac{z_1}{s} - \frac{z_2}{s}... - \frac{z_{n_z}}{s}\right)\right) = K^{\frac{1}{np-nz}}(-1)^{\frac{1}{np-nz}} \tag{5.165}$$

There are of course $(n_p - n_z)$ possible values for the $(n_p - n_z)$th root of (-1). Rearranging Equation (5.165) and expressing the various $(n_p - n_z)$th roots of (-1) in polar form:

$$s = \frac{p_1 + p_2... + p_{np} - z_1 - z_2... - z_{n_z}}{np - nz} + K^{\frac{1}{np-nz}}\left(\cos\left(\frac{(2k+1)\pi}{np - nz}\right) + j\sin\left(\frac{(2k+1)\pi}{np - nz}\right)\right) \tag{5.166}$$

where $k = 0 \ldots (n_p - n_z - 1)$. In other words, for large K, the roots of the characteristic equation tend asymptotically to large multiples of the $(n_p - n_z)$th root of -1, shifted by a term sometimes referred to as the centroid of the poles and zeros.

3. When K is neither very small nor very large, the roots of the characteristic equation will take some intermediate values which may be real or complex. The purpose of the root locus methods is to show graphically how the roots move between the open-loop poles and the open-loop zeros or the large complex values (asymptotes).

The rules for plotting the root locus of an arbitrary closed-loop transfer function were originally proposed in a paper by Evans (1948). They are given in detail, though in slightly different forms and orders, in any comprehensive text on classical control theory (e.g. Dorf and Bishop (2021), which use a slightly different convention for poles and zeros or Van de Vegte (1993)) and the reader is referred to such a text. A summary of the rules, with minor differences from the approach presented here, is given on the RoyMech website (Beardmore, 2022), along with a number of examples. The rules are summarised here (without further proof) to act as a starting point, noting that this version of the rules does not cover all contingencies.

1. The root locus plot is symmetric about the real axis because complex roots occur in conjugate pairs.
2. The number of paths on the root locus is equal to the number of open-loop poles n_p.
3. The paths start (for $K = 0$) at the open-loop poles and are normally marked on the complex plane with X symbols.
4. n_Z of these terminate at the open-loop zeros, which are marked on the complex plane with O symbols.
5. The remaining $n_P - n_Z$ paths do not terminate but tend towards asymptotes oriented at angles θ to the real axis where

$$\theta = \frac{(2k+1)\pi}{n_P - n_Z} \quad \text{where } k = 0, 1 \ldots (n_P - n_Z - 1) \tag{5.167}$$

so that (for example) if there are five poles and two zeros, then there will be three asymptotes oriented at $\pi/3$, π and $5\pi/3$ radians (60°, 180° and 300°) to the real axis (Figure 5.102a). Similarly, if there are five poles and one zero, there will be four asymptotes oriented at $\pi/4$, $3\pi/4$, $5\pi/4$ and $7\pi/4$ radians (45°, 135°, 225° and 315°) to the real axis (Figure 5.102b).

6. The asymptotes radiate outwards from the point on the real axis sometimes known as the centroid of the poles and zeros:

$$s = \frac{p_1 + p_2 \ldots + p_{n_P} - z_1 - z_2 \ldots - z_{n_Z}}{n_P - n_Z} \tag{5.168}$$

7. Root loci can be drawn on the real axis in the regions to the left of an odd number of poles and zeros. If there are an even number of poles and/or zeros on the real axis to the right of a given point, no root locus is present at that point.
8. Where paths come together along the real axis (or converge back onto the real axis), the values of s at the 'breakout' and 'breakin' points (collectively known as 'singular' points, where there are multiple roots at a single point) are given by rearranging Equation (5.152) to give K in terms of s and solving the equation:

$$\frac{dK}{ds} = 0 \tag{5.169}$$

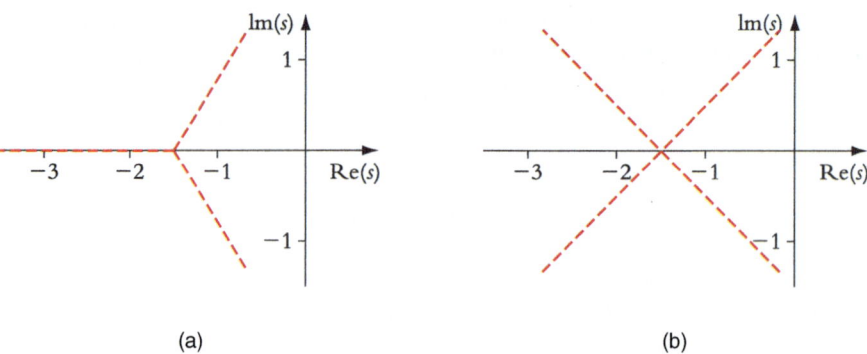

FIGURE 5.102   Typical asymptotes for (a) $n_P = 5$, $n_Z = 2$ and (b) $n_P = 5$, $n_Z = 1$.

Note that not all solutions to Equation (5.169) will necessarily correspond to breakaway (singular) points.

9. Where a path starts at a complex pole, the angle at which it leaves the pole, plus the sum of each of the angles to the real axis (horizontal) direction of the present pole from each of the other poles, adds up to an angle $(2k + 1)\pi$ radians where k is a real integer (in other words, adds up to an odd multiple of π). This is best illustrated via the example shown in Figure 5.103.

   The angles of arrival of the paths at complex zeros are found by a process that is effectively the opposite of the above (the angle of arrival at a zero, plus the angles from the other zeros, minus the angles from the poles adds up to an odd multiple of π).

10. The loci cross the imaginary axes at points that can be determined by replacing s with $j\omega$ and solving. It may or may not be necessary to find the limiting value of K for stability (for example, by applying the Routh–Hurwitz stability criterion) before solving to find the imaginary roots. This can be a little tedious and hence will not be explored in detail here, though the worked example includes a simple illustration.

11. Once we have plotted the root locus, the value of K can be chosen to ensure that the roots stay well away from the right-hand part of the complex plane. If a given point on the root

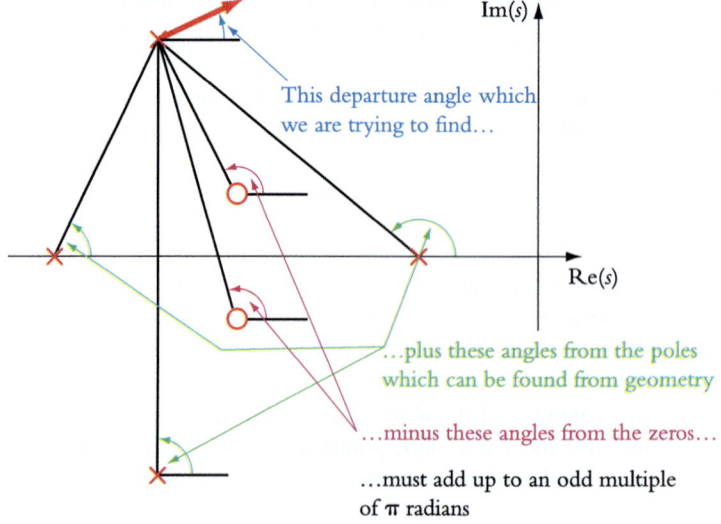

FIGURE 5.103   Finding the angle of departure from a complex pole.

locus is chosen, the corresponding value of K can be found as the product of the distances of that point from the poles divided by the product of the distances of that point from the zeros:

$$K = \frac{|s - p_1| \times |s - p_2| \times \ldots \times |s - p_{n_P}|}{|s - z_1| \times |s - z_2| \times \ldots \times |s - z_{n_Z}|} \quad (5.170)$$

If there are no zeros, the denominator is unity.

With the availability of computing packages such as MATLAB, root locus plots can be drawn automatically and accurately without the need to apply the rules given above. Various tools are available for plotting root loci from the transfer function, both for online use (for example that is available at the PUC-Rio website) and for download (e.g. the RootLocs application). It is nonetheless useful to be able to obtain a graphical visualisation of a system's behaviour without the need for computational effort. It is also very useful to have a good understanding of how the presence (and introduction) of poles and zeros within the transfer function of the controller can affect the shape of the root locus plot for the system, and hence how they can influence the behaviour of the system.

The power of the root locus method lies in the fact that an experienced control system designer can choose a suitable point on the root locus plot that will give a good compromise between stability and responsiveness; the corresponding value of K can then be calculated from rule 11 above.

WORKED EXAMPLE

Plot the root locus of a closed-loop system having the following open-loop transfer function:

$$G(s)H(s) = \frac{1}{(s^2 + 2s + 2)(s + 2)} \quad (5.171)$$

This can be expressed as

$$G(s)H(s) = \frac{1}{(s + 1 + j)(s + 1 - j)(s + 2)} \quad (5.172)$$

which has three poles at $s = (-1 + j)$, $s = (-1 - j)$ and s = -2, and no zeros. Following the rules above:

1. Symmetry about the real axis will be assumed and is satisfied by the open-loop poles.
2. There are three open-loop poles; so, there are three paths on the root locus plot.
3. The paths start at the open-loop poles $s = (-1 + j)$, $s = (-1 - j)$ and $s = -2$, which are marked on the diagram with X symbols (Figure 5.104(a))
4. There are no zeros to be marked on the plot, so all of the paths end in asymptotes rather than at defined positions.
5. $n_P = 3$, $n_Z = 0$; so, there are three asymptotes at angles $\pi/3$, π and $5\pi/3$ radians (or 60°, 180° and 300°) to the real axis (Figure 5.104(b))
6. The asymptotes radiate outwards from the point:

$$s = \frac{(-1 + j) + (-1 - j) + (-2)}{3} = -\frac{4}{3} \quad (5.173)$$

as shown in Figure 5.104(b).

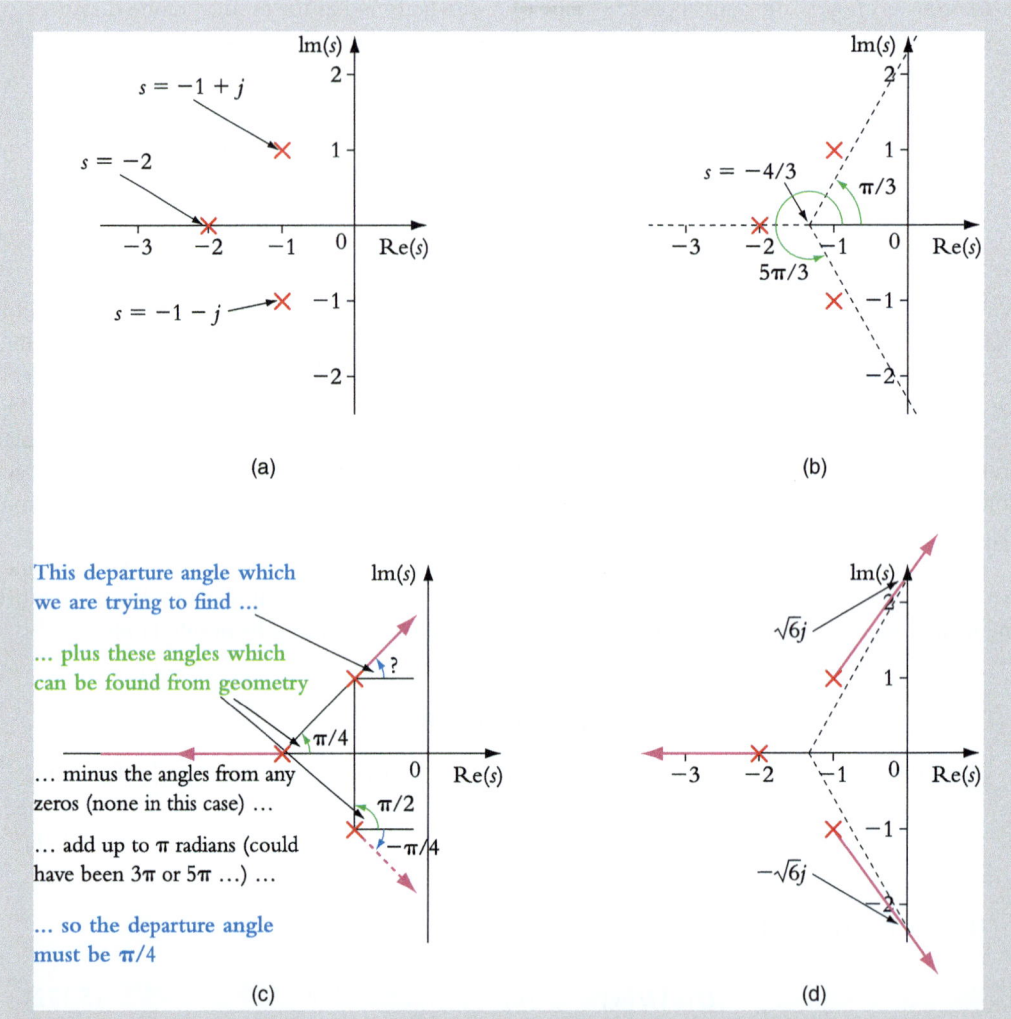

FIGURE 5.104 (a) Location of poles, (b) asymptotes, (c) angle of departure of the path from pole (−1 + j) and (d) completing the root locus plot including intersections with the imaginary axis.

7. The only point at which a path can be present on the real axis is to the left of the root $s = -2$, because only in this region is there an odd number of poles and zeros (in this case, only one pole and no zeros) to its right on the real axis. By enforcing this rule, it is seen that this particular path starts at the pole $s = -2$ and extends infinitely along the negative real axis (Figure 5.104(c)).

8. There is no breakout or breakin as the paths in this case either lie on the real axis or away from it.

9. The angle of departure from the pole $s = (-1 + j)$ (measured relative to the real axis), plus the total of the angles of that pole from the other poles, minus the total of the angles to that pole from the zeros, must add up to an odd multiple of π. It can be seen from Figure 5.104(c) how this is applied in the present case, with the total of the angles simply being π itself. The other angles are (in this example) simple to find geometrically, so the angle of departure is found as

$$\pi - \frac{\pi}{4} - \frac{\pi}{2} = \frac{\pi}{4}$$

By symmetry about the real axis, the angle of departure from the other pole must be $-\pi/4$.

10. Two of the paths cross the imaginary axis after leaving the complex poles. At this stage, it is noted that (by inserting the open-loop transfer function into (5.47), expanding the bracketed terms and rearranging) the characteristic equation of the closed-loop system can be expressed as

$$s^3 + 4s^2 + 6s + 4 + K = 0 \tag{5.174}$$

Making the substitution $s = j\omega$ gives

$$-j\omega^3 - 4\omega^2 + 6j\omega + 4 + K = 0 \tag{5.175}$$

Taking the real part of Equation (5.175) and rearranging gives

$$6\omega = \omega^3 \tag{5.176}$$

ω is therefore either zero (not a valid solution) or $\pm\sqrt{6}$ rad/s; so, the loci intersect the imaginary axis at the points $s = \pm\sqrt{6}j = \pm 2.44j$. Now taking the imaginary part of Equation (5.169), rearranging and inserting this value of ω gives the value of gain for limiting stability:

$$K = 4\omega^2 - 4 = 4 \times 6 - 4 = 20 \tag{5.177}$$

At this stage, the root locus plot can be completed as shown in Figure 5.104(d), with the two complex paths leaving the two complex poles at $\pm\pi/4$, crossing the imaginary axis at $\pm 2.44j$ and converging on the asymptotes at $\pi/3$ and $5\pi/3$. A control engineer may then choose (by eye and experience) a point on the plot for which the gain can then be calculated from the distances of that point to the poles and zeros. For example, if point $(-0.5 + 1.65j)$ on the locus is chosen by eye (Figure 5.105), then K is calculated as

$$K = \frac{\text{product of all distances from chosen } s \text{ to poles}}{\text{product of all distances from chosen } s \text{ to zeros}}$$

$$= \frac{|s+1-j| \times |s+1+j| \times |s+2|}{1} = \frac{|0.5+0.65j| \times |0.5+2.65j| \times |1.5+1.65j|}{1}$$

$$= 4.93 \tag{5.178}$$

noting that (in this case) the denominator is simply 1 as there are no zeros.

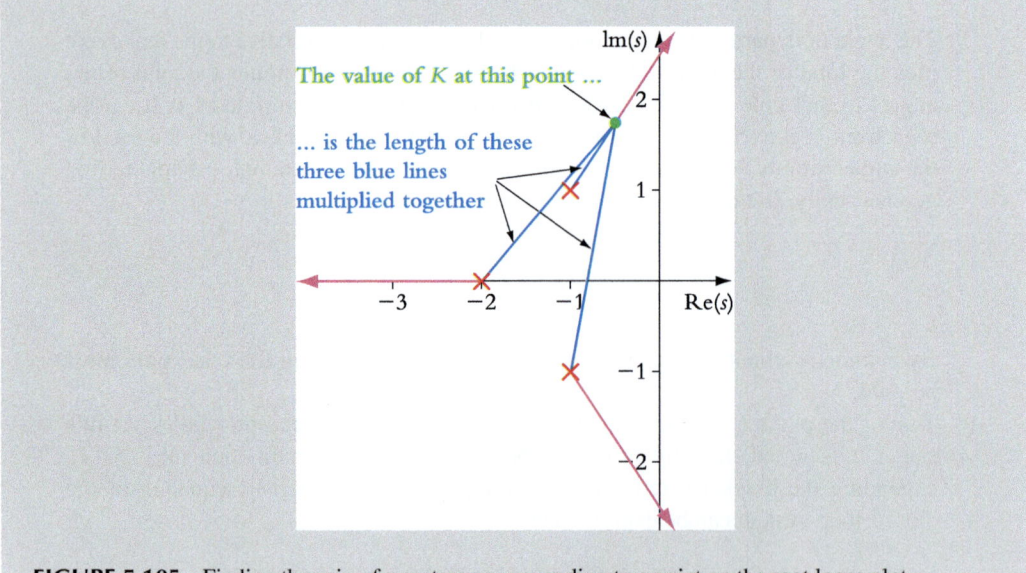

FIGURE 5.105   Finding the gain of a system corresponding to a point on the root locus plot.

5.14.1   Applications of the Root Locus Method

The root locus method can be used to explore further two of the examples from earlier in the present unit, namely the position controller and the antenna system, with particular reference to the stability and oscillatory response of the systems.

It has already been shown in Figure 5.60 that the open-loop transfer of the simplistic position controller system with a simple proportional amplifier is

$$\frac{V_P(s)}{V_D(s)} = \frac{K_a K_m K_s}{Js^2} \tag{5.179}$$

The root locus is very easy to plot (for varying amplifier gain) as we can immediately see:

- There are two poles, both of value $s = 0$; so, there are two paths
- As the poles are equal, they also form the breakout point
- There are no zeros; so, there are $2 - 0 = 2$ asymptotes
- The centroid of the poles is $s = 0$; so, the asymptotes radiate from the origin at an angle $\pm \pi/2$ to the real axis.

Without further ado, the root locus plot can be drawn as shown in Figure 5.106, showing the poles of the closed-loop transfer function as the amplifier gain is increased.

This illustrates that the root locus lies in the imaginary axis – in other words, the response is always oscillatory and is on the margin of stability, with any input resulting in a position that oscillates at a frequency dependent upon the amplifier gain, with the oscillations never dying away. However, if as before the simple amplifier is replaced with a PID controller/amplifier with gain K_c and proportional and integral time constants T_I and T_D, respectively, (and this time ignoring any additional external load), the open-loop transfer function now becomes

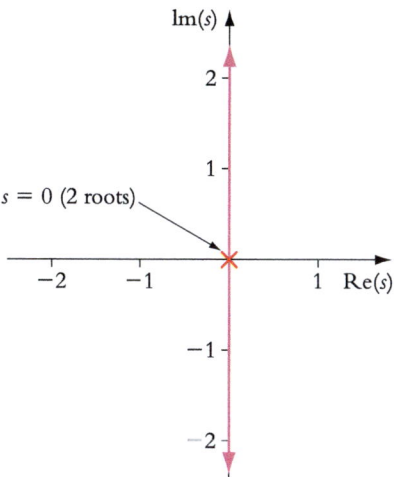

FIGURE 5.106   Root locus plot for a simplistic positioning system with proportional control.

$$\frac{V_P(s)}{V_D(s)} = \frac{K_c K_m K_s \left(1 + \dfrac{1}{T_I s} + T_D s\right)}{J s^2} = \frac{K_c K_m K_s T_D \left(s^2 + \dfrac{1}{T_D} s + \dfrac{1}{T_I T_D}\right)}{J s^3}$$

Once again, assuming that $T_D = 0.5$ s and $T_I = 1$ s, and assuming also the system parameters chosen in Section 5.11, the open-loop transfer function becomes:

$$\frac{V_P(s)}{V_D(s)} = \frac{K_c K_m K_s T_D \left(s^2 + 2s + 2\right)}{J s^3} = \frac{K_c K_m K_s T_D \left(s + 1 + j\right)\left(s + 1 - j\right)}{J s^3}$$

This system now has

- Three poles, all at $s = 0$
- Two zeros, at $s = (-1 + j)$ and $s = (-1 - j)$
- Two paths that arrive at the zeros at angles that can be found in a similar way to that for the angle of departure from poles: sum of angles from zeros – sum of angles from poles comes to $(2k + 1)\pi$ for integer k, that is odd multiples of π. In the case of the root at $s = (-1 - j)$, try $k = -1$; so, angle $= -\pi - (\pi/2 - 3 \times 3\pi/4) = 3\pi/4$, and by symmetry, the angle of arrival at the other complex zero is $-3\pi/4$. It can also be shown that the angles of departure from the three coincident poles at the origin are π and $\pm\pi/3$, though this is not obvious from the rules already described
- One asymptote at angle π radians, radiating from the point $s = 2$.

The root locus plot can now be drawn as shown in Figure 5.107.

It is seen that, as the gain is increased, the roots move into the right-hand half of the plane and the system becomes unstable; however, for larger values of K, the system becomes stable again as two of the closed-loop poles tend to the open-loop zeros in the left-hand half of the complex plane (indicating decaying oscillatory responses), and the third closed-loop pole tends to $-\infty$, indicating a rapidly decaying monotonic response. This example shows the usefulness of being able to visualise what is going on within a system that may initially seem to exhibit puzzling or counterintuitive behaviour, in this case showing unstable behaviour at small gains and becoming stable for larger

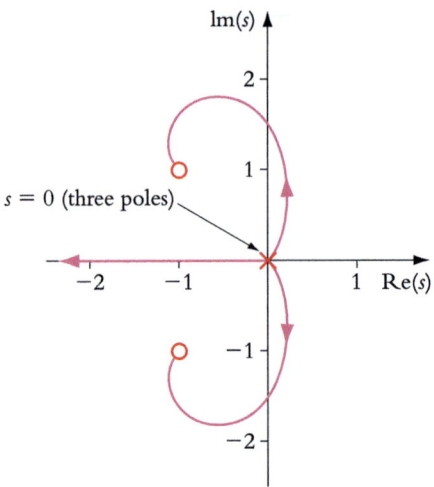

FIGURE 5.107 Root locus plot for a simplistic positioning system with PID control.

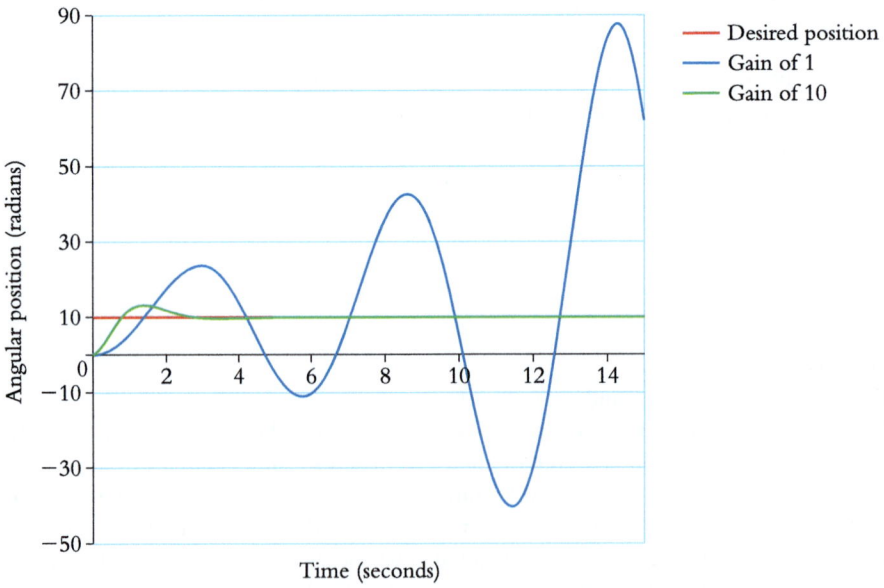

FIGURE 5.108 Response of positioning system with PID control for two values of gain.

gains. The practical implication of this is shown in Figure 5.108, which shows the response for gains of 1 and 10, respectively.

The root locus plot can also be drawn for the radar antenna problem. The block diagram in Figure 5.74 may be rearranged slightly by placing the common factor of K_t inside the feedback loop as shown in Figure 5.109.

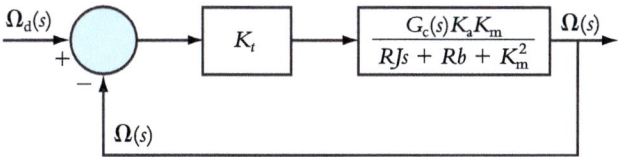

FIGURE 5.109   Rearranged block diagram for the complete antenna system.

After multiplying the terms on the forward path and substituting for proportional plus integral action, the following open-loop transfer function is obtained:

$$\text{Open loop transfer function} = \frac{G_c\left(s\right)K_t K_a K_m}{RJs + Rb + K_m^{\,2}}$$

$$= \frac{K_c\left(1 + 1/T_I s\right)K_t K_a K_m}{RJs + Rb + K_m^{\,2}} = \frac{K_c K_t K_a K_m\left(s + 1/T_I\right)}{RJs\left(s + \left(Rb + K_m^{\,2}\right)/RJ\right)}$$

This is now a second-order system with poles at $s = 0$ and $s = -(Rb + K_m^2)/RJ$, and a zero at $s = -1/T_I$. Using the same system constants as before, but with $T_I = 0.2$ s (as this involves more interesting behaviour than other values which are arguably more suitable), the poles are now at $s = 0$ and $s = -2$, and the zero is at $s = -5$. The root locus plot shown in Figure 5.110 can be drawn.

This shows that, as the gain is increased from zero, the behaviour is initially overdamped, then becomes (slightly) underdamped, and then becomes overdamped again. Such behaviour would not be obvious from examining the open-loop transfer function! The practical implication of this is shown via plots for $K_a = 0.1$ ($\zeta = 1.48$), 4 ($\zeta = 0.67$) and 20 ($\zeta = 1.1$) as shown in Figure 5.111.

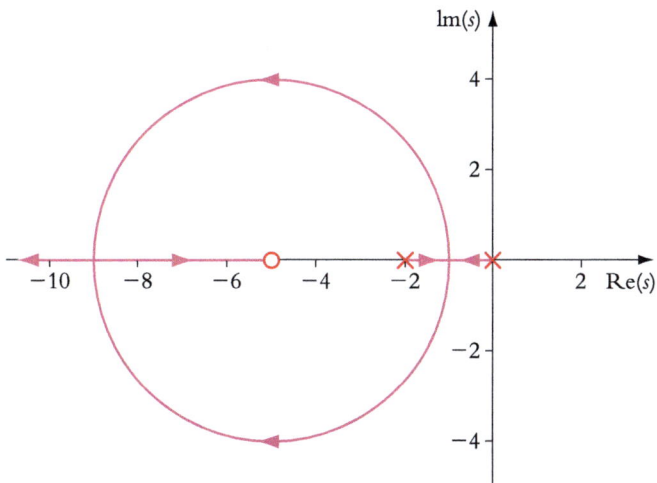

FIGURE 5.110   Root locus plot for the antenna system.

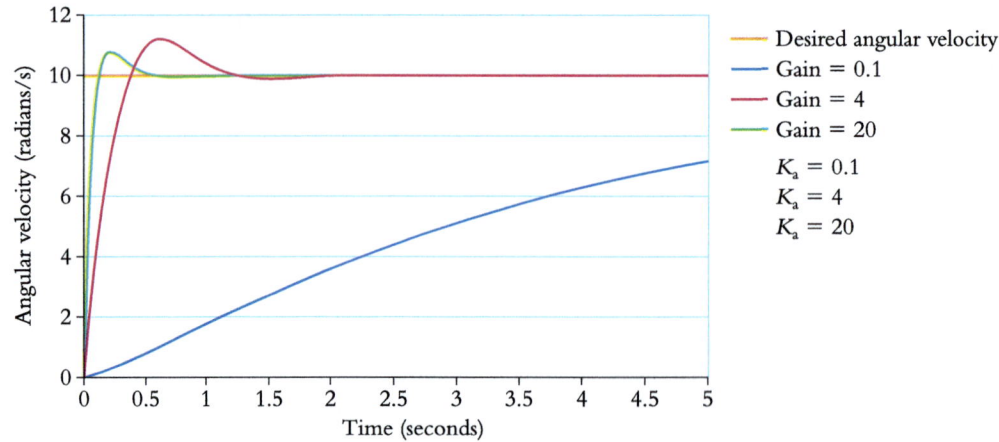

FIGURE 5.111 Responses of antenna system for various values of K_a.

LEARNING SUMMARY

At the end of this section, you should be able to:

✓ Draw root locus plots and interpret them in terms of system stability and response

---

ACKNOWLEDGEMENTS

The author is very grateful to a number of colleagues for their helpful input and suggestions on this unit, notably Alan Watson, Dimitrios Chronopoulos and David Branson III.

REFERENCES

Beardmore, R. http://www.roymech.org/Related/Control/root_locus.html accessed 16 August 2022.

Bennett, S., 1979, *A History of Control Engineering, 1800–1930*. Peter Peregrinus/IEE.

Bennett, S., 1993, *A History of Control Engineering, 1930–1955*, 1st ed. Peter Peregrinus.

Bolton, W., 1998, *Control Engineering*, 2nd ed., Pearson.

Brown, G. S. and Hall, A. C., 1946, "Dynamic Behavior and Design of Servomechanisms," *Trans. ASME*, 68, 503.

Dorf, R.C. and Bishop, R.H., 2021, *Modern Control Systems*, 14th ed., Pearson.

Evans, W.R., 1948, "Graphical analysis of control systems," *Transac. Amer. Inst. Electrical Engineers*, 67, 547–551.

Hall, A. C., 1943, The Analysis and Synthesis Of Linear Servomechanisms. Thesis, Massachusetts Institute of Technology.

Healey, M., 1967, *Tables of Laplace, Heaviside, Fourier and Z transforms*. Chambers.

Heaviside, O., 1893, "On Operators in Physical Mathematics", *Proc. Roy. Soc. London*, 52, 504–529.

"How Do Self Balancing Scooters Work?" https://bestelectrichoverboard.com/how-do-self-balancing-scooters-work/ accessed 16 August 2022.

Hurwitz, A., 1895, "Über die Bedingungen, unter welchen eine Gleichung nur Wurzeln mit negativen reellen Theilen besitzt," *Math. Ann.*, 46, 273–284.

Kreyszig, E. (2011), *Advanced engineering mathematics*, 10th ed. Wiley.

Lützen, J., 1979, "Heaviside's Operational Calculus and the Attempts to Rigorise It," *Arch. Hist. Exact Sci.*, 21, 161–200. https://doi.org/10.1007/BF00330405

Mathworks: Simulink™ product website Details can be found at: http://www.mathworks.com/products/simulink/ (accessed 16 August 2022).

McRuer, D.T., Graham, D. and Ashkenas, I., 1974, *Aircraft Dynamics and Automatic Control.* Princeton University Press.

Minorsky, N., 1922, "Directional stability of automatically steered bodies," *Journal of American Society of Naval Engineers*, 34, 280–309.

Nyquist, H., 1932, "Regeneration Theory," *Bell System Technical Journal*, 11, 126–147.

"Root Locus Plot Simulator", https://www.maxwell.vrac.puc-rio.br/26235/26235.PHP accessed 16 August 2022.

"RootLocs – plot Root Locus diagrams on your PC or Mac!" http://www.coppice.myzen.co.uk/RootLocs_Site/RootLocs.html (accessed 16 August 2022).

Routh, E.J., 1877, *A treatise on the stability of a given state of motion: Particularly steady motion.* Macmillan.

Spencer, A. J. M., Parker, D. F., Berry, D. S., England, A. E. H., [need other authors] (1977), *Engineering Mathematics*, Vol. 1. Van Nostrand Reinhold.

Van de Vegte, J., 1993 *Feedback Control Systems*, 3rd ed., Pearson.

6 Structural Vibration

Edward Williams and Alastair Campbell Ritchie
University of Nottingham, Nottingham, United Kingdom

6.1 INTRODUCTION

Vibration in a structure is characterized by a cyclic displacement, with a characteristic frequency and amplitude. This is sometimes desirable in a musical instrument or a loudspeaker, but in most engineering situations it should be minimized.

You will be familiar with the sensation of vibration from the action of a buzzing mobile phone. Musical instruments rely on vibrations of a particular frequency to produce the notes played by a musician, and less obviously, vibration produces the sounds in speakers or earphones. While these provide pleasure to the user, vibration in most engineering structures is something to be avoided.

This unit will describe some basic causes and effects of vibration and show you how to model the vibration of structures mathematically.

A recurring theme is **resonance**. If a structure is excited at a frequency close to one of its natural frequencies, significant vibration will occur.

Large alternating displacements often produce large alternating stresses. It is no surprise, therefore, that **resonance-induced fatigue** is a major cause of in-service failures – normally when the effects of vibration have not been considered adequately at the design stage.

In design, it is common to calculate the stresses due to static loads and to use a reserve factor to make allowance for effects that have been excluded from the analysis. An alternating force that induces resonance can easily produce stresses that are 100 times higher than a static force of the same magnitude. This compares to stress margins (the ratio of ultimate stress to expected in-service stresses) varying from 0.5 (aerospace) to 3 (bridges and buildings).

The most dramatic failure caused by resonant vibration was probably that of the Tacoma Narrows Bridge in Washington State (Figure 6.1). It collapsed in a truly spectacular fashion in 1940, after only four months of use. From its opening, bending waves were set up if the wind speed rose and people would drive across 'Galloping Gertie' just for the fun of the experience (Disney World didn't exist in 1940!). On the 7th of November, a wind speed of only 42 mph produced a severe torsional vibration, due to the shedding of vortices from the deck of the bridge. At its most extreme, the sides of the bridge were moving up and down by over 8 m. Take a look at the video at www.tinyurl.com/2ca5mh. Seeing is believing.

Designers now understand more about vertical bending and torsional vibration that can be caused by wind or by loads moving across bridges. Or at least they thought they did until the opening of the Millennium Bridge in London showed that there was still more to learn (Figure 6.2). It was built for pedestrians and cyclists, and the designers had taken account of the vertical forces from their feet as they walked across. A big crowd had gathered for the opening ceremony and when they started to walk across, the bridge began to sway from side-to-side. No structure is completely rigid, and the crowd found that the bridge swayed slightly from side to side. To help steady themselves, they moved their feet apart as they walked and in doing so exerted a small alternating horizontal force onto the bridge with each step. The problem was that in order to keep their balance, people on the bridge responded in the same way at the same time and the combined effect was sufficient to excite a horizontal bending mode of the bridge. Take a look at the video at www.tinyurl.com/33xeyle. The problem was cured by a combination of viscous dampers and tuned-mass absorbers.

DOI: 10.1201/9780429319495-6

FIGURE 6.1   Wind-induced torsional vibration of the Tacoma Narrows Bridge shortly before it collapsed.

FIGURE 6.2   The crowd of pedestrians who had gathered for the opening of the Millennium Bridge in London caused a pronounced side-to-side vibration when they began to walk across.

Section 6.2 explains how to calculate the resonant frequencies of structures and see how structures deform when they vibrate, termed their 'mode shapes'. Structures only vibrate if they are subjected to some form of excitation, and Section 6.3 considers the response of structures that exhibit only one resonant frequency. Section 6.4 extends this to structures with many resonant frequencies. Even the most sophisticated computer analysis has to make assumptions and simplifications that may not reflect the actual properties of the structure. Section 6.5 describes how a structure's resonant frequencies and mode shapes can be measured experimentally. At the preliminary stage of a design, approximate values can help determine whether vibration is likely to be an issue, and Section 6.6 introduces some useful techniques for obtaining these. Finally, Section 6.7 introduces two common techniques for controlling the potentially damaging effects of vibration.

6.2   NATURAL FREQUENCIES AND MODE SHAPES

6.2.1   Introduction

Cups and glasses will ring in response to a tap on the rim – and every cup has a characteristic resonant frequency. The sound heard is emitted at the dominant natural frequency of the structure. As noted in Section 6.1, resonance-induced fatigue is a major cause of premature structural failure, and this section discusses several ways of modelling structures so that their natural frequencies can be calculated.

Before doing so, we introduce the classification of structure in terms of ***degrees of freedom***. These can be thought of as 'possible motions'. The number of degrees of freedom for a structure:

- Is related to the number of resonant frequencies it has;
- Affects the number of motion coordinates needed to define its position.

A rigid body can have up to six degrees of freedom: three translations and three rotations. In addition to the six rigid body freedoms, a flexible body can have vibrational modes, such as bending and twisting, in addition to considerations due to flexible couplings between components.

A key step is to decide how many degrees of freedom are appropriate. Structures are generally constrained so that their freedom to move is limited, and even in vehicles, ships and aircraft, this assumption holds – vibrational motion will be relative to another body. Even so, this can still leave perhaps thousands of possible ways (***modes of vibration***) in which a structure may vibrate. In most cases, we are only interested in a few dominant modes, and we will look for a mathematical model that concentrates on these. For example, if we wanted to study the way an aircraft's wing vibrates when it hits an air pocket, we would make a model of the wing and its attachment to the aircraft, ignoring other effects that hitting the air pocket would have.

We will start by looking at structures that can be characterized by just one type of motion. We will then move on to those that can exhibit two or more types of motion.

The mathematical descriptions we will develop relate the motion of the structure to the forces that cause it. Central to this process is ***Newton's second law of motion***.

In linear motion (translation), this can be expressed as

$$\overrightarrow{\begin{array}{c}\text{Resultant force in the direction of}\\\text{the acceleration}\end{array}} = \text{Mass} \times \overrightarrow{\begin{array}{c}\text{Absolute acceleration of}\\\text{the centre of mass}\end{array}}$$

This will be familiar as

$$\boldsymbol{F} = m \times \boldsymbol{a}$$

where $\boldsymbol{F}$ is the force acting on a body of mass m and $\boldsymbol{a}$ is the resultant acceleration. Note that $\boldsymbol{F}$ and $\boldsymbol{a}$ are both vectors while the mass is a scalar quantity. A second important form of Newton's second law is for torsional forces:

$$\overrightarrow{\begin{array}{c}\text{Resultant torque}\left(\text{moment}\right)\\\text{about the axis in the}\\\text{direction of the acceleration}\end{array}} = \text{Moment of inertia about the axis} \times \overrightarrow{\begin{array}{c}\text{Absolute}\\\text{angular}\\\text{accelaration}\end{array}}$$

Or, mathematically,

$$\boldsymbol{T} = \boldsymbol{I} \times \boldsymbol{\alpha}$$

where T is the torque acting on the body, I is the moment of inertia about the axis of rotation, and α is the angular acceleration $\frac{d\omega}{dt}$. In this case, I is often expressed as a matrix.

It is important to note that 'the axis' about which the torque is taken *must* be one of the following.

- A fixed axis.
- An 'instantaneous centre of rotation' (this is a point on a body that is stationary for an instant, such as the point of contact of a tyre rolling along the road).
- The centre of mass.

No other axis is allowed.

For a fuller discussion of these equations, see Unit 6 in Part 1.

When solving problems, the most effective approach is usually to write it down as it's said, noting that forces and velocities are vector quantities. Put the forces (resolved into the vector direction that you have chosen for acceleration) on the left of the equals sign and the mass and acceleration terms on the right. Don't mix the two together. 'Mass × acceleration' is *not* itself a force. It describes the effect that the applied forces have on the mass.

6.2.2 STRUCTURES WITH ONE DOMINANT MODE OF VIBRATION

Initially, we look at structures that can be modelled with just one degree of freedom. This is the simplest model for a vibrating structure. The model normally takes the form of a rigid body connected to a fixed point by one or more springs of negligible mass. It works well for structures with one resonance mode or where one resonance mode dominates the vibration behaviour. It gives good insight into vibration behaviour and is often used as a first approximation for more complicated structures.

In this unit, we will look at a number of different examples where single-degree-of-freedom dynamic models are appropriate, but the general approach to the analysis will be the same in all cases and will involve the following steps:

1. Convert the physical structure into a dynamic mass–spring model. Drawing a diagram at this point will help with the next step.
2. Draw a free-body diagram. Displacing the body from its equilibrium position will create forces in the restraining springs that will try to return the body to equilibrium. The free-body diagram can be thought of as a 'snapshot' of the state of the system when it has moved away from equilibrium by a chosen amount. Drawing the free-body diagram is the key step in the analysis and should be tackled systematically. There are three stages:
     i.   Start with the element of interest in equilibrium and draw it as a free body. To create the 'free' body, draw it without any of the restraining springs. The forces exerted by the springs will be added in stage (iii).
     ii. Select a motion coordinate to describe how the system will deflect from its equilibrium position and mark it on the diagram.
     iii. Apply a positive deflection in the chosen motion coordinate, identify the forces (and/or moments) that result and draw them on the diagram. It is critical that the positive directions of the forces due to a positive deflection are shown correctly.
3. Apply the appropriate form of Newton's second law of motion to give the equation of motion for the system.

At this stage, we will concentrate on finding the ***natural frequency*** of the systems. This is the frequency at which a system will vibrate when displaced from equilibrium and then released. Later, we will consider the effects of external excitation and damping, but these are omitted for the moment.

When we look at the effects of excitation, we will find that, for most engineering structures, there is a maximum response if the excitation frequency coincides with the natural frequency (which is why we need to know what its value is). This effect is called ***resonance***, and the term ***resonant frequency*** is often used instead of natural frequency (although, strictly speaking, the two are different as we shall see later).

6.2.2.1 Example 1: Simple Mass–Spring System

If a mass m [kg] is suspended from a spring of stiffness k [N/m], it will move down (in the x direction) under the effect of gravity and stretch the spring by a distance x_{eq} before reaching its static equilibrium position (Figure 6.3).

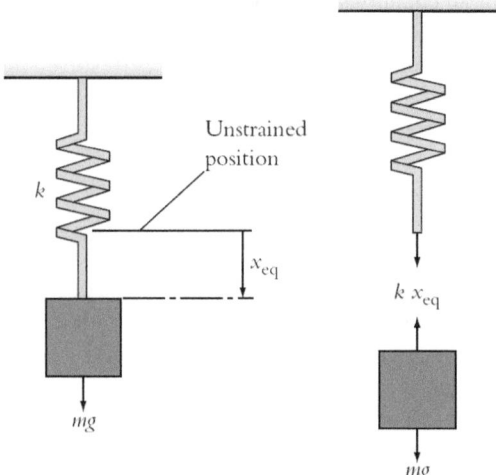

FIGURE 6.3 Schematic representation of mass-spring system.

Once it is in equilibrium, the force in the spring (given by the stiffness multiplied by the change of length) exactly balances the gravitational force on the mass and hence

$$mg = kx_{eq}$$

What then happens if the mass is given a *further* downward displacement x away from equilibrium and then released (Figure 6.4)?

The displacement x produces an additional tensile force kx in the spring as shown in Figure 6.4(b). Since, we know from the static equilibrium case that $mg = kx_{eq}$, the magnitude of the resultant force is kx as shown in Figure 6.4(c).

Remembering that the displacement and the force are vectors, we see that a displacement from equilibrium in the $+x$ direction produces an opposite resultant force in the $-x$ direction on the mass. Since the force will always act to return the mass to equilibrium, the term ***restoring force*** is often used.

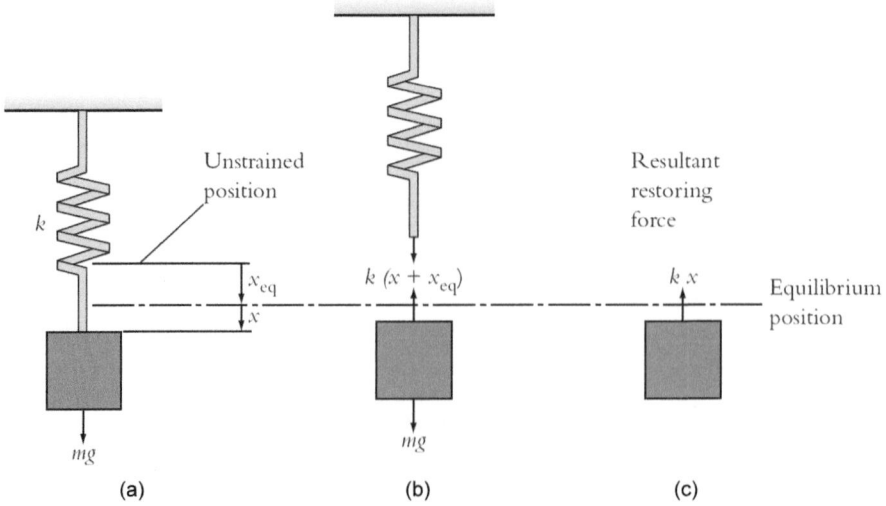

FIGURE 6.4   (a) Equilibrium position; (b) additional displacement; (c) free body diagram of mass for mass-spring system.

If the mass is displaced by a distance x at $t = 0$, then there will be a resultant force on the mass to return it to the equilibrium position

$$m\ddot{x} = -kx$$

The displacement and velocity are plotted against time in Figure 6.5 for a system that is displaced at $t = 0$ then released and allowed to oscillate freely. In this case, a positive value corresponds to the direction in Figure 6.4. As would be expected, the acceleration is proportional to the displacement and hence the maximum velocity magnitudes (A, C) correspond to the points at which the displacement is zero and the mass comes to a stop momentarily when the spring is at its maximum or minimum extent (B, D).

We see that the resultant force on the mass depends only on the displacement measured from the equilibrium position. Here, and in all other problems, the static forces in springs exactly balance any gravitational forces on the mass under equilibrium conditions. Because they always cancel each other, it is safe to ignore them and start at the equilibrium position and consider displacements away from that position.

Note: Throughout this unit, we assume that displacements are small and that the spring stiffnesses are constant. This means that the force in a spring varies linearly with displacement. Such systems are called **linear systems**. If displacements are not small, the stiffness may vary with displacement, but this is not considered here.

To create a mathematical model of the problem, the first step is to replace the physical system (Figure 6.6) with a dynamic mass–spring model In this simple case, the mass is represented by a square block and the coil spring by a zig-zag representation.

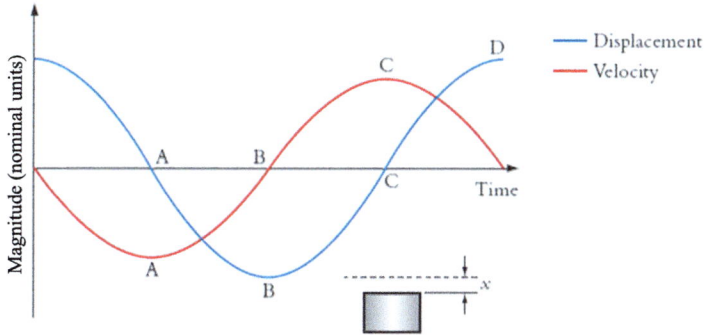

FIGURE 6.5   Velocity and displacement plotted against time for mass-spring system.

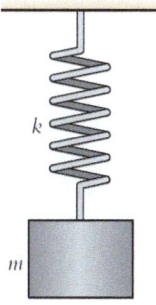

FIGURE 6.6   Physical system.

Step 1 Dynamic mass–spring model

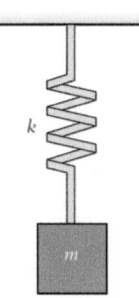

Step 2 The second stage is to create a ***free-body*** diagram for the element of interest in this system, which is the mass.

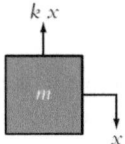

    (i)   Remove the spring to leave the mass by itself.
   (ii)   Mark the chosen positive direction for displacement.
  (iii)   If a positive displacement is applied to the mass, write down the expression for the magnitude of the force and add an arrow to give its direction.

Step 3 Equation of motion

Applying Newton's second law, the resultant force will cause the mass to accelerate. In this example, the tension in the spring is directed upwards in response to a downward (positive) displacement, hence:

$$-kx = m\ddot{x}$$

This is normally rearranged to give a second-order ordinary differential equation with constant coefficients:

$$m\ddot{x} + kx = 0 \tag{6.1}$$

Any sinusoidal function can satisfy this equation, but from the earlier description of what happens to the mass when it is displaced downwards and then released, the most suitable mathematical form would be $x(t) = X\cos(\omega t)$. This describes a sinusoidally varying displacement at frequency ω with magnitude X and a maximum value at $t = 0$. The frequency of the vibration is the ***natural frequency*** and this form of displacement is formally termed ***simple harmonic motion***.

For a displacement given by $x(t) = X\cos\omega t$, $\dot{x} = -\omega X\sin\omega t$ and $\ddot{x} = -\omega^2 X\cos\omega t$.

Substituting for $x(t)$ and its derivatives into the equation of motion (Equation 6.1) gives:

$$-m\omega^2 X\cos\omega t + kX\cos\omega t = 0$$

Cancelling $X \cos \omega t$,

$$-m\omega^2 + k = 0 \quad \omega^2 = \frac{k}{m}$$

The natural frequency ω for this system is therefore given by $\sqrt{\dfrac{k}{m}}$. The symbol ω_n is normally used for the natural frequency. When substituting numerical values into the equation of motion, ω must have the units of ***radians/s*** to make the equation consistent. However, the value would normally be quoted (in a report, for example) as a frequency in Hz (hertz, cycles per second). The two are related as follows:

$$f_n\left[\text{Hz}\right] = \frac{\omega_n}{2\pi}\left[\text{rad}/\text{s}\right]$$

You will find that other systems have different equations of motion but provided that we can assume that displacements remain small, all will have a similar form:

$$M\ddot{z} + Kz = 0$$

where z is the chosen motion coordinate. Following the above analysis, we find that the natural frequency is given by

$$\omega_n = \sqrt{\frac{K}{M}}\left[\text{rad}/\text{s}\right]$$

6.2.2.2   Example 2: Vertical Vibration of a Block on a Flexible Cantilever Beam

This system can be made to vibrate by lifting the block up (causing the supporting beam to bend) and then releasing it, resulting in an up-and-down oscillation. The block will be treated as a rigid mass and the supporting beam as a massless spring. The ***bending stiffness*** of the beam is given by $k_\text{B} = \dfrac{3EI_2}{L^3}$. Note that the symbol I_2 in this formula is the second moment of area for the beam. Since displacements are assumed to be small, we consider only the vertical translation of the mass (Figure 6.7 and Figure 6.8).

Step 1 Dynamic model
For a small displacement x from the equilibrium position, the force acting to restore the mass to its equilibrium position is $F = -k_b x$.
   Applying Newton's second law,

$$m\ddot{x} = -k_b x$$

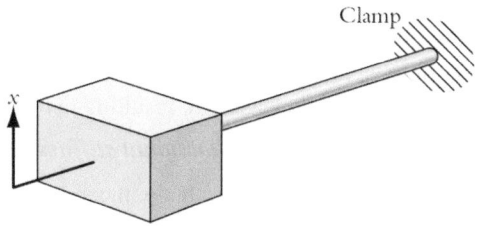

FIGURE 6.7   Physical system.

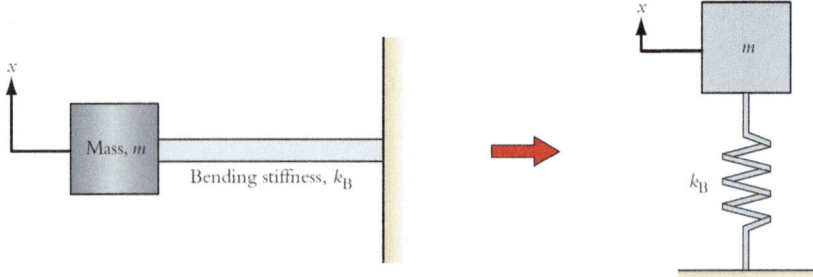

FIGURE 6.8 Model of vertical motion of supported mass for small displacements.

Hence, the equation of motion for the system will be given by:

$$m\ddot{x} + k_b x = 0$$

To solve this equation, a variety of methods can be used. To solve using the conventional general solution and particular integral method, the second order differential equation $m\ddot{x} + k_b x = 0$ has the general solution:

$$x(t) = A\sin(\omega_n t) + B\cos(\omega_n t)$$

where ω_n, the natural frequency is given by

$$\omega_n = \sqrt{\frac{k_B}{m}} \tag{6.2}$$

And particular integral

$$x(t) = 0$$

The initial velocity will be given by

$$\dot{x}(t) = A\omega_n \cos(\omega_n t) - B\omega_n \sin(\omega_n t)$$

From the boundary conditions for a small displacement δ at $t = 0$ and no initial velocity, $A = 0$ and $B = \delta$. Hence,

$$x(t) = \delta\cos(\omega_n t)$$

As can be seen, the natural frequency $\omega_n = \sqrt{k_B/m}$ is found rapidly – and often, in structural vibration, this may be the only result needed.

6.2.2.3   Example 3: Torsional Vibration of a Block on a Cantilever Beam

The block and beam from the previous example can also be made to vibrate by rotating the block about the axis of the support beam and then releasing it, resulting in torsional vibration. In this case, the beam can be modelled as a torsion spring with **torsional stiffness** $k_T = \dfrac{GJ}{L}$ (with units of Nm/rad). The block can again be modelled as a rigid body, but with a moment of inertia I about the beam axis, which we will assume to be a fixed axis. Methods to calculate the moment of inertia of a rigid body are covered in Unit 6 of Part 1.

Step 1 Dynamic model
It is convenient to use rotational vector notation in this example. As shown below figure the angular displacement vector acts along the axis of rotation in a direction given by the 'right-hand rule'.

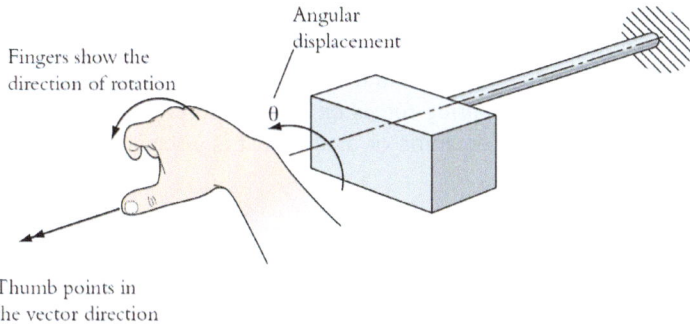

Step 2 Free-body diagram
If the block rotates by an angle θ to the position shown by the dashed lines in the figure below, then the induced torque in the support beam will be $k_T\theta$. This will act in a direction to oppose the imposed rotation and to return the block to its equilibrium position.

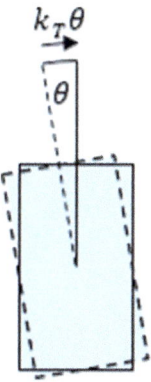

Step 3 Equation of motion
Applying Newton's second law for rotation $\left(\text{torque} = I\ddot{\theta}\right)$:

$$I\ddot{\theta} = -k_T\theta$$

Rearranging:

$$I\ddot{\theta} + k_T\theta = 0$$

From the coefficients in the equation of motion, the natural frequency for this system is therefore

$$\omega_n = \sqrt{\frac{k_T}{I}} \tag{6.3}$$

6.2.2.4 Example 4: Rocker System

This example consists of a rigid bar of negligible mass with a fixed pivot at one end and a large mass attached at the other. The rocking motion about the pivot is restrained by two springs, one attached to the mass and the other that is connected to the bar at point B.

Note: We will use the angular displacement of the bar about the fixed pivot as the motion coordinate and assume that this displacement is small. This makes the assumptions that $\cos\theta \approx 1$ and $\sin\theta \approx \tan\theta \approx \theta$ are valid. Taken together with the earlier assumption that stiffness values are constant, the linear dependence of the spring forces on the motion coordinate, θ, will be maintained. As a result, the model will give an equation of motion that is a second-order ordinary differential equation with constant coefficients.

Step 1 Dynamic model

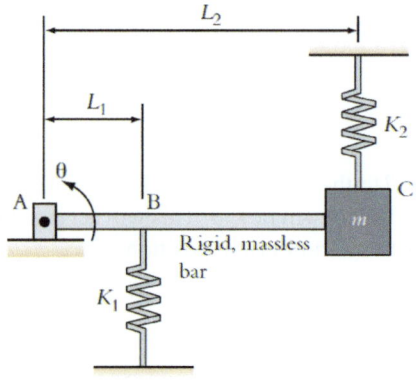

Step 2 Free–body diagram

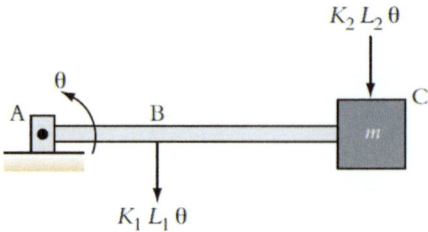

For a positive rotation of the bar, point B will move up by $L_1\theta$. This puts the spring K_1 into tension, resulting in a downward force on the bar. When point C moves up by $L_2\theta$, the spring K_2 is put into compression resulting in a downward force on the bar.

Step 3 Equation of motion
The moment of inertia of the mass about the pivot is given by

$$I = mL_2{}^2$$

The net moment about the fixed pivot A of the forces $M(A)$ is given by

$$M(A) \circlearrowright \quad M(A) = -K_1 L_1 \theta \times L_1 - K_2 L_2 \theta \times L_2$$

This moment will result in an angular acceleration $\ddot{\theta}$ such that

$$M(A) = I\ddot{\theta} = mL_2^2 \ddot{\theta} = -K_1 L_1^2 \theta - K_2 L_2^2 \theta$$

Rearranging,

$$mL_2^2 \ddot{\theta} + \left(K_1 L_1^2 + K_2 L_2^2 \right) \theta = 0$$

From the coefficients of $\ddot{\theta}$ and θ, the expression for the natural frequency is

$$\omega_n = \sqrt{\frac{K_1 L_1^2 + K_2 L_2^2}{mL_2^2}} \tag{6.4}$$

6.2.3 LUMPED MASS–SPRING SYSTEMS

Many structures exhibit more than one mode of vibration, each with a different frequency. For example, the Tacoma Narrows Bridge had become known for its bending vibration almost from the day it opened, but it was a torsional vibration excited by a particular speed of crosswind that caused its dramatic collapse.

Some structures can be approximated by several rigid bodies connected by massless springs. The analysis approach is an extension of the one used for single-degree-of-freedom systems, except that each of the rigid elements will have a separate free-body diagram. Each will also have its own equation of motion.

6.2.3.1 Example 1: Single-Axle Caravan

In this example, the body of the caravan is assumed to behave as a rigid mass and is connected to the axle (also assumed to be a rigid mass) by road springs in the suspension. The axle is separated from the road by flexible tyres. The mass of the tyres and the suspension springs is assumed to be negligible in comparison with the body and axle parts. A suitable dynamic model for studying vertical vibration is shown in Figure 6.9.

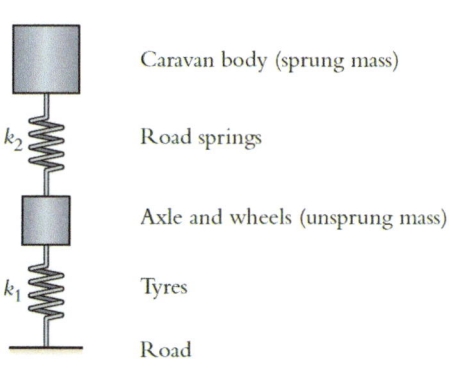

FIGURE 6.9 Physical system model of single axle caravan.

Since the body and the axle can move separately from each other, this model has two degrees of freedom (two independent possible motions). Two coordinates are needed to describe how the system moves: axle displacement (x_1) and body displacement (x_2).

This model will have two natural frequencies, each of which will have a characteristic pattern of displacement – its *mode shape*.

Step 1 Dynamic mass–spring model
Whenever a problem is presented, the first step is to draw a simplified diagram of the system – this will help to clarify the elements of the system, degrees of freedom and directions of forces and motion.

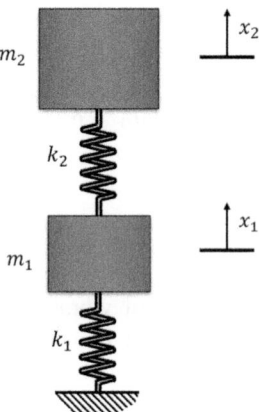

The system is modelled as a smaller mass m_1 (the wheels and axle) connected to the ground through a spring k_1 (representing the behaviour of the pneumatic tyres) and to the caravan body m_2 through a spring k_2 which represents the suspension system. At rest, the system is taken to be in equilibrium and so x_1 and x_2 are displacements from the equilibrium position.

To begin the analysis, we consider the forces acting on mass m_1. To find these, we can draw free-body diagrams of the two springs and two masses, as shown in Figures 6.10–6.12.

Note that in all of these cases, the springs are shown to be exerting a tensile force (pulling on the masses) – this will not always be the case, as the spring may also be in compression, but it is vital to be consistent in following the convention.

The equation of motion for mass m_1 is, therefore,

$$m_1\ddot{x}_1 = k_2 x_2 - (k_2 - k_1)x_1$$

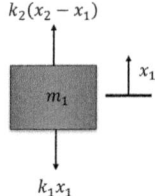

FIGURE 6.10 Free-body diagram for mass m_1.

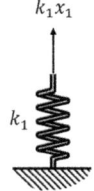

FIGURE 6.11   Free-body diagram for spring k_1.

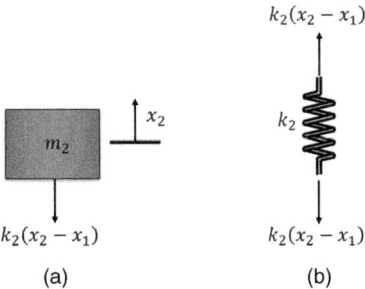

(a)   (b)

FIGURE 6.12   (a) Free-body diagram for mass m_2, (b) Free-body diagram for spring k_2.

And the equation of motion for mass m_2 will be given by

$$m_2\ddot{x}_2 = -k_2\left(x_2 - x_1\right)$$

Rearranging these equations gives

$$m_1\ddot{x}_1 + \left(k_2 - k_1\right)x_1 - k_2 x_2 = 0 \tag{6.5}$$

$$m_2\ddot{x}_2 + k_2\left(x_2 - x_1\right) = 0 \tag{6.6}$$

To express these in the matrix vector form, the acceleration vector is given by: $\{\ddot{x}\} = \begin{Bmatrix} \ddot{x}_1 \\ \ddot{x}_2 \end{Bmatrix}$ and the displacement vector is $\{x\} = \begin{Bmatrix} x_1 \\ x_2 \end{Bmatrix}$. The mass matrix is $[\mathrm{M}] = \begin{bmatrix} m_1 & 0 \\ 0 & m_2 \end{bmatrix}$ and the spring stiffness matrix is given by $[K] = \begin{bmatrix} (k_1 + k_2) & -k_2 \\ -k_2 & k_2 \end{bmatrix}$. Hence, Equations 6.5 and 6.6 become

$$\begin{bmatrix} m_1 & 0 \\ 0 & m_2 \end{bmatrix}\begin{Bmatrix} \ddot{x}_1 \\ \ddot{x}_2 \end{Bmatrix} + \begin{bmatrix} (k_1 + k_2) & -k_2 \\ -k_2 & k_2 \end{bmatrix}\begin{Bmatrix} x_1 \\ x_2 \end{Bmatrix} = \begin{Bmatrix} 0 \\ 0 \end{Bmatrix}$$

or

$$[\mathrm{M}]\{\ddot{x}\} + [K]\{x\} = \{0\}$$

At this point, it is possible to double-check the working – from the terms of the matrices:

1. The terms in the leading diagonals of $[M]$ and $[K]$ must always be *always positive*;
2. The off-diagonal terms may be positive or negative;
3. $[M]$ and $[K]$ are *often symmetric* about the leading diagonal.

It is also worth noting that the equations are **coupled**: each involves both x_1 and x_2. This makes sense in physical terms: if one mass moves, this will affect the springs and the other mass will also move. The motion of one mass cannot occur independently of the other, hence the equations must be solved simultaneously.

For free vibration of the system at one of its natural frequencies, ω, the motion of each mass will be sinusoidal. The system's vibration is given by $x_1(t) = X_1 \cos \omega t$ and $x_2(t) = X_2 \cos \omega t$ where X_1 and X_2 are the amplitudes of vibration of m_1 and m_2, respectively:

$$m_1 \ddot{x}_1 + (k_1 + k_2) x_1 - k_2 x_2 = -m_1 X_1 \omega^2 \cos \omega t + (k_1 + k_2) X_1 \cos \omega t - k_2 X_2 \cos \omega t = 0$$

$$m_2 \ddot{x}_2 - k_2 x_1 + k_2 x_2 = -m_2 X_2 \omega^2 \cos \omega t - k_2 X_1 \cos \omega t + k_2 X_2 \cos \omega t = 0$$

Hence

$$-m_1 X_1 \omega^2 + (k_1 + k_2) X_1 - k_2 X_2 = 0$$

and

$$-m_2 X_2 \omega^2 - k_2 X_1 + k_2 X_2 = 0$$

When this is presented in the matrix form,

$$\begin{bmatrix} (k_1 + k_2) - m_1 \omega^2 & -k_2 \\ -k_2 & k_2 - m_2 \omega^2 \end{bmatrix} \tag{6.7}$$

$$\begin{Bmatrix} X_1 \\ X_2 \end{Bmatrix} = \begin{Bmatrix} 0 \\ 0 \end{Bmatrix} \tag{6.8}$$

or

$$([K] - \omega^2 [M])\{X\} = \{0\} \tag{6.9}$$

Alternatively,

$$[Z]\{X\} = \{0\}$$

where

$$[Z] = [K] - C[M]$$

From Equation 6.9, it can be seen that there will be vectors $\{X\}$ which satisfy Equation (6.9) and values of ω^2 for which the determinant of the matrix $[Z]$ is zero, similar to an eigenvalue–eigenvector relationship. In practice, it's often easiest to use a computer to solve these problems using software such as MATLAB.

It is possible to solve these problems analytically if there are only two or three degrees of freedom: For a non-trivial solution of Equation (6.9), the determinant of matrix $|Z|$ must be zero[1]:

$$|Z| = \begin{vmatrix} (k_1 + k_2) - m_1 \omega^2 & -k_2 \\ -k_2 & k_2 - m_2 \omega^2 \end{vmatrix} = 0$$

Multiplying out the determinant in this case gives

$$|Z| = \left((k_1 + k_2) - m_1\omega^2\right)\left(k_2 - m_2\omega^2\right) - \left(-k_2\right)^2 = 0$$

$$m_1 m_2 \omega^4 - \left(m_1 k_2 + m_2\left(k_1 + k_2\right)\right)\omega^2 + k_1 k_2 = 0 \tag{6.10}$$

Equation (6.10) is known as the **Frequency equation**. For this problem, it is a quadratic in ω^2 and will have two roots, ω_{n1}^2 and ω_{n2}^2, where ω_{n1} and ω_{n2} are the two natural frequencies of the system. To find the corresponding mode shapes, we substitute each root back into Equation (6.7) or (6.8) to get the relationship between X_1 and X_2. Since Equations (6.7) and (6.8) form a pair of homogeneous equations, we cannot find unique solutions for X_1 and X_2 separately.

As we are interested in the relative magnitudes of X_1 and X_2, *we can set the* amplitude of one component of $\{X\}$ equal to unity and then find the other amplitude relative to this at the natural frequency ω_{n1}. For example, if we choose $X_2 = 1$, Equation (6.8) becomes:

$$\left(k_1 + k_2 - \omega_{n1}^2 m_1\right) X_1 - k_2 \cdot 1 = 0$$

Hence,

$$\begin{Bmatrix} X_1 \\ X_2 \end{Bmatrix} = \begin{Bmatrix} \dfrac{k_2}{k_1 + k_2 - \omega_{n1}^2 m_1} \\ 1.0 \end{Bmatrix} \tag{6.11}$$

The vector $\begin{Bmatrix} X_1 \\ X_2 \end{Bmatrix}$ is the **mode shape**, which tells us the relative amplitude and phase of the two masses for vibration at frequency ω_{n1}.

Numerical example
For the caravan system shown in Figure 6.15, let $m_1 = 70\,\text{kg}$, $m_2 = 350\,\text{kg}$, $k_1 = 400\,\text{Nm}^{-1}$ and $k_2 = 40\,\text{Nm}^{-1}$.

Natural frequencies
Putting these values into Equation (6.10) gives

$$24500\omega^4 - 156800\omega^2 + 16000 = 0$$

Solving for ω^2 gives two roots: $\omega_{n1}^2 = 0.104\,\text{s}^{-2}$ and $\omega_{n2}^2 = 6.296\,\text{s}^{-2}$. Hence, $\omega_{n1} = 0.322\,\text{rads}^{-1}$ (0.051 Hz) and $\omega_{n2} = 2.51\,\text{rads}^{-1}\left(0.4\,\text{Hz}\right)$

6.2.3.2 Mode Shapes
6.2.3.2.1 *Mode 1*
From Equation (6.11), for $\omega_{n1}^2 = 0.104\,\text{s}^{-2}$

$$\begin{Bmatrix} X_1 \\ X_2 \end{Bmatrix} = \begin{Bmatrix} \dfrac{k_2}{k_1 + k_2 - \omega_{n1}^2 m_1} \\ 1.0 \end{Bmatrix} = \begin{Bmatrix} \dfrac{40}{440 - 0.104 \times 70} \\ 1.0 \end{Bmatrix}$$

$$\begin{Bmatrix} X_1 \\ X_2 \end{Bmatrix} = \begin{Bmatrix} 0.092 \\ 1.0 \end{Bmatrix}$$

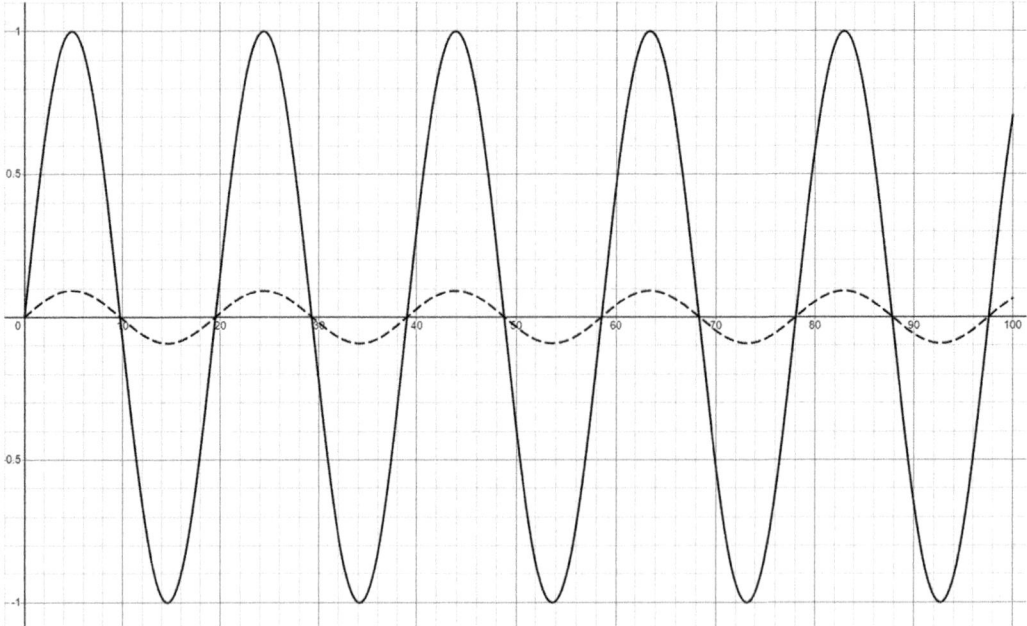

FIGURE 6.13 Plot of the relative displacement of the body X_2 (solid line) and the axle X_1 (dashed line) for mode 1. Time given in seconds.

In this mode, the axle and the body vibrate in phase with each other (X_1 and X_2 are both positive values) and the amplitude of the displacement of the body is about 10 times that of the axle, as can be seen from the plot of relative displacements in Figure 6.13. This can be attributed to the higher stiffness of the tyres compared to the springs in the suspension. This low-frequency vibration (0.05 Hz) is commonly called the 'bounce' mode of the vehicle.

Since the axle movement (X_1) is small compared with that of the body in this mode, we could have made the assumption that the tyres were effectively rigid, implying that the axle will not vibrate. This would have given a single-degree-of-freedom system with the body mass supported only by the suspension springs. The natural frequency for this system is 0.054 Hz, which is similar to the figure of 0.051 Hz obtained above.

6.2.3.2.2 Mode 2

From Equation (6.11), for $\omega_{n2}{}^2 = 6.296\,\mathrm{s}^{-2}$

$$\begin{Bmatrix} X_1 \\ X_2 \end{Bmatrix} = \begin{Bmatrix} \dfrac{k_2}{k_1 + k_2 - \omega_{n2}{}^2 m_1} \\ 1.0 \end{Bmatrix} = \begin{Bmatrix} \dfrac{40}{440 - 6.296 \times 70} \\ 1.0 \end{Bmatrix}$$

$$\begin{Bmatrix} X_1 \\ X_2 \end{Bmatrix} = \begin{Bmatrix} -55.56 \\ 1.0 \end{Bmatrix}$$

In this mode, the axle and the body vibrate out of phase with each other (X_1 and X_2 have opposite signs), and the amplitude of the displacement of the axle is over 50 times that of the body. This is known as a 'wheel-hop' mode and is at a much higher frequency (0.4 Hz) than the bounce mode. Figure 6.14 shows the relative displacements for mode 2.

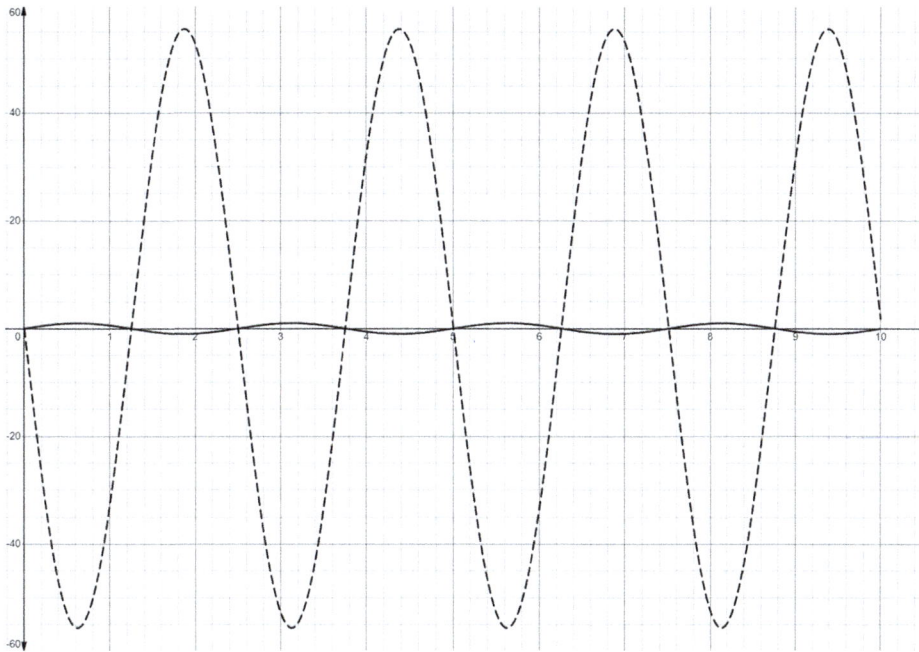

FIGURE 6.14 Plot of the relative displacement of the body X_2 (solid line) and the axle X_1 (dashed line) for mode 2. Time given in seconds.

6.2.3.3 Example 2: 2D Vehicle Model (Coupled Bounce and Pitch)

The previous worked example showed that the bounce mode of the caravan could be estimated by assuming that the tyres were rigid. This example further applies this observation to a four-wheeled vehicle (shown schematically in Figure 6.15) to study the combined effects of bounce and pitch.

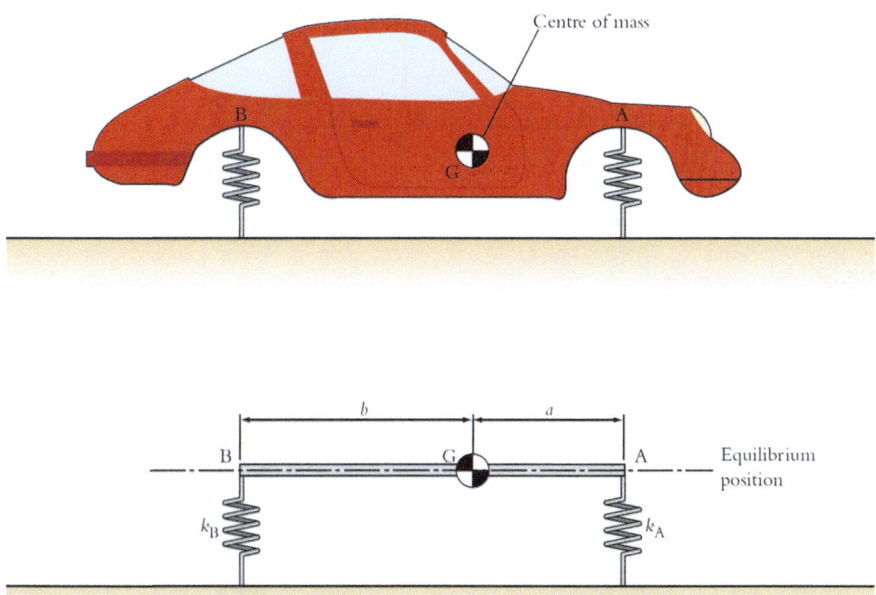

FIGURE 6.15 Two dimensional representation of four wheeled vehicle model.

Step 1 The following assumptions will be used in the dynamic model:

1. Roll motion is not considered – pitch and vertical translation only.
2. The body is rigid, with mass, m, and moment of inertia, I_m, about its centre of mass.
3. The tyres are very stiff, so that the axles do not move.
4. k_A and k_B are the combined stiffnesses for the front and rear springs, respectively.
5. The shock absorbers are ignored.

A and B (the attachment points for the suspension springs) along with the centre of mass, G, are marked on the dynamic model, which is shown in its equilibrium position.

We will use y_G (vertical displacement of G) together with θ (pitch angle) as motion coordinates. These are convenient since they link directly with the two equations of motion for the body. The vertical translation equation will need the absolute acceleration of the centre of mass and the rotational equation will use G as the axis (there is no fixed axis on the car body) and we will need the angular acceleration of the body.

Step 2 Free-body diagram
To draw the free-body diagram, the directions of z_G and θ are as given in Figure 6.23. For small θ, the spring extensions are given by

$$x_A = y_G - a\theta \qquad\qquad (6.12)$$

and

$$x_B = y_G + b\theta$$

With both springs in tension, the positive directions of the forces on the body are as shown:

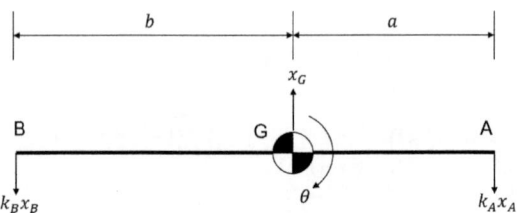

Step 3 Equations of motion
For vertical translation of the centre of mass

$$-k_A\left(y_G - a\theta\right) - k_B\left(y_G + b\theta\right) = m\ddot{y}_G$$

Rearranging gives

$$m\ddot{y}_G + \left(k_A + k_B\right)y_G + \left(bk_B - ak_A\right)\theta = 0$$

For rotation about the centre of mass

$$+k_A\left(y_G - a\theta\right)\cdot a - k_B\left(y_G + b\theta\right)\cdot b = I_G\ddot{\theta}$$

which becomes

$$I_G\ddot{\theta} + \left(bk_B - ak_A\right)y_G + \left(a^2k_A + a^2k_B\right)\theta = 0$$

In the matrix form,

$$\begin{bmatrix} m & 0 \\ 0 & I_G \end{bmatrix}\begin{Bmatrix} \ddot{y}_G \\ \ddot{\theta} \end{Bmatrix} + \begin{bmatrix} k_A + k_B & bk_B - ak_A \\ bk_B - ak_A & a^2k_A + b^2k_B \end{bmatrix}\begin{Bmatrix} y_G \\ \theta \end{Bmatrix} = \begin{Bmatrix} 0 \\ 0 \end{Bmatrix}$$

If we take the displacement to be of the form $z_G(t) = Y_G\cos\omega t$ and $\theta(t) = \Theta\cos\omega t$, then

$$\ddot{y}_G = -\omega^2 Y_G\cos\omega t \text{ and } \ddot{\theta} = -\omega^2\Theta\cos\omega t,$$

$$\begin{bmatrix} k_A + k_B - m\omega^2 & bk_B - ak_A \\ bk_B - ak_A & a^2k_A + b^2k_B - I_G\omega^2 \end{bmatrix}\begin{Bmatrix} Y_G \\ \Theta \end{Bmatrix} = \begin{Bmatrix} 0 \\ 0 \end{Bmatrix} \tag{6.13}$$

or

$$[Z]\begin{Bmatrix} Y_G \\ \Theta \end{Bmatrix} = \begin{Bmatrix} 0 \\ 0 \end{Bmatrix} \tag{6.14}$$

As in the previous example, we find the natural frequencies from the determinant of the matrix $[Z]$:

$$\det[Z] = |Z| = \left(\left(k_A + k_B - m\omega^2\right)\left(a^2k_A + b^2k_B - I_G\omega^2\right) - \left(bk_B - ak_A\right)^2\right) = 0$$

This gives

$$mI_G\omega^4 - \left(\left(a^2k_A + b^2k_B\right)m + \left(k_A + k_B\right)I_G\right)\omega^2 + k_Ak_B\left(a + b\right)^2 = 0$$

The mode shapes can be found by substituting the natural frequencies into either Equation (6.13) or (6.14). Using Equation (6.13), for example

$$\left(k_A + k_B - m\omega^2\right)X_G + \left(bk_B - ak_A\right)\Theta = 0$$

Let $\Theta = 1$ rad to give

$$\begin{Bmatrix} X_G \\ \Theta \end{Bmatrix} = \begin{Bmatrix} \left(ak_A - bk_B\right)/\left(k_A + k_B - m\omega^2\right) \\ 1.0 \end{Bmatrix} \tag{6.15}$$

WORKED EXAMPLE

m = 900 kg	I_G = 1000 kg m^2
k_A = 25 kN/m	k_B = 10 kN/m
a = 1 m	b = 2 m

Natural Frequencies

The frequency equation becomes

$$\begin{vmatrix} 35 \times 10^3 - 900\omega^2 & -5 \times 10^3 \\ -5 \times 10^3 & 25 \times 10^3 + 40 \times 10^3 - 100\omega^2 \end{vmatrix} = 0$$

or

$$\begin{vmatrix} 35 - 0.9\omega^2 & -5 \\ -5 & 65 - \omega^2 \end{vmatrix} = 0$$

Expanding,

$$0.9\omega^4 - 93.5\omega^2 + 2250 = 0$$

Roots are $\omega_{n1}^2 = 37.8\,\text{s}^{-2}$ and $\omega_{n2}^2 = 66.0\,\text{s}^{-2}$
    Hence, $\omega_{n1} = 0.98\,\text{Hz}$ and $\omega_{n2} = 1.29\,\text{Hz}$

Mode Shapes

Mode 1: Put $\omega_{n1}^2 = 37.8\,\text{s}^{-2}$ into Equation (6.15) to give $\begin{Bmatrix} x_G \\ \Theta \end{Bmatrix} = \begin{Bmatrix} 5.43 \\ 1.0 \end{Bmatrix} \begin{matrix} \text{m} \\ \text{rad} \end{matrix}$

To visualize the mode shape, $\begin{Bmatrix} X_A \\ X_B \end{Bmatrix}$ is more convenient. This ratio can be obtained by using Equations (6.12). Thus,

$$\begin{Bmatrix} X_A \\ X_B \end{Bmatrix} = \begin{Bmatrix} X_G - a\Theta \\ X_G + b\Theta \end{Bmatrix} = \begin{Bmatrix} 5.43 - 1 \times 1 \\ 5.43 + 2 \times 1 \end{Bmatrix} = \begin{Bmatrix} 4.43 \\ 7.43 \end{Bmatrix}$$

Normalizing to make $X_B = 1.0$, we get

$$\begin{Bmatrix} X_A \\ X_B \end{Bmatrix} = \begin{Bmatrix} 0.60 \\ 1.0 \end{Bmatrix}$$

as shown in Figure 6.16.

This 'snapshot' of the mode at its extreme position shows a predominant bounce motion, with a little pitching.

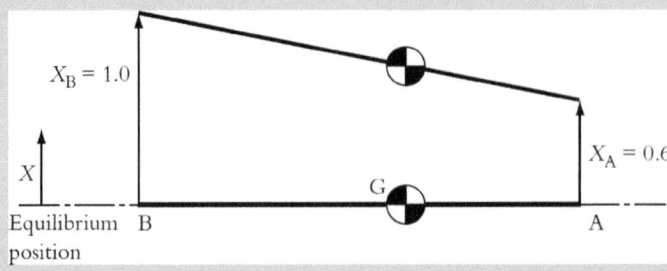

FIGURE 6.16   Mode 1: predominantly vertical displacement (bounce) with some axial rotation (pitch).

Mode 2: Put $\omega_{n2}^2 = 66.0\,\text{s}^{-2}$ into Equation (6.15) to give $\begin{Bmatrix} x_G \\ \Theta \end{Bmatrix} = \begin{Bmatrix} -0.205 \\ 1.0 \end{Bmatrix} \begin{matrix} \text{m} \\ \text{rad} \end{matrix}$

or from Equation (6.12), $\begin{Bmatrix} X_A \\ X_B \end{Bmatrix} = \begin{Bmatrix} -0.67 \\ 1.0 \end{Bmatrix}$.

In this case, we see mainly pitching motion, with only a small displacement at the centre of mass (Figure 6.17).

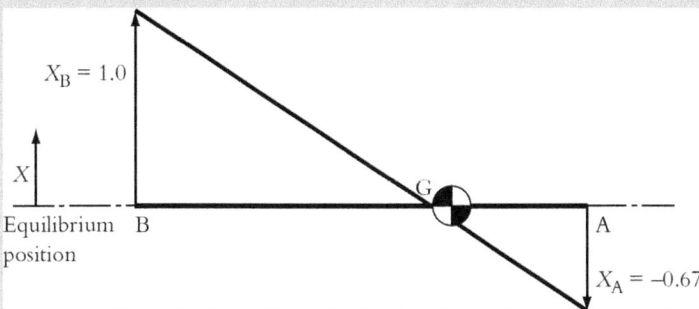

FIGURE 6.17 Mode 2: predominantly axial rotation (pitch) and only a small vertical displacement (bounce).

If the car is driven over a road with a gradually undulating profile (such as produced by ground subsidence), the car will tend to bounce up and down with mode 1 dominating the response. Alternatively, if the front wheels hit an obstruction in the road (e.g. a speed bump or pothole), a pitching response dominated by mode 2 would result. We will see in Section 6.4 that the response of a structure can be written as a combination of its modes of vibration, and most responses will be a mixture of modes 1 and 2.

6.2.4 TORSIONAL SYSTEMS

Most mechanical power is transmitted via rotating shafts. Examples include the turbine–alternator sets in power stations, the tail rotor drive in a helicopter and the propeller shaft in a ship's propulsion system. One of the problems for the vibration engineer is that power surges or sudden changes in load can induce transient torsional oscillations that are superimposed on uniform rotation. The drive torque may also contain fluctuating components that can cause steady-state torsional oscillations. In each case, the torsional natural frequencies and mode shapes are required.

6.2.4.1 Example: Main Drive Shaft of a Gas Turbine Engine

There are two drive shafts in the V2500 engine, which powers the majority of Airbus A320 aircraft (and several others) currently in service (Figure 6.18). The main shaft transmits power from a five-stage turbine at the rear to the fan and the first three compressor stages. The whole assembly rotates freely, supported by bearings. To study the torsional behaviour, the dynamic model will assume that the fan and turbine assemblies can be treated as rigid elements and that the mass of the shaft can be neglected.

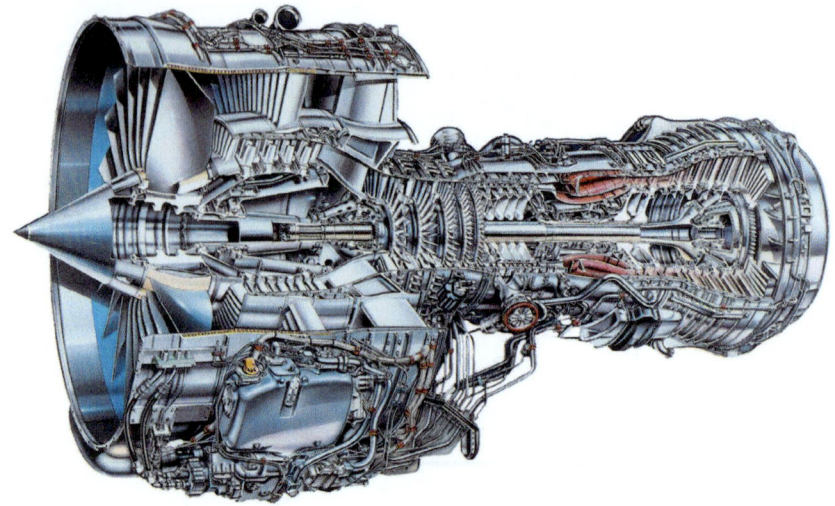

FIGURE 6.18   V2500 twin-shaft aero engine.

Step 1 Dynamic mass–spring model
The engine rotates anti-clockwise when viewed from the front, so this has been chosen for the positive direction of rotation for the two motion coordinates, θ_1 and θ_2.

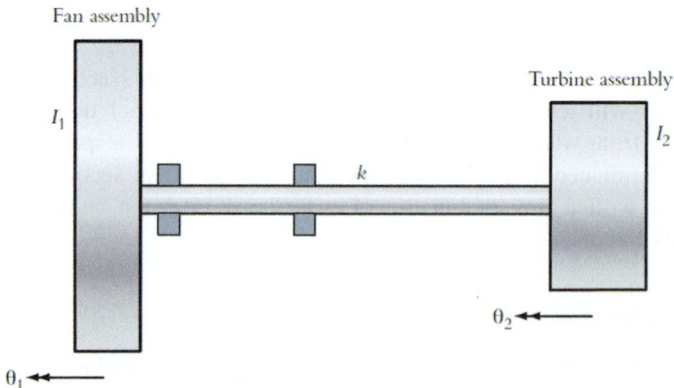

Step 2 Free-body diagrams
If both motion coordinates are given a positive rotation, any twisting of the shaft will be given by $(\theta_1 - \theta_2)$. Hence, the torque in the shaft will be given by $k(\theta_1 - \theta_2)$. This is positive for $\theta_1 > \theta_2$ and has the vector directions shown below.

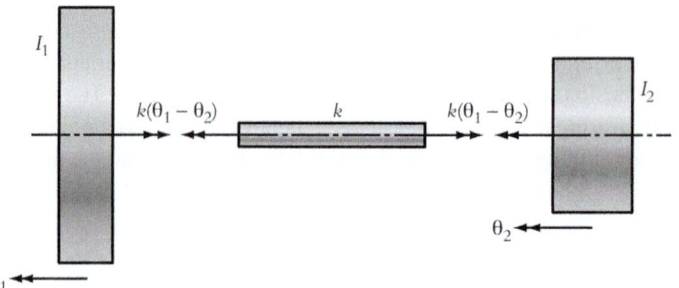

Step 3 Equations of motion

From the physical model, angular accelerations of the fan $\left(\ddot{\theta}_1\right)$ and turbine $\left(\ddot{\theta}_2\right)$ will be given by:

For the fan, $I_1\ddot{\theta}_1 = -k\left(\theta_1 - \theta_2\right)$

For the turbine, $I_2\ddot{\theta}_2 = k\left(\theta_1 - \theta_2\right)$

Hence, $I_1\ddot{\theta}_1 + k\theta_1 - k\theta_2 = 0$

And $I_2\ddot{\theta}_2 - k\theta_1 + k\theta_2 = 0$

This can be expressed in matrix form as

$$\begin{bmatrix} I_1 & 0 \\ 0 & I_2 \end{bmatrix} \begin{Bmatrix} \ddot{\theta}_1 \\ \ddot{\theta}_2 \end{Bmatrix} + \begin{bmatrix} k & -k \\ -k & k \end{bmatrix} \begin{Bmatrix} \theta_1 \\ \theta_2 \end{Bmatrix} = \begin{Bmatrix} 0 \\ 0 \end{Bmatrix}$$

As in the linear translation cases, this will have a general solution of the form $\theta(t) = \Theta\cos\omega t$ where Θ is the magnitude of the oscillation. If we substitute $\theta_1(t) = \Theta_1\cos\omega t$ and $\theta_2(t) = \Theta_2\cos\omega t$ into the equation of motion, then the matrix equation becomes

$$\begin{bmatrix} I_1 & 0 \\ 0 & I_2 \end{bmatrix} \begin{Bmatrix} -\omega^2\Theta_1 \\ -\omega^2\Theta_2 \end{Bmatrix} + \begin{bmatrix} k & -k \\ -k & k \end{bmatrix} \begin{Bmatrix} \Theta_1 \\ \Theta_2 \end{Bmatrix} = \begin{Bmatrix} 0 \\ 0 \end{Bmatrix}$$

Or

$$\begin{bmatrix} -\omega^2 I_1 & 0 \\ 0 & -\omega^2 I_2 \end{bmatrix} \begin{Bmatrix} \Theta_1 \\ \Theta_2 \end{Bmatrix} + \begin{bmatrix} k & -k \\ -k & k \end{bmatrix} \begin{Bmatrix} \Theta_1 \\ \Theta_2 \end{Bmatrix} = \begin{Bmatrix} 0 \\ 0 \end{Bmatrix}$$

Hence,

$$\begin{bmatrix} k - I_1\omega^2 & -k \\ -k & k - I_2\omega^2 \end{bmatrix} \tag{6.16}$$

$$\begin{Bmatrix} \Theta_1 \\ \Theta_2 \end{Bmatrix} = \begin{Bmatrix} 0 \\ 0 \end{Bmatrix} \tag{6.17}$$

To find the natural frequencies, we find the values where the determinant of the matrix is zero, as before

$$\left(k - I_1\omega^2\right)\left(k - I_2\omega^2\right) - \left(-k\right)^2 = 0$$

$$I_1 I_2\omega^4 - k\left(I_1 + I_2\right)\omega^2 = 0$$

The roots are therefore $\omega_{n1}{}^2 = 0$ and $\omega_{n2}{}^2 = \dfrac{k\left(I_1 + I_2\right)}{I_1 I_2}$

The mode shapes are found by substituting the roots into Equation (6.16) or (6.17). Using Equation (6.17) and with $\Theta_2 = 1$,

$$\begin{Bmatrix} \Theta_1 \\ \Theta_2 \end{Bmatrix} = \begin{Bmatrix} \left(k - I_2\omega^2\right)/k \\ 1.0 \end{Bmatrix} = \begin{Bmatrix} 1 - \left(I_2\omega^2\right)/k \\ 1.0 \end{Bmatrix} \tag{6.18}$$

For mode 1, $\omega_{n1} = 0$ and $\begin{Bmatrix} \Theta_1 \\ \Theta_2 \end{Bmatrix} = \begin{Bmatrix} 1 \\ 1 \end{Bmatrix}$ This describes a ***rigid body mode***, which in this case tells us that the rotor is able to rotate continuously in one direction, with no twisting of the shaft ($\Theta_1 = \Theta_2$). A rigid body mode is characterized by a natural frequency equal to zero and by a mode shape in which there is no deformation of any of the parts: the system moves as a single rigid body.

Any structure that is capable of moving without deformation (this is true of any structure not connected to the ground) *will* have one (or more) rigid body mode with $\omega_n = 0$. It follows that the frequency equation will not contain a constant term. Since you can tell in advance that this should be the case, it's a useful check that the frequency equation is correct.

For mode 2, $\omega_{n2} = \dfrac{\sqrt{k\left(I_1 + I_2\right)}}{I_1 I_2}$ and $\begin{Bmatrix} \Theta_1 \\ \Theta_2 \end{Bmatrix} = \begin{Bmatrix} 1 - \dfrac{I_2\left(I_1 + I_2\right)}{I_1 I_2} \\ 1.0 \end{Bmatrix} = \begin{Bmatrix} -\dfrac{I_2}{I_1} \\ 1.0 \end{Bmatrix}$

This mode has a non-zero frequency and the fan and turbine vibrate 180° out of phase with each other, resulting in significant alternating twisting of the shaft.

6.2.5 SHAFT WHIRL

Shaft whirl is a potentially destructive, self-sustaining flexural vibration observed in rotating shafts. It occurs when the rotational frequency of the shaft coincides with a resonant frequency: ***critical speeds***. A given shaft will be designed to operate with some maximum speed and if this maximum design speed is below the lowest critical speed, whirl will not be a problem. Unfortunately, this is not always possible, and it is vital to be able to calculate what the critical speeds will be. For example, the need to minimize the mass of the drive shafts in an aero engine and the very high rotational speeds of gas turbines means that they may have critical speeds within the operating range.

To examine the characteristics of shaft whirl behaviour, the shaft is modelled as a 'beam' with a circular cross-section to find its natural frequencies. The analysis below shows that shafts potentially have an infinite number of flexural natural frequencies, which means that they have an infinite number of critical speeds.

6.2.6 FLEXURAL VIBRATION OF UNIFORM BEAMS

Apart from shafts, many structures exhibit beam-like vibrational behaviour. Examples include aircraft wings, helicopter rotor blades and suspension bridge decks (all of which can vibrate in response to aerodynamic interaction) and tall buildings, which vibrate significantly during earthquakes. While these are more complex than uniform beams, they exhibit many of the same characteristics and this section will provide some insight into this behaviour.

The beam's mass and stiffness are distributed along the length, and so the analysis must begin by dividing the beam into a series of infinitesimal elements of length δx as shown in Figure 6.19.

The equations derived and used in this section depend on the use of the sign convention (shown in Figure 6.20) that defines the positive directions of the transverse displacement, $Y(x, t)$, the shear force S and bending moment, M. It is the same sign convention used in Unit 3.

From beam bending theory, the Bending Moment is related to the curvature of the beam, $\dfrac{\partial^2 y}{\partial x^2}$ by

$$M = -EI\frac{\partial^2 y}{\partial x^2} \tag{6.19}$$

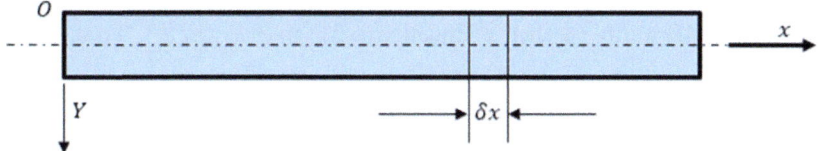

FIGURE 6.19 Mathematical model of beam in bending.

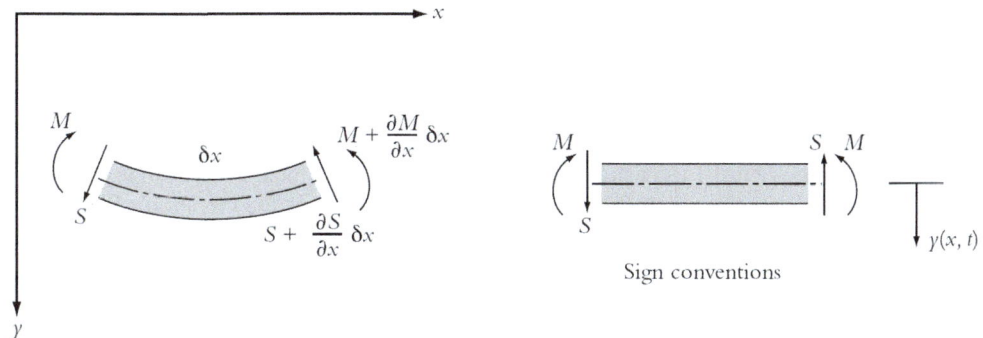

Sign conventions

FIGURE 6.20 Forces and moments on an element in the beam.

The net force on the element (neglecting gravity) is given by

$$S - \left(S + \frac{\partial S}{\partial x} \delta x \right)$$

If the mass of the element is $\rho A \delta x$, where A is the cross-sectional area of the beam and ρ is the density of the beam, then the equation of motion for linear displacement of the element is given by

$$S - \left(S + \frac{\partial S}{\partial x} \delta x \right) = \left(\rho A \delta x \right) \frac{\partial^2 y}{\partial t^2}$$

This gives

$$\frac{\partial S}{\partial x} = -\rho A \frac{\partial^2 y}{\partial t^2} \tag{6.20}$$

To simplify the analysis, we will neglect the rotational inertia of the element and shear deformation of cross-sectional planes. Both assumptions tend to give an overestimate of the natural frequencies of the beam. While this error is normally small for the first few modes, it increases progressively when higher frequencies are evaluated.

Considering rotational motion about an axis through the centre of mass of the element, the sum of moments on the element is given by

$$S \frac{\delta x}{2} + \left(S + \frac{\partial S}{\partial x} \delta x \right) \frac{\delta x}{2} - M + \left(M + \frac{\partial M}{\partial x} \delta x \right) = 0$$

Treating $\dfrac{\partial S}{\partial x} \dfrac{(\delta x)^2}{2}$ as negligible gives the result

$$S = -\frac{\partial M}{\partial x} \tag{6.21}$$

From Equation (6.19), $M = -EI \dfrac{\partial^2 y}{\partial x^2}$; hence, Equation (6.21) gives $S = EI \dfrac{\partial^3 y}{\partial x^3}$. From Equation (6.20),

$$\frac{\partial S}{\partial x} = -\rho A \frac{\partial^2 y}{\partial t^2}$$

Hence,

$$EI \frac{\partial^4 y}{\partial x^4} = -\rho A \frac{\partial^2 y}{\partial t^2} \tag{6.22}$$

This is the governing differential equation for free vibration of the beam. Note that the displacement y is a function of space (x) and time (t). The objective in solving Equation (6.22) is to find the natural frequencies and corresponding mode shapes.

For free vibration at one of the natural frequencies, the displacement of any point on the beam in the y-direction will oscillate sinusoidally, but the local amplitude of the vibration will vary along the length.

If we substitute

$$y(x,t) = Y(x)\cos \omega t$$

Then,

$$EI \frac{d^4 Y(x)}{dx^4}\cos \omega t = \rho A \omega^2 Y(x)\cos \omega t$$

$$\therefore \frac{d^4 Y(x)}{dx^4} = \frac{\rho A \omega^2}{EI} Y(x)$$

For a uniform cross-section, A and I will be constant. We define the **wavenumber**, λ, as

$$\lambda^4 = \frac{\rho A \omega^2}{EI} \tag{6.23}$$

Hence,

$$\frac{d^4 Y(x)}{dx^4} = \lambda^4 Y(x)$$

If $Y(x)$ is of the form,

$$Y(x) = Ae^{\alpha x}$$

Then,

$$\alpha^4 Ae^{\alpha x} = \lambda^4 Ae^{\alpha x}$$

or

$$\alpha^4 = \lambda^4$$

If $\alpha^4 = \lambda^4$, then remember that any real number λ^4 will have four roots with the same magnitude, one will be negative and two of these will be imaginary (for example, $i^4 = 1$). Hence,

$$\alpha = \pm\lambda \text{ or } \pm i\lambda$$

The complete solution for $Y(x)$ is therefore

$$Y_x = A_1 e^{\lambda x} + A_2 e^{-\lambda x} + A_3 e^{i\lambda x} + A_4 e^{-i\lambda x}$$

which may also be written as

$$Y(x) = C_1 \sin \lambda x + C_2 \cos \lambda x + C_3 \sinh \lambda x + C_4 \cosh \lambda x \qquad (6.24)$$

This is a *general equation* giving the deflected shape of any beam of the uniform cross-section. The constants C_1 to C_4 need to be determined from the boundary conditions at the ends of the beam. Four basic types of support are shown in Table 6.1 and other types of boundary conditions are considered later.

The following steps should be used to find the solutions for particular cases:

Step 1 Start by identifying the boundary conditions. These will set the values to be used for C_1 to C_4 in Equation (6.24) to give $Y(x)$ and then the displacement will be given by $y(x,t) = Y(x)\cos \omega t$.

TABLE 6.1

Boundary Conditions for Basic Types of Support

Descriptive Terms	Diagrammatic	Boundary Conditions
Cantilever beam: Built-in clamped encastré Constraints: No translation No rotation		$y = 0 \quad \dfrac{\partial y}{\partial x} = 0$
Simple support hinged pinned Constraints: No translation		$y = 0$ $M = 0 \therefore \dfrac{\partial^2 y}{\partial x^2} = 0$
Free Constraints: Nil		$M = 0 \therefore \dfrac{\partial^2 y}{\partial^2 x} = 0$ $S = 0 \therefore \dfrac{\partial^3 y}{\partial x^3} = 0$

Massless slider
Constraints:
No rotation
Translation in one direction only
  (perpendicular to the long axis of the
  beam)
		$\dfrac{\partial y}{\partial x} = 0$
		$S = 0 \therefore \dfrac{\partial^3 y}{\partial x^3} = 0$

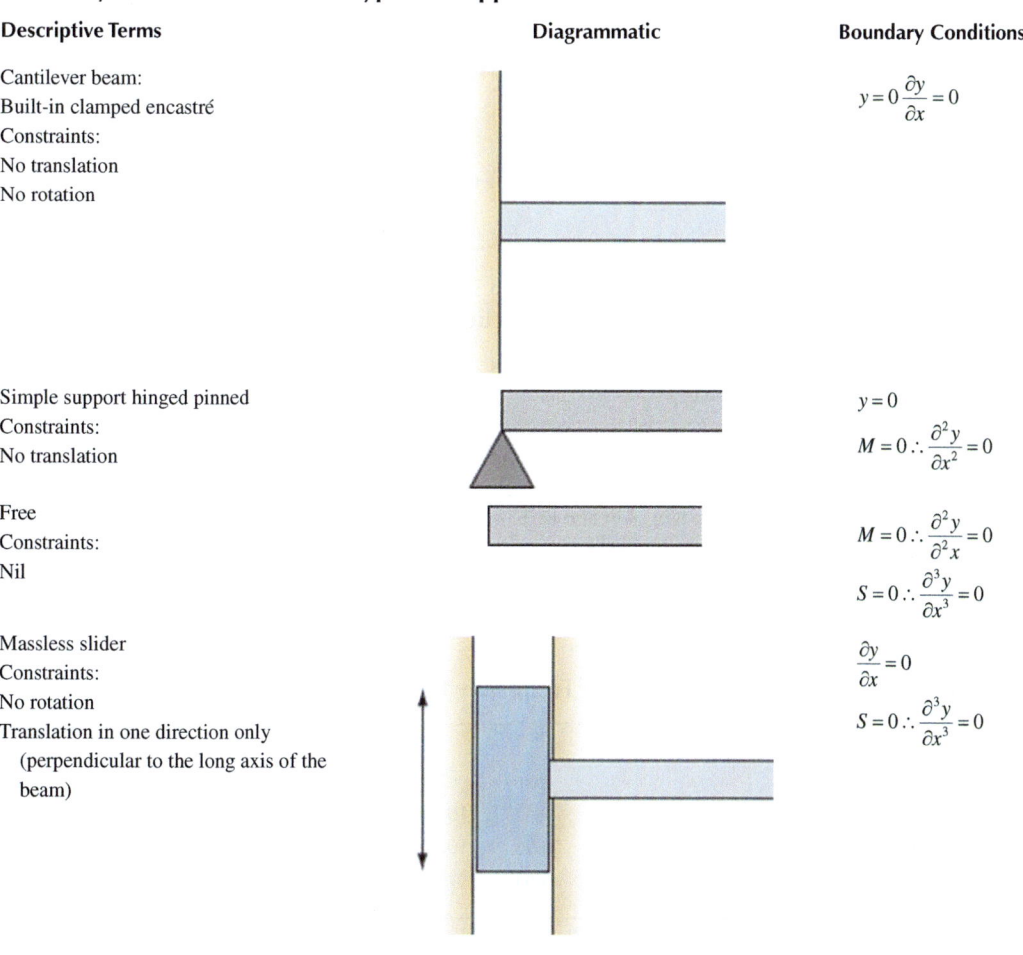

Step 2 Since each of the boundary condition equations depends on $C_1 - C_4$, each of these equations will take the form

$$[Z]C = 0 \qquad\qquad (6.25)$$

where $\{C\}$ is a vector of the constants $C_1 - C_4$ and $[Z]$ is a coefficient matrix.

Step 3 For a valid solution, $\det[Z] = 0$.

This gives the ***frequency equation*** and its roots will give the natural frequencies of the beam.

Step 4 When each root is substituted back into Equation (6.25), the solution vector $\{C\}$ will define the corresponding ***mode shape*** when the values are put into Equation (6.24). Selecting the values of $\lambda_r L$ from Table 6.3 for the beam of interest, the natural frequencies can be found from Equation (6.23). That is: $\omega_r = \left(\dfrac{\lambda_r L}{L}\right)^2 \sqrt{\dfrac{EI}{\rho A}}$

6.2.6.1 Example 1: Simply Supported Beam

Step 1

The boundary conditions at $x = 0$ and $x = L$ are $Y = 0$ and $M = 0$ (Figure 6.21); hence, $\dfrac{\partial^2 y}{\partial x^2} = 0$. (Eqn 6.19)

Since $y(x,t) = Y(x)\cos \omega t$, these boundary conditions become

$$Y = 0 \text{ and } \frac{d^2 Y}{dx^2} = 0$$

From Equation (6.24),

$$Y(x) = C_1 \sin \lambda x + C_2 \cos \lambda x + C_3 \sinh \lambda x + C_4 \cosh \lambda x$$

TABLE 6.2

Frequency Equation for Particular End Conditions

Pinned-Pinned	$\sin \lambda L = 0$
Clamped-clamped and free-free[a]	$\cos \lambda L \cosh \lambda L - 1 = 0$
Clamped-pinned and free-pinned[a]	$\tan \lambda L - \tanh \lambda L = 0$
Clamped-free	$\cos \lambda L \cosh \lambda L + 1 = 0$

[a] A free-free beam will also have two rigid body modes and a free-pinned beam one rigid body mode, each corresponding to $\lambda L = 0$.

TABLE 6.3

Numerical Values of Roots, $\lambda_r L$, of Frequency Equations

r	1	2	3	4	5	>5
Pinned-pinned	π	2π	3π	π	5π	$r\pi$
Clamped-clamped and free-free[a]	4.730	7.853	10.996	14.137	17.279	$\approx (r + 0.5)\,\pi$
Clamped-pinned and free-pinned[a]	3.927	7.096	10.210	13.351	16.493	$\approx (r + 0.25)\,\pi$
Clamped-free	1.875	4.694	7.855	10.996	14.137	$\approx (r - 0.5)\,\pi$

[a] A free-free beam will also have two rigid body modes and a free-pinned beam one rigid body mode, each corresponding to $\lambda L = 0$.

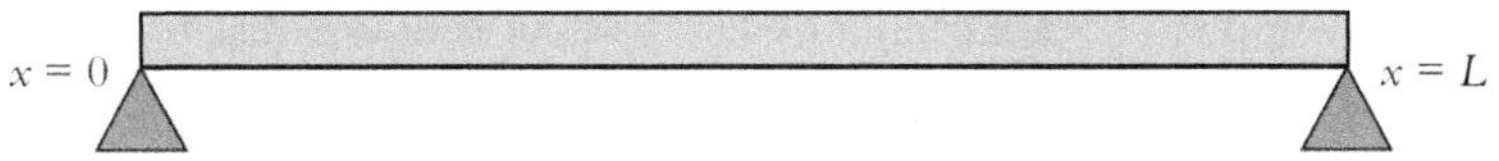

FIGURE 6.21   System model for simply supported beam.

$$\frac{d^2Y}{dx^2} = -\lambda^2 C_1 \sin \lambda x - \lambda^2 C_2 \cos \lambda x + \lambda^2 C_3 \sinh \lambda x + \lambda^2 C_4 \cosh \lambda x$$

Note that $\dfrac{d}{d\theta}\sin h\theta = \cos h\theta$ and $\dfrac{d}{d\theta}\cos h\theta = \sin h\theta$

Hence, at $x = 0$

$$Y(0) = C_1 \times 0 + C_2 \times 1 + C_3 \times 0 + C_4 \times 1 = 0$$

$$\left(\frac{d^2Y}{dx^2}\right)_{x=0} = -\lambda^2 C_1 \times 0 - \lambda^2 C_2 \times 1 + \lambda^2 C_3 \times 0 + \lambda^2 C_4 \times 1 = 0$$

and at $x = L$

$$Y(L) = C_1 \sin \lambda L + C_2 \cos \lambda L + C_3 \sin h\lambda L + C_4 \cos h\lambda L = 0$$

$$\frac{d^2Y}{dx^2} = -\lambda^2 C_1 \sin \lambda L - \lambda^2 C_2 \cos \lambda L + C_3 \sin h\lambda L + \lambda^2 C_4 \cosh \lambda L = 0$$

Step 2
The coefficient matrix $[Z]$ is given by

$$\begin{bmatrix} 0 & 1 & 0 & 1 \\ 0 & -\lambda^2 & 0 & \lambda^2 \\ \sin \lambda L & \cos \lambda L & \sin h\lambda L & \cos h\lambda L \\ -\lambda^2 \sin \lambda L & -\lambda^2 \cos \lambda L & \lambda^2 \sin h\lambda L & \lambda^2 \cos h\lambda L \end{bmatrix} \begin{Bmatrix} C_1 \\ C_2 \\ C_3 \\ C_4 \end{Bmatrix} = \begin{Bmatrix} 0 \\ 0 \\ 0 \\ 0 \end{Bmatrix} \qquad (6.26)$$

Step 3
This is the particular form of Equation (6.25) for a simply supported beam. The natural frequencies are found from the determinant of the coefficient matrix: Expanding the determinant of the coefficient matrix and equating it to zero gives

$$|Z| = -1 \times \begin{vmatrix} 0 & 0 & \lambda^2 \\ \sin \lambda L & \sin h\lambda L & \cos h\lambda L \\ -\lambda^2 \sin \lambda L & \lambda^2 \sin h\lambda L & \lambda^2 \cos h\lambda L \end{vmatrix} -1 \times \begin{vmatrix} 0 & -\lambda^2 & 0 \\ \sin \lambda L & \cos \lambda L & \sin h\lambda L \\ -\lambda^2 \sin \lambda L & -\lambda^2 \cos \lambda L & \lambda^2 \sin h\lambda L \end{vmatrix}$$

$$|Z| = -\lambda^2 \left(\sin \lambda L \times \lambda^2 \sin h\lambda L + \lambda^2 \sin \lambda L \sin h\lambda L\right) - \lambda^2 \left(\sin \lambda L \times \lambda^2 \sin h\lambda L + \lambda^2 \sin \lambda L \sin h\lambda L\right)$$

$$= -4\lambda^4 \sin \lambda L \sin h\lambda L = 0$$

This expression is termed the ***frequency equation*** – note that the wavenumber is directly related to the frequency (see Equation 6.23). To identify the roots of the frequency equation, it is important to note that a simply supported beam cannot deflect without bending, that is it has no freedom to move as a rigid body and therefore will not have a rigid body mode of vibration. From Equation (6.23), $\lambda^4 = \dfrac{\rho A \omega^2}{EI}$, it follows that $\lambda = 0$ will not be a root of the frequency equation. Not only is $\neq 0$, but $\sinh \lambda L \neq 0$

The frequency equation therefore reduces to

$$\sin \lambda L = 0$$

which has roots $\lambda_r L = r\pi$ for $r = 1, 2, 3, \ldots$

From Equation (6.23), $\omega^2 = \dfrac{EI\lambda^4}{\rho A} = \dfrac{EI}{\rho A}\left(\dfrac{r\pi}{L}\right)^4$, the natural frequencies for $r = 1, 2, 3, \ldots$ are

$$\omega_r = \left(\dfrac{r\pi}{L}\right)^2 \sqrt{\dfrac{EI}{\rho A}}$$

Step 4
To find the corresponding mode shapes, we substitute the roots into Equation (6.26) and solve for the constants $C_1 - C_4$.

From (6.26a), $\quad\quad\quad\quad\quad\quad\quad$ $C_2 + C_4 = 0$

From (6.26b), $\quad\quad\quad\quad\quad\quad\quad$ $\lambda_r^2\left(-C_2 + C_4\right) = 0$

Since $\lambda_r \neq 0$, it follows that $\quad\quad$ $C_2 = C_4 = 0$

From (6.26c), $\quad\quad\quad\quad\quad\quad\quad$ $\sin \lambda_r L \cdot C_1 + \sinh \lambda_r L \cdot C_3 = 0$

Since, $\sin \lambda_r L = 0$ (the frequency equation) and $\sinh \lambda_r L \neq 0$, $C_3 = 0$.

The only non-zero constant is C_1. Its value is arbitrary and we normally choose $C_1 = 1$. From Equation (6.24), the mode shape expression becomes

$$Y_r(x) = \sin \lambda_r x = \sin \dfrac{r\pi x}{L}, \text{ with } r = 1, 2, 3, \ldots$$

The first three mode shapes are as shown in Figure 6.22. These will be familiar as simple harmonic motion or as the deflected shapes of the strings in a musical instrument.

6.2.6.2 Example 2: Vibration of a Cantilever (Clamped-Free) Beam

In the next case, we will study a cantilever that is clamped at $x = 0$ and free at $x = L$. See Figure 6.23.

Step 1
The beam will have one end fixed – displacement at this point will be zero, and as the end is fixed, the gradient $\left(\dfrac{\partial y}{\partial x}\right)$ will also be zero. At the free end, the end can have any displacement or angle, but the moment and shear force at this point are zero – hence, the curvature $\dfrac{\partial^2 y}{\partial^2 x^2} = 0$. The boundary conditions will therefore be

$$x = 0 \quad\quad y = 0 \quad\quad \dfrac{\partial y}{\partial x} = 0$$

$$x = L \quad\quad M = 0 \quad\quad \dfrac{\partial^2 y}{\partial x^2} = 0$$

$$\quad\quad\quad\quad\quad S = 0 \quad\quad \dfrac{\partial^3 y}{\partial x^3} = 0$$

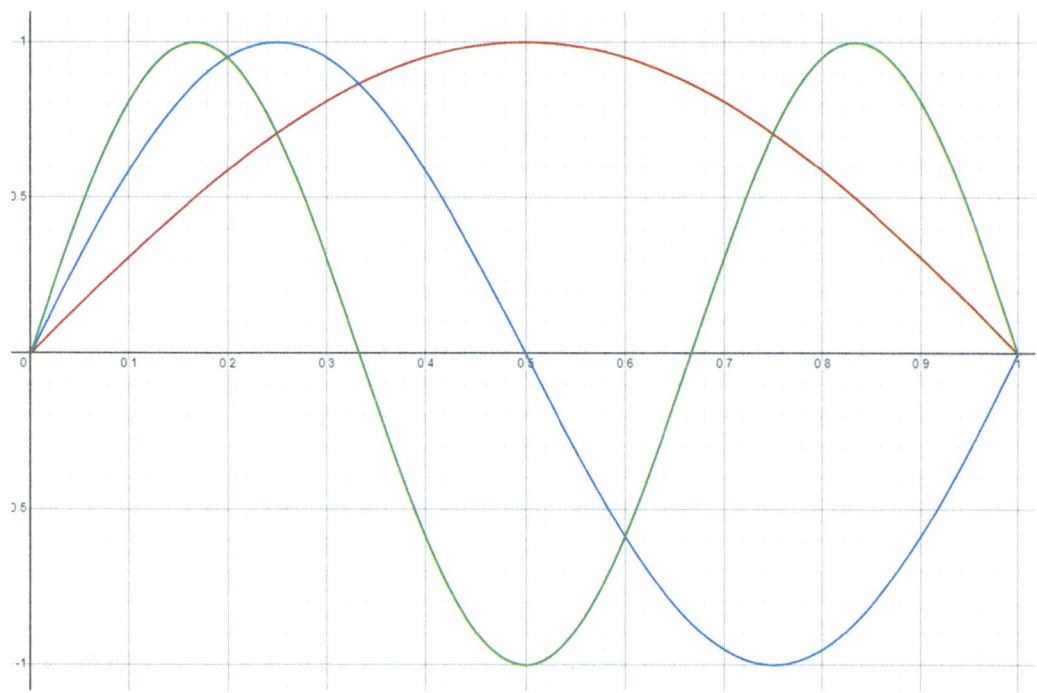

Mode #1 ($r = 1$) $Y_1(x) = \sin \dfrac{\pi x}{L}$

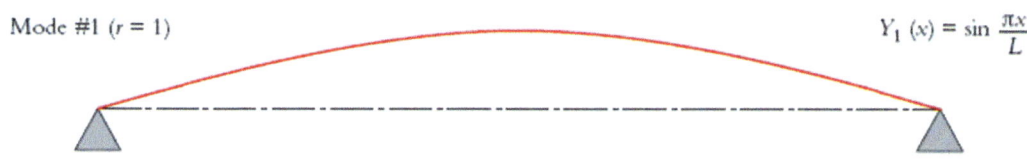

Mode #2 ($r = 2$) $Y_2(x) = \sin \dfrac{2\pi x}{L}$

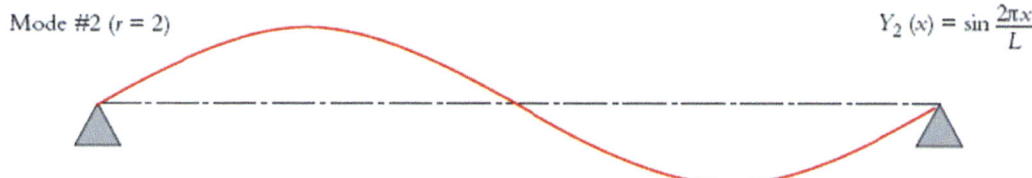

Mode #3 ($r = 3$) $Y_3(x) = \sin \dfrac{3\pi x}{L}$

FIGURE 6.22   Modes 1–3 for vibration of a simply supported beam.

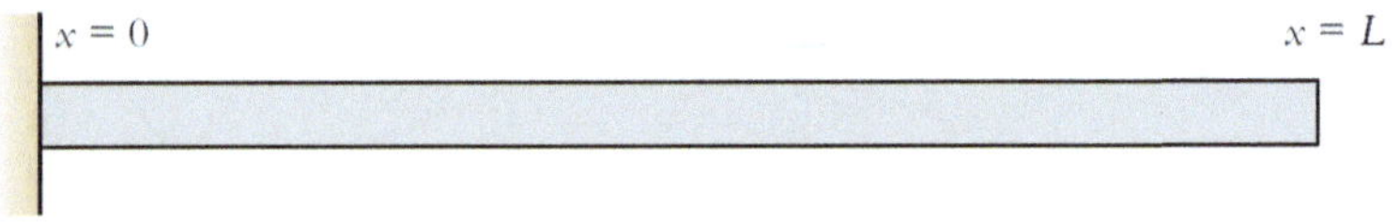

FIGURE 6.23 Boundary conditions for a cantilever beam with one free and one built in end.

Using the same form for $y(x,t)$ as before

$$y(x,t) = Y(x)\cos \omega t$$

$$Y(x) = C_1 \sin \lambda x + C_2 \cos \lambda x + C_3 \sinh \lambda x + C_4 \cosh \lambda x$$

From the non-zero terms in the boundary conditions,

$$\text{At } x = 0, y = 0 \text{ so } Y(x) = C_2 \cos \lambda x + C_4 \cosh \lambda x = C_2 + C_4 = 0$$

$$\text{At } x = 0, \frac{dY}{dx} = 0 \text{ so } \frac{dY}{dx} = C_1 \lambda \cos \lambda x + C_3 \lambda \cosh \lambda x = C_1 \lambda + C_3 \lambda = 0$$

$$\text{At } x = L, M = 0 \text{ so } \frac{d^2 y}{dx^2} = -\lambda^2 C_1 \sin \lambda L - \lambda^2 C_2 \cos \lambda L + \lambda^2 C_3 \sinh \lambda L + \lambda^2 C_4 \cosh \lambda L = 0$$

$$S = 0 \text{ so } \frac{d^3 y}{dx^3} = -\lambda^3 C_1 \cos \lambda L + \lambda^3 C_2 \sin \lambda L + \lambda^3 C_3 \cosh \lambda L + \lambda^3 C_4 \sinh \lambda L = 0$$

Step 2

This then gives the following coefficient matrix:

$$\begin{bmatrix} 0 & 1 & 0 & 1 \\ \lambda & 0 & \lambda & 0 \\ -\lambda^2 \sin \lambda L & -\lambda^2 \cos \lambda L & \lambda^2 \sinh \lambda L & \lambda^2 \cosh \lambda L \\ -\lambda^3 \cos \lambda L & \lambda^3 \sin \lambda L & \lambda^3 \cosh \lambda L & \lambda^3 \sinh \lambda L \end{bmatrix} \begin{Bmatrix} C_1 \\ C_2 \\ C_3 \\ C_4 \end{Bmatrix} = \begin{Bmatrix} 0 \\ 0 \\ 0 \\ 0 \end{Bmatrix} \begin{matrix} (a) \\ (b) \\ (c) \\ (d) \end{matrix} \qquad (6.27)$$

This is the particular version of Equation (6.25) for a cantilever beam.

Step 3

We find the ***frequency equation*** by finding when the determinant of the coefficient matrix is zero. The determinant of the coefficient matrix will simplify to give:

$$1 + \cos \lambda L \cosh \lambda L = 0$$

To find the solutions to this equation, values for λL *are obtained numerically*. Values for the first five wavenumbers are given in Table 6.3. From the wavenumbers, the natural frequencies can be found using Equation (6.23).

Step 4

The ***mode shapes*** are obtained by substituting $\lambda = \lambda_r$ into Equation (6.27) and solving for the constants $C_1 - C_4$.

From (6.27a) and (6.27b)

$$C_3 = -C_1 \text{ and } C_4 = -C_2$$

Thus, from (6.27c) or (6.27d)

$$C_2 = -\frac{\sin \lambda_r L + \sinh \lambda_r L}{\cos \lambda_r L + \cosh \lambda_r L} C_1$$
$$= \sigma_r C_1$$

This gives C_2, C_3 and C_4 in terms of C_1, an arbitrary constant.

If we choose $C_1 = 1$, the mode shape becomes

$$Y_r(x) = \sin \lambda_r x - \sinh \lambda_r x + \sigma_r \left(\cos \lambda_r x - \cosh \lambda_r x \right)$$

When each value of λ_r is used in this equation, a different deflected shape is obtained (Figure 6.24–6.26).

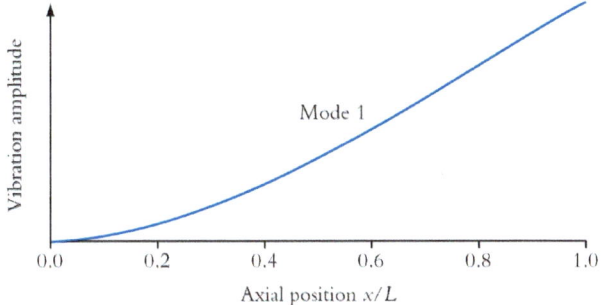

FIGURE 6.24   Mode 1.

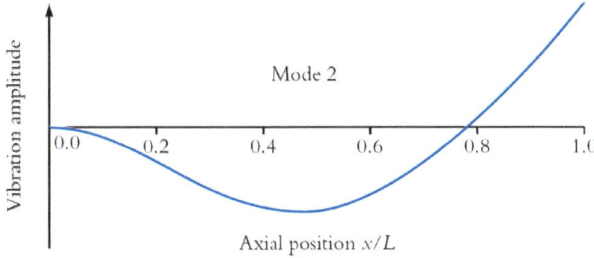

FIGURE 6.25   Mode 2.

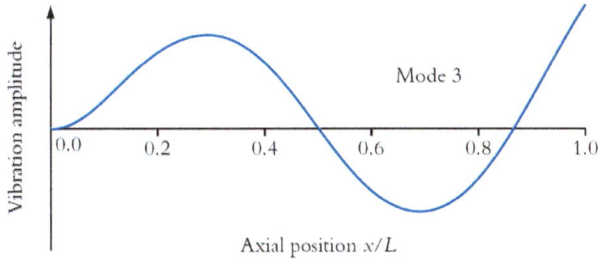

FIGURE 6.26   Mode 3.

6.2.7   OTHER BOUNDARY CONDITIONS

6.2.7.1   Example: Cantilever Beam with a Mass at the Free End

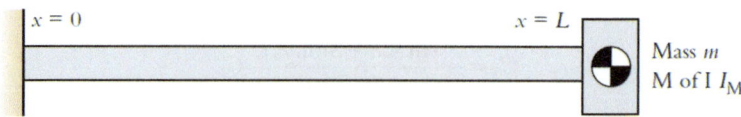

Step 1

The boundary conditions at the clamped end are identical to the previous example: $Y = 0$ and $\dfrac{dY}{dx} = 0$ at $x = 0$.

However, at $x = L$, the mass must be supported by a force and any moments due to the moment of inertia of the mass must be countered: hence $S \neq 0$ and $M \neq 0$. To look at the effect that the mass will have on the vibration of the beam, we use two of the basic principles of mechanics. These are:

1. Compatibility of displacements
2. Equilibrium of forces and moments.

Consider first the shear force reaction between the beam and the mass.

The principle of **compatibility** determines that the displacement at the end of the beam is the same as the displacement of the mass.

The principle of **equilibrium of forces** determines that the shear force on the beam is equal and opposite to the force on the mass, as the mass and the beam have a fixed connection.

The free-body diagram is

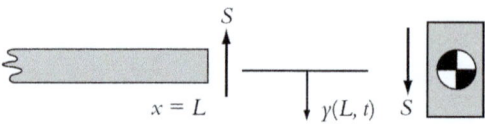

It is vital that the sign convention is consistent: note the positive directions for the displacement coordinate and the shear force on the beam. From the free-body diagram, equations of motion can be derived for the beam and the mass:

Beam	Mass
From Equations (6.19) and (6.21): $S(t) = EI\left(\dfrac{\partial^3 y}{\partial x^3}\right)_{x=L}$	From equilibrium, the acceleration of the mass must equal that of the beam at $x = L$ $S(t) = m\left(\dfrac{\partial^2 y}{\partial t^2}\right)_{x=L}$

Since the displacement takes the form $y(x,t) = Y(x)\cos\omega t$, it follows that:

$$S(t) = m \times \left(-\omega^2 Y(L)\cos\omega t\right)$$

And

$$S(t) = EI\left(\frac{d^3Y}{dx^3}\right)_{x=L} \cos \omega t$$

From Equation (6.23), $\omega^2 = \dfrac{EI\lambda^4}{\rho A}$, so equating the two expressions for $s(t)$ and substituting for ω^2:

$$EI\left(\frac{d^3Y}{dx^3}\right)_{x=L} \cos \omega t + mY(L)\cos \omega t \frac{EI\lambda^4}{\rho A} = 0$$

Eliminating EI and $\cos \omega t$ gives

$$\left(\frac{d^3Y}{dx^3}\right)_{x=L} + mY(L)\frac{(\lambda L)^4}{\rho A L^4} = 0 \tag{6.28}$$

To give the complete solution, we must also consider the bending moments.

> *Compatibility* states that the rotation of the mass must be equal to the slope at the end of the beam.
> Moment equilibrium states that the bending moment on the beam is equal and opposite to the bending moment on the mass.

The free-body diagram is

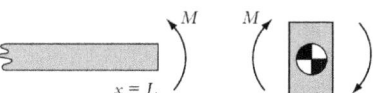

$x = L$

Consideration of moments for the beam and the mass give

Beam	Mass
From Equation (6.19), the bending moment at the end of the beam is: $M(t) = -EI\left(\dfrac{\partial^2 y}{\partial x^2}\right)_{x=L}$	Equation of motion for the mass in rotation: $M(t) = I_M\left(\dfrac{\partial^2 \theta}{\partial t^2}\right)_{x=L}$
As before, $y(x,t) = Y(x)\cos \omega t$ and so, $M(t) = -EI\left(\dfrac{d^2Y}{dx^2}\right)_{x=L} \cos \omega t$	Compatibility (rotation of mass = angle of beam) $\theta(t) = \dfrac{\partial y}{\partial x} = \dfrac{dY}{dx}\cos \omega t$ Hence, $\dfrac{\partial^2 \theta}{\partial t^2} = -\omega^2\left(\dfrac{dY}{dx}\right)\cos \omega t$

Therefore, the equation of motion for the mass becomes

$$M(t) = -I_M\omega^2\left(\frac{dY}{dx}\right)_{x=L} \cos \omega t$$

So:

$$M(t) = -EI\left(\frac{d^2Y}{dx^2}\right)_{x=L} \cos\omega t = -I_M\omega^2\left(\frac{dY}{dx}\right)_{x=L} \cos\omega t$$

Substituting for ω^2 from Equation (6.23) and eliminating gives

$$\left(\frac{d^2Y}{dx^2}\right)_{x=L} - \frac{I_M(\lambda L)^4}{\rho AL^4}\left(\frac{dY}{dx}\right)_{x=L} = 0 \qquad (6.29)$$

From these boundary conditions, we have the following four equations:

$$Y(0) = 0 \qquad\qquad (a)$$

$$\left(\frac{dY}{dx}\right)_{x=0} = 0 \qquad\qquad (b)$$

$$\left(\frac{d^3Y}{dx^3}\right)_{x=L} + \frac{m(\lambda L)^4}{\rho AL^4}Y(K) = 0 \qquad\qquad (c)$$

$$\left(\frac{d^2Y}{dx^2}\right)_{x=L} - \frac{I_M(\lambda L)^4}{\rho AL^4}\left(\frac{dY}{dx}\right)_{x=L} = 0 \qquad\qquad (d)$$

Step 2
To set up the coefficient matrix (Equation 6.25) for this system, substitute for $Y(x)$ and its derivatives from Equation (6.24).

Steps 3 and 4 then follow as in the previous examples.

6.2.7.2   Other Structures with Distributed Mass and Stiffness

The models presented in the earlier parts of this section are based on a range of assumptions that may or may not be applicable in particular cases. An aircraft wing, for example, will exhibit bending modes of vibration, so the study of uniform beams gives some insight into the expected behaviour. However, a wing does not have a uniform cross-section; so, the analysis will not be able to predict the natural frequencies of this structure accurately.

For this and many other structures, engineers will usually turn to the finite element method. There is an introduction to finite elements in Unit 3 related to obtaining displacements and stresses due to static loads. This involves creating and solving an equation of the form

$$[K]\{X\} = \{P\}$$

[K] is a global stiffness matrix formed by combining the individual stiffness matrices of the finite elements chosen to discretize the structure and {P} is a vector of applied loads. {X} is the solution vector giving the displacements at the nodes.

As well as solving static problems, most commercial finite element codes can also solve vibration problems by calculating a mass matrix in addition to the stiffness matrix. This is then used to set up and solve an eigenvalue problem of the form

$$\left([K] - \omega^2[M]\right)\{X\} = \{0\}$$

The solution will give the natural frequencies and mode shapes for the structure.

6.3 RESPONSE OF DAMPED SINGLE-DEGREE-OF-FREEDOM SYSTEMS

6.3.1 INTRODUCTION

So far, we have looked at the free vibration of systems and their natural frequencies and mode shapes. This section will analyse the response of single-degree-of-freedom systems to external excitation. This takes the form either of applied forces and/or moments or imposed displacements on parts of the system.

We will also introduce the effects of damping. Damping is the process by which an oscillatory motion will attenuate – as would be observed in any real-world situation. While there are a variety of mechanisms by which damping can occur, in this section, we will consider only viscous damping, where the force opposing the motion and causing attenuation is proportional to the velocity (angular velocity in a rotating system). This is the principle in the majority of deliberately damped systems, such as the shock absorbers in a car or the front forks of a mountain bike as shown in Figure 6.27.

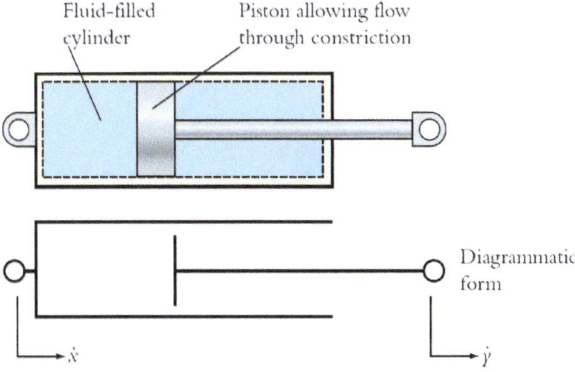

FIGURE 6.27 Schematic representation of damper/shock absorber.

In the viscous damping model, we assume that the force in the damper is proportional to the relative velocity between the ends and acts to oppose the imposed motion. The constant of proportionality is called the damping coefficient (normally given the symbol c) and has units of N/(m/s) [or Ns/m].

Hence, the force opposing the motion is $c\,(\dot{x} - \dot{y})$.

Note that dampers do not impose any stiffness on the structure; they only transmit a force if there is relative motion between the ends. If there is no motion, there is no force.

For most structural vibrations, there is no 'deliberate' damping as would be seen in a shock absorber and the level of damping is low. For this reason, any differences between the model and the actual mechanism are sufficiently low in magnitude for the error introduced to be negligible.

6.3.2 EQUATIONS OF MOTION

In the examples that follow, the steps leading to the equation of motion are the same as those presented in Section 6.2 with the addition of damping forces and external excitation. We will then look at the solution to the equation of motion for a few types of excitation.

6.3.2.1 Example 1: Mass–Spring–Damper System

Step 1 Dynamic mass–spring model

Figure 6.41 shows a simple mass–spring–damper system, in which the mass is coupled to a fixed base via a spring element with stiffness k and a damper with damping constant c.

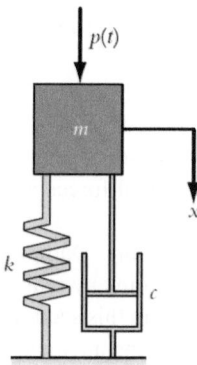

Step 2 Draw a free-body diagram for mass:

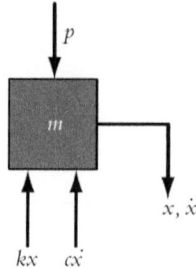

From the free-body diagram, the force p will cause a positive (downward) displacement of the mass from its equilibrium position. This will compress the spring, resulting in an upward (negative) force acting on the underside of the mass. The positive (downward) velocity of the mass will move the damper. The damper will oppose this motion, resulting in an upward force acting on the underside of the mass.

Step 3 Equation of motion

$$m\ddot{x} = p - kx - c\dot{x}$$

or

$$p(t) = m\ddot{x} + c\dot{x} + kx \tag{6.30}$$

6.3.2.2 Example 2: Rocker System
Step 1 Dynamic mass–spring model

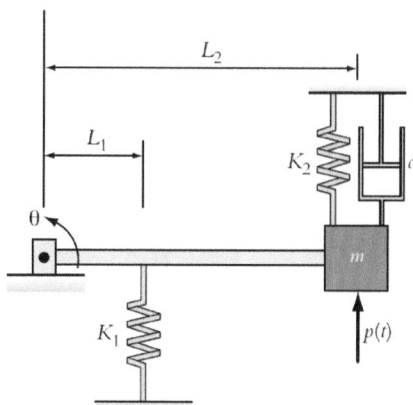

Step 2 Free-body diagram (for small displacements)
This is similar to the rocker system given as an example in Section 6.2, but with the addition of the damper and the applied force. In this case, a positive (anti-clockwise) angular velocity will produce upward velocity of the mass and cause the damper to compress. The damper will oppose this motion, resulting in a downward force acting on the top of the mass.

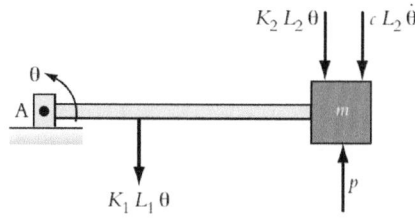

Step 3 Equation of motion

Taking moments about the fixed pivot at A, the equation of motion is

$$-K_1 L_1 \theta \cdot L_1 - K_2 L_2 \theta \cdot L_2 - c L_2 \dot{\theta} \cdot L_2 + p \cdot L_2 = I_A \ddot{\theta}$$

If the rocker is modelled as a point mass at the end of a massless bar, the moment of inertia of the rocker about the pivot is $I_A = m L_2^2$.

Hence, on rearranging,

$$m L_2^2 \ddot{\theta} + c L_2^2 \dot{\theta} + \left(K_1 L_2^2 + K_1 L_2^2 \right) \theta = p(t) L_2 \tag{6.31}$$

6.3.2.3 Example 3: Single-Axle Trailer with Sprung Suspension

We will make the following simplifying assumptions:

1. The tyres are very stiff compared to the suspension springs (typically they are about ten times stiffer).
2. The tyres do not lose contact with the road. Taken with (1), this means that the vertical motion of the axle will follow the road profile exactly.
3. The trailer body behaves as a rigid mass.
4. Only the vertical motion of the body is considered; pitching and rolling are ignored.

Step 1 Dynamic mass–spring model

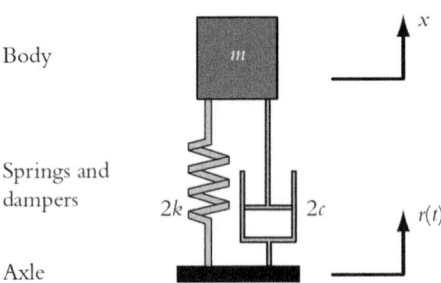

Step 2 Free-body diagram

In this case, the excitation takes the form of a defined **displacement** instead of a force. With the assumption that the tyres are effectively rigid, the road profile over which the trailer travels exactly defines the motion (displacement and velocity) of the axle. Any difference between the relative displacement or velocity of the body and axle will give rise to forces in the springs and dampers that will react with the underside of the body.

From the free-body diagram, using the sign convention given, the force in the spring is given by *Spring force = Stiffness × Change of length*. In this case, the change in length of the spring is the difference between the body and axle displacements, or $(r - x)$. This will give a positive vertical force when the axle displacement is greater than the body displacement. This corresponds to the spring being in compression.

For the damper, ***Damper force = Damping coefficient × Relative velocity***. The relative velocity is given by $\frac{d}{dt}(r - x)$ or $(\dot{r} - \dot{x})$. From the sign convention, the expression for the relative velocity will be positive if the axle velocity $\dot{r}$ is greater than the body velocity $\dot{x}$. The effect of the damper will always be in the opposite direction to the relative velocity; hence, a positive value for $(\dot{r} - \dot{x})$ gives an upward reaction force acting on the body of the trailer.

Step 2 Equation of motion
The net force acting on the body of the trailer is given by:

$$F = 2k(r - x) + 2c(\dot{r} - \dot{x})$$

Hence, from Newton's second law,

$$m\ddot{x} = 2k(r - x) + 2c(\dot{r} - \dot{x})$$

Thus,

$$m\ddot{x} + 2c\dot{x} + 2kx = 2c\dot{r}(t) + 2kr(t) \tag{6.32}$$

For any given vehicle speed, the shape of the road profile, $r(t)$, can be measured as an explicit function of time, so that we know exactly what the displacement of the axle will be. Differentiating the displacement gives the axle velocity $\dot{r}(t)$, again as an explicit function of time. As a result, the excitation terms on the right-hand side of Equation (6.23) are completely defined.

Collecting the equations of motion for the three examples, we have

Mass–spring–damper system	$m\ddot{x} + c\dot{x} + kx = P(t)$
Rocker system	$mL_2{}^2\ddot{\theta} + cL_2{}^2\dot{\theta} + (K_1L_1{}^2 + K_1L_2{}^2)\theta = L_2P(t)$
Single–axle trailer	$m\ddot{x} + 2c\dot{x} + 2kx = 2kx = 2c \cdot \dot{r}(t) + 2k \cdot r(t)$

While each of the equations is different in detail, you will see that they all share a common mathematical form, that of a **second-order ordinary differential equation with constant coefficients**. All linear, single-degree-of-freedom systems have this form, which can be written generically as

$$M\ddot{z} + C\dot{z} + Kz = F(t) \tag{6.33}$$

in which

z is the response coordinate
M is the coefficient of $\ddot{z}$ (normally mass, or moment of inertia in rotational systems)
C is the coefficient of $\dot{z}$ (damping factor)
K is the coefficient of z (force due to displacement from equilibrium)
$F(t)$ is the excitation function; independent of z.

The exact form of the three coefficients and the excitation function will depend on the system being analysed. The mathematical solution depends on the nature of the excitation function and the amount of damping in the system. Mathematics textbooks will provide a more comprehensive explanation of these problems, and Unit 4 presents the use of Laplace transforms for this purpose. In this unit, we will confine the analysis to three specific cases: 'free' vibration, harmonic (sinusoidal) and arbitrary periodic excitation.

6.3.3 FREE VIBRATION

If no externally applied forces or moments act on a structure, it can vibrate freely – hence the term 'free' vibration. We have previously used this to find the natural frequency of undamped systems. In real physical systems, damping will cause any vibration to stop once the stimulus is removed, and consequently, free vibration is also termed as the 'transient response'.

For $F(t) = 0$, we can use a solution of the form $z(t) = Ae^{\lambda t}$
Substituting into the equation of motion gives

$$M\lambda^2 Ae^{\lambda t} + C\lambda Ae^{\lambda t} + KAe^{\lambda t} = 0$$

For a non-trivial solution,

$$M\lambda^2 + C\lambda + K = 0$$

so that

$$\lambda_{1,2} = \frac{-C \pm \sqrt{C^2 - 4KM}}{2M} \qquad (6.34)$$

The complete solution is then

$$z(t) = A_1 e^{\lambda_1 t} + A_2 e^{\lambda_2 t} \qquad (6.35)$$

The magnitudes A_1 and A_2 are found from the initial conditions specified in the problem. These will normally be the displacement and velocity at $t = 0$.

It can be seen from Equation (6.34) that the roots $\lambda_{1,2}$ can be real, imaginary or complex, depending on the amount of damping present. While Equation (6.35) gives a full description of the response, other forms are easier to visualize. Four cases with different damping levels are considered.

6.3.4 ZERO DAMPING

For zero damping, the system will oscillate with a sinusoidal waveform, as the roots will be

$$\lambda_{1,2} = \pm\frac{\sqrt{-4KM}}{2M} = \pm i\sqrt{\frac{K}{M}} \text{ and } e^{i\sqrt{\frac{K}{M}}} = \cos\sqrt{\frac{K}{M}} + i\sin\sqrt{\frac{K}{M}}.$$

$\sqrt{\dfrac{K}{M}}$ is the **undamped natural frequency** and is given the symbol ω_n.

From Equation (6.34), the complex term is affected by the damping factor, and hence any under-damped system will have a frequency at which it will vibrate freely. For a damped system, the natural frequency will be $\dfrac{\sqrt{C^2 - 4KM}}{2M}$ and this is termed the damped natural frequency. Returning to the general case, Equation (6.35) becomes

$$z(t) = A_1 e^{i\omega_n t} + A_2 e^{-i\omega_n t}$$

Using Euler's formula, $e^{i\theta} = \cos\theta + i\sin\theta$, this becomes

$$z(t) = B\cos\omega_n t + C\sin\omega_n t \qquad (6.36)$$

The constants B and C are found from the initial conditions of the problem.

6.3.5 OVER DAMPING

If the damping level is such that $C^2 > 4KM$, the two roots will be **real** and **negative**. The response, as given by Equation (6.35), is the sum of two decaying exponential functions. The constants A_1 and A_2 are found from the initial conditions.

6.3.6 CRITICAL DAMPING

A special case for the roots of Equation (6.34) occurs if $C^2 = 4KM$. This value of damping is known as 'critical damping', which is thus given by

$$C_{crit} = 2\sqrt{KM} \qquad (6.37)$$

This is an important relationship and represents the boundary between two types of response. As we shall see in the next section, for subcritical damping the system will oscillate about its equilibrium position before coming to rest. For overdamping, the free response decays exponentially back to its equilibrium position without oscillating.

From Equation (6.34),

$$\lambda_1 = \lambda_2 = -\frac{C_{crit}}{2M} \equiv -\omega_n$$

In reality, the critical damping point is a singularity which cannot be achieved – all systems that appear to be critically damped will be very slightly under- or overdamped. From the overdamped case, if the two roots are $(\lambda + \delta)$ and $(\lambda - \delta)$, then for critical damping

$$z(t) = \lim_{\delta \to 0}\left(A_1 e^{-(\lambda+\delta)t} + A_2 e^{-(\lambda-\delta)t} \right) = \left(B_1 + B_2 t \right) e^{-\lambda t} \qquad (6.38)$$

Where A_1, A_2, B_1 and B_2 are constants dependent on the boundary conditions. Note the t in the $B_2 t$ term.

6.3.7 LIGHT DAMPING

Most engineering structures will be underdamped. From experience, we know that a structure will oscillate and eventually come to rest. This is described mathematically by the case where $C^2 < 4KM$ and the roots of Equation (6.34) are a complex conjugate pair:

$$\lambda_{1,2} = -\frac{C}{2M} \pm i\frac{\sqrt{4KM - C^2}}{2M} \qquad (6.39)$$

Here, we will introduce the **damping ratio**, $\gamma = \dfrac{C}{\text{critical damping}} = \dfrac{C}{2\sqrt{KM}}$

Using the expression for the undamped natural frequency, ω_n, Equation (6.39) becomes

$$\lambda_{1,2} = -\gamma\omega_n \pm i\omega_n\sqrt{1 - \gamma^2} \qquad (6.40)$$

Substitution into Equation (6.35) gives the solution

$$z(t) = A_1 e^{\left(-\gamma\omega_n + i\omega_n\sqrt{1-\gamma^2}\right)t} + A_2 e^{\left(-\gamma\omega_n - i\omega_n\sqrt{1-\gamma^2}\right)t} \qquad (6.41)$$

Again, making use of the complex exponential identities and the fact that A_1 and A_2 are a complex conjugate pair, Equation (6.41) can be rewritten as

$$z(t) = e^{-\gamma\omega_n t}\left(B_1 \cos \omega_n\sqrt{1 - \gamma^2}\, t + B_2 \sin \omega_n\sqrt{1 - \gamma^2}\, t \right) \qquad (6.42)$$

Equation (6.42) describes a sinusoidal waveform (indicated by the terms in the brackets) with an exponentially decaying term at the front that will cause the amplitude of the sinusoid to decrease over time, matching what we would expect to see in real life. B_1 and B_2 are both real amplitudes. An alternative form of Equation (6.42) is given by

$$z(t) = C_0 e^{-\gamma \omega_n t} \cos\left(\omega_n \sqrt{1-\gamma^2}\, t - \psi \right) \tag{6.43}$$

where ψ is the phase angle of the vibration. As with the previous cases, the two constants can be found from the initial conditions.

Equations (6.42) and (6.43) both give the frequency of vibration as $\omega_n\sqrt{1-\gamma^2}$. This is the **damped natural frequency** and is lower than the undamped natural frequency, ω_n.

WORKED EXAMPLE

When at rest in equilibrium, the mass in the diagram receives an impulse J of magnitude 5 Ns applied at time $t = 0$. Find the response for $t > 0$.

Data: $k = 500$ N/m　　　$c = 20$ Ns/m　　　$m = 10$ kg

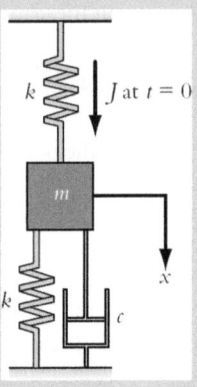

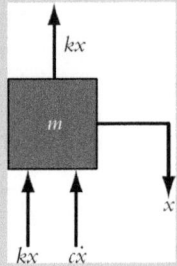

Equation of Motion

$$-2kx - c\dot{x} = m\ddot{x}$$

or

$$m\ddot{x} + c\dot{x} + 2kx = 0$$

Note that the impulse J does not appear on the free-body diagram or in the equation of motion as it does not exist for $t > 0$.

From the equation of motion, $\omega_n = \sqrt{\dfrac{2k}{m}} = 10\,\text{rad/s}$ and $\gamma = \dfrac{c}{2\sqrt{2km}} = 0.1$. Since the damping ratio is less than 1.0 in this problem, the system is underdamped. From Equation (6.42), the response is:

$$x(t) = e^{-\gamma \omega_n t}\left(B_1 \cos \Omega_n t + B_2 \sin \Omega_n t\right)$$

where $\Omega_n = \omega_n \sqrt{1-\gamma^2} = 9.9\,\text{rad/s}$.

To find the values of B_1 and B_2, we use the initial conditions for the system.

$$x = 0 \text{ at } t = 0; \text{ hence, } B_1 = 0;$$

This gives the solution as:

$$x(t) = B_2 e^{-\gamma \omega_n t} \sin \Omega_n t$$

The second initial condition is the impulse, which has a magnitude of 5Ns. The unit here indicates that the impulse is in the form of momentum, so the change in momentum of the mass is given by

$$J = m\left(\dot{x}_0 - 0\right)$$

Therefore,

$$\dot{x}_0 = \frac{J}{m}$$

From the general solution, $\dot{x}(t)$ is given by

$$\dot{x}(t) = B_2\left[\Omega_n e^{-\gamma \omega_n t} \cos \Omega_n t - \gamma \omega_n e^{-\gamma \omega_n t} \sin \Omega_n t\right]$$

From the boundary condition,

$$\dot{x}(0) = \frac{J}{m} = B_2\left[\Omega_n\right]$$

This gives

$$B_2 = \frac{J}{m\Omega_n} \text{ and } x(t) = \frac{J}{m\Omega_n} e^{-\gamma \omega_n t} \sin \Omega_n t$$

Substituting the numerical values gives $x(t) = 0.0505 e^{-0.99t} \sin 9.9t$

6.3.8 ESTIMATING DAMPING

While the mass and stiffness of a structure (and hence the undamped natural frequency) can normally be calculated, structural damping is very difficult to predict. Equations (6.42) and (6.43) show that the rate of decay of the free vibration of a structure depends directly on the damping ratio, and this will allow us to measure the damping ratio experimentally.

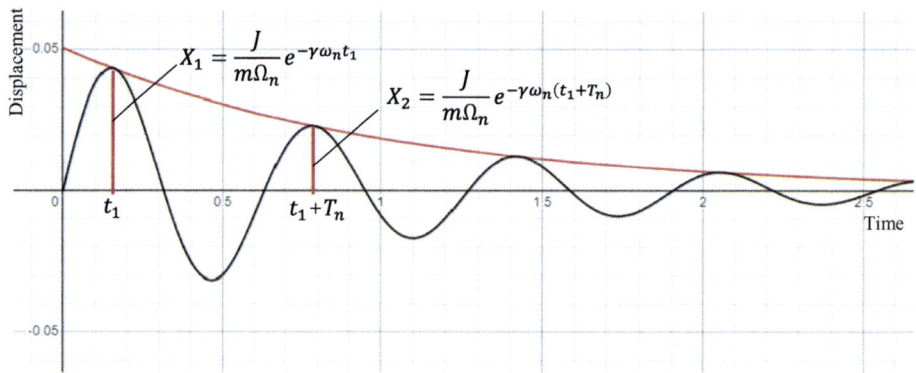

FIGURE 6.28 Response of a damped system to free vibration.

For the worked example above, if k and m are known, but c is unknown, we could design an experiment to measure the transient displacement from the impulse. We would expect to see a response waveform as shown in Figure 6.28.

Because the system is at rest at $t = 0$, the expression for the displacement is

$$x(t) = \frac{J}{m\Omega_n} e^{-\gamma\omega_n t} \sin\Omega_n t$$

Note that we do not yet know the damping ratio, and hence the relationship between Ω_n and ω_n $\Omega_n = \omega_n\sqrt{1-\gamma^2}$. The amplitude of the first peak at $t = t_1$ is

$$X_1 = \frac{J}{m\Omega_n} e^{-\gamma\omega_n t_1}$$

and the amplitude of the second peak, which occurs one period $\left(T_n = \dfrac{2\pi}{\omega_n\sqrt{1-\gamma^2}} \right)$ later, is

$$X_2 = \frac{J}{m\Omega_n} e^{-\gamma\omega_n (t_1+T_n)}$$

The ratio of the peaks, $\dfrac{X_1}{X_2} = e^{\gamma\omega_n T_n}$. Assuming that the system is lightly damped ($\gamma < 0.2$),

$$T_n = \frac{2\pi}{\omega_n\sqrt{1-\gamma^2}} \approx \frac{2\pi}{\omega_n}$$

Taking logs of both sides of the expression for the ratio of the peaks, we get

$$\ln\left(\frac{X_1}{X_2}\right) = \gamma\omega_n T_n = \frac{2\pi\gamma}{\sqrt{1-\gamma^2}} \approx 2\pi\gamma$$

From the graph, $X_1 = 0.0431\,\text{m}$ and $X_2 = 0.0229\,\text{m}$ so that $\gamma = 0.101$ and $c = 20.1\,\text{Nsm}^{-1}$, which compare well with the exact values. For best practice in determining the damping ratio experimentally, the ratio of any two successive peaks is a constant; so, several estimates of the damping ratio can be taken from a waveform with repeated peaks.

6.3.9   HARMONIC EXCITATION

Many structures are subject to harmonic excitation, such as vibration from a piece of machinery. If the structure is subjected to an excitation given by $f(t)$, the displacement of the structure $z(t)$ is the solution of:

$$M\ddot{z} + C\dot{z} + Kz = f(t)$$

$z(t)$ will consist of the general solution (the damped transient response derived above) and the particular integral, that will be dependent on the excitation $f(t)$. Figure 6.29 gives an example of a system's response to a sinusoidal excitation whose frequency differs significantly from the natural frequency of the system. As can be seen, the early influence of the transient response decays quickly until the total waveform is at the same frequency as the stimulus, but often out of phase.

This response (the blue line in Figure 6.29) is termed the **steady-state response** and is usually the most important consideration to the engineer.

6.3.9.1   Method 1: Direct Substitution

For the spring–damper system with the characteristic equation given by Equation (6.33),

$$M\ddot{z} + C\dot{z} + Kz = F(t)$$

If the harmonic excitation is given by $F(t) = F \cos \omega t$, we can expect the response of the system to be sinusoidal, and the frequency will be the same as the excitation. We can expect a phase difference (lag) between the excitation and the response (figure 6.30). The key information needed from the solution is the amplitude of the response and the phase relative to the excitation.

The response of the system (particular integral) can therefore be written as

$$z(t) = Z \cos(\omega t + \alpha) \tag{6.44}$$

where Z is the *amplitude* of the vibration and α is the *phase angle*.

To find Z and α, we substitute for $z(t)$ and its derivatives in the equation of motion:

$$-M\omega^2 Z \cos(\omega t + \alpha) - C\omega Z \sin(\omega t + \alpha) + KZ \cos(\omega t + \alpha) = F \cos \omega t$$

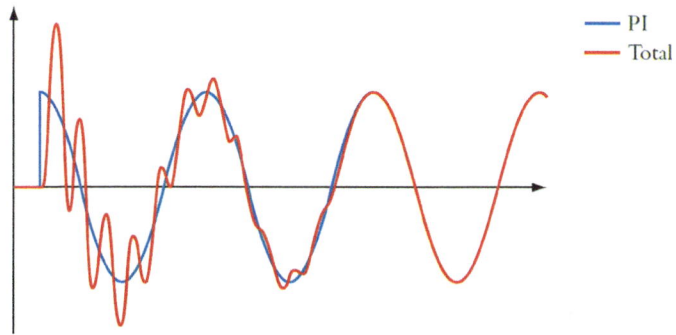

FIGURE 6.29   Response of the system to harmonic excitation.

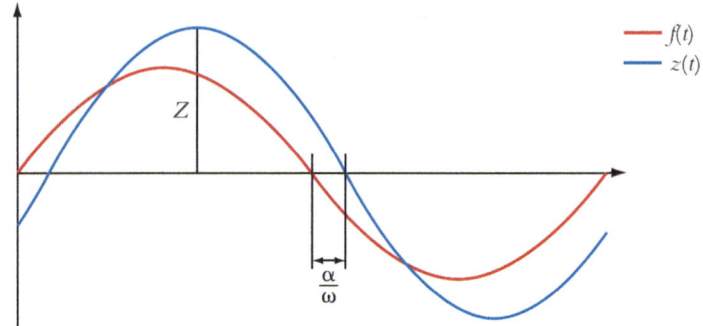

FIGURE 6.30   Relationship between excitation f(t) and response z(t).

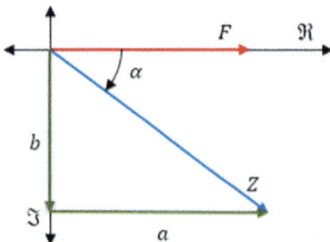

FIGURE 6.31   Phasor diagram of relationship between excitation F and response Z.

The next step is to find the magnitude of the response, Z. Using the trigonometric identities $\cos(\omega t + \alpha) = \cos \omega t \cos \alpha - \sin \omega t \sin \alpha$ and $\sin(\omega t + \alpha) = \sin \omega t \cos \alpha + \cos \omega t \sin \alpha$, all terms in $\cos \omega t$ on the left-hand side must be equal to F and all terms in $\sin \omega t$ must equal to zero. Solving for Z gives

$$Z = \frac{F}{\sqrt{\left(K - M\omega^2\right)^2 + \omega^2 C^2}} \tag{6.45}$$

And solving for α

$$\alpha = \tan^{-1}\left(\frac{-\omega C}{K - M\omega^2}\right) \tag{6.46}$$

Equation (6.46) shows that the phase angle α is negative, indicating that the response of a system will follow, or *lag* behind the excitation.

A convenient method to show the amplitude and phase of the response in comparison to the excitation is with a phasor diagram, as shown in Figure 6.31. As can be seen, the phase angle is accurately represented but the scales of excitation F (force) and response Z (displacement) will depend on the scale used.

6.3.9.2   Method 2: Complex Algebra

A second method is to use complex algebra, in particular the substitutions $f(t) = Fe^{i\omega t}$ for the excitation and $z(t) = Ze^{i(\omega t + \alpha)}$ for the response. This has the added advantage that the same mathematical approach can be extended to more complicated (and therefore more realistic) structures and problems.

Excitation

$$f(t) = Fe^{i\omega t}$$
(6.47)

Response

$$z(t) = Ze^{i(\omega t + \alpha)}$$
(6.48a)

Equation 6.48a can be further refined

$$z(t) = Ze^{i(\omega t + \alpha)} = Ze^{i\alpha}e^{i\omega t}$$

Hence, we can define

$$z(t) = Z^{\star}e^{i\omega t}$$
(6.48b)

where $Z^{\star}$ is a complex number: $Z^{\star} = \cos\alpha + i\sin\alpha$

The derivatives of Equation (6.48b) are $\dot{z} = i\omega Z^{\star}e^{i\omega t}$ and $\ddot{z} = -\omega^2 Z^{\star}e^{i\omega t}$.
When these are substituted into the equation of motion, we get

$$\left(-M\omega^2 + i\omega C + K\right)Z^{\star}e^{i\omega t} = Fe^{i\omega t}$$

Hence,

$$Z^{\star} = \frac{F}{\left(K - M\omega^2\right) + i\omega C}$$

or

$$Z^{\star} = \frac{F}{\left(K - M\omega^2\right)^2 + \omega^2 C^2}\left[\left(K - M\omega^2\right) - i\omega C\right]$$

Since $Z^{\star} = a + ib$, Z is the overall magnitude such that $Z = \sqrt{a^2 + b^2}$.
Hence,

$$Z = \frac{F}{\sqrt{\left(K - M\omega^2\right)^2 + \omega^2 C^2}}$$

and

$$\alpha = \tan^{-1}\left(\frac{-\omega C}{K - M\omega^2}\right)$$

Plotting $Z^{\star}$ and F on the complex (Argand) plane (Figure 6.31) gives the same phasor diagram discussed before. In practice, the imaginary part of $Z^{\star}$ is always negative.

6.3.10 FREQUENCY RESPONSE FUNCTION

To examine the relationship between the excitation frequency and the response, we will consider the **frequency response function** (FRF). This is defined as the **response magnitude/unit applied force**. Note that it is used exclusively for **force** excitation.

From the general form of the equation of motion,

$$M\ddot{z} + C\dot{z} + Kz = f(t)$$

Dividing by M and noting that $\dfrac{C}{M} = 2\gamma\omega_n$ and $\dfrac{K}{M} = \omega_n^2$, we get

$$\ddot{z} + 2\gamma\omega_n\dot{z} + \omega_n^2 z = \dfrac{f(t)}{M}$$

If $f(t) = Fe^{i\omega t}$ and $z(t) = Z^{*}e^{i\omega t}$, the equation of motion becomes:

$$Z^{*}e^{i\omega t}\left(-\omega^2 + 2i\gamma\omega_n\omega + \omega_n^2\right) = \dfrac{Fe^{i\omega t}}{M}$$

Hence,

$$Z^{*} = \dfrac{F}{M}\dfrac{1}{\left(\omega_n^2 - \omega^2\right) + i2\gamma\omega_n\omega}$$

The FRF is therefore

$$H(\omega) = \dfrac{Z^{*}}{F} = \dfrac{1}{M\left(\left(\omega_n^2 - \omega^2\right) + i2\gamma\omega_n\omega\right)} \tag{6.49a}$$

An alternative expression that emphasizes the frequency dependence (using the relationship that $\dfrac{K}{M} = \omega_n^2$) is:

$$H(\omega) = \dfrac{Z^{*}}{F} = \dfrac{1}{K}\dfrac{1}{\left(1 - \dfrac{\omega^2}{\omega_n^2}\right) + i2\gamma\dfrac{\omega}{\omega_n}} \tag{6.49b}$$

The magnitude is:

$$\left|H(\omega)\right| = \dfrac{1}{K}\dfrac{1}{\sqrt{\left(1 - \dfrac{\omega^2}{\omega_n^2}\right)^2 + 4\gamma^2\dfrac{\omega^2}{\omega_n^2}}} \tag{6.50}$$

and the phase angle is

$$\alpha = \tan^{-1}\left(\dfrac{-2\gamma\dfrac{\omega}{\omega_n}}{1 - \dfrac{\omega^2}{\omega_n^2}}\right) \tag{6.51}$$

It is clear from these expressions that the response of the structure (its amplitude and phase angle) is dependent on the ratio between the excitation frequency and the undamped natural frequency $\dfrac{\omega}{\omega_n}$ and the damping ratio γ.

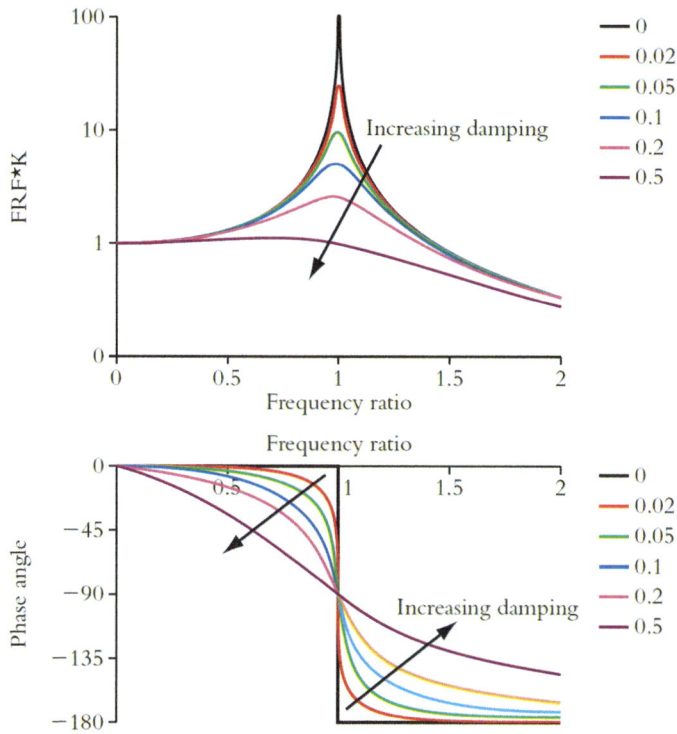

FIGURE 6.32   Amplitude (top) and phase angle (bottom) of FRF plotted against frequency ratio.

Figure 6.32 shows how the FRF amplitude and phase angle (y axis) vary with both the frequency ratio (x axis) and damping ratio (see legend).

As most structures have damping ratios of less than 0.1, the resonant peak (the maximum response near the frequency ratio of 1.0) will be large. Increasing damping will reduce the height of this resonant peak. In the phase angle plot, we see that for low values of the damping ratio, the phase angle is close to 0° (i.e. the response is nearly in phase with the excitation) for frequencies below the undamped natural frequency and close to −180° (nearly out-of-phase) above the undamped natural frequency.

The frequency giving the maximum response is the **resonant frequency**. With low damping, the resonant frequency, the undamped natural frequency and the damped natural frequency are all virtually identical. For higher damping (e.g. $\gamma = 0.2$), the resonant frequency is lower than the undamped natural frequency.

The relationship between the phase angle and the FRF is plotted in Figure 6.33. Each point on the plot corresponds to the phasor diagram in Figure 6.52 (the arrow illustrates one particular frequency). As the frequency ratio increases, the phase angle's magnitude will increase (note the negative value for α). When $\omega = 0$, the amplitude is 1.0 and the phase angle is 0°. When $\omega = \omega_n$, the phase angle is −90° and the response is purely imaginary. As ω tends to infinity, the amplitude tends to zero and the phase angle tends to −180°.

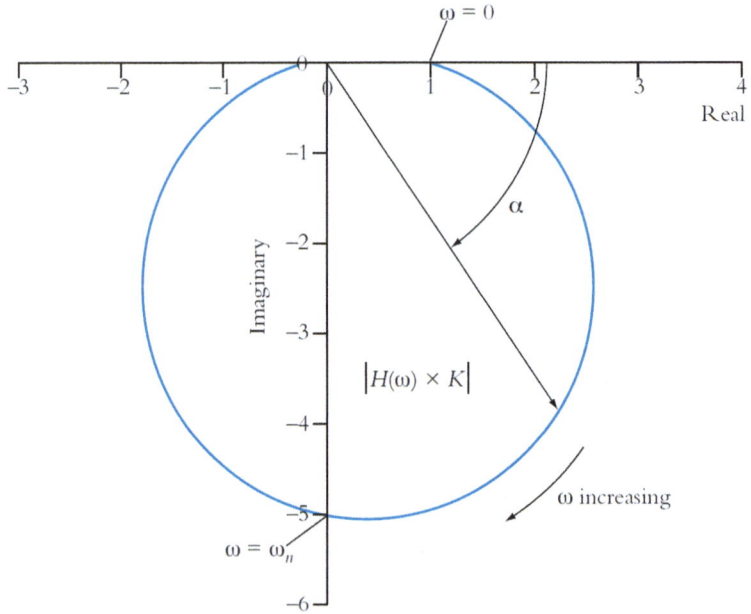

FIGURE 6.33 Relationship between the imaginary component of the FRF plotted and the real component for damping ratio $\gamma = 0.1$.

WORKED EXAMPLE – USE OF THE FRF

Find the waveform of the steady-state vertical displacement of a three-cylinder diesel generator (mass, $m = 4000$ kg) that operates at a crank speed, Ω, of 300 rpm. The machine is supported on a set of four resilient mounts that have a combined vertical stiffness of 2.5 MN/m and a damping ratio of 0.04. The generator can be modelled as a rigid mass and the mounts as a spring–damper combination as shown in Figure 6.34

The effect of the masses of the reciprocating pistons is to produce a vertical force on the compressor given by:

$$s(t) = 1.3\Omega^2 \cos\Omega t + 3.0\Omega^2 \cos 2\Omega t = S_1 \cos \omega_1 t + S_2 \cos \omega_2 t$$

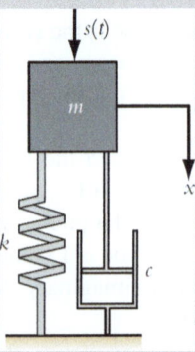

FIGURE 6.34 Schematic of the mass-spring-damper system.

Solution: Each sinusoidal term in the excitation will produce a steady-state response with the same frequency. Hence, $S_1 \cos \omega_1 t$ will produce a response in the form $X_1 \cos(\omega_1 t + \alpha_1)$. The magnitude and the phase angle can both be found from the FRF and as the two functions are linear and independent, the overall response will be the sum of the two responses.

The undamped natural frequency of the system is given by

$$\omega_n = \sqrt{\frac{K}{m}} = \sqrt{\frac{2.5 \times 10^6}{4 \times 10^3}} = 25 \, \text{rad s}^{-1}$$

Using the FRF, we use.

The effect of the masses of the reciprocating pistons is to produce a vertical force on the compressor given by:

$$s(t) = 1.3 \Omega^2 \cos \Omega t + 3.0 \Omega^2 \cos 2\Omega t = S_1 \cos \omega_1 t + S_2 \cos \omega_2 t$$

Solution: Each sinusoidal term in the excitation will produce a steady-state response with the same frequency. Hence, $S_1 \cos \omega_1 t$ will produce a response in the form $X_1 \cos(\omega_1 t + \alpha_1)$. The magnitude and the phase angle can both be found from the FRF and as the two functions are linear and independent, the overall response will be the sum of the two responses.

The undamped natural frequency of the system is given by

$$\omega_n = \sqrt{\frac{K}{m}} = \sqrt{\frac{2.5 \times 10^6}{4 \times 10^3}} = 25 \, \text{rad s}^{-1}$$

Using the FRF, we use Equation (6.50) to calculate the magnitude

$$\left| H(\omega_1) \right| = \frac{1}{K} \frac{1}{\sqrt{\left(1 - \dfrac{\omega_1^2}{\omega_n^2}\right)^2 + 4\gamma^2 \dfrac{\omega_1^2}{\omega_n^2}}}$$

We then use the information given in the question to calculate the numerical values – a similar process is used for ω_2. This gives the values

$$\left| H(\omega_1) \right| = 6.81 \times 10^{-4} \, \text{mm / N and} \left| H(\omega_2) \right| = 7.52 \times 10^{-5} \, \text{mm / N}$$

The phase angles are calculated from Equation (6.51)

$$\alpha = \tan^{-1} \left(\frac{-2\gamma \dfrac{\omega_1}{\omega_n}}{1 - \dfrac{\omega_1^2}{\omega_n^2}} \right)$$

So, $\alpha_1 = -170°$ and $\alpha_2 = -178°$

Thus, for the magnitudes given in the question (remember to convert Ω into radians/s),

$S_1 = 1283 \, \text{N}, X_1 = 0.87 \, \text{mm}$ with a phase angle of $-170°$

$S_2 = 2961 \, \text{N}, X_2 = 0.22 \, \text{mm}$ with a phase angle of $-178°$

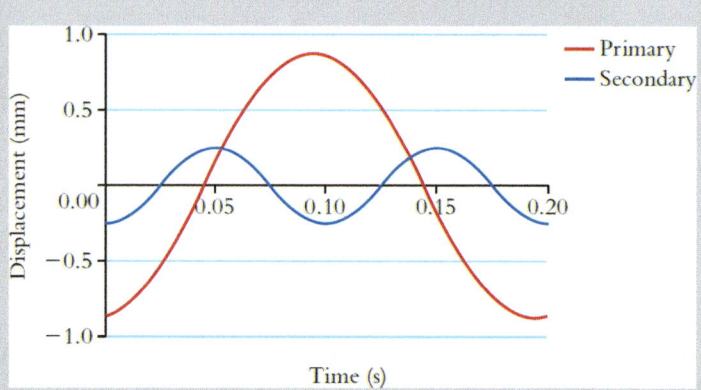

FIGURE 6.35 Vibration components: $X_1 \cos(\omega_1 t + \alpha_1)$ and $X_2 \cos(\omega_2 t + \alpha_2)$.

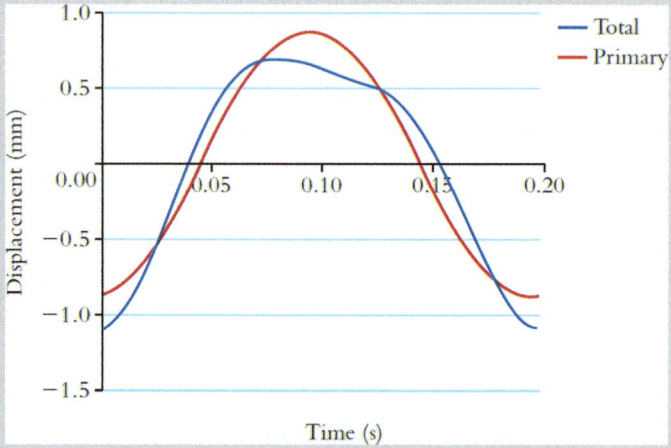

FIGURE 6.36 Total response (sum of the two waveforms).

Figure 6.35 plots the responses X_1 and X_2 against time and Figure 6.36 shows the overall response of the system.

The term 'primary' indicates that the vibration frequency ω_1 is equal to the rotational frequency of the crank. The 'secondary' component has a frequency that is twice the rotational frequency of the crank.

6.3.11 PERIODIC EXCITATION

Structures are rarely subjected to pure sinusoidal excitation in their operating environment. In many situations, *periodic* excitation is present. One example would be due to reciprocating machinery, such as the generator considered in Section 6.3.10.

If we consider the system shown in Figure 6.37, the normal mathematical approach for solving the differential equation

$$m\ddot{x} + c\dot{x} + kx = p(t)$$

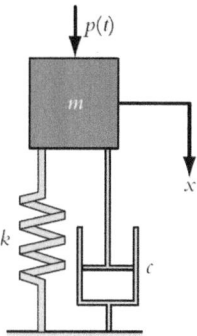

FIGURE 6.37   Mass-spring-damper system subjected to a periodic force p(t).

is to fit an appropriate expression for the particular integral. For a periodic waveform that repeats exactly every T seconds, the waveform can be represented by a Fourier series of the form:

$$p(t) = A_0 + \sum_{j=1}^{\infty} \left[A_j \cos(j\omega_0 t) + B_j \sin(j\omega_0 t) \right]$$

where the repetition frequency, $\omega_0 = \dfrac{2\pi}{T} \left[\text{rad}/\text{s} \right]$.

The steady-state response will also be sinusoidal with the same frequencies and we can calculate this using the FRF.

The process to calculate the response is as follows:

Use Fourier series analysis to decompose the excitation waveform into a set of sinusoids.

We can use the complex number representation of the waveforms to good effect here. For a typical pair of components we write

$$A_j \cos(j\omega_0 t) + B_j \sin(j\omega_0 t) = P_j^* e^{ij\omega_0 t} \text{ where } P_j^* = A_j + iB_j$$

(If the discrete Fourier transform is used, this complex number form is calculated directly.) The response to this excitation can be calculated from the FRF, so that

$$X_j^* = H(j\omega_0) \times P_j^*$$

The response term, X_j^*, will have the form $C_j + i\,D_j$ which can be interpreted as

$$C_j \cos(j\omega_0 t) + D_j \sin(j\omega_0 t)$$

Having done this for all frequencies in the excitation, the various frequency responses can be summed together.

As an example, Figure 6.38 shows a periodic excitation in the form of a series of rectangular force pulses with a repetition period of 0.25 s (i.e. a repetition frequency of 4 Hz). The response of a damped sprung system corresponding to Figure 6.37 is shown in Figure 6.39.

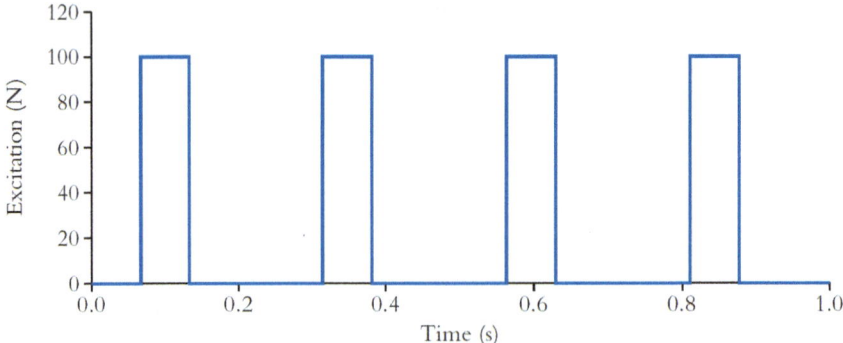

FIGURE 6.38 The 4 Hz force pulse waveform.

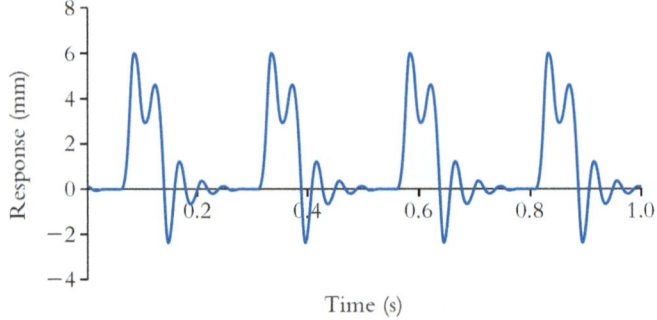

FIGURE 6.39 Response of single-degree-of-freedom system.

Figure 6.40 shows the first 12 Fourier coefficients calculated in Step 1 with cosine terms A_j (in the upper plot) and sine terms B_j (lower plot). Fourier coefficients are plotted against frequency. In addition to the A_0 at 0 Hz on the top graph, the components appear at 4 Hz, 8 Hz, 12 Hz, … (i.e. the $j\omega_0$ terms from the Fourier series, which are at integer multiples of the repetition frequency). In most cases, a 95%+ match can be achieved with 12 harmonics.

[Note that since $\cos(0) = 1$ and $\sin(0) = 0$, the A_0 term can be seen as a special case of the general summation when $j = 0$.]

Figure 6.41 shows the FRF for the system in the example, with the real and imaginary parts plotted against frequency. The system has an undamped natural frequency of 25 Hz and a damping ratio of 0.2.

In Step 2, we perform the complex multiplications between the FRF and excitation components. The FRF $H(\omega)$ at 20 Hz is calculated from Equation (6.49b) and the fifth Fourier component (at 20 Hz) is as follows:

$$P_5^* = -2.98 + i3.77\,\text{N}$$

$$H(20\,\text{Hz}) = 0.0629 - i0.0559\,\text{mm / N}$$

The FRF is shown in Figure 6.62. After complex multiplication, we get

$$P_5^* = -0.376 - i\,0.045\,\text{mm}$$

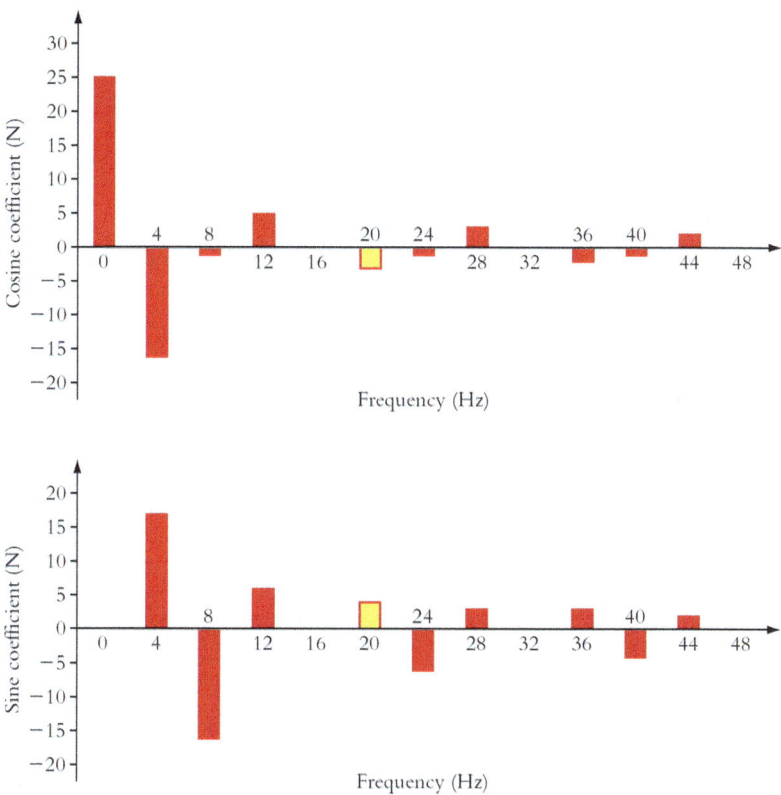

FIGURE 6.40 Cosine (top) and sine (bottom) coefficients of the Fourier series expansion of the excitation waveform up to $j = 12$.

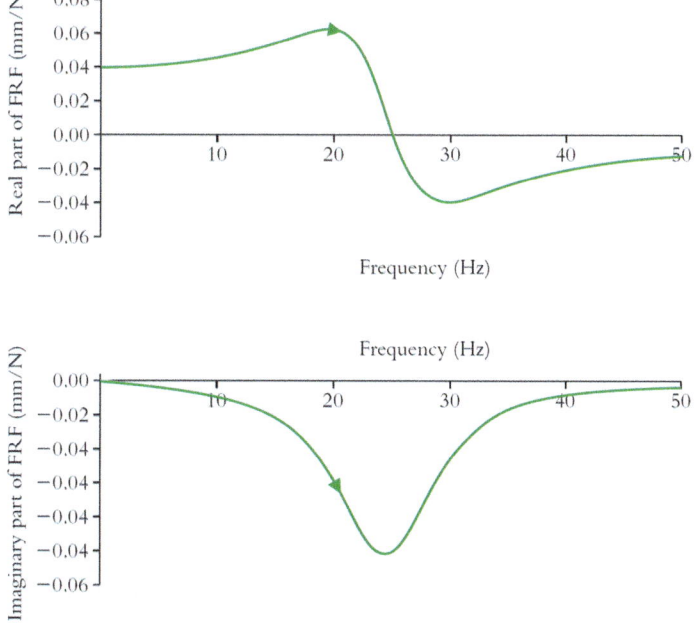

FIGURE 6.41 Real and imaginary parts of the FRF, natural frequency = 25 Hz, damping ratio = 0.2.

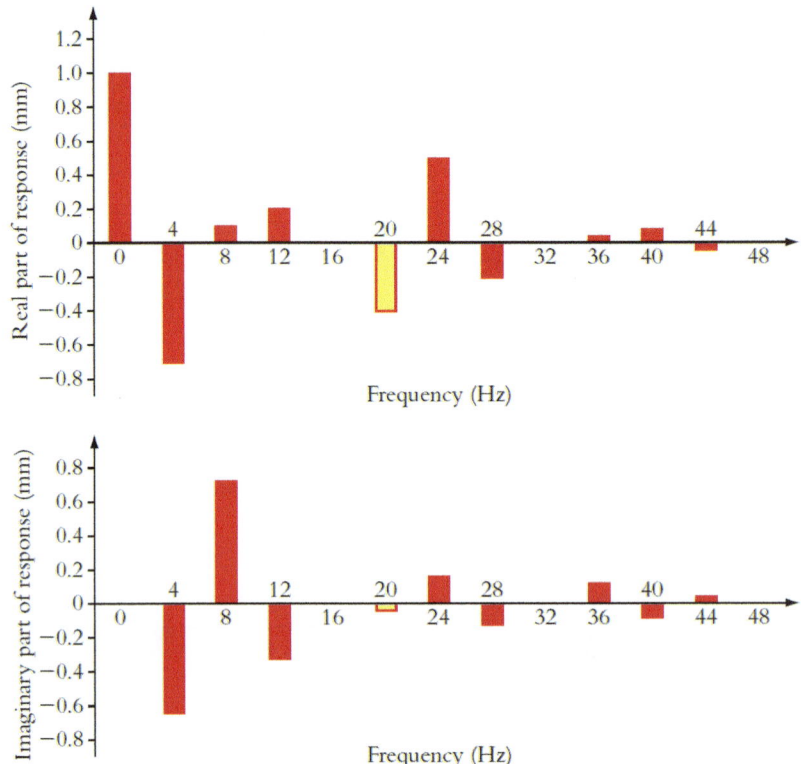

FIGURE 6.42   Real (cosine) and imaginary (sine) coefficients of the response of the system.

For a large number of components, this procedure would normally be done by software.

Component 5 is highlighted in Figure 6.42, which shows all of the C_j (top) and D_j (bottom) terms plotted against frequency. These are the cosine and sine coefficients of the response waveform, which are added together in Step 3. The resulting waveform is shown in Figure 6.43.

Figure 6.63 shows how the response waveform builds up as more terms are added. Figure 6.63(a) shows the constant term and the first three sinusoidal components (up to $j = 5$, or 20 Hz) and Figure 6.63(b) includes up to $j = 13$ (52 Hz). The waveform does not change much once the $j = 6$ term (24 Hz) has been added. This is because the excitation components are smaller above this frequency (Figure 6.60) and the FRF magnitude is also reducing (Figure 6.61).

Notes

1. The above procedure will work for any periodic excitation.
2. The FRF used here is the same expression that was derived using pure sinusoidal excitation. We are thus able to use the result from a mathematically convenient excitation case (but one which is *not* often experienced in practice) to find the response to a more complicated (and practical) excitation.
3. Most engineering structures have more than one resonant frequency. Provided the FRF can be calculated (as explained in Section 6.4) or measured (as in Section 6.5), the same procedure can be used.

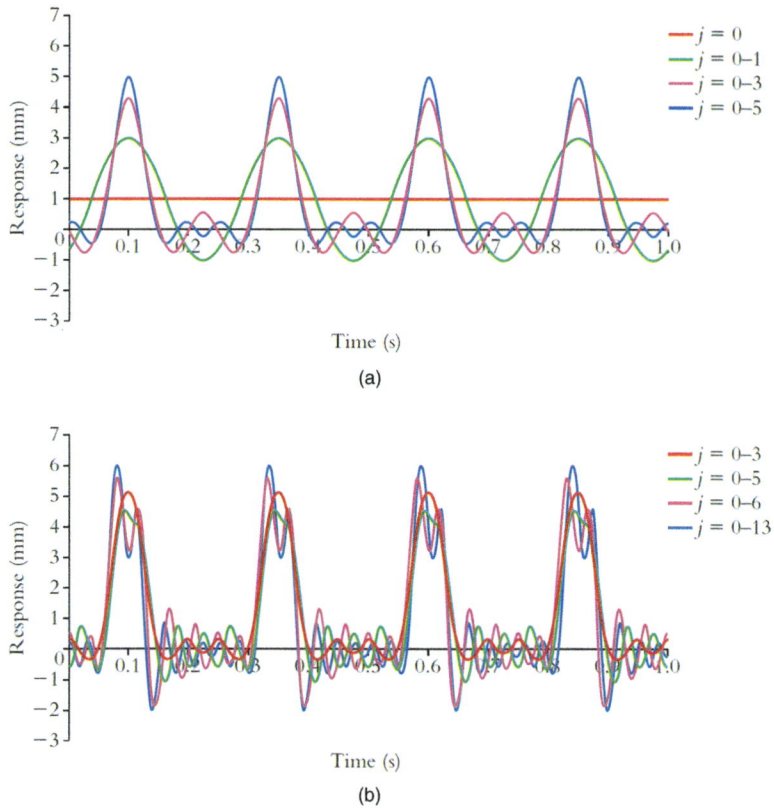

FIGURE 6.43   The summation process in Step 3: (a) terms up to $j = 5$ and (b) terms up to $j = 13$.

To illustrate further, Figure 6.44(a) shows the magnitude plot of the frequency response function of a structure that has five resonant frequencies in the range from 0 to 350 Hz. To find the response of this system to the same 4 Hz pulse sequence, the three-step procedure is repeated, using the new FRF. The result is shown in Figure 6.44(b).

As can be seen from Figure 6.44(b), there are significant similarities between the response of the 5 degrees of freedom system and the single-degree-of-freedom response waveform in Figure 6.59. This can be attributed to the original single-degree-of-freedom's resonant frequency being identical to the lowest natural frequency of the five-degrees-of-freedom system. The single-degree-of-freedom system is therefore a good approximation to the more complicated five-degrees-of-freedom structure. Section 6.6 will show how to create single-degree-of-freedom approximations for more complicated structures. The method is very powerful as it can be applied to any structure subjected to any periodic excitation.

LEARNING SUMMARY

By the end of this section, you should have learnt:

✓   To obtain the response of single-degree-of-freedom mass–spring–damper systems for the cases of 'free' vibration and for harmonic (sinusoidal) and arbitrary periodic excitation.

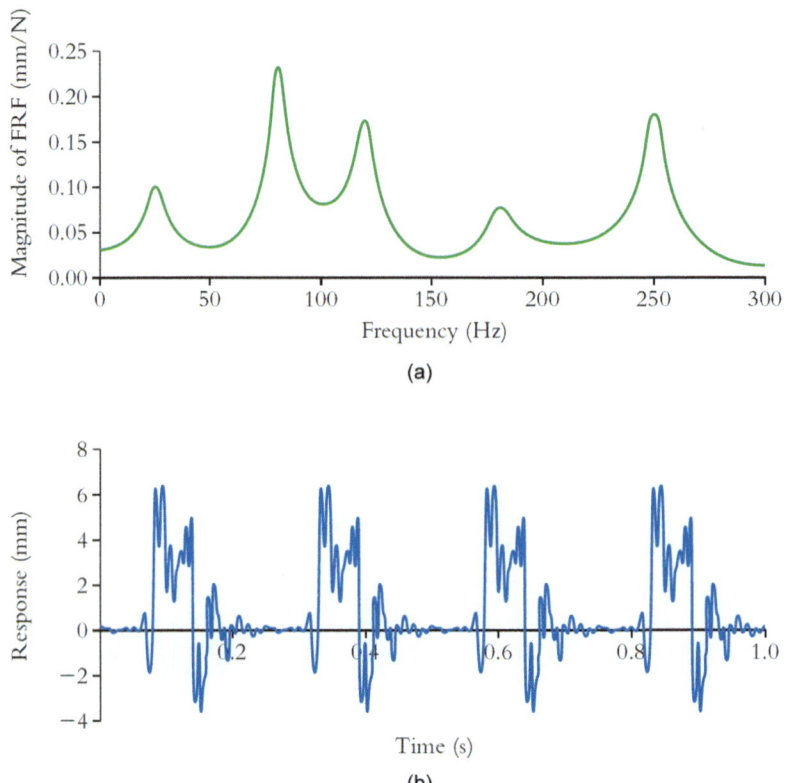

FIGURE 6.44 (a) Magnitude plot of FRF of a five-degrees-of-freedom system and (b) its response to 4 Hz force pulses.

6.4 RESPONSE OF DAMPED MULTI-DEGREE-OF-FREEDOM SYSTEMS

6.4.1 INTRODUCTION

Section 6.2 covered free, undamped vibration of multi-degree-of-freedom systems of the form:

$$[M]\{\ddot{x}\} + [K]\{x\} = \{0\}$$

These were solved with the substitution, $x(t) = \{X\}\cos\omega t$ and led to the eigenvalue problem:

$$\left([K] - \omega^2[M]\right)\{X\} = \{0\} \tag{6.52}$$

Solving this gave the natural frequency ω_r and modal vectors $\{X\}$ for each mode. Section 6.3 discussed the forced response of damped single-degree-of-freedom systems. This section extends this to damped systems with many degrees of freedom. The inclusion of damping and excitation requires a set of equations in the form:

$$[M]\{\ddot{x}\} + [C]\{\dot{x}\} + [K]\{x\} = \{p(t)\} \tag{6.53}$$

In order to solve these equations, we have to uncouple the equations to give a set of individual second-order ordinary differential equations with only one response variable. These can then be solved using the methods described in Section 6.3.

The steps involved are as follows:

Step 1 Uncouple the equations of motion to give a set of individual second-order ordinary differential equations.
Step 2 Solve each equation.
Step 3 Combine the solutions to give the full response of the system.

6.4.2 ORTHOGONALITY OF MODES

The modal vectors can be uncoupled because they are orthogonal, and hence the dot-product of these modal vectors will be zero (i.e. $\{X\}_n^T \cdot \{X\}_m = 0$ where n and m are mode numbers and $\{X\}_n^T$ is the transpose of $\{X\}_n$). For a typical mode, r, we can rearrange Equation (6.52) to give:

$$[K]\{X\}_r = \omega_r^2 [M]\{X\}_r$$

If we multiply by the transpose of a different mode shape vector, $\{X\}_s^T$ we get

$$\{X\}_s^T [K]\{X\}_r = \omega_r^2 \{X\}_s^T [M]\{X\}_r \qquad (6.54)$$

The corresponding expression for mode s is

$$\{X\}_r^T [K]\{X\}_s = \omega_s^2 \{X\}_r^T [M]\{X\}_s \qquad (6.55)$$

Since $[K]$ and $[M]$ are symmetric, it follows that

$$X_r^T [K] X_s = X_s^T [K] X_r \text{ and } X_r^T [M] X_s = X_s^T [M] X_r$$

Subtracting (6.55) from (6.54) gives

$$0 = \left(\omega_r^2 - \omega_s^2\right)\{X\}_r^T [M]\{X\}_s$$

Thus, provided that $\omega_r \neq \omega_s$

$$\{X\}_r^T [M]\{X\}_s = 0 \qquad (6.56)$$

from which it follows that

$$\{X\}_r^T [K]\{X\}_s = 0 \qquad (6.57)$$

For $r = s$,

$$\{X\}_r^T [M]\{X\}_r = M_r \qquad (6.58)$$

where M_r is the **modal mass** (a scalar) for mode r and

$$X_r^T [K] X_r = K_r \qquad (6.59)$$

K_r is the **modal stiffness** for mode r. Each individual mode shape is unique and independent of the others.

6.4.3 MODAL MATRIX AND MODAL SCALING

The **modal matrix**, [Φ], is obtained by assembling all of the modal vectors, column by column. That is,

$$[\Phi] = \left[\{X\}_1 \{X\}_2 \ldots X_r \ldots \right]$$

Thus, column r of the modal matrix is the modal vector for mode r, with each element giving the amplitude of the displacement in each coordinate for that mode.

$$\{X\}_r = \begin{Bmatrix} X_{1r} \\ X_{2r} \\ X_{3r} \\ \vdots \end{Bmatrix}$$

And so the full form of the matrix will be:

$$[\Phi] = \begin{bmatrix} X_{11} & X_{12} & X_{13} & \cdots \\ X_{21} & X_{22} & X_{23} & \cdots \\ X_{31} & X_{32} & X_{33} & \cdots \\ \vdots & \vdots & \vdots & \ddots \end{bmatrix}$$

It is important to note that X_{13} is *not* the same as X_{31}. In particular,

The first suffix (i.e. the row number) = response coordinate number
The second suffix (i.e. the column number) = mode number.

Thus, X_{13} is the response in coordinate 1 for the third mode of vibration and X_{31} is the response in coordinate 3 for the first mode of vibration. Following the results, for single mode vectors (Equations 6.56–6.59),

$$[\Phi]^T [M][\Phi] = \begin{bmatrix} \ddots & & \\ & M_r & \\ & & \ddots \end{bmatrix}$$

and

$$[\Phi]^T [K][\Phi] = \begin{bmatrix} \ddots & & \\ & K_r & \\ & & \ddots \end{bmatrix}$$

These are the **modal mass** and **modal stiffness matrices**, and they are **diagonal matrices**. These allow the equations of motion to be uncoupled.

Modal vectors have an arbitrary scaling. In Equation (6.11), for example, the mode shape vectors were found by setting the amplitude of one of the coordinates to be 1.0. The scaling adopted also affects the numerical values of the modal masses and stiffnesses. As a result, these will also contain an arbitrary factor. It is common to choose to scale the vectors so that all of the modal masses are equal to one. Starting with an arbitrarily scaled vector $\{X\}$ and the resulting modal mass M_r, the modal vector can be scaled (or **normalized**) to **unit modal mass** using the expression:

$$U_r = \frac{\{X\}_r}{\sqrt{M_r}} \tag{6.60}$$

One advantage of using this method of scaling is that, in addition to having all of the modal masses equal to one, the modal stiffness values are numerically equal to the square of the relevant natural frequency, that is, $M_R = 1.0$ and $K_R = \omega_r^2$ for all modes.

6.4.4   FORCED RESPONSE OF AN UNDAMPED MULTI-DEGREE-OF-FREEDOM STRUCTURE

Before considering a damped multi-degree-of-freedom structure, we will consider the equivalent undamped problem. The coupled equations of motion are

$$\left[M\right]\{\ddot{x}\} + \left[K\right]\{x\} = p(t) \tag{6.61}$$

where $p(t)$ is the stimulus applied to the structure. For step 1 of the solution process, we uncouple the equations by transforming the vector $\{x\}$ into the modal co-ordinate vector $\{q\}$ using the substitution

$$\{x\} = \left[\Phi\right]\{q\} \tag{6.62}$$

Substituting into Equation (6.61) and premultiplying by the transpose $[\Phi]^T$ gives

$$\left[\Phi\right]^T\left[M\right]\left[\Phi\right]\{\ddot{q}\} + \left[\Phi\right]^T\left[K\right]\left[\Phi\right]\{q\} = \left[\Phi\right]^T\{p(t)\}$$

Using the orthogonality properties for modal vectors and assuming that they have been scaled to unit modal mass gives

$$\left[I\right]\{\ddot{q}\} + \begin{bmatrix} \ddots & & \\ & \omega_r^2 & \\ & & \ddots \end{bmatrix}\{q\} = \left[\Phi\right]^T\{p(t)\} = \{f(t)\} \tag{6.63}$$

Just as Equation (6.61) describes the equations of motion in terms of the physical coordinates, Equation (6.63) describes the equations of motion in terms of the modal coordinates. Equation (6.62) can be seen mathematically as a vector space transformation, taking the equations from **physical space** (described by the physical displacement coordinates in $\{x\}$) to **modal space** (described by the modal coordinates $\{q\}$). This is analogous to stress transformations in solid mechanics, where a complex state of stress can be expressed as two or three principal stresses.

Because the two matrices on the left-hand side of Equation (6.63) are both diagonal matrices, the individual equations are *not* coupled and each involves only one mode. The equation for a typical mode r has the form

$$\ddot{q}_r + \omega_r^2 q_r = f_{r(t)} \tag{6.64}$$

Note that if the vectors have *not* been scaled to unit modal mass, the substitution of Equation (6.62) into (6.61) still results in uncoupled equations, but each then has the form

$$M_r\ddot{q}_r + K_r q_r = f_r(t)$$

In either case, there will be as many individual equations as there are modes of the structure, with each one involving only one of the modal coordinates.

Each equation describes a single mode of vibration and will tell us how much that mode contributes to the overall response. This has the natural frequency ω_r and is subjected to excitation defined by $f_r(t)$. The fact that these are modal rather than physical space equations does not affect the mathematics.

The solutions to the modal space equations are the responses of each individual mode of vibration expressed in terms of the modal coordinates, $q_r(t)$. Finding the expressions for these responses is step 2 of the solution process.

Once we have them, we can obtain the response in each of the physical coordinates by reusing Equation (6.62),

$$\{x\} = [\Phi]\{q\}$$

This is step 3 of the solution process.

Written out longhand, these equations are

$$\begin{Bmatrix} x_1(t) \\ x_2(t) \\ \vdots \\ x_j(t) \\ \vdots \end{Bmatrix} = \begin{bmatrix} u_{11} & u_{12} & \cdots & u_{12} & \cdots \\ u_{21} & u_{12} & \cdots & u_{12} & \cdots \\ \vdots & \vdots & \ddots & \vdots & \cdots \\ u_{j1} & u_{j2} & & u_{12} & \\ \vdots & \vdots & & \vdots & \end{bmatrix} \begin{Bmatrix} q_1(t) \\ q_2(t) \\ \vdots \\ q_r(t) \\ \vdots \end{Bmatrix}$$

The response in a typical coordinate, j, is therefore

$$x_j(t) = \sum_r u_{jr} q_r(t) \tag{6.65}$$

Equation (6.65) demonstrates that the response of the structure at any instant is a weighted sum of the individual modal vectors. The modal coordinates, $q_r(t)$ define how much each mode shape contributes to the structure's overall response at each instant.

WORKED EXAMPLE

Find the steady-state response of each mass in the two-degrees-of-freedom system shown in figure 6.45 to a sinusoidal force p_1 of amplitude 10 N and frequency 6 Hz applied to the 1 kg mass.

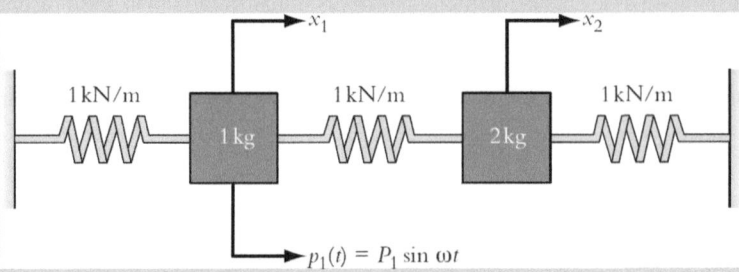

FIGURE 6.45   Two degree of freedom sprung mass system.

The equations of motion, natural frequencies and modal vectors are obtained first:

$$1\,\text{kg} \times \ddot{x}_1 + 1000x_1 + 1000(x_1 - x_2) = p_1(t)$$

$$2\,\text{kg} \times \ddot{x}_2 - 1000(x_1 - x_2) + 1000x_2 = 0$$

In matrix form, these become:

$$\begin{bmatrix} 1 & 0 \\ 0 & 2 \end{bmatrix} \begin{Bmatrix} \ddot{x}_1 \\ \ddot{x}_2 \end{Bmatrix} + \begin{bmatrix} 2000 & -1000 \\ -1000 & 2000 \end{bmatrix} \begin{Bmatrix} x_1 \\ x_2 \end{Bmatrix} = \begin{Bmatrix} p_1(t) \\ 0 \end{Bmatrix}$$

The natural frequencies from the determinant $\begin{vmatrix} 2000 - \omega^2 & -1000 \\ -1000 & 2000 - 2\omega^2 \end{vmatrix}$ (see Section 6.2 Equations 6.5–6.9) are $\omega_1^2 = 634\,\text{s}^{-2}$ and $\omega_2^2 = 2366\,\text{s}^{-2}$ giving natural frequencies of 4.0 Hz and 7.7 Hz, respectively.

The corresponding mode shape vectors from the solutions of $(2000 - \omega^2)x_1 - 1000x_2 = 0$ for $x_2 = 1.0$ and the natural frequencies ω_1 and ω_2 (Equation 6.8) are

$$\{X\}_1 = \begin{Bmatrix} X_{11} \\ X_{21} \end{Bmatrix} = \begin{Bmatrix} 0.732 \\ 1.0 \end{Bmatrix} \text{ and } \{X\}_2 = \begin{Bmatrix} X_{12} \\ X_{22} \end{Bmatrix} = \begin{Bmatrix} -2.732 \\ 1.0 \end{Bmatrix}$$

Using Equations (6.58) and (6.59) with the modal matrix $\begin{bmatrix} 0.732 & -2.732 \\ 1.0 & 1.0 \end{bmatrix}$, the modal mass and modal stiffness values are:

$$M_1 = 2.54, M_2 = 9.46, K_1 = 1608 \text{ and } K_2 = 22392$$

When the matrix given above is rescaled to give unit modal mass by dividing each column by the square root of its modal mass, for example $\left(\dfrac{0.732}{\sqrt{2.54}} = 0.460 \right)$ it becomes

$$[\Phi] = \begin{bmatrix} 0.460 & -0.888 \\ 0.628 & 0.325 \end{bmatrix}$$

It is worth noting that $[\Phi]^T [K][\Phi] = \begin{bmatrix} 634 & 0 \\ 0 & 2366 \end{bmatrix} = \begin{bmatrix} \omega_1^2 & 0 \\ 0 & \omega_2^2 \end{bmatrix}$, that is the transformed modal stiffnesses are equal to the corresponding natural frequencies.

From $\{f(t) = [\Phi]^T p(t)\}$, the modal forces are

$$\begin{Bmatrix} f_1(t) \\ f_2(t) \end{Bmatrix} = \begin{bmatrix} 0.460 & 0.628 \\ -0.888 & 0.325 \end{bmatrix} \begin{Bmatrix} p_1(t) \\ 0 \end{Bmatrix} = \begin{Bmatrix} 0.460 \times p_1(t) \\ -0.888 \times p_1(t) \end{Bmatrix}$$

The modal equation for mode 1 is

$$\ddot{q}_1 + \omega_1^2 q_1 = f_1(t)$$

or

$$\ddot{q}_1 + 634q_1 = 0.460p_1(t) \qquad (6.66)$$

To give the mode 1 steady-state response, we use $p_1(t) = 10e^{i\omega t}$ and $q_1(t) = Q_1^\star e^{i\omega t}$ in (6.66) with an excitation frequency of 6 Hz $\left(\omega = 12\pi = 37.7\,\mathrm{rad\,s^{-1}}\right)$ to give

$$Q_1^\star = \frac{4.60}{634 - (6 \times 2\pi)^2} = -0.00584$$

Similarly for mode 2,

$$Q_2^\star = \frac{-8.88}{2366 - (6 \times 2\pi)^2} = -0.009\,40$$

From Equation (6.65),

$$\begin{Bmatrix} X_1^\star e^{i\omega t} \\ X_2^\star e^{i\omega t} \end{Bmatrix} = \begin{bmatrix} 0.460 & -0.888 \\ 0.628 & 0.325 \end{bmatrix} \begin{Bmatrix} Q_1^\star e^{i\omega t} \\ Q_2^\star e^{i\omega t} \end{Bmatrix}$$

Hence,

$$X_1^\star = 0.460 \times Q_1^\star - 0.888 \times Q_2^\star = 5.66\,\mathrm{mm}$$

and

$$X_2^\star = 0.628 \times Q_1^\star + 0.325 \times Q_2^\star = -6.72\,\mathrm{mm}$$

Note that $X_1^\star$ is positive, showing that mass 1 moves in phase with the force applied to it, while the negative sign for $X_2^\star$ shows that mass 2 moves 180° out of phase with the applied force. When these modal responses are calculated across a range of frequencies, the modal responses (Q_n) and the displacements of the two masses (X_n) are as shown in Figures 6.46 and 6.47.

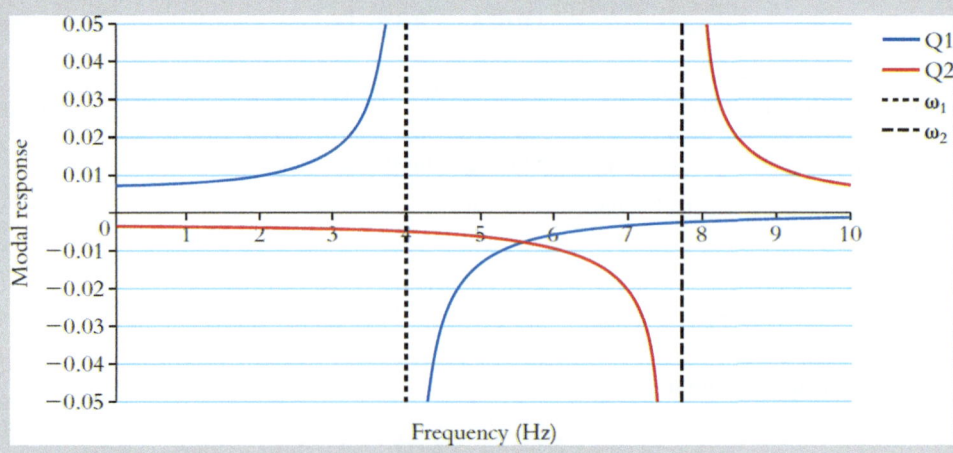

FIGURE 6.46   Modal responses.

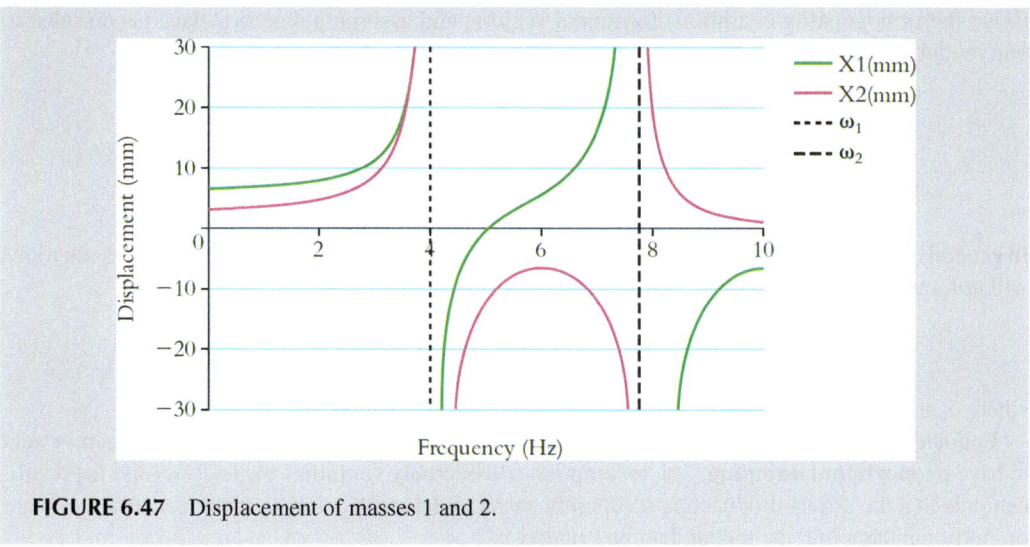

FIGURE 6.47   Displacement of masses 1 and 2.

We see from the modal responses that each produces an infinite response if the excitation frequency coincides with the relevant natural frequency (4.0 Hz and 7.7 Hz, in this case). At frequencies in the vicinity of each natural frequency, the relative displacements of the two masses will closely resemble the proportions given by the relevant mode shape. In the vicinity of ω_1, for example, $|Q_1^*| \gg |Q_2^*|$; so, the first mode will dominate the response. The chosen excitation frequency of 6 Hz is about midway between the two natural frequencies and we find that the two modes make similar contributions at this frequency.

At a frequency of approximately 5.5 Hz, mass 1 has a displacement amplitude of zero. This may seem surprising as the force is applied to this mass. The ability to design a dynamic system that has zero displacement at the point of force application is the basis for **tuned vibration absorbers**. These were used to help solve the resonant vibration problem on London's Millennium Bridge.

6.4.5   Forced Response of a Damped Structure

As with single-degree-of-freedom systems, we have to add further terms in $\dot{x}$ to take damping into account. The equations of motion are:

$$\left[M\right]\{\ddot{x}\}+\left[C\right]\{\dot{x}\}+\left[K\right]\{x\} = \{p(t)\}$$

As in the undamped case, the equations can be uncoupled by transforming into modal space using the modal matrix $\left[\Phi\right]$.

Step 1
Substitute $\left[\Phi\right]\{q\} = \{x\}$, where $\left[\Phi\right]$ is the modal matrix of the *undamped* system into the equation of motion and then pre-multiply by $\left[\Phi\right]^T$:

$$\left[\Phi\right]^T\left[M\right]\left[\Phi\right]\{\ddot{q}\}+\left[\Phi\right]^T\left[C\right]\left[\Phi\right]\{\dot{q}\}+\left[\Phi\right]^T\left[K\right]\left[\Phi\right]\{q\} = \left[\Phi\right]^T\{p(t)\}$$

Using the orthogonality conditions for modal vectors, and assuming that they have been scaled to unit modal mass, we get

$$[I]\{\ddot{q}\} + [\Phi]^T [C][\Phi]\{\dot{q}\} + \begin{bmatrix} \ddots & & \\ & \omega_r^2 & \\ & & \ddots \end{bmatrix} \{q\} = [\Phi]^T \{p(t)\} = \{f(t)\}$$

In general, the **modal damping matrix**, $[\Phi]^T[C][\Phi]$, will *not* be a diagonal matrix and the equations will *not* uncouple. The exception is if [C] can be written in the form

$$[C] = a_1 [M] + a_2 [K]$$

where a_1 and a_2 are constants. In this case, the equations *will* uncouple.

Engineers commonly assume that [C] has the above form, in which case the system can be said to have **proportional damping**. The assumption considerably simplifies the analysis and the justification is that the error introduced is acceptably small as damping is low in most real structures. For proportional damping, the modal damping matrix is

$$[\Phi]^T [C][\Phi] = \begin{bmatrix} \ddots & & \\ & 2\gamma_r \omega_r & \\ & & \ddots \end{bmatrix}$$

where γ_r is the damping ratio for mode r. The modal space equation for mode r is

$$\ddot{q}_r + 2\gamma_r \omega_r \dot{q}_r + \omega_r^2 q_r = f_r(t)$$

Steps 2 and 3 of the solution are as before.

6.4.6 FREQUENCY RESPONSE FUNCTION

The FRF for a multi-degree-of-freedom system can be calculated using a simple extension of the method used in Section 6.3 for a single-degree-of-freedom system. We will consider the case of sinusoidal excitation applied to coordinate k and find the response in coordinate j.

Section 6.4 showed that the equations of motion can be uncoupled to give a set of equations describing the behaviour of each mode. For mode r,

$$\ddot{q}_r + 2\gamma_r \omega_r \dot{q}_r + \omega_r^2 q_r = f_r(t)$$

The modal force vector, $\{f(t)\}$, is calculated from $[\Phi]^T \{p(t)\}$. For a situation where the only non-zero term is the kth coordinate, that is

$$\{p(t)\} = \begin{Bmatrix} 0 \\ \vdots \\ 0 \\ p_k(t) \\ 0 \\ \vdots \\ 0 \end{Bmatrix}$$

From $\{f(t)\} = [\Phi]^T \{p(t)\}$, $f_r(t) = u_{kr} p_k(t)$. If the input is a harmonic excitation, $f_r(t) = u_{kr} P_k e^{i\omega t}$ and $q_r(t) = Q_r^* e^{i\omega t}$. The modal equation of motion therefore becomes:

$$-\omega^2 Q_r^* e^{i\omega t} + 2i\gamma_r \omega_r \omega Q_r^* e^{i\omega t} + \omega_r^2 Q_r^* e^{i\omega t} = f_r(t) = u_{kr} P_k e^{i\omega t}$$

Eliminating $e^{i\omega t}$ and re-arranging gives

$$Q_r^* = \frac{u_{kr} P_k}{\left(\omega_r^2 - \omega^2\right) + i2\gamma_r \omega_r \omega}$$

or

$$q_r(t) = \frac{u_{kr} P_k e^{i\omega t}}{\left(\omega_r^2 - \omega^2\right) + i2\gamma_r \omega_r \omega}$$

Combining the individual modal responses gives the response in coordinate j

$$x_j(t) = X_j^* e^{i\omega t} \sum_r u_{jr} q_r(t)$$

$$= \sum_r \frac{u_{jr} u_{kr} P_k e^{i\omega t}}{\omega_r^2 - \omega^2 + i2\gamma_r \omega_r \omega}$$

The FRF giving the response in coordinate j due to a unit sinusoidal force in coordinate k is therefore

$$H_{jk}(\omega) = \frac{X_j^*}{P_k} = \sum_r \frac{u_{jr} u_{kr}}{\left(\omega_r^2 - \omega^2\right) + i2\gamma_r \omega_r \omega} \qquad (6.67)$$

LEARNING SUMMARY

By the end of this section, you should have learnt:

✓ To uncouple the equations of motion of a proportionally damped multi-degree-of-freedom structure by making use of the orthogonality properties of the modal matrix;

✓ To formulate response expressions, including the FRF.

6.5 EXPERIMENTAL MODAL ANALYSIS

6.5.1 INTRODUCTION

The methods for predicting the natural frequencies and mode shapes of different types of structures are shown in Section 6.2. In each case, assumptions and approximations were made. However, none of these models were able to predict the damping level in the structure.

Experimental modal analysis tests a structure as it is without the need for modelling approximations. For each mode of vibration within a chosen frequency range, it identifies the natural frequency, damping and mode shape. These are the modal parameters.

For example, Figure 6.48 shows the first bending and first torsional mode shapes obtained from a test on a beam that is free to vibrate. While many structures (such as helicopter rotor blades and

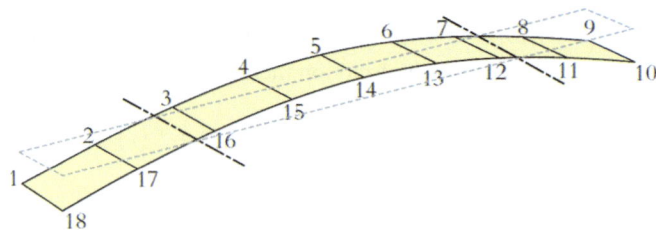

First bending mode

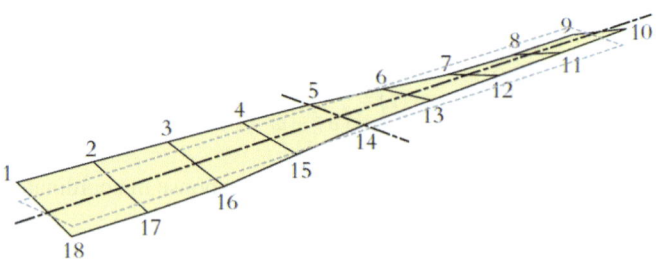

First torsional mode

FIGURE 6.48   First bending and torsional modes for a free beam.

lamp posts) have beam-like characteristics, few meet the idealized conditions incorporated into these theoretical models.

In this particular test, each mode shape has been defined in terms of the amplitude and relative phase of the vertical motion of 18 points on the beam (nine along each edge). Each point oscillates sinusoidally about the undisplaced (equilibrium) position, shown by the dotted lines in the figure. Figure 6.48 itself shows snapshots of the displaced beam when it is at one extreme of this motion. As can be seen in the figure, points on the beam along the chain-dashed lines are stationary. These are the nodal lines and are unique to each mode shape.

A common way of performing these tests is to excite the beam by means of a single impulse (usually a tap with a hammer), measure all resulting FRFs and then find the modal parameters using a curve fitting procedure.

6.5.2   FREQUENCY RESPONSE FUNCTION TESTING

To recap, the FRF defines the response per unit applied force across a range of frequencies (for further detail, see Section 6.3). The FRF for an excitation frequency ω was shown (Equation 6.49) to be

$$H(\omega) = \frac{Z^\star}{F} = \frac{1}{M}\frac{1}{\left(\omega_n^2 - \omega^2 i 2\gamma\omega_n\omega\right)} \tag{6.68}$$

where $Z^\star$ is a complex number containing amplitude and phase information about the response, F is the force amplitude, M is the coefficient of acceleration in the equation of motion and ω is the excitation frequency. The expression also contains two of the modal parameters; ω_n, which is the undamped natural frequency, and γ, the damping ratio.

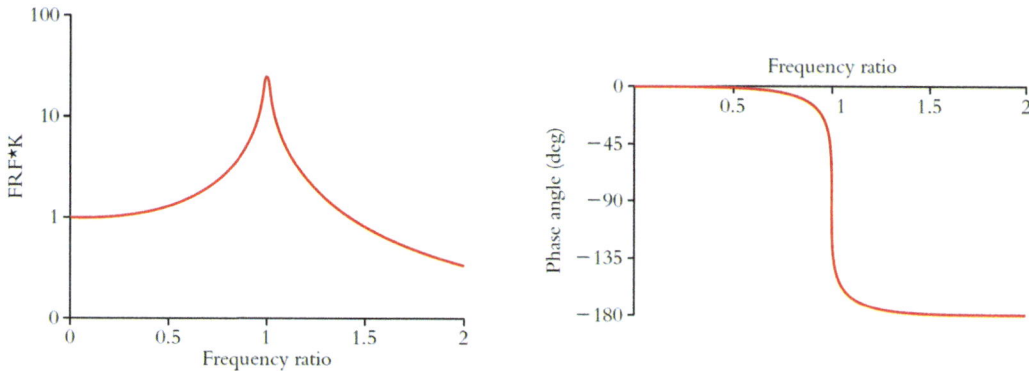

FIGURE 6.49 Amplitude and phase spectra plotted against frequency ratio $\frac{\omega}{\omega_n}$ for a single degree of freedom system.

The amplitude and phase spectra for a single degree of freedom system are shown in Figure 6.49, plotted against the frequency ratio, $\frac{\omega}{\omega_n}$:

If a sinusoidal force was applied to a structure and the steady-state amplitude and phase of the response were measured for that frequency, this would generate one point on each graph. A complete FRF plot could be produced by repeating this test at many different excitation frequencies. This is called a swept-sine test.

For a single-degree-of-freedom system, there will be one resonant peak. For a multi-degree-of-freedom system, a swept-sine test would find several resonant peaks.

Figure 6.50 shows the FRF of a multi-degree-of-freedom system. There are five resonant peaks in the range analysed, indicating five modes of vibration. Note that each peak has a related phase shift of 180°. This is the same characteristic seen in the FRF for the single-degree-of-freedom system in Figure 6.50.

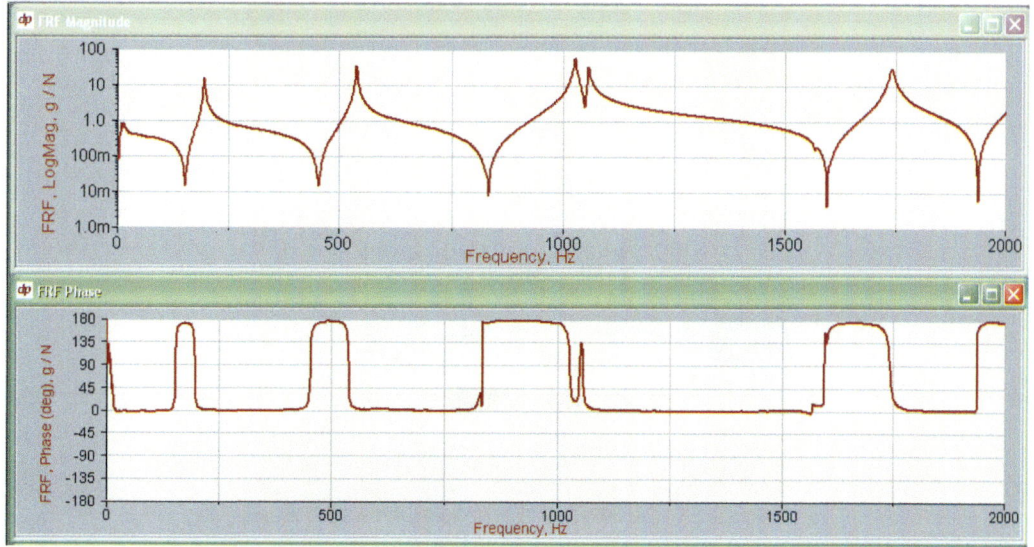

FIGURE 6.50 Experimentally determined frequency response function of a multi-degree of freedom system.

Since a single-degree-of-freedom system has only one coordinate to describe it, the response in that coordinate tells us all we need to know. For a multi-degree-of-freedom system, the displacement varies from point to point on the structure. These different mode shapes (bending and torsion) will vary between resonant frequencies. Equation (6.67) gives the FRF for a multi-degree-of-freedom system giving the response in coordinate j due to a unit sinusoidal force in coordinate k:

$$H_{jk}(\omega) = \frac{X_j^{\star}}{P_k} = \sum_r \frac{u_{jr}u_{kr}}{\left(\omega_r^2 - \omega^2\right) + i2\gamma_r\omega_r\omega} \tag{6.69}$$

where r is the mode number and ω_r and γ_r are the undamped natural frequency and damping ratio for mode number r. u_{jr} and u_{kr} are the mode shape values in coordinates j and k for mode r. This can also be given as

$$H_{jk}(\omega) = \sum_r \frac{\left(A_{jk}\right)_r}{\left(\omega_r^2 - \omega^2\right) + i2\gamma_r\omega_r\omega} \tag{6.70}$$

The terms $\left(A_{jk}\right)_r = u_{jr}u_{kr}$ are termed the *residues* and are simply the product of the mode shape values of the structure at the response and excitation positions for each mode.

Each of the five modes in Figure 6.51 is described mathematically by an expression similar to Equation (6.68). Note that if there was only one mode, the summation in Equation (6.70) would only have one term and the residue would be equal to 1 as the mode shapes were scaled to give modal masses equal to **1.0** in Equation (6.69).

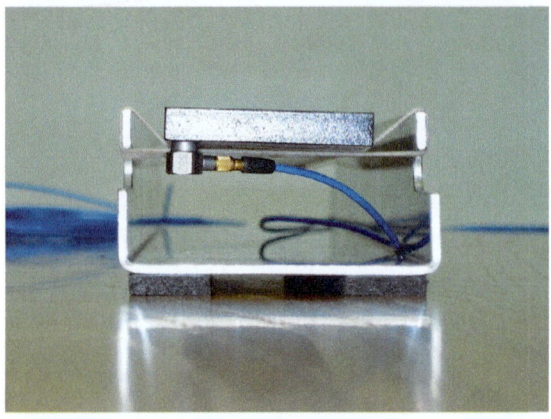

FIGURE 6.51   (Top) Beam supported in a cradle using rubber bands to simulate a free-free condition. An instrumented hammer is next to the beam; (Bottom) Accelerometer measures the vertical response at one corner (Point #1).

Measuring FRFs lies at the heart of experimental modal analysis. In principle, this can be done by applying a known force to the structure at point k, measuring its response at point j and then dividing one by the other. That is,

$$H_{jk}(\omega) = \frac{X_j^\star}{F_k} \tag{6.71}$$

An 'impact test' is a simple and effective technique for small, lightly damped structures. In a typical set-up for such a test, the structure is excited by a tap from a special instrumented hammer containing a force transducer. Accelerometers attached to the structure measure the response. A set-up for a test on a free-free beam is shown in Figure 6.51.

The impact excites a wide range of frequencies simultaneously and the resulting transient response is picked up by the accelerometer. Typical force and response signals are shown in Figure 6.52.

Each mode behaves as a separate single-degree-of-freedom system and gives a transient response, a decaying sinusoidal vibration at that mode's natural frequency. The overall response is the combination of the oscillations from all of the modes in the range. Provided that the structure behaves as a linear system, the principle of superposition applies and the individual modal responses will not interact.

The input and response are both time domain signals: The input (Figure 6.53, top left) is an impulse and the response (top and bottom right) can be seen to decay over time and to have several core frequencies. The response signals are filtered and analysed using the discrete Fourier transform

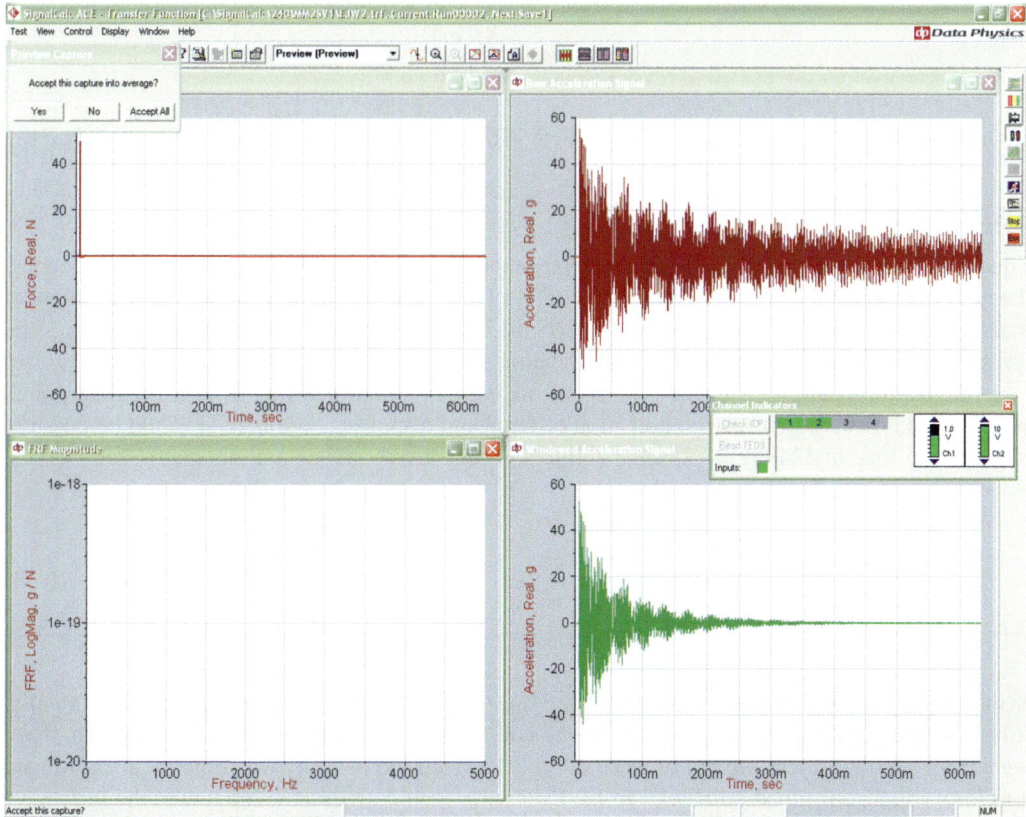

FIGURE 6.52 Typical force and response signals.

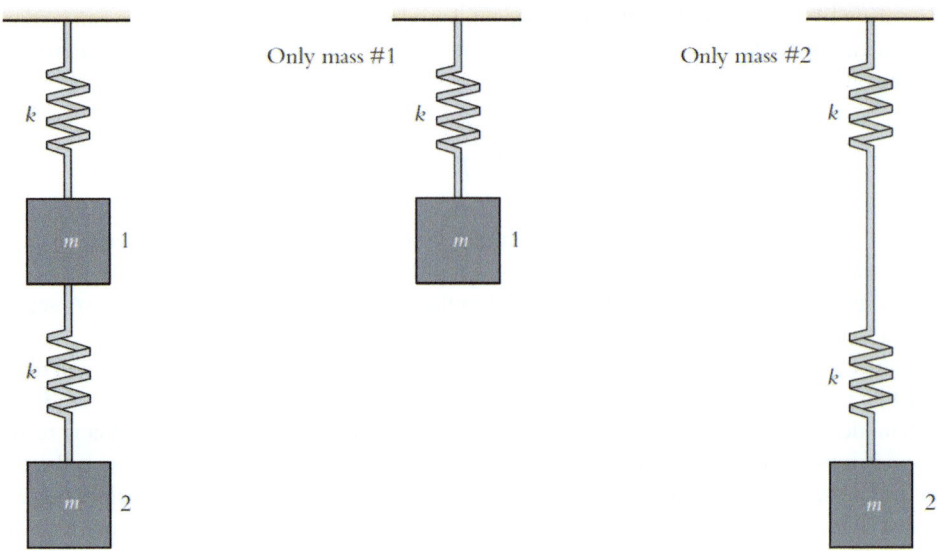

FIGURE 6.53   Two-degree-of-freedom system and its two subsystems.

to identify the individual frequency components. For each pair of frequency components, dividing response by force gives the FRF at that frequency.

$$H_{jk}(\omega) = \frac{X_j(\omega)}{F_k(\omega)} \tag{6.72}$$

Doing this for all frequencies gives the complete FRF. An example can be seen in Figure 6.51.

6.5.3 MODE IDENTIFICATION

Equation (6.70) shows that the FRF embodies the modal properties of the system (natural frequency, damping and mode shape). Once the values of $(A_{jk})_r$, ω_r and γ_r have been found for each mode to make Equation (6.70) match the measured FRF, the modal parameters can be identified. This is the basis for many commercial parameter identification programs.

In the earlier example of a free–free beam, the vertical coordinates at 18 measurement points were selected. If we used every combination of force and response coordinate, it would be possible to measure a total of 324 (18 × 18) separate FRFs. These tests could be put into matrix form as:

$$\begin{Bmatrix} x_1 \\ x_2 \\ \vdots \\ x_{18} \end{Bmatrix} = \begin{bmatrix} H_{1,1} & H_{1,2} & \cdots & H_{1,18} \\ H_{2,1} & H_{2,2} & \cdots & H_{2,18} \\ \vdots & \vdots & \ddots & \vdots \\ H_{18,1} & H_{18,2} & \cdots & H_{18,18} \end{bmatrix} \begin{Bmatrix} F_1 \\ F_2 \\ \vdots \\ F_{18} \end{Bmatrix} \tag{6.73}$$

In this matrix, $H_{1,2}$ is the response in coordinate 1 due to a force applied in coordinate 2, etc. Normally, it is only necessary to measure one row or one column of the matrix of FRFs. For example, if we were using an impact test, we might place the accelerometer to measure the response at coordinate 1 to forces applied from coordinates 1 to 18. Doing this would give us the top row of the matrix.

Equation (6.70) shows that each FRF can be computed if the residue, natural frequency and damping values are known for each mode. The right-hand side of Equation (6.70) can be used as a parametric fitting curve in which the modal parameters must be selected to give the best fit between the computed and measured FRFs. A variety of algorithms are used to do the curve fitting, but a description of these is outside the scope of this book.

The frequency and damping of a mode are global properties and are the same regardless of the measurement positions. However, the residue for each mode, $\left(A_{jk}\right)_r$, will be different for each of the FRFs.

Once the curve fitting has been done for all of the FRFs, there will be residue values for each of the measurement coordinates and each mode. In the beam example, if we had measured the top row of the FRF matrix, the curve-fitting process would have given us the following vector of residues for each mode:

$$\left\{A_{1,1} \quad A_{1,2} \quad \cdots \quad A_{1,18}\right\}_r \tag{6.74}$$

Since the residues are simply the products of mode shape values, Equation (6.74) can be rewritten as:

$$\left\{\left(u_1\right)_r \cdot \left(u_1\right)_r \quad \left(u_1\right)_r \cdot \left(u_2\right)_r \cdots \left(u_1\right)_r \cdot \left(u_{18}\right)_r\right\} \tag{6.75}$$

$\left(u_1\right)_r$ is a common factor in every term in this case. This value can be found provided that one of the measurements corresponds to the coordinate where the force is applied. In this example, this is coordinate #1. Once the value of $\left(A_{1,1}\right)_r$ is known, the common factor can be found in Equation (6.76).

$$\left(u_1\right)_r \cdot \left(u_1\right)_r = \left(A_{1,1}\right)_r$$

or

$$\left(u_1\right)_r = \sqrt{\left(A_{1,1}\right)_r} \tag{6.76}$$

Hence, the complete mode shape vector can be found from the corresponding residue vector using Equation (6.75).

$$\left(u_k\right)_r = \frac{\left(A_{1,k}\right)_r}{\sqrt{\left(A_{1,1}\right)_r}} \tag{6.77}$$

LEARNING SUMMARY

By the end of this section, you should have learnt:

- ✓ How the expression for the FRF for a multi-degree-of-freedom structure can be used to identify the natural frequencies, damping and mode shapes of structures from experimental tests.

6.6   APPROXIMATE METHODS

6.6.1   INTRODUCTION

In many cases, the lowest natural frequency is the most important. As an example, if the first critical speed of a shaft is outside its operating range, whirl will be avoided. It is often the case that the first mode gives the largest displacement for a given excitation.

This section presents two methods for estimating the lowest natural frequency and shows how single-degree-of-freedom approximations of more complex systems can be created.

6.6.2 DUNKERLEY'S METHOD

Stanley Dunkerley developed a method of estimating the lowest critical speed of shafts. His formula can be used to estimate the lowest natural frequency of any system.

$$\frac{1}{\omega_n^{\,2}} \approx \frac{1}{\omega_1^{\,2}} + \frac{1}{\omega_2^{\,2}} + \cdots + \frac{1}{\omega_r^{\,2}} + \cdots \tag{6.78}$$

where ω_r is the natural frequency of a subsystem that has only the rth mass present. The following examples show how the formula is used.

6.6.2.1 Example 1: Two-Degree-of-Freedom System

The system on the left of Figure 6.53 has two masses, so there will be two subsystems, one with only mass 1 and the other with only mass 2. The two subsystems are also shown in Figure 6.53.

> **Subsystem with only mass 1**: When we remove mass 2 to create this subsystem, we also ignore the lower spring, since this is assumed to be massless. The natural frequency of the subsystem is given by:

$$\omega_1^{\,2} = \frac{k}{m}$$

> **Subsystem with only mass 2**: Removing mass 1 leaves mass 2 supported by two springs in series, each of stiffness k. These are equivalent to a single spring of stiffness $\frac{k}{2}$. The natural frequency of the subsystem is given by

$$\omega_2^{\,2} = \frac{\dfrac{k}{2}}{m} = \frac{k}{2m}$$

With two subsystems, Dunkerley's formula will have two terms in the series on the right-hand side. Substituting the subsystem results into the formula, we get

$$\frac{1}{\omega_n^{\,2}} \approx \frac{m}{k} + \frac{2m}{k} = \frac{3m}{k}$$

Hence,

$$\omega_n = \sqrt{\frac{k}{3m}} \tag{6.79}$$

As a numerical example, if m = 2 kg and k = 200 N/m, Equation (6.79) gives the lowest natural frequency to be 0.918 Hz. The exact answer is 0.984 Hz.

6.6.2.2 Example 2: Shaft with Added Masses

The phenomenon of shaft whirl, and a method for calculating the whirl frequencies for plain shafts with uniform cross-sections, were presented in Section 6.2. While this could be a good approximation for the drive shaft on a vehicle, for example, it would not be appropriate for the steam turbine

FIGURE 6.54 Steam turbine shaft and discs.

shaft shown in Figure 6.54. This carries several large-bladed discs that add significant mass to the system and affect its natural frequencies and hence its whirl speeds. Whirl can be avoided if the rotational frequency of the shaft remains below its lowest natural frequency in flexure and Dunkerley's method is able to give an estimate of this.

We will look at a simplified example consisting of a uniform shaft carrying three added masses.

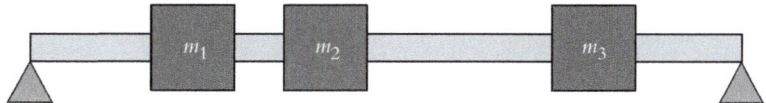

Subsystem with only mass 1 present is:

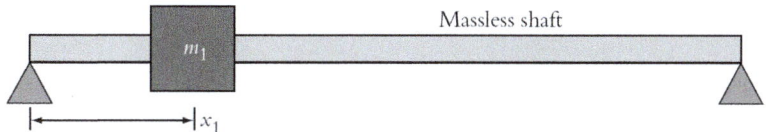

Since we are assuming that only mass 1 is present, we model the shaft as a massless spring. Hence, the natural frequency for the system can be written as:

$$\omega_1^2 = \left(\frac{\text{shaft stiffness at mass position, } x_1}{m_1} \right)$$

For a uniform, simply supported massless shaft, the stiffness at the position of mass 1 is

$$k_1 = \frac{3EIL}{x_1^2 \left(L - x_1 \right)^2}$$

Hence,

$$\omega_1{}^2 = \frac{3EIL}{m_1 x_1^2 \left(L - x_1\right)^2}$$

This can be repeated with mass 2 only and with mass 3 only to give ω_2^2 and ω_3^2. There is also a fourth system in which all three of the added masses are removed to leave only the mass of the shaft present. This is,

For a uniform, simply supported shaft, the first natural frequency is

$$\omega_0 = \left(\frac{\pi}{L}\right)^2 \sqrt{\frac{EI}{\rho A}}$$

Combining all of the frequencies using Dunkerley's formula gives:

$$\frac{1}{\omega_n{}^2} \approx \frac{1}{\omega_1{}^2} + \frac{1}{\omega_2{}^2} + \frac{1}{\omega_3{}^2} + \frac{1}{\omega_0{}^2}$$

For all systems, Dunkerley's formula always gives an **underestimate** of the true frequency. This is very useful, since it provides a conservative estimate for whirl calculations. Specifically, to avoid the possibility of whirl, the lowest critical speed (i.e. the lowest natural frequency) of the shaft must be greater than the top speed of the shaft. Therefore, if the top speed is below the value predicted by Dunkerley's formula, it should also be below the actual critical speed.

6.6.3 Rayleigh's Method

Lord Rayleigh observed that for an undamped system vibrating freely at one of its natural frequencies, energy is conserved so that

maximum kinetic energy = maximum strain energy

This is the basis for his method. The strain and kinetic energies can be found for any structure, provided that we know the deflected shape (i.e. the mode shape). Since we do not normally know the exact mode shape, it is necessary to make an estimate of it. This is an important step since the accuracy depends on making a good guess.

For systems with lumped mass and massless springs, the instantaneous kinetic energy for mass i is

$$\left(T\right)_{\text{mass } i} = \frac{1}{2} m_i \dot{x}_i^2$$

For sinusoidal vibration, $x_i(t) = X_i \sin \omega t$ and $\dot{x}_i(t) = \omega X_i \cos \omega t$. Hence the maximum kinetic energy for mass i is

$$\left(T_{\text{MAX}}\right)_{\text{mass } i} = \frac{1}{2} m_i \left(\omega X_i\right)^2 = \frac{1}{2} \omega^2 m_i X_i^2 \tag{6.80}$$

where X_i is the amplitude of vibration for mass i based on the assumed mode shape.

The maximum strain energy in the spring

$$\left(U_{\text{MAX}}\right)_{\text{spring } j} = \frac{1}{2} k_j \left(\text{maximum change of length}\right)^2 \tag{6.81}$$

The totals for the complete system are given by summing the contributions of all the masses and all the springs. Equating the strain and kinetic energies, and rearranging, gives an expression for ω^2. The result can also be calculated from the mass and stiffness matrices for the system. That is

$$\omega^2 = \frac{\{\phi\}^T [K]\{\phi\}}{\{\phi\}^T [M]\{\phi\}}$$

where $\{\phi\}$ is the assumed mode shape vector.

6.6.3.1 Example 1: Two-Degree-of-Freedom System

From Equation (6.78), the combined maximum kinetic energy for both masses (figure 6.55) is

$$T_{\text{MAX}} = \frac{1}{2}\omega^2 m_1 X_1^2 + \frac{1}{2}\omega^2 m_2 X_2^2$$

$$= \frac{1}{2}\omega^2 \left(m_1 X_1^2 + m_2 X_2^2\right)$$

From Equation (6.81), the combined maximum strain energy for the two springs is

$$U_{\text{MAX}} = \frac{1}{2}k_1 X_1^2 + \frac{1}{2}k_2\left(X_1 - X_2\right)^2$$

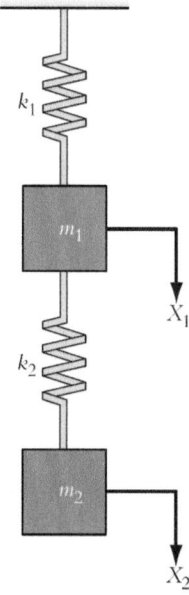

FIGURE 6.55 Springs, masses and displacements in two-degree-of-freedom system.

Equating T_{MAX} and U_{MAX}, we get

$$\omega_n^{\ 2} = \frac{k_1 X_1^{\ 2} + k_2 \left(X_1 - X_2 \right)^2}{m_1 X_1^{\ 2} + m_2 X_2^{\ 2}} \tag{6.82}$$

If we can estimate the mode shape, we will have values of X_1 and X_2 that can be substituted into this equation.

As a numerical example, take the case of $m_1 = m_2 = 2$ kg and $k_1 = k_2 = 200$ N/m. For the first mode of vibration, we would expect the two masses to vibrate in phase with each other and for $X_2 > X_1$. Given that the springs have equal stiffness, we might guess that they both stretch by the same amount, which would suggest that $X_1 = 1$ and $X_2 = 2$ would be a reasonable estimate.

Substituting these values into Equation (6.82) gives a value for ω_n of 1.007 Hz. The exact answer for this problem is 0.984 Hz, so the error is 2.3%.

The lowest mode shape often resembles the **static deflection shape**, so this is invariably a good 'guess'.

Here, noting that the bottom spring supports one mass, but the top spring carries the weight of both masses, the static extension of the top spring will be twice that of the lower spring. As a result, the static deflection shape is $X_1 = 2$ and $X_2 = 3$. This gives $\omega_n = 0.987$ Hz, which is an error of only 0.4%.

If we knew the exact mode shape, we would find that this gave the exact answer.

Because Rayleigh's method imposes a deflection shape on the system, it effectively constrains it to vibrate in a different way from the true mode shape. As a result, Rayleigh's method will always give an overestimate of the natural frequency, unless you happen to guess the exact mode shape by good fortune.

Thus,

$$\omega_{\text{Rayleigh}} \geqslant \omega_{\text{Exact}}$$

The technique to adopt is to try several possible mode shapes. The lowest of the predicted frequencies will be the most accurate.

6.6.4 RAYLEIGH'S METHOD FOR SHAFTS AND BEAMS

Rayleigh's method can also be applied to shafts and beams. In this case, the expressions for the maximum kinetic and strain energies are

$$T_{\text{MAX}} = \frac{1}{2} \omega^2 \int_0^L \rho A \left[Y \left(x \right) \right]^2 dx \tag{6.83}$$

$$U_{\text{MAX}} = \frac{1}{2} \int_0^L EI \left(\frac{\mathrm{d}^2 Y}{\mathrm{d}x^2} \right)^2 dx \tag{6.84}$$

where $Y(x)$ is the mode shape function, which defines the amplitude of vibration of the shaft/ beam along its length. Non-uniform cross-sections (where A and I are functions of x) can be analysed, as can systems of interconnected beams, and systems that include discrete masses and springs. In each case, we sum the contribution of each element to the overall strain and kinetic energies.

We need to guess $Y(x)$ in order to evaluate the integrals.

6.6.4.1 Example 1: Uniform Cantilever Beam

The exact answer to this problem can be found by substituting the first root of the frequency equation given in Table 6.3 into Equation (6.23). This gives

$$\omega_n = \frac{3.52}{L^2}\sqrt{\frac{EI}{\rho A}}$$

The main criterion for choosing the mode shape is to ensure that it satisfies the displacement and slope conditions at the ends of the beam/shaft.

For a cantilever beam, $Y = \dfrac{dY}{dx} = 0$ at $x = 0$. The exact mode shape (see Section 6.2) is:

$$Y_1(x) = \sin \lambda_1 x - \sin \lambda_1 x - \frac{\sin \lambda_1 L + \sin \lambda_1 L}{\cos \lambda_1 L + \cosh \lambda_1 L}(\cos \lambda_1 x - \cosh \lambda_1 x)$$

It would be impossible to 'guess' this function, particularly when we don't know the natural frequency (which is needed to work out the wavenumber, λ_1).

Choice 1: Quadratic function

$$Y(x) = Cx^2$$

This expression satisfies the displacement and slope conditions at the clamped end and gives a shape that is similar to the actual mode shape.

The maximum kinetic energy is given by

$$T_{MAX} = \frac{1}{2}\omega^2 \int_0^L \rho A \left[Cx^2\right]^2 dx$$

$$= \frac{1}{2}\omega^2 \rho A \times \frac{CL^5}{5}$$

$$= \omega^2 \frac{\rho A C^2 L^5}{5}$$

The maximum strain energy

$$U_{MAX} = \frac{1}{2}\int_0^L EI\left(\frac{d^2Y}{dx^2}\right) dx$$

$$= \frac{1}{2}\int_0^L EI(2C)^2 dx$$

$$= 2EIC^2L$$

Equating gives

$$\omega_n^2 = 20\frac{EI}{\rho A L^4}$$

Prediction is $\omega_n = \dfrac{4.47}{L^2}\sqrt{\dfrac{EI}{\rho A}}$, which is significantly (27%) higher than the exact value.

Choice 2: Static deflection shape $Y\{x\} = C\left[\left(\dfrac{x}{L}\right)^4 - 4\left(\dfrac{x}{L}\right)^3 + 6\left(\dfrac{x}{L}\right)^2\right]$

This expression gives: $\omega_n = \dfrac{3.54}{L^2}\sqrt{\dfrac{EI}{\rho A}}$, an error of less than 1%.

Choice 3: A trigonometric function

Suitable functions can often be found by inspection, avoiding the need to calculate the static deflection shape. For a cantilever beam, a suitable function would be:

$$Y(x) = 1 - \cos\left(\frac{\pi x}{2L}\right)$$

Using this, the prediction is $\omega_n = \dfrac{3.66}{L^2}\sqrt{\dfrac{EI}{\rho A}}$, which is 4% high.

Figure 6.56 compares each of the three approximate mode shapes with the exact expression. The quadratic function is the least accurate shape and gives the least accurate estimate of the natural frequency. Notice that there is greater bending in the quadratic function, resulting in an overestimate of the strain energy. Also, the displacement amplitudes are less, resulting in an underestimate of the kinetic energy. Since we effectively divide the strain energy by the kinetic energy (albeit without the ω^2 term), these two factors explain why the quadratic function gives such a high estimate. The static deflection shape is closest to the exact mode shape and this is why it gives the most accurate estimate of the natural frequency. The trigonometric function is a good compromise and can be found for different beams by inspection.

6.6.4.2 Example 2: Beams and Shafts with Added Masses

The added masses (Figure 6.57) don't change the strain energy, but they add extra kinetic energy.

For mass at $x = x_p$, the maximum velocity is $\omega Y(x_p)$

The contribution of mass p to the kinetic energy is $\dfrac{1}{2}m_p\left[\omega Y(x_p)\right]^2$

These contributions need to be added to the kinetic energy of the shaft itself. Hence, the total kinetic energy becomes

$$T_{MAX} = \frac{1}{2}\omega^2\int_0^L \rho A\left[Y(x)\right]^2 dx + \sum_{ALL\ MASSES}\frac{1}{2}m_r\omega^2\left[Y(x_p)\right]^2$$

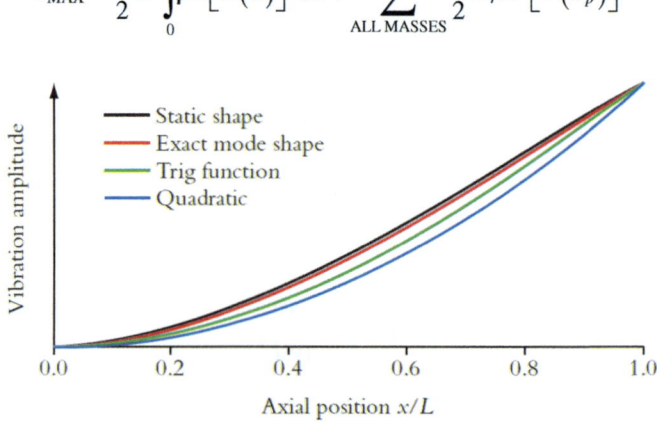

FIGURE 6.56 Comparison of mode shape estimate for a cantilever beam.

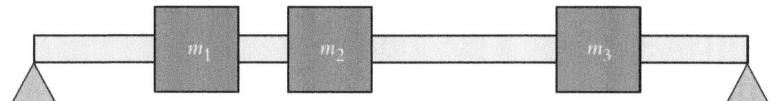

FIGURE 6.57 Simply supported beam with added masses. Note that masses are treated as "point masses" and do not contribute strain energy.

6.6.5 SINGLE-DEGREE-OF-FREEDOM DYNAMIC MODELS OF COMPLEX SYSTEMS

The approach is based on the observation that if *the total strain and kinetic energies of two different dynamic systems are identical, an exact analogue relationship will exist between the two systems.* Such systems are called **dynamically equivalent systems**.

In situations where the real system has many degrees of freedom, but where only one mode of vibration is of interest, the concept provides a powerful method for producing an approximate single-degree-of-freedom model to describe that mode of vibration. Not only are single-degree-of-freedom models easy to analyse, but examples in this section have confirmed that they can often give good predictions of the general behaviour of more complex systems. We will consider only undamped systems here.

The approximate model consists of a simple mass–spring system, in which the displacement of the mass represents the displacement of some chosen point on the real structure.

Using the concept of dynamically equivalent systems, the mass and spring stiffness of the approximate model are chosen so that the maximum strain and kinetic energies of real and model systems are the same.

To do this, we assume that the lowest mode of vibration is dominant and therefore defines the deformation pattern in the structure. This is a good assumption if the system is vibrating sinusoidally near its lowest mode of vibration, but can also work well in other cases. An example of the latter is shown in Figure 6.47 where the response of a five-degrees-of-freedom system to a sequence of rectangular pulses was shown for comparison with that of an equivalent single-degree-of-freedom model.

As with Rayleigh's method, the accuracy of the model relies on having a good estimate of the real structure's mode shape.

6.6.5.1 Example 1: Lumped Mass System

In this example, we will find an approximate single-degree-of-freedom model to analyse the motion of the top mass of the three-degree-of-freedom system shown in Figure 6.58.

The first step is to establish a link between the displacement of the mass in the approximate single-degree-of-freedom model and some chosen points in the real system. In this case, the obvious choice is to link the displacement of the approximate model to the displacement of the top mass, x, since it's the motion of this mass that we want to predict.

However, since we are assuming that the motion of the real system is given by our chosen mode shape, not only does the approximate model predict the behaviour of the coordinate used as the link with the real system but we can also work out the motion of the other coordinates since they are all linked by the assumed mode shape. In this example, it means that once we've found $x(t)$, we can get $y(t)$ and $z(t)$, since each will be related to $x(t)$ in proportion to the mode shape $\begin{Bmatrix} X \\ Y \\ Z \end{Bmatrix}$.

The approximate model is shown in Figure 6.59, with the motion coordinate chosen as x to match the coordinate of the top mass in the actual system.

To find the values for the mass and stiffness in the model, we equate the maximum kinetic and strain energies in the real and approximate systems.

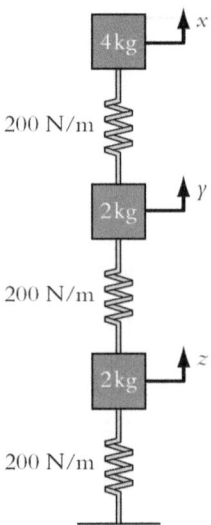

FIGURE 6.58 Three degree of freedom system.

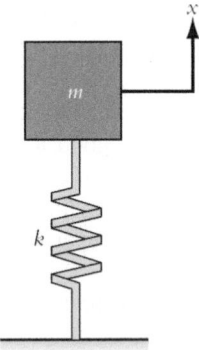

FIGURE 6.59 Approximate single degree of freedom model for three degree of freedom system.

Equating maximum kinetic energies in the real (LHS of the equation) and model systems (RHS of the equation), we get:

$$\frac{1}{2}4(\omega X)^2 + \frac{1}{2}2(\omega Y)^2 + \frac{1}{2}2(\omega Z)^2 = \frac{1}{2}m(\omega X)^2 \qquad (6.85)$$

Hence,

$$m = \frac{4X^2 + 2Y^2 + 2Z^2}{X^2}$$

Equating maximum strain energies in the real and model systems, we get:

$$\frac{1}{2}200(X-Y)^2 + \frac{1}{2}200(Y-Z)^2 + \frac{1}{2}200 \cdot Z^2 = \frac{1}{2}kX^2 \qquad (6.86)$$

Hence,

$$k = \frac{200\left[(X-Y)^2 + (Y-Z)^2 + Z^2\right]}{X^2}$$

To get values for m and k, we need an estimate for the mode shape. For the lowest mode shape, we expect all three masses in the real system to vibrate in phase with each other and, since the top mass is not connected to the ground, that $X > Y > Z$.

Choice 1: Since all springs have the same stiffness, we might guess that they all deflect by the same amount. This implies that the mode shape would be:

$$\left\{\begin{array}{c} X \\ Y \\ Z \end{array}\right\} = \left\{\begin{array}{c} 3 \\ 2 \\ 1 \end{array}\right\}$$

Substituting these values into Equations (6.83) and (6.84) gives, $m = 5.11$ kg and $k = 66.7$ N/m.

Choice 2: The static deflection shape for this system is $\left\{\begin{array}{c} X \\ Y \\ Z \end{array}\right\} = \left\{\begin{array}{c} 9 \\ 7 \\ 4 \end{array}\right\}$ and substituting these values

into Equations (6.83) and (6.84) gives, $m = 5.60$ kg and $k = 71.6$ N/m.

To decide which of these models is the most accurate, we can compare the natural frequencies predicted by each one.

Choice	m [kg]	k [N/m]	ω_n [Hz]
1	5.11	66.7	0.575
2	5.60	71.6	0.569

It can be seen that the second choice (the static deflection shape) is the more accurate since it gives a lower estimate of the natural frequency.

6.6.5.2 Example 2: Forced Response of a Cantilever Beam

In the second example, we will create a single-degree-of-freedom model of a uniform cantilever beam and use it to estimate the steady-state response at the free end due to a sinusoidal force having a frequency near the lowest natural frequency of the beam (figure 6.60).

There are two stages. First, we set up the approximate model and then we use it to do the steady-state response calculation.

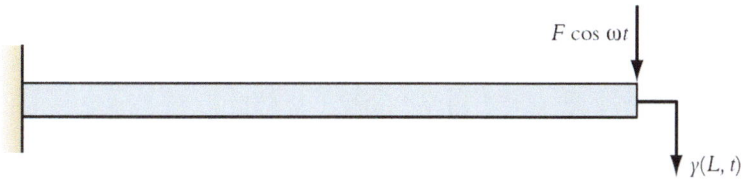

FIGURE 6.60   Cantilever beam subjected to periodic force.

In this case, we choose to link the displacement of the mass in the approximate model to the displacement at the free end of the cantilever. In terms of the chosen displacement variables,

$$Z(t) = \gamma(L,t)$$

For steady-state, sinusoidal vibration, the link can be written as

$$Z\cos\omega t = Y(L)\cos\omega t \text{ or } Z = Y(L)$$

To proceed, we need to choose an expression for $Y(x)$, the amplitude of vibration for the cantilever. Two possibilities are considered.

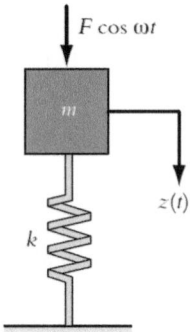

Choice 1: $Y(x) = Cx^2$

Linking this with the approximate model, $Z = Y(L) = CL^2$
Hence,

$$C = \frac{Z}{L^2} \text{ so that } Y(x) = \frac{Z}{L^2}x^2$$

This expression for $Y(x)$ is then used to calculate the maximum kinetic and strain energies in the beam. By equating these to the equivalent expressions for the single-degree-of-freedom model, we get the required mass and stiffness values.

For the kinetic energies,

$$T_{\text{MAX}} = \frac{1}{2}\omega^2 \int_0^L \rho A \left[Y(x)\right]^2 dx = \frac{1}{2}m\omega^2 Z^2$$

This gives the mass for the approximate model to be $m = 0.2\,\rho\,AL$. Note that $\rho\,AL$ is the mass of the beam.

For the strain energies,

$$U_{\text{MAX}} = \frac{1}{2}\int_0^L EI\left(\frac{d^2Y}{dx^2}\right)^2 dx = \frac{1}{2}kZ^2$$

This gives the spring stiffness for the model to be $k = 4\dfrac{EI}{L^3}$

Applying the natural frequency test to the approximate model, we find that $\omega_1 = \dfrac{4.47}{L^2}\sqrt{\dfrac{EI}{\rho A}}$, which is the same (poor) result obtained with Rayleigh's method using this choice for $Y(x)$.

Choice 2: Static deflection shape, $Y(x) = C\left[\left(\dfrac{x}{L}\right)^4 - 4\left(\dfrac{x}{L}\right)^3 + 6\left(\dfrac{x}{L}\right)^2\right]$

Linking this expression at $x = L$ with the coordinate for the approximate model, we get

$$Z = Y(L) = C\left[1^4 - 4 \times 1^3 + 6 \times 1^2\right] = 3C$$

Hence,

$$C = \frac{Z}{3} \text{ and } Y(x) = \frac{Z}{3}\left[\left(\frac{x}{L}\right)^4 - 4\left(\frac{x}{L}\right)^3 + 6\left(\frac{x}{L}\right)^2\right]$$

Using this expression to calculate the mass and stiffness values for the model gives the following:

$$m = 0.257\rho AL \text{ and } k = 3.20\frac{EI}{L^3}$$

The natural frequency test in this case gives $\omega_1 = \dfrac{3.53}{L^2}\sqrt{\dfrac{EI}{\rho A}}$, which is lower than the result from the first choice, confirming that the static deflection shape is the better approximation.

6.6.5.3 Forced Response Analysis

Having established the single-degree-of-freedom model, we can now use it to estimate the steady-state response at the free end of the beam due to a sinusoidal force. The equation of motion for the approximate model is:

$$m\ddot{z} + kz = F(t)$$

Making the standard substitutions, $F(t) = Fe^{i\omega t}$ and $z(t) = Z^* e^{i\omega t}$, we get

$$Z^* = \frac{F}{\left(k - m\omega^2\right)}$$

Note that since there is no damping in this model, the expression for Z^* is real.

This meets the objective of finding the steady-state amplitude of the deflection at the free end of the cantilever. However, since we have assumed that the deflected shape of the beam can be defined by the mode shape (that is, the function $Y(x)$), the expression for Z^* can also tell us the vibration amplitude at *any* point along its length. In the case of choice 2, this gives:

$$Y(x) = \frac{Z^*}{3}\left[\left(\frac{x}{L}\right)^4 - 4\left(\frac{x}{L}\right)^3 + 6\left(\frac{x}{L}\right)^2\right]$$

$$= \frac{F}{3\left(k - m\omega^2\right)}\left[\left(\frac{x}{L}\right)^4 - 4\left(\frac{x}{L}\right)^3 + 6\left(\frac{x}{L}\right)^2\right]$$

LEARNING SUMMARY

By the end of this section, you should have learnt:

✓ To use Dunkerley's method and Rayleigh's method to obtain estimates of the lowest natural frequency of structures;

✓ To create a single-degree-of-freedom approximation of a more complex structure and use it to estimate the structure's response

6.7   VIBRATION CONTROL TECHNIQUES

6.7.1   INTRODUCTION

Vibration isolators (also known as anti-vibration mounts) are used for reducing the vibration transmitted from a source. They work by introducing flexibility between a device and its support.

Case 1   In some cases, the source of vibration is within the device and *the objective is to minimize the force transmitted to the support*. Examples are a ship's engine that can transmit vibration to the deck structure or a passing train that can produce ground-borne vibration.

Case 2   In other cases, the source may be remote but causes the support for a device to vibrate. Here, *the objective is to minimize the displacement transmitted to the device*. Examples are a satellite mounted in a launch vehicle (figure 6.61) or the need to protect sensitive laser instruments from ground-borne vibration.

6.7.2   TYPES OF ISOLATOR

6.7.2.1   Elastomeric Isolator

This is the most common type of isolator. Elastomers can be moulded from many different combinations of many different materials, including natural rubber, neoprene, butyl and silicone. A typical mount made with these materials generally employs the elastomer in shear but many utilize compressive strain also. The mounts may employ the elastomer in a manner that provides both shear and compressive loading for effective isolation performance in both the horizontal as well as the vertical direction. It is relatively easy to design various degrees of damping, shape, load-deflection characteristics and transmissibility characteristics into elastomeric isolators. The inherent damping of elastomers is useful in preventing problems at resonance that would be difficult to restrain if coil springs were used.

For isolation from shocks, elastomers offer some significant advantages because of the fact that they can generally absorb more shock energy per unit weight than other forms of isolator systems.

6.7.2.2   Pneumatic Vibration Isolators

Pneumatic isolators are air-filled, reinforced rubber bellows with mounting plates on top and bottom. Isolators such as these can provide very low natural frequencies with small static deflections. To provide a 1 Hz natural frequency, a steel coil spring isolator would need to be about 600 mm long and capable of deflecting about 250 mm. It would thus be difficult to install and would also present some lateral stability problems. Unlike most isolators, which are passive devices, pneumatic isolators are also used with position feedback in active control systems. This is used, for example, to maintain the height of a table for mounting optical equipment that is sensitive to the slightest movement (figure 6.62).

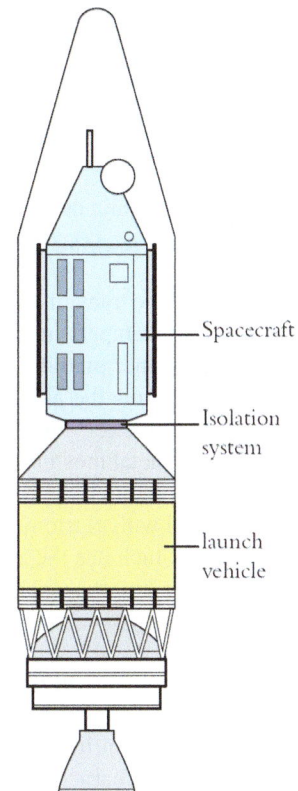

FIGURE 6.61   Satellite isolation system.

FIGURE 6.62   Table incorporating pneumatic isolators.

6.7.2.3 Coil Spring Isolators

Springs may be loaded in tension but it is frequently more convenient to load them in compression. Coil springs are used primarily for the isolation of low-frequency vibration. Consequently, they operate with a relatively large static deflection and lateral stability may be a problem. It can be shown that a coil spring will be stable if

$$\frac{\text{leteral stiffness}}{\text{axial stiffness}} > \frac{\text{static deflection}}{\text{working height}}$$

Coil springs possess practically no damping, and the transmissibility at resonance is extremely high. This can be overcome by the addition of friction dampers in parallel with the load-carrying spring and these types of isolators are widely used. Another method of adding damping to a spring is by the use of an air chamber with an orifice for metering the airflow. For applications where all metal isolators are desired because of temperature extremes or other environmental factors, damping can be added to a load-carrying spring by the use of metal mesh inserts.

High-frequency vibrations can be transmitted through the coils to the isolated unit. To overcome this, one or both ends of the spring can be fitted with elastomeric pads. Conventionally, the pad is attached to the bottom of the spring assembly, which has the added advantage of providing a non-slip surface that frequently eliminates the need to bolt the isolator to the floor.

6.7.2.4 Transmissibility Analysis

The isolators are invariably very much more flexible than the device they support, so the first approximation is to use a single-degree-of-freedom model in which the device to be isolated is treated as a rigid body and the isolators are represented by a spring–damper combination. The steady-state response to harmonic excitation provides a way of characterizing the isolation performance at different frequencies.

Case 1   Source of vibration within a device transmitting vibration to the support

We will assume that the device generates an excitation force, of amplitude P and frequency ω and use the method from Section 6.2.

Step 1 Dynamic mass–spring model

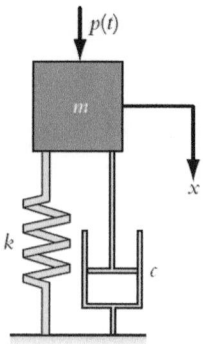

Step 2 Free-body diagram

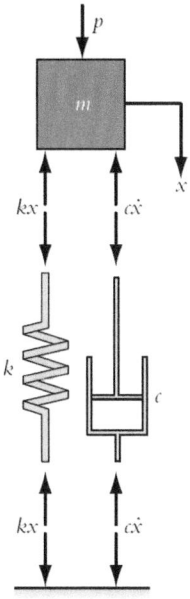

Step 3 Equations of motion

For the device:

$$p - kx - c\dot{x} = m\ddot{x}$$

or

$$m\ddot{x} + c\dot{x} + kx = p \tag{6.87}$$

The transmitted force is

$$q(t) = kx + c\dot{x} \tag{6.88}$$

Substitute $p(t) = Pe^{i\omega t}$, $q(t) = Q^\star$, $e^{i\omega t}$, and $x(t) = X^\star e^{i\omega t}$ into Equations (6.87) and (6.88) to give

$$X^\star = \frac{P}{\left(k - m\omega^2\right) + ic\omega} \text{ and } Q^\star = \left(k + ic\omega\right)X^\star$$

Eliminating $X^\star$ we get $\dfrac{Q^\star}{P} = \dfrac{\left(k + ic\omega\right)}{\left(k - m\omega^2\right) + ic\omega}$

Here, only the magnitude of the transmitted force is of interest and we can define **force transmissibility** as

$$T_F = \left|\frac{Q^\star}{P}\right| = \sqrt{\frac{\left(k^2 + c^2\omega^2\right)}{\left(k - m\omega^2\right)^2 + c^2\omega^2}} \tag{6.89}$$

Case 2   Vibration from the support transmitted to the device

The support vibration is defined by the displacement, $y(t) = Y \cos \omega t$.

Step 1 Dynamic mass–spring model

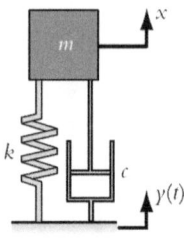

Step 2 Free-body diagram

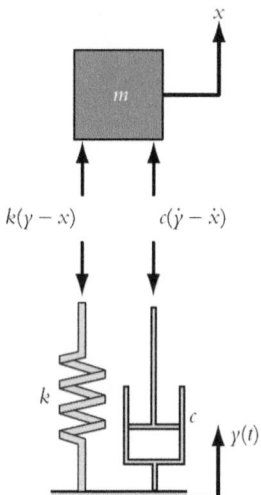

Step 3 Equations of motion

For the device:

$$k\left(y - x\right) + c\left(\dot{y} - \dot{x}\right) = m\ddot{x}$$

or

$$m\ddot{x} + c\dot{x} + kx = c\dot{y} + ky \qquad\qquad (6.90)$$

Substitute $y(t) = Ye^{i\omega t}$ and $x(t) = X^{\star} e^{i\omega t}$ into Equation (6.90) to give

$$X^{\star} = \frac{(k+ic\omega)Y}{(k-m\omega^2)+ic\omega}$$

Again, only the magnitude of the transmitted displacement is of interest and we can define **displacement transmissibility** as

$$T_D = \left|\frac{X^{\star}}{Y}\right| = \sqrt{\frac{(k^2+c^2\omega^2)}{(k-m\omega^2)^2+c^2\omega^2}} \tag{6.91}$$

Note that the force and displacement transmissibility expressions for this mass–spring–damper system are identical. The same is true of other physical systems when analysing the displacements or forces applied to a given pair of coordinates (the mass and the support in this case). Note also that Equations (6.89) and (6.91) only apply to this mass–spring–damper system. Other physical systems will have different transmissibility expressions.

The reader is left to confirm that Equations (6.89) or (6.91) can be rearranged by introducing the expressions for the natural frequency and damping ratio for the system, namely $\omega_n = \sqrt{\dfrac{k}{m}}$ and $\gamma = \dfrac{c}{2\sqrt{km}}$. The alternative expression for the transmissibility ratio is

$$T_{D,F} = \frac{\sqrt{1+4\gamma^2\dfrac{\omega^2}{\omega_n^2}}}{\sqrt{\left(1-\dfrac{\omega^2}{\omega_n^2}\right)^2+4\gamma^2\dfrac{\omega^2}{\omega_n^2}}} \tag{6.92}$$

This form of transmissibility expression emphasizes the importance of the ratio of the excitation frequency to the natural frequency in determining the effectiveness of the isolation system in reducing vibration.

The transmissibility graphs in Figure 6.63 are plotted for three different damping ratios. With zero damping, the transmissibility is infinite if the excitation frequency coincides with the natural frequency of the system. The main effect of increasing the damping ratio ($\gamma = 0.2$ is shown in

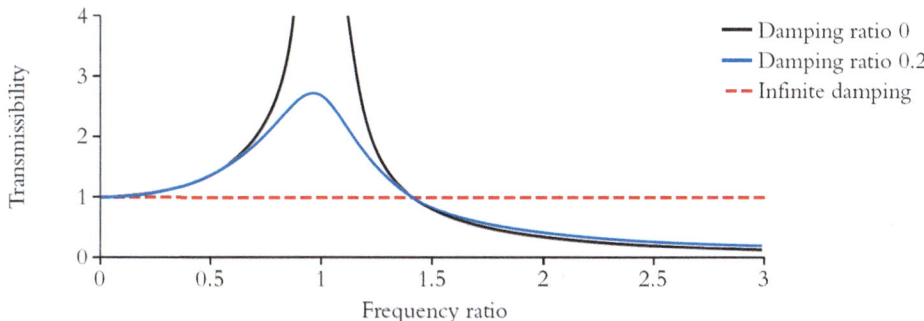

FIGURE 6.63 Transmissibility curves for a simple mass–spring–damper system.

Figure 6.63) is to reduce the maximum transmissibility in the vicinity of the natural frequency. At higher frequencies, the difference between the curves is small. Theoretically, infinite damping would result in no relative movement across the damper, effectively giving a rigid connection between the device and its support. This, of course, is the original situation without isolators.

It's easy to show from Equation (6.92) that $T = 1$ when $\omega / \omega_n = \sqrt{2}$ and that it is independent of the value of the damping ratio. This explains why the three curves intersect at this point on the graph.

Notice that if $\omega / \omega_n < \sqrt{2}$, the transmitted displacement (or force) is higher than the input.

In this case, it would be better to have no isolators and fix the device rigidly to the support! However, if $\omega / \omega_n > \sqrt{2}$, the transmissibility is less than 1.0, resulting in vibration reduction. The aim of selecting isolators is to ensure that the system operates in this **isolation region**.

It is convenient to define isolation efficiency to describe the reduction in displacement (or force) compared with the applied displacement (or force). This is:

$$\text{Isolation efficiency} = \frac{\text{Reduction in displacement}\left(\text{or force}\right)}{\text{Original displacement}\left(\text{or force}\right)} \times 100\% = \left(1 - T\right) \times 100\%$$

6.7.2.5 Isolator Selection

This section considers the task of selecting isolators to achieve a desired reduction in vibration. There are two constraints governing the selection: the lowest excitation frequency to be encountered, ω_{MIN}, and the maximum allowable transmissibility, T_{MAX}.

The combination of T_{MAX} and $\dfrac{\omega_{MIN}}{\omega_n}$ is marked in Figure 6.64 and represents the worst case in respect of isolation efficiency, since for any excitation frequency greater than ω_n, the transmissibility will be lower than T_{MAX}. It is apparent from Figure 6.64 that ω_n must be very much less than ω_{MIN} to ensure that T_{MIX} will be less than 1.0 for all excitation frequencies greater than or equal to ω_{MIN}.

Three variables affect the system's dynamics; m, k and c, and m and k together determine ω_n. The stiffness, k, is given by the set of isolators selected.

At first sight, the mass of the device may not seem to be a variable for this problem. However, m can be increased by mounting the machine on an inertia base. This will reduce ω_n and increase $\dfrac{\omega_{MIN}}{\omega_n}$ and have the beneficial effect of reducing the transmissibility. Figure 6.65 gives an example of an inertia base installation.

In most vibration situations, it is desirable to increase damping since this limits the amplifying effect of resonance. Here, since the system is designed to operate well above the resonant frequency, low damping is desirable since increasing γ will increase the transmissibility in the isolation region. Low damping is easy to achieve and most commercial isolators give a damping ratio that is less than 0.1.

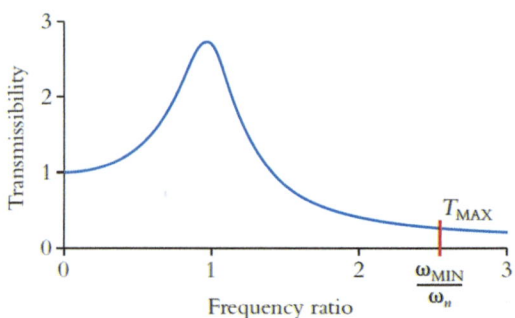

FIGURE 6.64 Design constraints for isolator selection.

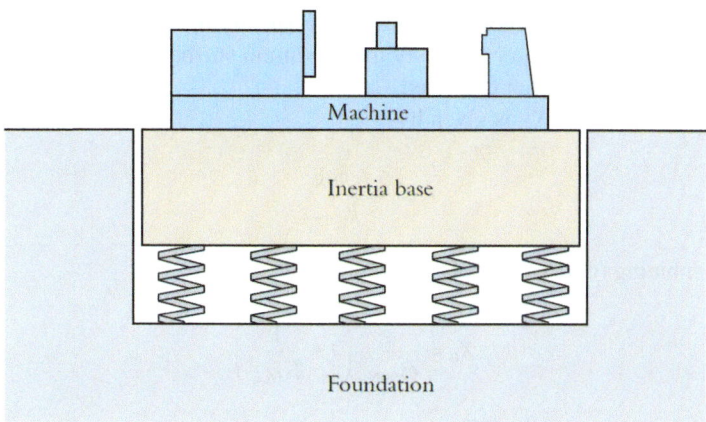

FIGURE 6.65 Example of inertia base used to increase the isolated mass.

It is normal to base the design on the assumption of zero damping and, as can be seen from the transmissibility graph in Figure 6.63, the error involved in the isolation region is small.

It is also normal to treat each isolator independently of the others. *In this case, m is the effective mass supported by the isolator in question.*

For the simple mass–spring model with zero damping,

$$T = \left| \frac{1}{1 - \dfrac{\omega^2}{\omega_n^2}} \right|$$

$$= \frac{1}{\dfrac{\omega^2}{\omega_n^2} - 1}$$

for the isolation region where $\omega > \omega_n$.

If T_{MAX} at $\omega = \omega_{MIN}$,

$$\omega_n^2 = \frac{T_{MAX}\omega_{MIN}^2}{1 + T_{MAX}}$$

Since $\omega_n^2 = \dfrac{k}{m}$, the required isolator stiffness is

$$k = m\omega_n^2 = \frac{mT_{MAX}\omega_{MIN}^2}{1 + T_{MAX}} \tag{6.93}$$

If selecting isolators from a manufacturer's catalogue, it is unlikely that one with precisely this stiffness will be found. The stiffness given by Equation (6.93) therefore gives the *maximum* value consistent with the design requirements.

It might appear from this that *any* isolator would be suitable provided its stiffness was less than this value. This is not the case, however, since there are also constraints imposed by static considerations. In the case of coil spring isolators, there could be installation, coil bottoming or lateral stability problems if the static deflection is too large. With elastomeric isolators, there are strength limitations under static load.

Manufacturers often express these constraints by specifying a **maximum static deflection**. Therefore, after selecting an isolator to satisfy the maximum stiffness limit, it is necessary to check that the static deflection limit is not exceeded.

The actual static deflection, X_0, is given by

$$X_0 = \frac{mg}{k_{\text{isolator}}} \tag{6.94}$$

Alternatively, combining (6.93) and (6.94) gives

$$X_0 = \frac{g}{\omega_{\text{MIN}}^2}\left(1 + \frac{1}{T_{\text{MAX}}}\right) \tag{6.95}$$

This represents the *minimum* static deflection consistent with the design requirements.

These equations can be incorporated into a procedure for selecting suitable isolators for an application. The procedure is as follows:

Step 1 Find the centre of mass of the machine.
Step 2 Select the number and position of attachment points for isolators.
Step 3 Estimate the load supported by each isolator.
Step 4 For each isolator position in turn:
    (a) calculate the maximum stiffness from Equation (6.93);
    (b) select an isolator with a lower stiffness;
    (c) check that this does not exceed any static deflection limit using Equation (6.94);
    (d) although this will give a satisfactory selection, it is often worth repeating (b) and
    (e) with other isolators having even lower stiffness; the lower the stiffness, the greater the isolation efficiency, so the limiting factor becomes the maximum allowable static deflection.

6.7.3 Tuned Vibration Absorbers

Tuned vibration absorbers were mentioned in Section 6.1 as part of the solution to the resonant vibration on the Millennium Bridge. They were also mentioned in Section 6.4 where it was observed that it was possible to design a dynamic system that has zero displacement at the point of force application. This section explains how this can be achieved.

Let's look at the case of a structure that is subjected to a sinusoidal force whose frequency is very close to a natural frequency and consider the worst-case scenario that the structure has no damping to limit the resonant amplitude. We can model the mode in question as the undamped single-degree-of-freedom system shown in Figure 6.66(a). For the original system without the tuned vibration absorber, the response amplitude is given by Equation (6.45) as

$$X_0 = \left|\frac{P_1}{k - m\omega^2}\right|$$

As a numerical example, we will take $k = 1$ MN/m, $m = 1000$ kg and $P_1 = 1$ kN to get the response spectrum shown in Figure 6.66(b). As expected, the model predicts a very large response when the excitation frequency is close to the natural frequency. The aim of adding the tuned vibration absorber is to reduce this response to zero.

The tuned absorber itself is a secondary mass–spring system in which the ratio of stiffness to mass is identical to the original system. This is attached to the original mass as shown in Figure 6.67(a). In

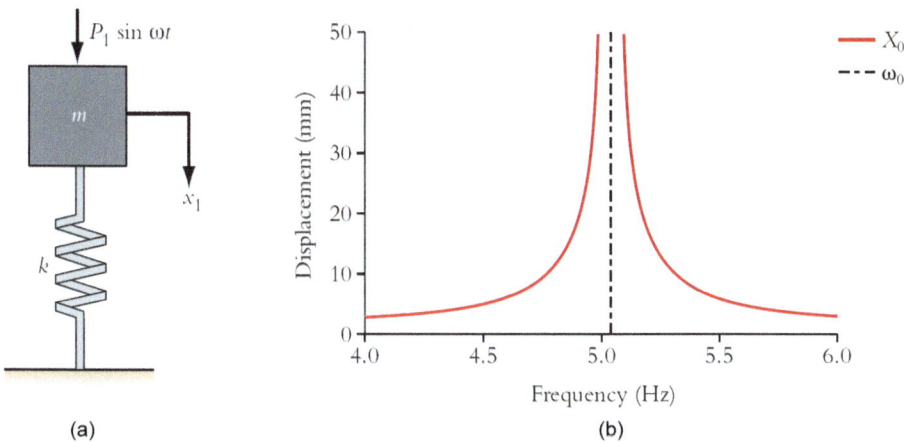

FIGURE 6.66   (a) Model of undamped single degree of freedom system; (b) Magnitude of response plotted against excitation frequency.

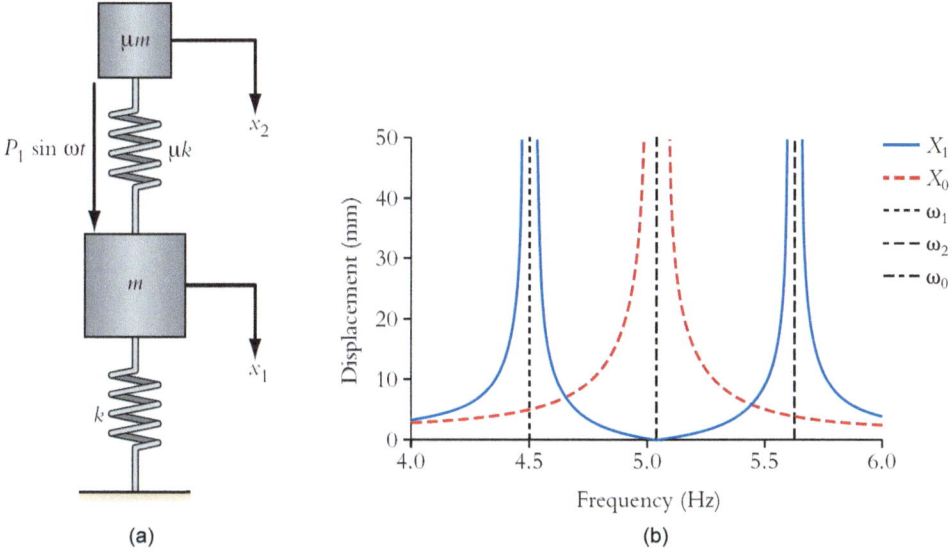

FIGURE 6.67   (a) Model of undamped two degree of freedom system; (b) Magnitude of responses plotted against excitation frequency.

the numerical example shown here, the ratio of the absorber mass to the original mass, μ, has been taken to be 0.01.

This can be analysed in exactly the same way as the worked example in Section 6.4 to give expressions for the response of each mass. It is seen from Figure 6.67(b) that the response amplitude X_1 is zero when the excitation frequency is equal to the original natural frequency. The effect of adding the tuned absorber is to create a two-degrees-of-freedom system whose natural frequencies lie on either side of the natural frequency of the original system. A limitation of this absorber is that it is only effective if the excitation frequency is constant and at a frequency very close to the natural frequency of the original system. If the excitation frequency deviates from this, there is a strong chance

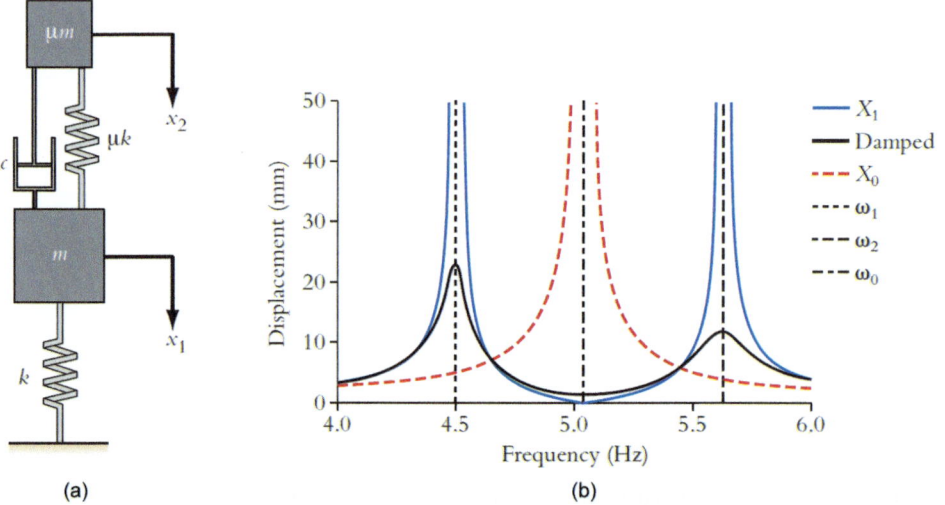

FIGURE 6.68 (a) Model of two degree of freedom system incorporating a damped vibration absorber; (b) Magnitude of responses plotted against excitation frequency.

of exciting one of the new natural frequencies, since it can be seen from Figure 6.67(b) that these are only about 0.5 Hz (10%) away from the original natural frequency. The separation of the new natural frequency can be increased by choosing a larger value for the mass ratio, μ. However, there may be limits to the amount of mass that can be added, for example in the case of the Millennium Bridge (Section 6.1).

This problem can be overcome by using a damped vibration absorber that uses a damper in parallel with the absorber spring, as in Figure 6.68(a). Note that the original structure remains undamped. The response spectrum with the damped absorber is shown in Figure 6.68(b) using a damping coefficient of 100 Ns/m. This has greatly reduced resonant peaks and still gives a very low response at the original natural frequency.

The modal approach to solving the equations of motion cannot be used for this problem since they cannot be uncoupled. The equations of motion are:

$$\begin{bmatrix} m & 0 \\ 0 & \mu m \end{bmatrix}\begin{Bmatrix} \ddot{x}_1 \\ \ddot{x}_2 \end{Bmatrix} + \begin{bmatrix} c & -c \\ -c & c \end{bmatrix}\begin{Bmatrix} \ddot{x}_1 \\ \ddot{x}_2 \end{Bmatrix} + \begin{bmatrix} k(1+\mu) & -\mu k \\ -\mu k & \mu k \end{bmatrix}\begin{Bmatrix} x_1 \\ x_2 \end{Bmatrix} = \begin{Bmatrix} p_1(t) \\ 0 \end{Bmatrix}$$

It will be seen that the damping matrix is not in the form $[C] = a_1[M] + a_2[K]$, meaning that we do not have proportional damping in this case. Since the equations of motion cannot be uncoupled, they must be solved as a simultaneous pair. For sinusoidal excitation, we can still make the usual substitutions, namely: $p_1(t) = P_1 e^{i\omega t}$, $x_1(t) = X_1^* e^{i\omega t}$ and $x_2(t) = X_2^* e^{i\omega t}$ This leads to the equations

$$\begin{bmatrix} \left(k(1+\mu)-m\omega^2\right)+i\omega c & -\mu k - i\omega c \\ -\mu k - i\omega c & \left(\mu k-\mu m\omega^2\right)+i\omega c \end{bmatrix}\begin{Bmatrix} X_1^* \\ X_2^* \end{Bmatrix} = \begin{Bmatrix} P_1 \\ 0 \end{Bmatrix}$$

The responses are then given by

$$\begin{Bmatrix} X_1^* \\ X_2^* \end{Bmatrix} = \begin{bmatrix} \left(k(1+\mu)-m\omega^2\right)+i\omega c & -\mu k - i\omega c \\ -\mu k - i\omega c & \left(\mu k-\mu m\omega^2\right)+i\omega c \end{bmatrix}^{-1}\begin{Bmatrix} P_1 \\ 0 \end{Bmatrix}$$

FIGURE 6.69   Tuned vibration absorber for overhead power lines (circled).

In practice, this problem would be solved using software such as MATLAB, which has the capability to invert complex matrices. A MATLAB script based on this example is included on the book's supporting website.

The most common application of tuned vibration absorbers is for overhead power lines, as shown in figure 6.69. When wind blows across the cables, vortex shedding can induce vertical oscillations, which can cause fatigue in the cable. This is most prevalent where the cable is supported by the transmission pylons, and vibration absorbers are attached here to suppress the oscillations.

The absorber consists of a pair of masses, each one on the end of a short cantilever made from cable. The bending of the cantilevers provides the required spring stiffness and friction between the cable strands provides damping.

LEARNING SUMMARY

By the end of this section, you should have learnt:

- ✓ How to derive the transmissibility expression that gives the ratio of transmitted force to applied force (or of transmitted displacement to applied displacement) as a function of excitation frequency;
- ✓ To select suitable isolators to achieve a given isolation efficiency;
- ✓ How tuned vibration absorbers can suppress resonant vibration in situations where sinusoidal excitation coincides with a natural frequency of the structure.

NOTE

1   For an explanation of why this must be the case, think about the implication of the determinant being zero: this means that non-zero terms for k_1, k_2, m_1 and m_2 that give a determinant of zero are the equivalents of the general solution for the second-order differential equation $\dfrac{d^2x}{dt^2} + b\dfrac{dx}{dt} + cx = 0$.

Questions

1 FLUID DYNAMICS

1. Show that the two-dimensional velocity field given by $u = ay$ and $v = bx$ satisfies the continuity equation. Using the Navier–Stokes equations, derive the expression for the pressure p.

2. The three-dimensional velocity profile for an incompressible flow is given by

$$u = x^3 + 2z^2 \text{ and } w = y^3 + 2yz$$

Derive a general form of the third velocity component, v, from the 3D continuity equation

$$\frac{\partial u}{\partial x} + \frac{\partial v}{\partial y} + \frac{\partial w}{\partial z} = 0.$$

3. By neglecting the viscous and gravity terms, derive the Bernoulli equation from the steady Navier–Stokes equations. (*Hint*: choose the x-axis along the streamline.)

4. For an axial flow inside a circular tube, the Reynolds number which causes the transition to turbulence is approximately 2300, based on the tube diameter and average flow velocity. If the tube diameter is 4 cm and the fluid is petrol at 20°C, then find the volumetric flow rate which causes the transition. The density and viscosity of the petrol are $\rho = 680$ kg/m³ and $\mu = 2.92 \times 10^{-4}$ kg/ms, respectively.

2 THERMODYNAMICS

2.1 REFRIGERATION

1. (a) What pressure is required in a freezer compartment evaporator coil using refrigerant R134a if the temperature is to be maintained at −20°C?
   (b) What is the enthalpy and entropy when R134a is:
      (i) all liquid?
      (ii) all vapour?
      (iii) What does the change in entropy between the two states tell you?
   (c) Assuming a mass flow rate of R134a at 20 g/s and making use of the SFEE, what heat transfer rate occurs in the evaporator if the refrigerant leaves at saturated gas condition (i.e. all vapour, with no superheat) and enters at saturated liquid condition (i.e. all liquid and no sub-cooling)?

2.2 AIR CONDITIONING

2. An air conditioning unit draws in 0.5 kg/s of atmospheric air at 30°C and specific humidity 0.012.
   (a) What is the mass flow rate of water vapour in the air?
   (b) Using the formula relating specific humidity to partial pressure of the water vapour and the atmospheric pressure and assuming one atmosphere atmospheric pressure, what is Lthe partial pressure of the water vapour?

DOI: 10.1201/9780429319495-7

(c) Find the p_g of water vapour. What is the relative humidity of the incoming air?

(d) Find the dew point corresponding to the pressure p_g. Sketch a T–v diagram showing the process that the water vapour in atmospheric air must go through in order to form dew. If the air is to be issued at 20°C and 50% humidity, determine the moisture removal rate and the heat transferred in the cooler and heater sections.

2.3 VAPOUR POWER CYCLES

3. A turbine receives steam at 550°C and 200 bar.
  (a) Assuming an isentropic expansion through the turbine to a pressure of 40 bar, what is the final temperature?
  (b) The turbine actually expands to 40 bar and 330°C. What is the isentropic efficiency of that turbine?

4. A power station uses the Rankine cycle between a boiler with feedwater at 200 bar and a condenser at a pressure corresponding to the saturated water vapour at 30°C.
  (a) What are the pressure and specific enthalpy in the condenser?
  (b) What are the temperature and specific enthalpy of the saturated steam?
  (c) If the mass flow rate of the water is 100 kg/s, what is the heat input to the steam in the boiler?
  (d) What is the work done in the pump?
  (e) The Rankine cycle is modified with a superheat steam pipe circuit prior to entering the high-pressure turbine. The cycle is also modified with reheat after exit from the high-pressure turbine up to 550°C before exhausting in the low-pressure turbine at the pressure of the condenser. Given the isentropic efficiency of the low-pressure turbine is 85%, what are:
    (i)   the specific enthalpy at turbine entry?
    (ii) the specific enthalpy at turbine exit?
    (iii) the steam quality at exit?
  (f) Knowing the pump work, the work produced by the high-pressure turbine and the low-pressure turbine work, what is:
    (i)   specific steam consumption?
    (ii) work ratio?

2.4 COMBUSTION CHEMISTRY

5. Write the reaction equation for the stoichiometric combustion of butane in oxygen.
  (a) If the mass of butane consumed is 20 g, how many moles of butane is this?
  (b) For this 20 g of butane, what mass of O_2 is consumed in the reaction?
  (c) How many moles of oxygen is this?
  (d) What are the masses of the gases in the products?

6. Write the stoichiometric reaction equation for butane in air.
  (a) What is the air-to-fuel ratio by volume and by mass?
  (b) What are the proportions by mass and volume of the product gases?
  (c) Determine how much air is required for the stoichiometric combustion of propane burning in air. Determine the AFR by mass and volume.
  (d) Propane is burned with 30% excess air. Determine the volume and mass fractions of the reactants and products.

2.5 HEAT TRANSFER

7. (a) What is the thermal resistance of a brick wall (k = 1.5 W/mK), 10-cm thick, on a room wall of height 2.3 m and width 3.1 m?

(b) Calculate the thermal resistance of the wall given that it has three layers: brick, insulation and brick. Each layer is 10-cm thick, k = 1.2 W/mK for the inner brick, 0.5 W/mK for the insulation and 2 W/mK for the outer brick.

(c) Given that the convective heat transfer to the wall's inner surface is 9 W/m²K, and the outer wall is 95 W/m²K, find the thermal resistance for the inside and outside convective conditions, and hence, for the overall wall.

(d) What is the heat transfer if the inner air is at 18°C and the outer air is at 1°C?

(e) Sketch the temperature profile through the wall.

(f) The ground at night has air blowing over it at 2°C, with a heat transfer coefficient of 20 W/m²K. It also sees the night sky, which is at −55°C. The view factor is 1 and the emissivity of the ground is 0.9. Given the ground has a temperature of 10°C at a depth of 3 m and conductivity of 2 W/m², calculate the surface temperature of the ground.

2.6 COMBUSTION ENERGY

8. Calculate the energy release by combusting 10 g of butane (C_4H_{10}), given that the enthalpy of formation of liquid butane at standard conditions is −147.6 kJ/mol. Assume the products contain water as vapour.

9. Butane is burned by a stove with 100% excess air at 25°C to heat up a pan of water. If the combusted gases escape at 800 K, how much heat is extracted from the flame per kmol of fuel?

10. What mass of butane is required to heat up 400 g of water from 25°C to boiling point?

2.7 HEAT EXCHANGERS

11. (a) A heat exchanger operates with an oil with c_p of 1.67 kJ/kgK and density 910 kg/m³ and water with c_p of 4.2 kJ/kgK and density 1000 kg/m³. The oil volume flow rate is 3158 l/h and the water flow rate is 2000 l/h. What are the capacity rates of the oil and water?

(b) The water enters at 50°C and leaves at 70°C and the oil enters at 120°C and leaves at 85°C. The heat exchanger is of the shell-and-tube type with counter-current flow. What is the logarithmic mean temperature difference?

(c) What is the surface area of the heat transfer surface within the heat exchanger if it has an overall heat transfer coefficient of 1100 W/m²K? If this is done with a tube having diameter 12 mm, what is the length of the tube in the exchanger?

12. A compact heat exchanger is constructed with a cross-flow matrix arrangement with unmixed streams for cooling oil in an air stream. The oil enters at 70°C and is required to leave at 30°C. The oil density is 700 kg/m³. The flow rate of oil is 5 litres per minute and has a specific heat capacity c = 1.7 kJ/kgK at 10°C and 2.5 kJ/kgK at 100°C.

(a) What is a suitable average specific heat capacity of the oil over the working range?

(b) What is the rate of heat transfer?

13. A compact heat exchanger has oil exchanging heat with water and is used to cool engine oil. The oil mass flow rate is 50 g/s and c_p = 1800 J/kgK. The water mass flow rate is 60 g/s. Given that the ingoing oil is at 130°C and the ingoing water is at 15°C,

(a) What are c_{min} and c_{max} for this situation?

(b) What is the maximum potential heat transfer in this situation, q_{max}?

(c) What is the NTU if the area is 0.5 m² and the overall heat transfer coefficient is 800 W/m²K?

(d) What is the effectiveness, ε, and hence, the actual heat transfer?

2.8 COMPRESSORS

14. A two-stage reciprocating air compressor, running at four cycles per second, delivers air at 18 bar from atmospheric air at 1.03 bar at 11°C. The dimensions of the first stage are 100 mm diameter and 100 mm stroke, with a clearance length of 5 mm. Taking the polytropic index as 1.25 for compression and expansion processes,
    (a) What are the swept volume and clearance volume of the first stage?
    (b) What is the intermediate pressure for minimum work?
    (c) What is the expansion of the clearance volume in the first stage?
    (d) What is the volumetric efficiency and hence the mass flow rate of air?

3   SOLID MECHANICS

1. A helicopter rotor shaft, 50 mm in diameter, transmits a torque of 2.4 kNm and an upward tensile lifting force of 125 kN. Determine the maximum tensile stress, maximum compressive stress and the maximum shear stress in the shaft.
2. For the purposes of analysis, a segment of a crankshaft in a vehicle is represented as shown in Figure Q3.1. The load is $P = 1$ kN and the dimensions are $b_1 = 80$ mm, $b_2 = 120$ mm and $b_3 = 40$ mm. The diameter of the shaft is $d = 20$ mm. Determine the maximum tensile, compressive and shear stress at point A, located on the surface of the shaft at the z axis.

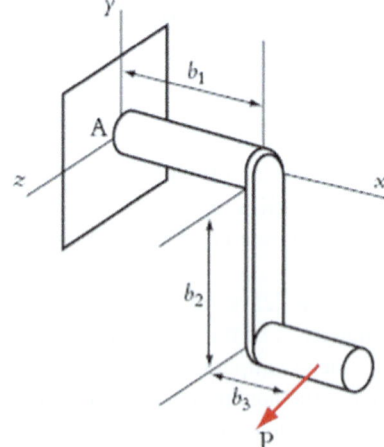

FIGURE Q3.1

3. Using the Tresca and von Mises yield criterion, calculate the pressure to cause yielding in a steel cylinder, which has an 80 mm diameter and 1 mm thickness. The cylinder is closed at each end; end effects should be neglected. Assume $\sigma_y = 250$ MPa.
4. What additional torque can be applied about the axis of the cylinder in Question 3 if yielding is to occur with an internal pressure of 4.0 MPa.
5. A steel beam of rectangular section 10 mm × 30 mm is subjected to pure bending in a plane parallel to the 30-mm faces. Ideal elastic-plastic behaviour may be assumed. Calculate the bending moment necessary for:
    (a) the onset of yield;
    (b) the complete yield through the section.
    Assume $\sigma_y = 250$ MPa.
6. The web and flanges of a straight I section steel beam are 80-mm wide and 10-mm thick. The beam is loaded in pure bending in the plane of the web until the whole of each flange

has yielded but the whole of the web remains elastic. Calculate the residual curvature in the unloaded beam. Assume ideal elastic-plastic behaviour. Assume $\sigma_y = 250$ MPa and $E = 200$ GPa.

7. Figure Q3.7 shows a simply supported beam carrying two concentrated loads at the positions indicated. Given that the beam has a rectangular cross-section as shown, calculate the deflection of the beam at a position 3 m from the left-hand end and at a position 5 m from the right-hand end. (E steel $= 200$ GPa.)

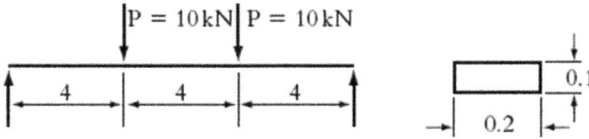

FIGURE Q3.7

8. Find the slope at point A and the deflection at point B of the beam shown in Figure Q3.8.

$$\left(EI = 4 \times 10^6 \, NM^2. \right)$$

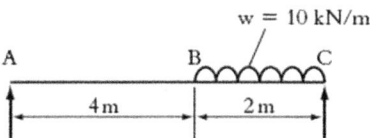

FIGURE Q3.8

9. A steel beam with both ends built in has an effective span of 7.5 m and carries a concentrated load of 60 kN at a point 4.5 m from the left-hand end. Calculate the position and magnitude of the maximum deflection due to this load, given that it occurs at a point between the left-hand end and the concentrated load. ($E = 208 \times 10^9 \, Nm^2$; $I = 85 \times 10^6 \, mm^4$.)

10. Find the critical load for a steel column of rectangular cross-section 50×25 mm and length 3 m for each of the following assumptions:
    (a) the ends are hinged;
    (b) the ends are built in;
    (c) one end is hinged and restrained from moving horizontally, the other end is clamped.

11. A new tripod for a surveying instrument will consist of three equal tubular legs, hinged at the top and resting on points at the bottom. When the instrument is set up for use, the top hinge points are equally spaced on a 100-mm diameter horizontal circle, while the pointed ends are equally spaced on a 1-m diameter circle. The top hinge points are 1.3 m above the level ground on which the pointed ends rest. The greatest expected instrument weight is 80 N. Assuming a factor of safety of 10 against elastic buckling, select the lightest safe aluminium tube ($E = 70$ GPa) from the following stock list.

Thickness (mm)	Outer Diameters Available (mm)
1	5, 10, 15, 20, 25 and 30
2	10, 15, 20, 25, 30 and 40
3	15, 20, 25, 30, 40 and 50

12. A hollow steel column, 9 m long and of circular cross-section 90 mm outer diameter and 75 mm inner diameter is subjected to an end thrust of 30 kN; the line of action of the thrust is parallel to the unstrained line of the strut but does not coincide with it. Under load, the maximum deviation of the strut from the straight is 80 mm. Find the eccentricity of the load and the maximum compressive and tensile stresses. The ends may be assumed to be hinged.

Assume $E = 200$ GPa

13. Figure Q3.13 shows the cross-section of a solid beam which carries a vertical shear force of 100 kN. Determine:
    (a) the shear stress just above and just below X–X;
    (b) the shear stress at the neutral axis of the section.
    (c) Sketch the shear stress distribution through the section and state where the maximum shear stress occurs.

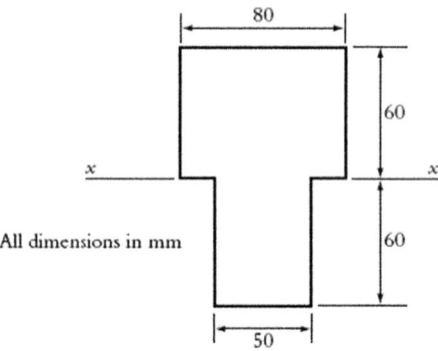

All dimensions in mm

FIGURE Q3.13

14. Show that the difference between the maximum and mean shear stress in the web of an i beam is $Sd^2/24I$, where d is the height of the web.

15. The outer dimensions of a channel girder section are 120 mm (web) × 50 mm (flanges); the web and flanges are 5 mm thick. Determine the position of the shear centre of the section.

16. Find the shear centre of the beam cross-section shown in Figure Q3.16 (it is much smaller than R)

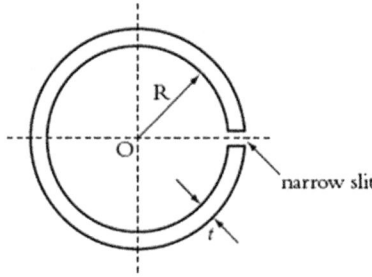

narrow slit

FIGURE Q3.16

17. A heat exchanger consists of a cylindrical vessel which contains 35 U-shaped tubes of 20 mm bore and 30 mm outside diameter. The ends of these tubes are welded to one of the flat end plates. The total length of the tubes within the vessel is 7 m. Calculate the hoop and axial stresses on the inside and outside of the straight parts of the tubes, remote from the

bend and the ends, due to pressures of 10 bar in the vessel (i.e. outside the tubes) and 100 bar inside the tubes.

18. A compound tube is made up of an inner steel tube, 4.5 in internal diameter and 6.5 in external diameter, on which is shrunk an outer steel tube that is 9 in in external diameter. If the radial pressure at the common surface is 4000 lbf/in^2, find the maximum and minimum circumferential stresses on both tubes due to shrinkage and plot the distribution of hoop stress across the wall of the compound tube.

19. A circular saw that is 5 mm thick and has a 900 mm diameter has a bore of 100 mm. The steel, of which the saw is made, has a density of 7800 kg m^{-3} and $\eta = 0.3$. Find the maximum speed permitted if the hoop stress is restricted to 240 MPa. What is the maximum value of the radial stress?

20. A bronze gyro-wheel has a moment of inertia of 0.4 kg m^2 and can be run with safety at 3000 rpm. A larger steel wheel of similar shape is required to give an angular momentum of 40×10^3 kg m^2 s^{-1} at 1800 rpm. In what ratio must all the linear dimensions be increased and what elastic limit is required for the steel so that both wheels have the same factor of safety. (Density of bronze = 8480 kg m^{-3}; Elastic limit of bronze = 78 MPa.)

21. Calculate the principal second moments of area and the directions of the principal axes for the section shown in Figure Q3.21.

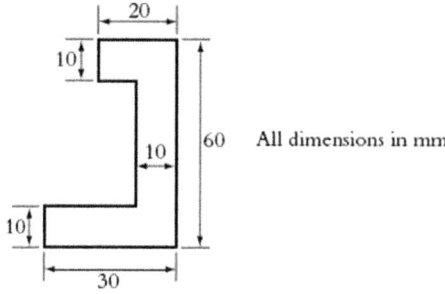

FIGURE Q3.21

22. A 50 mm × 30 mm × 5 mm angle is used as a cantilever of 500 mm, with the 30-mm leg horizontal and uppermost. A vertical load of 1000 N is applied at the free end. Determine the position of the neutral axis and the maximum tensile and compressive bending stresses.

23. Calculate the maximum tensile stress and the position of the neutral axis for the section shown in Figure Q3.23, when a bending moment of 225 Nm is applied about the x-axis in the sense shown.

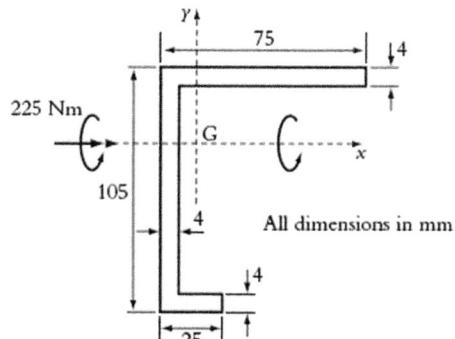

FIGURE Q3.23

24. Using strain energy, derive an expression for the deflection of the load point of the beam shown in Figure Q3.24.

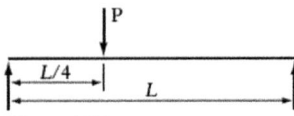

FIGURE Q3.24

25. Calculate the deflection beneath the force for the cantilevered bracket shown in Figure Q3.25. ($E = 200$ GPa; $G = 80$ GPa.)

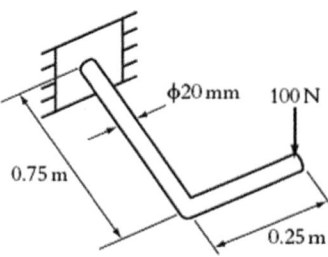

FIGURE Q3.25

26. Derive an expression for the increase in distance between the ends A and B of a thin bar of uniform cross-section consisting of a semi-circular portion CD and two straight portions AC and BD as shown in Figure Q3.26.

    If a bar is of diameter 6 mm, R is 40 mm and has a spring stiffness, $\dfrac{P}{\delta}$, of 1000 kg/m, show that the necessary length for L is approximately 210 mm. The bar is made from mild steel with Young's modulus $E = 210$ GPa.

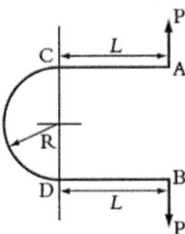

FIGURE Q3.26

27. A circular connecting rod is made from a steel having the following properties:
    - UTS: 950 MN/m²
    - YP: 800 MN/m²
    - Fatigue limit: 500 MN/m²

    The rod is subjected to a fully reversed axial load of 180 kN. Determine the minimum rod diameter, allowing a factory for safety of 2, if the rod end produces a fatigue strength reduction factor of 2.1, where the stress concentration factor is 2.5.

28. A mild steel cantilever beam of circular cross-section is subjected to a load at its free end which varies cyclically from P to $-3P$ as shown in Figure Q3.28. Determine the maximum value of P if the fatigue strength reduction factor for the fillet is 1.85 and a safety factor of 2.0 is assumed. (Hint: use the Goodman Diagram and apply the fatigue strength reduction factor to the stress amplitude only.)

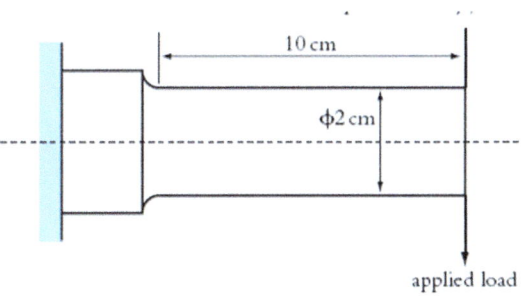

10 cm

φ2 cm

applied load

FIGURE Q3.28

29. A circular steel shaft having a transverse oil hole is subjected to a torsional load which varies from -100 Nm to 400 Nm (i.e. values in opposite senses). Determine the necessary shaft diameter, assuming that the hole causes a fatigue strength reduction factor of 1.75 and making use of a factor of safety of 1.5. Assume the following properties for the steel:
    - Ultimate tensile strength: 400 NM/m²
    - Fatigue endurance limit: 260 NM/m²
30. A rectangular section mild steel beam with the cross-sectional dimensions shown in Figure Q3.30 has a temperature given by

$$T = 50\cos\left(\frac{\pi y}{40}\right){}^\circ\text{C}$$

Find the factor of safety of the beam against yield, at a section remote from the ends. Assume $\sigma_y = 250$ MPa and $E = 200$ GPa.

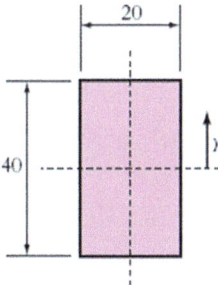

20

40

y

FIGURE Q3.30

31. A thin circular steel plate with a 400-mm outside diameter and a 200-mm insider diameter has a temperature distribution which varies approximately linearly with r. At the bore, $T = 100°C$, and at the outside $T = 30°C$. Find the growth of the bore and the growth of the outside diameter compared with their room temperature (20°C) values.

4 ELECTROMECHANICAL DRIVE SYSTEMS

1. A machine may be regarded as consisting of the system shown in Figure Q4.1, where $J_1 = J_2 = 0.1$ kg m², $n_1 = n_3 = 20$ and $n_2 = n_4 = 40$. Find the moment of inertia of the system referred to the input (bottom) shaft, ignoring the inertia of the gears themselves.

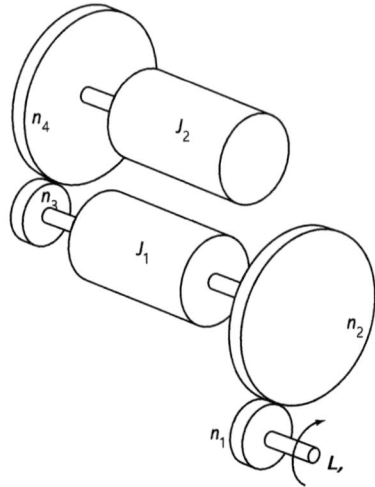

FIGURE Q4.1

(Hint: there are two ways to do this. Either work down the system, building up the contributions to the referred inertia for each axis or treat each inertia separately and refer it directly to the input shaft. Both methods are equivalent and give identical answers.)

2. Part of an x-y-z positioning system consists of a carriage which has a total mass of 6 kg and is carried on a sliding ball-assisted carriage (Figure Q4.2). There is no viscous friction and windage is negligible but the coefficient of (Coulomb) friction for the ball-slide may be regarded as 0.05. The carriage is attached to a continuous toothed belt which passes over two pulleys of effective (pitch) diameter 20 mm, one of which is connected directly to the motor which is used for positioning the carriage. The total mass of the belt is 50 g. Each pulley may be regarded as a solid cylinder with mass 10 g and diameter 18 mm.

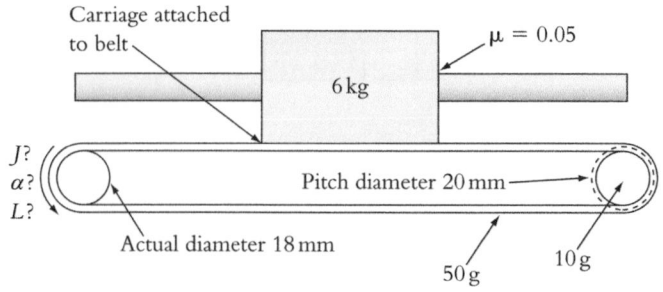

FIGURE Q4.2

(a) What is the moment of inertia of the system referred to the pulley that is connected to the motor?

(b) The carriage is required to move from one end of the 600-mm-long bed to the other in 1 second. Assuming that this is achieved by a period of constant acceleration over 300 mm, followed by constant deceleration over the remaining 300 mm, find the maximum linear acceleration of the carriage and hence find the maximum angular acceleration of the shaft.

(c) Find the frictional force (ignore the frictional effect of the weight of the belt), the frictional torque and the total torque required from the motor in order to achieve the necessary acceleration while overcoming the friction.

3. A mass of 100 kg is moved in a straight line using a lead screw of pitch 5 mm. As it is a ball-assisted high-quality leadscrew, its efficiency is high (95%), and it may be approximated as a solid cylinder of density 7800 kg m^{-3}, diameter 25 mm and length 2 m. What torque must be applied to the leadscrew to accelerate the mass from rest to 0.2 m/s over a distance of 0.5 m? Assume constant acceleration.

(Hint: Work out the linear acceleration of the mass using one of the kinematic formulae, and noting that 2π radians of rotation equates to 5 mm = 0.005 m of linear movement, work out the angular acceleration needed. Work out the moment of inertia of the screw itself and then use equation 4.26 to obtain the torque.)

FIGURE Q4.3

4. A 'Penny-farthing' (Figure Q4.3, a primitive type of bicycle, having a very large front wheel driven directly by the pedals) is being ridden by a person of mass 80 kg. The penny-farthing has a total mass of 20 kg of which 8 kg is concentrated in the rim of the front wheel. The front wheel has a radius of 0.65 m.

(a) Calculate the inertia of the system referred to the axis of the front wheel. Ignore the contribution to the rotational inertia of the front wheel spokes; also ignore the rotational inertia of the very small rear wheel.

(b) Calculate how much torque is required at the pedals to obtain an acceleration of 0.5 m/s^2

(i)   when the penny-farthing is being ridden on a level surface;

(ii) when it is climbing a gradient of 1 in 50 or 2% (equivalent to an angle of inclination of 0.02 radian).

5. A vehicle has a mass of 1600 kg. This total includes four wheels, each of mass 15 kg and diameter 0.7 m, which may be regarded for inertia purposes as solid discs.

(a) What is the moment of inertia of each wheel about its own axis?

(b) What is the moment of inertia of the vehicle referred to the axis of the wheels?

6. A mixing head for stirring a slurry has the following torque–speed characteristic when mixing is taking place:

$$L = 100 + \omega + 0.1\omega^2$$

where L is the torque in Nm and ω is the angular velocity in rad/s.

It is driven via a 50:1 worm gear of efficiency 45% from an induction motor with the following torque–speed characteristic in its operating region:

$$L' = 0.1(100\pi - \omega')$$

where L' and ω' are again in Nm and rad/s, respectively.
   (a) What is the torque–speed characteristic of the mixing head, referred to the axis of the motor, that is, expressed as L' vs ω'?
   (b) At what speed will the motor run when driving the mixing head? Express your answer in rad/s and rev/min.
   (c) At what speed (in rad/s and rev/min) will the mixing head itself rotate?
   (d) What power does the motor produce when driving the mixing head under load?

7. A floating container is pulled along a channel of water using a chain, which is hauled using a winch of diameter 300 mm. The drag force F on the container is related to its speed v by the following equation:

$$F = 500v^2$$

where F is in N and v is in m/s.

The winch is to be driven from a petrol engine with the following characteristics:

$$L' = 1 + 0.06\omega' - 0.0001\omega'^2 \text{ Nm}$$

which is valid within the range $100 < \omega < 450$ rad/s
   (a) Refer the force–speed characteristics of the container to the axis of the winch.
   (b) Hence, refer the force–speed characteristics of the container to the axis of the engine for the following combinations of gear ratio and efficiency:
       (i) 30:1, 65%
       (ii) 40:1, 55%
   (c) In both cases, find the combination of torque and angular velocity at which the engine will run when driving the winch. Hence, decide whether either of these combinations is feasible, and which will make the more effective use of the system.

8. A 36 V DC permanent magnet motor operates with an armature current of 1.5 A. The design constant k is 0.1 Vs (rad)$^{-1}$ and the armature resistance is 1.5 Ω.

Assuming that there are no losses, calculate the torque, speed in rev min^{-1} and output power.

The DC supply voltage is reduced to 16 V. The load torque is unchanged. Calculate the new speed and output power.

9. A 12 V DC permanent magnet motor has a no-load speed (ignoring frictional effects) of 9800 rev min^{-1}. When a load torque is applied, the motor speed drops to 8200 rev min^{-1} and the motor draws a current of 11 A. Calculate the motor design constant and armature resistance.

The supply voltage is reduced to 10 V when a load torque of 0.05 Nm is applied. Calculate the armature current, speed and mechanical output power.

10. A careful check of specification for the 12 V DC permanent magnet motor described in Question 9 reveals that when the motor runs on no-load (i.e. with no additional load torque beyond internal frictional effects) at a speed of 9800 rev min^{-1}, it draws a current

of 2 A. When a load torque is applied, the motor still runs at a speed of 8200 rev min⁻¹ and draws a current of 11 A. Calculate the revised figures for the motor design constant and armature resistance. (Hint: Using Equation 4.63, set up a pair of simultaneous equations for the two running conditions and solve them for the two unknowns, that is the motor design constant and the armature resistance).

Hence, calculate the internal frictional torque (which you may assume to be independent of speed over the range under consideration).

The supply voltage is reduced to 10 V and an external load torque of 0.05 Nm is applied. Calculate the new running speed.

11. A star-connected three-phase 415 V four-pole 50 Hz induction motor delivers a full load power of 9625 W when it runs at 1425 rev min⁻¹. Assuming that the torque *vs.* speed characteristic is linear over the motor's normal operating range, calculate the speed when the motor drives a load torque of 44.45 Nm.

12. A three-phase 25 kW, 415 V, 50 Hz and 1440 rev min⁻¹ cage induction motor is fed from a variable-frequency supply. The voltage and frequency are varied in proportion up to the rated frequency, above which the voltage is held constant. This arrangement is used to drive a blower which has a torque–speed characteristic given by $L = 6 \times 10^{-3} \omega^2$, where the torque, L, is in Nm and the angular velocity, ω, is in rad s⁻¹. Determine the frequency and line-to-line voltage at a blower speed of 1000 rev min⁻¹.

13. A star-connected motor is available with the following specification: 3000 W, 415 V, 50 Hz and 1450 rev min⁻¹.
    (a) What is its rated torque?
    (b) Assuming that the torque *vs.* speed characteristic is linear over the motor's normal operating range, at what speed will it run at its rated voltage if it is supplying a constant torque of 15 Nm?
    (c) If the supply voltage were to fall to 381 V line-to-line with the frequency remaining unchanged, at what speed would the motor run when supplying the same torque of 15 Nm?

14. (a) Explain the reasons why the use of a pneumatic motor may in some situations be preferable to the use of an electric motor. Illustrate your argument with proper graphs and examples where appropriate. Explain the shortcomings of pneumatic motors.
    (b) Draw a diagram illustrating the various features of a compressed air system, such as may be found in a typical factory. Briefly identify the function of each component and explain how the air pressure in the system is controlled.

5   FEEDBACK AND CONTROL THEORY

1. (a) Figure Q5.1(a) shows a multi-loop system. Determine the overall transfer function of the system shown in Figure Q5.1(a), relating the output $Y(s)$ to the input $X(s)$

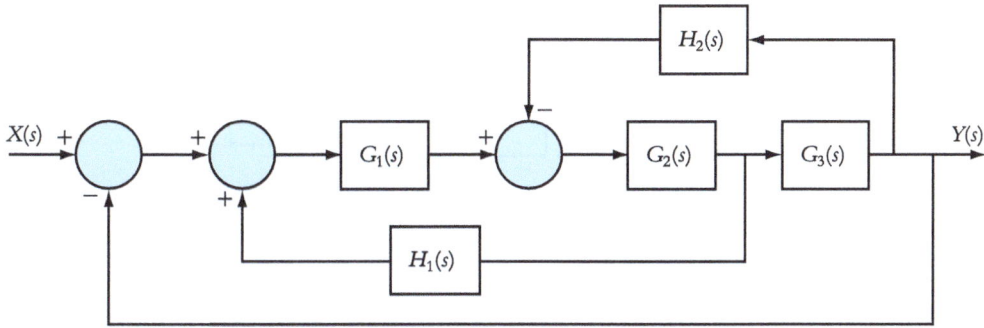

FIGURE Q5.1(A)

(b) Figure Q5.1(b) and Figure Q5.1(c) show two block diagrams for the same system. Determine the transfer functions $G(s)$ and $H(s)$ of the block diagram shown in Figure Q5.1(c) that are equivalent to those of the block diagram of Figure Q5.1(b).

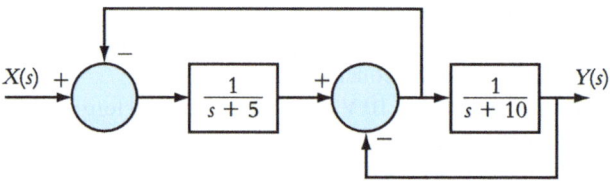

FIGURE Q5.1(B)

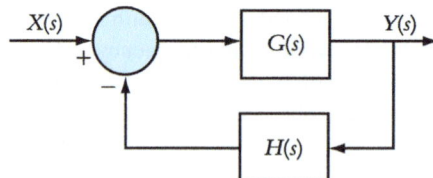

FIGURE Q5.1(C)

2. Figure Q5.2 shows a system for controlling the level of a liquid in a tank. The tank has a fixed cross-sectional area A, fixed linearised flow resistance R for the outflow (relating the outflow to the liquid level) and variable liquid level $h(t)$. For this system, the difference between the actual level h in the tank and the desired level h_i is used to form the error signal $\varepsilon(t)$, where the actual level h is measured by a transducer. This error signal is fed to a controller that drives a variable speed pump such that the volumetric inflow rate in the tank is given by $Q(s) = G_C(s)E(s)$, where $G_C(s)$ is the transfer function of the controller and $E(s)$ is the Laplace transform of the error signal $\varepsilon(t)$. In addition, there is an uncontrolled disturbance inflow to the tank which is given by $q_d(t)$.

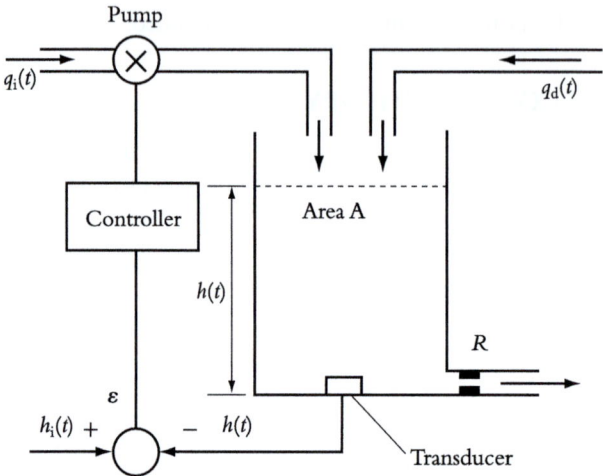

FIGURE Q5.2

For the case when the controller only contains proportional action with gain K:

(a) Draw the block diagram of the system, taking h_i to be the input and h to be the output.

(b) Determine the overall transfer function relating h to h_i and q_d and show that the system is of first order.

(c) Show the location of the closed loop pole of the system in the s-plane.

(d) With $q_d = 0$, if $A = 2$, $R = 10$ and $K = 0.3$ and a constant demand level $h_i = \bar{h}_i$, in consistent units, find the steady-state response in terms of $\bar{h}_i$.

3. Figure Q5.3 shows a schematic of a system for controlling the angular velocity Ω of a cable drum of radius R under conditions where the cable tension T is variable. The drum, which has a moment of inertia J_D is driven via an $n{:}1$ speed-reduction gearing by a servo-motor with the rotating parts having a moment of inertia J_M. The viscous friction coefficient C resists the rotation of the drum. The feedback signal V_o is derived from a tachogenerator and is subtracted from the demand signal $K_G\Omega_D$ to form the error signal. The controller is a simple proportional controller with gain K and delivers current I to the servo-motor that develops a drive torque L. The transfer functions for the tachogenerator and servo-motor, respectively, are as follows:

$$\frac{V_o(s)}{\Omega(s)} = \frac{K_G}{1+T_G s}, \frac{L(s)}{I(s)} = \frac{K_M}{1+T_M s}$$

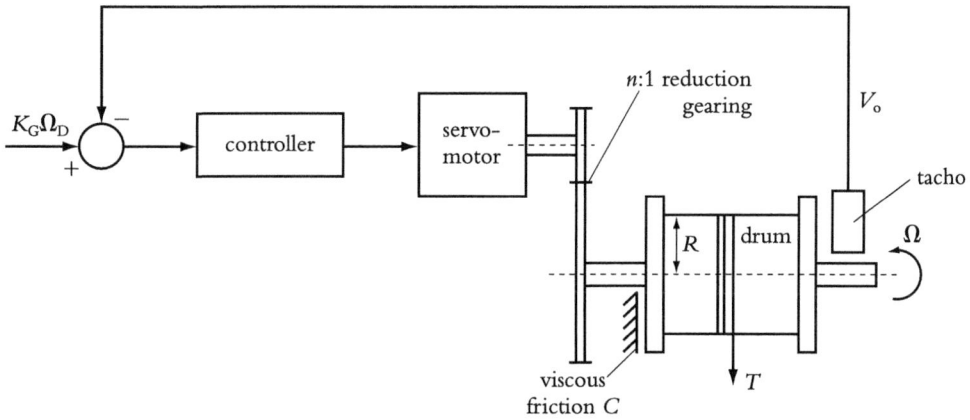

FIGURE Q5.3

(a) Draw a block diagram for the system and derive the overall transfer function relating the drum speed $\Omega(s)$ to the demand speed Ω_D, and cable tension T.

(b) For the case when $T_G = 0$, determine the order of the system and derive expressions for appropriate parameters which characterise completely the transient behaviour.

4. The block diagram in Figure Q5.4 represents a system for controlling the speed of a rotor that is driven by an electric motor. A tachogenerator is used to provide the feedback signal. T_M, T_L and T_T are time constants associated with the motor, rotor and tacho-generator, respectively, and K is a gain associated with the motor. For a particular system, the time constants have the following numerical values: $T_M = 0.5$, $T_L = 1$ and $T_T = 0.1$.

(a) Find the minimum value of K for which the steady-state error in rotor speed $(\Omega_i - \Omega_o)$ following a step change in input is 2% or less.

(b) Find the maximum value of K for which the system is stable. For this maximum value of K, determine the frequency at which the system would oscillate if disturbed.

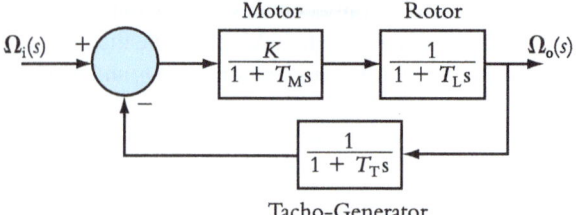

FIGURE Q5.4

5. The block diagram in Figure Q5.5 represents a system for controlling the rotational speed of an inertia load that is driven by a diesel engine. K and T_c are respectively the gain and time constant of the fuel injector. The demand speed is $\omega_i(t)$ and the actual load speed is $\omega_o(t)$ (these are expressed in the Laplace space as $\Omega_i(s)$ and $\Omega_o(s)$, respectively).

   (a) Derive an expression for the maximum value of K for which the closed-loop system is stable.

   (b) Show that the steady-state error in speed following a step change in demand is zero and derive an expression for the steady-state error when a ramp input is applied.

   (c) With reference to your answers to parts (a) and (b), explain whether it is better for the fuel injector time constant T_c to be shorter or longer.

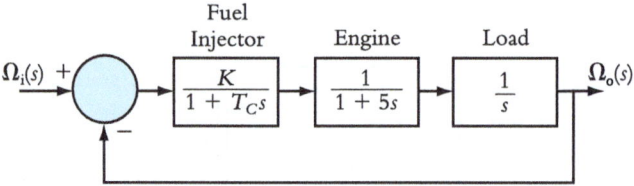

FIGURE Q5.5

6. (a) Figure Q5.6 shows the block diagram of a system for controlling the temperature of a liquid in a tank. For the case where the controller is a proportional controller with transfer function $G(s) = K$, draw the root locus plot for the system.

   (b) To reduce steady-state errors, the controller is modified to incorporate integral action so that the controller transfer function is now given by

$$G_C(s) = K\left(\frac{s+4}{s}\right)$$

Draw the root locus plot for the modified system and comment on the practical significance of the differences between the plots for the original and modified systems.

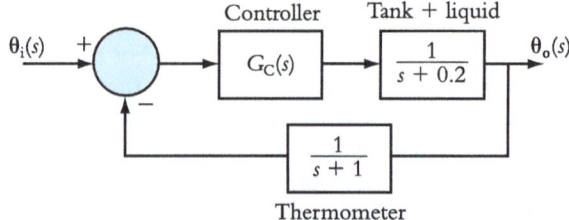

FIGURE Q5.6

7. Figure Q5.7 shows a schematic of a system for controlling the angular position (expressed in the Laplace domain) $\Theta_o(s)$ of a satellite dish. The dish assembly has a moment of inertia J_D and is driven via $n:1$ speed reduction gearing by a servo-motor with rotating parts having a moment of inertia J_M. A viscous friction of coefficient C resists the motion of the dish assembly. The feedback signal $V_o(s)$ is derived from a potentiometer and is subtracted from the demand signal $K_P\Theta_i(s)$ to form the error signal. The controller is a simple proportional controller with gain K and delivers current I to a servo-motor that develops a drive torque $L_M(s)$. The transfer functions for the potentiometer and servo-motor, respectively, are as follows:

$$\frac{V_o(s)}{\Theta_o(s)} = K_P; \quad \frac{L_M(s)}{I(s)} = \frac{K_M}{1+T_M s}$$

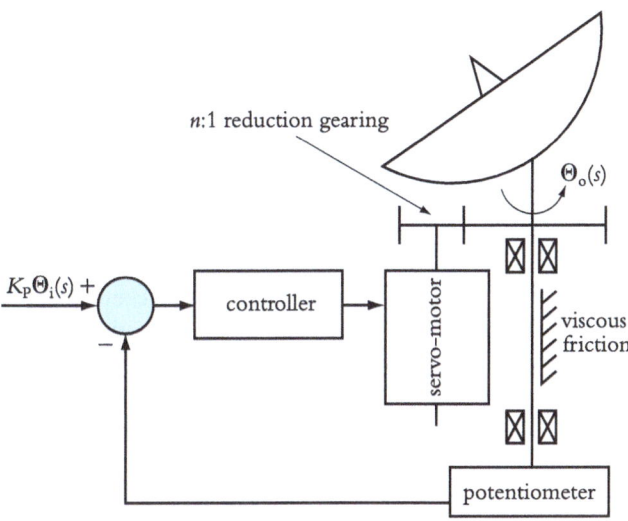

FIGURE Q5.7

(a) Draw a block diagram for the system and derive the overall transfer function relating the angular position of the dish $\Theta_o(s)$ to the demand rotation $\Theta_i(s)$

(b) For the case when $T_M = 0$, determine the order of the system and derive expressions for appropriate parameters which characterise completely the transient behaviour.

8. Find the numbers of positive roots of the following equations using the Routh-Hurwitz criterion:

(a) $s^3 + 4s^2 + s - 6 = 0$

(b) $s^4 - s^3 - 2s^2 - 2s + 4 = 0$

(c) $s^5 + 4s^4 - 14s^3 - 46s^2 + 25s + 150 = 0$

9. Plot the closed loop root loci for the situations where the open-loop transfer function is given by the following transfer functions combined with a variable gain K:

(a) $\dfrac{s-1}{(s^2+2s+2)(s+2)}$

(b) $\dfrac{s}{(s+1)(s+2)(s+3)(s+4)}$

6 STRUCTURAL VIBRATION

6.1 NATURAL FREQUENCIES AND MODE SHAPES

1. Derive the equation of motion and hence find the natural frequencies for the system shown in Figure Q6.1.

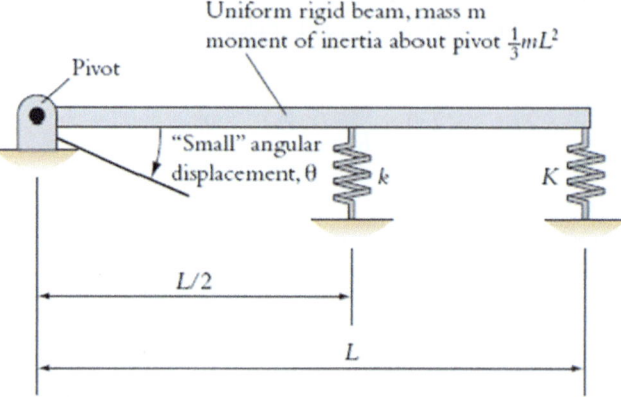

FIGURE Q6.1

2. A wheel (radius r, mass m and moment of inertia about its centre I) can roll without slipping on a horizontal plane. It is restrained by a horizontal spring (stiffness k) attached at one end to the centre of the wheel and at the other end to a rigid vertical wall as shown in Figure Q6.2. Derive the equation of motion and hence find the natural frequency for the system.

   What would the natural frequency be if there was no friction between the wheel and the plane?

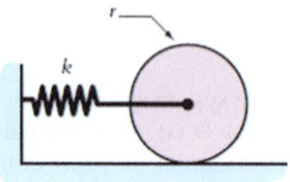

FIGURE Q6.2

3. Derive the equations of motion for the system shown in Figure Q6.3. Assume that all displacements are small.

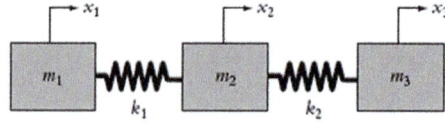

FIGURE Q6.3

4. Find the natural frequencies and mode shapes for the system shown in Figure Q6.3 when $k_1 = 10$ kN/m, $k_2 = 30$ kN/m, $m_1 = m_2 = 5$ kg and $m_3 = 10$ kg.
5. Derive the equations of motion for the system in Figure Q6.5. Assume that all displacements and angles are small.

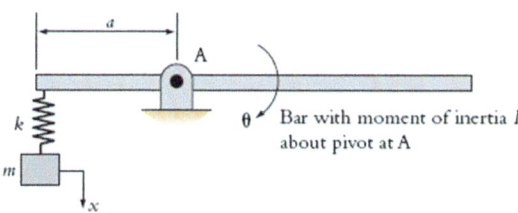

FIGURE Q6.5

6. Derive the frequency equation for flexural vibration of a uniform beam that is free at both ends and find an expression for the mode shape function.
7. A 25-mm diameter shaft, 1.5-m long, is held by two roller bearings at one end (giving a 'clamped' boundary condition) and by a self-aligning ball bearing at the other end (giving a 'pinned' boundary condition). Using the roots of the appropriate frequency equation given in Table 6.3 on page 398, find the first three critical speeds of the shaft.

6.2   RESPONSE OF DAMPED SINGLE-DEGREE-OF-FREEDOM SYSTEMS

8. If a heavily damped structure is given an initial displacement Z_0 and then released from rest, find the constants of integration and sketch the graph of $z(t)$ against time.
9. A critically damped structure is subjected to an impulse such that it acquires an instantaneous initial velocity, V_0, while the displacement remains zero. Show that the subsequent displacement is given by $z(t) = V_0 t e^{-\omega_n t}$.
10. The rigid beam shown in Figure Q6.10 has a moment of inertia of 10 kgm² about the pivot at A. End C is displaced downwards by 10 mm from its equilibrium position and then released from rest. Find the maximum upward displacement of C from its equilibrium position and the elapsed time at which this occurs.

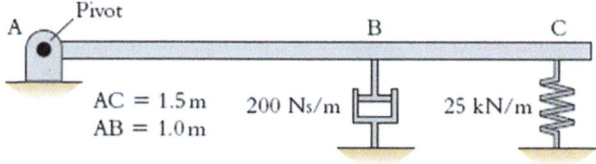

FIGURE Q6.10

11. Figure Q6.11 shows a rocker arm with moment of inertia I_O, about the pivot at O. Rubber blocks at ends A and B can each be modelled as a spring (stiffness, k) in parallel with a viscous damper (damping coefficient, c). The base of the block at A is attached to a rigid foundation. The base of the block at B is attached to a follower, which is driven by a cam that gives the follower a sinusoidal displacement of amplitude Y and frequency ω.

Derive an expression for the steady-state amplitude of the displacement at A, assuming that the angular displacement of the bar is small.

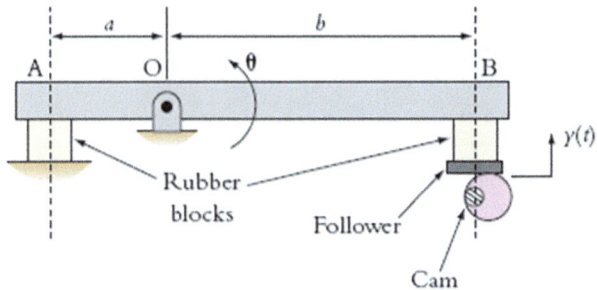

FIGURE Q6.11

12. For an undamped system with n degrees of freedom, show that the steady-state response in coordinate j due to a sinusoidal force of amplitude P and frequency ω applied in coordinate k is given by

$$x_j(t) = \left\{ \sum_{r=1}^{n} \frac{u_{jr}u_{kr}}{\omega r^2 - \omega^2} \right\} p \sin \omega t$$

6.3 APPROXIMATE METHODS

13. Use Dunkerley's Method and Rayleigh's Method to estimate the lowest natural frequency of the torsional system shown in Figure Q6.13.

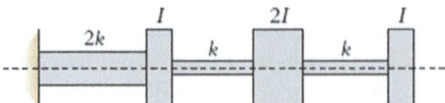

FIGURE Q6.13

14. A shaft with universal joints at each end has a length of 6 m, a second moment of area of 0.00025 m⁴ and a mass/unit length of 75 kg/m. It carries three discs, which can be regarded as point masses of 100, 150 and 200 kg located 1.2, 3 and 4.8 m from the left-hand end. Estimate the lowest critical speed using Dunkerley's and Rayleigh's Methods. Take $E = 207$ GN/m² and $\rho = 7800$ kg/m².

15. Find a single-degree-of-freedom approximate model to analyse the motion of the 1 kg mass in the system in Figure Q6.15 when it is vibrating near its lower natural frequency.

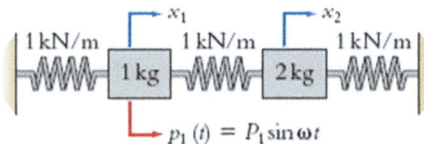

FIGURE Q6.15

16. Use the model from Question 15 to estimate the steady-state response of each mass in the two-degree-of-freedom system due to a sinusoidal force of amplitude 10 N and frequency 3.8 Hz applied to the 1-kg mass.

6.4 VIBRATION CONTROL TECHNIQUES

17. Derive the displacement transmissibility for the system in Figure Q6.17.

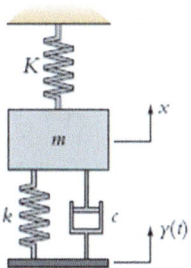

FIGURE Q6.17

18. A compressor of mass 300 kg is to be installed on four isolators, two at each end. The centre of mass is 0.4 m from end A and 0.2 m from end B. In the end elevation, the isolators are located symmetrically with respect to the centre of mass.

    Isolators are available with stiffnesses of 40, 70, 110, 180 and 290 kN/m and each has a maximum allowable static deflection of 10 mm. Select suitable isolators for the installation so that the isolation efficiency is at least 70% at the normal running speed of 870 rev/min. Estimate the actual isolation efficiency for each of the isolators you select. Neglect damping.

Index

Pages in *italics* refer to figures and pages in **bold** refer to tables.

Printed and bound by CPI Group (UK) Ltd, Croydon, CR0 4YY

06/11/2024

01784891-0002